AF840194

NERVOUS SYSTEM DRUG DELIVERY

NERVOUS SYSTEM DRUG DELIVERY

Principles and Practice

Edited by

Russell R. Lonser

Malisa Sarntinoranont

Krystof Bankiewicz

ELSEVIER

ACADEMIC PRESS
An imprint of Elsevier

Academic Press is an imprint of Elsevier
125 London Wall, London EC2Y 5AS, United Kingdom
525 B Street, Suite 1650, San Diego, CA 92101, United States
50 Hampshire Street, 5th Floor, Cambridge, MA 02139, United States
The Boulevard, Langford Lane, Kidlington, Oxford OX5 1GB, United Kingdom

© 2019 Elsevier Inc. All rights reserved.

No part of this publication may be reproduced or transmitted in any form or by any means, electronic or mechanical, including photocopying, recording, or any information storage and retrieval system, without permission in writing from the publisher. Details on how to seek permission, further information about the Publisher's permissions policies and our arrangements with organizations such as the Copyright Clearance Center and the Copyright Licensing Agency, can be found at our website: www.elsevier.com/permissions.

This book and the individual contributions contained in it are protected under copyright by the Publisher (other than as may be noted herein).

Notices
Knowledge and best practice in this field are constantly changing. As new research and experience broaden our understanding, changes in research methods, professional practices, or medical treatment may become necessary.

Practitioners and researchers must always rely on their own experience and knowledge in evaluating and using any information, methods, compounds, or experiments described herein. In using such information or methods they should be mindful of their own safety and the safety of others, including parties for whom they have a professional responsibility.

To the fullest extent of the law, neither the Publisher nor the authors, contributors, or editors, assume any liability for any injury and/or damage to persons or property as a matter of products liability, negligence or otherwise, or from any use or operation of any methods, products, instructions, or ideas contained in the material herein.

Library of Congress Cataloging-in-Publication Data
A catalog record for this book is available from the Library of Congress

British Library Cataloguing-in-Publication Data
A catalogue record for this book is available from the British Library

ISBN: 978-0-12-813997-4

For information on all Academic Press publications
visit our website at https://www.elsevier.com/books-and-journals

Publisher: Nikki Levy
Acquisition Editor: Melanie Tucker
Editorial Project Manager: Kristi Anderson
Production Project Manager: Bharatwaj Varatharajan
Cover Designer: Victoria Pearson

Typeset by SPi Global, India

ELSEVIER Book Aid International Working together to grow libraries in developing countries
www.elsevier.com • www.bookaid.org

Dedications

To my wife Carolyn, and to my daughters Hannah, Sarah, and Alicia, who support me in everything I do. And to Ed Oldfield, who always offered expert advice and thoughtful mentorship. *Russell R. Lonser, MD*

To my dad, who always gave the best advice, and to my mentors and students, who teach me new things every day. *Malisa Sarntinoranont, PhD*

To my wife Malina and to my daughter Olenka, both of whom made a great many sacrifices during my long working hours. *Krystof Bankiewicz, MD, PhD*

Contents

Chapter 12. Nanoparticles for Brain Tumor Delivery

TISTA ROY CHAUDHURI, ROBERT M. STRAUBINGER

Chapter 13. Solute Transport in the Cerebrospinal Fluid: Physiology and Practical Implications

MIKHAIL PAPISOV, VASILY BELOV

Chapter 14. Drug-Impregnated Polymer Delivery

RICCARDO SERRA, JOSHUA CASAOS, BETTY TYLER, HENRY BREM

Chapter 15. Immunomodulatory Methods

CYNTHIA KASSAB, BRITTANY PARKER KERRIGAN, HILLARY CARUSO, SANAALARAB AL ENAZY, AMY B. HEIMBERGER

Chapter 16. Convection-Enhanced Drug Delivery in the Central Nervous System

JERUSHA NAIDOO, MASSIMO FIANDACA, RUSSELL R. LONSER, KRYSTOF BANKIEWICZ

Chapter 17. Stem Cell Transplantation for Neurological Disease: Technical Considerations and Delivery Devices

ETHAN A. WINKLER, DANIEL A. LIM

Chapter 24. Central Nervous System Drug Delivery After Ischemic or Hemorrhagic Stroke

NGUYEN HOANG, SHAHID M. NIMJEE

Chapter 25. Intrathecal Drug Delivery for Cancer Pain

PHILIP S. KIM, PETER S. STAATS, TIMOTHY R. DEER, MICHAEL J. IADAROLA, ANDREW J. MANNES

Contributors

Sanaalarab Al Enazy Department of Neurosurgery, The University of Texas MD Anderson Cancer Center, Houston, TX, United States

Casey Anthony Department of Neurosurgery, Emory University School of Medicine, Atlanta, GA, United States

Isabelle Aubert Sunnybrook Research Institute, Sunnybrook Health Sciences Centre; Department of Laboratory Medicine and Pathobiology, University of Toronto, Toronto, Canada

Krystof Bankiewicz Interventional Neuro Center, Department of Neurological Surgery, University of California—San Francisco, San Francisco, CA, United States

Susan D. Bell Department of Neurological Surgery, The Ohio State University Wexner Medical Center, Columbus, OH, United States

Vasily Belov Massachusetts General Hospital, Harvard Medical School, and Shriners Hospitals for Children, Boston, MA, United States

Nicholas M. Boulis Department of Neurosurgery, Emory University School of Medicine; Department of Biomedical Engineering, Georgia Institute of Technology, Atlanta, GA, United States

Henry Brem Department of Neurosurgery and Hunterian Neurosurgical Research Laboratory, Johns Hopkins University School of Medicine, Baltimore, MD, United States

Matthew Campbell Smurfit Institute of Genetics, Trinity College Dublin, Dublin, Ireland

Hillary Caruso Department of Neurosurgery, The University of Texas MD Anderson Cancer Center, Houston, TX, United States

Joshua Casaos Department of Neurosurgery and Hunterian Neurosurgical Research Laboratory, Johns Hopkins University School of Medicine, Baltimore, MD, United States

Tista Roy Chaudhuri Department of Pharmaceutical Sciences, University at Buffalo, State University of New York, Buffalo, NY, United States

Aaron Dadas Flocel, Inc., Cleveland, OH, United States

Beverly L. Davidson The Raymond G. Perelman Center for Cellular and Molecular Therapeutics, The Children's Hospital of Philadelphia; The Perelman School of Medicine, The University of Pennsylvania, Philadelphia, PA, United States

Timothy R. Deer Center for Pain Relief, INC, Charleston, WV, United States

Long Di Department of Neurosurgery, Emory University School of Medicine, Atlanta, GA, United States

Nancy D. Doolittle Department of Neurology, Oregon Health & Science University, Portland, OR, United States

James B. Elder Department of Neurological Surgery, The Ohio State University Wexner Medical Center, Columbus, OH, United States

Rasha Elmghirbi Department of Physics, Oakland University, Rochester, MI, United States

James R. Ewing Department of Neurology, Henry Ford Hospital; Department of Neurology, Wayne State Medical School; Department of Physics, Oakland University, Rochester; Department of Neurosurgery, Henry Ford Hospital, Detroit, MI, United States

Massimo Fiandaca Brain Neurotherapy Bio, Inc., Oakland, CA, United States

Gerald Grant Department of Neurosurgery, Stanford University School of Medicine, Stanford, CA, United States

Chris Greene Smurfit Institute of Genetics, Trinity College Dublin, Dublin, Ireland

Kunal Gupta Department of Neurological Surgery, Oregon Health & Science University, Portland, OR, United States

Nathan Hardcastle Department of Neurosurgery, Emory University School of Medicine, Atlanta, GA, United States

Amy B. Heimberger Department of Neurosurgery, The University of Texas MD Anderson Cancer Center, Houston, TX, United States

Nguyen Hoang Department of Neurosurgery, The Ohio State University Wexner Medical Center, Columbus, OH, United States

Yuhao Huang Department of Neurosurgery, Stanford University School of Medicine, Stanford, CA, United States

Kristin Huntoon Department of Neurological Surgery, The Ohio State University Wexner Medical Center, Columbus, OH, United States

Sakibul Huq Department of Neurosurgery and Hunterian Neurosurgical Research Laboratory, Johns Hopkins University School of Medicine, Baltimore, MD, United States

Lee S. Hwang Department of Neurosurgery, The Cleveland Clinic, Cleveland, OH, United States

Michael J. Iadarola Department of Perioperative Medicine, National Institutes of Health, Bethesda, MD, United States

Damir Janigro Flocel, Inc.; Department of Physiology, Case Western Reserve University, Cleveland, OH, United States

Cynthia Kassab Department of Neurosurgery, The University of Texas MD Anderson Cancer Center, Houston, TX, United States

Brittany Parker Kerrigan Department of Neurosurgery, The University of Texas MD Anderson Cancer Center, Houston, TX, United States

Philip S. Kim Center for Interventional Pain & Spine, LLC, Helen Graham Center, Newark, DE, United States

Daria Krivosheya Department of Neurosurgery, The Cleveland Clinic, Cleveland, OH, United States

Paul Larson Department of Neurological Surgery, University of California-San Francisco, San Francisco, CA, United States

Daniel A. Lim Department of Neurological Surgery; Eli and Edythe Broad Center of Regeneration Medicine and Stem Cell Research, University of California San Francisco; Veterans Affairs Medical Center, San Francisco, CA, United States

Nir Lipsman Division of Neurosurgery; Sunnybrook Research Institute, Sunnybrook Health Sciences Centre, University of Toronto, Toronto, Canada

Russell R. Lonser Department of Neurological Surgery, The Ohio State University Wexner Medical Center, Columbus, OH, United States

Andrew J. Mannes Department of Perioperative Medicine, National Institutes of Health, Bethesda, MD, United States

Nicola Marchi Cerebrovascular Mechanisms of Brain Disorders Laboratory, Department of Neuroscience, Institute of Functional Genomics (INSERM), University of Montpellier, Montpellier, France

Bryn A. Martin Neurophysiological Imaging and Modeling Laboratory, Department of Biological Engineering, University of Idaho, Moscow, ID, United States

John M. McGregor Department of Neurological Surgery, The Ohio State University Wexner Medical Center, Columbus, OH, United States

Ying Meng Division of Neurosurgery; Sunnybrook Research Institute, Sunnybrook Health Sciences Centre, University of Toronto, Toronto, Canada

Karim Mithani Division of Neurosurgery, Sunnybrook Health Sciences Centre, University of Toronto, Toronto, Canada

Paul Morrison National Institute of Biomedical Imaging & Bioengineering, National Institutes of Health, Bethesda, MD, United States

Tavarekere Nagaraja Department of Neurosurgery, Henry Ford Hospital, Detroit, MI, United States

Jerusha Naidoo Interventional Neuro Center, Department of Neurological Surgery, University of California—San Francisco, San Francisco, CA, United States

Edward A. Neuwelt Departments of Neurology and Neurosurgery, Oregon Health & Science University, Portland, OR, United States

Shahid M. Nimjee Department of Neurosurgery, The Ohio State University Medical Center, Columbus, OH, United States

Claire O'Connor Smurfit Institute of Genetics, Trinity College Dublin, Dublin, Ireland

Soroush Heidari Pahlavian Laboratory of Functional MRI Technology, Stevens Neuroimaging and Informatics Institute, Department of Neurology, University of Southern California, Los Angeles, CA, United States

James Pan Department of Neurosurgery, Stanford University School of Medicine, Stanford, CA, United States

Mikhail Papisov Massachusetts General Hospital, Harvard Medical School, and Shriners Hospitals for Children, Boston, MA, United States

Kelly M. Poth Department of Neurosurgery, Emory University School of Medicine, Atlanta, GA, United States

Nirmala Ramanath Smurfit Institute of Genetics, Trinity College Dublin, Dublin, Ireland

R. Mark Richardson Department of Neurological Surgery, University of Pittsburgh, Pittsburgh, PA, United States

Kaitlin Sandor Department of Neurosurgery, Emory University School of Medicine, Atlanta, GA, United States

Malisa Sarntinoranont Department of Mechanical & Aerospace Engineering, University of Florida, Gainesville, FL, United States

Nathan R. Selden Department of Neurological Surgery, Oregon Health & Science University, Portland, OR, United States

Riccardo Serra Department of Neurosurgery and Hunterian Neurosurgical Research Laboratory, Johns Hopkins University School of Medicine, Baltimore, MD, United States

Bryan P. Simpson The Raymond G. Perelman Center for Cellular and Molecular Therapeutics, The Children's Hospital of Philadelphia; The Perelman School of Medicine, The University of Pennsylvania, Philadelphia, PA, United States

Peter S. Staats National Spine and Pain Centers, Shrewsbury, NJ, United States

Robert M. Straubinger Department of Pharmaceutical Sciences, University at Buffalo, State University of New York, Buffalo, NY; Department of Molecular and Cellular Biophysics and Biochemistry; Department of Pharmacology and Therapeutics, Roswell Park Comprehensive Cancer Center, Buffalo, NY, United States

Muhibullah S. Tora Department of Neurosurgery, Emory University School of Medicine; Department of Biomedical Engineering, Georgia Institute of Technology, Atlanta, GA, United States

Betty Tyler Department of Neurosurgery and Hunterian Neurosurgical Research Laboratory, Johns Hopkins University School of Medicine, Baltimore, MD, United States

Laura Vecchio Sunnybrook Research Institute, Sunnybrook Health Sciences Centre, University of Toronto, Toronto, Canada

Michael A. Vogelbaum Department of Neurosurgery, The Cleveland Clinic, Cleveland, OH, United States

Jolewis Washington Flocel, Inc., Cleveland, OH, United States

Ethan A. Winkler Department of Neurological Surgery, University of California San Francisco, San Francisco, CA, United States

Emily Youngers Department of Neurology, Oregon Health & Science University, Portland, OR, United States

Preface

New molecular insights into disorders of the nervous system have led to the development of therapeutic agents for ineffectively treated or untreatable diseases. Although these putative agents have shown benefit in in vitro studies and animal studies, they have not yet been effectively translated into successful clinical treatments. A major obstacle to the translation of potential therapeutic agents into clinical therapies is that these drugs cannot be reliably, efficiently delivered to the nervous system.

To overcome this limitation, researchers have been expanding their view, focusing on gaining greater insight into the biology of nervous system drug distribution. This broadening has resulted in new techniques for drug delivery, and the application of novel methods tailored to specific biological features across the spectrum of neurological diseases, which has demanded critical interdisciplinary efforts across basic, translational, and clinical teams.

The primary objective of this textbook is to provide readers with a comprehensive, detailed, state-of-the-art reference for nervous system drug delivery. Specifically, this volume provides the basic science background underlying each of the various drug delivery techniques, detailing the properties and features of each nervous system delivery method, and providing the foundational principles underlying the clinical application of drug delivery approaches to the targeted nervous system pathobiology. To accomplish the primary objective of the text and to facilitate reader understanding, this reference was designed to cover systematically all areas impacting drug delivery in three broad sections, including the *Physiology of Nervous System Drug Delivery* (Section I), *Nervous System Delivery Techniques* (Section II), and the *Clinical Application of Nervous System Drug Delivery* (Section III).

One of the more valuable contributions of this book is that it collates and communicates the knowledge of experts in the field of nervous system drug delivery and, through a deeper understanding across a variety of specialties, provides pathways that deepen interdisciplinary collaboration among basic, translational, and clinical drug delivery investigators. Because there are so many emerging areas in nervous system drug delivery spanning a wide range of disciplines (including engineering, immunology, pharmacology, neurology, surgical sciences, radiology, physics, biology, medical sciences, and gene therapy), the authors of this book appropriately represent a host of different industries. These experts provide clearly defined, up-to-date information framed by the historical context of early understanding of drug delivery in their respective fields. Each chapter was written to serve as a reference text for students and practiced experts across the wide spectrum of nervous system drug delivery research and clinical spheres.

It is only thanks to the hard work of an incredible group of individuals that this textbook was possible. We are deeply indebted to each of the expert authors. The efficient editorial staff members at Elsevier were likewise invaluable. We are deeply appreciative of the world-class editing and inestimable project management provided by Clare Sonntag. Finally, we remain ever-grateful for the unyielding support of our families and colleagues for the sacrifices they made as we saw this effort through.

Russell R. Lonser
Malisa Sarntinoranont
Krystof Bankiewicz

Acknowledgments

The editors wish to express their gratitude to Clare Sonntag, whose expert and indispensable editing was essential to the development, content, and completion of this textbook.

SECTION I

PHYSIOLOGY OF NERVOUS SYSTEM DRUG DELIVERY

CHAPTER

1

Fundamentals of Brain–Barrier Anatomy and Global Functions

Chris Greene, Matthew Campbell*, Damir Janigro*[†,‡]

[*]Smurfit Institute of Genetics, Trinity College Dublin, Dublin, Ireland [†]Flocel, Inc., Cleveland, OH, United States [‡]Department of Physiology, Case Western Reserve University, Cleveland, OH, United States

BARRIERS OF THE CENTRAL NERVOUS SYSTEM

The central nervous system (CNS) consists of the brain and spinal cord and controls all bodily activities. Central to this function is the neuron, an electrically excitable cell that requires a precise control of electrophysiological and chemical signals to function efficiently. As neurons lack regenerative capacities, it is vital to maintain a constant state of homeostasis in the CNS for the health and integrity of neurons. For efficient synaptic signaling between neurons, a tightly regulated control of the cerebral microenvironment is required to efficiently process the vast array of information received by the CNS and to synchronize its motor outputs. Indeed, the brain expends roughly 20% of the energy produced by the body, mostly through neural signaling, despite accounting for just 2% of bodily mass.[1] Therefore, to protect the homeostasis of this delicate biocomputing environment, it is vital to separate brain tissue from the peripheral circulation. It is important to remember, however, that the composition of peripheral blood is controlled in part by the brain itself (e.g., satiety and thirst centers).

In mammalians, this is achieved by three cellular barriers in the brain, which form an interface between the blood and neural tissue: the blood–cerebrospinal fluid (CSF) barrier, the arachnoid barrier, and the blood–brain barrier (BBB). The blood–CSF barrier is formed by epithelial cells of the choroid plexus.[2] The choroid plexus epithelium secretes CSF, which fills the cerebral and spinal subarachnoid spaces and ventricles, functioning as a buffer to protect the brain from injury and regulating cerebral blood flow (CBF) and molecular exchange with the brain. CSF is also the main component of interstitial fluid. A second barrier is formed by the arachnoid epithelium, an avascular membrane underlying the dura and completely enclosing the CNS. This forms a seal between the CSF and the extracellular fluids of the rest of the body.[3] The BBB, positioned along the blood vessels of the CNS, is a selective and tightly

Nervous System Drug Delivery
https://doi.org/10.1016/B978-0-12-813997-4.00001-3

© 2019 Elsevier Inc. All rights reserved.

regulated barrier that reflects the brain's critical roles in cognitive function and behavior, maintaining homeostasis and strictly coordinating the functions of peripheral organs.

The BBB is important not only in regulating the exchange of ions and nutrients between the blood and the brain, but also in protecting delicate neural tissue from potentially damaging blood- borne agents such as pathogens, immune cells, endogenous contaminants, and anaphylatoxins.[3] Additionally, the brain endothelium secretes approximately 200 mL of fresh interstitial fluid per day, creating an ideal ionic environment for neural function.[4] In fact, as the CNS has no local energy reserves, it requires a constant supply of glucose and oxygen delivered from the blood and is sensitive to changes in blood flow, which in turn is locally controlled by a process called *autoregulation*.[2] This energy need is fulfilled by the cerebral microvasculature. Microvessels in the brain have a combined surface area of 200 cm^2/g of tissue, with a capillary length of roughly 650 km, which accounts for more than 85% of total vessel length in the brain. This means that capillaries provide the largest surface area for molecular exchange of solutes between the blood and the brain.[5] Such is the extent of the cerebral vasculature that no neuron is farther than about 20 μm from a blood vessel.[6]

The BBB, formed by the endothelial cells (ECs) of the CNS, separates peripheral blood from brain tissue. Because of these specialized functions, CNS ECs are structurally and functionally different from ECs of the periphery. Notably, CNS ECs contain polarized expression of receptor proteins, regulating the entry and exit of material across cells (i.e., the transcellular pathway); highly electrical-resistant tight junctions (TJs), limiting the flux of material between ECs (i.e., the paracellular pathway); limited vesicular transport, preventing large hydrophilic molecules from entering the CNS; higher numbers of mitochondria, for greater energy expenditure; and an absence of fenestrations, preventing the rapid exchange of molecules between blood and tissue that is normally present in the periphery (Fig. 1).[3,7]

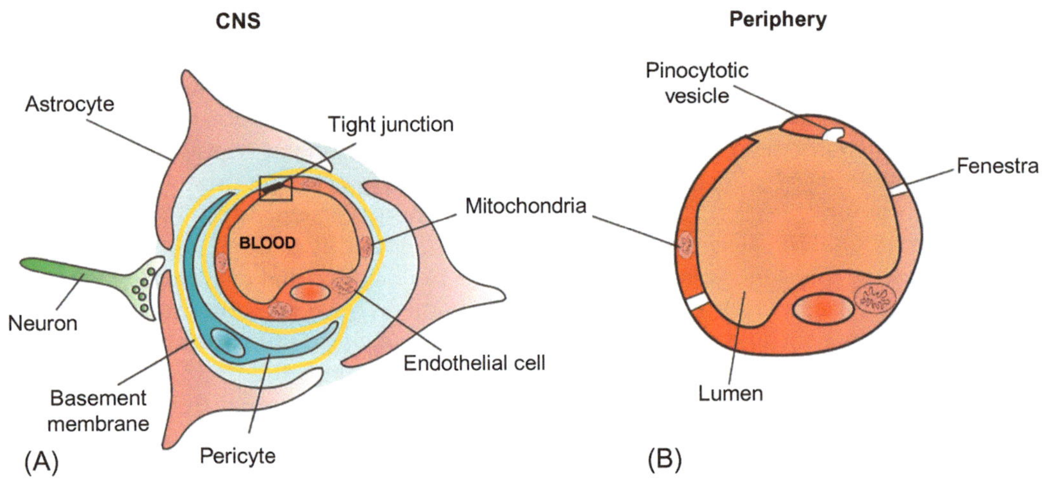

FIG. 1 Central nervous system (CNS) endothelial cells (ECs) vs. peripheral ECs. (A) CNS ECs are enveloped by pericytes, astrocytes, and a basement membrane in which bidirectional signaling of various signaling components enhances barrier integrity. CNS ECs contain more mitochondria for greater energy consumption, and restrict the movement of material from blood to brain and vice versa due to the presence of highly electrical-resistant tight junction (TJ) components located between adjacent ECs or, in capillaries, between opposing endings of the same EC. (B) ECs of the peripheral vasculature are characterized by the absence of TJs, fewer mitochondria, increased pinocytotic vesicles, and the presence of fenestra that allow the rapid exchange of material between the blood and the parenchyma.

THE NEUROVASCULAR UNIT

The neurovascular unit (NVU) is a mélange of cell types that interact and communicate with an intricate degree of cross talk to create a dynamic microenvironment (Fig. 2). Early transplantation experiments carried out by Stewart and Wiley showed that, after grafting of immature avascular brain tissue from embryonic quails into the coelom of chick embryos, the abdominal vessels vascularizing the grafted brain tissue formed structural and functional properties of the BBB, such as TJs and few pinocytotic vesicles. In contrast, when mesodermal tissue was transplanted into the brain, the capillaries in the grafts lacked barrier properties.[8] This pointed to yet-undefined cues from the neural microenvironment that were involved in developing barrier properties. Astrocytes are believed to be the inducing cell type.

Pericytes

Pericytes are specialized perivascular cells of mesoderm origin. They are embedded in the basement membrane (BM) that envelops blood capillaries. Pericytes are morphologically,

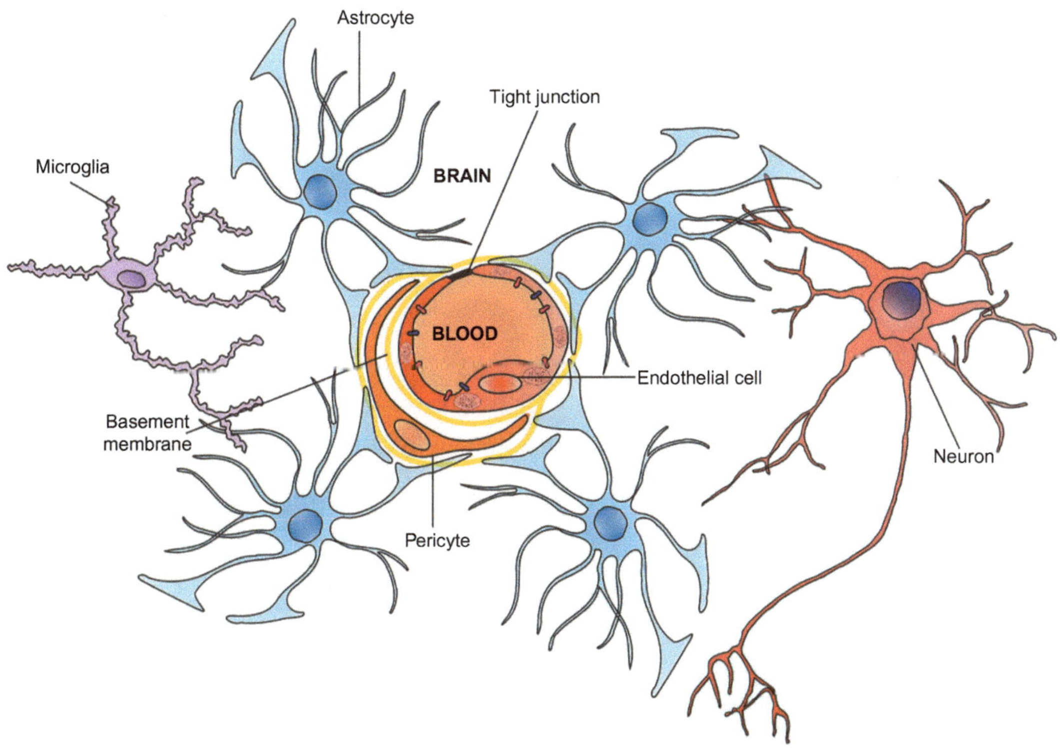

FIG. 2 The blood–brain barrier (BBB) and neurovascular unit (NVU). The NVU is an intricate milieu of endothelial cells, astrocytes, and pericytes that interacts with neurons, microglia, and other brain components to impart specific properties on the BBB. Pericytes partially surround the microvascular endothelium while astrocyte end-feet also surround the capillaries.

biochemically, and physiologically heterogeneous depending on vascular bed location, tissue type, and differentiation state. As a result, finding a pan-pericyte marker is extremely difficult.

Functionally, pericytes are key components of the BBB and the blood–retina barrier (BRB). They have an important role in regulating CBF and capillary diameter,[9] microvascular stabilization, and extracellular matrix protein secretion.[10] During early angiogenesis (driven primarily by vascular endothelial growth factor [VEGF] and Wnt/β-catenin signaling),[11-13] pericyte recruitment to cerebral blood vessels is key to BBB formation. Loss of pericyte coverage in platelet-derived growth factor or platelet-derived growth factor receptor-null mice has been shown to result in capillary microhemorrhages, TJ dysfunction, increased vascular permeability, failure to recruit pericyte precursor cells, and embryonic lethality.[11,14] Recruitment and attachment of pericytes are thought to be mediated by platelet-derived growth factor secretion from ECs and binding the platelet-derived growth factor receptor on pericytes.[15]

The barrier-promoting function of pericytes results from inhibition of molecules such as angiopoietin-2, PLVAP, and leukocyte adhesion molecules that promote vascular permeability and immune cell infiltration.[11] Pericytes are also critical to the integrity of the BBB during adulthood, as pericyte-deficient mice have less capillary coverage, reduced cerebral microcirculation, TJ dysfunction, and increased BBB permeability.[16,17] Armulik et al. also noted that pericytes guide astrocytic foot processes to the vessel wall via polarization of astrocyte end-feet and expression of cues that mediate attachment of astrocyte end-feet to the vessel wall.[17] Pericytes also express the contractile proteins α-smooth muscle actin, tropomyosin, and myosin[18]; furthermore, loss of pericyte coverage results in reduced CBF.[16] A recent study has highlighted the crucial role of pericytes in regulating CBF in which the neurotransmitter glutamate evokes the release of messengers, including prostaglandin E_2 and nitric oxide, which help dilate capillaries by actively relaxing pericytes.[19]

Neuroglia

Four principle glial cells reside in the brain: astrocytes, microglia, oligodendrocytes, and ependymal cells. Astrocytes are specialized glial cells derived from the developing neural tube. They protect neurons by regulating neurotransmitter levels and water and ion concentrations to maintain homeostasis of the neural microenvironment. Astrocytic end-feet envelop the abluminal surface of cerebral ECs.[2] These interactions are key in regulating brain water volume and synchronizing metabolite and ion levels with CBF and vasodilation in the adult brain. The most abundant water channel protein, aquaporin 4 (AQP4), is predominantly expressed in the end-feet surrounding cerebral vessels, and colocalizes with the inward rectifier potassium channel Kir4.1.[20]

Astrocytes maintain BBB integrity and express factors such as sonic hedgehog (Shh), which is known to upregulate claudin-5 and occludin levels in vitro.[21] Shh knockout mice exhibit embryonic lethality between E11 and E13.5. Although these mice showed normal vascular patterning, knockout mouse embryos had reduced levels of claudin-5 and occludin.[21] Additionally, conditional knockout of smoothened, a downstream signaling component of the Shh pathway, from ECs results in reduced TJ expression and increased extravasation of plasma proteins. Like pericytes, astrocytes express angiotensin-1,[22] which signals to Tie-2 receptors on ECs and leads to the development of more advanced TJs, inhibition of transcytosis, and downregulation of leukocyte adhesion molecules.[23] Taken together, these data suggest that astrocytes have key functions in maintaining cerebrovascular integrity. The finding that astrocytes are only present at the BBB postnatally in rats adds further weight to this argument.[11] Indeed, the cholesterol and phospholipid transporter molecule Apolipoprotein E (ApoE), which is produced by hepatocytes and astrocytes, signals through low-density lipoprotein receptor-related protein 1 on ECs to regulate TJ levels in the CNS. More recently, however, production and release of retinoic acid from radial glial cells (precursor astrocytes) have been shown to interact with retinoic acid receptor-β on developing ECs to induce barrier properties.[24]

Microglia are the resident immune cell of the CNS. They are derived from hematopoietic precursor cells that migrate from the embryonic yolk sac into the CNS.[25] In the developing brain, microglia are involved in engulfing and eliminating synapses in a process known as *synaptic pruning*.[26] Additionally, microglia secrete growth factors essential for neuronal survival.[27] Microglia contain highly motile processes and protrusions that constantly survey the neural microenvironment and interact with neurons, axons, and dendritic spines. Cumulative evidence shows that microglia are crucial regulators of synaptic plasticity,[28-30] neurogenesis,[31,32] learning, and memory.[29,33] As phagocytic cells, they survey the cerebral microenvironment and engulf and eliminate cellular debris and toxic proteins (i.e., amyloid plaques).

A hallmark characteristic of microglial response is the cells' ability to alter morphology. Classically, this altered morphology was associated with pathological transformation, but morphological changes only indicate that microglia have detected a change in homeostasis. In fact, transcriptome profiling of microglia in mice has shown that the phenotypic responses fail to conform to the M1 and M2 modes of activation.[34,35] Aberrations of normal microglial functions may contribute to disease processes in the CNS. In patients diagnosed with Alzheimer disease, microglia accumulate in senile plaques and failure of microglia to clear amyloid beta appears reduced, with disease progression with reduced expression of microglial amyloid beta phagocytic receptors in a mouse model of Alzheimer disease.[36,37] Genome-wide association studies have identified loci linked to Alzheimer disease that are expressed in microglia or myeloid cells. For example, individuals heterozygous for TREM2 variant R47H are at a significantly increased risk for Alzheimer disease.[38]

Basement Membrane

The often-overlooked component of the NVU is the acellular BM, yet it has a pivotal role in establishing and maintaining BBB properties (e.g., supporting pericytes and astrocytes). Structurally, BMs are a specialized layer of extracellular matrix proteins found basolateral to the endothelium and epithelium in all body tissues. Cells of the NVU synthesize and extracellularly secrete the proteins that form the BM. BMs are a heterogeneous mixture of proteins, the principal constituents of which are type IV collagen, laminin, nidogen/entactin, and perlecan[39] that form a layer approximately 20-200 nm in thickness. Astrocyte-specific deletion of laminin induces spontaneous hemorrhage in mice with impaired smooth muscle cell differentiation and loss of AQP4 and TJ proteins.[40] Furthermore, a subset of mice with a pericyte-specific deletion of laminin develop hydrocephalus, BBB breakdown, and loss of AQP4.[41] Today, BM components are routinely used in vitro in the culture of brain ECs.

Communication between cellular and acellular components of the NVU is critical for the health and integrity of the BBB, from embryogenesis into adulthood. Breakdown of this communication may contribute directly or indirectly to CNS disease pathogenesis. As will be discussed in the following sections, these cellular components of the NVU have important roles in the development of the BBB, as well as in forming transport routes across and between brain ECs. Disruption of these processes can result in impaired homeostasis and greater BBB permeability.

TRANSPORT ROUTES ACROSS THE BRAIN ENDOTHELIUM

Although oxygen and carbon dioxide can rapidly diffuse across the brain endothelium, only the smallest lipophilic molecules (i.e., <400 Da) that contain fewer than 8-10 hydrogen bonds can passively diffuse across the BBB (Fig. 3).[42] However, because the energy needs of the brain are met by the cerebral circulation, numerous protein transport systems are present on the luminal and abluminal surfaces of brain ECs to regulate the CNS entry and exit of molecules (Fig. 3).

Paracellular Pathway

The paracellular pathway comprises the inter-endothelial space between adjacent ECs positioned along cerebral blood vessels (Fig. 3A). The presence of TJ proteins limiting charge transfer in this space also restricts the flow of matter to all but the smallest solutes, ions, and lipid-soluble molecules. Many lipid-soluble molecules with molecular weights less than 400 Da can cross the BBB via diffusion through the endothelial lipid membranes (Fig. 3B). It has been estimated that all large molecule neurotherapeutics and up to 95% of small molecule drugs cannot circumvent the BBB.[43] Some substances can enter the brain through the paracellular pathway by deploying specific mechanisms. For example, leukocytes can gain entry to the brain in inflammatory situations. Leukocyte migration across the BBB is a multistep process that begins when leukocytes bind to P- and E-selectin on brain ECs to slow and bind to vascular cell adhesion molecule 1 and intercellular adhesion molecule 1 to attach to the ECs for subsequent transmigration (Fig. 3E).[44] Leukocyte adhesion molecules are typically expressed at low levels in the healthy brain microvasculature; however, they may become upregulated in inflammatory conditions.[44]

Transcellular Pathway

The transcellular pathway is the predominant pathway for the delivery of proteins, peptides, amino acids, ions, and carbohydrates to the brain. In addition, various therapeutic drugs have been delivered across the BBB by taking advantage of the molecular composition of various receptors expressed on the apical surface of brain ECs, such as the transferrin

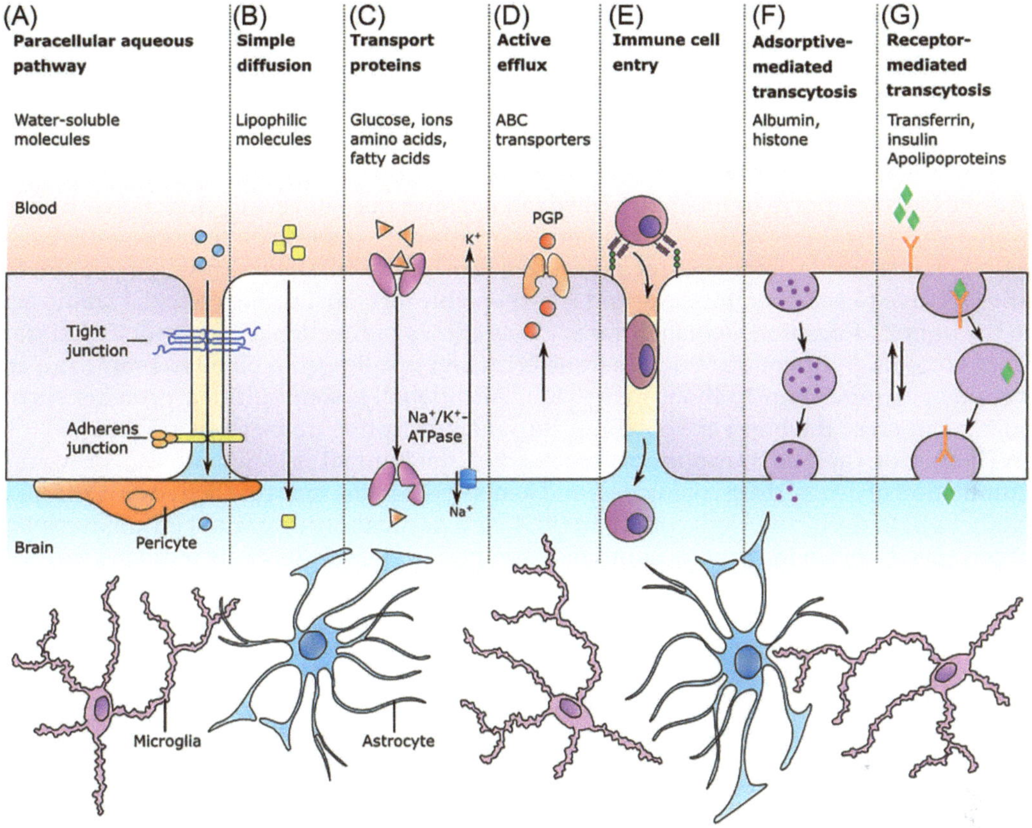

FIG. 3 Transport properties of the capillary endothelium. (A) Tight junction protein complexes seal the inter-endothelial space between neighboring endothelial cells and restrict the free movement of solutes. (B) O_2 and CO_2 cross the blood–brain barrier (BBB) by diffusion, as do lipophilic molecules with a molecular weight <400 Da and containing fewer than eight hydrogen bonds. (C) Protein transporters on the luminal surface of the endothelium facilitate the entry of glucose, amino acids, and nucleosides into the central nervous system. (D) ATP-binding cassette (ABC)-active efflux transporters limit entry of drugs and xenobiotics. (E) Immune cells bind to cell adhesion molecules such as P- and E-selectin to infiltrate the brain parenchyma via the paracellular or transcellular route. (F) Adsorptive-mediated transcytosis involves the cationization of plasma proteins such as albumin and subsequent binding of the positively charged protein to sites along the negatively charged plasma membrane. This induces internalization and transcytosis of the ligand across the cell within a vesicle. (G) Receptor-mediated transcytosis is used to transfer a variety of macromolecules, such as insulin, immunoglobulin G, and transferrin across the BBB. This process involves binding of a ligand to a receptor, endocytosis, and transport of the receptor-macromolecule complex within a vesicle, followed by dissociation of the complex and exocytosis of the ligand.

receptor. Ions and other small solutes can enter the brain via solute carriers on the apical and basal membrane, as can intracellular and extracellular enzymes. Ion regulation, which is critical for optimal synaptic signaling between neurons, is maintained by proteins such as the Na^+/K^+-ATPase pump on the abluminal membrane[45] and potassium channels (Fig. 3C). Carrier-mediated transport is the major route for the entry of essential nutrients such as glucose and amino acids. Glucose transporter 1, which is encoded by the SLC2A1 gene, is enriched in CNS ECs compared to peripheral tissues. Glucose transporter 1 facilitates the transport of glucose and has a profound impact on BBB and NVU integrity (Fig. 3C). Indeed, slc2a1$^{+/-}$ mice show extensive microvascular leakage to serum proteins as well as reduced TJ protein levels.[46]

The major facilitator superfamily domain containing 2a (mfsd2a) is specifically expressed in cerebral blood vessels. Mfsd2a is expressed during BBB formation and is the major route for docosahexaenoic uptake to the brain. Mfsd2a knockout mice have a dysfunctional BBB and elevated vesicular transcytosis.[47-49] Numerous transporters also facilitate amino acid entry to the CNS. For example, large neutral amino acid transporter is highly expressed in brain capillaries relative to peripheral tissues, and is responsible for transporting essential amino acids into the brain.[50] Receptor-mediated transcytosis allows bidirectional transendothelial transport of proteins and peptides, such as transferrin and insulin (from blood to brain) and apolipoproteins (from brain to blood) (Fig. 3G). Additionally, some plasma proteins such as albumin can cross the brain endothelium through adsorptive transcytosis (Fig. 3F).

ATP-binding cassette transporters expressed on the luminal side of brain ECs prevent the accumulation of drugs, drug conjugates, and xenobiotics in the brain via active efflux from the endothelium to the blood. P-glycoprotein (PGP) is a particularly important efflux transporter that protects the brain from many neurotoxic compounds by substantially reducing their CNS entry (Fig. 3D).[51] The functional importance of PGP at the BBB was investigated in mice deficient in Mdr1a and Mdr1b. Mdr1a knockout mice were found to have 10-fold increases in brain concentration of PGP substrates.[52]

JUNCTIONAL COMPLEXES OF THE BBB

Two major junctional complexes are present at the BBB: adherens junctions (AJs) and TJs. AJs are composed primarily of cadherin proteins that span the intercellular cleft and provide stability by linking to the cell cytoplasm by $\alpha/\beta/\gamma$ catenin proteins.[53] The precise role of AJs has yet to be resolved; however, it is thought that the molecular components play a key role in maintaining cellular polarity, providing stability, promoting EC survival, and responding to stimuli via interactions with cadherin proteins and the actin cytoskeleton. Evidence suggests that AJs are also essential for the formation of TJs. Unlike AJs, which are present in all vascular beds, TJs are enriched in the endothelium of the brain microvasculature. TJs appear as continuous, anastomosing, intramembranous networks of strands that interact with TJ proteins on the same cell or on adjacent ECs at so-called *kissing points* to eliminate the paracellular space[54] (Fig. 4). This fusion of TJs impedes the flow of solutes and ions from the blood to brain and vice versa, in turn creating a dynamic and highly regulatable barrier system.

Barrier properties at the BBB are conferred by highly electrical-resistant TJ proteins that limit the flux of all but the smallest molecules. TJs interact with TJs on adjacent ECs, as well as with intracellular scaffolding proteins that tether the TJs to cytoskeletal proteins. TJ protein complexes have two primary functions. First, TJ components significantly reduce BBB permeability to polar solutes and ions from the blood to brain and vice versa. This drastic reduction in the flow of ions across the BBB produces a high electrical resistance in vivo of roughly 1800 Ω cm^2.[55] Second, TJ components aid in the maintenance of cellular polarity

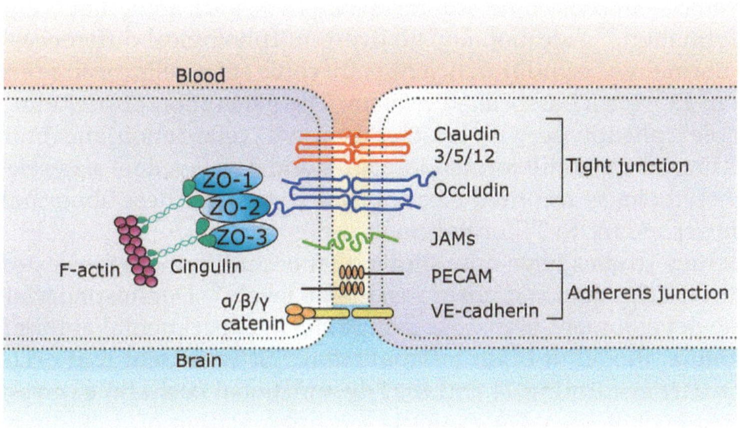

FIG. 4 Junctional complexes of the blood–brain barrier (BBB). Claudin-3, -5, and -12 and occludin are the major tight junction (TJ) components at the BBB, and are linked to the actin cytoskeleton via the Zonula Occludens family of scaffolding proteins and other intracellular proteins, such as cingulin. Other TJs, such as tricellulin and lipolysis-stimulated lipoprotein receptor, are enriched at three cell contacts (not shown here). Vascular endothelial-cadherin is a component of the adherens junction and is linked to the actin cytoskeleton by α/β/γ catenin proteins.

by preventing the lateral diffusion of membrane lipids and proteins between apical and basolateral compartments of ECs.[56] Although the BBB is a major obstacle for CNS entry, certain molecules and cells can use specific mechanisms to move between TJ complexes. For example, leukocytes can traverse the brain endothelium by binding to intracellular adhesion molecules (1 and 2) expressed on EC surfaces. This leads to transmigration across the ECs.[57]

The main TJ protein complexes are the claudins and occludin, which are enriched at two cell contacts. In addition, cytoskeletal scaffolding proteins called *zonula occludens* (ZO) interact with claudin and occludin on the intercellular domain of the plasma membrane to link the TJs to the actin cytoskeleton.[58-60] At points of three cell contacts, the TJ complexes tricellulin and lipolysis-stimulated lipoprotein receptor have been identified more recently as potential regulators of paracellular permeability.[61] The following TJs have been described in detail (Fig. 4).

Tight Junctions

Occludin

Occludin, part of the tetraspan family of integral membrane proteins, has four membrane-spanning domains and two extracellular loops. It is enriched at the TJ of epithelial cells and ECs.[62] The role of occludin was first elucidated after ectopic expression of chicken occludin in Sf9 insect cells, whereupon it induced the formation of TJ-like structures.[63] After this, it was found that a mutated occludin protein, introduced into Madin-Darby canine kidney (MDCK) cells, increased the paracellular leakage of MDCK cells to small tracers, implying a role in the barrier properties of TJs.[64] Subsequent investigations show that occludin-deficient embryonic stem cells could still form intact TJs, indicating that occludin is dispensable to barrier formation.[65] Additionally, no overt morphological differences were observed between TJs of normal vs occludin-deficient embryonic stem cells, or in normal localization and expression of ZO-1, a TJ-associated protein. The generation of occludin knockout mice revealed a complex phenotype with postnatal growth retardation and brain calcification. However, occludin null mice still formed intact TJs and displayed no size-selective loosening of the intestinal epithelia as recorded electrophysiologically. These abnormalities point to a role for occludin secondary to TJ formation.[66]

Indeed, numerous studies have now shown that occludin undergoes extensive modifications at the posttranscriptional and posttranslational levels.[67] Dephosphorylation of occludin occurs before the development of disease symptoms in experimental autoimmune encephalomyelitis, an animal model of brain inflammation.[68] It is evident that occludin has a role beyond that of a barrier-forming TJ and that dysfunctional occludin expression is involved in numerous neuropathologies. For example, phosphorylation of occludin at Ser-490 in response to VEGF leads to its ubiquitination and subsequent internalization. This process results in increased permeability to macromolecules and ions.[69] Furthermore, in a mouse model of neovascularization, occludin phosphorylation is required for VEGF-induced neovascularization and subsequent loss of BRB integrity.[70]

Claudins

Claudins are a multigene family of 20- to 24-kDa integral membrane proteins consisting of four membrane-spanning domains with a short N terminus, two extracellular loops, and a cytoplasmic tail. In all, 24 claudin proteins have been identified in humans,[71] with sequence analysis separating the family into two groups based on their sequence similarity and proposed function. Group one contains the classic claudins (i.e., 1-10, 14, 15, 17, 19) and group two contains the nonclassic claudins (i.e., 11-13, 16, 18, 20-24). Claudins are similar in structure to occludin despite having minimal sequence homology. Claudins are expressed in numerous tissues with claudins 3, 5, and 12 expressed in the brain endothelium; claudin-5 is the most enriched of these three.[72]

Claudins are the major component of the TJ, and a general role of claudins is the paracellular sealing function they perform, which limits the paracellular movement of material. The first extracellular domain (ECD) of claudins is known to be vital for the barrier properties of the TJs. Mutations to conserved cysteine residues in ECD1 of claudin-5 in MDCK cells results in increased paracellular permeability to mannitol and monosaccharides.[73] Claudins associate with claudin species on adjacent cells as well as form cis interactions on the same cell via their ECDs, providing structural integrity.[74,75] Claudins are a major structural component of the TJ and form the backbone of TJs through homotypic and heterotypic interactions via their ECDs.[74-77]

The spatial organization of claudin TJ strands is determined by the ZO scaffolding proteins. Most claudin species contain a C terminus PDZ-binding motif that can bind to PDZ motifs on the ZO proteins,[78] linking them to the actin cytoskeleton. In vitro models of the BBB as well as genetic mouse models have improved our understanding of the physiological roles of the claudin proteins. Claudin-1-deficient mice die within 1 day of birth from excessive skin dehydration, and were shown to have impaired barrier functions at the epidermis with increased permeation of a 600-Da tracer molecule.[79] Following transfection into TJ-free MDCK cells, claudin-5 formed stable TJ networks concurrent with a selective decrease in permeability to ions.[73] Claudin-5-deficient mice have an impaired BBB; electron microscopy reveals intact TJs at cell-to-cell contacts. Tracer molecule experiments revealed increased permeation of molecules up to approximately 800 Da in size. These mice also die within hours of birth from undefined causes.[80] The claudin proteins are expressed abundantly in numerous tissues and are central to the barrier function of tissues, such as the intestinal epithelia, inner ear, BBB, BRB, and blood–testis barrier. It is apparent that the claudins have an intrinsic role in regulating permeability. Manipulation of TJ components is a promising approach in improving drug delivery to the CNS, and may ultimately facilitate the development of novel therapeutic strategies for disorders of the CNS.[81]

Adherens Junctions

Like TJs, AJs associate with the actin cytoskeleton and are involved in the initiation and stabilization of cell-to-cell adhesion, regulation of the actin cytoskeleton, intracellular signaling, and transcriptional regulation. AJs are composed primarily of cadherin proteins (e.g., E-cadherin), which span the intercellular cleft and provide stability by initiating intercellular contacts through trans-pairing between cadherins on adjacent cells. Cadherins can also link directly to cell cytoplasm proteins, and the interactions between cadherins and catenins play a crucial role in the formation and function of AJs.[82] Catenin family members, including p120 catenin and β-catenin, can bind to classical cadherins to form a cadherin–catenin core complex that can subsequently bind to the F-actin cytoskeleton through α-catenin—a mechanism that is crucial for firm cell adhesion.[83] The molecular components of AJs have a key role in maintaining cellular polarity, providing stability, promoting EC survival, and responding to stimuli via interactions with cadherin proteins and the actin cytoskeleton.

AJs are involved in the regulation of endothelial permeability through dynamic opening and closing of cell-to-cell AJs as a result of the phosphorylation of vascular endothelial (VE)-cadherin and subsequent internalization.[84] Evidence also suggests that AJs are essential for the formation of TJs. VE-cadherin is an EC-specific AJ component that mediates the upregulation of claudin-5 through Akt-dependent phosphorylation of the forkhead box factor Fox01, which inhibits β-catenin translocation to the nucleus and represses claudin-5 transcription.[85]

Junctional Adhesion Molecules

Junctional adhesion molecules (JAMs), like TJ proteins, are integral membrane proteins that belong to the immunoglobulin superfamily. They consist of a single membrane-spanning domain, an ECD with an N terminus and a short cytoplasmic C terminus.[86] JAMs interact with scaffolding proteins such as ZO-1 and cingulin via a PDZ motif on the cytoplasmic C terminus.[87,88] JAMs also interact with JAM family members, as well as with other adhesion molecules on opposing ECs.[89] A function of JAM proteins is to promote cell polarity and localization of ZO-1 and occludin at cell contacts. This is achieved by JAM binding to Par3.[90] Emerging evidence has suggested a role for JAM family members in leukocyte migration across EC layers.[91,92]

Scaffolding Proteins

The membrane-associated guanylate-kinase protein family contains accessory elements for the transmembrane components of TJs. The ZO proteins (i.e., ZO-1, ZO-2, and ZO-3) are members of the membrane-associated guanylate-kinase protein family and link TJs to the actin cytoskeleton. ZO-1 is a 220-kDa protein essential for endothelial barrier formation, VE-cadherin-mediated cell tension, and actomyosin organization through its interaction with F-actin.[93] Mice deficient in ZO-1 are embryonic lethal by E10.5 with embryonic and extraembryonic defects, including impaired angiogenesis and increased apoptosis in the neural tube and notochord.[94] ZO-2 is a 160-kDa protein that is also critical for embryonic development with mice deficient for ZO-2, being embryonic lethal due to a loss of cell proliferation and induction of apoptosis between E6.5 and E7.5.[95] In contrast, mice deficient in ZO-3, a 130-kDa protein, developed normally and had no apparent phenotype.[95,96] The ZO proteins contain a PDZ motif on the C terminus to link ZO proteins with transmembrane proteins or with PDZ motifs on other proteins. ZO-1 binds to the claudins, occludin, and JAM via PDZ motifs as well as with various components of the cytoskeleton.[87,88,93] Knockdown of ZO-1 in MDCK cells delays TJ assembly, whereas knockout of ZO-1 in MDCK cells by TALEN-mediated gene targeting results in alterations in myosin organization at cell-to-cell contacts, as well as disruption of the localization of TJ proteins.[97,98]

Cingulin is a cytoplasmic protein localized on the cytoplasmic face of TJs.[99] It is a 140-kDa protein that interacts with several TJ species. In vitro studies have identified ZO-1, ZO-2, ZO-3, myosin, and AF-6 interacting with the N terminus of cingulin.[100] Investigations in MDCK cells have shown that cingulin is not involved in the basic structure or the function of TJs.[101] Subsequent work demonstrated that cingulin is involved in the regulation of cell proliferation and gene expression through RhoA signaling.[102]

OPENING THE BBB FOR THERAPEUTIC CONSIDERATIONS

Numerous studies have shown that it is possible to specifically target small interfering RNA (siRNA) to brain capillary ECs for targeted suppression of BBB-specific components. Delivery of siRNA to mouse brain capillary ECs in vivo could efficiently suppress the organic ion transporter 3 and reduce brain-to-blood transport of benzyl penicillin.[103] More recently, RNA interface (RNAi)-based suppression of claudin proteins has been used in numerous preclinical models of neurological disorders to improve drug penetration to the brain or to remove neurotoxic metabolites from the brain, such as amyloid-β (Aβ). Using systemically injected siRNA targeting claudin-5, Campbell et al. demonstrated the first RNAi-based modulation of the BBB in mice.[104] Levels of claudin-5 mRNA were reduced between 24 and 48 hours post-siRNA injection, with maximum suppression of claudin-5 occurring at 48 hours postinjection compared to non-targeting siRNA-injected mice. This facilitated a size-selective loosening of the BBB to molecules up to 1 kDa in size, including the contrast agent gadolinium (Gd-DTPA, 742 Da), while restricting a molecule 4.4 kDa in size. Importantly, this process was also reversible, with levels of claudin-5 returning to normal 72 hours post-siRNA injection, with BBB integrity also being restored.[104]

This technique was also adapted to successfully modulate the molecularly homologous inner blood–retina barrier (iBRB) in animal models of retinopathies.[105] Using this approach, it was possible to improve visual function in mouse models of retinitis pigmentosa (RP) and a light-induced retinal degeneration model. IMPDH knockout mice are a model of autosomal recessive RP. These mice lack an enzyme involved in the de novo synthesis of guanosine triphosphate (molecular weight [MW]: 523 Da), which is essential for visual transduction. Through targeted suppression of claudin-5 in the neural retina, systemic injection of guanosine triphosphate could bypass the iBRB and improve retinal function. Similarly, in BalB/c mice with light-induced retinopathy, systemic injection of the calpain inhibitor *N*-acetyl-Leu-Leu-Met-CHO (MW: 401 Da) could readily diffuse across the modulated iBRB and reduce photoreceptor cell death, a hallmark of light-induced retinopathy.

To adapt this process for the treatment of chronic diseases such as RP and age-related macular degeneration, Campbell and colleagues developed a doxycycline-inducible small hairpin RNA (shRNA) system for the transient knockdown of claudin-5 levels. The doxycycline-inducible shRNA sequence was inserted into the genome of an adeno-associated virus-2/9 vector that can persist in retinal or brain ECs after a single localized injection. This approach achieved therapeutic benefit in a laser-induced model of choroidal neovascularization, a hallmark of wet age-related macular degeneration. With this approach, it was possible to systemically deliver 17-AAG (MW: 585 Da) and Sunitinib malate (MW: 532 Da), two well-characterized VEGF inhibitors, across the BRB[106] and significantly reduce the volume of choroidal neovascularization lesions compared to the NT shRNA-treated contralateral eye. Importantly, this approach specifically downregulated claudin-5 at the BBB/BRB, whereas expression patterns of other TJs remained at normal physiological levels. It is also possible to modulate TJs in specific brain regions (e.g., by direct stereotaxic injection of adeno-associated virus vectors).

In the cold-induced model of traumatic brain injury, application of siRNAs targeting claudin-5 can transiently open the BBB, reduce cerebral edema, and improve neurological function.[107] Claudin-3, which is also expressed in brain capillary ECs, is overexpressed in 90% of ovarian cancers. RNAi suppression of claudin-3 reduces tumor cell proliferation and growth, tumor burden, and metastasis in several ovarian cancer models.[108] RNAi has provided a robust method for the transient and reversible modulation of the BBB. More recently, sequential delivery of siRNAs targeting claudin-5 and occludin transcripts has been shown to suppress both proteins and modulate the BBB to molecules up to 3 kDa in size. In Tg2576 mice, a murine model of familial Alzheimer disease, suppression of claudin-5 and occludin improved cognitive function in tandem with reduced brain levels of Aβ(1-40) and increased serum levels of Aβ(1-40), indicating that it is possible to remove pathogenic agents from the brain to the blood.[109] In summary, targeted suppression of TJs at the BBB/BRB increases paracellular permeability and enhances targeted drug delivery to neuronal regions. Through use of an inducible system, it is also possible to reverse BBB permeability by withdrawing the inducing agent.

In addition to these genetic methods of opening the BBB, various chemical agents have also been used to deliver drugs to the brain, with some success.[81] Intra-arterial administration of hyperosmotic mannitol (1.6 M) can induce BBB dysfunction (measured by contrast-enhanced computed tomography) to improve the CNS penetration of chemotherapies. Further methods have also been developed that specifically target TJs through the use of peptides.

Drug enhancer peptides have been designed based on the first extracellular loop of claudin-5. Data from in vitro investigations revealed that these peptides could drastically increase the permeability of brain ECs by downregulating claudin-5 and promoting redistribution away from cell-to-cell contacts. Subsequent in vivo studies revealed enhanced brain penetration of the contrast agent gadolinium within 4 hours of peptide treatment.[110]

Additional peptides targeting other TJ structures have also been described. A synthetic peptide derived from the second extracellular loop of tricellulin has been demonstrated to increase ion permeability in vitro in human epithelial colorectal adenocarcinoma cells, as well as increase leakage of small and large tracer molecules up to 10 kDa in size.[111] Although significant attention has been devoted to the therapeutic potential of modulation of the BBB in the treatment of neurological disorders, few studies have assessed the neurobehavioral consequences of specifically targeting BBB EC components to modulate BBB permeability. Recently we have shown that chronic suppression of claudin-5 induces numerous behavioral impairments in mice, including deficits in sensorimotor gating, analogous to those observed in psychiatric disorders.[112] Additionally, loss of claudin-5 in the nucleus accumbens has been shown to induce various depressive-like behaviors in a chronic social defeat model of depression.[113]

This chapter sought to establish the fundamental aspects of the BBB and enumerate strategies to bypass and modulate this obstacle to therapeutic intervention. The following chapter will discuss strategies and methods for modeling the BBB in physiological and pathological conditions.

References

1. Raichle ME, Gusnard DA. Appraising the brain's energy budget. *Proc Natl Acad Sci U S A*. 2002; 99(16):10237-10239.
2. Abbott NJ, Ronnback L, Hansson E. Astrocyte-endothelial interactions at the blood-brain barrier. *Nat Rev Neurosci*. 2006;7(1):41-53.
3. Abbott NJ, Patabendige AA, Dolman DE, Yusof SR, Begley DJ. Structure and function of the blood-brain barrier. *Neurobiol Dis*. 2010;37(1):13-25.
4. Hladky SB, Barrand MA. Fluid and ion transfer across the blood-brain and blood-cerebrospinal fluid barriers; a comparative account of mechanisms and roles. *Fluids Barriers CNS*. 2016;13(1):19.
5. Pardridge WM. Targeted delivery of protein and gene medicines through the blood-brain barrier. *Clin Pharmacol Ther*. 2015;97(4):347-361.
6. Schlageter KE, Molnar P, Lapin GD, Groothuis DR. Microvessel organization and structure in experimental brain tumors: microvessel populations with distinctive structural and functional properties. *Microvasc Res*. 1999; 58(3):312-328.
7. Saunders NR, Liddelow SA, Dziegielewska KM. Barrier mechanisms in the developing brain. *Front Pharmacol*. 2012;3:46.
8. Stewart PA, Wiley MJ. Developing nervous tissue induces formation of blood-brain barrier characteristics in invading endothelial cells: a study using quail-chick transplantation chimeras. *Dev Biol*. 1981;84(1):183-192.
9. Peppiatt CM, Howarth C, Mobbs P, Attwell D. Bidirectional control of CNS capillary diameter by pericytes. *Nature*. 2006;443(7112):700-704.
10. Winkler EA, Bell RD, Zlokovic BV. Central nervous system pericytes in health and disease. *Nat Neurosci*. 2011; 14(11):1398-1405.
11. Daneman R, Zhou L, Kebede AA, Barres BA. Pericytes are required for blood-brain barrier integrity during embryogenesis. *Nature*. 2010;468(7323):562-566.
12. Liebner S, Corada M, Bangsow T, et al. Wnt/beta-catenin signaling controls development of the blood-brain barrier. *J Cell Biol*. 2008;183(3):409-417.

13. Obermeier B, Daneman R, Ransohoff RM. Development, maintenance and disruption of the blood-brain barrier. *Nat Med*. 2013;19(12):1584-1596.
14. Lindahl P, Johansson BR, Leveen P, Betsholtz C. Pericyte loss and microaneurysm formation in PDGF-B-deficient mice. *Science*. 1997;277(5323):242-245.
15. Hellstrom M, Kalen M, Lindahl P, Abramsson A, Betsholtz C. Role of PDGF-B and PDGFR-beta in recruitment of vascular smooth muscle cells and pericytes during embryonic blood vessel formation in the mouse. *Development*. 1999;126(14):3047-3055.
16. Bell RD, Winkler EA, Sagare AP, et al. Pericytes control key neurovascular functions and neuronal phenotype in the adult brain and during brain aging. *Neuron*. 2010;68(3):409-427.
17. Armulik A, Genove G, Mae M, et al. Pericytes regulate the blood-brain barrier. *Nature*. 2010;468(7323):557-561.
18. Bandopadhyay R, Orte C, Lawrenson JG, Reid AR, De Silva S, Allt G. Contractile proteins in pericytes at the blood-brain and blood-retinal barriers. *J Neurocytol*. 2001;30(1):35-44.
19. Hall CN, Reynell C, Gesslein B, et al. Capillary pericytes regulate cerebral blood flow in health and disease. *Nature*. 2014;508(7494):55-60.
20. Higashi K, Fujita A, Inanobe A, et al. An inwardly rectifying K(+) channel, Kir4.1, expressed in astrocytes surrounds synapses and blood vessels in brain. *Am J Phys Cell Phys*. 2001;281(3):C922-C931.
21. Alvarez JI, Dodelet-Devillers A, Kebir H, et al. The Hedgehog pathway promotes blood-brain barrier integrity and CNS immune quiescence. *Science*. 2011;334(6063):1727-1731.
22. Milsted A, Barna BP, Ransohoff RM, Brosnihan KB, Ferrario CM. Astrocyte cultures derived from human brain tissue express angiotensinogen mRNA. *Proc Natl Acad Sci U S A*. 1990;87(15):5720-5723.
23. Lee SW, Kim WJ, Choi YK, et al. SSeCKS regulates angiogenesis and tight junction formation in blood-brain barrier. *Nat Med*. 2003;9(7):900-906.
24. Mizee MR, Wooldrik D, Lakeman KA, et al. Retinoic acid induces blood-brain barrier development. *J Neurosci*. 2013;33(4):1660-1671.
25. Ginhoux F, Greter M, Leboeuf M, et al. Fate mapping analysis reveals that adult microglia derive from primitive macrophages. *Science*. 2010;330(6005):841-845.
26. Paolicelli RC, Bolasco G, Pagani F, et al. Synaptic pruning by microglia is necessary for normal brain development. *Science*. 2011;333(6048):1456-1458.
27. Ueno M, Fujita Y, Tanaka T, et al. Layer V cortical neurons require microglial support for survival during postnatal development. *Nat Neurosci*. 2013;16(5):543-551.
28. George J, Cunha RA, Mulle C, Amedee T. Microglia-derived purines modulate mossy fibre synaptic transmission and plasticity through P2X4 and A1 receptors. *Eur J Neurosci*. 2016;43(10):1366-1378.
29. Rogers JT, Morganti JM, Bachstetter AD, et al. CX3CR1 deficiency leads to impairment of hippocampal cognitive function and synaptic plasticity. *J Neurosci*. 2011;31(45):16241-16250.
30. Sipe GO, Lowery RL, Tremblay ME, Kelly EA, Lamantia CE, Majewska AK. Microglial P2Y12 is necessary for synaptic plasticity in mouse visual cortex. *Nat Commun*. 2016;7:10905.
31. Bachstetter AD, Morganti JM, Jernberg J, et al. Fractalkine and CX 3 CR1 regulate hippocampal neurogenesis in adult and aged rats. *Neurobiol Aging*. 2011;32(11):2030-2044.
32. Sierra A, Encinas JM, Deudero JJ, et al. Microglia shape adult hippocampal neurogenesis through apoptosis-coupled phagocytosis. *Cell Stem Cell*. 2010;7(4):483-495.
33. Parkhurst CN, Yang G, Ninan I, et al. Microglia promote learning-dependent synapse formation through brain-derived neurotrophic factor. *Cell*. 2013;155(7):1596-1609.
34. Holtman IR, Raj DD, Miller JA, et al. Induction of a common microglia gene expression signature by aging and neurodegenerative conditions: a co-expression meta-analysis. *Acta Neuropathol Commun*. 2015;3:31.
35. Ransohoff RM. A polarizing question: do M1 and M2 microglia exist? *Nat Neurosci*. 2016;19(8):987-991.
36. Hickman SE, Allison EK, Khoury JE. Microglial dysfunction and defective β-amyloid clearance pathways in aging Alzheimer's disease mice. *J Neurosci*. 2008;28(33):8354-8360.
37. Ransohoff RM, Khoury JE. Microglia in Health and Disease. *Cold Spring Harb Perspect Biol*. 2016;8(1):a020560.
38. Colonna M, Wang Y. TREM2 variants: new keys to decipher Alzheimer disease pathogenesis. *Nat Rev Neurosci*. 2016;17(4):201-207.
39. LeBleu VS, Macdonald B, Kalluri R. Structure and function of basement membranes. *Exp Biol Med (Maywood)*. 2007;232(9):1121-1129.
40. Chen ZL, Yao Y, Norris EH, et al. Ablation of astrocytic laminin impairs vascular smooth muscle cell function and leads to hemorrhagic stroke. *J Cell Biol*. 2013;202(2):381-395.

41. Gautam J, Zhang X, Yao Y. The role of pericytic laminin in blood brain barrier integrity maintenance. *Sci Rep*. 2016;6:36450.
42. Pardridge WM. Alzheimer's disease drug development and the problem of the blood-brain barrier. *Alzheimers Dement*. 2009;5(5):427-432.
43. Pardridge WM. The blood-brain barrier: bottleneck in brain drug development. *NeuroRx*. 2005;2(1):3-14.
44. Takeshita Y, Ransohoff RM. Inflammatory cell trafficking across the blood-brain barrier: chemokine regulation and in vitro models. *Immunol Rev*. 2012;248(1):228-239.
45. Hawkins BT, Davis TP. The blood-brain barrier/neurovascular unit in health and disease. *Pharmacol Rev*. 2005; 57(2):173-185.
46. Winkler EA, Nishida Y, Sagare AP, et al. GLUT1 reductions exacerbate Alzheimer's disease vasculo-neuronal dysfunction and degeneration. *Nat Neurosci*. 2015;18(4):521-530.
47. Andreone BJ, Chow BW, Tata A, et al. Blood-brain barrier permeability is regulated by lipid transport-dependent suppression of caveolae-mediated transcytosis. *Neuron*. 2017;94(3). 581-594.e585.
48. Ben-Zvi A, Lacoste B, Kur E, et al. Mfsd2a is critical for the formation and function of the blood-brain barrier. *Nature*. 2014;509(7501):507-511.
49. Nguyen LN, Ma D, Shui G, et al. Mfsd2a is a transporter for the essential omega-3 fatty acid docosahexaenoic acid. *Nature*. 2014;509(7501):503-506.
50. Verrey F, Closs EI, Wagner CA, Palacin M, Endou H, Kanai Y. CATs and HATs: the SLC7 family of amino acid transporters. *Pflugers Arch*. 2004;447(5):532-542.
51. Lin JH, Yamazaki M. Role of P-glycoprotein in pharmacokinetics: clinical implications. *Clin Pharmacokinet*. 2003;42(1):59-98.
52. Schinkel AH, Smit JJ, van Tellingen O, et al. Disruption of the mouse mdr1a P-glycoprotein gene leads to a deficiency in the blood-brain barrier and to increased sensitivity to drugs. *Cell*. 1994;77(4):491-502.
53. Wolburg H, Lippoldt A. Tight junctions of the blood-brain barrier: development, composition and regulation. *Vasc Pharmacol*. 2002;38(6):323-337.
54. Tsukita S, Furuse M, Itoh M. Multifunctional strands in tight junctions. *Nat Rev Mol Cell Biol*. 2001;2(4):285-293.
55. Butt AM, Jones HC, Abbott NJ. Electrical resistance across the blood-brain barrier in anaesthetized rats: a developmental study. *J Physiol*. 1990;429:47-62.
56. van Meer G, Simons K. The function of tight junctions in maintaining differences in lipid composition between the apical and the basolateral cell surface domains of MDCK cells. *EMBO J*. 1986;5(7):1455-1464.
57. Steiner O, Coisne C, Cecchelli R, et al. Differential roles for endothelial ICAM-1, ICAM-2, and VCAM-1 in shear-resistant T cell arrest, polarization, and directed crawling on blood-brain barrier endothelium. *J Immunol*. 2010;185(8):4846-4855.
58. Fanning AS, Jameson BJ, Jesaitis LA, Anderson JM. The tight junction protein ZO-1 establishes a link between the transmembrane protein occludin and the actin cytoskeleton. *J Biol Chem*. 1998;273(45):29745-29753.
59. Fanning AS, Little BP, Rahner C, Utepbergenov D, Walther Z, Anderson JM. The unique-5 and -6 motifs of ZO-1 regulate tight junction strand localization and scaffolding properties. *Mol Biol Cell*. 2007;18(3):721-731.
60. Hartsock A, Nelson WJ. Adherens and tight junctions: structure, function and connections to the actin cytoskeleton. *Biochim Biophys Acta*. 2008;1778(3):660-669.
61. Ikenouchi J, Furuse M, Furuse K, Sasaki H, Tsukita S, Tsukita S. Tricellulin constitutes a novel barrier at tricellular contacts of epithelial cells. *J Cell Biol*. 2005;171(6):939-945.
62. Furuse M, Hirase T, Itoh M, et al. Occludin: a novel integral membrane protein localizing at tight junctions. *J Cell Biol*. 1993;123(6 Pt 2):1777-1788.
63. Furuse M, Fujimoto K, Sato N, Hirase T, Tsukita S, Tsukita S. Overexpression of occludin, a tight junction-associated integral membrane protein, induces the formation of intracellular multilamellar bodies bearing tight junction-like structures. *J Cell Sci*. 1996;109(Pt 2):429-435.
64. Balda MS, Whitney JA, Flores C, Gonzalez S, Cereijido M, Matter K. Functional dissociation of paracellular permeability and transepithelial electrical resistance and disruption of the apical-basolateral intramembrane diffusion barrier by expression of a mutant tight junction membrane protein. *J Cell Biol*. 1996;134(4):1031-1049.
65. Saitou M, Fujimoto K, Doi Y, et al. Occludin-deficient embryonic stem cells can differentiate into polarized epithelial cells bearing tight junctions. *J Cell Biol*. 1998;141(2):397-408.
66. Saitou M, Furuse M, Sasaki H, et al. Complex phenotype of mice lacking occludin, a component of tight junction strands. *Mol Biol Cell*. 2000;11(12):4131-4142.
67. Cummins PM. Occludin: one protein, many forms. *Mol Cell Biol*. 2012;32(2):242-250.

68. Morgan L, Shah B, Rivers LE, et al. Inflammation and dephosphorylation of the tight junction protein occludin in an experimental model of multiple sclerosis. *Neuroscience*. 2007;147(3):664-673.
69. Murakami T, Felinski EA, Antonetti DA. Occludin phosphorylation and ubiquitination regulate tight junction trafficking and vascular endothelial growth factor-induced permeability. *J Biol Chem*. 2009;284(31):21036-21046.
70. Liu X, Dreffs A, Díaz-Coránguez M, et al. Occludin S490 phosphorylation regulates vascular endothelial growth factor-induced retinal neovascularization. *Am J Pathol*. 2016;186(9):2486-2499.
71. Morita K, Furuse M, Fujimoto K, Tsukita S. Claudin multigene family encoding four-transmembrane domain protein components of tight junction strands. *Proc Natl Acad Sci U S A*. 1999;96(2):511-516.
72. Daneman R, Zhou L, Agalliu D, Cahoy JD, Kaushal A, Barres BA. The mouse blood-brain barrier transcriptome: a new resource for understanding the development and function of brain endothelial cells. *PLoS One*. 2010;5(10).
73. Wen H, Watry DD, Marcondes MC, Fox HS. Selective decrease in paracellular conductance of tight junctions: role of the first extracellular domain of claudin-5. *Mol Cell Biol*. 2004;24(19):8408-8417.
74. Blasig IE, Winkler L, Lassowski B, et al. On the self-association potential of transmembrane tight junction proteins. *Cell Mol Life Sci*. 2006;63(4):505-514.
75. Furuse M, Sasaki H, Tsukita S. Manner of interaction of heterogeneous claudin species within and between tight junction strands. *J Cell Biol*. 1999;147(4):891-903.
76. Krause G, Winkler L, Mueller SL, Haseloff RF, Piontek J, Blasig IE. Structure and function of claudins. *Biochim Biophys Acta*. 2008;1778(3):631-645.
77. Krause G, Winkler L, Piehl C, Blasig I, Piontek J, Muller SL. Structure and function of extracellular claudin domains. *Ann N Y Acad Sci*. 2009;1165:34-43.
78. Itoh M, Furuse M, Morita K, Kubota K, Saitou M, Tsukita S. Direct binding of three tight junction-associated MAGUKs, ZO-1, ZO-2, and ZO-3, with the COOH termini of claudins. *J Cell Biol*. 1999;147(6):1351-1363.
79. Furuse M, Hata M, Furuse K, et al. Claudin-based tight junctions are crucial for the mammalian epidermal barrier: a lesson from claudin-1-deficient mice. *J Cell Biol*. 2002;156(6):1099-1111.
80. Nitta T, Hata M, Gotoh S, et al. Size-selective loosening of the blood-brain barrier in claudin-5-deficient mice. *J Cell Biol*. 2003;161(3):653-660.
81. Greene C, Campbell M. Tight junction modulation of the blood brain barrier: CNS delivery of small molecules. *Tissue Barriers*. 2016;4(1).
82. Takeichi M. Dynamic contacts: rearranging adherens junctions to drive epithelial remodelling. *Nat Rev Mol Cell Biol*. 2014;15(6):397-410.
83. Hansen SD, Kwiatkowski AV, Ouyang CY, et al. alphaE-catenin actin-binding domain alters actin filament conformation and regulates binding of nucleation and disassembly factors. *Mol Biol Cell*. 2013;24(23):3710-3720.
84. Dejana E, Orsenigo F, Lampugnani MG. The role of adherens junctions and VE-cadherin in the control of vascular permeability. *J Cell Sci*. 2008;121(Pt 13):2115-2122.
85. Taddei A, Giampietro C, Conti A, et al. Endothelial adherens junctions control tight junctions by VE-cadherin-mediated upregulation of claudin-5. *Nat Cell Biol*. 2008;10(8):923-934.
86. Martin-Padura I, Lostaglio S, Schneemann M, et al. Junctional adhesion molecule, a novel member of the immunoglobulin superfamily that distributes at intercellular junctions and modulates monocyte transmigration. *J Cell Biol*. 1998;142(1):117-127.
87. Bazzoni G, Martinez-Estrada OM, Orsenigo F, Cordenonsi M, Citi S, Dejana E. Interaction of junctional adhesion molecule with the tight junction components ZO-1, cingulin, and occludin. *J Biol Chem*. 2000;275(27):20520-20526.
88. Ebnet K, Schulz CU, Meyer Zu Brickwedde MK, Pendl GG, Vestweber D. Junctional adhesion molecule interacts with the PDZ domain-containing proteins AF-6 and ZO-1. *J Biol Chem*. 2000;275(36):27979-27988.
89. Weber C, Fraemohs L, Dejana E. The role of junctional adhesion molecules in vascular inflammation. *Nat Rev Immunol*. 2007;7(6):467-477.
90. Itoh M, Sasaki H, Furuse M, Ozaki H, Kita T, Tsukita S. Junctional adhesion molecule (JAM) binds to PAR-3: a possible mechanism for the recruitment of PAR-3 to tight junctions. *J Cell Biol*. 2001;154(3):491-497.
91. Aurrand-Lions M, Lamagna C, Dangerfield JP, et al. Junctional adhesion molecule-C regulates the early influx of leukocytes into tissues during inflammation. *J Immunol*. 2005;174(10):6406-6415.
92. Bradfield PF, Nourshargh S, Aurrand-Lions M, Imhof BA. JAM family and related proteins in leukocyte migration (Vestweber series). *Arterioscler Thromb Vasc Biol*. 2007;27(10):2104-2112.
93. Tornavaca O, Chia M, Dufton N, et al. ZO-1 controls endothelial adherens junctions, cell-cell tension, angiogenesis, and barrier formation. *J Cell Biol*. 2015;208(6):821-838.

94. Katsuno T, Umeda K, Matsui T, et al. Deficiency of zonula occludens-1 causes embryonic lethal phenotype associated with defected yolk sac angiogenesis and apoptosis of embryonic cells. *Mol Biol Cell.* 2008; 19(6):2465-2475.
95. Xu J, Kausalya PJ, Phua DC, Ali SM, Hossain Z, Hunziker W. Early embryonic lethality of mice lacking ZO-2, but Not ZO-3, reveals critical and nonredundant roles for individual zonula occludens proteins in mammalian development. *Mol Cell Biol.* 2008;28(5):1669-1678.
96. Adachi M, Inoko A, Hata M, et al. Normal establishment of epithelial tight junctions in mice and cultured cells lacking expression of ZO-3, a tight-junction MAGUK protein. *Mol Cell Biol.* 2006;26(23):9003-9015.
97. Tokuda S, Higashi T, Furuse M. ZO-1 knockout by TALEN-mediated gene targeting in MDCK cells: involvement of ZO-1 in the regulation of cytoskeleton and cell shape. *PLoS One.* 2014;9(8).
98. McNeil E, Capaldo CT, Macara IG. Zonula occludens-1 function in the assembly of tight junctions in Madin-Darby canine kidney epithelial cells. *Mol Biol Cell.* 2006;17(4):1922-1932.
99. Citi S, Amorosi A, Franconi F, Giotti A, Zampi G. Cingulin, a specific protein component of tight junctions, is expressed in normal and neoplastic human epithelial tissues. *Am J Pathol.* 1991;138(4):781-789.
100. Cordenonsi M, D'Atri F, Hammar E, et al. Cingulin contains globular and coiled-coil domains and interacts with ZO-1, ZO-2, ZO-3, and myosin. *J Cell Biol.* 1999;147(7):1569-1582.
101. Guillemot L, Citi S. Cingulin regulates claudin-2 expression and cell proliferation through the small GTPase RhoA. *Mol Biol Cell.* 2006;17(8):3569-3577.
102. Citi S, Paschoud S, Pulimeno P, et al. The tight junction protein cingulin regulates gene expression and RhoA signaling. *Ann N Y Acad Sci.* 2009;1165:88-98.
103. Hino T, Yokota T, Ito S, et al. In vivo delivery of small interfering RNA targeting brain capillary endothelial cells. *Biochem Biophys Res Commun.* 2006;340(1):263-267.
104. Campbell M, Kiang AS, Kenna PF, et al. RNAi-mediated reversible opening of the blood-brain barrier. *J Gene Med.* 2008;10(8):930-947.
105. Campbell M, Nguyen AT, Kiang AS, et al. An experimental platform for systemic drug delivery to the retina. *Proc Natl Acad Sci U S A.* 2009;106(42):17817-17822.
106. Campbell M, Humphries MM, Kiang AS, et al. Systemic low-molecular weight drug delivery to pre-selected neuronal regions. *EMBO Mol Med.* 2011;3(4):235-245.
107. Campbell M, Hanrahan F, Gobbo OL, et al. Targeted suppression of claudin-5 decreases cerebral oedema and improves cognitive outcome following traumatic brain injury. *Nat Commun.* 2012;3:849.
108. Huang YH, Bao Y, Peng W, et al. Claudin-3 gene silencing with siRNA suppresses ovarian tumor growth and metastasis. *Proc Natl Acad Sci U S A.* 2009;106(9):3426-3430.
109. Keaney J, Walsh DM, O'Malley T, et al. Autoregulated paracellular clearance of amyloid-beta across the blood-brain barrier. *Sci Adv.* 2015;1(8).
110. Dithmer S, Staat C, Muller C, et al. Claudin peptidomimetics modulate tissue barriers for enhanced drug delivery. *Ann N Y Acad Sci.* 2017;1397(1):169-184.
111. Cording J, Arslan B, Staat C, et al. Trictide, a tricellulin-derived peptide to overcome cellular barriers. *Ann N Y Acad Sci.* 2017;1405(1):89-101.
112. Greene C, Kealy J, Humphries MM, et al. Dose-dependent expression of claudin-5 is a modifying factor in schizophrenia. *Mol Psychiatry.* 2017.
113. Menard C, Pfau ML, Hodes GE, et al. Social stress induces neurovascular pathology promoting depression. *Nat Neurosci.* 2017;20(12):1752-1760.

CHAPTER

2

Blood–Brain Barrier in Disease States

Aaron Dadas, Jolewis Washington*, Nicola Marchi*[†], *Damir Janigro**[,‡]

[*]Flocel, Inc., Cleveland, OH, United States [†]Cerebrovascular Mechanisms of Brain Disorders Laboratory, Department of Neuroscience, Institute of Functional Genomics (INSERM), University of Montpellier, Montpellier, France [‡]Department of Physiology, Case Western Reserve University, Cleveland, OH, United States

BLOOD–BRAIN BARRIER RESEARCH

Only recently has the blood–brain barrier (BBB) reached acceptance as a mechanism of neurological diseases.[1-6] Fig. 1 shows the increased number of hits on PubMed for the terms "blood–brain barrier" and "blood–brain barrier AND (neurological OR disease)" over the past five decades. Note how the contribution of neurological diseases to the overall number of hits lagged behind the almost-exponential increase in BBB hits (Fig. 1C). Reluctance to link the BBB to neurological conditions was perhaps due to the neuro-centric view of the brain before the year 2000. It also appears that not all neurological diseases experience the same degree of association with the BBB (Fig. 2). The fact that so many pathologies implicate the BBB is due to its association with multidrug resistance, especially in regard to chemotherapeutic or antineoplastic agents.

With this in mind, we summarize current knowledge on BBB pathophysiology, based on the data shown in Figs. 1 and 2. We then briefly discuss the normal BBB, describe multidrug resistance in greater detail, and conclude by reviewing how the BBB affects the etiology of neurological diseases such as epilepsy, Alzheimer disease, and traumatic brain injury (TBI).

BUILDING A BBB FOR EXPERIMENTAL STUDIES

Several review articles[5, 7, 8] and even the preceding chapter in this very book describe the cellular and molecular building blocks of the human BBB. We will instead focus on models of

https://doi.org/10.1016/B978-0-12-813997-4.00002-5
© 2019 Elsevier Inc. All rights reserved.

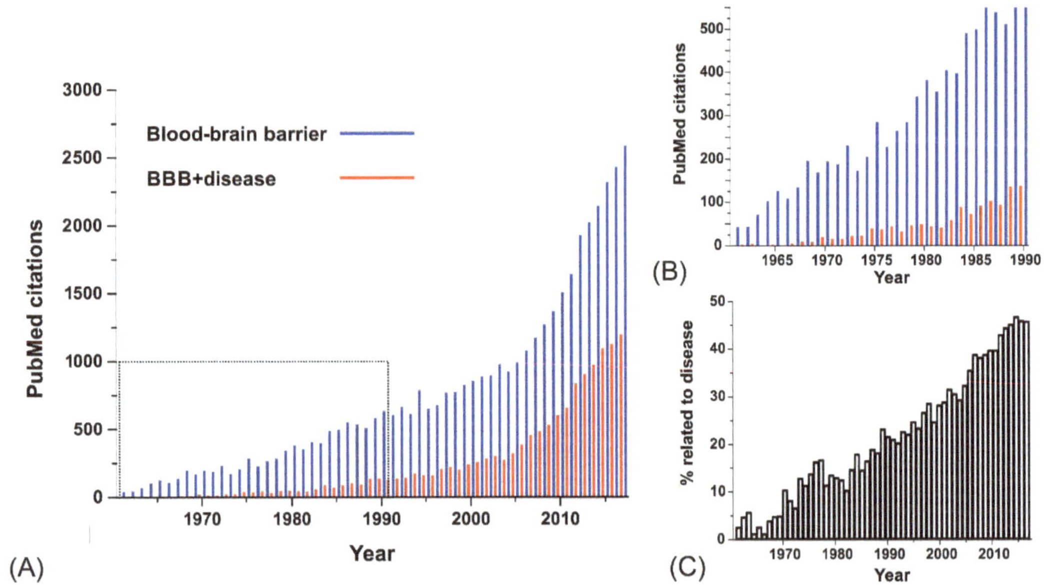

FIG. 1 Increased awareness of the role of the blood–brain barrier (BBB) in human disease is evidenced by the number of hits in PubMed searches. (A) Near-exponential changes in return of the search "blood–brain barrier." When mining the literature for "blood–brain barrier and neurological diseases," the number of hits decreases substantially across the years. (B) However, by narrowing the search to a specific period of time (i.e., 1960-1990) it becomes apparent that the popularity of "blood–brain barrier and neurological diseases" significantly lagged behind the use of the abbreviation "BBB"; this is also underscored in (C), wherein the ratio of "blood–brain barrier"/"blood–brain barrier and neurological diseases" is plotted vs. time. Today, nearly half of BBB-related publications deal with neurological diseases; this is a large increase compared to <10% in the 1960s and <20% before the 1990s.

the human BBB and the usefulness of these models for studying the BBB under physiological and pathological conditions. The architecture and the physiology of cerebral vessels vary from venules to BBB capillaries. The BBB consists of microvascular endothelial cells that line the brain microvessels together with the closely associated astrocytic end-feet processes. The capillary endothelium is characterized by the presence of tight junctions, lack of fenestrations, and minimal pinocytotic vesicles. In particular, tight junctions between cerebral endothelial cells form a diffusion barrier, which selectively excludes most blood-borne substances from entering the brain and vice versa, protecting it from systemic influences mediated by substances of all sizes or by polar molecules, such as water-soluble compounds. The BBB also plays a crucial role in determining neurotoxicity and in carrying out specific transport mechanisms to prevent it. However, as mentioned further, the same efflux transport systems may contribute to multidrug resistance during treatment of many disorders of the central nervous system (CNS), such as drug-refractory epilepsy or pharmaco-resistant brain tumors.

In addition, the BBB, along with factors produced by resident neuroepithelial cells, plays a crucial role in regulating the trafficking, survival, and effector functions of leukocytes during inflammatory states. Venular endothelial cells are exposed to a significantly lower level of shear stress (up to 4-5 dynes/cm^2) than those that form the BBB (up to 20 dynes/cm^2), which

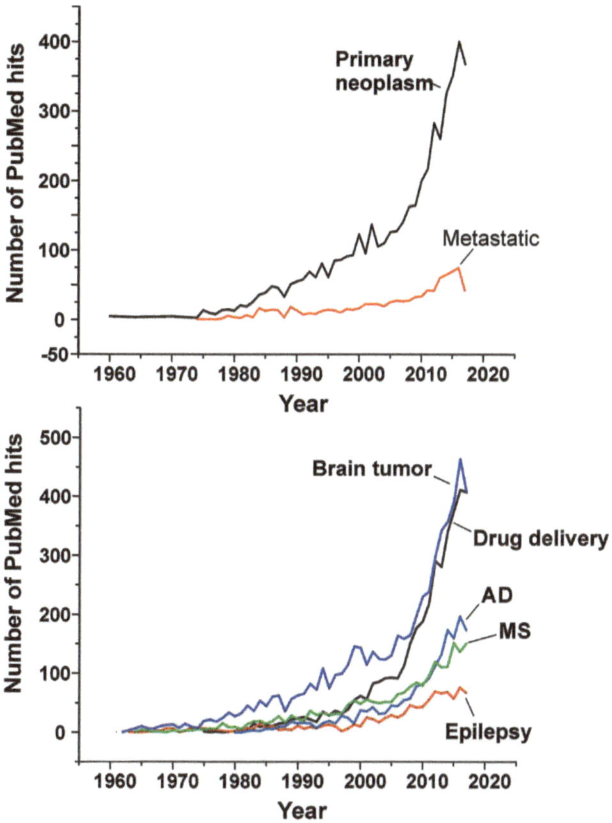

FIG. 2 Literature searches aimed at unveiling the most common topics of preclinical blood–brain barrier (BBB) research reveal a considerable preference for drug delivery and neoplasm-related topics. Articles addressing the BBB as an etiological mechanism of multiple sclerosis (MS), Alzheimer disease (AD), or seizure disorders lag behind both historic onset of awareness of the BBB, as well as absolute number of citations, shown in Fig. 1.

in turn affects the physiological and functional properties of the vascular endothelium. Thus, postcapillary blood vessels (i.e., venules) allow more blood-borne substances to access the subendothelial space. Such access is provided by more permissive endothelial junctions, fenestrations, discontinuities in the capillary walls, or pinocytotic vesicles. Perivascular pericytes also play an important role in cerebrovascular physiology.[9] An ideal cerebrovascular model should be able to reproduce these largely vessel-specific hemodynamic and cellular characteristics, as well as the fluid properties of the intracranial space, if possible.

In vivo rodent models of the BBB have been used for decades. By taking advantage of modern molecular technologies, mice have been used with variable success to reproduce human pathology.[10] For example, a size-selective loosening of the BBB in claudin-5-deficient mice[11] and a dose-dependent expression of claudin-5 were found to be modifying factors in schizophrenia.[12] The use of larger animal models is gaining momentum, and porcine (and "minipig"[13]) models have been used to study, for example, the effects of TBI on the BBB,[14, 15] cell trafficking,[16] and acute seizures.[17] Pig models do not allow for molecular screening based on transgene or knockout technology, but facilitate a translational comparison to the human brain, owing to similar skull characteristics and a non-lissencephalic brain. In addition, pig models are amenable for imaging by routine clinical tools, such as magnetic resonance

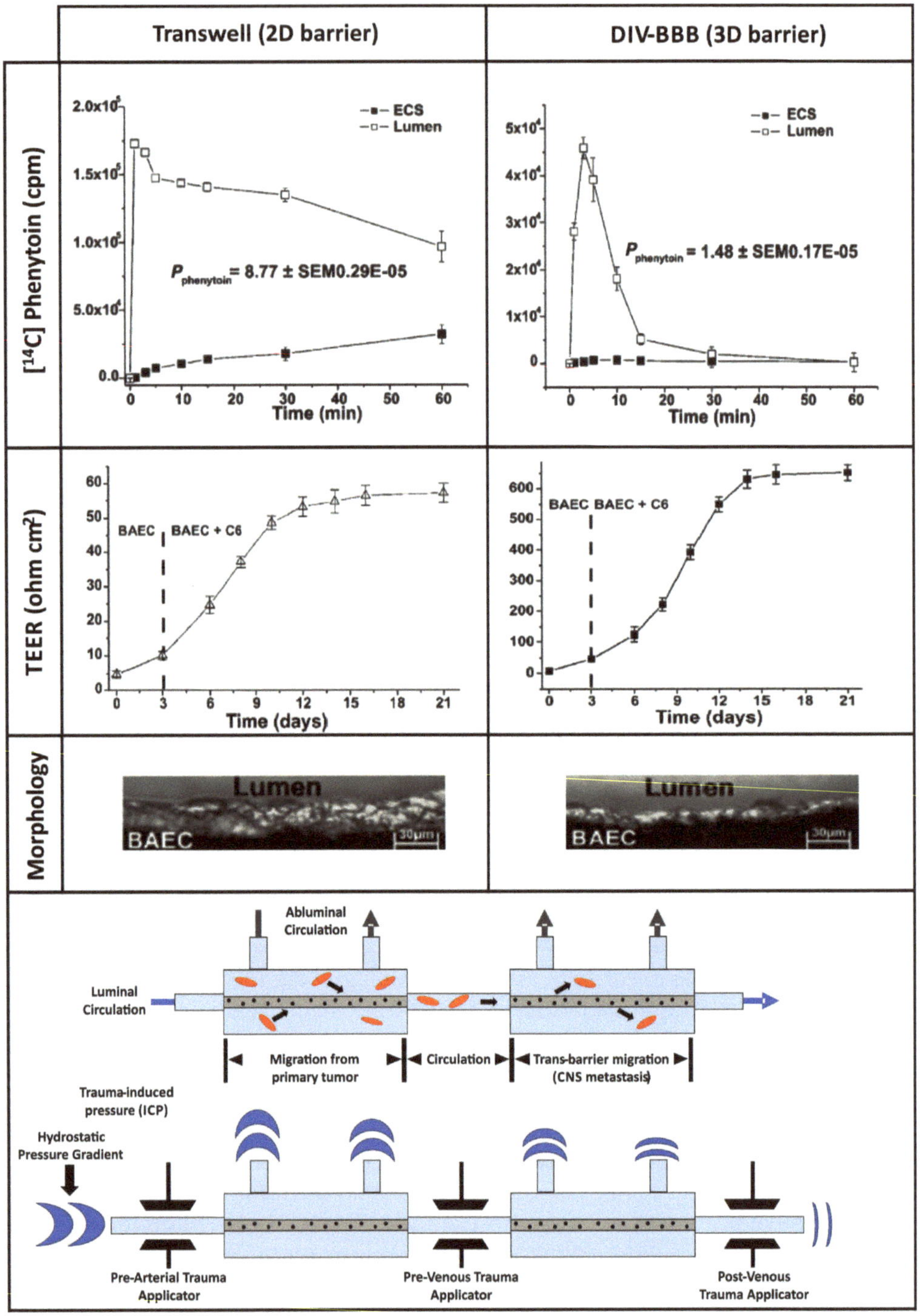

FIG. 3

See figure legend on opposite page

imaging (MRI) or computed tomography.[16, 17] Finally, nonhuman primates have been used as a surrogate of clinical studies in many laboratories.[18, 19]

In vitro cell barrier culture systems allow us to examine the physiology and mechanisms of the human BBB in a more direct, less expensive, and possibly high-throughput manner. An in vitro cell barrier model provides a scaffold upon which the user can incorporate many types of human endothelial cells, cell lines, cocultures further containing neuroglial cells, and establish a "control" brain environment to assess biological functions of the BBB under homeostatic conditions. These testing environments come in two varieties.

The first approach—and, until recent years, the predominant method—is a two-dimensional (2D) cell culture plate (e.g., the Transwell or Netwell plates [Sigma-Aldrich]). The cell monolayer is grown on a semipermeable support made of glass or polystyrene and is immersed in a static fluid environment, with the media requiring manual replacement at intervals throughout the duration of the experiment. Although this method allows for high-throughput evaluation of barrier properties (up to 96-well plates, in many cases), there are some notable disadvantages to a 2D cell culture. As previously noted, the luminal boundary of in vivo vasculature is subject to continual blood flow; therefore, there is a constant shear stress on the apical surface of these cells—a force that varies between arterial and venous fluid compartments. In 2D cellular arrays, there is no fluid motion, and the cells take on a considerably different morphology, tight junction integrity, and spatial orientation of transporters, receptors, and genes than one would expect in vivo.[20] Unsurprisingly, these changes to the cell barrier microenvironment result in experimental data that are often not reflective of in vivo BBB physiology or biological function, and these setbacks must be addressed (Fig. 3).[21, 22]

A second approach to cell culture is the use of three-dimensional (3D) cell constructs, such as cell spheroids[21, 23] or vessel-shaped structures (e.g., the dynamic in vitro BBB,[22] microfluidic conduits,[24] etc.). Both of these methods are more accurate than 2D cultures in terms of cell morphology and cell-to-cell interactions; however, the vessel-shaped constructs further provide isolated luminal/abluminal compartments as well as luminal fluid perfusion. This incorporation of dynamic fluid perfusion—along with vessel-shaped spatial orientation of human endothelial cells and microglia—confers one of the most physiologically

FIG. 3 **Cont'd,** Structural and functional differences between 2D (Transwell) and 3D (dynamic in vitro blood–brain barrier [DIV-BBB]) cell barrier culture systems. Row 1 of the chart reveals that a fluid-static 2D culture fails to properly regulate drug permeability at rates expected in vivo (0.0001 cm/s for phenytoin), whereas the 3D culture approaches a physiologically accurate rate very quickly. Row 2 further delineates this structural disparity by showing a much lower transendothelial electrical resistance (TEER) value across the Transwell cell barrier than the DIV-BBB cell barrier, with the latter approaching TEER levels recorded in vivo. The microscopic images in Row 3 highlight a lack of spatial orientation and cell-to-cell interaction between cells of the 2D culture, as they instead form a multilayer with little tight junction organization, as opposed to the uniform monolayer in the 3D vessel scaffold. Lastly, the bottom diagrams depict two functional applications of a 3D cell culture model that are not currently possible with 2D systems, such as tumor metastasis between different tissues (diagram, top) and multicompartment analysis of hemodynamic properties or differential response in fluid compartments to trauma (diagram, bottom). *Reproduced with permission from Santaguida S, Janigro D, Hossain M, Oby E, Rapp E, Cucullo L: Side by side comparison between dynamic versus static models of blood–brain barrier in vitro: A permeability study.* Brain Res. *2006;1109:1-13.*

representative models of the human in vivo vasculature to date (Fig. 3) and is a major step forward in allowing us to replicate human BBB physiology in vitro.

THE BBB AS AN IMPEDIMENT TO THERAPEUTIC INTERVENTIONS FOR NEUROLOGICAL DISEASE

Rational CNS drug design cannot entirely and exclusively rely upon the physical–chemical properties of putative neurotherapeutics, as lipophilicity alone is a poor predictor for drug penetration in the CNS. Pharmacotherapy for brain disorders depends on the pharmacokinetic and pharmacodynamic properties of the drug in question. Pharmacokinetic properties result from the interaction of the drug with multidrug transporters.[25, 26] As drugs required to act in the brain (e.g., chemotherapeutics and antiepileptic drugs [AEDs]) have become more lipophilic, the potency of active efflux mechanisms has become apparent.[27] It is true that lipophilic drugs with low molecular weight are (in theory) able to cross the brain endothelium via the lipid membranes. Such lipophilic drugs, however, are potential substrates for efflux carriers of the BBB,[28] particularly P-glycoprotein (P-gp, a product of the MDR1 gene), which is predominantly located on the endothelial luminal membrane[29-31] but is also expressed in astrocytes and other parenchymal cells.[32, 33] Up to 50% of drug candidates may be substrates for drug transporters, leading to a phenomenon known as *pharmacokinetic multidrug resistance*.[29, 30, 33-36] In the case of epilepsy, multidrug resistance is thought by some to be a distinct pathology from epilepsy itself, and therefore may require separate treatment.[37]

In addition to transporters, metabolizing enzymes (e.g., the CYP family) are functionally active at the BBB[38, 39] and can convert a therapeutic agent into a neurotoxic one,[40] thereby controlling both pharmacokinetics and pharmacodynamics of a given therapeutic agent (Fig. 4). Whether brain-specific enzymatic machinery is responsible for the metabolism of AEDs is unclear. It is often assumed that most of the metabolic burden of drug metabolism is hepatic, but increasing evidence points to the CNS as a drug-metabolizing organ.[38, 39, 45] The concomitant CNS expression of MDR1 and P450 enzymes in the drug-resistant epileptic brain suggests that the bioavailability of AEDs is dictated by mechanisms similar to those used by the principal drug-metabolizing organ, the liver.[38] The presence of a BBB machinery to manage drug fate is an important finding, as a local metabolic process may allow the formation and uncontrolled accumulation of molecules in the brain, with unpredictable or even toxic results. These complicating factors further confound brain pharmacology in the setting of pathological conditions with multidrug resistance.[30, 46, 47]

As underscored in Fig. 2, brain neoplasms have been the primary foci of preclinical and clinical BBB research efforts. Somewhat surprisingly, most of the work in this field focuses on relatively uncommon types of primary tumors of the CNS (e.g., glioblastoma), whereas the types of tumors that are more common (e.g., metastatic, secondary) have been largely ignored. Thanks in part to the pioneering work by Ed Neuwelt at Oregon Health Science University, however, the brain tumor and BBB research efforts enjoyed a "human model" to study drug extravasation and chemotherapeutic efficacy.[48] A treatment called blood–brain barrier disruption (BBBD)[49-51] has been used throughout the past decades as a last-resort treatment for primary CNS lymphomas and other brain tumors. In this protocol, intra-arterial hyperosmotic mannitol (1.4 M) is administered via the carotid or vertebral artery, and the

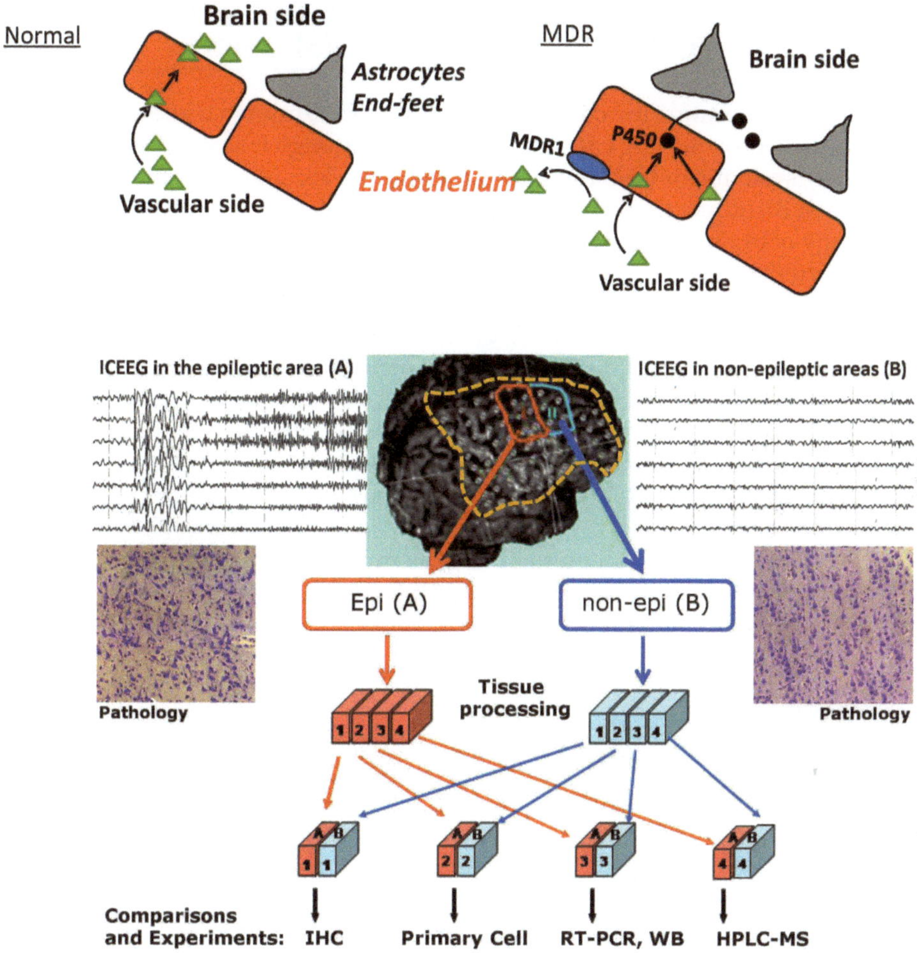

FIG. 4 Translational approaches to blood–brain barrier (BBB) studies have become quite sophisticated, as shown by Marchi and Ghosh.[38, 41–44] The approach translates in vitro findings using cells from human epileptic brain resections that have been characterized as *epileptogenic* (or "silent") on intraoperative electrocorticography. This approach also allows researchers to measure and compare levels of antiepileptic drugs in vivo and in vitro, and facilitates the study of the contribution of transporters and metabolizing enzymes. By using this approach, a role for drug-metabolizing enzymes (e.g., P450) expressed at the BBB has been discovered to act synergistically with multidrug transporters. In addition, molecular screening and/or BBB cells isolated from patients allow researchers to pinpoint patient-specific defects in drug pharmacokinetics (for earlier research using this approach, see Dombrowski et al.[30]).

level of BBBD is confirmed by contrast-enhanced computed tomography immediately after chemotherapy treatment. This allows measurement of the effects of chemotherapy on brain tumors surrounded by a transiently permeable BBB. This approach also revealed that disruption of the BBB was seizure-promoting, which was one of the first reports of a clear causal relationship between BBBD and a disease or symptoms.[17]

THE BBB AS A TARGET FOR THERAPEUTIC INTERVENTIONS FOR NEUROLOGICAL DISEASES

In attempting to determine the contribution of the BBB to human disease, one immediately faces a classic chicken-or-the-egg causality dilemma. Although it is accepted that the BBB fails because of many systemic or CNS disorders, it is far less accepted that the BBB plays a causative role in such disorders (Fig. 5).[3, 4, 52-55] The first publications implicating the BBB in the pathophysiology of brain edema were published in the 1960s[56, 57]; its role in autoimmune diseases was reported in 1961.[58] Today, the number of pathologies implicating the BBB has grown to encompass virtually every major condition affecting the brain (Table 1). In several monogenic diseases (e.g., De Vivo disease, GLUT1 deficiency syndrome), the role of the BBB is implicit to the disease itself.[71] GLUT1 (a product of the human gene SLC2A1) is expressed at high levels by brain endothelial cells. Patients with GLUT1 deficiency syndrome have a defect in the BBB GLUT1 glucose transporter, causing decreased glucose concentration in the CNS. This syndrome is part of a spectrum of neurological phenotypes that result from glucose-transport deficiency. Similar to GLUT1 deficiency, a handful of mono- or polygenic "BBB diseases" have been reported.[8, 72]

In diseases triggered by a small number of gene mutations/deletions, identifying the role of a given cell type in which these genes are expressed is relatively easy. However, a causal relationship becomes problematic when several etiological mechanisms act concurrently to

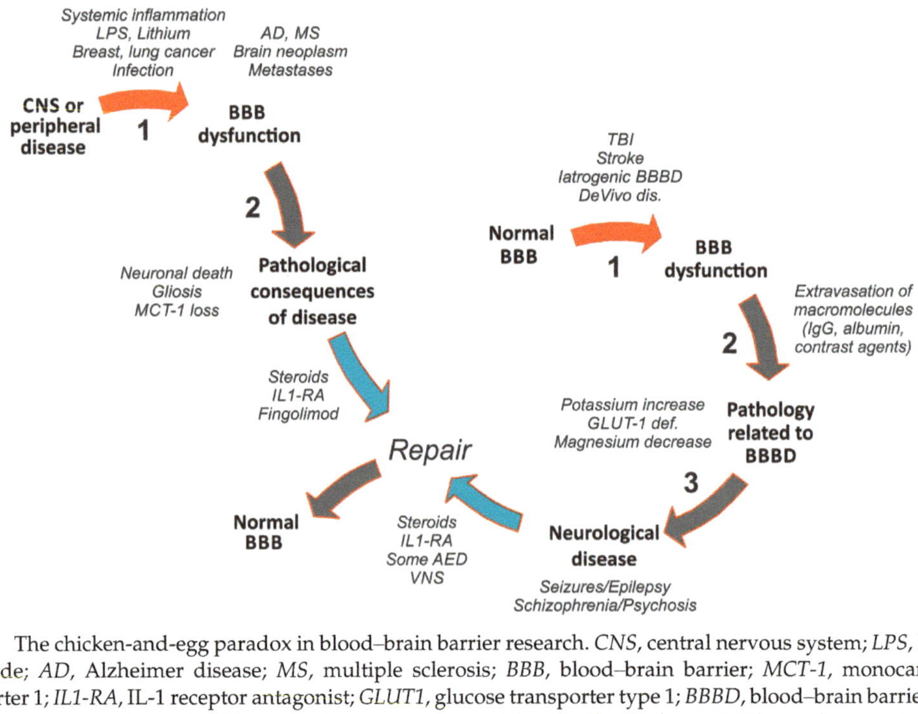

FIG. 5 The chicken-and-egg paradox in blood–brain barrier research. *CNS*, central nervous system; *LPS*, lipopolysaccharide; *AD*, Alzheimer disease; *MS*, multiple sclerosis; *BBB*, blood–brain barrier; *MCT-1*, monocarboxylate transporter 1; *IL1-RA*, IL-1 receptor antagonist; *GLUT1*, glucose transporter type 1; *BBBD*, blood–brain barrier disruption; *IgG*, immunoglobulin G; *TBI*, traumatic brain injury; *AED*, antiepileptic drugs; *VNS*, vagal nerve stimulator.

TABLE 1 Short List of Neurological Diseases With Demonstrated or Suspected Involvement of the Blood–Brain Barrier

Disease process, reference	Mechanism involved	Pathology and symptoms	Treatment	Direct role for BBB?
Genetic, monogenic				
De Vivo disease[59]	GLUT1 deficiency	Seizures, mental retardation	Ketogenic diet	Yes, GLUT1 is expressed by BBB ECs
CNS hemorrhage, calcification[60]	Occludin, JAM-C mutations	Whole-brain hemorrhage and calcification	None specific for disease	Yes, tight junctions are involved in disease etiology
Cavernous malformation[61]	CCM 1-3	Abnormal vascular structure, vessels prone to rupture	None specific for disease	Yes, CCM is expressed by ECs
CADASIL[62]	Notch-3	Stroke, migraine, cognition, psychiatric	Only symptomatic (triptans and NSAIDs)	Yes, gene controls vasculogenesis in the brain
Acquired				
Brain edema[63-65]	Multifactorial	Cognitive decline, motor, sensory deficits	Mannitol, diuretics	Yes, but other factors play a role
Epilepsy[3, 66, 67]	Inflammation; increased extracellular potassium	Seizures and comorbidities of epilepsy	AEDs, anti-inflammatory, IL1-RA; steroids	Yes, but only for a subset of seizure type
Autoimmune CNS disease[68]	IgG and leukocytes in CNS	Seizures, MS; AD?	Anti-inflammatory; IL1-RA; steroids; IV IgG	Yes, but other factors play a role
Stroke	Inflammation and disruption of energy metabolism; glutamate	Reperfusion injury, sensory motor, aphasia, cognition	tPA; anti-inflammatory	Yes, but other factors play a role
SVID[69, 70]	Genetic/acquired	Vascular dementias	Management of risk factors; statins	Yes, but other factors play a role
Alzheimer disease[6]	Genetic/acquired	Dementia	None available	Yes, but other factors play a role

BBB, *blood–brain barrier;* ECs, *endothelial cells;* CNS, *central nervous system;* JAM-C, *junctional adhesion molecule C;* CCM, *cerebral cavernous malformation;* CADASIL, *cerebral autosomal dominant arteriopathy with subcortical infarcts and leukoencephalopathy;* Notch-3, *neurogenic locus notch homolog protein 3;* NSAIDs, *nonsteroidal anti-inflammatory drugs;* AED, *antiepileptic drugs;* IL1-RA, *interleukin-1 receptor antagonists;* IgG, *immunoglobulin G;* AD, *Alzheimer disease;* IV IgG, *intravenous immunoglobulin G;* tPA, *tissue plasminogen activator;* SVID, *small vessel ischemic disease, also referred to as white matter hyperintensities.*

TABLE 2 Environmental Factors Affecting Integrity of the Blood–Brain Barrier

Agent, reference	Effect on BBB	Disease
Agents of structural change		
Aluminum[73, 74]	Increases the rate of transmembrane diffusion and selectively changes saturable transport systems without disrupting the integrity of the membranes or altering CNS hemodynamics	Alzheimer disease, progression of stroke, ischemia/reperfusion injury, hypoxia/reoxygenation, and vascular dysfunction
Ammonia[75]	Alters the passage of different molecules across the BBB, both transcellularly and paracellularly	Vasogenic brain edema, secondary brain damage
Arsenic[76, 77]	Acts as a cytoskeletal interference by inhibiting neurofilament transport	Parkinson disease, neurotoxicity, nephrotoxicity
Mercury[78]	Inhibits amino acid and glutamate transport	Damage to CNS
Nitric oxide[79]	Induces the breakdown of ECs, causing substances to enter the brain passively	Vasogenic brain edema, secondary brain damage
Agent of ion substitution		
Lead[80]	Can substitute for calcium ions and pass through calcium channels in barrier	Brain damage, mental retardation, behavioral problems, nerve damage; possibly Alzheimer disease, Parkinson disease, and schizophrenia

BBB, *blood–brain barrier;* CNS, *central nervous system;* ECs, *endothelial cells*

result in a pathology. This grows even more complicated when environmental factors are considered (Table 2). One obstacle is the lack of highly predictive, noninvasive markers of BBB integrity. Three methods have been used to noninvasively determine BBB integrity: (1) MRI with or without contrast,[81-83] (2) sampling of cerebrospinal fluid (CSF) and blood to determine albumin coefficient,[84] and (3) peripheral blood analysis for biomarkers such as S100B.[85, 86] These methods of detecting a disrupted BBB are conceptually identical, because they measure passage of molecules across a "leaky" BBB. When contrast MRI is the selected method, the tracer gadolinium is most commonly used to perform dynamic contrast-enhanced MRI (known as DCE-MRI).[87] When contrast is not used, a fluid-attenuated inversion recovery (or FLAIR) MRI[88] is used to reveal regions of disrupted BBB.

The albumin ratio is the oldest and most invasive of the three methods; it is based on measuring albumin permeability by "tapping" either the lumbar sac or ventricular CSF.[89] The least invasive, least expensive method measures protein normally retained in the CNS by the BBB after these appear in blood or other peripheral body fluids.[83, 90-93] Excluding CSF sampling, the remaining methods (i.e., blood biomarkers or MRI) are both common, but each has its own advantages and drawbacks. Blood markers are extremely convenient and easy to obtain, but offer no information on the location of permeability defects. MRI, on the other hand, provides a superb topographic view of BBB leakage, but is costly and mostly unusable for time-course studies or population analyses. Nonetheless, these methods have made it possible for researchers to recognize a causative effect of BBBD in seizures,[3, 90, 91] TBI,[94, 95] and small-vessel ischemic disease,[96, 97] among others.

HOW DOES THE BBB CONTROL NEURONAL ACTIVITY?

An excellent example of how the BBB directly affects neuronal firing is the role it plays in acute and delayed consequences of TBI. Early posttraumatic seizures occur immediately *after* TBI and are *caused* by TBI. Exacerbation of preexisting seizures, however, is not a good clinical example of post-TBI seizures caused by the TBI. Furthermore, the temporal relationship between the traumatic event and seizures is a key factor in the underlying mechanisms of ictogenesis, as early posttraumatic seizures (2%-5% of mild TBI cases, 10%-15% of severe TBI cases)[98-102] are different in mechanism from late seizures. In addition, late seizures are most common after penetrating head wound, such as one would expect in war-related injury. One study assessed 421 Vietnam veterans with penetrating head wounds; 53% of these veterans went on to develop posttraumatic epilepsy, and 15 years later half of them were still experiencing seizures.[103]

The BBB maintains ion gradients specific to the cerebral environment.[104] Of particular importance to the discussion that follows are the ion gradients that, when increased due to disruption of the BBB, lead to neuronal excitability. BBB disruption and the accompanying neuronal hyperexcitability promote seizures.[17, 105-108] Furthermore, direct increase of potassium levels in the brain has also been shown to cause seizures.[109] The concentration of the gradients that exist on either side of the BBB and their effects on neuronal excitability have been previously described.[104] What follows is a brief discussion of the regulation of ions and molecules relevant to TBI and seizure in the context of brain edema.

Potassium (K^+) regulation is critical to the proper function of all excitable cells, including neurons.[110] K^+ in the brain extracellular space is maintained at approximately 3 mM, regardless of serum potassium levels (which are usually around 5 mM). Intracellular K^+ levels (approximately 60 mM) are kept high compared to the extracellular compartment. This dramatic difference in concentration allows for the rapid repolarization required of CNS neurons after depolarization. There are therefore two distinct K^+ gradients: the gradient across cell membranes and the gradient across the BBB. K^+ homeostasis is achieved via a glial buffering system in addition to the energy-dependent neuronal mechanism.[111]

Glia, and in particular astroglia, play the largest role in K^+ buffering by two mechanisms: K^+ uptake and spatial buffering. In the K^+ uptake mechanism, astroglia remove K^+ in the extracellular space through a glial-specific Na^+/K^+/ATPase better suited for K^+ buffering than the neuronal isotype.[112, 113] Astrocytes also possess a $Na^+/K^+/Cl^-$ cotransporter (NKCC1) that has recently been implicated in potassium buffering.[114] Using these channels, astrocytes are able to temporarily sequester excess extracellular K^+ and release it back into the extracellular space when K^+ levels drop. The second mechanism, spatial buffering, is a means by which astrocytes can remove K^+ from areas of high concentration and release it elsewhere, in areas of comparably lower concentration. This is achieved by a syncytium of astroglia connected by gap junctions that directly link neighboring cells.[115] In areas of high K^+ concentration, K^+ enters the cell primarily through potassium inward rectifier (Kir) channels, specifically Kir4.1.[116] Kir4.1 channels localize at astrocyte end-foot processes.[117] These channels move K^+ into astrocytes, and are unique in that they have higher conductance at negative membrane potentials.[118] K^+ entry triggers a depolarization that travels through the astroglial network, resulting in net outward movement of K^+. In this way, astrocytes are able to spread K^+ across a large area while causing only a transient increase in local intracellular K^+ concentrations.[111]

The movement of water across the BBB and within the brain parenchyma follows osmotic gradients. In the brain, water moves through aquaporin channels—specifically, aquaporin 1 and aquaporin 4 (AQP4). For the purpose of this discussion, we focus only on the role of AQP4, as it is the primary water channel expressed on astrocytes and co-localizes with Kir4.1 channels.[119]

Astrocytes express high levels of AQP4 on end-foot processes surrounding barrier capillaries, and their proximity suggests a role in the regulation of water movement into and out of the brain parenchyma.[119] However, AQP4 can be found on end-foot processes of astrocytes at synapses, suggesting an additional role in neuronal water uptake. It is interesting to note that immunocytochemical experiments do not detect the presence of the same aquaporin on neurons, indicating that astrocytes may be responsible for water homeostasis in the brain. AQP4 co-localization with Kir4.1 channels suggests the coupling of K^+ and water movement within the brain as well as across the BBB.[120] The juxtaposition of these channels has led to the hypothesis that, in addition to the spatial buffering of potassium, the astroglial syncytium may redistribute water throughout the cortex and, in particular, the perivascular areas thought to be sinks for excess water and K^+.[121] Experiments have shown that an increase in extracellular K^+ leads to a decrease in extracellular space, suggesting a link between K^+ and water movement.[122] Additionally, it has been demonstrated that Kir4.1 channels and AQP4 channels may be associated extracellularly by the dystrophin–glycoprotein complex.[123, 124]

Glutamate is an excitatory neurotransmitter, and its concentration is tightly controlled across the BBB and in the extracellular space of the brain.[125] Concentrations of glutamate are considerably higher in the blood than in the brain; extracellular cerebral glutamate levels are maintained at approximately 1 μM, whereas plasma glutamate levels are approximately 50 μM.[126] Similar to K^+, extracellular glutamate concentrations are regulated by astrocytes. Released at the synaptic cleft, excess glutamate is taken up by astrocytic processes that surround the synaptic cleft and are enriched with high-affinity glutamate transporter 1 and glutamate aspartate transporter 1 receptors.[127, 128] Both transporters require the cotransport of Na^+ with H^+, as well as the movement of one K^+ and one bicarbonate (HCO_3^-) or OH^- out of the cell. Once in the cytoplasmic compartment of astroglia, glutamine synthase catalyzes the condensation of glutamate with ammonia to form glutamine. Glutamine is then released by the astrocyte and taken up by neurons that convert it back to glutamate.

CONCLUSIONS

Research on the BBB has reached new heights—the role of BBB endothelial cells in promoting neurological diseases or facilitating recovery has become textbook material. Future directions include an in-depth genetic analysis of BBB polymorphisms in humans, as well as the development of novel tools to image BBB function. Large animal models and sophisticated in vitro replicas of the BBB will likely fill in the gap between clinical and experimental research.

References

1. Jiang X, Andjelkovic AV, Zhu L, et al. Blood-brain barrier dysfunction and recovery after ischemic stroke. *Prog Neurobiol*. 2018;163-164:144-171.

2.. Marchi N, Granata T, Alexopoulos A, Janigro D. The blood-brain barrier hypothesis in drug resistant epilepsy. *Brain*. 2012;135(Pt 4).
3. Marchi N, Granata T, Janigro D. Inflammatory pathways of seizure disorders. *Trends Neurosci*. 2014;37(2):55-65.
4. Oby E, Janigro D. The blood-brain barrier and epilepsy. *Epilepsia*. 2006;47(11):1761-1774.
5. Ochocinska MJ, Zlokovic BV, Searson PC, et al. NIH workshop report on the trans-agency blood-brain interface workshop 2016: exploring key challenges and opportunities associated with the blood, brain and their interface. *Fluids Barriers CNS*. 2017;14(1):12.
6. Sweeney MD, Sagare AP, Zlokovic BV. Blood-brain barrier breakdown in Alzheimer disease and other neurodegenerative disorders. *Nat Rev Neurol*. 2018;14(3):133-150.
7. Haddad-Tovolli R, Dragano NRV, Ramalho AFS, Velloso LA. Development and function of the blood-brain barrier in the context of metabolic control. *Front Neurosci*. 2017;11:224.
8. Zhao Z, Nelson AR, Betsholtz C, Zlokovic BV. Establishment and dysfunction of the blood-brain barrier. *Cell*. 2015;163(5):1064-1078.
9. Giannoni P, Badaut J, Dargazanli C, et al. The pericyte-glia interface at the blood-brain barrier. *Clin Sci (Lond)*. 2018;132(3):361-374.
10. Sohet F, Daneman R. Genetic mouse models to study blood-brain barrier development and function. *Fluids Barriers CNS*. 2013;10(1):3.
11. Nitta T, Hata M, Gotoh S, et al. Size-selective loosening of the blood-brain barrier in claudin-5-deficient mice. *J Cell Biol*. 2003;161(3):653-660.
12. Greene C, Kealy J, Humphries MM, et al. Dose-dependent expression of claudin-5 is a modifying factor in schizophrenia. *Mol Psychiatry*. 2018;23:2156-2166.
13. Hildebrand F, Andruszkow H, Huber-Lang M, Pape HC, van Griensven M. Combined hemorrhage/trauma models in pigs-current state and future perspectives. *Shock*. 2013;40(4):247-273.
14. Morganti-Kossmann MC, Yan E, Bye N. Animal models of traumatic brain injury: is there an optimal model to reproduce human brain injury in the laboratory? *Injury*. 2010;41(Suppl 1):S10-S13.
15. Shridharani JK, Wood GW, Panzer MB, et al. Porcine head response to blast. *Front Neurol*. 2012;3:70.
16. Marchi N, Teng Q, Nguyen MT, et al. Multimodal investigations of trans-endothelial cell trafficking under condition of disrupted blood-brain barrier integrity. *BMC Neurosci*. 2010;11:34.
17. Marchi N, Angelov L, Masaryk T, et al. Seizure-promoting effect of blood-brain barrier disruption. *Epilepsia*. 2007;48(4):732-742.
18. Kamimura HA, Flament J, Valette J, et al. Feedback control of microbubble cavitation for ultrasound-mediated blood-brain barrier disruption in non-human primates under magnetic resonance guidance. *J Cereb Blood Flow Metab*. 2018: 271678X17753514.
19. League-Pascual JC, Lester-McCully CM, Shandilya S, et al. Plasma and cerebrospinal fluid pharmacokinetics of select chemotherapeutic agents following intranasal delivery in a non-human primate model. *J Neuro-Oncol*. 2017;132(3):401-407.
20. Edmondson R, Broglie JJ, Adcock AF, Yang L. Three-dimensional cell culture systems and their applications in drug discovery and cell-based biosensors. *Assay Drug Dev Technol*. 2014;12(4):207-218.
21. Chitcholtan K, Sykes PH, Evans JJ. The resistance of intracellular mediators to doxorubicin and cisplatin are distinct in 3D and 2D endometrial cancer. *J Transl Med*. 2012;10:38.
22. Santaguida S, Janigro D, Hossain M, Oby E, Rapp E, Cucullo L. Side by side comparison between dynamic versus static models of blood-brain barrier in vitro: a permeability study. *Brain Res*. 2006;1109(1):1-13.
23. Kenny PA, Lee GY, Myers CA, et al. The morphologies of breast cancer cell lines in three-dimensional assays correlate with their profiles of gene expression. *Mol Oncol*. 2007;1(1):84-96.
24. Cho H, Seo JH, Wong KH, et al. Three-dimensional blood-brain barrier model for in vitro studies of neurovascular pathology. *Sci Rep*. 2015;5:15222.
25. Bauer B, Yang X, Hartz AM, et al. In vivo activation of human pregnane X receptor tightens the blood-brain barrier to methadone through P-glycoprotein up-regulation. *Mol Pharmacol*. 2006;70(4):1212-1219.
26. Marchi N, Hallene KL, Kight KM, et al. Significance of MDR1 and multiple drug resistance in refractory human epileptic brain. *BMC Med*. 2004;2:37.
27. Rambeck B, Jurgens UH, May TW, et al. Comparison of brain extracellular fluid, brain tissue, cerebrospinal fluid, and serum concentrations of antiepileptic drugs measured intraoperatively in patients with intractable epilepsy. *Epilepsia*. 2006;47(4):681-694.

28. Nikolic K, Mavridis L, Djikic T, et al. Drug design for CNS diseases: polypharmacological profiling of compounds using cheminformatic, 3D-QSAR and virtual screening methodologies. *Front Neurosci.* 2016;10:265.
29. Abbott NJ, Khan EU, Rollinson CM, et al. Drug resistance in epilepsy: the role of the blood-brain barrier. *Novartis Found Symp.* 2002;243:38-47. discussion 47-53, 180-185.
30. Dombrowski SM, Desai SY, Marroni M, et al. Overexpression of multiple drug resistance genes in endothelial cells from patients with refractory epilepsy. *Epilepsia.* 2001;42(12):1501-1506.
31. Schinkel AH. P-Glycoprotein, a gatekeeper in the blood-brain barrier. *Adv Drug Deliv Rev.* 1999;36(2-3):179-194.
32. Marroni M, Agrawal ML, Kight K, et al. Relationship between expression of multiple drug resistance proteins and p53 tumor suppressor gene proteins in human brain astrocytes. *Neuroscience.* 2003;121(3):605-617.
33. Marroni M, Marchi N, Cucullo L, Abbott NJ, Signorelli K, Janigro D. Vascular and parenchymal mechanisms in multiple drug resistance: a lesson from human epilepsy. *Curr Drug Targets.* 2003;4(4):297-304.
34. Loscher W, Potschka H. Role of multidrug transporters in pharmacoresistance to antiepileptic drugs. *J Pharmacol Exp Ther.* 2002;301(1):7-14.
35. Marroni M, Kight KM, Hossain M, Cucullo L, Desai SY, Janigro D. Dynamic in vitro model of the blood-brain barrier. Gene profiling using cDNA microarray analysis. *Methods Mol Med.* 2003;89:419-434.
36. Potschka H, Fedrowitz M, Loscher W. Multidrug resistance protein MRP2 contributes to blood-brain barrier function and restricts antiepileptic drug activity. *J Pharmacol Exp Ther.* 2003;306(1):124-131.
37. Granata T, Marchi N, Carlton E, et al. Management of the patient with medically refractory epilepsy. *Expert Rev Neurother.* 2009;9(12):1791-1802.
38. Ghosh C, Gonzalez-Martinez J, Hossain M, et al. Pattern of P450 expression at the human blood-brain barrier: roles of epileptic condition and laminar flow. *Epilepsia.* 2010;51(8):1408-1417.
39. Ghosh C, Marchi N, Desai NK, et al. Cellular localization and functional significance of CYP3A4 in the human epileptic brain. *Epilepsia.* 2011;52(3):562-571.
40. Ghosh C, Hossain M, Spriggs A, et al. Sertraline-induced potentiation of the CYP3A4-dependent neurotoxicity of carbamazepine: an in vitro study. *Epilepsia.* 2015;56(3):439-449.
41. Boussadia B, Gangarossa G, Mselli-Lakhal L, et al. Lack of CAR impacts neuronal function and cerebrovascular integrity in vivo. *Exp Neurol.* 2016;283(Pt A):39-48.
42. Ghosh C, Hossain M, Puvenna V, et al. Expression and functional relevance of UGT1A4 in a cohort of human drug-resistant epileptic brains. *Epilepsia.* 2013;54(9):1562-1570.
43. Ghosh C, Hossain M, Solanki J, Dadas A, Marchi N, Janigro D. Pathophysiological implications of neurovascular P450 in brain disorders. *Drug Discov Today.* 2016;21(10):1609-1619.
44. Ghosh C, Marchi N, Hossain M, et al. A pro-convulsive carbamazepine metabolite: quinolinic acid in drug resistant epileptic human brain. *Neurobiol Dis.* 2012;46(3):692-700.
45. Ghosh C, Puvenna V, Gonzalez-Martinez J, Janigro D, Marchi N. Blood-brain barrier P450 enzymes and multidrug transporters in drug resistance: a synergistic role in neurological diseases. *Curr Drug Metab.* 2011;12(8):742-749.
46. Loscher W, Potschka H. Drug resistance in brain diseases and the role of drug efflux transporters. *Nat Rev Neurosci.* 2005;6(8):591-602.
47. Marchi N, Gonzalez-Martinez J, Nguyen MT, Granata T, Janigro D. Transporters in drug-refractory epilepsy: clinical significance. *Clin Pharmacol Ther.* 2010;87(1):13-15.
48. Kroll RA, Neuwelt EA. Outwitting the blood-brain barrier for therapeutic purposes: osmotic opening and other means. *Neurosurgery.* 1998;42(5):1083-1099. discussion 1099-1100.
49. Angelov L, Doolittle ND, Kraemer DF, et al. Blood-brain barrier disruption and intra-arterial methotrexate-based therapy for newly diagnosed primary CNS lymphoma: a multi-institutional experience. *J Clin Oncol.* 2009; 27(21):3503-3509.
50. Greig NH. Brain tumours and the blood-brain barrier. In: Neuwelt EA, ed. *Implications of the Blood-Brain Barrier and Its Manipulation.* New York-London: Plenum; 1989:77-106.
51. Roman-Goldstein S, Clunie DA, Stevens J, et al. Osmotic blood-brain barrier disruption: CT and radionuclide imaging. *AJNR Am J Neuroradiol.* 1994;15(3):581-590.
52. Neuwelt EA, Bauer B, Fahlke C, et al. Engaging neuroscience to advance translational research in brain barrier biology. *Nat Rev Neurosci.* 2011;12(3):169-182.
53. Shlosberg D, Benifla M, Kaufer D, Friedman A. Blood-brain barrier breakdown as a therapeutic target in traumatic brain injury. *Nat Rev Neurol.* 2010;6(7):393-403.

54. Tagge CA, Fisher AM, Minaeva OV, et al. Concussion, microvascular injury, and early tauopathy in young athletes after impact head injury and an impact concussion mouse model. *Brain*. 2018;141(2):422-458.
55. Tomkins O, Feintuch A, Benifla M, Cohen A, Friedman A, Shelef I. Blood-brain barrier breakdown following traumatic brain injury: a possible role in posttraumatic epilepsy. *Cardiovasc Psychiatry Neurol*. 2011;2011:765923.
56. Aird RB, Becker RA. The blood-brain barrier in clinical disease: a review. *J Nerv Ment Dis*. 1963;136:517-526.
57. Katzman R, Gonatas N, Levine S. Electrolytes and fluids in experimental focal leukoencephalopathy. *Arch Neurol*. 1964;10:58-65.
58. Luse SA, McDougal Jr DB. Electron microscopic observations on allergic encephalomyelitis in the rabbit. *J Exp Med*. 1960;112:735-742.
59. Janigro D. Blood-brain barrier, ion homeostatis and epilepsy: possible implications towards the understanding of ketogenic diet mechanisms. *Epilepsy Res*. 1999;37(3):223-232.
60. Mochida GH, Ganesh VS, Felie JM, et al. A homozygous mutation in the tight-junction protein JAM3 causes hemorrhagic destruction of the brain, subependymal calcification, and congenital cataracts. *Am J Hum Genet*. 2010;87(6):882-889.
61. Fischer A, Zalvide J, Faurobert E, Albiges-Rizo C, Tournier-Lasserve E. Cerebral cavernous malformations: from CCM genes to endothelial cell homeostasis. *Trends Mol Med*. 2013;19(5):302-308.
62. Zhu S, Nahas SJ. CADASIL: imaging characteristics and clinical correlation. *Curr Pain Headache Rep*. 2016; 20(10):57.
63. Campbell M, Hanrahan F, Gobbo OL, et al. Targeted suppression of claudin-5 decreases cerebral oedema and improves cognitive outcome following traumatic brain injury. *Nat Commun*. 2012;3:849.
64. Iffland P, Grant GG, Janigro D. Mechanisms of cerebral edema leading to early seizures after traumatic brain injury. In: Lo EH, ed. *Vascular Mechanisms in CNS Trauma*. New York: Springer Science+Business Media; 2014:29-45.
65. Verkman AS. Aquaporins in clinical medicine. *Annu Rev Med*. 2012;63:303-316.
66. Marchi N, Granata T, Freri E, et al. Efficacy of anti-inflammatory therapy in a model of acute seizures and in a population of pediatric drug resistant epileptics. *PLoS One*. 2011;6(3).
67. Vezzani A, Granata T. Brain inflammation in epilepsy: experimental and clinical evidence. *Epilepsia*. 2005; 46(11):1724-1743.
68. Obermeier B, Verma A, Ransohoff RM. The blood-brain barrier. *Handb Clin Neurol*. 2016;133:39-59.
69. Ding J, Sigurethsson S, Jonsson PV, et al. Large perivascular spaces visible on magnetic resonance imaging, cerebral small vessel disease progression, and risk of dementia: the age, gene/environment susceptibility-reykjavik study. *JAMA Neurol*. 2017;74(9):1105-1112.
70. Iadecola C. The pathobiology of vascular dementia. *Neuron*. 2013;80(4):844-866.
71. De Vivo DC, Trifiletti RR, Jacobson RI, Ronen GM, Behmand RA, Harik SI. Defective glucose transport across the blood-brain barrier as a cause of persistent hypoglycorrhachia, seizures, and developmental delay. *N Engl J Med*. 1991;325(10):703-709.
72. Rosenberg GA. Neurological diseases in relation to the blood-brain barrier. *J Cereb Blood Flow Metab*. 2012;32 (7):1139-1151.
73. Banks WA, Niehoff ML, Drago D, Zatta P. Aluminum complexing enhances amyloid beta protein penetration of blood-brain barrier. *Brain Res*. 2006;1116(1):215-221.
74. Chen L, Yokel RA, Hennig B, Toborek M. Manufactured aluminum oxide nanoparticles decrease expression of tight junction proteins in brain vasculature. *J NeuroImmune Pharmacol*. 2008;3(4):286-295.
75. Skowronska M, Albrecht J. Alterations of blood brain barrier function in hyperammonemia: an overview. *Neurotox Res*. 2012;21(2):236-244.
76. DeFuria J, Shea TB. Arsenic inhibits neurofilament transport and induces perikaryal accumulation of phosphorylated neurofilaments: roles of JNK and GSK-3beta. *Brain Res*. 2007;1181:74-82.
77. Tyler CR, Allan AM. The effects of arsenic exposure on neurological and cognitive dysfunction in human and rodent studies: a review. *Curr Environ Health Rep*. 2014;1:132-147.
78. Aschner M, Aschner JL. Mercury neurotoxicity: mechanisms of blood-brain barrier transport. *Neurosci Biobehav Rev*. 1990;14(2):169-176.
79. Thiel VE, Audus KL. Nitric oxide and blood-brain barrier integrity. *Antioxid Redox Signal*. 2001;3(2):273-278.
80. Sanders T, Liu Y, Buchner V, Tchounwou PB. Neurotoxic effects and biomarkers of lead exposure: a review. *Rev Environ Health*. 2009;24(1):15-45.

81. Tomkins O, Shelef I, Kaizerman I, et al. Blood-brain barrier disruption in post-traumatic epilepsy. *J Neurol Neurosurg Psychiatry*. 2008;79(7):774-777.
82. Weissberg I, Veksler R, Kamintsky L, et al. Imaging blood-brain barrier dysfunction in football players. *JAMA Neurol*. 2014;71(11):1453-1455.
83. Zhang J, Puvenna V, Janigro D. Biomarkers of traumatic brain injury and their relationship to pathology. In: Laskowitz D, Grant G, eds. *Translational Research in Traumatic Brain Injury*. Boca Raton (FL): CRC Press/Taylor and Francis Group; 2016.
84. Blyth BJ, Farahvar A, He H, et al. Elevated serum ubiquitin carboxy-terminal hydrolase L1 is associated with abnormal blood-brain barrier function after traumatic brain injury. *J Neurotrauma*. 2011;28(12):2453-2462.
85. Marchi N, Cavaglia M, Fazio V, Bhudia S, Hallene K, Janigro D. Peripheral markers of blood-brain barrier damage. *Clin Chim Acta*. 2004;342(1-2):1-12.
86. Marchi N, Rasmussen P, Kapural M, et al. Peripheral markers of brain damage and blood-brain barrier dysfunction. *Restor Neurol Neurosci*. 2003;21(3-4):109-121.
87. Chai WY, Chu PC, Tsai CH, et al. Image-guided focused-ultrasound CNS molecular delivery: an implementation via dynamic contrast-enhanced magnetic-resonance imaging. *Sci Rep*. 2018;8(1):4151.
88. Weier K, Fluri F, Kos S, Gass A. Postcontrast flair MRI demonstrates blood-brain barrier dysfunction in PRES. *Neurology*. 2009;72(8):760-762.
89. Reiber H. Dynamics of brain-derived proteins in cerebrospinal fluid. *Clin Chim Acta*. 2001;310(2):173-186.
90. Bargerstock E, Puvenna V, Iffland P, et al. Is peripheral immunity regulated by blood-brain barrier permeability changes? *PLoS One*. 2014;9(7).
91. Marchi N, Bazarian JJ, Puvenna V, et al. Consequences of repeated blood-brain barrier disruption in football players. *PLoS One*. 2013;8(3).
92. O'Connell B, Kelly AM, Mockler D, et al. Use of blood biomarkers in the assessment of sports-related concussion-A systematic review in the context of their biological significance. *Clin J Sport Med*. 2018;28:561-571.
93. Puvenna V, Brennan C, Shaw G, et al. Significance of ubiquitin carboxy-terminal hydrolase L1 elevations in athletes after sub-concussive head hits. *PLoS One*. 2014;9(5).
94. Dadas A, Janigro D. The role and diagnostic significance of cellular barriers after concussive head trauma. Concussion 2018. 3
95. Dadas A, Washington J, Marchi N, Janigro D. Improving the clinical management of traumatic brain injury through the pharmacokinetic modeling of peripheral blood biomarkers. *Fluids Barriers CNS*. 2016;13(1):21.
96. Choi H, Puvenna V, Brennan C, et al. S100B and S100B autoantibody as biomarkers for early detection of brain metastases in lung cancer. *Transl Lung Cancer Res*. 2016;5(4):413-419.
97. Vogelbaum MA, Masaryk T, Mazzone P, et al. S100beta as a predictor of brain metastases: brain versus cerebrovascular damage. *Cancer*. 2005;104(4):817-824.
98. Annegers JF, Grabow JD, Groover RV, Laws Jr ER, Elveback LR, Kurland LT. Seizures after head trauma: a population study. *Neurology*. 1980;30(7 Pt 1):683-689.
99. Annegers JF, Grabow JD, Kurland LT, Laws Jr. ER. The incidence, causes, and secular trends of head trauma in Olmsted County, Minnesota, 1935-1974. *Neurology*. 1980;30(9):912-919.
100. Hauser WA, Annegers JF, Kurland LT. Incidence of epilepsy and unprovoked seizures in Rochester, Minnesota: 1935-1984. *Epilepsia*. 1993;34(3):453-468.
101. Langendorf FG, Pedley TA, Temkin NR. Posttraumatic seizures. In: Engel G, Pedley TA, eds. *Epilepsy: A Comprehensive Textbook*. 2nd ed. Philadelphia: Lippincott Williams & Wilkins; 2008:2537-2542.
102. Temkin NR, Dikmen SS, Wilensky AJ, Keihm J, Chabal S, Winn HR. A randomized, double-blind study of phenytoin for the prevention of post-traumatic seizures. *N Engl J Med*. 1990;323(8):497-502.
103. Salazar AM, Jabbari B, Vance SC, Grafman J, Amin D, Dillon JD. Epilepsy after penetrating head injury. I. Clinical correlates: a report of the Vietnam Head Injury Study. *Neurology*. 1985;35(10):1406-1414.
104. Janigro D. Are you in or out? Leukocyte, ion, and neurotransmitter permeability across the epileptic blood-brain barrier. *Epilepsia*. 2012;53(Suppl 1):26-34.
105. Marchi N, Fan Q, Ghosh C, et al. Antagonism of peripheral inflammation reduces the severity of status epilepticus. *Neurobiol Dis*. 2009;33(2):171-181.
106. Marchi N, Oby E, Batra A, et al. In vivo and in vitro effects of pilocarpine: relevance to ictogenesis. *Epilepsia*. 2007;48(10):1934-1946.

107. Marchi N, Teng Q, Ghosh C, et al. Blood-brain barrier damage, but not parenchymal white blood cells, is a hallmark of seizure activity. *Brain Res*. 2010;1353:176-186.
108. Seiffert E, Dreier JP, Ivens S, et al. Lasting blood-brain barrier disruption induces epileptic focus in the rat somatosensory cortex. *J Neurosci*. 2004;24(36):7829-7836.
109. Janigro D, Gasparini S, D'Ambrosio R, McKhann 2nd G, DiFrancesco D. Reduction of K+ uptake in glia prevents long-term depression maintenance and causes epileptiform activity. *J Neurosci*. 1997;17(8):2813-2824.
110. Misonou H. Homeostatic regulation of neuronal excitability by K(+) channels in normal and diseased brains. *Neuroscientist*. 2010;16(1):51-64.
111. Kofuji P, Newman EA. Potassium buffering in the central nervous system. *Neuroscience*. 2004;129(4):1045-1056.
112. Gloor SM. Relevance of Na,K-ATPase to local extracellular potassium homeostasis and modulation of synaptic transmission. *FEBS Lett*. 1997;412(1):1-4.
113. Sweadner KJ. Two molecular forms of (Na+ + K+)-stimulated ATPase in brain. Separation, and difference in affinity for strophanthidin. *J Biol Chem*. 1979;254(13):6060-6067.
114. Kahle KT, Simard JM, Staley KJ, Nahed BV, Jones PS, Sun D. Molecular mechanisms of ischemic cerebral edema: role of electroneutral ion transport. *Physiology (Bethesda)*. 2009;24:257-265.
115. Pannasch U, Vargova L, Reingruber J, et al. Astroglial networks scale synaptic activity and plasticity. *Proc Natl Acad Sci U S A*. 2011;108(20):8467-8472.
116. Ohno Y, Hibino H, Lossin C, Inanobe A, Kurachi Y. Inhibition of astroglial Kir4.1 channels by selective serotonin reuptake inhibitors. *Brain Res*. 2007;1178:44-51.
117. Higashi K, Fujita A, Inanobe A, et al. An inwardly rectifying K(+) channel, Kir4.1, expressed in astrocytes surrounds synapses and blood vessels in brain. *Am J Phys Cell Phys*. 2001;281(3):C922-C931.
118. Olsen ML, Sontheimer H. Functional implications for Kir4.1 channels in glial biology: from K+ buffering to cell differentiation. *J Neurochem*. 2008;107(3):589-601.
119. Medici V, Frassoni C, Tassi L, Spreafico R, Garbelli R. Aquaporin 4 expression in control and epileptic human cerebral cortex. *Brain Res*. 2011;1367:330-339.
120. Nielsen S, Nagelhus EA, Amiry-Moghaddam M, Bourque C, Agre P, Ottersen OP. Specialized membrane domains for water transport in glial cells: high-resolution immunogold cytochemistry of aquaporin-4 in rat brain. *J Neurosci*. 1997;17(1):171-180.
121. Rash JE, Yasumura T, Hudson CS, Agre P, Nielsen S. Direct immunogold labeling of aquaporin-4 in square arrays of astrocyte and ependymocyte plasma membranes in rat brain and spinal cord. *Proc Natl Acad Sci U S A*. 1998;95(20):11981-11986.
122. Dietzel I, Heinemann U, Hofmeier G, Lux HD. Transient changes in the size of the extracellular space in the sensorimotor cortex of cats in relation to stimulus-induced changes in potassium concentration. *Exp Brain Res*. 1980;40(4):432-439.
123. Ahn AH, Kunkel LM. Syntrophin binds to an alternatively spliced exon of dystrophin. *J Cell Biol*. 1995; 128(3):363-371.
124. Connors NC, Kofuji P. Dystrophin Dp71 is critical for the clustered localization of potassium channels in retinal glial cells. *J Neurosci*. 2002;22(11):4321-4327.
125. O'Kane RL, Martinez-Lopez I, DeJoseph MR, Vina JR, Hawkins RA. Na(+)-dependent glutamate transporters (EAAT1, EAAT2, and EAAT3) of the blood-brain barrier. A mechanism for glutamate removal. *J Biol Chem*. 1999;274(45):31891-31895.
126. Hawkins RA. The blood-brain barrier and glutamate. *Am J Clin Nutr*. 2009;90(3):867S-874S.
127. Storck T, Schulte S, Hofmann K, Stoffel W. Structure, expression, and functional analysis of a Na(+)-dependent glutamate/aspartate transporter from rat brain. *Proc Natl Acad Sci U S A*. 1992;89(22):10955-10959.
128. Tanaka K, Watase K, Manabe T, et al. Epilepsy and exacerbation of brain injury in mice lacking the glutamate transporter GLT-1. *Science*. 1997;276(5319):1699-1702.

CHAPTER

3

Pharmacokinetics of Systemic Drug Delivery

Claire O'Connor[a], *Nirmala Ramanath*[a], *Matthew Campbell*
Smurfit Institute of Genetics, Trinity College Dublin, Dublin, Ireland

PHARMACOKINETICS AND THE BLOOD–BRAIN BARRIER

Pharmacokinetics, the branch of pharmacology concerned with the movement of drugs within the body, determines the fate of a drug after administration.[1] Systemic delivery is often the favored method, due to ease of administration. However, despite major advances in drug development for disorders of the central nervous system (CNS), the efficacy of treatments delivered to the CNS is still suboptimal.[2,3] This inefficacy is due to the presence of the blood–brain barrier (BBB), a physiological barrier that restricts the movement of material from the peripheral system to the brain and vice versa. The BBB limits the delivery of many approved and clinically enabled drugs to the brain.[4,5] In fact, it completely prevents the passage of large-molecule neurotherapeutics, and allows just 5% of small-molecule therapeutics access to the brain.[5]

Although the BBB represents a major obstacle in systemic administration and delivery of targeted drugs to the CNS, crossing the BBB remains a favored route for drug delivery to the brain. Every neuron is within approximately 25 μm of a capillary; therefore, once the BBB is successfully bypassed, the diffusion distance to the site of action is very short.[6,7] Additionally, the large surface area of the CNS vascular network (150-200 cm^2/g tissue) provides significant access to the brain parenchyma.[8] These properties make the BBB an obstacle worth overcoming. Major advances in our understanding of BBB components and of the neurovascular unit (NVU) as a whole have allowed us to manipulate this dynamic structure and to develop approaches that may facilitate CNS-targeted drug delivery via the systemic circulation. This chapter discusses the various methods by which BBB penetration may be analyzed

[a] These authors contributed equally to this work.

https://doi.org/10.1016/B978-0-12-813997-4.00003-7

© 2019 Elsevier Inc. All rights reserved.

and assesses the methods currently being developed to bypass the BBB to ultimately improve treatment of CNS disorders.

THE NEUROVASCULAR UNIT AND THE BLOOD–BRAIN BARRIER

The CNS controls vital processes such as cognition and the coordination of peripheral organ function. As a result, the vasculature of the CNS is both structurally and functionally distinct from the vasculature of the rest of the body.[9,10] For reliable neural signaling through chemical and electrical signals, the ionic microenvironment surrounding the synapses must be maintained.[11,12] The significant metabolic and energy demands of the CNS must be met, and the CNS must be protected from substances that trigger uncontrolled activation of inflammatory or neuronal responses to maintain the integrity of the delicate brain network. Positioned along the blood vessels of the CNS is the BBB, a highly selective and tightly regulated barrier that reflects the vital functions of the brain. The BBB is important in controlling the uptake of ions and nutrients between the blood and brain, and in protecting neural tissue from potentially damaging agents such as pathogens, immune cells, anaphylatoxins, drugs, and drug conjugates. Lining the blood vessels of the CNS are endothelial cells (ECs), which are the principal components of the BBB.[13]

The ECs of the CNS have developed numerous mechanisms that limit the transcellular and paracellular movement of material from blood to brain and vice versa. The presence of BBB-specific receptor and transport proteins, in addition to the absence of fenestrae and the low rate of fluid-phase transcytosis noted in ECs, results in limited transcellular transport.[9,14] Paracellular transport is highly restricted due to the presence of tight junctions (TJs), which create extremely high electrical resistance between ECs.[15] TJs are dynamic structures that link adjacent ECs, and are composed of a number of transmembrane proteins and membrane-associated cytoplasmic proteins. Transmembrane proteins include occludin and the claudins, both of which are tetra-spanning proteins.[16] These proteins function as dimers, in which the two extracellular loops of each monomer interact with a counterpart on the neighboring ECs (Fig. 1).[17] Membrane-associated cytoplasmic proteins, such as zonula occludens-1 (ZO-1), connect the transmembrane proteins to the actin cytoskeleton, providing structural support.[18] Together, these properties limit the transcellular and paracellular pathways so that only small, nonpolar, lipophilic molecules can pass through the BBB.[19] However, owing to the exceedingly high energy demands of the brain, the brain endothelium contains numerous transport systems for the uptake of ions, amino acids, proteins, and carbohydrates.[9] Even if molecules do cross the BBB, they may not last long in the brain microenvironment, as abluminal transporters on ECs ensure rapid efflux of material, while various enzymes actively degrade foreign molecules.[20,21]

Cerebral ECs are known for their unique nature, morphology, and functionality; however, the famous quail chick transplantation experiments demonstrated that the properties of the BBB are not intrinsic.[22] In fact, it is the interactions between developing vascular tissue and the other components of the NVU (e.g., nervous tissue) that are critical in the development of the BBB.[22] Therefore, the BBB cannot function independently; instead, it requires complex cooperative interactions with the other cells of the NVU, both during development and in its maintenance. These other cell types that constitute the NVU are pericytes, astrocytes,

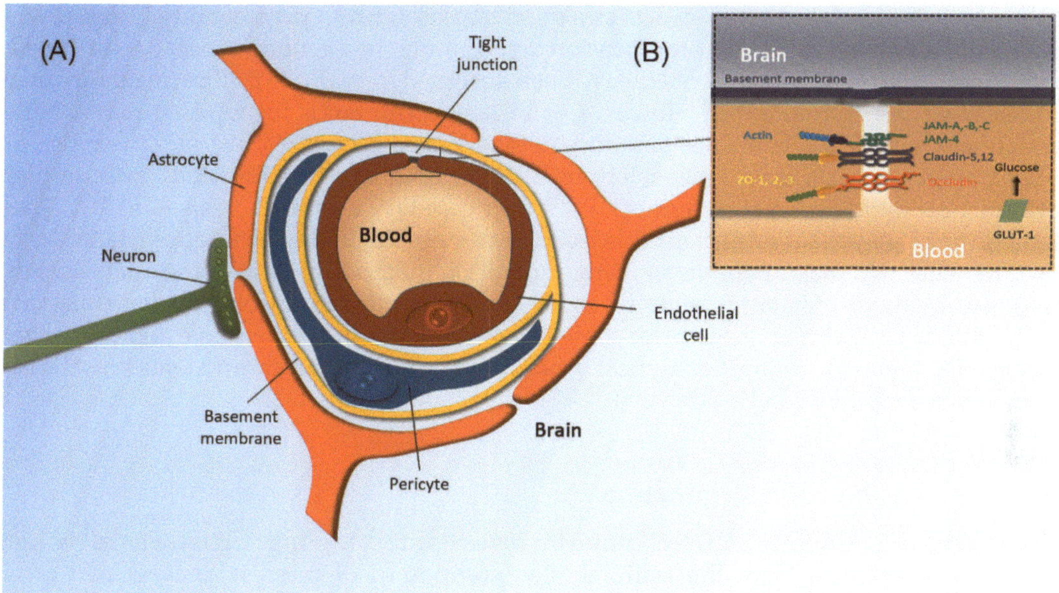

FIG. 1 Endothelial cells, pericytes, astrocytes, microglia, interneurons, and neurons make up a complex composition of cell types that interact closely to form the neurovascular unit. Inset: Tight junction components create a paracellular barrier between adjacent endothelial cells, separating the blood and brain microenvironment.

microglia, neurons, interneurons, and the noncellular basement membrane (Fig. 1).[14] Each component of the NVU is closely and reciprocally linked, which results in the formation of an anatomical, functional unit. In essence, the BBB is a highly efficient system of regulating blood flow and ensures that the requirements of the brain are met without undue risk.

PHARMACOKINETIC ANALYSIS METHODS

Clearly, the BBB presents many logistical problems in terms of delivering drugs to the CNS. Evaluating BBB penetration is an essential task in the development of effective CNS-targeted drugs, as it is critical to know to what extent a drug crosses the BBB from the systemic circulation.[23] Pharmacokinetics, as previously mentioned, is concerned with the movement of drugs within the body. More specifically, it is a mathematical analysis of the rates of absorption, distribution, metabolism, and excretion, otherwise known as *the ADME processes*, and of concentration–time relationships.[1] When delivering drugs to the brain, not only it is essential that the drug crosses the BBB but also that it is retained in the brain long enough to produce the desired effect.[1,24] Various methods of pharmacokinetics analysis can help evaluate the delivery of a molecule from the systemic circulation to the CNS, providing important information on BBB penetration and the therapeutic potential of the drug.[25]

When discussing drug delivery, the success of delivery of the solute is described as the percentage of the overall injected dose delivered per gram of brain.[26] The success depends on two factors: permeability–surface area product (PS), and plasma area under the curve at a given time after injection.[27] PS acts as a good marker for BBB permeability,[28] and is what many of the methods in this chapter focus on determining. In a general sense, the established in vivo methods previously described have been designed to evaluate BBB permeability using radiolabelled test compounds.[29] However, with new compounds for which a radiolabelled version may not be available, a separation method is also required.[27]

Separation methods used extensively in BBB permeability investigations include high-performance liquid chromatography (HPLC) and capillary electrophoresis (CE).[30-35] These methods have played an important role as adjunct methods to the techniques described below. There are two major benefits of using separation methods as part of the analysis. First, separation methods enhance selectivity, as the analyte of interest is isolated from substances that may interfere with analysis prior to determination. Second, when investigating a large number of analytes via a single analytical procedure, the individual components must be separated out before they can be analyzed.[27]

Brain Uptake Index Determination

The brain uptake index (BUI) technique involves tissue sampling.[29] First, a small volume of solution containing both the radiolabelled compound of interest as well as a radiolabelled diffusible reference compound, which is often an internal standard such as 3H-water, is injected as a bolus dose into the carotid artery.[36] Within 5 to 15 seconds of the injection, the animal is decapitated. The brain is then homogenized and the radioactivity in the brain as well as in injected solution is analyzed.[37,38] The BUI is calculated using the following formula:

$$\mathrm{BUI} = \frac{Et}{Er} \times 100 \tag{1}$$

where Et stands for the unidirectional extraction of the test and Er represents the reference compound.[39,40] Once the cerebral blood flow (F) value is known, the Renkin-Crone equation can be used to convert the Et value into BBB PS products, as seen in Eq. (2).[27,39,41]

$$Et = 1 - e^{(-\mathrm{PS}/F)} \tag{2}$$

As mentioned, the tracer technique is useful; however, more recently, separation methods such as HPLC have been used to determine BUI and PS product.[31,42]

In Situ Brain Perfusion

This method of analysis is similar to BUI determination, in that the tracer molecule is injected through the internal carotid artery of the model, with brain perfusion being stopped on decapitation of the animal, after which the brain is removed for ex vivo analysis.[43] In this case, the tracer is added to a suitable perfusate solution and it is perfused via the carotid artery at a rate of 2.5-10 mL/min.[44,45] The various distribution measurements obtained from the ex vivo analysis are then used to quantify the BBB PS product, as described in Eq. (3).

$$PS = \frac{[Vd - Vo]Cp(T)}{AUC|_0^t} \tag{3}$$

where *Vd* represents brain volume of distribution of the test compound, *Vo* describes the plasma volume of distribution of the marker, *Cp*(*T*) is the terminal plasma concentration, and AUC is the plasma area under the curve at a given time after injection.[27]

The quantification of solute distribution in brain tissue was often measured using techniques such as autoradiography; however, this method cannot distinguish parent compounds from any possible metabolites.[46] Therefore, again, separation methods such as HPLC or mass spectrometry are preferred to ensure the integrity of tracer.[33]

This method offers several advantages compared to the BUI determination method. First, due to the extended exposure, this method is more pertinent for solutes with poor BBB permeability.[27] Second, it allows both passive and carrier-mediated transport to be evaluated.[39] Third, this method does not disrupt the BBB, which provides the opportunity to evaluate the ability of a substance to pass across an undisturbed BBB, which is a key factor in the development of CNS-targeted therapeutics.[43] Finally, compared with the BUI method, in situ brain perfusion is more sensitive due to the long experimental time period.[39]

Microdialysis Method and Improvements Using Cassette Dosing

Intracerebral microdialysis is a method that can be used to assess BBB permeability. This method requires only a small volume for analysis with minimal preparation, can be done in vivo, and the animal need not be euthanized. Intracerebral microdialysis can also determine brain concentrations of drug over time.[23,27] However, this method involves insertion of a probe into the brain of a living animal, which could disrupt or alter tissue behavior, affecting permeability readouts.[23,47,48]

The general principle of intracerebral microdialysis involves stereotactic implantation of a microdialysis probe—which consists of a semipermeable membrane that allows passage of molecules, depending on their molecular weight—into a specific region of the brain; a certain molecular weight cutoff allows passage of select molecules through the membrane. These molecules diffuse across an osmotic gradient.[27,49] As a result, the concentration of free drug in the brain can be assessed and calculated using the formula for in vivo concentration recovery, which tracks the transport of a molecule from brain tissue to the dialysate (i.e., the fraction that passes through the membrane).

$$E_d = (C_{\text{out}} - C_{\text{in}})/(\text{brain}_{\text{EcF}} - C_{\text{in}}) = 1 - \exp\{-1/[F(R_d + R_m + R_e)]\} \tag{4}$$

where E_d = dialysate extraction fraction, C_{out} = dialysate concentration, C_{in} = perfusate concentration, $\text{brain}_{\text{EcF}}$ = molecule concentration in brain extracellular fluid surrounding probe (periprobe concentration), F = flow rate of the perfusate, R = mass transfer resistance for—R_d = dialysate; R_m = membrane; and R_e = external medium.[50,51]

Because the permeability of drugs across the BBB can be low, it is necessary to use a method of analysis that is highly sensitive for drug permeation. In this regard, HPLC is an ideal method.[27,52] In addition to HPLC, CE is an extremely useful means of separation and quantification in microdialysis sampling. The application of a high applied direct current voltage is what provides CE with its characteristic ability to resolve complex aqueous samples with high resolution. Microdialysis sampling can only produce small sample volumes; therefore, a particular benefit of CE as an adjunct method of analysis for microdialysis is that CE requires only small volumes for highly precise analysis.[53] An additional benefit to CE is that the voltametric characterization allows for detection and identification of unknown analytes in the microdialysate.[27] Simultaneous administration of several compounds can be challenging and identification of multiple compounds can be time consuming, particularly if they have different physiological properties.[54] However, simultaneous administration of drugs is sometimes necessary to conserve animal usage, to improve treatment through multiple drug–drug interactions in the specimen, or to increase the throughput of microdialysis studies.[55,56]

It is critical to validate the cassette before using it for high-throughput studies. Such validation can be done by comparing the $K_{p,\text{brain}}$ (ratio of total brain to total plasma drug concentrations) for cassette delivery and individual delivery in the specimen, after which the cassette can be used freely.[56]

Pharmacological Inhibition Assay

Another strategy for improving drug delivery is inhibition of certain transporters, such as P-glycoprotein (P-gp).[57] This type of delivery can be analyzed through designed inhibition assays that use probe substrates, for example, using [^{3}H]vinblastine for P-gp, with concurrent administration of P-gp inhibitors, where substrate concentration can be measured in the brain and plasma after a designated time interval.[58,59]

$$E = E_0 + (E_{max} - E_0)\,(C^\gamma/IC_{50}{}^\gamma + C^\gamma) \tag{5}$$

can be used, where E is the relative change in the accumulation of substrate seen in the presence of inhibitor relative to the control without inhibitor, E_0 is the accumulation of probe in the absence of inhibitor normalized to unity, E_{max} is the maximum increase in accumulation of the probe, IC_{50} is the concentration of inhibitor at which half-maximal effect is seen, C is the concentration of inhibitor, and γ is the factor that determines the slope of the curve.[58] Plasma and brain specimens from in vivo studies can also be analyzed through liquid chromatography–mass spectrometry.[58]

Flux Assay

The transport of drugs across the BBB can also be measured through flux assays: radiolabeled inhibitor can be measured in transwells from the apical to basal, or from the basal to apical direction through the BBB. Permeability can be calculated using the following equation:

$$P_{\text{app}} = (dQ/dt)/(A \times C_0) \tag{6}$$

where P_{app} is the apparent permeability, dQ/dt is the mass transport rate obtained from the slope of the plot for amount transported vs time, A is the apparent surface area of the monolayer of cells, and C_0 is the initial concentration of drug treatment.[58] The efflux ratio is the P_{app}

from the basal to apical direction over the P_{app} in the apical to basal direction and can be useful for efflux protein-mediated assays.[58]

PHARMACOKINETIC METHODS FOR OVERCOMING THE BLOOD–BRAIN BARRIER

Peptide Mimics

Peptide mimics, or *peptidomimetics*, are synthesized short-protein-like chains that have been designed to mimic an endogenous peptide or a component of that endogenous protein.[60] Molecular properties, such as biological activity of these peptide mimics, are modified so that when the mimic interacts with the same protein as their endogenous counterpart, it may alter the properties of its binding partner, promoting degradation or resulting in its inactivation.[60,61] Peptidomimetics, therefore, have the potential to modulate BBB permeability and to improve CNS-targeted drug delivery. With respect to BBB modulation, many of the peptides synthesized mimic an extracellular loop domain of one of the many TJ proteins.

For example, occludin, a predominant TJ protein, has been used to demonstrate the use of synthetic peptides in modulating BBB permeability. The peptide mimic corresponding to the second extracellular domain of occludin was capable of reversibly reducing the protein levels of occludin present at TJs. This resulted in both a decrease in transepithelial electrical resistance while simultaneously increasing paracellular permeability. Highly selective alteration was noted at the TJ, with only occludin affected, whereas other components of TJs (e.g., ZO-1, ZO-2) were unaffected. This highlights the specificity of this treatment, a property that may prove to be a major benefit in regard to developing drug delivery strategies.[62]

Peptide mimics have also been designed to target another predominant TJ protein group/family, the claudins. C1C2 is a claudin-1 targeted peptidomimetic, emulating the first extracellular loop domain of claudin 1 protein. In a proof-of-principle experiment demonstrating how a peptide mimic could be used in conjunction with a therapeutic to enhance efficacy, C1C2 was administered to Wistar rats, which resulted in the transient modulation of the blood–nerve barrier through the reduction of claudin-1 expression, which in turn enhanced the analgesic properties of opioid-based agents.[63] Additionally, the C1C2 peptide mimic has been noted to affect the localization of other claudins (claudin-1–claudin-5) as well as occludin, redistributing these proteins from junctions to the cytosol, which results in a reversible increase in permeability of the various barriers in vivo.[64]

Peptide mimic zonula occludens toxin, an enterotoxin produced by *Vibrio cholerae*, has been shown to enhance drug delivery through the paracellular pathway by reversibly opening TJs.[65,66] Additionally, zonula occludens toxin was shown to be nontoxic and to increase permeability in a time- and dosage-dependent manner, which are important properties to consider if zonula occludens toxin is to be developed as an adjunct therapy.[66] Furthermore, ΔG, the biologically active 12 kDa fragment of zonula occludens toxin, acts as an absorption enhancer, improving bioavailability of therapeutic agents across TJ-mediated barriers.[67] This has been useful in enhancing delivery of anticancer drugs, such as methotrexate and paclitaxel. These drugs are hydrophilic and lipophilic, respectively; these properties result in poor brain distribution without the help of ΔG.[67] The coadministration of ΔG with a CNS-targeted compound shows great therapeutic potential.

RNA Interference

RNA interference (RNAi) is a process that allows for specific downregulation of a distinct mRNA transcript through an endogenous degradation system.[68,69] Synthetic small interfering RNA (siRNA) is relatively novel technology. It has been less than 20 years since proof-of-principle experiments demonstrated the efficacy of RNAi for highly specific gene knockdown; however, its potential has resulted in it becoming a multibillion-dollar industry.[70] Due to its high level of specificity, RNAi has been used to degrade disease-causing proteins such as age-related macular degeneration with great success, and some clinical trials examining RNAi have reached phase III.[71-73] RNAi is a promising technology, with potential medical applications not only in the direct destruction of disease-causing proteins, but also in other areas, including the targeted disruption of TJ proteins of the BBB and modulation of BBB integrity to facilitate drug delivery.

As previously described, claudin-5 is a predominant TJ protein, and the most highly enriched in the TJs of the BBB. siRNA targeting this protein has shown successful reduction in the claudin-5 protein level and transient modulation of the BBB to molecules of up to 1 kDa in size.[74] This example demonstrates the ability of these TJ-protein-targeting siRNAs to alter the size selectivity of the TJ without affecting the overall constitution of the structure, as none of the other TJ proteins are affected. Therefore, the application of RNAi techniques to alter BBB permeability may improve the pharmacokinetics of drugs whose site of action lies behind these barriers.

Campbell et al.[75] used RNAi to target claudin-5 in the TJs of the blood–retina barrier of a rodent model of autosomal recessive retinitis pigmentosa (IMPDH −/−). This model lacks an essential enzyme required for the synthesis of GTP, a factor required for visual phototransduction. However, systemic replacement of GTP was not previously possible due to its molecular weight (i.e., 523 Da), which prevents passive diffusion through the barrier. RNAi treatment and the resulting transient size-selective modulation of the barrier allowed systemic administration of GTP to bypass the blood–retina barrier and improve visual function. This adjunct therapy provided therapeutic effects with limited side effects, and provides a platform for further development of novel therapeutics for disorders that currently have few treatment options, if any.[75]

Intra-Arterial Hyperosmolar Infusion

The infusion of hyperosmolar solutions into the internal carotid artery was one of the earliest methods of BBB disruption, used to improve bioavailability of drugs in the CNS.[76] This temporary disruption of the BBB is the result of hyperosmolar solutions such as mannitol, which shrinks ECs and creates gaps in what were once "tight" junctions. This method has been shown effective in both rats and humans.[77]

Additionally, a lower concentration of mannitol administered intravenously has shown promise in enhancing delivery of chemotherapeutics to brain tumors.[78] Despite the potential of mannitol infusion for BBB disruption (and thus its ability to enhance drug delivery), a number of concerns surround the use of this method, including the complex surgical procedures required for its administration and a significant risk of focal seizure development as a major adverse effect.[79,80] Overall, this method carries a very heavy risk burden, but with enhanced knowledge and development of better technology, less invasive methods with better safety profiles may be on the horizon.

Focused Ultrasound

Focused ultrasound (FUS), coupled with circulating microbubbles, is similar to mannitol conceptually: that is, altered BBB integrity is the result of physical disruption to the barrier. Microbubbles are composed of air or perfluorocarbon enclosed in a lipid/albumin carrier and are administered intravenously.[81] The ultrasound is applied transcranially.[2] Although the exact mechanism by which FUS induces reversible opening of the BBB is still under investigation, the likely explanation is that microbubbles expand and oscillate when they enter the FUS field, resulting in alterations to the blood vessel walls, such as widening TJs and increasing BBB permeability overall.[2,82] These microbubbles are not necessarily essential, but their presence allows use of a lower-intensity ultrasound. Additionally, as the name suggests, the treatment is focused; in other words, it affects only precise neural regions of interest. The lower intensity and the specificity of FUS reduce potential harm, highlighting the major technological advances that have occurred since the mannitol disruption method was first put forward. A range of CNS-specific therapeutics that would usually be blocked by the BBB has been delivered successfully to their site of action following FUS application, including chemotherapeutics and stem cells.[83,84] With FUS, BBB disruption lasts between 4 and 24 hours.[85] FUS is still in its nascent stages and its precise mechanism of action still needs to be elucidated. Therefore, further study of its mechanisms and any adverse effects must be carried out before FUS begins to achieve any clinically relevant results.

Molecular Trojan Horse Drug Delivery Across The BBB

The molecular Trojan horse (MTH) approach involves using natural transport mechanisms native to the BBB. This is a particularly effective drug delivery system for larger molecules that have difficulty crossing the BBB, especially compared to other techniques, such as FUS.[40,86] MTH technology often involves endogenous receptor-mediated transport (RMT) mechanisms, such as those described below, that take up ligands such as growth factors, enzymes, and circulating plasma proteins into the brain.[87,88]

Insulin Receptors

Brain ECs possess insulin receptors, which aggregate in "coated pits" on the plasma membrane and can be used to deliver therapeutics. As insulin molecules in the circulation bind to these receptors, the pits invaginate and form coated vesicles. The endosome then acidifies and the insulin dissociates from the receptor and traverses the membrane.[87,89] This efficient transport is evidenced by the fact that insulin mRNA is not present in the brain, but insulin itself is present in the brain and in the periphery derived from blood.[90-92]

Ligands acting similarly to endogenous insulin, such as genetically engineered chimeric or humanized variation murine monoclonal antibodies against human insulin receptor (HIRMAb), can be used to cross from the luminal to the abluminal space of the BBB.[92,93] Other nanoparticles targeting this receptor also offer possible treatment strategies,[94] such as ^{125}I-Abeta1-40 conjugated to 83-14 monoclonal antibody (mAb), which is used to diagnose Alzheimer disease.[87,95,96] However, this pathway must be used with caution, as a change in insulin metabolism may affect endogenous glucose levels in the brain as well, producing adverse effects.[92,97] In patients with existing disease conditions such as diabetes mellitus or Alzheimer disease, the levels of endogenous insulin receptor may need to be factored into strategies for drug delivery.[94,98]

Transferrin Receptor

Transferrin receptor (TfR) expression is regulated by the iron regulatory protein system.[94,99] Iron accumulation has been associated with cell damage, so the TfR-mediated transcytosis of transferrin-bound iron through the BBB is pertinent to drug delivery for many neurodegenerative disorders.[92,100] TfR mediates bidirectional transport of holo-transferrin from blood to brain, and the reverse process of apo-transferrin from brain to blood.[40,101] Transferrin itself is an endogenous peptide that can be used via the MTH approach to cross the BBB.[40] Alternatively, a mAb such as OX26 can enter the brain using this receptor pathway, as it is able to bind to the TfR.[2,102]

However, OX26, along with other mAbs such as R17-217 and 8D3, bind to an alternative binding site on the same receptor.[87,103] Drugs attached to carriers mimicking the transferrin substrate can be used to provide treatment through this particular transcytosis mechanism,[2] and this approach has been successfully used to deliver 5-flurouracil through coupling the drug with Tf.[104] Cationic liposomes coated with transferrin also resulted in an increase in luciferase gene expression in C6 glioma cells.[87,105] With certain conditions (e.g., brain injury), TfR levels noticeably increase and therefore can be used to target for drug delivery.[87]

Limitations of this method include the fact that the endogenous plasma concentration of transferrin is 25 μmol/L, which is very high, and has a tendency to saturate the Tf-binding sites on the TfR.[92,106] It may be preferable to use mAbs that can bind to available epitopes on receptors that may not be the endogenous ligand binding site.[92]

Lipoprotein Receptors

Low-density lipoprotein receptor (LDLR) and lipoprotein receptor-related protein (LRP) types 1 and 2 are related and are present on the BBB.[87] LRP type 1 is a scavenger receptor that can bind to several ligands and mediates endocytosis of ligands into cells (apolipoprotein E, alpha-2macroglobulin, and amyloid beta precursor protein).[107] Proteins targeting LRP1 are taken up at slightly lower rates than proteins that target transcytosis receptors (e.g., TfR and insulin receptor).[87,92] Similarly, polysorbate-80-coated poly(butylcyanoacrylate) nanoparticles, which access this pathway, allow drugs that do not typically penetrate the BBB (e.g., doxorubicin, loperamide, and tubocurarine) to get through.[87,108]

LRP is expressed in diseases involved with inflammatory dysregulation and inhibits inflammation in diseases such as Parkinson disease, Alzheimer disease, multiple sclerosis, and several others, and can be a useful mechanism as part of MTH.[87,109] Apolipoprotein E is an excellent Trojan horse molecule, as it aids in transportation of lipids to the CNS and can be used to coat nanoparticles.[2,110] Nanoparticles like siRNA can also potentially mimic low-density lipoproteins and interact with this pathway, and become endocytosed into the brain.[111]

Diphtheria Toxin Receptor

Diphtheria toxin receptor is a precursor of heparin-binding epidermal growth factor, and this transport receptor endocytoses molecules from the blood into the brain across the BBB.[87,112] It has no endogenous ligands and is not involved with transporting nutrients across the BBB. However, this receptor is also involved in inflammation in brain diseases, much like the LDLR. Therefore, due to the prevalence of diphtheria toxin receptor in disease conditions and because of its lack of involvement in essential functions, it is useful for MTH therapeutic molecules to deliver agents into the brain using mAbs.[87,113,114]

Overall, it can be productive to target various RMT systems with Trojan horse molecules and to use endogenous pathways to deliver drugs across the BBB, depending on the expression levels of certain receptors based on the disease state.

Immunoglobulin G Fusion Proteins and Molecular Trojan Horse Technology

Immunoglobulin G (IgG) molecules do not reach the brain unless the BBB is disrupted, or unless they are specifically engineered to do so.[5] An IgG molecule can be genetically engineered with various fusion proteins to serve as MTH molecules and use RMT to enter the brain. This is done by fusing the therapeutic protein to the carboxyl terminus of the IgG molecule heavy chain, which can consist of neurotrophin, a decoy receptor, lysosomal enzymes, or a therapeutic antibody.[92,115] The IgG domain is a mAb that is fused to the therapeutic antibody and is then free to target the endogenous BBB receptor (e.g., TfR, insulin receptor), serving as the MTH vessel to breach the BBB.[116,117] For example, a HIRMAb-IDS (iduronate 2-sulfatase, a recombinant lysosomal enzyme) fusion protein was created, and relative to IDS delivery into the brain on its own, the fusion protein uptake was increased 34-fold.[116] Cationized immunoglobulins are also useful as drug carriers.[2]

Avidin-Biotin Technology in Molecular Trojan Horse Technology

Peptide pharmaceuticals can also be reengineered with avidin-biotin technology, through the creation of IgG-avidin fusion proteins.[92,118] One such molecule, avidin, has a high binding affinity to biotin, whose dissociation constant K_D is10^{-15}M, the dissociation $t_{1/2}$ is 89 days, which is the time at which the receptor is half occupied,[119,120] which is long and ideal for delivering therapeutics to the brain.

Pharmaceuticals such as siRNAs are monobiotinylated. They can then be conjugated to a streptavidin-conjugated mAb and are made up in separate vials before being mixed and administered intravenously.[92] For example, HIRMAb and TfRMAb were conjugated in a 1:1 ratio with streptavidin through thioether linkage, which then forms a complex with the biotinylated peptide.[120] This can be used as an MTH through the RMT pathways discussed, and these streptavidin- or avidin-conjugated IgG antibodies can enter the brain via transferrin or insulin transport, with the therapeutic peptide.

Efflux Transporter Inhibitors

Sometimes even highly lipophilic molecules, which should have no problem crossing the BBB, cannot accumulate in the brain.[2,19] This can be partly attributed to efflux transporters such as P-gp, which is present in high concentrations on the luminal side of ECs of the BBB, and binds with weak cations and large lipophiles.[2,121] Efflux transporters are important in preventing entry of toxins and drugs across the BBB, and also in transporting molecules out of the brain and into the bloodstream.[87]

In some diseases, such as epilepsy and focal cerebral ischemia, P-gp is overexpressed and is thought to contribute to drug resistance.[87,122] Efflux transport by P-gp can therefore hinder successful drug delivery, and hence should be targeted by inhibitors such as verapamil, clarithromycin, erythromycin, or ritonavir.[57] Inhibitors block the substrate-binding site, changing the permeability of the BBB.[2] Alternatively, drugs can be chemically modified to avoid the substrate binding site on the P-gp transporter.[87] The P-gp inhibitor approach is most useful for delivering drugs in maximal concentrations for short durations.[87,123] However, inhibiting P-gp gives rise to some safety concerns, such as respiratory depression[57] or the entry of toxins into the brain.[2] These potential adverse effects must be taken into consideration when drug delivery options are assessed.

Viral Vectors

Viral vectors such as lentivirus, herpes simplex virus, adenovirus, and adeno-associated virus (AAV) can be used to deliver gene therapies into the brain through their natural ability to transcribe nucleic acids in infected cells.[94] Some limitations of this method include manufacturing costs and the safety of viral vectors—currently, AAVs are the safest to use clinically, but AAV delivery requires direct injection into the cerebrospinal fluid or stereotaxic injections into the brain, which can be difficult to perform.[94,124]

AAV6 and 8 showed relatively inadequate levels of entry into the CNS in neonate or adult mice.[125] Many other serotypes of AAV have been developed, of which AAV9 has had very high success bypassing the BBB. AAV9 vectors injected intravenously have a unique ability to be transduced in the CNS, whereas AAV9 infects astrocytic-perivascular-endfeet surrounding the capillary ECs, and may require RMT, transport proteins, or other mechanisms to cross the BBB.[125-127] In one study, investigators injected the tail veins of adult mice with doses of self-complementary AAV9-CB-GFP ranging from 4×10^{11} to 4×10^{12} particles, and demonstrated persistence of AAV9 vector even after 7 weeks postinjection.[125] Various AAVs have successfully been used to treat rodents, but more work is required to translate this success to human patients.[94,128]

Liposome Delivery of Therapeutics

Liposomes are effective vehicles for delivery of drugs into the brain. However, conventional liposomes, which are formed through naturally occurring phospholipids and cholesterol, are associated with high systemic plasma clearance—this is often a pharmacokinetic barrier to successful drug delivery.[129,130] To overcome this hurdle, conventional liposomes are coated with polyethylene glycol (PEG), which helps prevent binding of opsonins from the plasma and helps liposomes avoid detection by phagocytic cells, increasing persistence of these sterically stabilized PEG liposomes, which have a half-life of up to 90 hours in humans.[130-132] This makes these liposomes highly advantageous for clinical applications.

PEGylated liposomes can be used to target the brain for delivery of therapeutics through conjugation with a suitable targeting component. A special class of PEGylated immunoliposomes (PILs), which have been conjugated with mAbs, allow the entire complex to undergo receptor-mediated transcytosis across the BBB.[130,133] Using such mAbs coupled to the PEG chains on PILs, targeting the transferrin and insulin receptors, can act as MTH liposomes.[120,133] These sophisticated PILs have been shown to successfully deliver plasmid DNA encoding short hairpin RNAs against human epidermal growth factor receptor, using mouse TfR and human insulin receptor mAbs. Administering PILs weekly to mice with intracranial human brain cancer led to 90% survival increase through successful delivery of short hairpin RNAs targeting epidermal growth factor receptor.[120,134]

Specifically, other targeting vectors to the brain that can be coupled with PILs are cationized albumin, or the OX26 mAb specifically to the rat TfR135—transcytosis of OX26 immunoliposomes has successfully been shown in in vitro models of the BBB based in immortalized RBE4 rat brain capillary ECs.[130,135] Another PEGylated phospholipid method of delivery to the brain involved use of biotinylated PEGylated liposomes and a streptavidin-conjugated antibody of choice.[130,136] The biotin–streptavidin interaction has a dissociation constant of approximately 1.3×10^{-15} M, and has a strong noncovalent bond to deliver therapeutic agents.[130]

CONCLUSIONS

As innovative methods for enhanced drug delivery across the BBB continue to develop, methodological approaches must similarly adapt to the quantitative needs of pharmacokinetic analyses. In time, these novel approaches to surmounting the challenge of BBB drug delivery should lead to major advances in understanding, treatment, and management of neural malignancies.

References

1. Alavijeh MS, Chishty M, Qaiser MZ, Palmer AM. Drug metabolism and pharmacokinetics, the blood-brain barrier, and central nervous system drug discovery. *NeuroRx*. 2005;2(4):554-571.
2. Kasinathan N, Jagani HV, Alex AT, Volety SM, Rao JV. Strategies for drug delivery to the central nervous system by systemic route. *Drug Deliv*. 2015;22(3):243-257.
3. Serwer L, Hashizume R, Ozawa T, James CD. Systemic and local drug delivery for treating diseases of the central nervous system in rodent models. *J Vis Exp*. 2010;(42).

4. Azad TD, Pan J, Connolly ID, Remington A, Wilson CM, Grant GA. Therapeutic strategies to improve drug delivery across the blood-brain barrier. *Neurosurg Focus*. 2015;38(3):E9.
5. Pardridge WM. The blood-brain barrier: bottleneck in brain drug development. *NeuroRx*. 2005;2(1):3-14.
6. Abbott NJ, Friedman A. Overview and introduction: the blood-brain barrier in health and disease. *Epilepsia*. 2012;53(Suppl 6):1-6.
7. Schlageter KE, Molnar P, Lapin GD, Groothuis DR. Microvessel organization and structure in experimental brain tumors: microvessel populations with distinctive structural and functional properties. *Microvasc Res*. 1999; 58(3):312-328.
8. Abbott NJ, Ronnback L, Hansson E. Astrocyte-endothelial interactions at the blood-brain barrier. *Nat Rev Neurosci*. 2006;7(1):41-53.
9. Abbott NJ, Patabendige AA, Dolman DE, Yusof SR, Begley DJ. Structure and function of the blood-brain barrier. *Neurobiol Dis*. 2010;37(1):13-25.
10. Engelhardt B, Liebner S. Novel insights into the development and maintenance of the blood-brain barrier. *Cell Tissue Res*. 2014;355(3):687-699.
11. Hansen AJ. Effect of anoxia on ion distribution in the brain. *Physiol Rev*. 1985;65(1):101-148.
12. Saunders NR, Liddelow SA, Dziegielewska KM. Barrier mechanisms in the developing brain. *Front Pharmacol*. 2012;3:46.
13. Luissint AC, Artus C, Glacial F, Ganeshamoorthy K, Couraud PO. Tight junctions at the blood brain barrier: physiological architecture and disease-associated dysregulation. *Fluids Barriers CNS*. 2012;9(1):23.
14. Obermeier B, Daneman R, Ransohoff RM. Development, maintenance and disruption of the blood-brain barrier. *Nat Med*. 2013;19(12):1584-1596.
15. Keaney J, Campbell M. The dynamic blood-brain barrier. *FEBS J*. 2015;282(21):4067-4079.
16. Haseloff RF, Dithmer S, Winkler L, Wolburg H, Blasig IE. Transmembrane proteins of the tight junctions at the blood-brain barrier: structural and functional aspects. *Semin Cell Dev Biol*. 2015;38:16-25.
17. Tsukita S, Furuse M, Itoh M. Multifunctional strands in tight junctions. *Nat Rev Mol Cell Biol*. 2001;2(4):285-293.
18. Zlokovic BV. The blood-brain barrier in health and chronic neurodegenerative disorders. *Neuron*. 2008; 57(2):178-201.
19. Banks WA. Characteristics of compounds that cross the blood-brain barrier. *BMC Neurol*. 2009;9(Suppl 1):S3.
20. Upadhyay RK. Drug delivery systems, CNS protection, and the blood brain barrier. *Biomed Res Int*. 2014;2014.
21. Urquhart BL, Kim RB. Blood-brain barrier transporters and response to CNS-active drugs. *Eur J Clin Pharmacol*. 2009;65(11):1063-1070.
22. Stewart PA, Wiley MJ. Developing nervous tissue induces formation of blood-brain barrier characteristics in invading endothelial cells: a study using quail–chick transplantation chimeras. *Dev Biol*. 1981;84(1):183-192.
23. de Lange EC, de Boer BA, Breimer DD. Microdialysis for pharmacokinetic analysis of drug transport to the brain. *Adv Drug Deliv Rev*. 1999;36(2-3):211-227.
24. Kassel DB. Applications of high-throughput ADME in drug discovery. *Curr Opin Chem Biol*. 2004;8(3):339-345.
25. Bickel U, Yoshikawa T, Pardridge WM. Delivery of peptides and proteins through the blood-brain barrier. *Adv Drug Deliv Rev*. 2001;46(1-3):247-279.
26. Falcone JA, Salameh TS, Yi X, et al. Intranasal administration as a route for drug delivery to the brain: evidence for a unique pathway for albumin. *J Pharmacol Exp Ther*. 2014;351(1):54-60.
27. Dash AK, Elmquist WF. Separation methods that are capable of revealing blood-brain barrier permeability. *J Chromatogr B Anal Technol Biomed Life Sci*. 2003;797(1-2):241-254.
28. Weidman EK, Foley CP, Kallas O, et al. Evaluating permeability surface-area product as a measure of blood-brain barrier permeability in a murine model. *AJNR Am J Neuroradiol*. 2016;37(7):1267-1274.
29. Oldendorf WH. Measurement of brain uptake of radiolabeled substances using a tritiated water internal standard. *Brain Res*. 1970;24(2):372-376.
30. Doležal R, Karásková N, Musil K, et al. Characterization of the Penetration of the Blood–Brain Barrier by High-Performance Liquid Chromatography (HPLC) Using a Stationary Phase with an Immobilized Artificial Membrane. *Anal Lett*. 2018;51(15):2401-2414.
31. Heyn H, McCarthy DJ, Curry SH, Eisman MS, Anders MW. Brain uptake and biotransformation of remacemide hydrochloride, a novel anticonvulsant. *Drug Metab Dispos*. 1994;22(3):443-446.
32. Kaul S, Faiman MD, Lunte CE. Determination of GABA, glutamate and carbamathione in brain microdialysis samples by capillary electrophoresis with fluorescence detection. *Electrophoresis*. 2011;32(2):284-291.

33. Killian DM, Chikhale PJ. A bioreversible prodrug approach designed to shift mechanism of brain uptake for amino-acid-containing anticancer agents. *J Neurochem*. 2001;76(4):966-974.
34. Smith NW, Evans MB. Capillary zone electrophoresis in pharmaceutical and biomedical analysis. *J Pharm Biomed Anal*. 1994;12(5):579-611.
35. Zhang YN, Yang YF, Yang XW. Blood-brain barrier permeability and neuroprotective effects of three main alkaloids from the fruits of Euodia rutaecarpa with MDCK-pHaMDR cell monolayer and PC12 cell line. *Biomed Pharmacother*. 2018;98:82-87.
36. Oldendorf WH, Stoller BE, Tishler TA, Williams JL, Oldendorf SZ. Transient blood-brain barrier passage of polar compounds at low pH. *Am J Phys*. 1994;267(6 Pt 2):H2229-H2236.
37. Bronzino JD, Molecular PDR. *Cellular, and Tissue Engineering*. 1st ed. Boca Raton, FL: CRC Press; 2018.
38. Oldendorf WH. Carrier-mediated blood-brain barrier transport of short-chain monocarboxylic organic acids. *Am J Phys*. 1973;224(6):1450-1453.
39. Bickel U. How to measure drug transport across the blood-brain barrier. *NeuroRx*. 2005;2(1):15-26.
40. Pardridge WM. Molecular Trojan horses for blood-brain barrier drug delivery. *Discov Med*. 2006;6(34):139-143.
41. Renkin EM. Transport of potassium-42 from blood to tissue in isolated mammalian skeletal muscles. *Am J Phys*. 1959;197:1205-1210.
42. Wang W, Luo S, Chen Y, Li B, Hattori M. Effective separation and simultaneous determination of corynoxeine and its metabolites in rats by high-performance liquid chromatography with tandem mass spectrometry and application to pharmacokinetics and in vivo distribution in main organs. *Anal Sci*. 2016;32(6):705-707.
43. Takasato Y, Rapoport SI, Smith QR. An in situ brain perfusion technique to study cerebrovascular transport in the rat. *Am J Phys*. 1984;247(3 Pt 2):H484-H493.
44. Allen DD, Smith QR. Characterization of the blood-brain barrier choline transporter using the in situ rat brain perfusion technique. *J Neurochem*. 2001;76(4):1032-1041.
45. Dagenais C, Rousselle C, Pollack GM, Scherrmann JM. Development of an in situ mouse brain perfusion model and its application to mdr1a P-glycoprotein-deficient mice. *J Cereb Blood Flow Metab*. 2000;20(2):381-386.
46. Solon EG, Schweitzer A, Stoeckli M, Prideaux B. Autoradiography, MALDI-MS, and SIMS-MS imaging in pharmaceutical discovery and development. *AAPS J*. 2010;12(1):11-26.
47. Benveniste H, Drejer J, Schousboe A, Diemer NH. Regional cerebral glucose phosphorylation and blood flow after insertion of a microdialysis fiber through the dorsal hippocampus in the rat. *J Neurochem*. 1987;49 (3):729-734.
48. Westerink BH, De Vries JB. Characterization of in vivo dopamine release as determined by brain microdialysis after acute and subchronic implantations: methodological aspects. *J Neurochem*. 1988;51(3):683-687.
49. Bungay PM, Wang T, Yang H, Elmquist WF. Utilizing transmembrane convection to enhance solute sampling and delivery by microdialysis: theory and in vitro validation. *J Membr Sci*. 2010;348(1-2):131-149.
50. Bungay PM, Morrison PF, Dedrick RL. Steady-state theory for quantitative microdialysis of solutes and water in vivo and in vitro. *Life Sci*. 1990;46(2):105-119.
51. de Lange EC, de Boer AG, Breimer DD. Methodological issues in microdialysis sampling for pharmacokinetic studies. *Adv Drug Deliv Rev*. 2000;45(2-3):125-148.
52. de Lange ECM, de Boer AG, Breimer DD. Intracerebral microdialysis. In: Pardridge WM, ed. *Introduction to the Blood-Brain Barrier: Methodology, Biology and Pathology*. Cambridge: Cambridge University Press; 1998:94-112.
53. Nandi P, Lunte SM. Recent trends in microdialysis sampling integrated with conventional and microanalytical systems for monitoring biological events: a review. *Anal Chim Acta*. 2009;651(1):1-14.
54. Durk MR, Deshmukh G, Valle N, Ding X, Liederer BM, Liu X. Use of subcutaneous and intraperitoneal administration methods to facilitate cassette dosing in microdialysis studies in rats. *Drug Metab Dispos*. 2018; 46(7):964-969.
55. Deshmukh G, Sun K, Liederer BM, Ding X, Liu X. Use of cassette dosing to enhance the throughput of rat brain microdialysis studies. *Drug Metab Dispos*. 2015;43(7):1123-1128.
56. Gustafsson S, Lindstrom V, Ingelsson M, Hammarlund-Udenaes M, Syvanen S. Intact blood-brain barrier transport of small molecular drugs in animal models of amyloid beta and alpha-synuclein pathology. *Neuropharmacology*. 2018;128:482-491.
57. Finch A, Pillans P. P-glycoprotein and its role in drug-drug interactions. *Aust Prescr*. 2014;37:137-139.
58. Agarwal S, Sane R, Gallardo JL, Ohlfest JR, Elmquist WF. Distribution of gefitinib to the brain is limited by P-glycoprotein (ABCB1) and breast cancer resistance protein (ABCG2)-mediated active efflux. *J Pharmacol Exp Ther*. 2010;334(1):147-155.

59. Agarwal S, Sane R, Ohlfest JR, Elmquist WF. The role of the breast cancer resistance protein (ABCG2) in the distribution of sorafenib to the brain. *J Pharmacol Exp Ther*. 2011;336(1):223-233.
60. Vagner J, Qu H, Hruby VJ. Peptidomimetics, a synthetic tool of drug discovery. *Curr Opin Chem Biol*. 2008; 12(3):292-296.
61. Qvit N, Rubin SJS, Urban TJ, Mochly-Rosen D, Gross ER. Peptidomimetic therapeutics: scientific approaches and opportunities. *Drug Discov Today*. 2017;22(2):454-462.
62. Wong V, Gumbiner BM. A synthetic peptide corresponding to the extracellular domain of occludin perturbs the tight junction permeability barrier. *J Cell Biol*. 1997;136(2):399-409.
63. Sauer RS, Krug SM, Hackel D, et al. Safety, efficacy, and molecular mechanism of claudin-1-specific peptides to enhance blood-nerve-barrier permeability. *J Control Release*. 2014;185:88-98.
64. Staat C, Coisne C, Dabrowski S, et al. Mode of action of claudin peptidomimetics in the transient opening of cellular tight junction barriers. *Biomaterials*. 2015;54:9-20.
65. Fasano A, Fiorentini C, Donelli G, et al. Zonula occludens toxin modulates tight junctions through protein kinase C-dependent actin reorganization, in vitro. *J Clin Invest*. 1995;96(2):710-720.
66. Karyekar CS, Fasano A, Raje S, Lu R, Dowling TC, Eddington ND. Zonula occludens toxin increases the permeability of molecular weight markers and chemotherapeutic agents across the bovine brain microvessel endothelial cells. *J Pharm Sci*. 2003;92(2):414-423.
67. Menon D, Karyekar CS, Fasano A, Lu R, Eddington ND. Enhancement of brain distribution of anticancer agents using DeltaG, the 12 kDa active fragment of ZOT. *Int J Pharm*. 2005;306(1-2):122-131.
68. Kim D, Rossi J. RNAi mechanisms and applications. *Biotechniques*. 2008;44(5):613-616.
69. O'Keeffe E, Campbell M. Modulating the paracellular pathway at the blood-brain barrier: current and future approaches for drug delivery to the CNS. *Drug Discov Today Technol*. 2016;20:35-39.
70. Greene C, Campbell M. Tight junction modulation of the blood brain barrier: CNS delivery of small molecules. *Tissue Barriers*. 2016;4(1).
71. Chakraborty C, Sharma AR, Sharma G, Doss CGP, Lee SS. Therapeutic miRNA and siRNA: Moving from bench to clinic as next generation medicine. *Mol Ther Nucleic Acids*. 2017;8:132-143.
72. Garba AO, Mousa SA. Bevasiranib for the treatment of wet, age-related macular degeneration. *Ophthalmol Eye Dis*. 2010;2:75-83.
73. Kaiser PK, Symons RC, Shah SM, et al. RNAi-based treatment for neovascular age-related macular degeneration by Sirna-027. *Am J Ophthalmol*. 2010;150(1):33-39 e32.
74. Campbell M, Kiang AS, Kenna PF, et al. RNAi-mediated reversible opening of the blood-brain barrier. *J Gene Med*. 2008;10(8):930-947.
75. Campbell M, Nguyen AT, Kiang AS, et al. An experimental platform for systemic drug delivery to the retina. *Proc Natl Acad Sci U S A*. 2009;106(42):17817-17822.
76. Rapoport SI. Effect of concentrated solutions on blood-brain barrier. *Am J Phys*. 1970;219(1):270-274.
77. Cosolo WC, Martinello P, Louis WJ. Christophidis N. Blood-brain barrier disruption using mannitol: time course and electron microscopy studies. *Am J Phys*. 1989;256(2 Pt 2):R443-R447.
78. Wakai A, McCabe A, Roberts I, Schierhout G. Mannitol for acute traumatic brain injury. *Cochrane Database Syst Rev*. 2013;(8).
79. Fieschi C, Lenzi GL, Zanette E, Orzi F, Passero S. Effects on EEG of the osmotic opening of the blood-brain barrier in rats. *Life Sci*. 1980;27(3):239-243.
80. Marchi N, Angelov L, Masaryk T, et al. Seizure-promoting effect of blood-brain barrier disruption. *Epilepsia*. 2007;48(4):732-742.
81. Fan C-H, Yeh C-K. Microbubble-enhanced focused ultrasound-induced blood-brain barrier opening for local and transient drug delivery in central nervous system disease. *J Med Ultrasound*. 2014;22(4):183-193.
82. Sheikov N, McDannold N, Vykhodtseva N, Jolesz F, Hynynen K. Cellular mechanisms of the blood-brain barrier opening induced by ultrasound in presence of microbubbles. *Ultrasound Med Biol*. 2004;30(7):979-989.
83. Burgess A, Ayala-Grosso CA, Ganguly M, Jordao JF, Aubert I, Hynynen K. Targeted delivery of neural stem cells to the brain using MRI-guided focused ultrasound to disrupt the blood-brain barrier. *PLoS ONE*. 2011;6(11).
84. Park J, Zhang Y, Vykhodtseva N, Jolesz FA, McDannold NJ. The kinetics of blood brain barrier permeability and targeted doxorubicin delivery into brain induced by focused ultrasound. *J Control Release*. 2012;162(1):134-142.
85. Burgess A, Hynynen K. Noninvasive and targeted drug delivery to the brain using focused ultrasound. *ACS Chem Neurosci*. 2013;4(4):519-526.
86. Deng CX. Targeted drug delivery across the blood-brain barrier using ultrasound technique. *Ther Deliv*. 2010; 1(6):819-848.

87. Chen Y, Liu L. Modern methods for delivery of drugs across the blood-brain barrier. *Adv Drug Deliv Rev*. 2012; 64(7):640-665.
88. Hymel D, Peterson BR. Synthetic cell surface receptors for delivery of therapeutics and probes. *Adv Drug Deliv Rev*. 2012;64(9):797-810.
89. Fan JY, Carpentier JL, Gorden P, et al. Receptor-mediated endocytosis of insulin: role of microvilli, coated pits, and coated vesicles. *Proc Natl Acad Sci U S A*. 1982;79(24):7788-7791.
90. Coker 3rd GT, Studelska D, Harmon S, Burke W, O'Malley KL. Analysis of tyrosine hydroxylase and insulin transcripts in human neuroendocrine tissues. *Brain Res Mol Brain Res*. 1990;8(2):93-98.
91. Duffy KR, Pardridge WM. Blood-brain barrier transcytosis of insulin in developing rabbits. *Brain Res*. 1987; 420(1):32-38.
92. Pardridge WM. Drug transport across the blood-brain barrier. *J Cereb Blood Flow Metab*. 2012;32(11):1959-1972.
93. Coloma MJ, Lee HJ, Kurihara A, et al. Transport across the primate blood-brain barrier of a genetically engineered chimeric monoclonal antibody to the human insulin receptor. *Pharm Res*. 2000;17(3):266-274.
94. Dong X. Current strategies for brain drug delivery. *Theranostics*. 2018;8(6):1481-1493.
95. Wu D, Yang J, Pardridge WM. Drug targeting of a peptide radiopharmaceutical through the primate blood-brain barrier in vivo with a monoclonal antibody to the human insulin receptor. *J Clin Invest*. 1997;100(7):1804-1812.
96. Zhang Y, Schlachetzki F, Pardridge WM. Global non-viral gene transfer to the primate brain following intravenous administration. *Mol Ther*. 2003;7(1):11-18.
97. Xie L, Helmerhorst E, Taddei K, Plewright B, Van Bronswijk W, Martins R. Alzheimer's beta-amyloid peptides compete for insulin binding to the insulin receptor. *J Neurosci*. 2002;22(10):RC221.
98. Ho L, Yemul S, Knable L, et al. Insulin receptor expression and activity in the brains of nondiabetic sporadic Alzheimer's disease cases. *Int J Alzheimers Dis*. 2012;2012.
99. Jeong SY, Crooks DR, Wilson-Ollivierre H, et al. Iron insufficiency compromises motor neurons and their mitochondrial function in Irp2-null mice. *PLoS ONE*. 2011;6(10).
100. Ali SA, Joao HC, Hammerschmid F, Eder J, Steinkasserer A. Transferrin trojan horses as a rational approach for the biological delivery of therapeutic peptide domains. *J Biol Chem*. 1999;274(34):24066-24073.
101. Skarlatos S, Yoshikawa T, Pardridge WM. Transport of [125I]transferrin through the rat blood-brain barrier. *Brain Res*. 1995;683(2):164-171.
102. Huwyler J, Wu D, Pardridge WM. Brain drug delivery of small molecules using immunoliposomes. *Proc Natl Acad Sci U S A*. 1996;93(24):14164-14169.
103. Rocha S. Targeted drug delivery across the blood brain barrier in Alzheimer's disease. *Curr Pharm Des*. 2013; 19(37):6635-6646.
104. Soni V, Kohli DV, Jain SK. Transferrin coupled liposomes as drug delivery carriers for brain targeting of 5-florouracil. *J Drug Target*. 2005;13(4):245-250.
105. da Cruz MT, Simoes S, de Lima MC. Improving lipoplex-mediated gene transfer into C6 glioma cells and primary neurons. *Exp Neurol*. 2004;187(1):65-75.
106. Qian ZM, Li H, Sun H, Ho K. Targeted drug delivery via the transferrin receptor-mediated endocytosis pathway. *Pharmacol Rev*. 2002;54(4):561-587.
107. Shibata M, Yamada S, Kumar SR, et al. Clearance of Alzheimer's amyloid-ss(1-40) peptide from brain by LDL receptor-related protein-1 at the blood-brain barrier. *J Clin Invest*. 2000;106(12):1489-1499.
108. Schroder U, Sabel BA. Nanoparticles, a drug carrier system to pass the blood-brain barrier, permit central analgesic effects of i.v. dalargin injections. *Brain Res*. 1996;710(1-2):121-124.
109. Dehouck B, Fenart L, Dehouck MP, Pierce A, Torpier G, Cecchelli R. A new function for the LDL receptor: transcytosis of LDL across the blood-brain barrier. *J Cell Biol*. 1997;138(4):877-889.
110. Zensi A, Begley D, Pontikis C, et al. Albumin nanoparticles targeted with Apo E enter the CNS by transcytosis and are delivered to neurones. *J Control Release*. 2009;137(1):78-86.
111. Mathupala SP. Delivery of small-interfering RNA (siRNA) to the brain. *Expert Opin Ther Pat*. 2009;19(2):137-140.
112. Mekada E, Okada Y, Uchida T. Identification of diphtheria toxin receptor and a nonproteinous diphtheria toxin-binding molecule in Vero cell membrane. *J Cell Biol*. 1988;107(2):511-519.
113. Gaillard P, Brink A, de Boer AG. Diphtheria toxin receptor-targeted brain delivery. *Int Congr Ser*. 2005;1277:185-198.
114. Lobeck K, Drevet P, Leonetti M, et al. Towards a recombinant vaccine against diphtheria toxin. *Infect Immun*. 1998;66(2):418-423.

115. Boado RJ, Zhang Y, Zhang Y, Xia CF, Pardridge WM. Fusion antibody for Alzheimer's disease with bidirectional transport across the blood-brain barrier and abeta fibril disaggregation. *Bioconjug Chem*. 2007;18(2):447-455.
116. Boado RJ, Hui EK, Lu JZ, Sumbria RK, Pardridge WM. Blood-brain barrier molecular trojan horse enables imaging of brain uptake of radioiodinated recombinant protein in the rhesus monkey. *Bioconjug Chem*. 2013; 24(10):1741-1749.
117. Boado RJ, Hui EK, Lu JZ, Zhou QH, Pardridge WM. Selective targeting of a TNFR decoy receptor pharmaceutical to the primate brain as a receptor-specific IgG fusion protein. *J Biotechnol*. 2010;146(1-2):84-91.
118. Penichet ML, Kang YS, Pardridge WM, Morrison SL, Shin SU. An antibody-avidin fusion protein specific for the transferrin receptor serves as a delivery vehicle for effective brain targeting: initial applications in anti-HIV antisense drug delivery to the brain. *J Immunol*. 1999;163(8):4421-4426.
119. Magalhaes ML, Czekster CM, Guan R, Malashkevich VN, Almo SC, Levy M. Evolved streptavidin mutants reveal key role of loop residue in high-affinity binding. *Protein Sci*. 2011;20(7):1145-1154.
120. Xia CF, Zhang Y, Zhang Y, Boado RJ, Pardridge WM. Intravenous siRNA of brain cancer with receptor targeting and avidin-biotin technology. *Pharm Res*. 2007;24(12):2309-2316.
121. Bendayan R, Ronaldson PT, Gingras D, Bendayan M. In situ localization of P-glycoprotein (ABCB1) in human and rat brain. *J Histochem Cytochem*. 2006;54(10):1159-1167.
122. Spudich A, Kilic E, Xing H, et al. Inhibition of multidrug resistance transporter-1 facilitates neuroprotective therapies after focal cerebral ischemia. *Nat Neurosci*. 2006;9(4):487-488.
123. Begley DJ. ABC transporters and the blood-brain barrier. *Curr Pharm Des*. 2004;10:1295-1312.
124. Gray SJ, Woodard KT, Samulski RJ. Viral vectors and delivery strategies for CNS gene therapy. *Ther Deliv*. 2010;1 (4):517-534.
125. Foust KD, Nurre E, Montgomery CL, Hernandez A, Chan CM, Kaspar BK. Intravascular AAV9 preferentially targets neonatal neurons and adult astrocytes. *Nat Biotechnol*. 2009;27(1):59-65.
126. Shen F, Kuo R, Milon-Camus M, et al. Intravenous delivery of adeno-associated viral vector serotype 9 mediates effective gene expression in ischemic stroke lesion and brain angiogenic foci. *Stroke*. 2013;44(1):252-254.
127. Vagner T, Dvorzhak A, Wojtowicz AM, Harms C, Grantyn R. Systemic application of AAV vectors targeting GFAP-expressing astrocytes in Z-Q175-KI Huntington's disease mice. *Mol Cell Neurosci*. 2016;77:76-86.
128. Natarajan G, Leibowitz JA, Zhou J, et al. Adeno-associated viral vector-mediated preprosomatostatin expression suppresses induced seizures in kindled rats. *Epilepsy Res*. 2017;130:81-92.
129. Frank MM. The reticuloendothelial system and bloodstream clearance. *J Lab Clin Med*. 1993;122(5):487-488.
130. Schnyder A, Huwyler J. Drug transport to brain with targeted liposomes. *NeuroRx*. 2005;2(1):99-107.
131. Ishii T, Asai T, Oyama D, et al. Amelioration of cerebral ischemia-reperfusion injury based on liposomal drug delivery system with asialo-erythropoietin. *J Control Release*. 2012;160(1):81-87.
132. Moghimi SM, Patel HM. Opsonophagocytosis of liposomes by peritoneal macrophages and bone marrow reticuloendothelial cells. *Biochim Biophys Acta*. 1992;1135(3):269-274.
133. Huwyler J, Yang J, Pardridge WM. Receptor mediated delivery of daunomycin using immunoliposomes: pharmacokinetics and tissue distribution in the rat. *J Pharmacol Exp Ther*. 1997;282(3):1541-1546.
134. Zhang Y, Zhang YF, Bryant J, Charles A, Boado RJ, Pardridge WM. Intravenous RNA interference gene therapy targeting the human epidermal growth factor receptor prolongs survival in intracranial brain cancer. *Clin Cancer Res*. 2004;10(11):3667-3677.
135. Cerletti A, Drewe J, Fricker G, Eberle AN, Huwyler J. Endocytosis and transcytosis of an immunoliposome-based brain drug delivery system. *J Drug Target*. 2000;8(6):435-446.
136. Pardridge WM. shRNA and siRNA delivery to the brain. *Adv Drug Deliv Rev*. 2007;59(2-3):141-152.

CHAPTER

4

Pharmacokinetics of Drug Delivery Past the Blood–Brain Barrier

John M. McGregor, Nancy D. Doolittle†, Emily Youngers†, Susan D. Bell*, Edward A. Neuwelt‡*

***Department of Neurological Surgery, The Ohio State University Wexner Medical Center, Columbus, OH, United States †Department of Neurology, Oregon Health & Science University, Portland, OR, United States ‡Departments of Neurology and Neurosurgery, Oregon Health & Science University, Portland, OR, United States**

INTRODUCTION

The blood–brain barrier (BBB) and the blood–cerebrospinal fluid barrier are the anatomic and physiologic structures that protect the sensitive central nervous system (CNS) homeostasis from exposure to molecules that are toxic or that would result in adverse neurologic effects. The protective functions of the BBB have hampered the ability of therapeutics to cross into the CNS and effect the treatment of various diseases, including degenerative disease, trauma, ischemia, and tumors. As further studies have elucidated the anatomy and functional capacities of the BBB, characteristics have emerged that have enabled the creation of therapeutic molecules that may be able to concentrate more easily in the CNS, perform beneficial effects, and, with appropriate design, may minimize CNS or systemic toxicities.

Drugs that can bypass the BBB may have a unique biochemical design of therapeutic molecules that allows them to cross the BBB and remain in therapeutic concentrations in the CNS. These therapies may circumvent the BBB through temporary opening of the barrier, or they may bypass the BBB via direct administration into the CNS. As therapeutic agents are developed, their benefits must be tested—namely, their ability to enter and remain in the CNS, and their potential toxic and adverse effects on the CNS and other systemic systems (given that they are designed to move higher concentrations of substance into tissue). In this chapter, we review the anatomy that makes up the BBB, describe methods available to measure pharmacokinetics across the BBB, and discuss various methods for delivering therapeutics across the

Nervous System Drug Delivery
https://doi.org/10.1016/B978-0-12-813997-4.00004-9

© 2019 Elsevier Inc. All rights reserved.

BBB, including the most clinically characterized method: Osmotic BBB disruption. As a more complete understanding of BBB physiology evolves and as further strategies for BBB manipulation are uncovered, advances in drug and molecule delivery will result.

ANATOMY OF THE BBB

The BBB consists of over 400 miles of capillaries lined by endothelial cells, which themselves are surrounded by astrocytes, pericytes, and basal lamina. Altogether, these elements form the neurovascular unit (NVU).[1,2] This anatomic arrangement creates the functional BBB that protects the CNS from unwanted exposure. The concept of a BBB was first identified based on experiments performed in 1909 by Goldman, who noted that dye injected intravenously in the rat stained every other organ and system except the brain and the spinal cord. Alternately, dye injected into the CNS via the cerebrospinal fluid did not escape the CNS, staining only the brain and spinal cord.[3-5]

The BBB is a variably permeable layer. The endothelial cells of CNS capillaries form tight junctions to limit permeability. Transmembrane proteins claudin,[6] occludin,[7] junctional adhesion molecules,[8] and zona occludins,[9] among others, assemble across the adjacent endothelial cells to form the tight junctions. These molecules limit the diffusion of substances across the endothelial layer between cells. The endothelial cells of the CNS do not have fenestrations or pores typical of endothelial cells in the periphery. They have few vesicles.[1,10] As a result, while small lipophilic molecules less than 400 Da can cross the BBB passively, larger and polar molecules cannot.[11]

Still, certain specific molecules larger than 400 Da and specific polar molecules do cross the BBB by means of select transport mechanisms across the NVU. Several types of systems allow transport of molecules across the BBB. Endothelial cells do admit certain cationic species via transcytosis, so-called adsorptive-mediated transport. Additionally, active transport systems exist in the endothelial cells but are comparatively limited to only the passage of important endogenous molecules, such as amino acids, glucose, and vitamins. The molecules that cross the BBB in this manner are subject to specific transport-protein-mediated influx.[2,11,12] These proteins mediate the binding of the intended molecules at the vascular side, and they are transported across the endothelial cell to the abluminal CNS side for release.[2,13,14] These molecules can be divided into two general categories of transport-protein-mediated influx: carrier- and receptor-mediated transports.

Carrier-mediated transports use a specific surface carrier molecule that binds and admits its specific ligand into the cell cytoplasm, where it diffuses across to the abluminal surface and releases the ligand to the CNS. Receptor-mediated systems use a surface receptor to bind to the specific ligand, at which point the whole structure undergoes receptor-induced endocytosis, transportation of the vesicle across the cytoplasm, exocytosis on the abluminal side, and ultimately, release of the ligand. These receptor-mediated transport proteins include the insulin and the transferrin receptor systems.[2,13,14] There is a limited number of certain peptides, called peptide vectors, which seem to have the ability to cross the endothelial cell layers in mechanisms still to be elucidated. Finally, a few cells themselves (e.g., monocytes, neutrophils, and macrophages) have demonstrated the ability to traverse the BBB both in healthy subjects and in disease states. These transport systems are much more limited in the CNS than in the systemic circulation. These features allow the NVU to maintain a protected environment for the CNS, and as a result restrict influx from the serum of almost all larger molecules, and 98% of even small molecules.[11,15]

In addition to influx regulation, the BBB has efflux transport proteins that further limit the bioavailability of molecules. Proteins such as p-glycoprotein, breast cancer resistance protein, and multidrug resistance proteins remove molecules as they enter the brain. Even many lipophilic molecules may cross the BBB easily only to be transported back across by efflux proteins, thereby limiting their ability to affect the CNS.[2,11,16,17] In addition, in disease states, the BBB may be structurally and functionally altered, making it variable in its permeability. This only adds to the challenges of predicting the behavior of therapeutic molecules and their potential to cross the BBB in disease states.[2]

METHODS OF MEASURING PHARMACOKINETICS ACROSS THE BBB

One of the first steps in identifying substances that might be useful in treating various conditions in the CNS from degenerative disease, infection, trauma, or tumor, is to identify which are able to reach the CNS effectively. Both the ability to cross the BBB and the ability to remain in the CNS in concentrations sufficient to provide action are required. Many methods exist for analyzing the ability of therapeutics to cross the BBB from the blood stream, measuring their resulting concentrations in the CNS, and thus anticipating their functional capacity. In vitro and in vivo measurement techniques have limitations; therefore, they have complementary roles in helping identify likely beneficial molecules. These techniques are reviewed in detail by Bickel,[18] and are summarized further in this chapter.

Classic in vivo experimental studies test the concentration of a substance in the brain tissue compared to the vascular concentration at either a single point or at various time points. This requires tissue sampling and measurement of CNS drug levels in tissue samples. These measures assume that there is no change in concentration or effect after a substance crosses the BBB (e.g., efflux, metabolism, or binding to non-effector molecules). The vascular compartment must either go through a washout phase prior to sampling to ensure accurate measuring of tissue concentrations, or a non-absorbed substance is used and measured in comparison to estimate the fraction of drug absorbed vs that which remains intravascularly. In the single-sample setting, brain tissue concentration is

$$C_{\mathrm{br}} = K_{\mathrm{in}} \times \mathrm{AUC}^{T}{}_{0} \tag{1}$$

where C_{br} is brain tissue concentration corrected for vascular concentration and K_{in} is a value for tissue clearance and AUC (area under the curve) is the integral of plasma concentration (C_{p}) from time$=0$ to time$=T$ as the drug is being absorbed.[19]

When measured in comparison to a non-absorbed substance, a total plasma vascular volume (V_0) is obtained for the tissue sample. In this setting, brain tissue concentration is

$$C_{\mathrm{br}} = (V_{\mathrm{D}} - V_0)C_{\mathrm{p}}(T) \tag{2}$$

where V_{D} is the volume of distribution of the drug in the total brain at the time T the sample was acquired.

The volume of distribution is defined as

$$V_D = C_{br}/C_p \tag{3}$$

Based on these experimentally measured values, the tissue clearance becomes

$$K_{in} = (V_D - V_0)C_p(T)/\mathrm{AUC}^T{}_0 \tag{4}$$

The ability of a substance to cross the BBB (i.e., permeability) is also related to the capillary transport, as identified in the Renkin-Crone equation:

$$K_{in} = F\left(1 - e^{-PS/F}\right) \tag{5}$$

where PS is the product of permeability (P) and capillary surface area (S), and F is a measure of blood flow.[18-22] The value for permeability of a substance is then defined by experimentally identified values and set parameters.

If multiple tissue samplings are used, a series of concentrations are obtained over time.[23] In the analysis, a linear absorption across the BBB is assumed. In this setting, the increasing tissue concentration over time extrapolated back to the point of injection should lead to a non-zero y-intercept at $T = 0$, which would indicate the vascular distribution at time zero (i.e., the intravascular volume). In this setting, then in theory, a non-absorbed vascular marker concentration is not required, since the graph of the data should provide the vascular volume just prior to the entry across the BBB. The relationship of the concentrations is defined by

$$V_D = C_{br(total)}(T)/Cp(T) = K_{in}\left(\mathrm{AUC}^T{}_0/C_p(T)\right) + V_i \tag{6}$$

where K_{in} is the slope of the regression line and V_i is the y-intercept that determines the initial plasma volume of distribution before a substance can cross the BBB.[24] Care must be taken to interpret these values, knowing the typical intravascular volume of the brain is between 10 and 12 μL/g. Abnormally high V_i should be a cause for concern, as it may indicate that efflux or metabolic effects are taking place.[25]

Models that account for these additional drug delivery behaviors become substantially more complex. Experimental measurements that attempt to quantify efflux across the BBB are available, but they require studies more complex than tissue sampling (e.g., endothelial cell culture experiments). Analysis of distribution of substances and various concentrations in compartments of the CNS can determine the amount of substance in the extracellular, intracellular, and cerebrospinal fluid compartments, allowing for additional information about potential drug mechanisms of action in the CNS beyond just crossing of the BBB alone.[26,27] The calculation of brain tissue concentration (C_{br}), then, is adjusted by the rate of elimination (k_{out}) in the equation:

$$C_{br}(t) = K_{in}\sum_{i=1}^{n} C_i\left(e^{-k\mathrm{out}(t)} - e^{\lambda i(t)}\right)/(\lambda_i - k_{out}) \tag{7}$$

where C_i and λ_i are determined from the serum plasma exponential disposition function.[28] These equations all assume a linear influx and efflux. Other conditions, such as saturation kinetics and variable transport, require further modifications.

One such modification method is a single-pass brain perfusion to assess degree of movement or substance across the BBB. The brain uptake index is a measure of extraction of a substance across the BBB. Isolated brain perfusion models can measure brain uptake experimentally, at the same time controlling for multiple systemic parameters that influence the amount of drug that crosses the BBB. In these models, the makeup of the serum perfusing the brain is varied according to parameters such as osmotic pressure, electrolyte concentrations, pH, additional binding proteins in serum, etc. The perfusate is rapidly infused into the carotid artery in a single bolus without allowing serum perfusion. Immediate tissue sampling is performed and concentrations are analyzed. Any effects of systemic circulation and perfusion washout are minimized.

Other brain perfusion studies include those in which the carotid circulation is isolated and a continuous perfusion is performed. These experiments can better characterize slower transport systems, such as those that use transcytosis. The fraction of substance that persists in the endothelial cells (either bound or metabolized there) can be measured by cell separation centrifugation, which isolates the endothelial cells from the rest of the tissue. This method gives an indication of brain uptake into the CNS, but unfortunately only analyzes vascular concentration at one moment in time.[18,29-31] Spatial information can be obtained by perfusion techniques that use radiolabeled substrates. Concentrations and their locations in tissues can be performed by microscopic analysis of the CNS tissues after perfusion and sacrifice using various imaging techniques. These methods can identify the cellular location of materials down to a micrometer level.[18,32]

Other experimental options allow for assessing substance transport across the BBB with spatial characterization and multiple real time in vivo analyses. Several available imaging techniques can be noninvasive and offer real-time assessments of BBB permeability. These include single-photon emission computed tomography, magnetic resonance imaging, and positron emission tomography.[18,33,34] Concentrations in brain extracellular space can be measured by micro-dialysis techniques. In these experiments, microcatheters are inserted in situ into the CNS stereotactically, and continuous sampling of the extracellular fluid provides a measure of concentration of substance in the CNS of the awake animal. Limitations of this method include the inability to measure directly the intracellular concentrations, where the likely effect of many CNS agents is to be encountered, and the potential overestimation of drug influx secondary to focal BBB damage at the site of catheter placement.[18,35,36]

Cell culture models of endothelial cells lead to additional in vitro options for experimental analysis of the BBB. Transport protein can be isolated and studied, the underlying gene identified, and the processes of binding, transmembrane transport, and ultimate efflux across the endothelial layer can be elucidated for specific molecules. Assays of binding can be carried out in cellular suspensions, which do not differentiate between luminal and abluminal surface binding to endothelial cells.[37,38] Analysis of fluorescein-labeled ligands using confocal microscopy can assess both the binding of tagged substances to endothelial cell surfaces specifically and can monitor the efflux of tagged cotransported proteins for those molecules that are not fluorescein labeled.[39]

Several versions of monolayer endothelial cell cultures that attempt to reproduce the physiology of the BBB have been created, each with its own approximations and limitations. As the knowledge of the complexity and function of the BBB has grown, the influence of the adjacent CNS anatomy at the level of the endothelial cell has come into play. Attempts to reproduce the complex NVU, complete with influence of astrocytes, pericytes, basement membrane, etc., are ongoing. The complex microenvironment of the NVU has specific structural features that need to be reproduced for utility in BBB modeling. Key features for replication include proper geometry, including shear forces and ultrastructural anatomy; the surrounding microenvironment, including the basement membrane and extracellular matrix; the functional barrier as measured by transendothelial cell electrical resistance, permeabilities, and efflux transport; cellular functions, including BBB markers, proteins, and their turnover; and the ability to achieve cocultures with the additional CNS cell types involved in the NVU, including pericytes and astrocytes.[40-43] Each of these in vitro methods help to model the BBB. They tend to be complementary—they each add features to the analysis of the pharmacokinetics of a substance to help predict its clinical behavior in the CNS when administered systemically.

METHODS OF ENHANCING DRUG DELIVERY TO THE CNS

As the understanding of the features of the BBB and their anatomy and biochemistry has grown, the limitations of the BBB (which keep molecules from entering the CNS) prove a challenge for the development of therapeutics. However, with further understanding of various mechanisms of entrance across the BBB by certain specific molecules and as the details of the behavior of the various components of the NVU come to light, new opportunities for drug delivery become possible.[11,17]

Transcytosis

Cationic molecules that present themselves to the luminal surface of the endothelial cells will bind to the surface via electrostatic forces. In this position, they are enfolded into vesicles, transported across the cell, and released on the abluminal surface. Cationic substances such as bovine serum albumin traverse the BBB by means of transcytosis. They have been used as carriers for therapeutic molecules that otherwise do not readily cross the BBB. Designing cationic substances that bind to the gangliosphingolipid GM1 on endothelial cell surface membranes have been shown to enhance delivery of therapeutic agents across the BBB by means of trancytosis.[11,17,44]

Carrier- and Receptor-Mediated Transport

Molecules critical to brain health include glucose, vitamins, amino acids, and many peptides. These substances cross the BBB by means of specific carrier transmembrane transport proteins. Carrier systems utilize carrier proteins that bind the ligand on the surface and transport it across to the abluminal side. Receptor systems first actively bind the ligand to a cell surface protein. Active endocytosis occurs, and then transport across the endothelial cell takes place to move molecules to the abluminal surface, where the ligand is released. Receptor transports are designed for specific molecules. Transferrin, insulin, glucose, and low-density lipoprotein transporters have also been identified. Designing molecules that take advantage of these ligand-specific carriers is a method for enhancing the crossing of macromolecules that otherwise could not cross the intact BBB.

Glucose transporters are carrier systems. GLUT1 is a member of the solute carrier group. Antiepileptic prodrugs have been developed that can use this pathway to transport molecules across the BBB.[45,46] Levodopa crosses the BBB by means of an amino-acid-specific carrier transporter molecule.[47]

Receptor transport systems include transferrin transport, which has been shown to be a potentially useful transport system with which to couple transferrin with a therapeutic molecule to traverse the BBB.[48-50] Delivery of small interfering RNA across the BBB has been shown to be facilitated by transferrin receptor protein pathways.[51,52] Folate receptors are another system used to transport specific therapeutic substances (in this case, cytochrome c protein and certain nanocrystals) across the BBB.[53,54] Other transport receptors used to attempt therapeutic drug delivery include the lipoprotein receptor,[55] interleukin-13 receptor alpha2,[56] glutamate receptor,[57] scavenging receptor SRA1 and SRB1,[58] and the insulin receptor.[59]

Peptides have been identified to provide a vector for transport of therapeutic molecules across the BBB. Apolipoprotein E (ApoE) is a protein molecule that can target the low-density lipoprotein receptor (LDL-R) and trigger endocytosis, and can potentially transport a bound ligand across the endothelial layer. ApoE-I and ApoE-II and angiopep-2 are other developed molecules. These peptide-mediated transport molecules also use the LDL-R-related protein system to deliver targeted substances even more efficiently to the CNS.[11,60-62]

In addition to transport-mediated influx, the endothelial layer has efflux proteins that limit the ability to concentrate molecules in the CNS. Multidrug-resistant proteins and breast cancer resistance proteins are efflux pumps that can remove cationic amphiphilic drugs and neutral compounds from the CNS across the BBB, back into the vascular lumen. P-glycoprotein is another ATP linked efflux pump that is able to effectively eliminate lipophilic molecules that would otherwise pass easily into the CNS across the BBB.[17,63,64] Strategies that block these efflux mechanisms have been useful in increasing the concentration of therapeutic substances entering the CNS.

Molecules can be engineered to take advantage of the above-described pathways. They have been shown at least in experimental models to increase the concentration of target ligands in the CNS. Increased toxicity in the brain and in other end organs must be considered when transcellular transporter proteins are targeted. Competitive inhibition of influx transport of endogenous substances can be detrimental to the CNS and to other end organ metabolism, and blockade of efflux transport can have unexpected toxicities both in the CNS and in vital systemic organs.

Designer Molecules and Manufactured Delivery Systems

Certain engineered substances may help increase the delivery of targeted therapeutics across the BBB and into the CNS. These molecules are built to take advantage of the biochemical and physical characteristics of the BBB in an effort to maximize transport into the CNS. The development of nanoparticle techniques has allowed the creation of multiple options for enhanced delivery, and various surface molecules and core substances have been designed with specific target molecules in mind. Although these have the potential to increase CNS concentrations, they can also increase concentration in other end organs, bringing potential unplanned toxicities, as noted with the above protein-mediated transport blockades.

However, increased therapeutic benefit may be achieved with lower administered concentrations using designed delivery systems.

Liposomes are phospholipid bilayers of phosphatidylcholine, sphingomyelin, or glycerophospholipids manufactured into spherical shapes that can enclose both hydrophilic and hydrophobic compounds. They can be manufactured to have a solid lipid core, the so-called *lipid nanoparticle*. These compounds then allow their incorporated therapeutic molecules to be delivered in higher concentrations across the BBB than could be achieved by direct serum administration alone. Many therapeutic agents are amenable to transport by way of these liposomes. These are created in the nanoparticle sizes, from 1 to 1000 nm, while capillary minimal sizes are on the order of 6 μm.[17,65]

Chitosan, a natural polysaccharide with antifungal and antibiotic properties, has been used as a building block for a nanoparticle carrier. It has potential as an oral delivery agent, as it can cross the gastrointestinal tract without degradation and can cross CNS endothelial cells to deposit carrier molecules (e.g., neuropeptides) in therapeutic concentrations.[11,66-68] Modifications to nanoparticle surfaces (e.g., polyethylene glycol, polysorbate 80, and poloxamer 188), and creation of a more cationic surface charge have been strategies used to increase concentrations of incorporated molecules into the CNS.[11]

Dendrimers and nanogels are assembled biopolymers and synthetic polymers that incorporate bound therapeutic molecules into nanoparticle structures that can then cross the BBB by nature of the surface properties built into their designs. Nanogels are made of individual polymers designed with specific properties that allow assembly into a spherical core and surface structure carrier that can store therapeutic molecules in the core. The surface properties of the assembled polymer allow it to cross the BBB. The internal cross-linking properties are engineered to disassemble once the target environment of the CNS is encountered, releasing the therapeutic ligand.[69,70] Dendrimers are biologic and synthetic polymers that are synthesized with a central core and highly branched surface molecules. They produce complex three-dimensional shapes, allowing for packing among the branches target ligands, including small molecule drugs and genes. All features of the dendrimer nanoparticle are modifiable for delivery purposes, including the core, the branch pattern, the surface characteristics, and the binding and releasing characteristics to target ligands. Many different types of dendrimers have been developed, including polyamidoamine dendrimers, phosphorus-containing dendrimers, polylysine dendrimers, and carbosilane dendrimers.[71,72]

Iron Particles: Ultrasmall Iron Oxide Particles

In addition to molecules that can be used as therapeutic agents, agents used to enhance imaging of the CNS play a critical role in assessment of any therapy. To be beneficial, these molecules must be able to identify disease across the BBB and to monitor ongoing treatment. Diseases in neurooncology tend to be infiltrative; therefore, one challenge of imaging in this field is measuring the degree of involvement of tumor within the CNS. Measuring the actual tumor volume, both initially and during treatment, remains a critical and controversial feat. Ultrasmall iron oxide particles have been developed and have demonstrated value in neuroimaging. These molecules consist of superparamagnetic iron oxide surrounded by a carbohydrate shell. In this form, the nanoparticle has been shown to improve imaging of tumors and tumor response to treatment. At the time of initial infusion, these agents tend not to have the degree of leakage noted with gadolinium, and therefore may act as an improved blood pool agent. This characteristic may help to further characterize cerebral blood volume with less initial leakage across the BBB, and to give an improved analysis of blood volume in relation to brain not infiltrated by tumor. With subsequent delayed imaging, the iron agent appears across the BBB in areas of tumor infiltration. These particles seem to be able to cross the BBB in association with tumor-activated macrophages. This characteristic may prove be beneficial in the assessment of disease response to treatment.[73,74]

METHODS THAT BYPASS THE BBB

The CNS extracellular space is a viable entry point for drug delivery to the CNS, but must be accessed directly to bypass the BBB. By placing therapeutic molecules in the brain extracellular space, the barrier to vascular delivery of substances is bypassed. This has been shown to be an effective method to introduce agents into the CNS. Drug-eluting wafers placed at the time of surgery for brain tumors are one option, and these have been shown to deliver BCNU in therapeutic concentrations.[75] The technique of convection-enhanced delivery involves placement of catheters stereotactically into the brain and the infusion of drug directly into the extracellular space. The bulk movement of fluid through the extracellular space allows for diffuse increases in concentration in the regions around the catheter insertion sites. Access via the vascular system is not required.[76]

Intranasal delivery of some substances has been shown to be a feasible method for drug delivery, bypassing the BBB at the NVU. Drugs administered in this manner can be adsorbed in the nasal mucosa, and utilizing both the olfactory nerves and the trigeminal nerves as pathways, can migrate into the CNS. These are fairly focal delivery methods in that their distributions are limited and degradation can readily occur. Techniques to extend the exposure and increase the concentrations in the CNS are necessary.[77]

METHODS THAT OPEN THE BBB

Focused ultrasound (FUS) is a relatively new technique to alter the BBB in a way that allows for molecules to cross the NVU and enter the CNS. Ultrasound is focused stereotactically at a target in the brain, and intravenous microbubbles are then circulated into the vascular system. As they encounter the endothelial cells under FUS, pores in the BBB are opened temporarily and systemic molecules are able to enter. The process is monitored in real time under magnetic resonance imaging. This is a fairly noninvasive method to deliver agents focally to lesions within the brain.[78,79] There are technical issues that must be considered that relate to anatomy (e.g., skull thickness). The technique has certain parameters that best provide focal BBB opening without focal tissue injury or inflammatory response.[80,81]

Certain chemicals have been shown to disrupt the BBB and to allow for entrance of molecules into the CNS. Bradykinin, histamine, and adenosine have been shown to act on tight junctions, temporarily opening them and allowing substances to diffuse. Bradykinin and its analogs, including receptor-mediated permeabilizer act on the kinin endothelial receptors B2 and B1, which then increases the permeability across tight junctions in the CNS.[82,83] Adenosine acts on the adenosine 2A receptor, a G-protein-coupled receptor found on endothelial cell membranes. Activation via agonists leads to increased permeability across the endothelial tight junctions.[84] The design of the agonists may be able to modulate the characteristics of the opening of the BBB including its duration.

Alkylglycerols are substances whose chemical makeups include varying lengths of alkyl and glycerol groups. These molecules are shown to have a temporary and reversible effect on BBB permeability. Their mechanism of action alters the cell morphology and the distribution of junctional proteins, resulting in increased permeability. These substances are thought to act at the level of the endothelial cytoskeleton.[85] Lysophosphatidic acid and borneol are two additional substances that can reversibly and temporarily alter the permeability of the BBB; lysophosphatidic acid through disruption of the structural integrity of tight junction proteins including claudin-5, occludin, and ZO-1; and borneol as an antagonist of multidrug-resistant proteins Mrp1, Mrd1a, and Mrd1b.[86-88]

OSMOTIC DISRUPTION OF THE BBB

Of all the clinical techniques identified as potential therapeutic methods for delivering therapeutic agents across the BBB into the CNS, osmotic BBB opening using hyperosmolar intra-arterial mannitol has been the longest consistently used and best-characterized technique in terms of safety and efficacy. The multicenter Blood–Brain Barrier Disruption Consortium was developed to administer osmotic BBB disruption with intra-arterial mannitol beginning in the 1990s. Over the years, Consortium members have included The Ohio State University, University of Oklahoma Health Science Center, University of Minnesota, Cleveland Clinic Foundation, University of Kentucky, Hadassah-Hebrew University Medical Center, Centre Hospital Universitaire de Sherbrooke, and Oregon Health & Science University (the coordinating center).

The Consortium has developed guidelines for the application of BBB disruption that include proper anesthesia, techniques for placement of a transfemoral arterial catheter, a grading system for measuring the degree of BBB disruption on postdisruption head computed tomograms, radiographic assessment of tumor response, guidelines for the technique and rate of mannitol and chemotherapy infusion, and guidelines for the care of patients before, during, and after the procedures. These guidelines have been developed over the course of decades of use of the technique in the treatment of neuro-oncologic patients. The consortium continues to hold annual meetings, partially funded by an NIH R13 grant. The meeting provides an opportunity for participating centers to discuss patient care, guideline development, recent advances in BBB science, and future directions. The proceedings are summarized by the participants and submitted for publication.[89-91]

The osmotic BBB disruption technique aims to maximize the delivery of therapeutic agents to the brain that otherwise are limited in their abilities to affect conditions in the CNS space with a minimum of potential systemic toxicities, all the while providing the best possible preservation of neurocognitive function and quality of life. BBB disruption is especially important in, and capable of, increasing the delivery of high-molecular-weight agents, such as proteins, antibodies, immunoconjugates, and viral vectors.[92-94]

The osmotic BBB disruption technique is accomplished through intra-arterial administration of a hyperosmolar solution of mannitol into the carotid or vertebral artery. Hyperosmolar mannitol causes an opening of the tight junctions at the endothelial cells through osmotic shrinkage. Therapeutic molecules have a transient temporal window in which they can cross into the CNS by way of simple diffusion, measured as outlined above. Calcium influx and nitrous oxide mediators, as well as cytoskeletal changes, are cellular messenger systems that likely facilitate the transient BBB opening. The clinical results of BBB disruption as a mechanism of delivery of high concentrations of therapeutic drug across the BBB have been most promising against the more chemosensitive brain tumors, such as primary CNS lymphoma.

Prior to use in the clinical setting, toxicity studies are conducted by the BBB preclinical and clinical teams in animal models and in Phase 1 studies, to identify agents with acceptable toxicity and safety profiles that may be used during BBB disruption. CNS and systemic toxicities must be assessed. Extensive preclinical animal studies are necessary and have proven beneficial prior to initiation of Phase 1 studies. When doxorubicin (intra-arterial),[95] cisplatin (intra-arterial) or 5-FU (intra-arterial)[96] were administered as single agents following osmotic BBB disruption, severe neurotoxicity was noted in preclinical studies. Etoposide phosphate administered in combination with melphalan (intra-arterial), methotrexate (intra-arterial), or carboplatin (intra-arterial) after BBB disruption was also associated with neurotoxicity if the anesthetic propofol was used.[97]

We have reported our consortium results in the treatment of primary CNS lymphoma.[98] A total of 149 patients with newly diagnosed primary CNS lymphoma were treated with BBB disruption in conjunction with intra-arterial methotrexate-based chemotherapy without the use of whole-brain radiation therapy. The protocol consisted of intra-arterial administration to two separate circulations (i.e., carotid and vertebrobasilar) on successive days, repeated monthly over the course of 1 year. A complete response of 56% and a partial response rate of 24% were noted. The overall response rate was 82%. The median overall survival was 3.1 years. There was a 25% estimated survival at 8.5 years. The median progression-free survival was 1.8 years. The 5- and 7-year progression-free survivals were 31% and 25%, respectively. Based on Karnofsky performance scale scores, risk groups were stratified. The low-risk group included patients younger than 60 years and with Karnofsky performance scale scores of 70 or higher. Median overall survival in that group was 14 years, with a survival plateau noted at 8 years.[98]

One stated effort of BBB-disruption chemotherapy is avoidance of neurocognitive decline, seen in patients who have undergone whole-brain radiation therapy. Long-term results in cognitive capabilities in our patient population have been evaluated.[99] Neurocognitive examinations were performed in the patient population that had achieved complete response. Examinations were performed between 2 and 26 years after treatment. The median time to evaluation was 12 years. Comparisons were made to pretreatment cognitive evaluations. Improvement in attention/executive function and improvement in verbal memory were all identified as significant compared to pretreatment evaluations.[99]

In an additional assessment of safety of the osmotic BBB disruption protocol, adverse events in patients undergoing repeat osmotic BBB disruption treatments were evaluated. We reported outcomes of three patients with recurrent primary CNS lymphoma who underwent a second year-long BBB disruption protocol, with a carboplatin regimen in conjunction with rituximab, a monoclonal antibody that reacts specifically with the B-cell antigen

CD20, induces apoptosis, and may synergize with platinum drugs. When added to the regimen, rituximab is infused 12 hours prior to day 1 of monthly BBB disruption, and is thus available to cross into the CNS when the barrier is opened. The initial results have been encouraging, and the safety profile is reasonable. These patients underwent 152 disruptions over the course of their combined treatments. There was no increase in adverse events noted in the repeat procedures, suggesting that osmotic BBB disruption presents additional opportunities for therapeutic drug delivery for previously treated patients as future advances in targeted molecules are made available.[100,101]

CONCLUSIONS

The BBB is a complex anatomic and physiologic structure that serves its function well to protect the sensitive CNS from molecules with potential toxic or unwanted neurologic effects. These protective functions, while critical, have limited our ability to introduce necessary therapeutic drugs to patients with diseases of the CNS. As we learn more about the BBB, however, great strides have been made in development of drugs that may traverse this natural barrier. Efforts to accurately measure the abilities of medications to cross the NVU into positions of therapeutic benefit are varied and are, in many ways, complementary. Certain drug modifications opened the door to enhanced therapeutics, but also to unexpected CNS toxicities and unanticipated concentrations and toxicities in systemic organ systems. Osmotic BBB disruption with intra-arterial mannitol has been well characterized and clinically applied for years as a successful method for delivery of CNS therapeutics across the NVU. Preclinical analysis, clinical applications of drug delivery and safety, and therapeutic benefit with minimal toxicities have evolved over the years thanks to preclinical and translational studies of the effects of BBB disruption. Future developments will shed light on opportunities for targeted drug delivery using known pathways through the NVU, with continued emphasis on specificity, safety, and therapeutic benefit.

References

1. Bhaskar S, Tian F, Stoeger T, et al. Multifunctional Nanocarriers for diagnostics, drug delivery and targeted treatment across blood-brain barrier: perspectives on tracking and neuroimaging. *Part Fibre Toxicol*. 2010;7:3.
2. Weiss N, Miller F, Cazaubon S, Couraud PO. The blood-brain barrier in brain homeostasis and neurological diseases. *Biochim Biophys Acta*. 2009;1788(4):842-857.
3. Alavijeh MS, Chishty M, Qaiser MZ, Palmer AM. Drug metabolism and pharmacokinetics, the blood-brain barrier, and central nervous system drug discovery. *NeuroRx*. 2005;2(4):554-571.
4. Clarke E, O'Malley CD. The human brain and spinal cord. In: *A historical study illustrated by writings from antiquity to the twentieth century*. Berkeley, CA: University of California Press; 1968.
5. Goyal K, Koul V, Singh Y, Anand A. Targeted drug delivery to central nervous system (CNS) for the treatment of neurodegenerative disorders: trends and advances. *Cent Nerv Syst Agents Med Chem*. 2014;14(1):43-59.
6. Günzel D, Yu AS. Claudins and the modulation of tight junction permeability. *Physiol Rev*. 2013;93(2):525-569.
7. Cummins PM. Occludin: one protein, many forms. *Mol Cell Biol*. 2012;32(2):242-250.
8. Ebnet K, Brinkmann BF, Kummer D, Misselwitz S, Peddibhotla SS, Tuncay H. Tight junctions, junctional adhesion molecules (jams), and the blood brain barrier. In: *Tight Junctions in Cancer Metastasis*. Springer; 2013:119-129.
9. Engelhardt B. Development of the blood-brain interface. In: *Blood-Brain Barriers: From Ontogeny to Artificial Intelligence*; 2006:9-39. vol. 1, 2006.

10. Kanwar JR, Sriramoju B, Kanwar RK. Neurological disorders and therapeutics targeted to surmount the blood-brain barrier. *Int J Nanomedicine*. 2012;7:3259-3278.
11. Crawford L, Rosch J, Putnam D. Concepts, technologies, and practices for drug delivery past the blood-brain barrier to the central nervous system. *J Control Release*. 2016;240:251-266.
12. Lu CT, Zhao YZ, Wong HL, Cai J, Peng L, Tian XQ. Current approaches to enhance CNS delivery of drugs across the brain barriers. *Int J Nanomedicine*. 2014;9:2241-2257.
13. Bicker J, Alves G, Fortuna A, Falcão A. Blood-brain barrier models and their relevance for a successful development of CNS drug delivery systems: a review. *Eur J Pharm Biopharm*. 2014;87(3):409-432.
14. Abbott NJ, Patabendige AA, Dolman DE, Yusof SR, Begley DJ. Structure and function of the blood-brain barrier. *Neurobiol Dis*. 2010;37(1):13-25.
15. Pardridge WM, Boado RJ, Kang YS. Vector-mediated delivery of a polyamide ("peptide") nucleic acid analogue through the blood-brain barrier in vivo. *Proc Natl Acad Sci USA*. 1995;92:5592-5596.
16. Pardridge WM, Eisenberg J, Yang J. Human blood-brain barrier transferrin receptor. *Metabolism*. 1987;36:892-895.
17. Patel MM, Patel BM. Crossing the blood-brain barrier: recent advances in drug delivery to the brain. *CNS Drugs*. 2017;31(2):109-133.
18. Bickel U. How to measure drug transport across the blood–brain barrier. *NeuroRx*. 2005;2(1):15-26.
19. Ohno K, Pettigrew KD, Rapoport SI. Lower limits of cerebrovascular permeability to nonelectrolytes in the conscious rat. *Am J Physiol (Lond)*. 1978;235:H299-H307.
20. Crone C. The permeability of capillaries in various organs as determined by use of the 'indicator diffusion' method. *Acta Physiol Scand*. 1963;58:292-305.
21. Renkin EM. Transport of potassium-42 from blood to tissue in isolated mammalian skeletal muscles. *Am J Physiol (Lond)*. 1959;197:1205-1210.
22. Fenstermacher JD, Blasberg RG, Patlak CS. Methods for quantifying the transport of drugs across brain barrier systems. *Pharmacol Ther*. 1981;14:217-248.
23. Patlak CS, Blasberg RG, Fenstermacher JD. Graphical evaluation of blood-to-brain transfer constants from multiple-time uptake data. *J Cereb Blood Flow Metab*. 1983;3:1-7.
24. Blasberg RG, Fenstermacher JD, Patlak CS. Transport of aminoisobutyric acid across brain capillary and cellular membranes. *J Cereb Blood Flow Metab*. 1983;3:8-32.
25. Smith QR. A review of blood-brain barrier transport techniques. *Methods Mol Med*. 2003;89:193-208.
26. Collins JM, Dedrick RL. Distributed model for drug delivery to CSF and brain tissue. *Am J Physiol (Lond)*. 1983;245:R303-R310.
27. Tetsuya T, Ohtsuki S. Brain-to-blood transporters for endogenous substrates and xenobiotics at the blood-brain barrier: an overview of biology and methodology. *NeuroRx*. 2005;2:63-71.
28. Bickel U, Thomsen T, Fischer JP, Weber W, Kewitz H. Galanthamine:pharmacokinetics, tissue distribution and cholinesterase inhibition in brain of mice. *Neuropharmacology*. 1991;30:447-454.
29. Oldendorf WH. Clearance of radiolabeled substances by brain after arterial injection using a diffusible internal standard. In: Marks N, Rodnight R, eds. *Research Methods in Neurochemistry*. New York: Plenum Publishing Co.; 1981:91-112.
30. Smith QR, Allen DD. In situ brain perfusion technique. *Methods Mol Med*. 2003;89:209-218.
31. Triguero D, Buciak J, Pardridge WM. Capillary depletion method for quantification of blood-brain barrier transport of circulating peptides and plasma proteins. *J Neurochem*. 1990;54:1882-1888.
32. Asotra K, Ningaraj N, Black KL. Measurement of blood-brain and blood-tumor barrier permeabilities with [14C]-labeled tracers. *Methods Mol Med*. 2003;89:177-190.
33. Huang SC, Hoh C, Barrio JR, Phelps ME. Measurement of bloodbrain barrier permeability in humans with positron emissiontomography. In: Pardridge WM, ed. *Introduction to the Blood-Brain Barrier: Methodology, Biology and Pathology*. Cambridge, UK: Cambridge University Press; 1998:122-146.
34. Tofts PS, Kermode AG. Measurement of the blood-brain barrier permeability and leakage space using dynamic MR imaging. 1. Fundamental concepts. *Magn Reson Med*. 1991;17:357-367.
35. Deguchi Y, Morimoto K. Application of an in vivo brain microdialysis technique to studies of drug transport across the blood-brain barrier. *Curr Drug Metab*. 2001;2:411-423.
36. Elmquist WF, Sawchuk RJ. Application of microdialysis in pharmacokinetic studies. *Pharm Res*. 1997;14:267-288.
37. Joo F. The blood-brain barrier in vitro: the second decade. *Neurochem Int*. 1993;23:499-521.

38. Pardridge WM. Isolated brain capillaries: and in vitro model of blood-barin barrier research. In: Pardridge WM, ed. *Introduction to the Blood-Brain Barrier: Methodology, Biology and Pathology*. Cambridge, UK: Cambridge University Press; 1998:49-61.
39. Miller DS, Nobmann SN, Gutmann H, Toeroek M, Drewe J, Fricker G. Xenobiotic transport across isolated brain microvessels studied by confocal microscopy. *Mol Pharmacol*. 2000;58:1357-1367.
40. Reichel A, Begley DJ, Abbott NJ. An overview of in vitro techniques for blood-brain barrier studies. *Methods Mol Med*. 2003;89:307-324.
41. Wevers NR, Kasi DG, Gray T, et al. A perfused human blood-brain barrier on-a-chip for high-throughput assessment of barrier function and antibody transport. *Fluids Barriers CNS*. 2018;15(1):23.
42. DeStefano JG, Jamieson JJ, Linville RM, Searson PC. Benchmarking in vitro tissue-engineered blood-brain barrier models. *Fluids Barriers CNS*. 2018;15(1):32.
43. Wolff A, Antfolk M, Brodin B, Tenje M. In vitro blood–brain barrier models—an overview of established models and new microfluidic approaches. *J Pharm Sci*. 2015;104:2727-2746.
44. Lu W, Zhang Y, Tan YZ, Hu KL, Jiang XG, Fu SK. Cationic albumin-conjugated pegylated nanoparticles as novel drug carrier for brain delivery. *J Control Release*. 2005;107:428-448.
45. Bonina FP, Arenare L, Palagiano F, et al. Synthesis, stability, and pharmacological evaluation of nipecotic acid prodrugs. *J Pharm Sci*. 1999;88:561-567.
46. Bonina FP, Arenare L, Ippolito R, et al. Synthesis, pharma- cokinetics and anticonvulsant activity of 7-chlorokynurenic acid prodrugs. *Int J Pharm*. 2000;202:79-88.
47. Mena I, Cotzias GC. Protein intake and treatment of Parkinson's disease with levodopa. *N Engl J Med*. 1975;292:181-184.
48. Qian ZM, Li H, Sun H, Ho K. Targeted drug delivery via the transferrin receptor- mediated endocytosis pathway. *Pharmacol Rev*. 2002;54:561-587.
49. Lee H, Engler JA, Collawn JF, Moore BA. Receptor mediated uptake of peptides that bind the human transferrin receptor. *Eur J Biochem*. 2001;268:2004-2012.
50. Johnsen KB, Moos T. Revisiting nanoparticle technology for blood–brain barrier transport: unfolding at the endothelial gate improves the fate of transferrin receptor-targeted liposomes. *J Control Release*. 2016;222:32-46.
51. Youn P, Chen Y, Furgeson DY. A myristoylated cell-penetrating peptide bearing a transferrin receptor-targeting sequence for neuro-targeted siRNA delivery. *Mol Pharm*. 2014;11:486-495.
52. Wei L, Guo XY, Yang T, et al. Brain tumor-targeted therapy by systemic delivery of siRNA with transferrin receptor-mediated core-shell nanoparticles. *Int J Pharm*. 2016;510:394-405.
53. Morales-Cruz M, Cruz-Montañez A, Figueroa CM, et al. Combining stimulus-triggered release and active targeting strategies improves cytotoxicity of cytochrome c nanoparticles in tumor cells. *Mol Pharm*. 2016;13:2844-2854.
54. Dong S, Cho HJ, Lee YW, Roman M. Synthesis and cellular uptake of folic acid-conjugated cellulose nanocrystals for cancer targeting. *Biomacromolecules*. 2014;15:1560-1567.
55. Xiao G, Gan LS. Receptor-mediated endocytosis and brain delivery of therapeutic biologics. *Int J Cell Biol*. 2013.
56. Madhankumar AB, Slagle-Webb B, Wang X, et al. Efficacy of interleukin-13 receptor-targeted liposomal doxorubicin in the intracranial brain tumor model. *Mol Cancer Ther*. 2009;8:648-654.
57. Webster CI, Caram-Salas N, Haqqani AS, et al. Brain penetra- tion, target engagement, and disposition of the blood–brain barrier-crossing bispecific antibody antagonist of metabotropic glutamate receptor type 1. *FASEB J*. 2016;30:1927-1940.
58. Srimanee A, Regberg J, Hällbrink M, et al. Role of scavenger receptors in peptide-based delivery of plasmid DNA across a blood–brain barrier model. *Int J Pharm*. 2016;500:128-135.
59. Dieu LH, Wu D, Palivan CG, et al. Polymersomes conjugated to 83-14 monoclonal antibodies: in vitro targeting of brain capillary endothelial cells. *Eur J Pharm Biopharm*. 2014;88:316-324.
60. Re F, Cambianica I, Zona C, Sesana S, Gregori M, Rigolio R. Functionalization of liposomes with ApoE-derived peptides at different density affects cellular uptake and drug transport across a blood–brain barrier model. *Nanomed Nanotechnol Biol Med*. 2011;7:551-559.
61. Re F, Cambianica I, Sesana S, et al. Functionalization with ApoE-derived peptides enhances the interaction with brain capillary endothelial cells of nanoliposomes binding amyloid-beta peptide. *J Biotechnol*. 2010;156:341-346.
62. Wang D, El-Amouri SS, Dai M, et al. Engineering a lysosomal enzyme with a derivative of receptor-binding domain of apoE enables delivery across the blood–brain barrier. *Proc Natl Acad Sci U S A*. 2013;110:2999-3004.

63. Chen Y, Liu L. Modern methods for delivery of drugs across the blood–brain barrier. *Adv Drug Deliv Rev*. 2012;64:640-665.
64. Fromm MF. P-glycoprotein: a defense mechanism limiting oral bioavailability and CNS accumulation of drugs. *Int J Clin Pharmacol Ther*. 2000;38:69-74.
65. Li X, Tsibouklis J, Weng T, et al. Nano carriers for drug transport across the blood brain barrier. *J Drug Target*. 2016;19:1-12.
66. Gan Q, Wang T. Chitosan nanoparticle as protein delivery carrier-Systematic examination of fabrication conditions for efficient loading and release. *Colloids Surf B: Biointerfaces*. 2007;59:24-34.
67. Bowman K, Leong KW. Chitosan nanoparticles for oral drug and gene delivery. *Int J Nanomedicine*. 2006;1:117-128.
68. Lalatsa A, Garrett NL, Ferrarelli T, Moger J, Schätzlein AG, Uchegbu IF. Delivery of peptides to the blood and brain after oral uptake of quaternary ammonium palmitoyl glycol chitosan nanoparticles. *Mol Pharm*. 2012;9:1764-1774.
69. Soni KS, Desale SS, Bronich TK. Nanogels: An overview of properties, biomedical applications and obstacles to clinical translation. *J Control Release*. 2016;240:109-126.
70. Aderibigbe BA, Naki T. Design and efficacy of nanogels formulations for intranasal administration. *Molecules*. 2018;23(6).
71. Somani S, Dufès C. Applications of dendrimers for brain delivery and cancer therapy. *Nanomedicine*. 2014;9:2403-2414.
72. Florendo M, Figacz A, Srinageshwar B, et al. Use of polyamidoamine dendrimers in brain diseases. *Molecules*. 2018;23(9).
73. Varallyay P, Nesbit G, Muldoon LL, et al. Comparison of two super paramagnetic viral-sized iron oxide particles ferumoxides and ferumoxtran-10 with a gadolinium chelate in imaging intracranial tumors. *AJNR Am J Neuroradiol*. 2002;23:510-519.
74. Neuwelt EA, Varallyay P, Bago A, Muldoon LL, Nesbit G, Nixon R. Imaging of iron oxide nanoparticles by MR and light microscopy in patients with malignant brain tumours. *Neuropathol Appl Neurobiol*. 2004;30(1):70.
75. Westphal M, Hilt DC, Bortey E, et al. A phase 3 trial of local chemotherapy with biodegradable carmustine (BCNU)wafers (Gliadel wafers) in patients with primary malignant glioma. *Neuro-Oncology*. 2003;5:79-88.
76. Lonser RR, Sarntinoranont M, Morrison PF, Oldfield EH. Convection-enhanced delivery to the central nervous system. *J Neurosurg*. 2015;122(3):697-706.
77. Hanson LR, Frey WH. Intranasal delivery bypasses the blood–brain barrier to target therapeutic agents to the central nervous system and treat neurodegenerative disease. *BMC Neurosci*. 2008;9:S5.
78. Hynynen K, McDannold N, Vykhodtseva N, Jolesz FA. Noninvasive MR imaging guided focal opening of the blood–brain barrier in rabbits. *Radiology*. 2001;220:640-646.
79. Aryal M, Arvanitis CD, Alexander PM, McDannold N. Ultrasound-mediated blood–brain barrier disruption for targeted drug delivery in the central nervous system. *Adv Drug Deliv Rev*. 2014;72:94-109.
80. Marquet F, Tung YS, Teichert T, Ferrera VP, Konofagou EE. Noninvasive, transient and selective blood–brain barrier opening in non-human primates in vivo. *PLoS ONE*. 2011;6:1-7.
81. McDannold N, Arvanitis CD, Vykhodtseva N, Livingstone MS. Temporary disruption of the blood–brain barrier by use of ultrasound and microbubbles: Safety and efficacy evaluation in rhesus macaques. *Cancer Res*. 2012;72:3652-3663.
82. Matsukado K, Inamura T, Nakano S, et al. Enhanced tumor uptake of carboplatin and survival in glioma-bearing rats by intracarotid infusion of the bradykinin analog, RMP-7. *Neurosurgery*. 1996;39:125-133.
83. Cote J, Savard M, Neugebauer W, et al. Dual kinin B1 and B2 receptor activation provides enhanced blood–brain barrier permeability and anticancer drug delivery into brain tumors. *Cancer Biol Ther*. 2013;14:806-811.
84. Kim DG, Bynoe MS. A2A adenosine receptor regulates the human blood–brain barrier permeability. *Mol Neurobiol*. 2015;52:664-678.
85. Hulper P, Veszelka S, Walter FR, et al. Acute effects of shortchain alkylglycerols on blood–brain barrier properties of cultured brain endothelial cells. *Br J Pharmacol*. 2013;169:1561-1573.
86. On NH, Savant S, Toews M, Miller DW. Rapid and reversible enhancement of blood–brain barrier permeability using lysophosphatidic acid. *J Cereb Blood Flow Metab*. 2013;33:1-11.
87. Masago K, Kihara Y, Yanagida K, et al. Lysophosphatidic acid receptor, LPA6, regulates endothelial blood-brain barrier function: Implication for hepatic encephalopathy. *Biochem Biophys Res Commun*. 2018;501(4):1048-1054.

88. Yu B, Ruan M, Dong X, Yu Y, Cheng H. The mechanism of the opening of the blood-brain barrier by borneol: a pharmacodynamics and pharmacokinetics combination study. *J Ethnopharmacol*. 2013;150(3):1096-1108.
89. Doolittle ND, Anderson CP, Bleyer WA, et al. for the Blood-Brain Barrier Disruption Consortium. Importance of dose intensity in neuro-oncology clinical trials: summary report of the sixth annual meeting of the Blood-Brain Barrier Disruption Consortium. *Neuro-Oncology*. 2001;3:46-54.
90. Doolittle ND, Abrey LE, Ferrari N, et al. Targeted delivery in primary and metastatic brain tumors: summary report of the seventh annual meeting of the Blood-Brain Barrier Disruption Consortium. *Clin Cancer Res*. 2002;8:1702-1709.
91. Doolittle ND, Abrey LE, Bleyer WA, et al. New frontiers in translational research in neuro-oncology and the blood-brain barrier: report of the tenth annual Blood-Brain Barrier Disruption Consortium Meeting. *Clin Cancer Res*. 2005;11(2 Pt 1):421-428.
92. Neuwelt EA. Mechanisms of disease: the blood-brain barrier. *Neurosurgery*. 2004;54:131-142.
93. Doolittle ND, Muldoon LL, Culp AY, Neuwelt EA. Delivery of chemotherapeutics across the blood-brain barrier: challenges and advances. *Adv Pharmacol*. 2014;71:203-243.
94. Neuwelt EA, Barnett PA, Hellstrom KE, et al. Delivery of melanoma-associated specific immunoglobulin monoclonal antibody and Fab fragments to normal brain utilizing osmotic blood-brain barrier disruption. *Cancer Res*. 1988;48:4725-4729.
95. Neuwelt EA, Pagel M, Barnett P, Glasberg M, Frenkel EP. Pharmacology and toxicity of intracarotid adriamycin administration following osmotic blood-brain barrier modification. *Cancer Res*. 1981;41:4466-4470.
96. Neuwelt EA, Barnett PA, Glasberg M, Frenkel EP. Pharmacology and neurotoxicity of cis-Diamminedichloroplatinum, bleomycin, 5-fluorouracil, and cyclophosphamide administration following osmotic blood-brain barrier modification. *Cancer Res*. 1983;43:5278-5285.
97. Fortin D, McCormick CI, Remsen LG, Nixon R, Neuwelt EA. Unexpected neurotoxicity of etoposide phosphate administered in combination with other chemotherapeutic agents after blood-brain barrier modification to enhance delivery, using propofol for general anesthesia, in a rat model. *Neurosurgery*. 2000;47:199-207.
98. Angelov L, Doolittle ND, Kraemer DF, et al. Blood-brain barrier disruption and intra-arterial methotrexate-based therapy for newly diagnosed primary CNS lymphoma: a multi-institutional experience. *J Clin Oncol*. 2009;27(21):3503-3509.
99. Doolittle ND, Dósa E, Fu R, et al. Preservation of cognitive function in primary CNS lymphoma survivors a median of 12 years after enhanced chemotherapy delivery. *J Clin Oncol*. 2013;31(31):4026-4027.
100. McGregor J, Bourekas E, Bell S. Repeat blood-brain barrier disruption in patients with recurrent primary central nervous system lymphoma. *Neuro Oncol*. 2008;10(5):759-915. ST-17.
101. Doolittle ND, Fu R, Muldoon LL, Tyson RM, Lacy C, Neuwelt EA. Rituximab in combination with methotrexate-based chemotherapy with blood–brain barrier disruption in newly diagnosed primary CNS lymphoma. *Hematol Oncol*. 2013;31(S1):179.

CHAPTER

5

Anatomy and Physiology of Cerebrospinal Fluid Dynamics

*Bryn A. Martin**, *Soroush Heidari Pahlavian*[†]

***Neurophysiological Imaging and Modeling Laboratory, Department of Biological Engineering, University of Idaho, Moscow, ID, United States [†]Laboratory of Functional MRI Technology, Stevens Neuroimaging and Informatics Institute, Department of Neurology, University of Southern California, Los Angeles, CA, United States**

INTRODUCTION

Understanding the anatomy and physiology of cerebrospinal fluid (CSF) dynamics is an essential element for investigating many diseases of the central nervous system (CNS) and potential disease therapeutics. CSF is a clear, colorless fluid with a viscosity similar to that of water at body temperature,[1] and constitutes a major part of the extracellular fluid of the CNS. CSF serves many physiological functions, including structural protection, metabolic homeostasis, and immunological support of the CNS. Abnormalities in CSF production, absorption, and/or dynamics are thought to play an important role in several CNS disorders such as Alzheimer disease,[2] hydrocephalus,[3, 4] normal-pressure hydrocephalus,[4] Chiari malformation,[5-7] and syringomyelia.[8, 9] A number of CSF-based therapeutics are presently under investigation, including intrathecal drug delivery (IDD),[10-12] CSF filtration,[13, 14] CSF cooling,[15, 16] and CSF pulse control.[17] This chapter provides a description of CSF function in the CNS and the physiological aspects of its pulsation and circulation throughout the cranial and spinal subarachnoid space (SAS).

SUMMARY OF CEREBROSPINAL FLUID SPACE ANATOMY

Cerebrospinal fluid resides in two primary regions that are located within the intracranial space and spine (Fig. 1). Total CSF volume is estimated to range from 250 to 400 mL in healthy adult humans,[18-22] with approximately 75% of that volume located intracranially. Intracranial

Nervous System Drug Delivery
https://doi.org/10.1016/B978-0-12-813997-4.00005-0

© 2019 Elsevier Inc. All rights reserved.

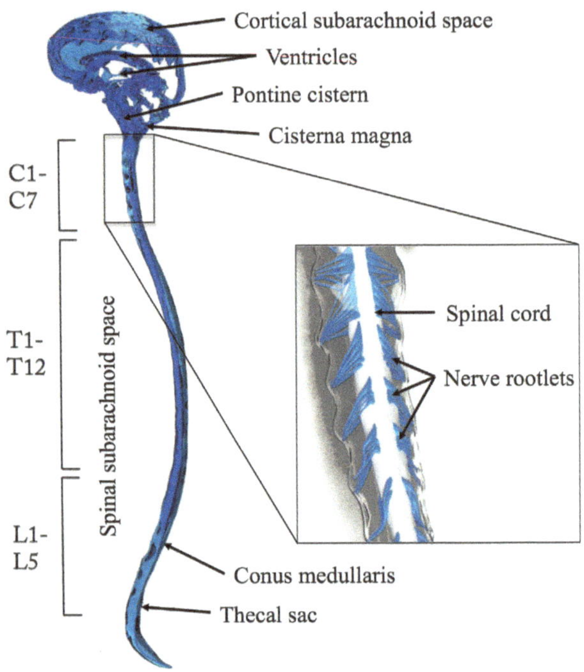

FIG. 1 Three-dimensional reconstruction of the complete cerebrospinal fluid system based on subject-specific, T2-weighted magnetic resonance imaging of a healthy 23-year-old female patient. *Based on Sass LR, Khani M, Natividad GC, Tubbs RS, Baledent O, Martin BA. A 3D subject-specific model of the spinal subarachnoid space with anatomically realistic ventral and dorsal spinal cord nerve rootlets.* Fluids Barriers CNS *2017;14(1):36.*

CSF spaces can be roughly divided into the cortical SAS, cisterns, and ventricles that contain about 180, 77, and 23 mL of CSF, respectively. The spinal CSF space has a volume of approximately 97 mL.[23] The intracranial ventricles consist of the lateral ventricles (~8.5 mL each), the third ventricle (~2.2 mL), and the fourth ventricle (~3.6 mL). These volumes are provided based on a single subject-specific model of the CSF system developed at University of Idaho using high-resolution T2-weighted magnetic resonance imaging (MRI), and can vary by subject and by disease state. It should be noted that the total CSF volume reported above is higher than the volume traditionally cited in medical textbooks[18, 24, 25] and review articles (i.e., ~150 mL).[26, 27] In fact, the reported value of ~150 mL total CSF volume has become so standard that an empirical reference is often omitted. Early studies reporting total CSF volume used relatively crude and invasive techniques [28] and have since been determined to be prone to error.[29, 30] The difference in total CSF volumes reported likely stems from the vastly different measurement methods. In general, more recent methods have yielded larger CSF volumes when noninvasive, high-resolution MRI techniques are used. More study is needed to understand CSF volume variation in healthy individuals and those diagnosed with CNS disease.

Designation of CSF-containing compartments including ventricles, cisterns, and SASs is based on their anatomical position (i.e., intra- vs extraparenchymal) and size (e.g., cisterns are dilated SASs). The intracranial ventricles are enveloped entirely by brain tissue. The SAS, both cranial and spinal, lies between pia and arachnoid mater. Within the spinal SAS, 31 pairs of dorsal and ventral nerve rootlets and denticulate ligaments are present

(Fig. 1). The spinal cord terminates at the conus and divides into many nerve segments called the *filum terminale* or "horse's tail."

Several openings connect the ventricular CSF spaces. The intraventricular foramina, also called the foramina of Monro, are openings (left and right) that connect the third ventricle to the lateral ventricles. A relatively long and narrow tube-like canal called the Aqueduct of Sylvius connects the third and fourth ventricles with a minimum diameter of approximately 1-3 mm. In addition to the Aqueduct of Sylvius, the fourth ventricle has three other narrow openings called the median aperture or foramen of Magendie, and the lateral apertures (left and right) or foramen of Luschka. The foramen of Magendie and Luschka are the only direct fluidic connections between the intraventricular CSF spaces and the cortical and spinal SAS.

The extracranial CSF spaces form a complex, interconnected region that includes the cortical SAS and a number of CSF cisterns that can be stratified into 7 "paired" (left and right) and 10 "unpaired" cisterns. Briefly, paired cisterns include the cerebellomedullary, Sylvian, trigeminal, cerebellopontine, carotid, crural, and ambient cisterns. Unpaired cisterns include the cisterna magna, cerebellar, superior cerebellar, quadrigeminal, interpeduncular, pericallosal, carotid, chiasmatic, prepontine, and premedullary cistern. Based on the University of Idaho model, the largest single cistern in terms of volume is the cerebellar cistern, which holds approximately 23 mL of CSF, followed by the Sylvian cistern, with a CSF volume of roughly 8.5 mL on each side (left and right). In total, there are 31 distinct CSF regions in terms of SASs, cisterns, and ventricles (Fig. 2 shows connectivity between regions). Note that

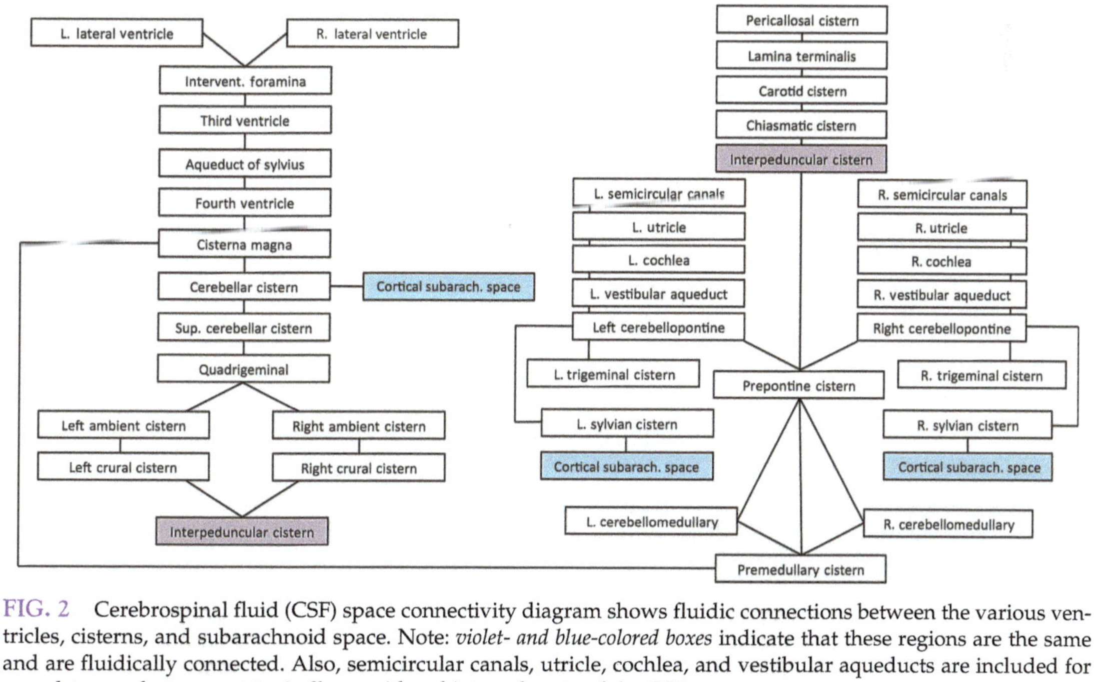

FIG. 2 Cerebrospinal fluid (CSF) space connectivity diagram shows fluidic connections between the various ventricles, cisterns, and subarachnoid space. Note: *violet- and blue-colored boxes* indicate that these regions are the same and are fluidically connected. Also, semicircular canals, utricle, cochlea, and vestibular aqueducts are included for completeness, but are not typically considered integral parts of the CSF system.

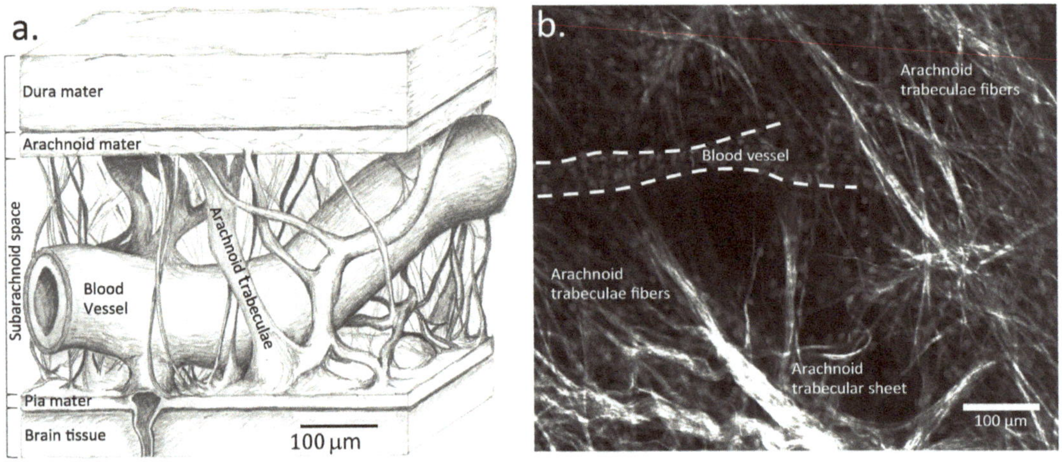

FIG. 3 (A) Anatomy of the subarachnoid space with detail of the location and structure of the arachnoid trabeculae, blood vessels, pia mater, arachnoid mater, dura mater, and brain tissue. (B) Second-harmonic image generation of ovine cortical subarachnoid space shows the collagen microstructure within the arachnoid trabeculae in terms of sheets, pillars, and blood vessel walls. Samples were stained with 4,6-diamidino-2-phenylindole (DAPI) to visualize cell nuclei. Ovine trabecular fiber diameter ranges from approximately 5 to 30 μm, and shows both straight and crimped morphology.

some texts may divide CSF into greater or fewer regions; this chapter highlights the regions most commonly described in the literature.

Cerebrospinal fluid system anatomy has important features that range in size over four orders of magnitude (10 μm-10 cm). Arterial and venous blood vessels traverse the CSF spaces to the brain and spinal cord. In addition, tiny web-like fibers measuring ~15 μm in diameter,[31] called *arachnoid trabeculae*, stretch from the arachnoid to the pia mater (Fig. 3). These fibers are distributed nonuniformly throughout the CSF system[32, 33] and can have a sheet-like morphology that contains holes.[34] Microstructural characterization of the pia arachnoid complex using optical scanning tomography has estimated arachnoid trabeculae volume fraction to be roughly 30% in a pig brain postmortem.[32] Many thin arachnoid membranes are also present in the CSF system, such as the dorsal and dorsolateral septum within the spinal SAS,[35] which may perturb distribution of tracers and affect CSF flow dynamics.[36]

BIOLOGICAL SIGNIFICANCE OF CSF

The presence of CSF within and around the brain and spinal cord is advantageous to the CNS in several ways. First, CSF provides a mechanical support system for the brain. The average adult brain weighs approximately 1400 g,[37] but when the brain is floating in CSF, buoyancy forces reduce its effective weight to roughly 50 g, according to Archimedes' law. CSF acts as a cushion that protects the brain from hitting the skull when the head moves abruptly or sustains mild trauma. Another protective function of CSF is its role in regulating intracranial pressure (ICP) in response to sudden fluctuations caused by obstructed venous outflow and/or expanding mass lesion. This function is described in the Munro–Kellie doctrine,[38] which states that the total volume of the cranial cavity is constant and equals the sum of brain, intracranial blood, and intracranial CSF. In essence, alterations in intracranial brain or blood volume are accommodated by a change in the intracranial CSF volume, thereby maintaining normal ICP.

A second important function of CSF is to sustain metabolic homeostasis of the CNS by supplying nutrients to neural and glial cells, removing metabolic by-products generated by the activities of these cells, and transporting neurotransmitters, hormones, and other biologically active substances throughout the CNS.[39, 40] According to a classical concept initially articulated by Davson and Segal,[41-43] CSF circulation and reabsorption serve as a sink for metabolic wastes that cannot be easily eliminated across the blood–brain barrier (BBB). More recently, the glymphatic system has been termed as a potential pathway for the exchange of CSF and interstitial fluid within the CNS tissue.[44-46] This process is thought to facilitate clearance of potentially neurotoxic waste products (e.g., β-amyloid) and its dysfunction was postulated to contribute to the progression of Alzheimer disease.[47] However, it is yet unclear how a flow of CSF to the interstitial spaces in the brain, as proposed in the glymphatic concept, could occur.[48]

CSF CIRCULATION

According to a classical hypothesis,[41, 49-51] CSF is primarily produced within the brain ventricles by the highly vascular ingrowth of ependymal linings, known as the *choroid plexus*, and circulates among ventricles, cisterns, and the SAS to be ultimately reabsorbed into the blood at the arachnoid granulations of the superior sagittal sinus and at other locations within the CSF system. According to this hypothesis, CSF is produced by the epithelial cells of the choroid plexus of the brain ventricles and flows from the lateral ventricles through the foramina of Monro to the third ventricle. From there, the CSF flows through the aqueduct of Sylvius into the fourth ventricle. CSF exits the ventricular system through the median aperture and the two lateral apertures of Luschka, and enters the cisterns surrounding the cerebellum in the inferior cranial SAS. The CSF continues to move superiorly and inferiorly to the cranial and spinal SAS, respectively.

The main postulates of the traditional hypothesis regarding the choroidal production and circulation of CSF were recently questioned.[52] Experimental measurement of aqueductal CSF flow in animals has demonstrated no CSF secretion inside the brain ventricles in physiological conditions.[53] Furthermore, recent MRI studies using the time-spatial labeling inversion pulse (Time-SLIP) technique[54] disputed the notion that CSF flows from sites of production to sites of absorption, as this imaging modality indicated a bidirectional flow between the lateral and third ventricles and an absence of CSF displacement toward the superior sagittal sinus. The aforementioned findings are in line with the older studies that reported no change in CSF formation rate or chemical composition after choroid plexectomy (i.e., removal) in animals.[55-59]

According to a new hypothesis, CSF is consistently produced and absorbed throughout the whole CSF system as a consequence of filtration and reabsorption of water through the capillary walls into the surrounding CNS tissues. Its volume is regulated by the hydrostatic and osmotic forces between CSF and neural tissues.[52] As such, a significant portion of CSF formation is thought to occur outside the brain ventricles, through filtration of water from blood to interstitial fluid and eventually CSF at the arterial capillaries (high capillary pressure). Based on this model, CNS venous capillaries that are under low hydrostatic pressure are the main spots for CSF reabsorption due to their large surface area. Meanwhile, arachnoid granulations and paraneural sheaths of the cranial and spinal nerves serve as accessory pathways for CSF absorption, which becomes significant when substances are infused into CSF under high experimental pressure.[60] Further research is needed to confirm the source(s) of CSF production and absorption.

ORIGINS OF CSF PULSATIONS

Cerebrospinal fluid pulsations and their sensitivity to the cardiac cycle and venous pressure have been documented since the earliest monomeric evaluation of CSF[61]; however, the mechanisms that drive pulsatile CSF flow are still not fully comprehended. The development of phase-contrast MRI techniques over the past several decades has allowed study of the pulsatile nature of CSF flow, which appears to be craniocaudally oriented during cardiac systole and in the reverse direction during diastole.[62-65] Some have suggested that CSF pulsations are caused by a change in brain volume (resulting from alterations in intracranial blood perfusion during cardiac and respiratory cycles).[66, 67]

Brain volume changes due to the phase difference between the inflow of arterial blood and outflow of venous blood from the cranium.[68, 69] The total cerebral blood volume is reported to change the same amount as the CSF exchange between the cranial and spinal SAS (0.5–2.0 mL per cardiac cycle).[70] Based on the Munro–Kellie hypothesis described earlier in this chapter, the volume change in intracranial blood must be compensated by CSF flow to and from the brain ventricles and cranial SAS. As such, a net increase in cranial blood volume during cardiac systole and the accompanying increase in ICP result in a craniocaudally directed CSF flow. Likewise, a decrease in intracranial blood volume during cardiac diastole leads to CSF flow in the caudocranial direction. CSF system compliance in healthy humans (i.e., cranial and spinal combined) has been estimated to range from roughly 0.5 to 2.0 mL/mmHg.[71]

Cranial venous return changes in response to respiration are thought to be an additional source underlying CSF pulsation. Williams documented the impact of respiration on CSF dynamics via thoracic pressure changes using invasive manometer recordings.[72] Recent noninvasive quantification of CSF flow using Time-SLIP[73] and real-time multislice echo planar imaging MRI techniques[74-76] have revealed a significant contribution of respiration on CSF pulsations in the ventricular system and SAS. These modalities found the magnitude of CSF movement during deep inhalation and exhalation to be greater than that of cardiac-generated pulsations. Alterations in thoracic pressure are transmitted to the intracranial space through nonvalved venous channels (i.e., the vertebrobasilar system and longitudinal vertebral veins) and thoracic CSF,[77] and are believed to affect cranial venous return and ICP.[73, 78] The increased cephalad CSF flow observed during deep inhalation could be explained by reduced cranial venous blood volume, which is caused by diminished thoracic pressure during deep inhalation. Similarly, during deep exhalation, an increase in thoracic pressure reduces the venous return from the brain and results in caudally directed CSF flow to compensate for increased intracranial venous blood volume. Other actions such as coughing,[79] sneezing, the Valsalva maneuver, or application of continuous positive airway pressure[80] could cause complicated pressure environments that are also likely to affect CSF pulse magnitude and flow direction.

Based on MRI measurements, the initiating point of CSF velocity pulsation has been estimated to be located at the foramen magnum near the base of the brain, and is associated with blood volume fluctuations in the nearby highly vascularized cerebellar tonsils.[68] From that location, the CSF pulse spreads outward in both cranial and caudal directions.[81, 82] Using high-speed phase-contrast MRI, CSF pulse wave velocity in the human cervical spine has been measured to be 4.6 m/s in the craniocaudal direction during systole.[83] Cortical CSF pulse wave velocity has not been measured.

IN VIVO ASSESSMENT OF CSF DYNAMICS

CSF dynamics (e.g., velocity, stroke volume, flow rate) have been quantified using MRI. These dynamics have been studied with the hope of elucidating the onset and progression of CNS disorders such as Type 1 Chiari malformation, hydrocephalus, normal-pressure hydrocephalus, and syringomyelia. The earliest MRI visualization of CSF flow was reported by Bradley et al. as a distinct MR artifact, known as flow void, caused by the diminishing signal due to the fluid motion.[3] Over the past decades, several studies have used different MRI-based flow-imaging techniques to evaluate the physiology and pathophysiology of CSF dynamics. The most commonly applied technique is cardiac-gated phase-contrast MRI,[84] which allows noninvasive measurement of CSF velocity profiles over the cardiac cycle. This technique has improved over the years in terms of spatial and temporal resolution. Current methods include four-dimensional phase-contrast MRI, which allows measurement of CSF velocity field in three directions over an entire volume (Fig. 4).[85, 86] In addition, real-time phase-contrast MRI has allowed assessment of CSF dynamics that are not cardiac-related and are caused by respiration and other maneuvers.[74, 76, 87]

The region of the CSF system with the highest CSF velocities in healthy people and patients is typically the aqueduct of Sylvius, with peak velocity values ranging from roughly 4 to 20 cm/s in the caudal direction during cardiac systole.[4, 88] Within the spinal SAS, peak CSF velocities are highest in the cervical spine, with a peak value from 2 to 20 cm/s and stroke volume of about 0.5-2 mL per CSF flow cycle.[85, 89] CSF stroke volume during each cardiac cycle has a decreasing trend, moving caudally along the spinal SAS, and is zero at the spinal SAS termination. This decreasing trend is attributed to distributed compliance along the spine, associated with blood vessels within the spine and the distensible dura and surrounding epidural space.[90]

NUMERICAL ASSESSMENT OF CSF DYNAMICS

Several researchers have implemented computational techniques as an engineering approach to help understand the nature of CSF dynamics and CSF alterations in patients with CNS disorders. These methods have helped to noninvasively investigate flow parameters, which can be difficult to measure using phase-contrast MRI or invasive modalities.

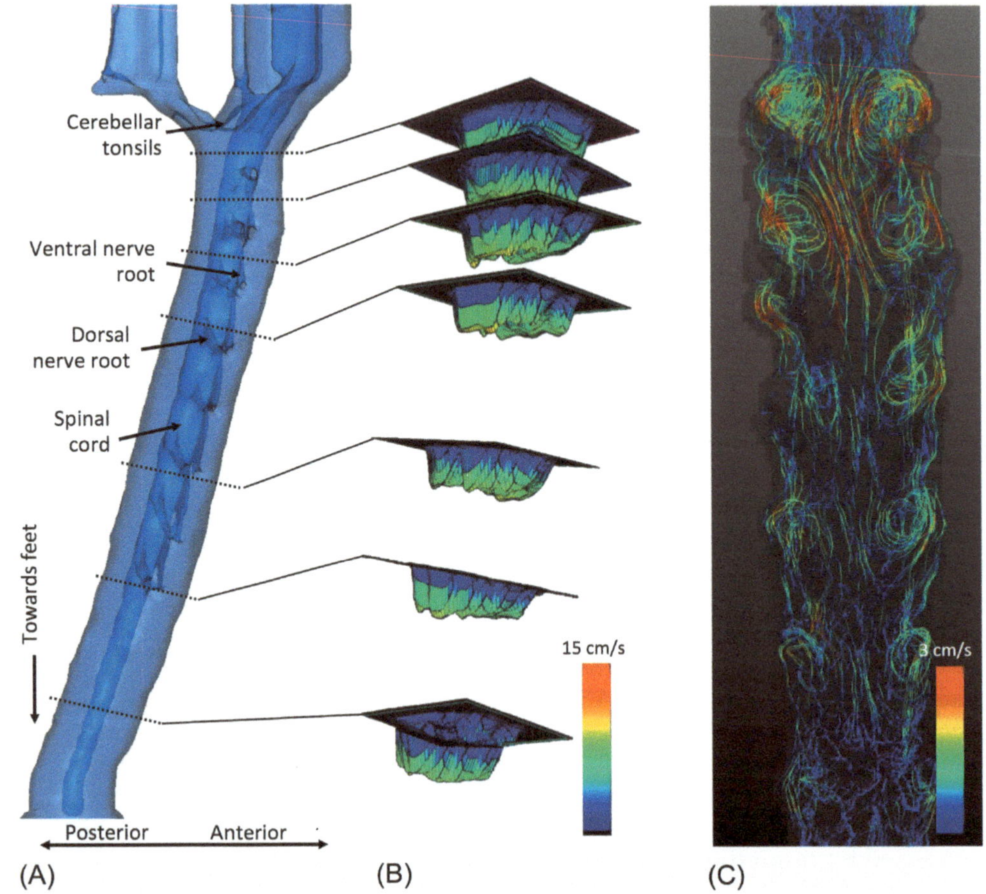

FIG. 4 Cerebrospinal fluid (CSF) velocity measurement in an in-vitro model of the craniovertebral junction using four-dimensional phase-contrast magnetic resonance imaging (4D Flow MRI). (A) Three-dimensional visualization of in vitro model. (B) CSF velocity profiles visualized by 4D Flow MRI. (C) CSF vortical flow structures around dorsal and ventral spinal cord nerve roots visualized by 4D Flow MRI-based streamlines.

The first numerical simulation of CSF motion in the spinal SAS was carried out by Loth et al.[91] who used an idealized annular geometry based on anatomical data from the Visible Human Project. In this study, CSF velocity and pressure fields were obtained by solving the two-dimensional governing equations of fluid dynamics with the assumption of rigid walls. The inertial effects were shown to dominate the flow field under normal physiological flow rates, particularly in the cervical and lower lumbar regions. More recently, Helgeland et al.[92] studied CSF flow in the spinal SAS using a direct numerical simulation model based on a subject-specific geometry of a patient diagnosed with a Type 1 Chiari malformation, and observed flow features similar to those observed in a transient flow. This similarity in features suggested the possible existence of transient velocity fluctuation in patients with extreme SAS geometrical abnormalities. It is unclear whether these velocity fluctuations exist in healthy subjects.

Numerical modeling has also been used to investigate physiological CSF hydrodynamics in the ventricles of the brain and cranial SAS. Kurtcuoglu et al.[88] investigated CSF motion in the third ventricle and in the aqueduct of Sylvius using a computational fluid dynamics model based on subject-specific geometry, flow rate, and wall motion. Their results demonstrated the existence of fluid jets exiting the aqueduct and recirculation zones superior and inferior to the observed jet. Subsequently, Gupta et al.[93, 94] characterized CSF motion in the cranial SAS and superior spinal SAS using subject-specific computational fluid dynamics models. These studies quantified the complex CSF flow patterns, including the large variation in the spatial distribution of velocities, and found the velocity profile of CSF entering the spinal SAS to be blunt or plug-like, which was consistent with in vivo observations.[89, 95]

In addition to their utility in detailing characteristics of CSF flow, numerical simulations provide a powerful tool for variational analysis to determine the importance of various anatomical features and physiological functions on CSF dynamics. These include spinal cord nerve roots,[96] denticulate ligaments,[97, 98] frequency and magnitude of CSF pulsations,[99] presence of arachnoid trabeculae,[98] focal spinal arachnoiditis,[100] tonsillar descent in patients with Chiari malformations,[6, 101] and properties of the spinal cord and dura.[102] In addition to producing new engineering-based parameters, these studies are necessary for evaluating the potential of such parameters for clinical use. Fig. 5 shows an example of numerical simulation of intrathecal CSF dynamics conducted by Khani et al.[96]

SOLUTE TRANSPORT WITHIN THE CSF

A solute injected into the CSF in the spinal SAS mixes with the CSF,[103] spreads throughout the CSF system, passes through the pia mater, and is then taken up into the brain parenchyma, perhaps via the paravascular route.[104] Unlike solutes injected into the blood, this process bypasses the BBB (and the endothelial tight junctions), which allows for delivery of many molecules to the brain that may not be deliverable through the vascular system.[105, 106] These range from small molecules to biologics, including protein and cell-based and gene therapies involving tropic factors to stimulate dying neurons.[107, 108] Moreover, direct delivery of therapeutic agents that bypass the BBB may allow lower doses, decreasing the risk of potential toxic effects, compared to therapeutic agents that are systemically administered.[109, 110] Several therapies have shown promise in animal studies[111]; others have been shown to be safe in human clinical trials.[112] In addition, CSF-based delivery is a relatively simple surgical intervention with a potentially lower risk to the patient than other surgical interventions, such as convection-enhanced drug delivery, and can be safer than other therapeutic options such as deep brain stimulation.

CSF solute transport has been investigated in several studies. Early clinical studies in humans showed that intrathecal and intracerebroventricular administration of small, lipid-soluble molecules did not result in efficient CNS tissue penetration.[113, 114] However, recent studies have demonstrated strong evidence that IDD might be ideal for the delivery of therapeutic proteins with large biomolecules.[115] Numerous animal studies have shown that leveraging the CSF to CNS pathway can result in widespread distribution of therapeutic agents in brain parenchyma.[115-119] Pharmacokinetics and bioavailability of therapeutic

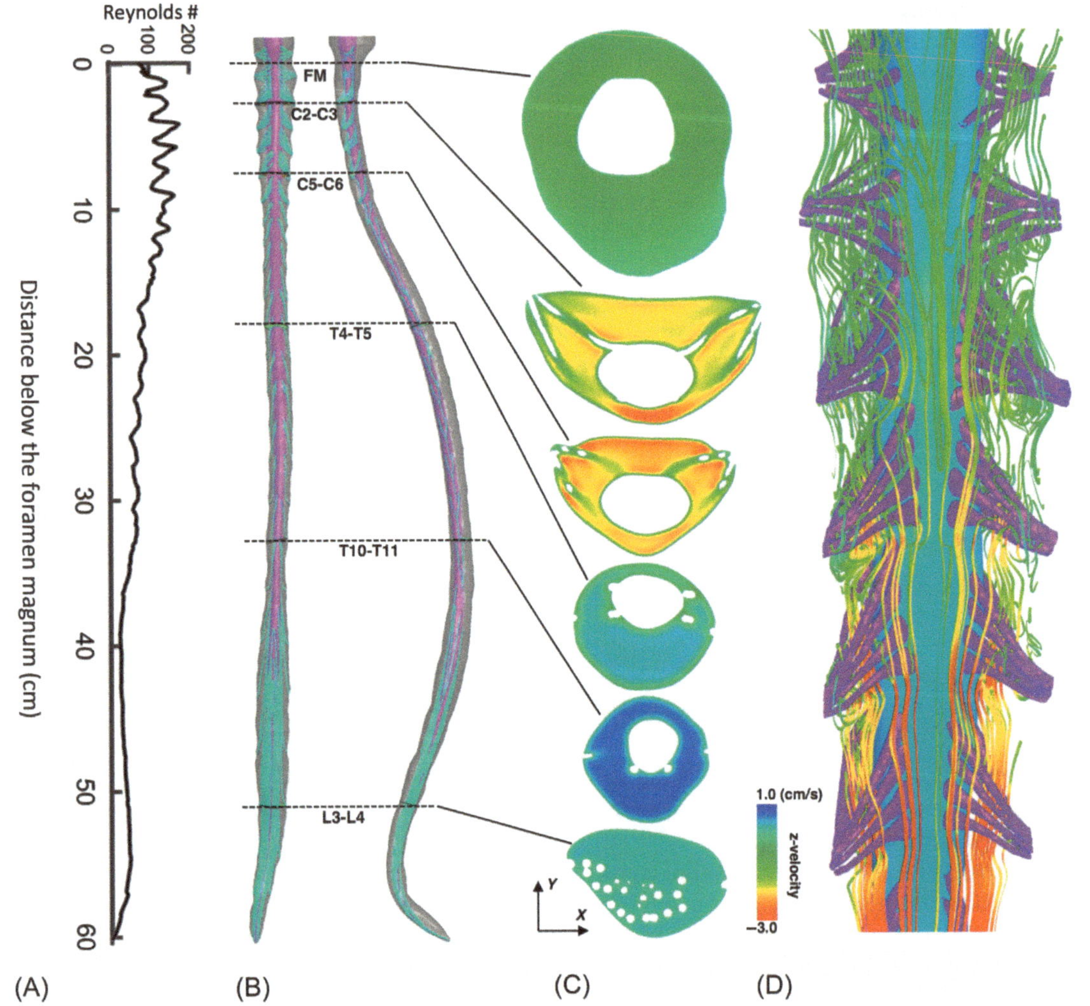

FIG. 5 Numerical simulation of intrathecal cerebrospinal fluid (CSF) dynamics by Khani et al.[96] (A) Reynolds number, computed based on hydraulic diameter along the model. (B) Three-dimensional geometric visualization of intrathecal space with 31 pairs of anatomically realistic dorsal and ventral nerve root pairs. (C) Numerical prediction of CSF velocity profiles around the spinal cord at 90 ms after cardiac systole (note: CSF flow is nonuniform along the spinal SAS). (D) Visualization of velocity streamlines in the cervical spine show vortices that form around spinal cord nerve roots during CSF flow reversal.

enzymes administered intrathecally have also been evaluated in experimental studies through direct measurements and noninvasive imaging techniques.[120]

Numerical modeling has helped elucidate how a net motion of particle tracers can occur along the spine while the CSF time-averaged flow (cyclic average flow rate at any cross section) can be zero. This net motion of particles has been reported in radionuclide tracer studies by Di Chiro[121] and others. In order to investigate this phenomenon, Sanchez et al.[122] formulated an anatomically idealized numerical model of the intrathecal space and identified that bulk CSF circulation along the spine, or localized "steady-streaming" CSF motion, can result due to convective acceleration. Local steady-streaming CSF velocities on one side of the spinal canal were found to reach up to $\sim$1 mm/s, depending on eccentricity of the spinal cord. Khani et al.[96] also corroborated these findings in a numerical model of the intrathecal space with anatomically realistic spinal cord geometry and nonuniform CSF pulsation. They found similar degrees of steady-streaming CSF velocities, with a complex steady-streaming velocity profile that was affected by the presence of spinal cord nerve roots and spinal curvature. Together, these numerical studies provide a mechanistic explanation for radionuclide tracer studies that indicate net CSF flow direction within the CSF system.

Variational analysis of different parameters involved in IDD has been carried out in multiple experimental and numerical studies. Using an in vitro model, Nelissen[123] investigated CSF dynamics around a catheter tip, and reported the adjacent flow vortices to be the main drivers of drug movement. The first numerical model of IDD was developed by Myers,[124] based on a three-dimensional geometry of the spinal SAS. This study reported that lower values for the ratio of the cross-sectional dimension of the spinal SAS to the diameter of the catheter can result in more uniform drug distribution. More recent simulation studies with varying degrees of anatomical detail and physiological accuracy have evaluated the impact of various parameters on IDD. These parameters include injection rate,[10, 125] catheter geometry and position,[10] CSF flow stroke volume and frequency,[126] anatomical microstructures,[127] and drug diffusion properties.[11] Numerical models were also implemented to evaluate the drug transport in CNS tissue.[128-130]

Cerebrospinal fluid dynamics are spatially and temporally complex due to the plethora of underlying physiologic and anatomic factors that govern its motion. Noninvasive MRI has been applied in many studies in an effort to objectively quantify volumes and flow velocities in the CSF system, but much remains to be known and confirmed under both physiological and pathophysiological conditions. Multidisciplinary research aimed to deepen our understanding of CSF, in combination with technological advances, will inevitably help develop new treatment strategies for patients suffering from the many pathologies in which CSF alterations are considered an important factor.

Acknowledgment

The authors thank Mohammadreza Khani, Lucas Sass, Nathan Schiele, Sophia Theodossiou, Claire Majors, Gabryel Conley Natividad, Ann Norton, and Suraj Thyagaraj for assistance with images used in figures.

Funding statement

This work was supported by an Institutional Development Award (IDeA) from the National Institute of General Medical Sciences (NIGMS) of the National Institutes of health (NIH) under Grant #P20GM1033408 and #4U54GM104944-04TBD, The National Institute of Mental Health Grant #1R44MH112210-01A1, University of Idaho Vandal Ideas Project, and American Syringomyelia and Chiari Alliance Project.

References

1. Bloomfield IG, Johnston IH, Bilston LE. Effects of proteins, blood cells and glucose on the viscosity of cerebrospinal fluid. *Pediatr Neurosurg*. 1998;28(5):246-251.
2. Weller RO. Pathology of cerebrospinal fluid and interstitial fluid of the CNS: significance for Alzheimer disease, prion disorders and multiple sclerosis. *J Neuropathol Exp Neurol*. 1998;57(10):885-894.

3. Bradley Jr. WG, Kortman KE, Burgoyne B. Flowing cerebrospinal fluid in normal and hydrocephalic states: appearance on MR images. *Radiology*. 1986;159(3):611-616.
4. Bradley Jr. WG, Scalzo D, Queralt J, Nitz WN, Atkinson DJ, Wong P. Normal-pressure hydrocephalus: evaluation with cerebrospinal fluid flow measurements at MR imaging. *Radiology*. 1996;198(2):523-529.
5. Bunck AC, Kroger JR, Juttner A, et al. Magnetic resonance 4D flow characteristics of cerebrospinal fluid at the craniocervical junction and the cervical spinal canal. *Eur Radiol*. 2011;21(8):1788-1796.
6. Shaffer N, Martin BA, Loth F. Cerebrospinal fluid hydrodynamics in type I Chiari malformation. *Neurol Res*. 2011;33(3):247-260.
7. Shaffer N, Martin BA, Rocque B, et al. Cerebrospinal fluid flow impedance is elevated in type I Chiari malformation. *J Biomech Eng*. 2013;136(2) 021012.
8. Clarke EC, Stoodley MA, Bilston LE. Changes in temporal flow characteristics of CSF in Chiari malformation type I with and without syringomyelia: implications for theory of syrinx development clinical article. *J Neurosurg*. 2013;118(5):1135-1140.
9. Stoodley MA, Gutschmidt B, Jones NR. Cerebrospinal fluid flow in an animal model of noncommunicating syringomyelia. *Neurosurgery*. 1999;44(5):1065-1075 [discussion 1075-1066].
10. Haga PT, Pizzichelli G, Mortensen M, et al. A numerical investigation of intrathecal isobaric drug dispersion within the cervical subarachnoid space. *PLoS One*. 2017;12(3).
11. Pizzichelli G, Kehlet B, Evju O, et al. Numerical study of intrathecal drug delivery to a permeable spinal cord: effect of catheter position and angle. *Comput Methods Biomech Biomed Engin*. 2017;1-10.
12. Tangen KM, Leval R, Mehta AI, Linninger AA. Computational and in vitro experimental investigation of intrathecal drug distribution: parametric study of the effect of injection volume, cerebrospinal fluid pulsatility, and drug uptake. *Anesth Analg*. 2017;124(5):1686-1696.
13. Tangen K, Narasimhan NS, Sierzega K, Preden T, Alaraj A, Linninger AA. Clearance of subarachnoid hemorrhage from the cerebrospinal fluid in computational and in vitro models. *Ann Biomed Eng*. 2016;44(12):3478-3494.
14. Bechter K, Schreiner V, Herzog S, et al. Cerebrospinal fluid filtration as experimental therapy in therapy refractory psychoses in Borna disease virus seropositive patients. Therapeutic effects, findings. *Psychiatr Prax*. 2003; 30(suppl 2):S216-S220.
15. Meylaerts SA, Kalkman CJ, de Haan P, Porsius M, Jacobs MJ. Epidural versus subdural spinal cord cooling: cerebrospinal fluid temperature and pressure changes. *Ann Thorac Surg*. 2000;70(1):222-227 [discussion 228].
16. Goetz T, Romero-Sierra C, Ethier R, Henriksen RN. Modeling of therapeutic dialysis of cerebrospinal fluid by epidural cooling in spinal cord injuries. *J Neurotrauma*. 1988;5(2):139-150.
17. Luciano MG, Dombrowski SM, Qvarlander S, et al. Novel method for dynamic control of intracranial pressure. *J Neurosurg*. 2017;126(5):1629-1640.
18. Levi Chazen J, Dyke JP, Holt RW, et al. Automated segmentation of MR imaging to determine normative central nervous system cerebrospinal fluid volumes in healthy volunteers. *Clin Imaging*. 2017;43:132-135.
19. Hodel J, Lebret A, Petit E, et al. Imaging of the entire cerebrospinal fluid volume with a multistation 3D SPACE MR sequence: feasibility study in patients with hydrocephalus. *Eur Radiol*. 2013;23(6):1450-1458.
20. Courchesne E, Chisum HJ, Townsend J, et al. Normal brain development and aging: quantitative analysis at in vivo MR imaging in healthy volunteers. *Radiology*. 2000;216(3):672-682.
21. Coffey CE, Lucke JF, Saxton JA, et al. Sex differences in brain aging: a quantitative magnetic resonance imaging study. *Arch Neurol*. 1998;55:169. Erratum in: Arch Neurol 1998 May;55(5):627.
22. Pfefferbaum A, Mathalon DH, Sullivan EV, Rawles JM, Zipursky RB, Lim KO. A quantitative magnetic resonance imaging study of changes in brain morphology from infancy to late adulthood. *Arch Neurol*. 1994; 51(9):874-887.
23. Sass LR, Khani M, Natividad GC, Tubbs RS, Baledent O, Martin BA. A 3D subject-specific model of the spinal subarachnoid space with anatomically realistic ventral and dorsal spinal cord nerve rootlets. *Fluids Barriers CNS*. 2017;14(1):36.
24. Guyton AC, Hall JE. *Textbook of Medical Physiology*. 9th ed. Philadelphia, PA: W.B. Saunders; 1996.
25. Davson H, Segal MB. *Physiology of the CSF and Blood-Brain Barriers*. Boca Raton, FL: CRC Press; 1996.
26. Johanson CE, Duncan III JA, Klinge PM, Brinker T, Stopa EG, Silverberg GD. Multiplicity of cerebrospinal fluid functions: new challenges in health and disease. *Cerebrospinal Fluid Res*. 2008;5:10.
27. Sakka L, Coll G, Chazal J. Anatomy and physiology of cerebrospinal fluid. *Eur Ann Otorhinolaryngol Head Neck Dis*. 2011;128(6):309-316.

28. Last RJ, Tompsett DH. Casts of the cerebral ventricles. *Br J Surg*. 1953;40(164):525-543.
29. Wyper DJ, Pickard JD, Matheson M. Accuracy of ventricular volume estimation. *J Neurol Neurosurg Psychiatry*. 1979;42(4):345-350.
30. Grant R, Condon B, Lawrence A, et al. Human cranial CSF volumes measured by MRI: sex and age influences. *Magn Reson Imaging*. 1987;5(6):465-468.
31. Killer HE, Laeng HR, Flammer J, Groscurth P. Architecture of arachnoid trabeculae, pillars, and septa in the subarachnoid space of the human optic nerve: anatomy and clinical considerations. *Br J Ophthalmol*. 2003; 87(6):777-781.
32. Scott G, Coats B. Microstructural characterization of the pia-arachnoid complex using optical coherence tomography. *IEEE Trans Med Imaging*. 2015;34(7):1452-1459.
33. Scott GG, Margulies SS, Coats B. Utilizing multiple scale models to improve predictions of extra-axial hemorrhage in the immature piglet. *Biomech Model Mechanobiol*. 2016;15(5):1101-1119.
34. Mortazavi MM, Quadri SA, Khan MA, et al. Subarachnoid Trabeculae: a comprehensive review of their embryology, histology, morphology and surgical significance. *World Neurosurg*. 2017;111:279-290.
35. Nauta HJ, Dolan E, Yasargil MG. Microsurgical anatomy of spinal subarachnoid space. *Surg Neurol*. 1983; 19(5):431-437.
36. Gottschalk A, Schmitz B, Mauer UM, et al. Dynamic visualization of arachnoid adhesions in a patient with idiopathic syringomyelia using high-resolution cine magnetic resonance imaging at 3 T. *J Magn Reson Imaging*. 2010;32(1):218-222.
37. Bradbury MWB. *The Concept of a Blood-Brain Barrier*. Chichester, NY: Wiley; 1979.
38. Mokri B. The Monro–Kellie hypothesis applications in CSF volume depletion. *Neurology*. 2001;56 (12):1746-1748.
39. Johnston M, Zakharov A, Koh L, Armstrong D. Subarachnoid injection of microfil reveals connections between cerebrospinal fluid and nasal lymphatics in the non-human primate. *Neuropathol Appl Neurobiol*. 2005; 31(6):632-640.
40. Knigge K, Joseph S, Schock D, et al. Role of the ventricular system in neuroendocrine processes: synthesis and distribution of thyrotropin releasing factor (TRF) in the hypothalamus and third ventricle. *Can J Neurol Sci*. 1974;1(01):74-84.
41. Davson H, Segal MB. *Physiology of the CSF and Blood Brain Bariers*. CRC-Press; 1995.
42. Segal M. Extracellular and cerebrospinal fluids. *J Inherit Metab Dis*. 1993;16(4):617-638.
43. Segal M, Pollay M. The secretion of cerebrospinal fluid. *Exp Eye Res*. 1977;25:127-148.
44. Iliff JJ, Wang M, Zeppenfeld DM, et al. Cerebral arterial pulsation drives paravascular CSF-interstitial fluid exchange in the murine brain. *J Neurosci*. 2013;33(46):18190-18199.
45. Murlidharan G, Crowther A, Reardon RA, Song J, Asokan A. Glymphatic fluid transport controls paravascular clearance of AAV vectors from the brain. *JCI Insight*. 2016;1(14):e88034.
46. Simon MJ, Iliff JJ. Regulation of cerebrospinal fluid (CSF) flow in neurodegenerative, neurovascular and neuroinflammatory disease. *Biochim Biophys Acta*. 2016;1862(3):442-451.
47. Iliff JJ, Lee H, Yu M, et al. Brain-wide pathway for waste clearance captured by contrast-enhanced MRI. *J Clin Invest*. 2013;123(3):1299-1309.
48. Abbott NJ, Pizzo ME, Preston JE, Janigro D, Thorne RG. The role of brain barriers in fluid movement in the CNS: is there a 'glymphatic' system? *Acta Neuropathol*. 2018;135(3):387-407.
49. Cserr HF. Role of secretion and bulk flow of brain interstitial fluid in brain volume regulation. *Ann N Y Acad Sci*. 1988;529(1):9-20.
50. Cushing H. Studies on the cerebro-spinal fluid: I. Introduction. *J Med Res*. 1914;31(1):1.
51. Dandy WE. Experimental hydrocephalus. *Ann Surg*. 1919;70(2):129.
52. Orešković D, Klarica M. The formation of cerebrospinal fluid: nearly a hundred years of interpretations and misinterpretations. *Brain Res Rev*. 2010;64(2):241-262.
53. Oreskovic D, Klarica M, Vukic M. Does the secretion and circulation of the cerebrospinal fluid really exist? *Med Hypotheses*. 2001;56(5):622-624.
54. Yamada S, Kelly E. Cerebrospinal fluid dynamics and the pathophysiology of hydrocephalus: new concepts. *Semin Ultrasound CT MR*. 2016;37(2):84-91.
55. Hammock MK, Milhorat TH. Recent studies on the formation of cerebrospinal fluid. *Dev Med Child Neurol*. 1973;15(s29):27-34.

56. Milhorat T, Hammock M, Davis D, Fenstermacher J. Choroid plexus papilloma. *Pediatr Neurosurg*. 1976; 2(5):273-289.
57. Milhorat TH. Choroid plexus and cerebrospinal fluid production. *Science*. 1969;166(3912):1514-1516.
58. Milhorat TH. The third circulation revisited. *J Neurosurg*. 1975;42(6):628-645.
59. Milhorat TH, Hammock MK, Chien T, Davis DA. Normal rate of cerebrospinal fluid formation five years after bilateral choroid plexectomy: case report. *J Neurosurg*. 1976;44(6):735-739.
60. Bulat M, Klarica M. Recent insights into a new hydrodynamics of the cerebrospinal fluid. *Brain Res Rev*. 2011; 65(2):99-112.
61. O'Connell JE. The vascular factor in intracranial pressure and the maintenance of the cerebrospinal fluid circulation. *Brain*. 1943;66(3):204-228.
62. Enzmann D, Pelc N. Normal flow patterns of intracranial and spinal cerebrospinal fluid defined with phase-contrast cine MR imaging. *Radiology*. 1991;178(2):467-474.
63. Greitz D, Franck A, Nordell B. On the pulsatile nature of intracranial and spinal CSF-circulation demonstrated by MR imaging. *Acta Radiol*. 1993;34(4):321-328.
64. Nitz W, Bradley Jr W, Watanabe A, et al. Flow dynamics of cerebrospinal fluid: assessment with phase-contrast velocity MR imaging performed with retrospective cardiac gating. *Radiology*. 1992;183(2):395-405.
65. Scroth G, Klose U. Cerebrospinal fluid flow. I. Physiology of cardiac-related pulsation. *Neuroradiology*. 1992; 35(1):9.
66. Avezaat C, Van Eijndhoven J, Wyper D. Cerebrospinal fluid pulse pressure and intracranial volume-pressure relationships. *J Neurol Neurosurg Psychiatry*. 1979;42(8):687-700.
67. Portnoy HD, Chopp M. Cerebrospinal fluid pulse wave form analysis during hypercapnia and hypoxia. *Neurosurgery*. 1981;9(1):14-27.
68. Henry–Feugeas MC, Idy–Peretti I, Baledent O, et al. Origin of subarachnoid cerebrospinal fluid pulsations: a phase-contrast MR analysis. *Magn Reson Imaging*. 2000;18(4):387-395.
69. Wåhlin A, Ambarki K, Hauksson J, Birgander R, Malm J, Eklund A. Phase contrast MRI quantification of pulsatile volumes of brain arteries, veins, and cerebrospinal fluids compartments: repeatability and physiological interactions. *J Magn Reson Imaging*. 2012;35(5):1055-1062.
70. Freund M, Adwan M, Kooijman H, et al. Measurement of CSF flow in the spinal canal using MRI with an optimized MRI protocol: experimental and clinical studies. *RoFo: Fortschritte auf dem Gebiete der Rontgenstrahlen und der Nuklearmedizin*. 2001;173(4):306-314.
71. Marmarou A, Shulman K, LaMorgese J. Compartmental analysis of compliance and outflow resistance of the cerebrospinal fluid system. *J Neurosurg*. 1975;43(5):523-534.
72. Williams B. Simultaneous cerebral and spinal fluid pressure recordings. I. Technique, physiology, and normal results. *Acta Neurochir*. 1981;58(3–4):167-185.
73. Yamada S, Miyazaki M, Yamashita Y, et al. Influence of respiration on cerebrospinal fluid movement using magnetic resonance spin labeling. *Fluids Barriers CNS*. 2013;10(1):36.
74. Yildiz S, Thyagaraj S, Jin N, et al. Quantifying the influence of respiration and cardiac pulsations on cerebrospinal fluid dynamics using real-time phase-contrast MRI. *J Magn Reson Imaging*. 2017;46(2):431-439.
75. Dreha-Kulaczewski S, Joseph AA, Merboldt KD, Ludwig HC, Gartner J, Frahm J. Inspiration is the major regulator of human CSF flow. *J Neurosci*. 2015;35(6):2485-2491.
76. Chen L, Beckett A, Verma A, Feinberg DA. Dynamics of respiratory and cardiac CSF motion revealed with real-time simultaneous multi-slice EPI velocity phase contrast imaging. *NeuroImage*. 2015;122:281-287.
77. Guerci AD, Shi A-Y, Levin H, Tsitlik J, Weisfeldt ML, Chandra N. Transmission of intrathoracic pressure to the intracranial space during cardiopulmonary resuscitation in dogs. *Circ Res*. 1985;56(1):20-30.
78. Yannopoulos D, McKnite SH, Metzger A, Lurie KG. Intrathoracic pressure regulation for intracranial pressure management in normovolemic and hypovolemic pigs. *Crit Care Med*. 2006;34(12):S495-S500.
79. Bezuidenhout AF, Khatami D, Heilman CB, et al. Relationship between cough-associated changes in CSF flow and disease severity in chiari I malformation: an exploratory study using real-time MRI. *AJNR Am J Neuroradiol*. 2018;39(7):1267-1272.
80. Yiallourou TI, Daners MS, Kurtcuoglu V, et al. Continuous positive airway pressure alters cranial blood flow and cerebrospinal fluid dynamics at the craniovertebral junction. *Interdiscip Neurosurg Adv Tech Case Manage*. 2015; 2(3):152-159.

81. Yatsushiro S, Sunohara S, Takizawa K, Matsumae M, Kajihara N, Kuroda K. Characterization of cardiac- and respiratory-driven cerebrospinal fluid motions using correlation mapping with asynchronous 2-dimensional phase contrast technique. *Conf Proc IEEE Eng Med Biol Soc*. 2016;2016:3867-3870.
82. Takizawa K, Matsumae M, Sunohara S, Yatsushiro S, Kuroda K. Characterization of cardiac- and respiratory-driven cerebrospinal fluid motion based on asynchronous phase-contrast magnetic resonance imaging in volunteers. *Fluids Barriers CNS*. 2017;14(1):25.
83. Kalata W, Martin BA, Oshinski JN, Jerosch-Herold M, Royston TJ, Loth F. MR measurement of cerebrospinal fluid velocity wave speed in the spinal canal. *IEEE Trans Biomed Eng*. 2009;56(6):1765-1768.
84. Naidich TP, Altman NR, Gonzalez-Arias SM. Phase contrast cine magnetic resonance imaging: normal cerebrospinal fluid oscillation and applications to hydrocephalus. *Neurosurg Clin N Am*. 1993;4(4):677-705.
85. Bunck AC, Kroeger JR, Juettner A, et al. Magnetic resonance 4D flow analysis of cerebrospinal fluid dynamics in Chiari I malformation with and without syringomyelia. *Eur Radiol*. 2012;22(9):1860-1870.
86. Heidari Pahlavian S, Bunck AC, Thyagaraj S, et al. Accuracy of 4D flow measurement of cerebrospinal fluid dynamics in the cervical spine: an in vitro verification against numerical simulation. *Ann Biomed Eng*. 2016; 44(11):3202-3214.
87. Bhadelia RA, Patz S, Heilman C, et al. Cough-associated changes in CSF flow in Chiari I malformation evaluated by real-time MRI. *AJNR Am J Neuroradiol*. 2016;37(5):825-830.
88. Kurtcuoglu V, Soellinger M, Summers P, et al. Computational investigation of subject-specific cerebrospinal fluid flow in the third ventricle and aqueduct of Sylvius. *J Biomech*. 2007;40(6):1235-1245.
89. Quigley MF, Iskandar B, Quigley ME, Nicosia M, Haughton V. Cerebrospinal fluid flow in foramen magnum: temporal and spatial patterns at MR imaging in volunteers and in patients with Chiari I malformation. *Radiology*. 2004;232(1):229-236.
90. Martin BA, Reymond P, Novy J, Baledent O, Stergiopulos N. A coupled hydrodynamic model of the cardiovascular and cerebrospinal fluid system. *Am J Physiol Heart Circ Physiol*. 2012;302(7):H1492-H1509.
91. Loth F, Yardimci MA, Alperin N. Hydrodynamic modeling of cerebrospinal fluid motion within the spinal cavity. *J Biomech Eng*. 2001;123(1):71-79.
92. Helgeland A, Mardal KA, Haughton V, Reif BA. Numerical simulations of the pulsating flow of cerebrospinal fluid flow in the cervical spinal canal of a Chiari patient. *J Biomech*. 2014;47(5):1082-1090.
93. Gupta S, Soellinger M, Boesiger P, Poulikakos D, Kurtcuoglu V. Three-dimensional computational modeling of subject-specific cerebrospinal fluid flow in the subarachnoid space. *J Biomech Eng*. 2009;131(2).
94. Gupta S, Soellinger M, Grzybowski DM, et al. Cerebrospinal fluid dynamics in the human cranial subarachnoid space: an overlooked mediator of cerebral disease. I. Computational model. *J R Soc Interface*. 2010; 7(49):1195-1204.
95. Iskandar BJ, Quigley M, Haughton VM. Foramen magnum cerebrospinal fluid flow characteristics in children with Chiari I malformation before and after craniocervical decompression. *J Neurosurg*. 2004;101(Suppl 2): 169-178.
96. Khani M, Sass L, Xing T, Sharp MK, Balédent O, Martin B. Anthropomorphic model of intrathecal cerebrospinal fluid dynamics within the spinal subarachnoid space: spinal cord nerve roots increase steady-streaming. *J Biomech Eng*. 2018.
97. Pahlavian SH, Loth F, Luciano M, Oshinski J, Martin BA. Neural tissue motion impacts cerebrospinal fluid dynamics at the cervical medullary junction: a patient-specific moving-boundary computational model. *Ann Biomed Eng*. 2015;43(12):2911-2923.
98. Stockman HW. Effect of anatomical fine structure on the flow of cerebrospinal fluid in the spinal subarachnoid space. *J Biomech Eng*. 2006;128(1):106-114.
99. Linge SO, Haughton V, Lovgren AE, Mardal KA, Helgeland A, Langtangen HP. Effect of tonsillar herniation on cyclic CSF flow studied with computational flow analysis. *AJNR Am J Neuroradiol*. 2011; 32(8):1474-1481.
100. Bertram C, Bilston L, Stoodley M. Tensile radial stress in the spinal cord related to arachnoiditis or tethering: a numerical model. *Med Biol Eng Comput*. 2008;46(7):701-707.
101. Martin BA, Kalata W, Shaffer N, Fischer P, Luciano M, Loth F. Hydrodynamic and longitudinal impedance analysis of cerebrospinal fluid dynamics at the craniovertebral junction in type I Chiari malformation. *PLoS One*. 2013;8(10).

102. Cirovic S. A coaxial tube model of the cerebrospinal fluid pulse propagation in the spinal column. *J Biomech Eng.* 2009;131(2).
103. Papisov MI, Belov VV, Gannon KS. Physiology of the intrathecal bolus: the leptomeningeal route for macromolecule and particle delivery to CNS. *Mol Pharm.* 2013;10(5):1522-1532.
104. Iliff JJ, Wang M, Liao Y, et al. A paravascular pathway facilitates CSF flow through the brain parenchyma and the clearance of interstitial solutes, including amyloid beta. *Sci Transl Med.* 2012;4(147):147ra111.
105. Lu CT, Zhao YZ, Wong HL, Cai J, Peng L, Tian XQ. Current approaches to enhance CNS delivery of drugs across the brain barriers. *Int J Nanomedicine.* 2014;9:2241-2256.
106. Patel T, Zhou JB, Piepmeier JM, Saltzman WM. Polymeric nanoparticles for drug delivery to the central nervous system. *Adv Drug Deliv Rev.* 2012;64(7):701-705.
107. Deepa P, Shahani N, Alladi PA, et al. Down regulation of trophic factors in neonatal rat spinal cord after administration of cerebrospinal fluid from sporadic amyotrophic lateral sclerosis patients. *J Neural Transm (Vienna).* 2011;118(4):531-538.
108. Watanabe Y, Kazuki Y, Kazuki K, et al. Use of a human artificial chromosome for delivering trophic factors in a rodent model of amyotrophic lateral sclerosis. *Mol Ther Nucleic Acids.* 2015;4.
109. Bottros MM, Christo PJ. Current perspectives on intrathecal drug delivery. *J Pain Res.* 2014;7:615-626.
110. Bhatia G, Lau ME, Koury KM, Gulur P. Intrathecal drug delivery (ITDD) systems for cancer pain. *F1000Res.* 2013;2:96.
111. Chen L, Jiang M, Pei L. Comparison of three methods of drug delivery in the rat lumbar spinal subarachnoid space. *Anat Rec (Hoboken).* 2012;295(7):1212-1220.
112. Miller TM, Pestronk A, David W. An antisense oligonucleotide against SOD1 delivered intrathecally for patients with SOD1 familial amyotrophic lateral sclerosis: a phase 1, randomised, first-in-man study. *Lancet Neurol.* 2013;12:435. Erratum in: Lancet Neurol: 2013;12(5):423.
113. Greene HL, Hug G, Schubert WK. Metachromatic leukodystrophy: treatment with arylsulfatase-a. *Arch Neurol.* 1969;20(2):147-153.
114. Ghersi-Egea JF, Gorevic PD, Ghiso J, Frangione B, Patlak CS, Fenstermacher JD. Fate of cerebrospinal fluid-borne amyloid β-peptide: rapid clearance into blood and appreciable accumulation by cerebral arteries. *J Neurochem.* 1996;67(2):880-883.
115. Calias P, Banks WA, Begley D, Scarpa M, Dickson P. Intrathecal delivery of protein therapeutics to the brain: a critical reassessment. *Pharmacol Ther.* 2014;144(2):114-122.
116. Kakkis E, McEntee M, Vogler C, et al. Intrathecal enzyme replacement therapy reduces lysosomal storage in the brain and meninges of the canine model of MPS I. *Mol Genet Metab.* 2004;83(1–2):163-174.
117. Calias P, Papisov M, Pan J, et al. CNS penetration of intrathecal-lumbar idursulfase in the monkey, dog and mouse: implications for neurological outcomes of lysosomal storage disorder. *PLoS One.* 2012;7(1).
118. Dickson P, McEntee M, Vogler C, et al. Intrathecal enzyme replacement therapy: successful treatment of brain disease via the cerebrospinal fluid. *Mol Genet Metab.* 2007;91(1):61-68.
119. Hemsley KM, Norman EJ, Crawley AC, et al. Effect of cisternal sulfamidase delivery in MPS IIIA huntaway dogs—a proof of principle study. *Mol Genet Metab.* 2009;98(4):383-392.
120. Xie H, Chung JK, Mascelli MA, McCauley TG. Pharmacokinetics and bioavailability of a therapeutic enzyme (idursulfase) in cynomolgus monkeys after intrathecal and intravenous administration. *PLoS One.* 2015;10(4).
121. Dichiro G. Movement of the cerebrospinal fluid in human beings. *Nature.* 1964;204:290-291.
122. Sanchez AL, Martinez-Bazan C, Gutierrez-Montes C, et al. On the bulk motion of the cerebrospinal fluid in the spinal canal. *J Fluid Mech.* 2018;841:203-227.
123. Nelissen RM. *Fluid Mechanics of Intrathecal Drug Delivery;* 2008. 158 pp.
124. Myers MR. A numerical investigation into factors affecting anesthetic distribution during spinal anesthesia. *J Biomech.* 1996;29(2):139-149.
125. Kuttler A, Dimke T, Kern S, Helmlinger G, Stanski D, Finelli LA. Understanding pharmacokinetics using realistic computational models of fluid dynamics: biosimulation of drug distribution within the CSF space for intrathecal drugs. *J Pharmacokinet Pharmacodyn.* 2010;37(6):629-644.
126. Hsu Y, Hettiarachchi HD, Zhu DC, Linninger AA. The frequency and magnitude of cerebrospinal fluid pulsations influence intrathecal drug distribution: key factors for interpatient variability. *Anesth Analg.* 2012; 115(2):386-394.

127. Tangen KM, Hsu Y, Zhu DC, Linninger AA. CNS wide simulation of flow resistance and drug transport due to spinal microanatomy. *J Biomech.* 2015;48(10):2144-2154.
128. Linninger AA, Somayaji MR, Mekarski M, Zhang L. Prediction of convection-enhanced drug delivery to the human brain. *J Theor Biol.* 2008;250(1):125-138.
129. Sarntinoranont M, Banerjee RK, Lonser RR, Morrison PF. A computational model of direct interstitial infusion of macromolecules into the spinal cord. *Ann Biomed Eng.* 2003;31(4):448-461.
130. Støverud KH, Darcis M, Helmig R, Hassanizadeh SM. Modeling concentration distribution and deformation during convection-enhanced drug delivery into brain tissue. *Transp Porous Media.* 2012;92(1):119-143.

CHAPTER

6

Pharmacokinetics of Polymeric Drug Delivery

Riccardo Serra, Sakibul Huq, Henry Brem, Betty Tyler

Department of Neurosurgery and Hunterian Neurosurgical Research Laboratory, Johns Hopkins University School of Medicine, Baltimore, MD, United States

DRUG DELIVERY TO THE CENTRAL NERVOUS SYSTEM

Optimizing drug delivery is a central challenge for modern clinicians and scientists treating myriad medical conditions, particularly in the field of oncology. The tolerability and efficacy of therapies depend on several factors, including the concentration of a drug that can be achieved inside a given tissue or a single cell. Although most organs and systems are not protected by a complex interface and are therefore readily accessed via systemic administration of drugs, a few are efficiently separated from the general circulation by physical and functional barriers. These barriers are composed of cellular elements and extracellular matrix, active pumps, passive channels, and interstitial fluids. They may attenuate the power of many chemotherapeutics, as they prevent these drugs from reaching their targets at clinically meaningful concentrations. A trade-off is then made between maximizing the concentration of the drug in these hard-to-reach tissues and minimizing toxicities and adverse events.

Perhaps the most efficient, articulated, and challenging of these interfaces is the blood–brain barrier (BBB). The BBB constitutes an anatomical and functional interface that allows for protection and homeostasis of the central nervous system (CNS) through the coordination of many elements. The implication for drug delivery is that the BBB is highly selective of the chemicals and foreign molecules that can permeate the CNS. Under normal conditions, the BBB performs a function essential to the brain's survival and physiology; however, in pathological states, its tight regulation becomes a significant, sometimes insurmountable obstacle to effective CNS drug delivery. This is especially highlighted in the treatment of brain tumors.

Brain tumors, particularly those of glial origin, are highly heterogeneous and often profoundly resistant to chemo- and radiotherapy, historically posing a daunting challenge for physicians.[1] Gliomas are especially marked by this cellular heterogeneity, invasiveness,

Nervous System Drug Delivery
https://doi.org/10.1016/B978-0-12-813997-4.00006-2

© 2019 Elsevier Inc. All rights reserved.

pronounced stemness, and high frequency of tumor recurrence. These aggressive factors, combined with the low concentration of therapeutics that can be achieved in the brain parenchyma, have contributed decisively to the poor prognosis of these tumors.[2] The BBB plays a pivotal role in these cases. Its low permeability prevents chemotherapy molecules from diffusing through the basement membrane of intratumoral capillaries, and even when these structures are damaged by radiation or other treatments, the efficiency of the BBB is rarely impaired to the point of permitting the accumulation of consistent amounts of chemotherapeutics.[3, 4]

The concept of local drug delivery was developed and optimized for the specific purpose of bypassing the BBB and increasing the intraparenchymal concentrations of drug molecules, while simultaneously reducing systemic toxicity. Local drug delivery using polymer matrices began in the 1980s with the ideation of using biocompatible polymeric matrices loaded with chemotherapeutics that could be surgically placed directly into the brain parenchyma at the time of tumor resection. A polymeric mixture was created and loaded with carmustine; these Gliadel® wafers (Eisai Inc. for Arbor Pharmaceuticals) ultimately became the only interstitial therapy for malignant glioma approved by the Food and Drug Administration.[5-8] The approval and use of Gliadel® led to further investigations and increased research efforts involving intracranially targeted chemotherapy. Over the past decade, researchers have made advancements in biocompatible materials regarding the use of polymers for nanocarriers such as nanoparticles and micelles.[9] Polymeric matrices have become important assets for the local delivery of drugs thanks to their tolerability, biocompatibility, and ability to bypass or directly cross the BBB and release their drug payload intraparenchymally.

BRAIN TUMORS

Astrocytic brain tumors have been the primary target of local drug delivery efforts over the past several decades due to their aggressive behavior, tendency to recur following total or near-total resection, and genetic heterogeneity.[10-12] These factors, combined with the impermeability of the BBB, massively delayed any significant advances in treatments, resulting in poor prognosis for patients with these tumors. Glioblastoma multiforme (GBM), a grade IV astrocytic tumor, is the most aggressive and lethal malignancy of the CNS. It is most commonly localized in the cerebral hemispheres and is very rarely diagnosed in regions such as the cerebellum or the spinal cord.[2] Its incidence rate has increased steadily over the past several years,[1] with 3.2 new cases per 100,000 people per year.[13] The median age at diagnosis is 64 years, and the male-to-female ratio is approximately 1.6:1.0. Ethnicity is also a predisposing factor, with incidence of GBMs being nearly twice as high in white patients relative to African Americans and Native Americans.[14]

Current therapy for patients with GBM includes surgical resection and radiotherapy plus adjuvant chemotherapy. Therapeutic treatment frequently includes chemotherapy wafers implanted at the time of surgery and orally delivered temozolomide, due to the significant survival benefit it confers with moderate toxicity, as well as radiation and anti-angiogenesis therapy (bevacizumab).[15] However, recurrence is still the most prevalent cause of death in patients with GBM, and therefore this type of tumor is a key research focus. These tumors can aggressively infiltrate the brain parenchyma and sometimes invade regions as remote as the contralateral hemisphere, making total eradication of the tumor mass a significant challenge.[16] The treatment of GBM therefore poses a constellation of molecular and clinical problems that contribute to its historically poor prognosis: high intercellular genetic heterogeneity, immune suppression, regeneration of the tumor mass mediated by glioblastoma stem-like cells, BBB-mediated shielding from systemically administered drugs, hyperproliferation of the tumor, and drug resistance.

THE ROLE OF THE BLOOD–BRAIN BARRIER IN DRUG DELIVERY

Polymeric local delivery of therapeutics was originally conceived as a method to circumvent and bypass the obstacles posed by the BBB. Cytotoxic drugs could be directly shuttled inside the tumor tissue without undergoing systemic clearance, thus reaching higher concentration levels. However, the BBB is a complex structure made of different components, and each of these parts has a particular function that actively contributes to the protection and homeostasis of the CNS.[17]

Proceeding from the capillary lumen outward, the first physical barrier is formed by tightly interconnected pericytes (also called brain microvascular endothelial cells). These cells contain adherens and tight junctions on their membranes, thereby conferring an almost-complete impermeability to paracellular flow to this compartment.[18] The only transport systems present in endothelial cells are mediated by pinocytic vesicles and transporters (e.g., GLUT-1) or ATP-dependent pumps (e.g., P-glycoprotein). Given these energy demands, brain pericytes contain a uniquely high number of mitochondria.[19]

The second component of the BBB is the basement membrane, which contains a matrix rich in collagen-IV that promotes the expression and maintenance of tight juctions,[20] in addition to providing a scaffold for pericyte proliferation. Its alteration in pathological states is strongly associated with increased permeability of the BBB.[21, 22] This state has been exploited to improve the local delivery of drugs and therefore elevate the achievable intraparenchymal concentrations of therapeutics.[23, 24] On the other hand, an increase in BBB permeability also implies the diffusion of a wide range of molecules and substances that can cause significant brain damage, potentially leading to brain edema.[25]

Another key element of the BBB involves the association of astrocytes with the basement membrane and brain microvascular endothelial cells, ultimately enveloping more than 99% of the BBB endothelium.[26, 27] Astrocytic pedicles are a vital part of the neurovascular unit, and their homeostatic role is essential for proper neuronal functioning and cellular homeostasis.[28] The BBB is thus a complex biological phenomenon with clear clinical implications. In general, only highly lipophilic and small molecules (with a molecular weight <400 Da) are capable of freely crossing the BBB, largely limiting effective targeting of CNS pathologies.[29, 30]

A greater understanding has been built around the state of the BBB during malignancy, with consensus forming around the concept of an impaired and more permeable blood–brain tumor barrier (BBTB).[31] The BBTB reflects the perturbations of the BBB in the setting of malignancy. In these cases, when blood vessels are contained inside the malignancy, the BBB becomes leaky and develops structural and functional weaknessses.[31-33] Therapeutics can thus penetrate slightly deeper into the tumor parenchyma when the BBTB is present.

However, cytotoxic drugs still only rarely reach concentrations high enough to efficiently kill malignant cells and prolong patient survival.[34]

POLYMERIC DRUG DELIVERY TO THE BRAIN

To address the concerns and the challenges that stem from the impermeability of the BBB, chemotherapeutics were loaded into a biocompatible, biodegradable polymeric matrix. This implantable wafer was capable of delivering a considerable dose of a cytotoxic agent (in this case, 1,3-bis(2-chloroethyl)-1-nitrosourea, also known as carmustine or BCNU) directly into the brain parenchyma.[35] This innovation helped reduce the influence of the BBB on local concentrations of therapeutics, increase the drug's efficiency, and limit the magnitude and extent of systemic toxicities.

The main components of the polymer matrix are poly[bis(p-carboxyphenoxy)propane] anhydride (p(CPP)) and sebacic acid (SA), two elements with notably different rates of spontaneous degradation. Release kinetics of the components as well as the effective payload can be fine-tuned by varying the relative proportions of the p(CPP) and the SA. Drug-release kinetics are extremely sensitive to these proportions. In tailoring p(CPP) and SA proportions, a compromise is made between preservation of the embedded drug and its release inside the surgical resection cavity. A 50:50 wafer has been shown to resist degradation for periods of weeks to months (Fig. 1).[36-38]

Polymer Degradation in Animal and in Vitro Models

The final ratio of p(CPP) and SA was calculated based on their effects on the erosion process. p(CPP) degrades over a period of months to years, whereas SA is a highly hydrophobic molecule that rapidly undergoes hydrolysis. The 50:50 p(CPP):SA matrix was capable of

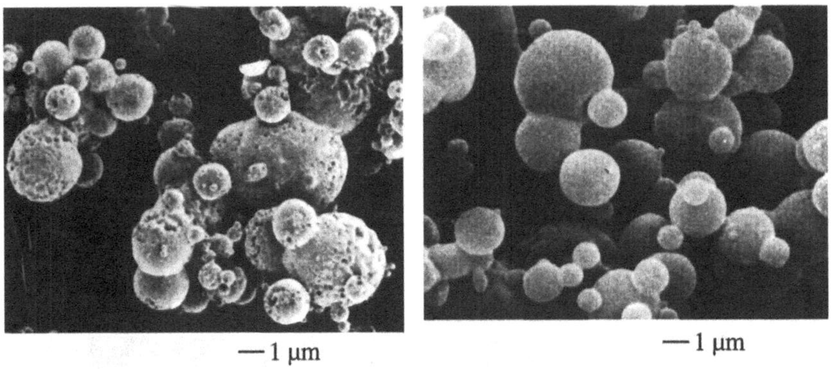

FIG. 1 Electron micrography of dried p(CPP:SA, 20:80) microspheres. (Left) Blank spheres are globular in shape and have a porous surface. (Right) When these are impregnated with BCNU, the surface tends to become smoother and porosity decreases. BCNU is usually homogeneously distributed in particles. *Reproduced with permission from Dang W, Daviau T, Brem H. Morphological characterization of polyanhydride biodegradable implant Gliadel during in vitro and in vivo erosion using scanning electron microscopy. Pharm Res 1996;13(5):683–691.*

resisting degradation for a longer period than the 20:80 wafers.[36, 37, 39] However, the 20:80 p(CPP):SA copolymer rendered a balanced rate of erosion, because adding SA to a p(CPP) matrix increased its degradation rate more than 800-fold.[39] The final formulation was confirmed after testing a wide variety of copolymers.

The hydrophobic properties of p(CPP) and SA are also important for protecting carmustine from spontaneous degradation by hydrolysis. Carmustine is sensitive to aqueous solutions, with a brief half-life of 45 minutes when in contact with phosphate-buffered saline (PBS) at pH 7.4 and 37°C.[40] However, when carmustine is enclosed in the protective matrix of the p(CPP):SA copolymer, the drug remains intact for a longer period, as it is shielded from the surrounding aqueous environment.

The rate of polymer degradation can be estimated both qualitatively and quantitatively, by means of scanning electron microscopy (SEM)[41] and precise measurements of polymer mass,[42] rate of release of low-molecular-weight oligomers, and chromatographic analysis.[37] Dang et al.[42] showed that polymer degradation was accomplished in two phases. First, water slowly permeates the superficial layers of the polymeric matrix, hydrolyzing the bonds between p(CPP) and SA, thereby disrupting its microscopic structure. This induction period lasts for approximately 10 hours, when the wafer loses a small fraction of its weight and the copolymer is fragmented into pieces with a molecular weight of about 10,000 Da. When this threshold is surpassed, p(CPP):SA molecules become soluble in water, and the erosion process enters its second phase.[42]

In the second phase, water rapidly permeates the polymeric matrix and brings it to complete dissolution over a period of 3–4 weeks. Fragments of the copolymer are therefore released at a significantly higher rate than during the first phase, and an exponential decrease in overall mass of the wafer is clearly detectable. Electron microscopy scanning shows how the polymer surface, uniformly smooth at the beginning of the SEM evaluation, appears more irregular and asymmetrical over time. These irregularities reflect the small gaps carved between the carmustine-loaded particles. These microscopic grooves illustrate the uniform distribution of the drug-containing microspheres and the different steps of wafer degradation in vivo.[41] As the wafers undergo hydrolysis, these gaps become larger and surface porosity increases, until water permeates the mass and rapidly hydrolyzes the polymer layer by layer, proceeding from the outermost regions to the inner core. This water infiltration is extensive enough that overall polymer weight noticeably increases during the second phase (Fig. 2).[42]

Single monomers start to release from the disc during this phase as a direct consequence of the hydrolytic breakdown of polymeric chains. The sensitivity of SA to hydrolysis is responsible for the rapid breakdown of SA chains, whereas p(CPP) tends to degrade more slowly and gradually, likely due to the stability of its anhydride bonds. SA diacids are also more water-soluble than CPP-CPP dimers, rendering another contribution to the observed differences in release kinetics.[43]

Polymeric wafers and their chemical properties were extensively evaluated using in vitro and in vivo approaches. Wu et al.[37] tested the effects of pH, mixing, and water on wafer degradation by placing empty p(CPP:SA) discs in solutions of 0.1 M PBS at 37°C at varying pH levels. Samples were taken at specific times and analyzed using high-performance liquid chromatography. Additional in vivo studies were performed with various formulations of empty and carmustine-loaded wafers in rat brains. Animals implanted with 20:80

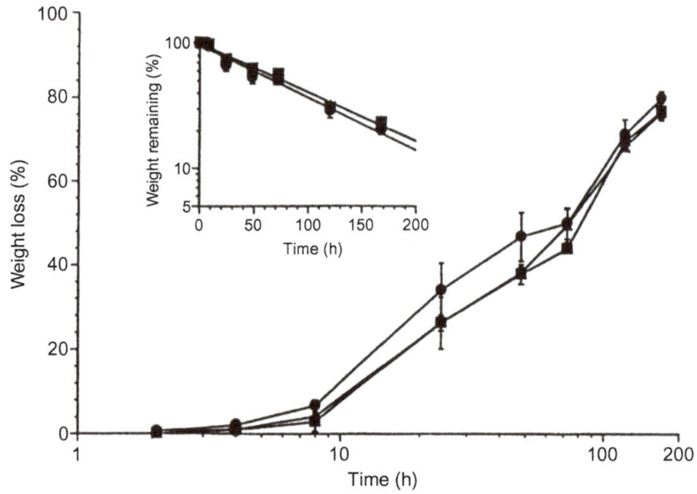

FIG. 2 Degradation in PBS 0.1 M (pH 7.4 and 37°C) of three wafers with different weights [(48 kDa M_w, •), (72 kDa M_w, ▪),(110 kDa M_w, ▲] over a period of 200 hours. *Reproduced with permission from Dang W, Daviau T, Ying P, et al. Effects of GLIADEL wafer initial molecular weight on the erosion of wafer and release of BCNU. J Control Release 1996;42(1):83–92.*

p(CPP:SA) discs received 20% carmustine wafers, 20% carmustine wafers 4 days after 9L gliosarcoma implantation, empty polymers, or empty polymers combined with a piece of an absorbable hemostatic material (Surgicel, Ethicon). Polymeric wafers in PBS were found to be more sensitive to hydrolysis and degraded faster, whereas those placed in rat brains were more resistant to degradation and tended to degrade over a longer period of time. However, erosion profiles were notably similar in in vitro and in vivo systems, suggesting that a common mechanism serves as a basis for wafer dissolution. Furthermore, the availability of polymeric molecules was shown to influence the erosion kinetics in vitro: monomers tend to saturate the space that surrounded the degrading wafer, slowing down the release of additional molecules from the remaining polymeric chains. This mechanism could explain the difference in the degradation rate that was observed between wafers placed in PBS and those implanted in rat brains.[37] Although the presence of a certain degree of variability has not been understood completely, it is generally accepted that polymeric wafers are constantly eroded in animal and in vitro systems within 6–8 weeks. The persistence of small remnants may therefore be imputable to the interposition of microscopic air bubbles or to the irregular contact between the brain and the polymeric matrix (Fig. 3).[37]

Release of Carmustine from The Wafer

In vitro carmustine release from the polymeric matrix is a gradual process that takes place over approximately 120 hours. Although the release of carmustine might seemingly not be influenced by the erosion kinetics of p(CPP:SA), it has been shown that in the first 24 hours following implantation, approximately 70% of the total payload is dispersed in the aqueous environment that surrounds the disc.[42] This burst release is therefore due not only to layer-wise degradation of the polymer, but also to the direct diffusion of carmustine through the polymeric matrix. After the first phase, a more gradual and controlled process occurs,

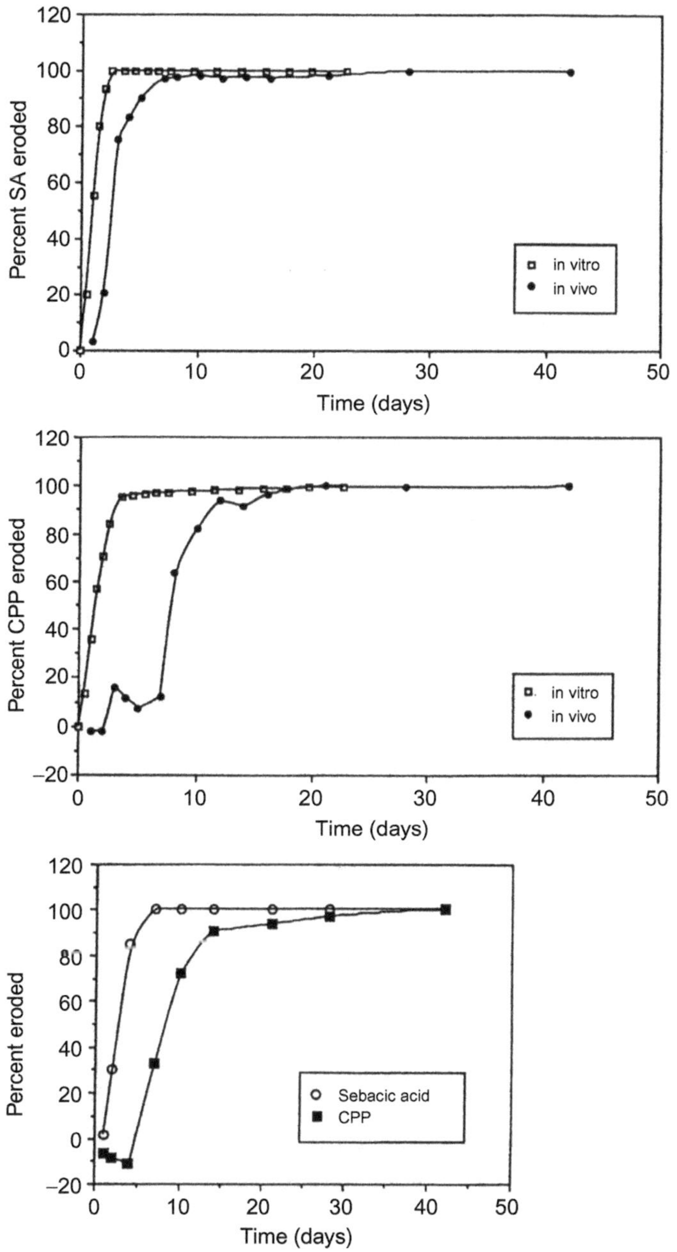

FIG. 3 Erosion of p(CPP:SA, 20:80) empty and 20% BCNU polymeric discs in healthy and 9L-implanted rat brains compared to PBS 0.1 M (pH 7.4 and 37°C). Sebacic acid and p(CPP) show different erosion kinetics but similar profiles; no visible remnant was observed at day 42 postimplantation. *Reproduced with permission Wu MP, Tamada JA, Brem H, Langer R. In vivo vs in vitro degradation of controlled release polymers for intracranial surgical therapy. J Biomed Mater Res A. 1994;28(3):387–395.*

with p(CPP:SA) degrading layer by layer, slowly diffusing the drug out of the polymer and into the tumor bed over a period of 5 days.[44]

Metabolic Processing of Sebacic Acid, p(CPP), and Carmustine in Animal Models

Metabolic disposition, distribution, and elimination of carmustine, SA, and p(CPP) were serially studied in two experiments with ^{14}C radio-labeled molecules. First, Domb et al.[45] implanted 50 adult Sprague-Dawley rats with intracranial radiolabeled polymeric wafers—either empty wafers or BCNU-loaded wafers. Samples of urine, feces, exhaled carbon dioxide, brain, kidney, liver, spleen, lung, fat, carcass, plasma, and polymeric disc remnants were collected and analyzed through a liquid scintillation system. Animals were sacrificed to examine the biodistribution and excretion of ^{14}C at 1, 4, and 7 days postimplantation. Levels of radioactivity were markedly different among the three components of polymeric wafers. Each warrants its own discussion and are described in the following sections.[45, 46]

Sebacic Acid

Sebacic acid is mainly eliminated through the respiratory system as carbon dioxide; the exhaled fraction is approximately 40% of the administered dose. Secondary and tertiary routes of excretion are the urinary system and gastrointestinal system, with 10% and 2% of total dose found in the urine and feces, respectively. Additionally, after 7 days, water collected at the implantation site contained 8% of the original radioactivity. Brain tissue retained around 1% of radiolabeled SA, with the rest of the analyzed organs containing approximately the same. Interestingly, the BCNU-loaded discs also demonstrated a slightly lower rate of SA excretion compared to the placebo wafers.[45, 46]

1,3-Bis-(p-Carboxyphenoxy)propane (p-CPP)

1,3-Bis-(p-carboxyphenoxy)propane [p(CPP)] is notably less sensitive to hydrolysis than SA. At days 4 and 7 postimplantation, approximately 96% of the original mass of p(CPP) was found in the recovered carmustine-loaded disc. It was also found that empty wafers were more likely to quickly release p(CPP) monomers into the surrounding aqueous environment. Placebo discs achieved a brain level of ^{14}C-p(CPP) of about 7%, while day 4 and 7 radioactivity in the brain was equal to only 1% of total p(CPP) contained in carmustine wafers. Regarding p(CPP) excretion, the gastrointestinal system was the primary route, with 2.5% of the p(CPP) found in the feces vs. 1% found in the urine. Total excretion after 7 days was a mere 10% of the p(CPP) mass, and marked inter-individual differences were recorded.[45, 46]

Carmustine

Carmustine distribution and excretion were substantially different from that of SA and p(CPP), with 60% of the initial carmustine dose found in the urine, whereas only 5% was exhaled as radiolabeled carbon dioxide (this fraction reached a peak during the first day of postimplantation, then slowly but constantly decreased throughout the length of the study). Interestingly, the percentage of carmustine bound to the wafer started decreasing immediately after polymer implantation: Only 50% of the initial dose was still contained inside the disc 48 hours postimplantation, quickly falling to 5% after 1 week. High-performance liquid chromatography and scintillation studies demonstrated similar levels of carmustine and radioactivity, but the amount of active drug was estimated to be two to five times lower than that of radiolabeled ^{14}C. This difference was likely due to local degradation and hydrolysis of carmustine. In this case, although the level of radioactivity at the implant site remained constant, the amount of active drug available in the wafer was greatly reduced.[45]

Carmustine Distribution in Animal Models and Humans

Drug biodistribution has a central role in determining the efficacy and tolerability of proposed treatments. To understand these parameters appropriately, many experiments and models were used to define BCNU distribution and elimination. Grossman et al.[47] used a New Zealand white rabbit model to assess the intracerebral distribution of carmustine released by implantable polymeric wafers. A fraction equal to 2.5%, 5%, and 10% of the final disc weight of tritiated (^{3}H) BCNU was added to the p(CPP:SA) 20:80 polymeric matrix and implanted in the left hemisphere of experimental subjects. Control animals received inulin-loaded polymeric wafers intracranially, and a third group underwent direct administration of intracerebral ^{3}H-carmustine via convection-enhanced delivery. Animals were euthanized after 3, 7, 14, or 21 days, and tissue samples were prepared for radiography. Autoradiographies were taken and analyzed by means of a digital system that converted the specific activity of ^{3}H-BCNU in pmol of carmustine/mg of tissue. Direct injection by means of convection-enhanced delivery of the labeled drug ensured rapid diffusion of ^{3}H-carmustine throughout the ipsilateral hemisphere and rapid clearance from the brain. Up to 15% of the sampled area was exposed to carmustine after 24 hours, and an even lower fraction was measured at 72 hours (Fig. 4).[47]

The distribution and clearance of ^{3}H-carmustine were quite different when the drug was administered by means of an implantable polymeric wafer. In these cases, 72 hours after intracranial placement, more than 40% of the initial dose was still traceable in the sampled slices, and the decrease was significantly slower and more gradual. The 15% threshold was reached between 180 and 350 hours postimplantation. The decrease in size of the area

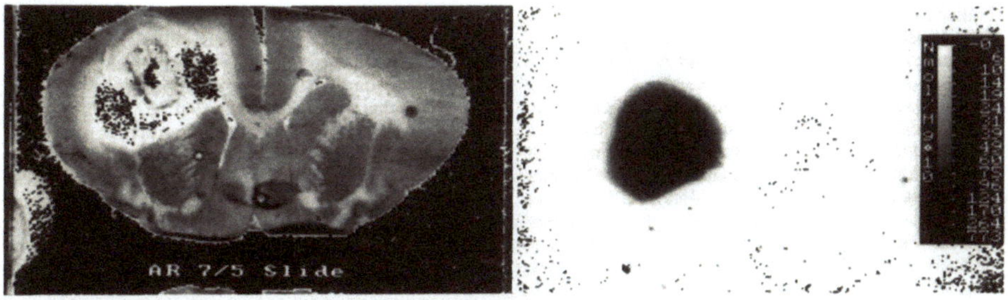

FIG. 4 Histological section and digitized autoradiograph of a rabbit brain implanted with a polymeric disc containing 1200 μg of ^{3}H-BCNU: images were taken at day 7 after wafer placement and are at the same magnification. *Reproduced with permission from Grossman SA, Reinhard C, Colvin OM, et al. The intracerebral distribution of BCNU delivered by surgically implanted biodegradable polymers. J Neurosurg 1992;76(4):640–647.*[47]

of interest explains this apparent surge in drug content, as total brain volume exposed to ^{3}H-carmustine shrinks with time, and the local concentration of carmustine seems to increase with time. These initial studies were hampered by difficulty in measuring low but significant concentrations of drug within the brain.

A subsequent study by Fung et al.[48] using ^{3}H-carmustine to track its degradation in vivo in F344 rats over a prolonged period of time showed a similar release pattern. A group of animals received a ^{3}H-carmustine solution 5 days after tumor implantation by means of a 5-μL microsyringe, while a second cohort was treated with intracranial administration of p(CPP:SA) discs impregnated with ^{3}H-carmustine. Animals were sacrificed after 6 hours in the microinjection cohort and at several prespecified time points (1, 3, 7, 14, and 30 days postimplantation) in the polymer group. Autoradiographies were obtained from tissue slides, and the data were analyzed by integrating the local concentrations measured for each pixel or region of interest. ^{3}H-carmustine was measured and showed an almost instantaneous release of drug when the polymeric wafer was immersed in PBS. Almost 50% of the carmustine load diffused through the disc in the first 8 hours, reaching a total of 84% after 24 hours. The release kinetics appeared to be linear with respect to the square root of time on the first day.

Drug concentrations were also extensively studied by Fung et al.[49] in another experiment on cynomolgus monkeys (*Macaca fascicularis*). The authors encapsulated ^{3}H-carmustine, 4-hydroperoxycyclophosphamide, or paclitaxel in polyanhydride pellets and implanted the wafers intracranially in 17 animals for a period of 30 days. Brain, blood, and CSF drug concentrations were evaluated by means of quantitative autoradiography, thin-layer chromatography, and scintillation counting, showing significant drug concentrations (0.4 μg for carmustine) in regions of the brain as distant as 5 cm from the implant site. Carmustine release in PBS was used as an in vitro control, and tritium activity in the ipsilateral and contralateral hemispheres was measured and compared to carmustine concentrations (obtained by thin-layer chromatography and scintillation counting) in different regions of the monkey brain. The researchers found that 30 days postimplantation, carmustine levels were still as high as 0.5 μM at a distance of 5 cm from the polymeric wafer, allowing effective targeting of malignant elements and cell death (Tables 1 and 2).[49]

TABLE 1 Tritium Activity in Ipsilateral and Contralateral Hemispheres, Blood, and Cerebrospinal Fluid

Time following implantation (days)	^{3}H activity in ipsilateral hemisphere (% dose/kg)	^{3}H activity in contralateral hemisphere (% dose/kg)	^{3}H activity in blood (% dose/kg)	^{3}H activity in CSF (% dose/kg)	^{3}H activity in CSF ÷ [^{3}H] activity in ipsilateral hemisphere
1	14±13	13±14	86±100	100±130	7.1
3	16±13	9.8±4	14±2.7	17±16	1.1
7	20±34	1.1±1.7	12±0.6	5.6±0.6	0.3
10–14	11±16	1±0.5	77±69	130±130	11
21–30	16±25	4.3±5.7	2, 15	11, 8.2	0.6

Tritium activity in ipsilateral and contralateral hemisphere, blood, CSF. Radioactivity was measured by means of scintillation counting. Data extrapolated from Fung LK, Ewend MG, Sills A, et al. Pharmacokinetics of interstitial delivery of carmustine, 4-hydroperoxycyclophosphamide, and paclitaxel from a biodegradable polymer implant in the monkey brain. Cancer Res 1998;58(4):672–684.

TABLE 2 Concentration of ^{3}H-BCNU in monkey brains at different intervals

Distance from the tip of the brain (cm)	Regional carmustine concentration in ipsilateral coronal sections (μM)			
	Day 1	Day 7	Day 14	Day 30
Implant center	350	30	29	19
	390	34	33	21
1.0–1.3	7.7	5.0	1.3	1.2
	7.5	5.6	1.3	1.2
1.3–1.5	1.9	1.9	0.4–0.3	0.3
	1.0	2.1		0.2
1.5–2.0	1.7	0.9	0.1–0	0.06
	0.8	1.0		0
2.0–3.0	1.7	0.13	<0.004	NA
	0.8	0.1	ND	NA
3.0–4.0	1.8	0.05	<0.004	0.05
	0.9	0	ND	0
4.0–5.0	1.4	0.04	<0.004	0.05
	0.4	0	ND	0

Concentrations of ^{3}H-BCNU in the monkey brains at different intervals. Radiation measurements were obtained by thin-layer chromatography and scintillation counting. For each entry, the value in the upper row represents the total regional concentration in a coronal section; the value in the lower row represents the drug concentration in the brain parenchyma of the coronal section, calculated by the equation $X_{parenchyma} = \frac{X_{total} - f_{blood} \cdot X_{blood} - f_{CSF} \cdot X_{CSF}}{f_{parenchyma}}$.

Reproduced with permission from Fung LK, Ewend MG, Sills A, et al. Pharmacokinetics of interstitial delivery of carmustine, 4-hydroperoxycyclophosphamide, and paclitaxel from a biodegradable polymer implant in the monkey brain. Cancer Res 1998;58(4):672–684.

Edema and Bulk-Flow Contribute to Interstitial Drug Delivery

In the same study, concentration profiles from the first 24 hours were also examined and found to be significantly different from those collected in subsequent days.[49] Carmustine appeared capable of diffusing further from the polymer immediately after its placement. This could be explained by the presence of a physiologic degree of vasogenic edema. In this process, wafer implantation may lead to an increase in vessel permeability, the development of a hydrostatic pressure differential between the tissue surrounding the implantation site and the cerebrospinal fluid (CSF), and the onset of a directional flow of fluids and solutes. This study also defined the depth of BCNU penetrance in the brain parenchyma around the wafer. The penetration distance for the drug was 1.1–1.3 mm for the first 3 days in rabbit brains.[47, 50] Another experiment yielded a different outcome in rats: carmustine was observed to reach a depth of 4.7 mm after 24 hours, with a sudden drop at days 3, 7, and 14 (1.6, 1.3, and 1.1 mm, respectively). This may be explained by the higher sensitivity of rats to local edema and surgical procedures compared to rabbits.[48]

The velocity of drug penetration in this work was calculated to be approximately 4.6 ± 2.5 mm/hours in experimental rats, decreasing progressively after the first 24 hours, and reaching a plateau at day 3 postimplantation.[48] Fung et al.[49] confirmed the existence of a similar mechanism in monkeys, in which they showed a pronounced sensitivity to polymer implantation (similar to results observed in rodents). The effusion of fluids in the extracellular space (ECS) and the consequent increase in interstitial pressure created a pressure gradient that promoted directional flow of edema into the CSF compartment. As a highly soluble and rapidly released compound, carmustine permeated the brain parenchyma more efficiently when transported by this directional stream of interstitial fluid.[49]

Additional Variables in Drug Penetration

Studies in nonhuman primates with the implanted carmustine wafer showed remarkable drug penetration in the brain throughout the first 30 days.[49] Furthermore, the concentrations of carmustine found in the region of the polymer wafers were very high as compared with the in vitro therapeutic threshold deemed necessary. Both the 7- and 30-day delivery of carmustine from the intracranially implanted wafer (~8 mg/kg) provided a higher area under the curve than intravenous delivery of a higher dose of carmustine (~10 mg/kg). To obtain the same concentration of carmustine that was achieved by polymeric delivery at distant locations, the required intravenous dose would have had to be four times higher (>40 mg/kg; Fig. 5).[49] Another interesting finding is the sustained release of carmustine during the first 30 days: The brain parenchyma enclosed in a radius of 1–2 cm from the polymeric implant is in fact exposed to BCNU concentrations of about 0.1–7.5 μM for almost a month after wafer placement.[49]

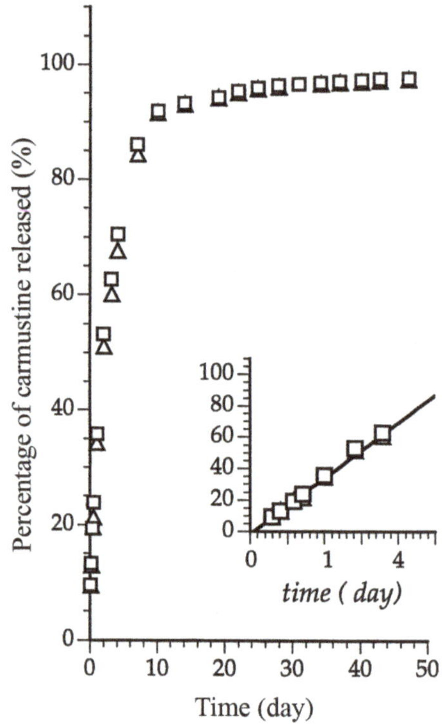

FIG. 5 The area under the curve in a nonhuman primate brain for carmustine (7-day [●] and 30-day [▪] exposure), 4-HC (◆), and paclitaxel (▲) delivered by intracranially implanted polymers (8 mg/kg) compared to the area under the curve for carmustine delivered by the internal carotid artery with hemoperfusion (10 mg/kg) and intravenous administration (10 mg/kg). *Reproduced with permission from Fung LK, Ewend MG, Sills A et al. Pharmacokinetics of interstitial delivery of carmustine, 4-hydroperoxycyclophosphamide, and paclitaxel from a biodegradable polymer implant in the monkey brain. Cancer Res 1998;58 (4):672–684.*

The presented findings are the result of the interactions of many different factors that affect drug diffusion and biodistribution. In addition to local delivery of carmustine to the surroundings of the implant site, BCNU is also redistributed to other regions of the body and the CNS due to its high vascular permeability, marked lipophilicity, and high penetration coefficient.[49]

PHARMACOKINETIC MODEL

A model that summarized the movement of carmustine and other compounds across the interstitium and the brain ECS was developed to determine drug biodistribution throughout the brain and its compartments. A set of equations has been extensively used for various delivery systems, many of which were based on biocompatible and biodegradable polymeric matrices.[49-61] Key brain parenchyma elements included in these equations are the cell membrane (CM), the ECS, and the intracellular space (ICS). These compartments are closely interrelated, with significant movement of solutes and water between them. Through these equations, it is assumed that a highly lipophilic drug such as BCNU would begin releasing from the polymer as soon as the wafer is implanted in the parenchyma.

A fraction of the total dose of carmustine diffuses in the extracellular fluid and permeates the cell membrane, reaching the ICS. The remaining carmustine, assuming that the brain is in close contact with a high number of vessels, crosses the semipermeable tissue capillaries and enters the systemic circulation, thus being removed from the implantation site and surrounding brain parenchymal area. In this case, the total amount of unbound carmustine contained inside the brain is determined by the following equation:

$$C = \alpha C^0_{ecs} + \beta C^0_{ics} + (1 - \alpha - \beta) C^0_{cm} \tag{1}$$

where α and β correspond to the percent of the total brain volume attributable to the ECS and ICS compartments. The volume of the CM is indirectly calculated by knowing the remaining two percentages, while the number of moles of BCNU per unit of volume can be represented by C^0_{ecs}, C^0_{ics}, and C^0_{cm}. A drug such as BCNU can be transported by diffusion or convection, eliminated through capillaries/metabolism/degradation, or bound to the brain parenchyma and its cellular elements. Taking into account these three mechanisms, local concentration of carmustine can be described by the equation of Patlak and Fenstermacher[60, 61] as a function of time and position (in the region that surrounds the polymeric implant):

$$\frac{\partial C}{\partial t} + \overline{v}\overline{V}C = D_b \overline{V}^2 C + R_e(C) - \frac{\partial B}{\partial t} \tag{2}$$

where C represents the concentration of the diffusible drug in the parenchyma that surrounds the implant (g/cm^3 tissue), v is the fluid velocity in the ECS (cm/s), D is the diffusion coefficient of the drug in the parenchyma (cm^2/s), R_e(C) is the rate of drug removed from the ECF by local capillaries (g/s cm^3 tissue), B is the concentration of drug bound or internalized in cells (g/cm^3 tissue), and t is the time following implantation.[49, 56]

From this equation, a more complex formula was derived that accounts for a series of variables and coefficients:

$$\frac{\partial C}{\partial t} = -\overline{V}*\left[-\alpha^* D_{ecs}\overline{V}C^0_{ecs} + \alpha^*\overline{V}_r C^0_{ecs}\right] - \left[\alpha^* k_{ecs}{}^* C^0_{ecs} + \beta^* k_{ics}{}^* C^0_{ics} + (1-\alpha-\beta)^* k_{cm}{}^* C^0_{cm}\right] - \frac{\partial B}{\partial t} \quad (3)$$

where V_r is the vector that describes the fluid velocity; t is the time after implantation; D_{ecs} is the diffusion coefficient for BCNU in the ECS; and k_{ics}, k_{ecs}, and k_{cm} are the elimination constants for each compartment.[49]

Bound BCNU is described by an equation that is similar to that of unbound carmustine:

$$B = \alpha^* B^0_{ecs} + \beta^* B^0_{ics} + (1-\alpha-\beta)^* B^0_{cm} \quad (4)$$

In this case, local equilibration of carmustine concentrations is reached in the ECS–ICS–CM system, given the following assumptions: carmustine is neither eliminated nor bound to CMs, the concentration is proportional to that of the free molecule in the ECS and ICS, and BCNU equilibrates between the different compartments. Given these premises, and considering the brain to be isotropic, the Patlak and Fenstermacher equation can be expressed in these terms:

$$\frac{\partial C^0_{ecs}}{\partial t} + \left(\frac{\alpha}{\alpha^*}\right)\overline{V}_r{}^*\overline{V}C^0_{ecs} = \left(\frac{\alpha}{\alpha^*}\right)D_{ecs}\overline{V}^2 C^0_{ecs} - \left(\frac{\alpha^* k_{ecs} + \beta^* k_{ics}{}^* P_{i:e}}{\alpha^*}\right)C^0_{ecs} \quad (5)$$

with:

$$\alpha^* = \alpha(1+k_{ecs}) + \beta^* P_{i:e}{}^*(1+k_{ics}) + (1-\alpha-\beta)P_{m:e} \quad (6)$$

where k_{ecs} and k_{ics} represent the binding constants between the bound/unbound state for carmustine in the ECS and ICS. These parameters are as follows:

$$k_{ecs} = \frac{B^0_{ecs}}{C^0_{ecs}} \quad (7)$$

$$k_{ics} = \frac{B^0_{ics}}{C^0_{ics}} \quad (8)$$

where $P_{i:e}$ and $P_{m:e}$ are the partition coefficients between ICS-ECS and ECS-CM:

$$P_{i:e} = \frac{C^0_{ics}}{C^0_{ecs}} \quad (9)$$

$$P_{m:e} = \frac{C^0_{cm}}{C^0_{ecs}} \quad (10)$$

Using the partition coefficients ($P_{i:e}$ and $P_{m:e}$), Eq. (1) can be expressed as the diffusion/convection/elimination equation[62]:

$$\frac{\partial C}{\partial t} + v^*\overline{V}C = D\overline{V}^2 C - kC \quad (11)$$

In this equation, v is the apparent velocity in the ECS, D the apparent diffusion coefficient in the brain parenchyma for carmustine, and k is the elimination constant $[(\alpha k_{ecs} + \beta k_{ics} \cdot P_{i:e})/\alpha^*]$. This equation can be further simplified by excluding convection, usually not preponderant in the brain, and considering diffusion as the only mechanism of transport active in the brain. The diffusion/elimination equation can therefore be written in the following terms:

$$\frac{\partial C}{\partial t} = D\overline{V}^2 C - kC \tag{12}$$

These equations have been extensively used to calculate the concentration of carmustine at the polymer-brain interface in animals implanted with various polymer geometries and percentages of carmustine. Different experimental conditions and polymer-brain interfaces have been used to represent a number of unique patterns of drug biodistribution in the CNS. To match these variables appropriately, a large set of mathematical models has been developed and used to describe the distribution and elimination kinetics of carmustine in the brain parenchyma.

In all these animals models, the release of carmustine appeared to be extremely rapid, with greater than 95% of the initial load diffused in the first 2 days.[48, 49] As a consequence of this mechanism and limited drug permeability in the brain parenchyma, local concentrations of BCNU tend to be extremely high in the first few hours postimplantation, suddenly dropping after an initial spike and remaining constant for the following 2–3 weeks.

All the approaches used for these experiments were standardized using thick coronal sections and quantitative autoradiography with ^{3}H-carmustine. This method allowed researchers to carefully determine the amount of radiolabelled drug in the regions surrounding the polymeric implant. Equations were appropriately tailored to the animal species used for each experiment, and experimental data matched those predicted by mathematical models.

CLINICAL USE OF LOCAL CHEMOTHERAPY IN HUMANS

The preclinical development of chemotherapy wafers involved a multitude of in vitro and in vivo studies, including experiments on rats, rabbits, and nonhuman primates that elucidated key aspects of intracranial biodistribution, polymer and drug behavior and interaction, as well as safety and efficacy. Wafer biocompatibility and safety studies were first conducted on rodents and rabbits. These studies showed that p(CPP):SA copolymers, in a 20:80 ratio, were safe and capable of degrading entirely, without leaving behind non-erodible residues in the resection cavity.[63-66] Efficacy studies were then performed using animal models of gliosarcoma. Rats previously implanted with intracranial tumor and treated with interstitial carmustine:p(CPP):SA polymer wafers survived significantly longer than those that received intraperitoneal injections of carmustine.[67] Untreated control animals showed significantly shorter median and overall survivals than the groups that underwent treatment. Similar results were obtained from rats implanted with subcutaneous 9L gliosarcomas. These studies confirmed that local chemotherapy administered in close proximity to the resection margin provided a significant advantage in terms of improved survival over systemically administered cytotoxic treatments.[67] Safety was also assessed in an experiment with 18 cynomolgus macaques. With the exception of peri-implant inflammation and edema, local and systemic adverse events were not recorded, paving the way for clinical investigation.[68]

For clinical wafer formation, p(CPP):SA is mixed with carmustine, dissolved in dichloromethane, and spray dried. This process yields spherical particles that are physically compressed to form the final Gliadel® wafer, a disc comprising microspheres with diameters between 1 and 20 μm.[41] Carmustine accounts for 3.85% of the wafer's final weight, which is approximately 200 mg. This amount of carmustine was chosen after an initial Phase I-II trial in humans demonstrated superior efficacy and low incidence of adverse effects.[5] Chemotherapy wafers are surgically implanted inside the resection cavity in close contact with tumor margins. Up to eight wafers are commonly used, allowing the drug concentration in the surrounding tissues to reach levels cytotoxic to malignant cells. Copolymeric wafers then degrade over a period of 5–6 weeks, releasing a total dose of 60 mg of carmustine to the brain. The localized placement and drug release from these wafers remove the burden of systemic side effects and toxicities caused by systemically administered BCNU (and other nitrosoureas). No hepatic, renal, or hematologic toxicities were reported in clinical trials.[5-8, 69] BCNU-loaded polymeric wafers were approved by the Food and Drug Administration in 1996 under the commercial name Gliadel®, leading to major advancements in prolonging survival for patients with either recurrent[6] or primary[7, 8, 69] GBM. Multiple meta-analyses have been conducted that show Gliadel® to be significantly beneficial for patients with either newly diagnosed or recurrent GBM.[70-72]

CONCLUSIONS

Experimental measurements and mathematical models of carmustine penetration and biodistribution have shown that the local intracranial strategy is capable of precisely targeting malignant brain tumors with high concentrations of chemotherapeutics while delivering and maintaining therapeutic levels of these compounds. The safety and efficacy of the polymer-drug combination made local delivery of BCNU a viable alternative to systemic therapy, reducing the magnitude of adverse effects and optimizing the targeting of malignant cells. The advent of local polymeric drug delivery in the brain began a new chapter for the treatment of GBM. By introducing a tool capable of circumventing the BBB without causing significant toxicities and side effects, Gliadel® reflected the potential of collaborative efforts among bioengineers, pharmacologists, oncologists, and neurosurgeons—and ultimately, resulted in improved therapeutic options for patients.

References

1. Alifieris C, Trafalis DT. Glioblastoma multiforme: pathogenesis and treatment. *Pharmacol Ther*. 2015;152:63-82.
2. Thakkar JP, Dolecek TA, Horbinski C, et al. Epidemiologic and molecular prognostic review of glioblastoma. *Cancer Epidemiol Biomarkers Prev*. 2014;23(10):1985-1996.
3. Wang Z, Sun H, Sebastian Yakisich J. Overcoming the blood-brain barrier for chemotherapy: limitations, challenges and rising problems. *Anticancer Agents Med Chem*. 2014;14(8):1085-1093.
4. Drapeau A, Fortin D. Chemotherapy delivery strategies to the central nervous system: neither optional nor superfluous. *Curr Cancer Drug Targets*. 2015;15(9):752-768.
5. Brem H, Mahaley Jr MS, Vick NA, et al. Interstitial chemotherapy with drug polymer implants for the treatment of recurrent gliomas. *J Neurosurg*. 1991;74(3):441-446.

6. Brem H, Piantadosi S, Burger PC, et al. Placebo-controlled trial of safety and efficacy of intraoperative controlled delivery by biodegradable polymers of chemotherapy for recurrent gliomas. *Lancet.* 1995;345(8956):1008-1012.
7. Westphal M, Hilt DC, Bortey E, et al. A phase 3 trial of local chemotherapy with biodegradable carmustine (BCNU) wafers (Gliadel wafers) in patients with primary malignant glioma. *Neuro-Oncology.* 2003;5(2):79-88.
8. Westphal M, Ram Z, Riddle V, Hilt D, Bortey E. Group ECotGS. Gliadel® wafer in initial surgery for malignant glioma: long-term follow-up of a multicenter controlled trial. *Acta Neurochir.* 2006;148(3):269-275.
9. Kumari A, Yadav SK, Yadav SC. Biodegradable polymeric nanoparticles based drug delivery systems. *Colloids Surf B: Biointerfaces.* 2010;75(1):1-18.
10. Thuy MN, Kam JK, Lee GC, et al. A novel literature-based approach to identify genetic and molecular predictors of survival in glioblastoma multiforme: analysis of 14,678 patients using systematic review and meta-analytical tools. *J Clin Neurosci.* 2015;22(5):785-799.
11. Uribe D, Torres Á, Rocha JD, et al. Multidrug resistance in glioblastoma stem-like cells: role of the hypoxic microenvironment and adenosine signaling. *Mol Asp Med.* 2017;55:140-151.
12. Persano L, Rampazzo E, Basso G, Viola G. Glioblastoma cancer stem cells: role of the microenvironment and therapeutic targeting. *Biochem Pharmacol.* 2013;85(5):612-622.
13. Ostrom QT, Gittleman H, Liao P, et al. CBTRUS Statistical Report: primary brain and other central nervous system tumors diagnosed in the United States in 2010–2014. *Neuro-Oncology.* 2017;19(Suppl_5):v1-v88.
14. Ostrom QT, Gittleman H, Farah P, et al. CBTRUS statistical report: primary brain and central nervous system tumors diagnosed in the United States in 2006-2010. *Neuro-Oncology.* 2013;15(Suppl_2):ii1-ii56.
15. Stupp R, Mason WP, Van Den Bent MJ, et al. Radiotherapy plus concomitant and adjuvant temozolomide for glioblastoma. *N Engl J Med.* 2005;352(10):987-996.
16. Xie Q, Mittal S, Berens ME. Targeting adaptive glioblastoma: an overview of proliferation and invasion. *Neuro-Oncology.* 2014;16(12):1575-1584.
17. Woodworth GF, Dunn GP, Nance EA, Hanes J, Brem H. Emerging insights into barriers to effective brain tumor therapeutics. *Front Oncol.* 2014;4:126.
18. Vorbrodt AW, Dobrogowska DH. Molecular anatomy of intercellular junctions in brain endothelial and epithelial barriers: electron microscopist's view. *Brain Res Rev.* 2003;42(3):221-242.
19. Oldendorf WH, Cornford ME, Brown WJ. The large apparent work capability of the blood-brain barrier: a study of the mitochondrial content of capillary endothelial cells in brain and other tissues of the rat. *Ann Neurol.* 1977; 1(5):409-417.
20. Savettieri G, Di Liegro I, Catania C, et al. Neurons and ECM regulate occludin localization in brain endothelial cells. *Neuroreport.* 2000;11(5):1081-1084.
21. Rascher G, Fischmann A, Kröger S, Duffner F, Grote E-H, Wolburg H. Extracellular matrix and the blood-brain barrier in glioblastoma multiforme: spatial segregation of tenascin and agrin. *Acta Neuropathol.* 2002;104(1):85-91.
22. Liu KJ, Rosenberg GA. Matrix metalloproteinases and free radicals in cerebral ischemia. *Free Radic Biol Med.* 2005;39(1):71-80.
23. Cosolo WC, Martinello P, Louis WJ, Christophidis N. Blood-brain barrier disruption using mannitol: time course and electron microscopy studies. *Am J Phys Regul Integr Comp Phys.* 1989;256(2):R443-R447.
24. Shaltiel-Karyo R, Frenkel-Pinter M, Rockenstein E, et al. A blood-brain barrier (BBB) disrupter is also a potent α-synuclein (α-syn) aggregation inhibitor a novel dual mechanism of mannitol for the treatment of parkinson disease (PD). *J Biol Chem.* 2013;288(24):17579-17588.
25. Obermeier B, Daneman R, Ransohoff RM. Development, maintenance and disruption of the blood-brain barrier. *Nat Med.* 2013;19(12):1584.
26. Hawkins BT, Davis TP. The blood-brain barrier/neurovascular unit in health and disease. *Pharmacol Rev.* 2005; 57(2):173-185.
27. Persidsky Y, Ramirez SH, Haorah J, Kanmogne GD. Blood–brain barrier: structural components and function under physiologic and pathologic conditions. *J NeuroImmune Pharmacol.* 2006;1(3):223-236.
28. Abbott NJ, Rönnbäck L, Hansson E. Astrocyte–endothelial interactions at the blood–brain barrier. *Nat Rev Neurosci.* 2006;7(1):41.
29. Smith QR, Ziylan Y, Robinson PJ, Rapoport SI. Kinetics and distribution volumes for tracers of different sizes in the brain plasma space. *Brain Res.* 1988;462(1):1-9.
30. Ghose AK, Viswanadhan VN, Wendoloski JJ. A knowledge-based approach in designing combinatorial or medicinal chemistry libraries for drug discovery. 1. A qualitative and quantitative characterization of known drug databases. *J Comb Chem.* 1999;1(1):55-68.

31. Gao X, Li C. Nanoprobes visualizing gliomas by crossing the blood brain tumor barrier. *Small*. 2014;10(3):426-440.
32. Papadopoulos C, SS DCD, BA Bell M. Emerging molecular mechanisms of brain tumour oedema. *Br J Neurosurg*. 2001;15(2):101-108.
33. Liebner S, Fischmann A, Rascher G, et al. Claudin-1 and claudin-5 expression and tight junction morphology are altered in blood vessels of human glioblastoma multiforme. *Acta Neuropathol*. 2000;100(3):323-331.
34. Van Tellingen O, Yetkin-Arik B, De Gooijer M, Wesseling P, Wurdinger T, De Vries H. Overcoming the blood–brain tumor barrier for effective glioblastoma treatment. *Drug Resist Updat*. 2015;19:1-12.
35. Leong K, D'Amore P, Marletta M, Langer R. Bioerodible polyanhydrides as drug-carrier matrices. II. Biocompatibility and chemical reactivity. *J Biomed Mater Res A*. 1986;20(1):51-64.
36. Akbari H, D'emanuele A, Attwood D. Effect of fabrication technique on the erosion characteristics of polyanhydride matrices. *Pharm Dev Technol*. 1998;3(2):251-259.
37. Wu MP, Tamada JA, Brem H, Langer R. In vivo versus in vitro degradation of controlled release polymers for intracranial surgical therapy. *J Biomed Mater Res A*. 1994;28(3):387-395.
38. Kumar N, Langer RS, Domb AJ. Polyanhydrides: an overview. *Adv Drug Deliv Rev*. 2002;54(7):889-910.
39. Leong K, Brott B, Langer R. Bioerodible polyanhydrides as drug-carrier matrices. I: characterization, degradation, and release characteristics. *J Biomed Mater Res A*. 1985;19(8):941-955.
40. Chae GS, Lee JS, Kim SH, et al. Enhancement of the stability of BCNU using self-emulsifying drug delivery systems (SEDDS) and in vitro antitumor activity of self-emulsified BCNU-loaded PLGA wafer. *Int J Pharm*. 2005; 301(1-2):6-14.
41. Dang W, Daviau T, Brem H. Morphological characterization of polyanhydride biodegradable implant Gliadel® during in vitro and in vivo erosion using scanning electron microscopy. *Pharm Res*. 1996;13(5):683-691.
42. Dang W, Daviau T, Ying P, et al. Effects of GLIADEL® wafer initial molecular weight on the erosion of wafer and release of BCNU. *J Control Release*. 1996;42(1):83-92.
43. Tamada J, Langer R. The development of polyanhydrides for drug delivery applications. *J Biomater Sci Polym Ed*. 1992;3(4):315-353.
44. Uhrich KE, Cannizzaro SM, Langer RS, Shakesheff KM. Polymeric systems for controlled drug release. *Chem Rev*. 1999;99(11):3181-3198.
45. Domb AJ, Rock M, Schwartz J, et al. Metabolic disposition and elimination studies of a radiolabelled biodegradable polymeric implant in the rat brain. *Biomaterials*. 1994;15(9):681-688.
46. Domb A, Rock M, Perkin C, Yipchuck G, Broxup B, Villemure J. Excretion of a radiolabelled anticancer biodegradable polymeric implant from the rabbit brain. *Biomaterials*. 1995;16(14):1069-1072.
47. Grossman SA, Reinhard C, Colvin OM, et al. The intracerebral distribution of BCNU delivered by surgically implanted biodegradable polymers. *J Neurosurg*. 1992;76(4):640-647.
48. Fung LK, Shin M, Tyler B, Brem H, Saltzman WM. Chemotherapeutic drugs released from polymers: distribution of 1, 3-bis (2-chloroethyl)-l-nitrosourea in the rat brain. *Pharm Res*. 1996;13(5):671-682.
49. Fung LK, Ewend MG, Sills A, et al. Pharmacokinetics of interstitial delivery of carmustine, 4-hydroperoxycyclophosphamide, and paclitaxel from a biodegradable polymer implant in the monkey brain. *Cancer Res*. 1998;58(4):672-684.
50. Strasser J, Fung L, Eller S, Grossman S, Saltzman W. Distribution of 1, 3-bis (2-chloroethyl)-1-nitrosourea and tracers in the rabbit brain after interstitial delivery by biodegradable polymer implants. *J Pharmacol Exp Ther*. 1995;275(3):1647-1655.
51. Morrison P, Dedrick R. Transport of cisplatin in rat brain following microinfusion: an analysis. *J Pharm Sci*. 1986; 75(2):120-128.
52. Nicholson C. Interaction between diffusion and Michaelis-Menten uptake of dopamine after iontophoresis in striatum. *Biophys J*. 1995;68(5):1699-1715.
53. Jain RK. Transport of molecules in the tumor interstitium: a review. *Cancer Res*. 1987;47(12):3039-3051.
54. Saltzman WM, Radomsky ML. Drugs released from polymers: diffusion and elimination in brain tissue. *Chem Eng Sci*. 1991;46(10):2429-2444.
55. Dang W, Saltzman WM. Dextran retention in the rat brain following release from a polymer implant. *Biotechnol Prog*. 1992;8(6):527-532.
56. Mak M, Fung L, Strasser JF, Saltzman WM. Distribution of drugs following controlled delivery to the brain interstitium. *J Neuro-Oncol*. 1995;26(2):91-102.
57. Krewson CE, Klarman ML, Saltzman WM. Distribution of nerve growth factor following direct delivery to brain interstitium. *Brain Res*. 1995;680(1-2):196-206.

58. Krewson CE, Saltzman WM. Transport and elimination of recombinant human NGF during long-term delivery to the brain. *Brain Res*. 1996;727(1-2):169-181.
59. Mahoney MJ, Saltzman WM. Controlled release of proteins to tissue transplants for the treatment of neurodegenerative disorders. *J Pharm Sci*. 1996;85(12):1276-1281.
60. Fenstermacher J, Patlak C. The movements of water and solutes in the brains of mammals. In: *Dynamics of Brain Edema*. Springer; 1976:87-94.
61. Fenstermacher J, Patlak C, Blasberg R. Transport of material between brain extracellular fluid, brain cells and blood. *Fed Proc*. 1974;33:2070-2074.
62. Fleming AB, Saltzman WM. Pharmacokinetics of the carmustine implant. *Clin Pharmacokinet*. 2002;41(6):403-419.
63. Brem H, Kader A, Epstein JI, et al. Biocompatibility of a biodegradable, controlled-release polymer in the rabbit brain. *Sel Cancer Ther*. 1989;5(2):55-65.
64. Laurencin C, Domb A, Morris C, et al. Poly (anhydride) administration in high doses in vivo: studies of biocompatibility and toxicology. *J Biomed Mater Res A*. 1990;24(11):1463-1481.
65. Tamargo RJ, Epstein JI, Reinhard CS, Chasin M, Brem H. Brain biocompatibility of a biodegradable, controlled-release polymer in rats. *J Biomed Mater Res A*. 1989;23(2):253-266.
66. Brem H, Domb A, Lenartz D, Dureza C, Olivi A, Epstein JI. Brain biocompatibility of a biodegradable controlled release polymer consisting of anhydride copolymer of fatty acid dimer and sebacic acid. *J Control Release*. 1992; 19(1-3):325-329.
67. Tamargo RJ, Myseros JS, Epstein JI, Yang MB, Chasin M, Brem H. Interstitial chemotherapy of the 9L gliosarcoma: controlled release polymers for drug delivery in the brain. *Cancer Res*. 1993;53(2):329-333.
68. Brem H, Tamargo RJ, Olivi A, et al. Biodegradable polymers for controlled delivery of chemotherapy with and without radiation therapy in the monkey brain. *J Neurosurg*. 1994;80(2):283-290.
69. Valtonen S, Ul T, Toivanen P, et al. Interstitial chemotherapy with carmustine-loaded polymers for high-grade gliomas: a randomized double-blind study. *Neurosurgery*. 1997;41(1):44-49.
70. W-K X, Shao C, Qi Z-Y, Yang C, Wang Z. The role of Gliadel wafers in the treatment of newly diagnosed GBM: a meta-analysis. *Drug Des Devel Ther*. 2015;9:3341.
71. Chowdhary SA, Ryken T, Newton HB. Survival outcomes and safety of carmustine wafers in the treatment of high-grade gliomas: a meta-analysis. *J Neuro-Oncol*. 2015;122(2):367-382.
72. Attenello FJ, Mukherjee D, Datoo G, et al. Use of Gliadel (BCNU) wafer in the surgical treatment of malignant glioma: a 10-year institutional experience. *Ann Surg Oncol*. 2008;15(10):2887.

CHAPTER

7

Pharmacokinetic Models of Convection-Enhanced Drug Delivery

Malisa Sarntinoranont, Paul Morrison†*

*Department of Mechanical & Aerospace Engineering, University of Florida, Gainesville, FL, United States †National Institute of Biomedical Imaging & Bioengineering, National Institutes of Health, Bethesda, MD, United States

INTRODUCTION

The brain offers substantial obstacles to successful drug delivery, the most substantial of which is the blood–brain barrier, which limits penetration of the majority of systemically delivered compounds. This is especially a barrier for delivery of large-molecular–weight compounds such as biologics. Of the few local drug delivery methods available for the CNS, convection-enhanced delivery (CED) continues to be a promising method for providing controlled, large-volume coverage.[1, 2] With CED, solutions of drug are infused directly into brain parenchyma through infusion cannulas that are placed at the target site. Infusions drive the flow of infusate directly through the interstitial space. In this way, CED overcomes slow diffusion limits encountered by macromolecular therapeutic agents.

Precise delivery of infusate requires a noncompliant infusion system that does not expand with pressurization.[3] Also, brain tissues are ultrasoft and require special considerations during infusion in order to provide predictable distributions within surrounding tissues. For example, controlled infusion rates and specialized tips are usually required to avoid backflow along the cannula track.[4-8] Induced flows are much larger than endogenous flow within the brain, and this can result in shunting of infusate along paths of least resistance to fissures, ventricles, along perivascular spaces, and other brain surfaces. Because of the potential for diverting flow, visualization of CED distribution through a co-infusion of tracer has proven to be an important check. As a result, CED imaging studies have provided a basis for quantifying CED within the CNS.

Another key means of quantifying CED is predictive transport modeling. For CED, these models need to provide detailed spatial information. Distributed models developed for CED combine both fluid and mass transport considerations to predict spatial and temporal drug

Nervous System Drug Delivery
https://doi.org/10.1016/B978-0-12-813997-4.00007-4

© 2019 Elsevier Inc. All rights reserved.

history. Fluid dynamic models for infusate flow through interstitial tissue are based on porous media transport theory. These range from simple one-dimensional, spherical infusion simulations to fully three-dimensional (3D) models that account for realistic anatomy and variation in underlying tissue structure. Predicted infusate penetration depends on the infusion protocol, catheter tip design, and targeted tissue region of interest. Furthermore, pharmacokinetic analysis requires consideration of diffusion, binding, and metabolism characteristics that depend on tissue-drug interactions, and are specific to the drug and tissue region considered. Because of these considerations, pharmacokinetic analysis of CED requires a more complicated formalism than required when using a compartmental modeling approach, where the agent is distributed homogeneously and only time-dependent concentrations, $C(t)$, are considered. Partial differential equations are solved to determine velocity flow fields, $\mathbf{v}(\mathbf{x}, t)$, and resulting spatial drug concentrations, $C(\mathbf{x}, t)$ as a function of spatial coordinate, $\mathbf{x}$.

A number of pharmacokinetic studies have been conducted as CED has progressed to the level of clinical trials. Pharmacokinetic analyses of quinolinic acid (QA) and GDNF have been quantified for the potential treatment of Parkinson disease (PD) and substance-P neurotoxin for the treatment of chronic pain. These have been conducted within different animal models ranging from rodents to primates. When considering pharmacokinetic modeling, the choice of CED model depends on the tissue region, infusion protocol, drug compound, and catheter design. In this chapter, we focus on CED pharmacokinetic models that incorporate tissue flow calculations into distributed transport models to predict $C(\mathbf{x},t)$. First, we summarize major parameters needed to quantify flow and pharmacokinetics of CED. Next, we review three special cases that provide a range of CED examples for: (i) distribution of a small-molecular–weight compound (QA) in gray matter regions of the brain characterized by isotropic transport properties; (ii) distribution of a macromolecular biologic (substance P-diphtheria toxin) within white matter regions of the spinal cord characterized by tissue anisotropy, realistic anatomical boundaries, and intracellular trafficking kinetics; and (iii) distribution of a macromolecule (GDNF) to gray matter regions where back flow and cylindrical distribution may arise from variations in catheter geometry and infusate flow. Lastly, we provide an overview and comment on additional considerations when developing pharmacokinetic models of CED.

MAJOR TISSUE FLOW AND PHARMACOKINETIC PARAMETERS

While simple in concept, CED drug coverage is dependent on a number of variables governing flow, drug deposition, and clearance in CNS tissues. Because of its infusate flow dependence, CED analysis must consider parameters governing infusate flow through interstitial or parenchymal spaces. Pharmacokinetic analysis of CED also requires an understanding of drug-specific parameters. As infused molecules move through the interstitial space, local concentration is determined by a combination of convection, diffusion, binding, and metabolism.

ϕ (Fluid Volume Fraction)

Tissue can be considered a porous or multiphase medium consisting of a solid matrix saturated with fluid that occupies an interconnected pore space. In CNS tissues, the solid matrix consists of the collective of cells, extracellular matrix, and vasculature. The remaining space is the interstitial space through which infusate can flow. This fluid-filled space has been measured in diffusion experiments to be approximately 20% of brain tissue volume using a real-time iontophoretic tetramethylammonium method.[9] Fluid volume fraction, ϕ, in tissue is also referred to as the pore space, porosity, or extracellular volume fraction.

K (Hydraulic Conductivity or Permeability)

Fluid flow through porous media is governed by Darcy's law

$$\mathbf{v} = -\mathbf{K} \cdot \nabla p \tag{1}$$

where the volume averaged fluid velocity vector, $\mathbf{v}$, in tissue is proportional to the pressure gradient, ∇p, across the tissue.The hydraulic conductivity or permeability tensor, $\mathbf{K}$, is a measure of the conductance of the material to fluid flow. In Eq. (1), vector and tensor quantities provide a 3D representation of Darcy's law. Volume average velocity is calculated over both solid and fluid phases, and this term is useful when accounting for mass fluid balances. Fluid flow only occurs through interconnected pore space and this local, interstitial (pore) velocity vector is given by

$$\mathbf{v}_i = \frac{1}{\phi} \mathbf{v} \tag{2}$$

This velocity is useful when considering convective transport through tissue.

Hydraulic conductivity is the main property governing pressure-driven flow through tissue. It is dependent on the characteristic geometry of the pore space; that is, pore size and connectivity and interstitial fluid viscosity. Large values correspond to low tissue resistance to flow, and smaller values require larger pressure gradients to maintain flow. During CED, the high pressures that are generated at the infusion site are based on this property. The pressure gradients that develop drive infusate penetration into surrounding tissues. Hydraulic conductivity is considered to be isotropic in gray matter[7, 10] and highly anisotropic in white matter because of the presence of the aligned nerve fibers.[1, 11] Regional differences in hydraulic conductivity such as between white matter and gray matter will also determine flow patterns. Regions of low conductivity, such as along white matter tracks or perivascular spaces, act as paths of least resistance for diverting flow.

D (Diffusivity)

Diffusional transport in tissue is driven by concentration gradients. Because it occurs within the interstitial space, it is also dependent on the structural geometry of this space and size of the molecule. Because diffusion is generally hindered within tissue, the effective diffusion tensor, D_e, of the molecule through tissue is less than that measured in free water, D_{free}. D_e has been determined in brain tissue slices using techniques such as the iontophoretic tetramethylammonium method which measures ϕ and effective tortuosity, $\tau = \sqrt{\frac{D_{free}}{D_e}}$.[9]

Within isotropic media where diffusivity is a scalar value, D_e of small molecular weight species can be related using the following relation[12]

$$D_e = \tau^{-2} D_{free}^{37°C} \propto \tau^{-2} (MW)^{-0.5} \tag{3}$$

where MW is the molecular weight of the diffusing species. By using this relation, the diffusion coefficient for an arbitrary molecular weight species can be determined with respect to a reference substance using the following relation[12]

$$\frac{D_e}{D_{e\_ref}} = \left(\frac{MW_{ref}}{MW}\right)^{0.5} \tag{4}$$

For macromolecules (MW > 40 kDa), the exponent increases from 0.5 to nearly unity, e.g., albumin.

Ps (Product of Permeability Coefficient and Surface Area Per Unit Volume)

During and following CED, infused drugs can be taken up by vasculature and be removed from tissue by this route. The *Ps* term accounts for passive movement of species from the extracellular space across microvasculature. *P* is the vascular permeability coefficient that is a measure of the drug exchange across microvascular walls. It is dependent on pore density and pore channel structure in microvessel walls. For small molecules, the permeability is also dependent on diffusion and other considerations, such as lipophilicity and charge. For large molecules, permeability is additionally dependent on filtration (convection across the vessel wall) as described by Starling's equation. The term *s* refers to the surface area of the entire microvascular bed per unit volume of tissue.

Ps may also be obtained from MW scaling relationships with known reference values using a similar approach as described for diffusivity.[13, 14] For small molecular weight species,[12]

$$\frac{Ps}{Ps_{ref}} = \left(\frac{MW_{ref}}{MW}\right)^{0.63} \tag{5}$$

where Ps_{ref} and MW_{ref} are the known values of the reference species in the tissue of interest.

R (Whole Tissue to Extracellular Space Partitioning Parameter)

If we consider the total concentration of bound plus free drug to be denoted by *C* and the free extracellular (interstitial) drug is C_e, then these concentrations are related by

$$C = RC_e \tag{6}$$

where *R* is the whole tissue to extracellular partitioning parameter. For nonbinding substances distributed only in the extracellular space, then *R* is equal to the extracellular volume fraction, ϕ. For nonbinding substances that distribute equally within intracellular and extracellular tissue spaces, $R = 1$. Small neutral molecules (generally lipophilic) can often be assumed to equilibrate quickly across cell membranes and reach equilibrium with nonspecific binders associated with the cell.

Small fully charged molecules are lipophobic and are excluded from the intracellular space when there are no specific membrane transporters present in the cell membrane.

If transporters (e.g., organic acid transporters) are present, these molecules can access the intracellular space. However, intracellular concentrations may not be equal to extracellular concentrations based on different influx and efflux transporter rates. To determine this, experimental transporter data must be available from the literature.

Macromolecules are too large to cross cell membranes unless specific uptake mechanisms are present. Without cell uptake, their intracellular concentrations are zero (e.g., albumin). With cell uptake present, independent experimental data must be available from literature to quantify rates of cell membrane binding and uptake. An example of species binding, uptake, and recycling is described in our case study for Substance-P diphtheria toxin, which is taken up by the SP-mediated endocytotic pathway.[15]

k (Reaction Constant)

Many molecules are subject to metabolic degradation or transformation in the interstitial and intracellular spaces, with the vast majority occurring in the intracellular space. Compound interactions with cell or extracellular matrix binding sites, cell uptake, and metabolism can be included through reaction terms. Estimates for reaction rate constants for these reactions generally come from whole tissue pharmacokinetic data. Introduced compounds can also interact with cell or extracellular matrix binding sites. Less frequently, rate constants can be estimated from cell or tissue culture measurements. However, care should be taken when using these values, since cofactors and cellular interactions may be missing in vitro. These rate constants are specific to each molecular agent.

CASE MODELS OF CONVECTION-ENHANCED DELIVERY

Different pharmacokinetic modeling approaches can be undertaken to capture local distributions and clearance during and following CED within the CNS. We highlight three major examples: QA,[16] substance P-diphtheria toxin,[15] and GDNF.[17]

Case i: Quinolinic Acid

Convection-enhanced delivery provides a means to inhibit deep brain regions in a reproducible and targeted manner. Lonser et al.[16] investigated CED of the excitotoxin QA to selectively ablate regions of the globus pallidus internus (GPi) in a primate model of PD. Hyperactivity of the subthalamic nucleus and the GPi is associated with neuronal dysregulation and results in many of the disorders of movement connected to PD and failure of medical treatment.[18-22] The goal of this study was to determine if selective lesioning of neuronal subpopulations in the GPi by CED provided improvement in PD symptoms, including reduction in movement disturbances. For successful treatment, CED of QA must be reproducibly targeted. Modeling was used to assess whether CED could be sufficiently localized to allow chemical pallidotomy within the GPi while avoiding unwanted toxicity in neighboring nuclei.

Quinolinic acid ($C_7H_5NO_4$) is an endogenous excitotoxin that is normally present in nanomolar concentration.[23, 24] It is a small molecular compound (MW 167 Da) for which diffusion can play a significant transport role. Accordingly, a CED model was chosen to model the GPi gray matter as a homogenous region characterized by isotropic transport and mechanical properties. Simulation revealed that for a given low convective inflow, an infusate concentration could be identified that would balance the outward convective-diffusion flow against the relatively rapid microvascular clearance of this small molecule to assure that the range of neurotoxic exposure could indeed be limited to the tissue volume of the GPi.

The developed model thus represents an early use of isotropic CED modeling to estimate therapeutic infusate concentration. The CED model assumed spherically symmetric distribution around the cannula tip infused at a low 0.1 μL/min flow rate (the largest flow rate to avoid backflow along the cannula shaft) and a total infusate volume of 5 μL (equivalent to filling the extracellular space of the GPi in 50 min). The inner boundary of the infused GPi was prescribed as the outer radius of the infusion cannula embedded within tissue. For the primate, the GPi outer radius was set to 1.5 mm, the outer radius of the largest inscribed sphere. QA concentrations relative to the infusate concentration were computed for each radial distance, r, from the point of infusion; that is, the tip of the infusion cannula. Determining from this the relative QA concentration at the GPi outer boundary at 50 minutes, it needed only to be divided into the known neurotoxic threshold value of 1.8 mM[25] to obtain the estimate for infusate concentration, 50 mM, required to restrict ablation to the GPi.

Interstitial tissue distributions of QA are governed by a convection–diffusion-clearance formulation previously developed by Morrison et al.[25]

$$R\frac{\partial C_e}{\partial t} = -\nabla \cdot (\phi \mathbf{v_i} C_e) + \phi D_e \nabla^2 C_e + k_{pex}\phi C_p - \left[k_{epx}\phi + k_{irr}\right] C_e \tag{7}$$

C_e is the extracellular (interstitial) concentration of QA, and R is 1.34.[26] This equation accounts for transport of QA with time through the interstitial space of tissue considered to be a porous medium with a fluid volume fraction of ϕ. The first term on the right accounts for convection of QA by infusate flow with interstitial velocity, $\mathbf{v_i}$. For a constant volumetric in flow rate, Q, interstitial radial velocity only varies with radial location based on the scalar relation

$$v_i(r) = \frac{Q}{4\pi r^2 \phi} \tag{8}$$

The second term is the diffusion of QA, which is driven by concentration gradients, ∇C_e, in the interstitial space. The use of effective diffusivity accounts for volume averaged tissue diffusion. The third term accounts for uptake from vasculature. The fourth term accounts for clearance of QA due to transport across microvasculature and metabolism, respectively. Rate constants, k_{pex} and k_{epx}, are the passive influx and active efflux rate constants on an interstitial basis. C_p is the plasma QA concentration. k_{irr}/R is the first-order degradation constant on a total tissue basis. Parameter estimates for diffusivity and reaction constants were taken from literature.[16]

Eq. (7) is a partial differential equation for which boundary and initial conditions are also needed in order to solve for $C_e(r,t)$. For the innermost boundary, the mass rate of delivery of QA from the cannula, QC_{inf}, was taken as equal to the convective rate of mass entering interstitial space across a spherical surface centered on the cannula tip; that is, $4\pi r_i^2 \phi v_i(r_i) C_e(r_i,t)$. Concentrations are solved for regions beyond the GPi boundary, and the outer boundary at $r = r_o$ must be far enough out so that the concentration reduces to normal endogenous tissue values, $C_e(r_o,t) = C_{enorm}$. Since endogenous QA is assumed to be negligible, QA concentrations in tissue are zero everywhere before the start of infusion. Computational software is used to determine numerical solutions and predict $C_e(r,t)$.

During CED, infusate displaces existing interstitial fluid and begins to fill the interstitial space with QA. A QA concentration profile then develops as the compound flows radially outward from the cannula tip, declining in value somewhat slowly in the high-convective region near the tip, then more rapidly with progressing radius as the interstitial velocity slows and microvascular clearance has more time to act over each increment of radius. Gradually, because of the dropping radial velocity and the relatively large diffusion coefficient of a small molecule like QA (relative to macromolecules), convective contributions to the mass transport of QA decline and diffusional transport becomes dominant. After the end of infusion at $t=t_f$, spread of QA continues by diffusion alone, further reducing tissue concentrations over time until all the QA is cleared from tissue. Tissue concentration after the end of CED is also solved by Eq. (7) with the convective transport term set to zero. For this case, the boundary condition at the inner boundary is no flux, $\nabla C_e=0$, and the boundary condition at the outermost boundary is the same as before. The concentration profile right at the end of CED infusion, $C_e(r, t_f)$, is used as the initial condition.

For a given set of infusion rates and transport properties, interstitial concentration profiles can be solved for QA over time. In this primate model, predicted concentration profiles were used to determine an optimal QA infusate concentration of 50 mM and infusate volume of 5 μL for excitotoxic lesioning of neurons. Fig. 1 shows concentration profiles for QA

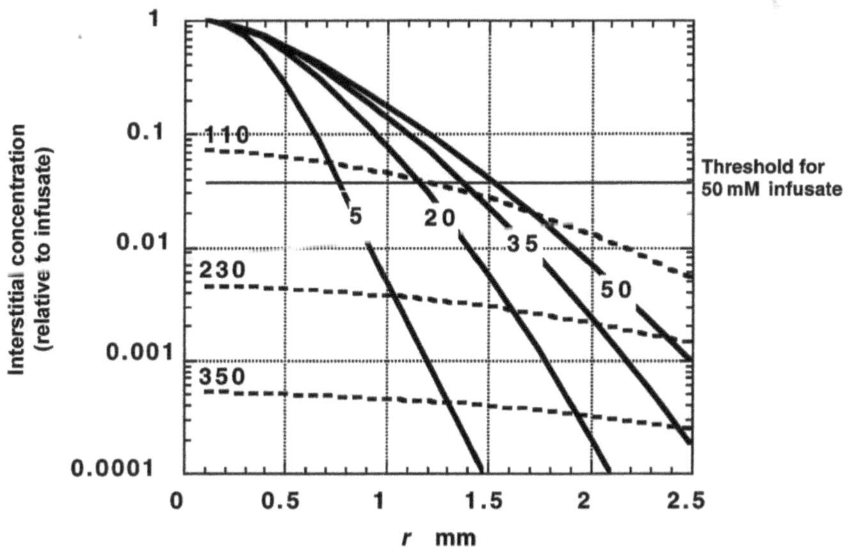

FIG. 1 Simulated QA concentration profiles in brain tissue during CED at 0.1 μL/min. Normalized profiles, C_e/C_{inf}, where C_{inf}=50 mM QA are given at 5, 20, 35, and 50 min (heavy solid lines) with radial distance from the cannula tip. At the end of CED infusion (50 min), the distance with QA distributed above the threshold value of 1.8 mM QA reaches 1.5 mm, which is the distance of the nearest GPi border. After the end of infusion, diffusion-only spread is given by dashed lines at different times 110, 230, and 350 minutes. *Used with permission from Lonser RR, Corthesy ME, Morrison PF, Gogate N, Oldfield EH. Convection-enhanced selective excitotoxic ablation of the neurons of the globus pallidus internus for treatment of parkinsonism in nonhuman primates. J Neurosurg 1999;91(2):294-302.*

distributed by convection–diffusion on tissue length scales on the order of millimeters. At the end of infusion, the concentration of QA relative to infusate concentration was computed as 0.036 at the outer GPi boundary of 1.5 mm. The threshold value of 1.8 mM QA divided by 0.036 yielded the QA infusate concentration of 50 mM. After the end of infusion, QA tissue concentrations reduced with time. Continued diffusional spread was limited by metabolism and uptake and did not extend significantly beyond the GPi border. In this way, neuronal loss was predicted to be primarily limited to the region of CED distribution. Delivery parameters predicted by the model were confirmed through CED of biotinylated albumin, which confirmed filling of the GPi with minimal entry into the neighboring globus pallidus externus. QA treatment of primates exhibiting characteristic parkinsonian features resulted in substantial improvement of symptoms, whereas untreated animals remained unchanged. This model helped to establish that CED could provide a sufficient range for QA in the GPi, despite its small molecular size, to make it a candidate for effective ablation of the GPi in advanced PD.

Case ii: SP-DT Neurotoxin

In this study, CED of substance P (SP) nocitoxins to the spinal cord was under consideration for the treatment of chronic pain.[15] 3D computational modeling was used to assess whether SP-diphtheria toxin fusion protein (SP-DT), delivered by CED to the white matter of the dorsal horn of the spinal cord, could be targeted to lamina I and II of the horn to achieve ablation of pain-transmitting neurons while avoiding toxicity to sensitive anterior motor neurons. Sensitivity of neuronal cell kill to the placement of the cannula, infusate concentration, and uncertainties in receptor density were assessed by combining a computational fluid dynamic CED model with an SP-DT dose–response model. This model provided the first use of macromolecular CED modeling to describe effective drug volumes in the CNS while accounting for (a) tissue anisotropy, (b) realistic anatomical boundaries, and (c) intracellular trafficking kinetics coupled to response predictions of cytotoxicity in NK1 receptor-expressing neurons. This model provides a potential guide for the CNS delivery of many of the new biologics.

Substance P-diphtheria toxin fusion protein (~47,000 MW) is a COOH-terminal truncated diphtheria toxin that retains the transmembrane and catalytic domains of diphtheria toxin but replaces the native binding domain with SP[27] or an analog of SP stabilized against peptidase activity.[28] In the treatment of persistent pain, the guiding idea was to take advantage of the SP-mediated endocytotic pathway, using it as a transport system to deliver cytotoxic fusion proteins to neurokinin type 1 receptor (NK1 R)-expressing neurons located within the dorsal horn of the spinal cord. SP-DT exotoxin conjugates were identified as capable of selectively lesioning spinal cord neurons, thereby blocking nociceptive transmission mediated by small-diameter fibers.[27, 29-32]

When solving for realistic anatomy, the resulting 3D structures are no longer represented by idealized planar, cylindrical, or spherical shapes. A computational modeling approach usually needs to be undertaken to solve the partial differential equations governing flow and transport within these spaces. This SP-DT CED model extended a previous computational fluid dynamic model for CED in a segment of a rat spinal cord[33] that captured albumin transport in gray and anisotropic white matter regions. The human spinal cord geometry was created for a single transverse section of the spinal cord at the C7 level, on which white and gray matter regions, including the different lamina, were delineated (Fig. 2). To create a 3D

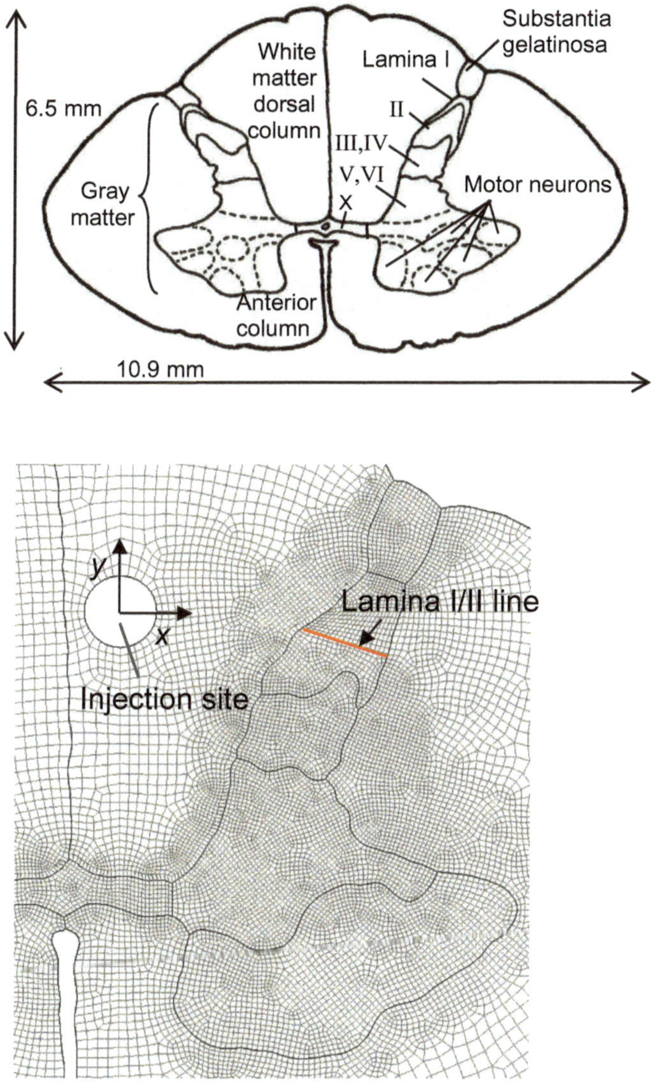

FIG. 2 The human spinal cord model at the C7 level. The top figure shows a transverse section with gray and white matter regions of interest. The bottom figure shows a portion of the corresponding finite element mesh used in the computational model. A CED infusion site is introduced into the dorsal column. The line intersecting the lamina I/II region was selected for viewing concentrations along this cross section. *Used with permission from Sarntinoranont M, Iadarola MJ, Lonser RR, Morrison PF. Direct interstitial infusion of NK(1)-targeted neurotoxin into the spinal cord: a computational model. Am J Phys Regul Integr Comp Phys 2003;285 (1):R243-R254.*

model, segmented outlines were projected (20 mm) in the craniocaudal (z) direction, creating separate white and gray matter volumes. A spherical infusion site (radius = 0.306 mm) was embedded into the white matter dorsal column.

The principle that fluid flows along path of least resistance also holds true within tissues. For CED in regions within or near regions of embedded axonal fibers like white matter, flows and drug distributions are no longer radially symmetric and preferential flow occurs in the interstitial space parallel to aligned nerve fibers.[1, 11] In the spinal cord, Lonser et al.[11] and Wood and colleagues[34] showed that direct infusion into the white matter dorsal column provides a basis for safe distribution of macromolecules that distribute axially over significant volumes of white matter. To account for flow patterns that lead to this distribution, 3D steady-state flow within porous media is predicted by solving the continuity relation, which is an expression of conservation of fluid mass within the tissue

$$\nabla \cdot \mathbf{v} = 0 \tag{9}$$

and Darcy's law as expressed in Eq. (1). Hydraulic conductivity is assumed to be isotropic in gray matter[7, 10] and highly anisotropic in white matter. More specifically, transport is considered to be transversely isotropic along the axial direction of the spinal cord such that hydraulic conductivity is greater in the z-direction, and has the same properties in the transverse plane corresponding to x- and y-directions such that $K_{wm\_z} > K_{wm\_x}$ and $K_{wm\_x} = K_{wm\_y}$. In tensor form this becomes

$$\mathbf{K} = \begin{bmatrix} K_{wm-x} & 0 & 0 \\ 0 & K_{wm-y} & 0 \\ 0 & 0 & K_{wm-z} \end{bmatrix} \tag{10}$$

Within brain tissue where axonal fiber direction is not necessarily in the z-direction, we modified this formulation to assign transversely isotropic $\mathbf{K}$ at each point in tissue based on a known fiber direction, $\mathbf{v}_3$[35]

$$\mathbf{K} = \mathbf{V} \begin{bmatrix} K_{trans} & 0 & 0 \\ 0 & K_{trans} & 0 \\ 0 & 0 & K_{axial} \end{bmatrix} \mathbf{V}^{\mathbf{T}} \tag{11}$$

where

$$\mathbf{V} = [\mathbf{v}_1 \; \mathbf{v}_2 \; \mathbf{v}_3].$$

$\mathbf{V}$ is a coordinate transformation tensor that assigns K_{axial} to the fiber direction given by the unit vector, $\mathbf{v}_3$. In this equation, $\mathbf{v}_1$ and $\mathbf{v}_2$ are arbitrary orthogonal unit vectors in the transverse plane. Fiber direction at each point in brain tissue can be calculated from diffusion tensor imaging data sets that use the effective diffusion of water to calculate directions of maximum diffusivity.[36, 37]

Boundary conditions are also needed to solve Eqs. (1) and (9) to determine the tissue velocity field, $\mathbf{v}(\mathbf{x},t)$. A volumetric infusion rate of 0.1 μL/min was applied at the infusion site boundary. Zero fluid pressure conditions were applied at the external spinal cord-CSF boundaries. Resulting pressure fields within tissue were relative to this boundary value. Along symmetrical planes, symmetry boundary conditions were assumed $-\mathbf{n} \cdot \mathbf{v} = 0$, where $\mathbf{n}$ is the unit normal vector to the surface.

As the SP-DT complex moves through interstitial space, it is subject to protein degradation and cellular interaction. These phenomena can be captured in a kinetic model that can be incorporated into the 3D transport model. We used literature data of Garland et al.,[38] Bowden et al.,[39] and Grady et al.[40] to develop a simple kinetic model of SP binding and intracellular trafficking[41] that incorporates the principal trafficking steps (shown in Fig. 3), including

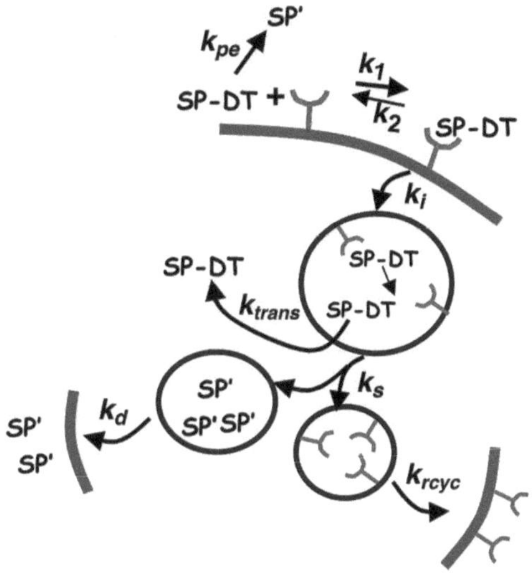

FIG. 3 Schematic representation of substance P trafficking outside and within a cell. SP can be partially degraded to SP′ outside the cell. SP can also be bound to receptors on the cell surface and internalized into endosomes. From this location, SP is transferred to cytosol or sorted and recycled to the cell membrane surface. *Used with permission from Sarntinoranont M, Iadarola MJ, Lonser RR, Morrison PF. Direct interstitial infusion of NK(1)-targeted neurotoxin into the spinal cord: a computational model. Am J Phys Regul Integr Comp Phys 2003;285(1):R243-R254.*

surface binding between SP and NK1R, clathrin-mediated endocytosis followed by spatial translation to a perinuclear endosome, where SP is sorted from its receptor, rapid degradation of SP once sorted to the late endosomes/early lysosomes, and return of sorted receptor to plasma membrane via recycling endosomes. In addition, the model follows previous research allowing for a very low rate of transfer of the diphtheria toxin moiety from early sorting endosomes to cytosol.[42] Transport equations were derived by applying mass conservation to the following SP-DT and the NK1R compartments:

SP_e SP-DT in the extracellular space
B bound SP-DT on the cell surface
R free NK1R on the cell surface
SP_s early/sorting endosomal SP-DT
R_s NK1R in early/sorting endosome
R_{post} NK1R in recycling endosomes, and
C_{cyt} SP-DT/diphtheria toxin in cytosol

In this case, concentrations of each compartment are expressed on a tissue-averaged basis (rather than the interstitial concentrations of Eq. (7) for Case (i). Of these quantities, SP_e is freely moving by convection and diffusion only through the extracellular space ($R=\phi$). Concentration is governed by

$$\frac{\partial SP_e}{\partial t} = -\nabla \cdot (\mathbf{v_i} SP_e) + \nabla \cdot (\mathbf{D_e} \cdot \nabla SP_e) - k_m SP_e - k_1 SP_e R + k_2 B \quad (12)$$

where

k_m rate of degradation by peptidases
k_1 rate SP-DT association to the receptor
k_2 rate of dissociation from the receptor

Effective diffusivity was considered isotropic in gray matter regions (D_{gm}=constant). Anisotropy of $\mathbf{D_e}$ was implemented in white matter, again assuming transverse isotropy

$$\mathbf{D_e} = \begin{bmatrix} D_{wm-x} & 0 & 0 \\ 0 & D_{wm-y} & 0 \\ 0 & 0 & D_{wm-z} \end{bmatrix} \tag{13}$$

where effective tissue diffusion was greater in the axial direction compared to the transverse direction, $D_{wm\text{-}z} > D_{wm\text{-}x}$ and $D_{wm\text{-}x} = D_{wm\text{-}y}$. Diffusivity values were estimated from measurements of anisotropic diffusion in the spinal cord.[43] Competitive binding with native SP was neglected because of the low endogenous levels present (~0.4 nM). Transport of SP_e across the external spinal cord/CSF surfaces was assumed to be dominated by convection, with negligible loss due to diffusion, $-\phi D_e \nabla SP_e \cdot \boldsymbol{n} = 0$. Along the symmetrical plane, flux of SP_e normal to the plane surface was set to zero, $\nabla SP_e \cdot \boldsymbol{n} = 0$.

Intracellular trafficking constants are as follows:

k_i rate of SP-DT internalization
k_{trans} rate of SP-DT transfer from endosome to cytosol
k_s rate of SP-DT and NK1R intracellular sorting
k_{rcyc} rate of NK1R transfer from the recycling endosomes to the cell surface

As SP-DT moves through the extracellular space, it is subject to cell binding and uptake. We apply a kinetic model to track these quantities at each point in tissue space. Equations associated with binding and cellular trafficking within NK1R-expressing neurons are given by

$$\frac{dB}{dt} = k_1 SP_e R - k_2 B - k_i B \tag{14}$$

$$\frac{dSP_s}{dt} = k_i B - k_s SP_s \tag{15}$$

$$\frac{dC_{cyt}}{dt} = k_{trans} SP_s \tag{16}$$

$$\frac{dR}{dt} = -k_1 SP_e R + k_2 B + k_{rcyc} R_{post} \tag{17}$$

$$\frac{dR_s}{dt} = k_i B - k_s R_s \tag{18}$$

$$\frac{dR_{post}}{dt} = k_s R_s - k_{rcyc} R_{post} \tag{19}$$

Because NK1R are conserved

$$B_{max} = B + R + R_s + R_{post} \tag{20}$$

where B_{max} is the NK1R capacity and varies within different laminae of the spinal cord. Incorporation of this trafficking model allows for estimation of the cytosolic diphtheria toxin concentration, which is ultimately responsible for cell kill. To determine the temporal and spatial distribution of these SP-DT complexes, Eqs. (12) and (14)-(20) were solved over the spinal cord volume using the previously determined flow field, $\mathbf{v}_i(\mathbf{x},t)=\frac{1}{\phi}\mathbf{v}(\mathbf{x},t)$, in the convective term. For the initial condition before infusion, concentration of membrane-bound NK1R was set equal to the corresponding B_{max}. All other concentrations were initially zero. Concentration of SP_e at the infusion site boundary was set equal to the tissue average value, ϕC_{inf}, and C_{inf} is the infusate concentration, a constant value (77, 385, and 770 nM).

In model predictions, CED in white matter dorsal column resulted in large-scale distribution of SP-DT along the craniocaudal direction. Corresponding transverse transport from the white matter into the adjacent gray matter is predicted to be more limited. However, it provides an ideal configuration for targeting neurons within gray matter regions of laminae I, II, and X, which are immediately adjacent. Approximate contours of cellular uptake (Fig. 4) and cytotoxicity (Fig. 5) were predicted. Cytotoxicity was predicted by locating the tissue regions where NK1R expressing cells contained ≥10 molecules of the diphtheria toxin moiety in the cytosol.

For constant CED infusion in the dorsal horn, the cytotoxic border predicted for NK1 receptor-expressing neurons was predicted to reach an asymptotic limit. This was estimated to be ~60% ablation of target neurons in lamina I/II. These predictions were sensitive to certain model parameters, NK1 receptor density, and white matter protease activity. The computational model was also used to vary certain drug and infusion parameters, and these model studies showed treatment was improved by a stable SP analog (longer half-life), high infusate concentration, and careful catheter placement adjacent to target lamina I/II.

This model was helpful in guiding development of CED treatment for localized chronic pain by assessing whether the drug was delivered sufficiently well to kill sensory neurons in the spinal cord that are responsible for the transmission of nociceptive signals. Similar CED models can be developed that combine mass transport, intracellular trafficking, and dose–response relations to predict the extent of therapeutic response for other agents with knowledge of certain tissue transport and kinetic parameters. Potential pharmacological agents with distribution that may be so predicted include biologics for treatment of neurodegenerative disorders. This model shows the ability of pharmacokinetic CED models to provide information on cell response, not just drug coverage.

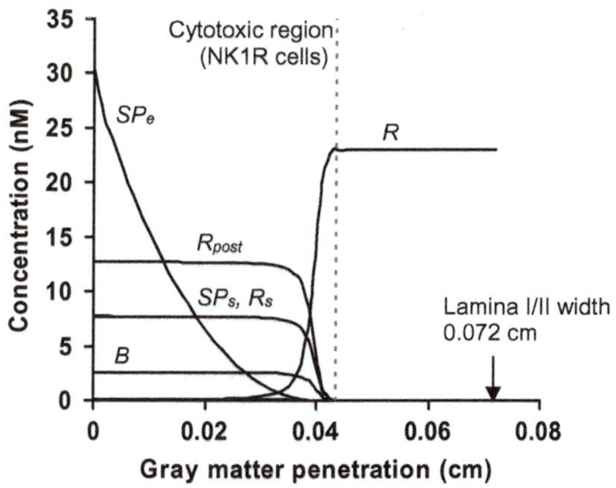

FIG. 4 Predicted concentrations within tissue based on the SP-DT kinetic model. Concentrations after 48-μL infusion are given at the level of the cannula tip along a line of interest through lamina I/II. SP_e is transported by convection and diffusion. All other quantities are based on local cellular binding, uptake, and sorting. *Used with permission from Sarntinoranont M, Iadarola MJ, Lonser RR, Morrison PF. Direct interstitial infusion of NK(1)-targeted neurotoxin into the spinal cord: a computational model. Am J Phys Regul Integr Comp Phys 2003;285(1):R243-R254.*

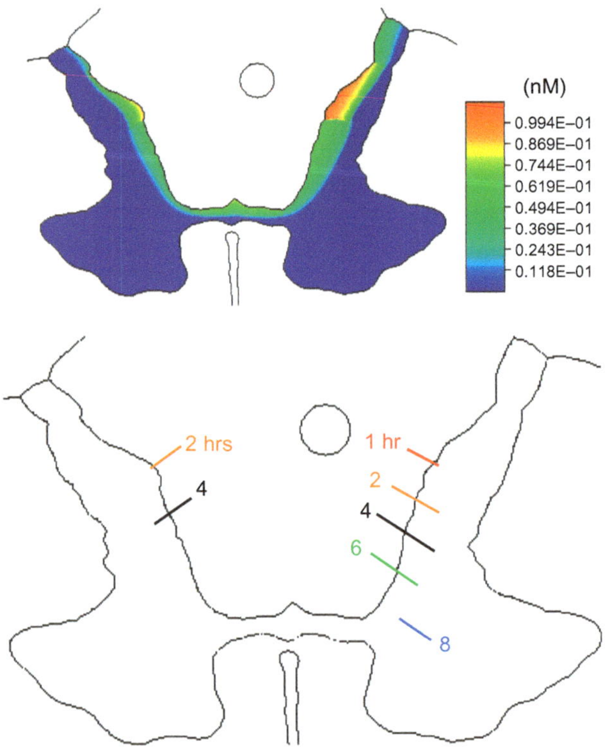

FIG. 5 Predicted concentrations at the level of the cannula tip for *(left)* cytosol concentration of SP-DT/diphtheria toxin, C_{cyst}, after 48-μL infusion and *(right)* regions of cytotoxicity where lines corresponded to the cytotoxic border for different times after the start of infusion. *Used with permission from Sarntinoranont M, Iadarola MJ, Lonser RR, Morrison PF. Direct interstitial infusion of NK(1)-targeted neurotoxin into the spinal cord: a computational model. Am J Phys Regul Integr Comp Phys 2003;285(1):R243-R254.*

Case iii: GDNF

In this study by Morrison et al.,[17] CED models were developed for glial cell line-derived neurotrophic factor (GDNF) delivered into the human striatum. GDNF is a macromolecule (MW 30.1 kDa) explored as a regenerative therapy for PD. Once CED of GDNF into the striatum was found to reverse parkinsonian symptoms in animal models of PD,[44, 45] these studies were followed up by clinical trials of GDNF delivered to the putamen.[46-48] Clinical trials resulted in divergent conclusions on significance of treatment, for which differences in drug delivery were implicated as a potential cause. This study presents the use of CED modeling to assess whether just the use of differing cannula protocols in trial treatments was sufficient to account for the different clinical responses. This model provided first use of macromolecular CED modeling to estimate drug distributions for a variety of cannula designs subject to (a) retrograde flow back along the cannula body,[8] and (b) the presence of extrachoroidal fluid production.

In particular, effects of different cannula geometry (single-hole cannula and multiport cannula) with different outer diameters, infusion rates (0.0333–10.92 μL/min), infusion concentrations (0.0226–0.226 μg/μL), and cannula placement (posterior vs mid-putamen) were simulated using a computational modeling approach. GDNF distributions were predicted in an isotropic cylindrical tissue volume based on human brain atlas data sets.[49, 50] The volume was equivalent to combined regions of human putamen, globus pallidus externus, and GPi.

The model accommodated different cannula designs resulting in different infusion site considerations for geometry and flow. Simulations also accounted for timing of infusions to include both continuous and repeat bolus infusions. High infusion pressures can result in flow back along the cannula track between the cannula wall and damaged tissue track, which will change the infusion site profile. The extent of this backflow profile depends on numerous factors, including infusion pressure, cannula diameter, elastic modulus and hydraulic permeability, as well as the cut surface profile of the surrounding tissue. In this model, the geometry of backflow was idealized to be a tapered fluid annulus connected to a fluid-filled cavity at the cannula tip (Fig. 6). A simple scaling formula was used to estimate backflow distance, x_m, as a function of cannula radius, r_c, and infusion rate, Q

$$x_m = 3.95\left[\ln\left(\frac{L}{r_c}\right)\right]^{0.8} Q^{0.6}\, {r_c}^{0.8} \tag{21}$$

where x_m, r_c, and L are given in centimeters and Q is in μL/min. L is a characteristic dimension for the tissue volume being infused, where backflow distances are generally insensitive to

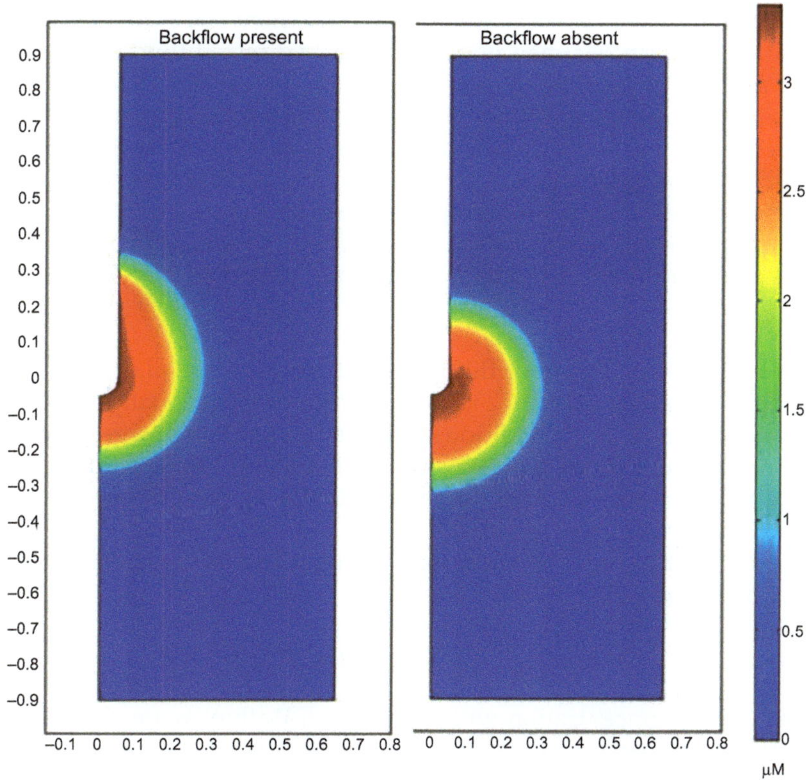

FIG. 6 Images revealing magnitude of backflow effect on GDNF concentration distribution. *(Left)* Computed distribution (half plane) when backflow is present. *(Right)* Computed distribution when backflow is absent. The outside diameter of the cannula was 1.04 mm and volumetric inflow rate was 0.1 μL/min. Numbers on the x- and y-axes represent distance coordinates in centimeters. Concentrations are expressed in μM units. *Used with permission from Morrison PF, Lonser RR, Oldfield EH. Convective delivery of glial cell line-derived neurotrophic factor in the human putamen. J Neurosurg 2007;107(1):74-83.*

small changes in L. In this model, an annulus with length x_m was placed concentrically around the cannula. The annulus tapered from a width of h_o at the cannula tip to zero at x_m. A fluid-filled hemisphere ($r=r_c+h_o$) was placed at the tip of the cannula.

Infusate enters tissue from the fluid–tissue interface at the tip of the cannula and along the fluid–tissue interface along the backflow annulus. Fluid exited the cannula tip with a parabolic Poiseuille velocity profile consistent with the infusion rate. Within fluid-filled spaces at the tip and within the backflow annulus, flow is governed by Navier–Stokes equations

$$\rho(\mathbf{v}_f \cdot \nabla)\mathbf{v}_f = -\nabla p + \nabla \cdot \left[\mu\left(\nabla \mathbf{v}_f + \nabla \mathbf{v}_f{}^T\right)\right] \tag{22}$$

where $\mathbf{v}_f$ is the fluid velocity vector and p is the fluid pressure. ρ is the density of fluid, and μ is the fluid viscosity. This set of equations accounts for viscous flow patterns that develop as fluid exits the cannula tip and fills annular spaces. At the fluid–tissue interface, fluid pressure within fluid and porous media spaces were equal. Also, fluid flux at the fluid–tissue interface was continuous, $-\mathbf{n}\cdot\mathbf{v}_f = -\mathbf{n}\cdot\mathbf{v}_t$ where $\mathbf{v}_t$ is the volume averaged fluid velocity in tissue. Resulting entry tissue flows are higher near the cannula tip and taper off with backflow distance.

The multiport cannula had side port openings that were spaced closely enough together that inflow was modeled as uniform flux across an equivalent cylindrical surface. The boundary condition at this surface assumed uniform, radially directed fluid entry, $\mathbf{v}_t = Q/(2\pi r_c l)\ \mathbf{e}_r$ where $\mathbf{e}_r$ is the radial unit vector and l is the length of the multiport section.

The computational model is simplified in that only the tissue regions surrounding the cannula are modeled. Flow into surrounding gray matter tissue was modeled as isotropic porous media using Darcy's law as previously described by Eq. (1). Because of the slow CED rates and long time periods simulated, interstitial flow was considered to be driven by both CED infusion and endogenous flows within tissues. To account for extrachoroidal fluid production,[51] a source term is added to the continuity equation

$$\nabla \cdot \mathbf{v}_t = F \tag{23}$$

where F is the extrachoroidal rate of fluid production, e.g., 0.18 μL/g/min.[51] Flow within fluid, Eq. (22), and tissue, Eqs. (1) and (23), were solved computationally using a multiphysics software package. For boundary conditions, symmetry conditions were used along the cannula axis, and no-slip conditions ($\mathbf{v}_f = \mathbf{0}$) were applied along the outside surface of the cannula. Also, the model assumed zero pressure relative to the CSF pressure at the outer tissue cylinder boundaries.

GDNF transport within fluid-filled regions surrounding the cannula tip is governed by convection and diffusion with no reactions

$$\frac{\partial C_a}{\partial t} = -\nabla \cdot \left(\mathbf{v}_f C_a\right) + D_{free} \nabla^2 C_a \tag{24}$$

where C_a is the GDNF concentration within the fluid. Within tissue, GDNF was subject to convection, diffusion, slow clearance across microvasculature, dilution by extrachoroidal fluid production, metabolism, and binding to heparin sulfate moieties in the interstitial space.[52] Corresponding tissue transport is given by[25]

$$R\frac{\partial C_e}{\partial t} = -\nabla \cdot (\mathbf{v}_t C_e) + \phi D_e \nabla^2 C_e - PsC_e - \left[\frac{k_{int} B_{max}}{K_d + C_e}\right] C_e \tag{25}$$

where C_e is the free interstitial concentration of GDNF. Ps is the product of permeability coefficient and surface area per unit volume which accounts for slow efflux of GDNF across capillaries. k_{int} is the rate constant of GDNF cell internalization and metabolic loss from tissue. K_d is the GDNF-heparin dissociation constant. For GDNF bound to heparin sulfate

$$R = B_{max} \frac{K_d}{(K_d + C_e)^2} + \phi \tag{26}$$

where B_{max} is the average tissue-binding capacity. B_{max} can be estimated from short-term V_d/V_i ratios assuming complete titration of binding sites within the distribution volume.[53] To determine temporal and spatial distribution of GDNF, Eqs. (24) and (25) are solved using the previously determined flow field, $\mathbf{v}_t(\mathbf{x},t)$, in the convective term. There is a continuity of flux and concentration at fluid–tissue interfaces. Additional boundary conditions include impermeability at solid cannula surfaces and free flow at outer cylindrical tissue boundaries.

Convection-enhanced delivery infusion protocols for different clinical trials were simulated and in each, distribution of GDNF was computed taking into account the cannula parameters (diameter, positioning, steady or interrupted flow rate, and number of ports) and the possibility and extent of retrograde flow along the needle track (Fig. 7). Simulations ultimately showed that (a) very little difference in treatment volume existed between the two trials[46, 48] using end-hole catheters when the catheters were positioned on their (nearly identical) target coordinates, yet these trials reached opposite conclusions as to efficacy (differences in catheter dimensions and backflow distribution played only minor roles in the predicted distribution); (b) treatment volume for the trial using an on-target multiport

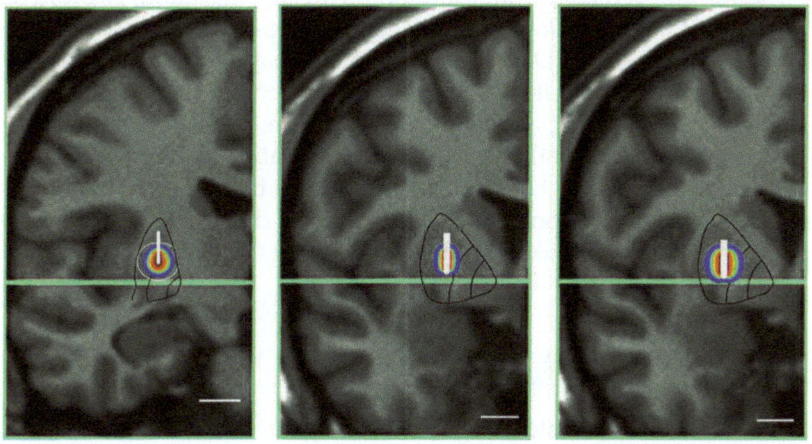

FIG. 7 Predicted CED distributions overlaid on coronal MR images of the human putamen and globus pallidus. Steady-state distributions of GDNF computed *(Left)* for two clinical trial protocols using the end-hole catheter, and a clinical trial using a multiport catheter *(Center)* for continuous basal inflow of 0.033 μL/min (equivalent to loss of all bolus flows to backflow) and *(Right)* where both basal and 6-hour bolus inflows (10.9 μL/min over 117 s) are operative and no fluid is lost to backflow. Color plots show distribution out to a putative minimum effective concentration (3.3 nM). *Black outline* indicates putamen and globus pallidus. Bars = 1 cm. *Adapted with permission from Morrison PF, Lonser RR, Oldfield EH. Convective delivery of glial cell line-derived neurotrophic factor in the human putamen. J Neurosurg 2007;107(1):74-83.*

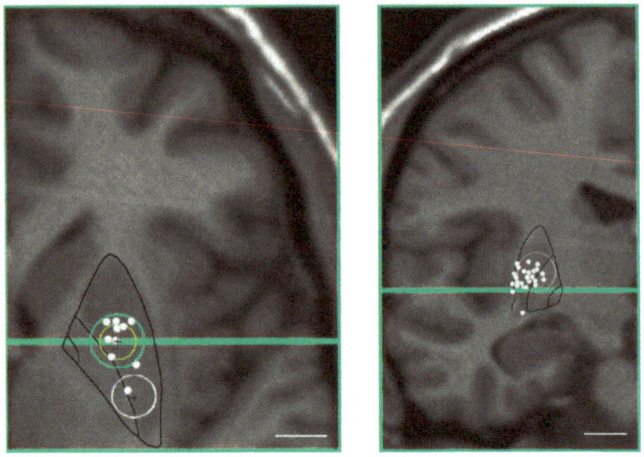

FIG. 8 Images showing comparative spreads of GDNF and experimental cannula placement distributions, according to clinical study protocol. *(Left)* Axial view of putative minimum effective concentration (3.3 nM) contour for the AmgenUT (*white* circle) and 12- to 18-month Bristol (*white* circle) trials that used end-hole cannulas and 16- to 24-week Kentucky protocols (*yellow* and *blue* circles) that used a multiport cannula. *Yellow* circle denotes Kentucky contour for basal inflow alone (equivalent to loss of all bolus flows to backflow); *blue* circle denotes the same contour if basal as well as bolus flows are operative and no fluid is lost to backflow. Upper cross denotes the mean observed catheter tip location in the Kentucky protocol, whereas the lower cross denotes the catheter tip target for the AmgenUT/Bristol protocols. White dots represent the individual catheter placements observed in the Kentucky trial. Distance between the horizontal blue line and bottom of image is the intercommissural distance. *(Right)* Coronal view through targeted cannula site for AmgenUT/Bristol protocols. White dots indicate the individual cannula placements observed in this plane in this clinical trial. Bars = 1 cm. *Adapted with permission from Morrison PF, Lonser RR, Oldfield EH. Convective delivery of glial cell line-derived neurotrophic factor in the human putamen. J Neurosurg 2007;107(1):74-83.*

catheter[47] with superimposed continuous and bolus dosing was about twofold larger relative to the end-hole cases if backflow losses in the high-flow portion are absent, smaller if backflow losses are operative as the CED model predicts. Just based on the volumes of distribution in these on-target computations, it is thus difficult to account for the pattern of efficacies reported by the trial investigators, and other reasons were sought. One possibility stemmed from the observation that actual catheter placements in patients were variable and often near a white-gray tissue boundary, where the likelihood of backflow redistribution increases (Fig. 8). This placement distribution suggests that all trial data be reanalyzed after excluding all such "near-boundary" points to see if efficacy findings become consistent.

OVERVIEW AND ADDITIONAL CONSIDERATIONS

When modeling CED, distributed models that capture local drug distributions require careful formulation and design. Drug size, binding, and metabolism influence drug penetration characteristics. Transport characteristics of the targeted tissue region are important for determining whether idealized geometries are sufficient or if anatomically realistic 3D models are needed. In addition, CED apparatus and infusion protocols may play significant roles in distributing the drug. These factors influence local delivery to varying extents, and pharmacokinetic experimental data can help to determine which factors are essential to include and which factors may be neglected. For example, fully 3D models are not needed for infusion into large gray matter regions, as shown in the case of CED of QA into the GPi.

Once governing equations are formulated, drug and tissue transport properties must also be determined. Ideally, this information will be available in the form of independent experimental data, either in vivo or in vitro, from literature. However, quantitative data needed for these properties are often lacking due to experimental limitations. For example, few measures of hydraulic conductivity are available since in vivo flow and pressure gradients are difficult to measure within the brain interstitium. Parameter analysis can help to provide a property estimate, as has been done to determine anisotropy of hydraulic conductivity in white matter (ratios for K_{axial}/K_{trans}).[33] Ultimately, pharmacokinetic CED models are experimentally driven, thus improved means to measure model parameters is important for the future development of these models.

All the presented models assumed a continuous medium approach, which is suitable for CED models where drug molecules can freely transport within the extracellular space (estimated to be on the order of 20-60 nm[54]). Large agents may encounter hindrance effects once they can no longer freely move through this space. Another consideration is that experimental studies show large agents such as liposomes, viral vectors, and some proteins to localize along blood vessels, suggesting that they accumulate along perivascular spaces.[55-57] These spaces may also provide a rapid convective transport route[58] and considerations for these transport routes may be incorporated into future CED models. An additional factor that was not presented includes mechanical deformations. CED models that account for tissue swelling with pressurization can model the effect of local pore expansion on CED distributions.[59-62] Tissue deformation considerations are also important when considering backflow, and more advanced fluid–tissue interaction models focusing on predicting the extent of backflow have also been developed.[63, 64]

Magnetic resonance-based CED models that account for realistic anatomy and anisotropic transport also continue to be investigated[65-68] along with models that account for other neurological disease states (e.g., brain tumors).[69, 70] The necessity of incorporating pathological changes highlights the need to develop improved experimental methods for measuring heterogeneous drug and transport properties. As the pharmacokinetic principles of CED are better understood and confirmed under a variety of pathological conditions, predictive modeling should be feasible across a wide spectrum of neurological disease states. The ability to accurately predict drug distributions based on clinical imaging will be important for drug trial development.

References

1. Bobo RH, Laske DW, Akbasak A, Morrison PF, Dedrick RL, Oldfield EH. Convection-enhanced delivery of macromolecules in the brain. *Proc Natl Acad Sci U S A*. 1994;91(6):2076-2080.
2. Lonser RR, Sarntinoranont M, Morrison PF, Oldfield EH. Convection-enhanced delivery to the central nervous system. *J Neurosurg*. 2015;122(3):697-706.
3. Nduom EK, Walbridge S, Lonser RR. Comparison of pulsed versus continuous convective flow for central nervous system tissue perfusion Laboratory investigation. *J Neurosurg*. 2012;117(6):1150-1154.
4. Croteau D, Walbridge S, Morrison PF, et al. Real-time in vivo imaging of the convective distribution of a low-molecular-weight tracer. *J Neurosurg*. 2005;102(1):90-97.
5. Krauze MT, Saito R, Noble C, et al. Reflux-free cannula for convection-enhanced high-speed delivery of therapeutic agents - Technical note. *J Neurosurg*. 2005;103(5):923-929.
6. Oh S, Odland R, Wilson SR, et al. Improved distribution of small molecules and viral vectors in the murine brain using a hollow fiber catheter. *J Neurosurg*. 2007;107(3):568-577.
7. Lieberman DM, Laske DW, Morrison PF, Bankiewicz KS, Oldfield EH. Convection-Enhanced Distribution of Large Molecules in Gray-Matter during Interstitial Drug Infusion. *J Neurosurg*. 1995;82(6):1021-1029.

8. Morrison PF, Chen MY, Chadwick RS, Lonser RR, Oldfield EH. Focal delivery during direct infusion to brain: role of flow rate, catheter diameter, and tissue mechanics. *Am J Phys Regul Integr Comp Phys*. 1999;277(4): R1218-R1229.
9. Sykova E, Nicholson C. Diffusion in brain extracellular space. *Physiol Rev*. 2008;88(4):1277-1340.
10. Chen MY, Lonser RR, Morrison PF, Governale LS, Oldfield EH. Variables affecting convection-enhanced delivery to the striatum: a systematic examination of rate of infusion, cannula size, infusate concentration, and tissue-cannula sealing time. *J Neurosurg*. 1999;90(2):315-320.
11. Lonser RR, Gogate N, Morrison PF, Wood JD, Oldfield EH. Direct convective delivery of macromolecules to the spinal cord. *J Neurosurg*. 1998;89(4):616-622.
12. Morrison PF. Distributed models of drug kinetics. In: Atkinson AJ, ed. *Principles of Clinical Pharmacology*. 3rd ed; 2012:117-138.
13. Rapoport SI, Ohno K, Pettigrew KD. Drug entry into the brain. *Brain Res*. 1979;172(2):354-359.
14. Dedrick RL, Flessner MF. Pharmacokinetic considerations on monoclonal antibodies. In: Mitchell MS, ed. *Progress in Clinical and Biological Research*. New York: Alan Liss; 1989:429–438. Vol. 288.
15. Sarntinoranont M, Iadarola MJ, Lonser RR, Morrison PF. Direct interstitial infusion of NK(1)-targeted neurotoxin into the spinal cord: a computational model. *Am J Phys Regul Integr Comp Phys*. 2003;285(1):R243-R254.
16. Lonser RR, Corthesy ME, Morrison PF, Gogate N, Oldfield EH. Convection-enhanced selective excitotoxic ablation of the neurons of the globus pallidus internus for treatment of parkinsonism in nonhuman primates. *J Neurosurg*. 1999;91(2):294-302.
17. Morrison PF, Lonser RR, Oldfield EH. Convective delivery of glial cell line-derived neurotrophic factor in the human putamen. *J Neurosurg*. 2007;107(1):74-83.
18. Albin RL, Young AB, Penney JB. The functional-anatomy of basal ganglia disorders. *Trends Neurosci*. 1989; 12(10):366-375.
19. Alexander GE, Crutcher MD. Functional architecture of basal ganglia circuits—neural substrates of parallel processing. *Trends Neurosci*. 1990;13(7):266-271.
20. Bergman H, Wichmann T, Delong MR. Reversal of experimental parkinsonism by lesions of the subthalamic nucleus. *Science*. 1990;249(4975):1436-1438.
21. Delong MR. Primate models of movement-disorders of basal ganglia origin. *Trends Neurosci*. 1990;13(7):281-285.
22. Graybiel AM. Neurotransmitters and neuromodulators in the basal ganglia. *Trends Neurosci*. 1990;13(7):244-254.
23. Moroni F, Lombardi G, Carla V, Moneti G. The excitotoxin quinolinic acid is present and unevenly distributed in the rat-brain. *Brain Res*. 1984;295(2):352-355.
24. Wolfensberger M, Amsler U, Cuenod M, Foster AC, Whetsell WO, Schwarcz R. Identification of quinolinic acid in rat and human-brain tissue. *Neurosci Lett*. 1983;41(3):247-252.
25. Morrison PF, Laske DW, Bobo H, Oldfield EH, Dedrick RL. High-flow microinfusion—tissue penetration and pharmacodynamics. *Am J Physiol*. 1994;266(1):R292-R305.
26. Beagles KE, Morrison PF, Heyes MP. Quinolinic acid in vivo synthesis rates, extracellular concentrations, and intercompartmental distributions in normal and immune-activated brain as determined by multiple-isotope microdialysis. *J Neurochem*. 1998;70(1):281-291.
27. Fisher CE, Sutherland JA, Krause JE, Murphy JR, Leeman SE, vanderSpek JC. Genetic construction and properties of a diphtheria toxin-related substance P fusion protein: In vitro destruction of cells bearing substance Preceptors. *Proc Natl Acad Sci U S A*. 1996;93(14):7341-7345.
28. Eison AS, Iversen SD, Sandberg BEB, Watson SP, Hanley MR, Iversen LL. Substance-P analog, dime-C7—evidence for stability in rat-brain and prolonged central actioNS. *Science*. 1982;215(4529):188-190.
29. Benoliel R, Eliav E, Mannes AJ, Caudle RM, Leeman S, Iadarola MJ. Actions of intrathecal diphtheria toxin substance P fusion protein on models of persistent pain. *Pain*. 1999;79(2-3):243-253.
30. Mantyh PW, Rogers SD, Honore P, et al. Inhibition of hyperalgesia by ablation of lamina I spinal neurons expressing the substance P receptor. *Science*. 1997;278(5336):275-279.
31. Nichols ML, Allen BJ, Rogers SD, et al. Transmission of chronic nociception by spinal neurons expressing the substance P receptor. *Science*. 1999;286(5444):1558-1561.
32. Wiley RG, Lappi DA. Destruction of neurokinin-1 receptor expressing cells in vitro and in vivo using substance P-saporin in rats. *Neurosci Lett*. 1997;230(2):97-100.
33. Sarntinoranont M, Banerjee RK, Lonser RR, Morrison PF. A computational model of direct interstitial infusion of macromolecules into the spinal cord. *Ann Biomed Eng*. 2003;31(4):448-461.
34. Wood JD, Lonser RR, Gogate N, Morrison PF, Oldfield EH. Convective delivery of macromolecules into the naive and traumatized spinal cords of rats. *J Neurosurg*. 1999;90(1):115-120.

35. Sarntinoranont M, Chen XM, Zhao JB, Mareci TH. Computational model of interstitial transport in the spinal cord using diffusion tensor imaging. *Ann Biomed Eng*. 2006;34(8):1304-1321.
36. Basser PJ, Pajevic S, Pierpaoli C, Duda J, Aldroubi A. In vivo fiber tractography using DT-MRI data. *Magn Reson Med*. 2000;44(4):625-632.
37. Conturo TE, Lori NF, Cull TS, et al. Tracking neuronal fiber pathways in the living human brain. *Proc Natl Acad Sci U S A*. 1999;96(18):10422-10427.
38. Garland AM, Grady EF, Payan DG, Vigna SR, Bunnett NW. Agonist-induced internalization of the substance-P (NK1) receptor expressed in epithelial-cells. *Biochem J*. 1994;303:177-186.
39. Bowden JJ, Garland AM, Baluk P, et al. Direct observation of substance-p-induced internalization of neurokinin-1 (NK1) receptors at sites of inflammation. *Proc Natl Acad Sci U S A*. 1994;91(19):8964-8968.
40. Grady EF, Garland AM, Gamp PD, Lovett M, Payan DG, Bunnett NW. delineation of the endocytic pathway of substance-P and its 7-transmembrane domain NK1 receptor. *Mol Biol Cell*. 1995;6(5):509-524.
41. Sarntinoranont M, Iadarola MJ, Morrison PF. A kinetic analysis of substance P trafficking. *J Pharm Sci*. 2003; 92(2):232-243.
42. Yazdi PT, Murphy RM. Quantitative-analysis of protein-synthesis inhibition by transferrin-toxin conjugates. *Cancer Res*. 1994;54(24):6387-6394.
43. Prokopova S, Vargova L, Sykova E. Heterogeneous and anisotropic diffusion in the developing rat spinal cord. *Neuroreport*. 1997;8(16):3527-3532.
44. Grondin R, Zhang ZM, Yi A, et al. Chronic, controlled GDNF infusion promotes structural and functional recovery in advanced parkinsonian monkeys. *Brain*. 2002;125:2191-2201.
45. Kordower JH, Emborg ME, Bloch J, et al. Neurodegeneration prevented by lentiviral vector delivery of GDNF in primate models of Parkinson's disease. *Science*. 2000;290(5492):767-773.
46. Lang AE, Gill S, Patel NK, et al. Randomized controlled trial of intraputamenal glial cell line-derived neurotrophic factor infusion in Parkinson disease. *Ann Neurol*. 2006;59(3):459-466.
47. Patel NK, Bunnage M, Plaha P, Svendsen CN, Heywood P, Gill SS. Intraputamenal infusion of glial cell line-derived neurotrophic factor in PD: A two-year outcome study. *Ann Neurol*. 2005;57(2):298-302.
48. Slevin JT, Gerhardt GA, Smith CD, Gash DM, Kryscio R, Young B. Improvement of bilateral motor functions in patients with Parkinson disease through the unilateral intraputaminal infusion of glial cell line-derived neurotrophic factor. *J Neurosurg*. 2005;102(2):216-222.
49. Schaltenbrand G, Wahren W, Schaltenbrand G, Wahren W. *Introduction to Stereotaxis With an Atlas of the Human Brain*. Springer-Verlag; 1977.
50. Talairach J, Tournoux P, Talairach J, Tournoux P. *Co-planar Stereotaxic Atlas of the Human Brain. 3-Dimensional Proportional System: An Approach to Cerebral Imaging*. Thieme Medical Publishers; 1988.
51. Abbott NJ. Evidence for bulk flow of brain interstitial fluid: significance for physiology and pathology. *Neurochem Int*. 2004;45(4):545-552.
52. Salvatore MF, Ai Y, Fischer B, et al. Point source concentration of GDNF may explain failure of phase II clinical trial. *Exp Neurol*. 2006;202(2):497-505.
53. Hamilton JF, Morrison PF, Chen MY, et al. Heparin coinfusion during convection-enhanced delivery (CED) increases the distribution of the glial-derived neurotrophic factor (GDNF) ligand family in rat striatum and enhances the pharmacological activity of neurturin. *Exp Neurol*. 2001;168(1):155-161.
54. Thorne RG, Nicholson C. In vivo diffusion analysis with quantum dots and dextrans predicts the width of brain extracellular space. *Proc Natl Acad Sci U S A*. 2006;103(14):5567-5572.
55. Hadaczek P, Yamashita Y, Mirek H, et al. The "perivascular pump" driven by arterial pulsation is a powerful mechanism for the distribution of therapeutic molecules within the brain. *Mol Ther*. 2006;14(1):69-78.
56. Krauze MT, Saito R, Noble C, et al. Effects of the perivascular space on convection-enhanced delivery of liposomes in primate putamen. *Exp Neurol*. 2005;196(1):104-111.
57. Mamot C, Nguyen JB, Pourdehnad M, et al. Extensive distribution of liposomes in rodent brains and brain tumors following convection-enhanced delivery. *J Neuro-Oncol*. 2004;68(1):1-9.
58. Foley CP, Nishimura N, Neeves KB, Schaffer CB, Olbricht WL. Real-time imaging of perivascular transport of nanoparticles during convection-enhanced delivery in the rat cortex. *Ann Biomed Eng*. 2012;40(2):292-303.
59. Basser PJ. Interstitial pressure, volume, and flow during infusion into brain tissue. *Microvasc Res*. 1992;44:143-165.
60. Brady M, Raghavan R, Chen ZJ, Broaddus WC. Quantifying fluid infusions and tissue expansion in brain. *IEEE Trans Biomed Eng*. 2011;58(8).
61. Chen XM, Sarntinoranont M. Biphasic finite element model of solute transport for direct infusion into nervous tissue. *Ann Biomed Eng*. 2007;35(12):2145-2158.

62. Smith JH, Garcia JJ. A nonlinear biphasic model of flow-controlled infusion in brain: Fluid transport and tissue deformation analyses. *J Biomech*. 2009;42(13):2017-2025.
63. Garcia JJ, Molano AB, Description SJH. Validation of a finite element model of backflow during infusion into a brain tissue phantom. *J Comput Nonlinear Dyn*. 2013;8(1).
64. Raghavan R, Mikaelian S, Brady M, Chen Z-J. Fluid infusions from catheters into elastic tissue: I. Azimuthally symmetric backflow in homogeneous media. *Phys Med Biol*. 2010;55(1):281-304.
65. Stoverud KH, Darcis M, Helmig R, Hassanizadeh SM. Modeling concentration distribution and deformation during convection-enhanced drug delivery into brain tissue. *Transp Porous Media*. 2012;92(1):119-143.
66. Rosenbluth KH, Eschermann JF, Mittermeyer G, Thomson R, Mittermeyer S, Bankiewicz KS. Analysis of a simulation algorithm for direct brain drug delivery. *NeuroImage*. 2012;59(3):2423-2429.
67. Kim JH, Astary GW, Kantorovich S, Mareci TH, Carney PR, Sarntinoranont M. Voxelized computational model for convection-enhanced delivery in the rat ventral hippocampus: comparison with in vivo mr experimental studies. *Ann Biomed Eng*. 2012;40(9):2043-2058.
68. Linninger AA, Somayaji MR, Mekarski M, Zhang LB. Prediction of convection-enhanced drug delivery to the human brain. *J Theor Biol*. 2008;250(1):125-138.
69. Magdoom KN, Pishko GL, Rice L, Pampo C, Siemann DW, Sarntinoranont M. MRI-based computational model of heterogeneous tracer transport following local infusion into a mouse hind limb tumor. *PLoS One*. 2014;9(3).
70. Zhan WB, Arifin DY, Lee TKY, Wang CH. Mathematical modelling of convection enhanced delivery of carmustine and paclitaxel for brain tumour therapy. *Pharm Res*. 2017;34(4):860-873.

CHAPTER

8

Mechanisms and Clinical Applications of Stem Cell Therapy

Muhibullah S. Tora[*,†], *Kelly M. Poth*[*], *Nathan Hardcastle*[*], *Kaitlin Sandor*[*], *Casey Anthony*[*], *Long Di*[*], *Nicholas M. Boulis*[*,†]

[*]Department of Neurosurgery, Emory University School of Medicine, Atlanta, GA, United States
[†]Department of Biomedical Engineering, Georgia Institute of Technology, Atlanta, GA, United States

STEM CELL TECHNOLOGIES

Stem cell therapies have been a source of excitement in the field of regenerative medicine, especially in the context of disorders of the central nervous system (CNS), which are associated with limited regenerative potential. Stem cell technologies have been applied in preclinical animal disease models and clinical trials, many with promising data regarding safety and efficacy.[1] The increasing number of clinical trials for stem cell therapy for various indications has resulted in the creation of guidelines for defining and regulating these technologies in the preclinical setting.[2, 3] In this chapter, we will provide an overview of the common categories of stem cells, therapeutic mechanisms and clinical applications, and recent clinical trials for indications of the CNS.

Stem cell therapies in the CNS operate through a variety of mechanisms. These include (but are not limited to) neuroprotection, inflammatory modulation, and cell replacement.[4-8] Each of these effects is largely dependent on the specific type of stem cell technology applied, and each carries its own advantages, limitations, and practical implications. Commonly used stem cells for the CNS in laboratory and clinical settings include human pluripotent stem cells (hPSCs), mesenchymal stem cells (MSCs), and fetal-derived neural progenitor stem cells (fNPCs). The prominent features of these types of stem cells are summarized in Table 1.

https://doi.org/10.1016/B978-0-12-813997-4.00008-6
© 2019 Elsevier Inc. All rights reserved.

TABLE 1 Features of Common Stem Cell Technologies

Stem cell type	Source	Clinical applications	Neurosurgical delivery	Allogeneic or autologous	Advantages	Limitations
Human pluripotent stem cells						
Human embryonic stem cell	Embryonic	Cell replacement	Intraparenchymal	Allogeneic	Gold standard of differentiability	- Limited availability - Regulatory and ethical barriers - Tumorigenicity - Immunogenicity
Human-induced pluripotent stem cell	Somatic	Cell replacement	Intraparenchymal	Both	Adult somatic source	- Transcriptional and epigenetic aberrations - Immunogenicity - Tumorigenicity
Neural progenitor stem cell	Fetal, adult	Cell replacement	Intraparenchymal	Allogeneic	Robust direction to central nervous system cell fates	- Limited availability - Mixed cell populations - Immunogenicity
Mesenchymal stem cell	*Fetal* - Umbilical cord - Amniotic fluid *Adult* - Bone marrow - Adipose - Synovium - Skeletal muscle	Cell replacement *Neuroregulatory secretome:* - Neuroprotection - Neurogenesis - Immunomodulation	Intra-arterial Intravenous Intraparenchymal Intrathecal Intranasal	Both	- Lesion homing - Migration - Neuroregulatory secretome - Numerous adult sources	- Controversial molecular markers - Phenotypic differences between sources - Unclear migration mechanisms

Human Pluripotent Stem Cells

Overview

Since the first in vitro derivation of hPSCs in 1998, these stem cells have been the subject of active research in preclinical studies. However, they have failed to live up to expectations in the translational setting.[9] Nevertheless, hPSCs continue to be the subject of study due to their ability to differentiate into any human cell type, offering paradigms for cell replacement therapy in the CNS, with putative approaches to implementation summarized in Fig. 1. hPSCs still have a long way to go before they can be safely sourced and used; for this reason, few recent trials have implemented hPSCs. The two most studied categories of hPSCs are human

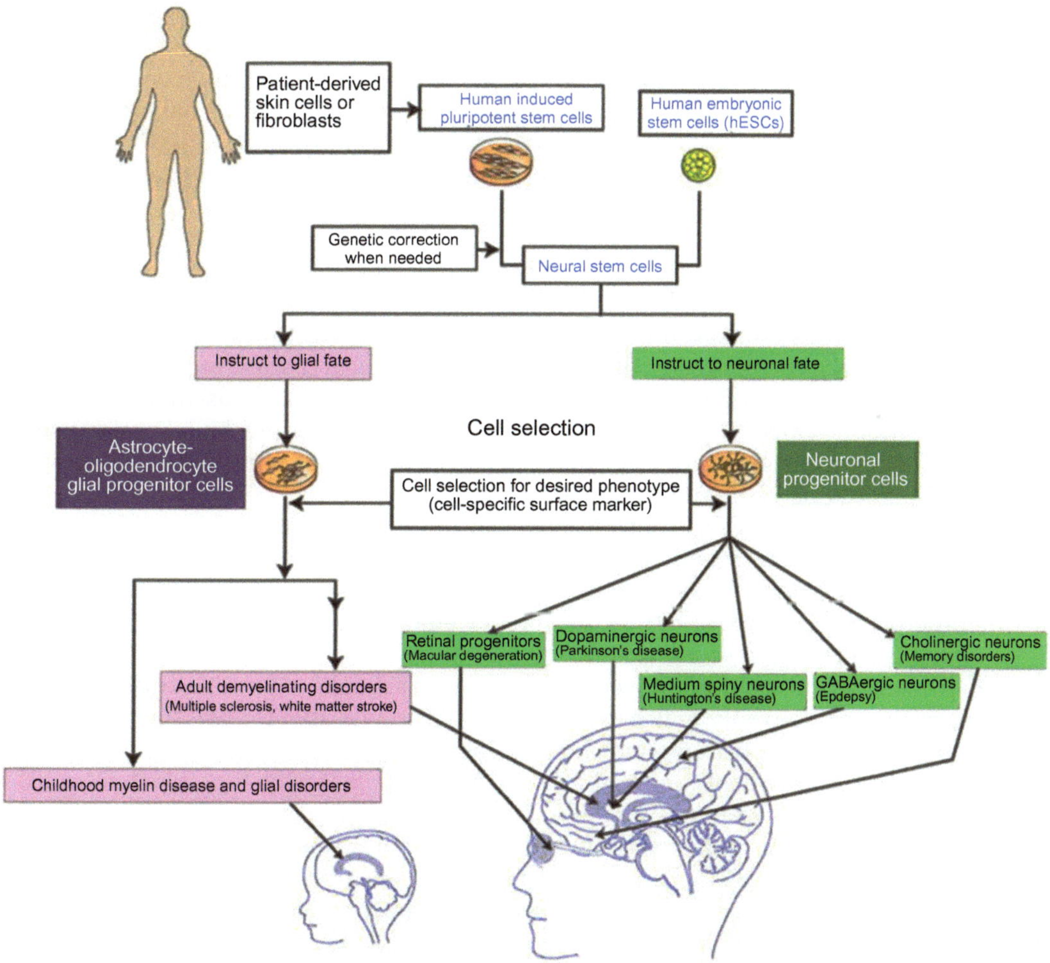

FIG. 1 Human embryonic stem cells (hESCs) and human induced pluripotent stem cells (hiPSCs) can be directed to terminal fates and can theoretically be employed for cell replacement in several disorders of the central nervous system, as seen here. The approaches largely require direct intraparenchymal injection to regions of pathology. *Reproduced with permission from Fox IJ, Daley GQ, Goldman SA, Huard J, Kamp TJ, Trucco M. Stem cell therapy. Use of differentiated pluripotent stem cells as replacement therapy for treating disease.* Science. *2014;345(6199):1247391.*

embryonic stem cells (hESCs) and human induced pluripotent stem cells (hiPSCs). Other types of hPSCs include parthogenetic[10] and human nuclear transfer embryonic stem cells[11]; present research on these types is primarily restricted to the preclinical setting.

Human embryonic stem cells are derived from human blastocysts and although they are considered the gold standard of differentiability, they are allogeneic to patients and can be immunologically rejected.[12] Furthermore, the requirement of embryo destruction for blastocyst harvesting leads to a practical barrier in the number of embryos required to treat each patient. Embryologic sourcing is controversial and has been restricted by funding restrictions, legal battles, and ethical concerns, all of which limit the application and development of hESC technology in the United States and the international community.[13, 14]

Human induced pluripotent stem cells circumvent many of the challenges faced by hESCs by induction of somatic cells to a pluripotent state. The first successful example of hiPSC production was in 2006, using retroviral delivery of transcription factors Oct4, Sox2, c-Myc, and Klf4 to mouse fibroblasts.[15] More efficient viral delivery using nonintegrative vectors, mRNA, chemical, and/or small-molecule induction of various adult somatic cells have proven to be crucial refinements in the production of hiPSCs, among other improvements.[16, 17] Means of production, classification, and epigenomic considerations of hiPSCs have been extensively discussed elsewhere.[12, 18]

Clinical Application and Therapeutic Mechanisms of Human Pluripotent Stem Cells

Human pluripotent stem cells are used extensively in the preclinical setting, where both hESCs and hiPSCs are considered powerful tools for preclinical disease modeling, including Rett Syndrome, Huntington disease, Parkinson disease (PD), spinal muscular atrophy, and pediatric gliomas.[19-22] hiPSCs are currently more commonly used in the laboratory setting than hESCs, as there is significant interest in using hiPSCs for in vitro modeling as well as translational paradigms using differentiated cells.[23]

Direct implantation of hPSCs is not a feasible paradigm, due to the unknown effects of the microenvironment following transplantation, the high risk of tumorigenicity, and inherent accumulated genetic defects introduced during expansion of undifferentiated lines.[18] The primary therapeutic mechanism for hPSCs is cell replacement therapy, where hPSCs are used to differentiate into a cell type of interest to ameliorate disease-related damage. The ideal approach for broad clinical application would require several banks of terminally differentiated, purified, and immunologically matched cell lines readily available for use, and either autologous or human-leukocyte-antigen matched allogeneic cells.

Under different culture conditions, hESCs can successfully derive functional neurons, oligodendrocytes, neural progenitors, and glial progenitors. Several examples of hESC-derived cell lines used in animal disease models include dopaminergic neurons for PD,[24] neural stem cells for spinal cord injury (SCI),[25] and oligodendrocyte precursors for remyelination and for SCI.[26] Notably, in rodent models of SCI, the intraparenchymal transplantation of hESC-derived oligodendrocyte and motor-neuron progenitors has been shown to promote robust functional recovery.[25, 27] In the clinical setting, one trial currently underway is using hESC-derived neural precursor cells for PD treatment.[28] In addition, a few clinical trials have been conducted for macular degeneration using hESC-derived retinal pigment epithelium.[29, 30] Overall, despite positive data in preclinical models and a handful of clinical trials for some indications, the clinical picture of hESC-derived cell lines is currently limited due to many of the aforementioned barriers.

The translational paradigm of hiPSCs is also primarily implemented through derivation of cell lines and the goal of banking human leukocyte antigen-matched cell populations for targeted intraparenchymal cell replacement therapy.[23] In rodent and nonhuman primate models of SCI, iPSC-derived cell lines have demonstrated successful restoration of function without tumorigenicity.[23] Notably, hiPSC-derived neural stem cells promoted functional recovery without tumorigenicity up to 12 weeks after transplantation in a marmoset model of SCI.[31] In hiPSCs, successful vector- and culture-driven induction methods have generated cell types for planned clinical trials, including neural stem cells, glial progenitor cells, oligodendrocyte precursors, retinal pigment epithelium, corneal endothelial and epithelial cells, and dopaminergic neurons.[23, 32-36] This is demonstrative of the rapid advancement of hiPSC technology.

Despite significant interest and increasing use of hiPSCs, there are still limitations in production technology[12, 23] and concerns of immunogenicity and tumorigenicity due to limited quantity and quality of data.[37-40] It is possible that with the continued rapid improvement of hiPSC technologies, there will be an increase in the number of clinical trials involving hiPSC-derived cell lines for cell replacement therapy in the near future.

Fetal-Derived Neural Progenitor Cells

Overview

Fetal-derived neural progenitor cells are resident cells that can be found in the fetal brain and spinal cord. These cells have been the subject of interest in the CNS due to their capacity to differentiate into CNS cell fates by using protocols that mimic developmental conditions with precise culture conditions.[41] Thanks to efficient production methodologies, including dual SMAD inhibition,[42] fNPCs have been directed toward neurons, glia, and several neuroectodermal cell types.[41, 43-46] Adult sources of neural progenitors are also available, but these are currently limited due to difficulty in culturing and characterization, albeit with rising improvement.[47] For this reason, the following discussion will focus on fNPCs and their more present clinical application.

It should be first noted that a major drawback of fNPCs is their limited availability. Even once fetal tissue is obtained, some translational paradigms require up to 10 aborted fetuses per patient.[41, 48] Compounding this limitation is the fact that no current production methodology can direct fNPCs to a purified cell fate, limiting the yield of a desired cell type. Indeed, the majority of production methodologies result in mixed populations, regardless of intended cell fate, which requires further purification.[49-51] Additionally, harvested fetal neural tissue is composed of a heterogeneous group of cells, of which as few as 2% of cells can be used for fNPC applications.[41] One method of circumventing this limited availability is conditional immortalization and expansion of lines, which have been used in industry-driven clinical trials (e.g., Neuralstem, Inc.; ReNeuron, StemCell Technologies).[41] The production, limitations, and refinement of fNPCs have been well described elsewhere.[41, 52] The take-home message regarding fNPCs is that they are a limited source of cells, with further refinement needed. Nevertheless, this has not prevented significant preclinical and clinical implementation.

Therapeutic Mechanisms and Clinical Applications of Fetal-Derived Neural Progenitor Cells

The translational paradigm for fNPCs is primarily focused on cell replacement therapy. Although uncommon, fNPCs can be genetically modified to increase therapeutic gene expression, thereby generating a more potent therapeutic effect.[53] fNPCs have been delivered through intraparenchymal injection in both rat and primate models of SCI.[54-56] These studies have shown demonstrable motor improvement, acceptable safety profiles, and immunohistochemical evidence of nerve fiber growth across the injured region. In rodent models of amyotrophic lateral sclerosis (ALS), fNPCs have been used for cell replacement and with genetic modifications for secretion of brain-derived neurotrophic factor. In these studies, rodents demonstrated survival of motor neurons, albeit with mixed results regarding motor improvement and mortality.[57, 58] Similar studies have been implemented in rodent and primate models of PD,[59, 60] stroke,[61-63], traumatic brain injury (TBI),[64, 65] and Huntington disease[66] with mixed but overall promising results.

The preclinical landscape has paved the way for clinical trials with fNPC-derived cell lines, especially for indications in which a focal region is a feasible site for intraparenchymal injection. Over the past 5 years, clinical trials have focused on indications such as SCI,[67] stroke,[68, 69] and ALS.[70, 71] Although therapeutic efficacy of fNPCs varied in these studies, one of the primary outcomes was the demonstration of the safety of intraparenchymal injection. For the spinal cord in particular, researchers were concerned that focal intraparenchymal injection could cause untenable injury, inflammation, and other adverse effects. In the first human clinical trial for stem cell transplantation in 12 patients with ALS, injection of fNPCs to the ventral horn was deemed safe in all patients, with no detectible motor function impairment.[72]

For further implementation, there are several areas in which fNPC technology could improve. First and foremost, the dose and quality of manufacturing practices in these studies varied widely. This calls for refined, standardized methodologies. Second, these clinical trials do not address the issue of immunogenicity due to the allogeneic nature of fNPCs. These studies have used different drug regimens and have discontinued immunosuppression at different time points due to drug-related toxicity. This has likely affected the efficacy of these trials and has highlighted the importance of establishing a clear immunosuppressive regimen for CNS implantation—keeping in mind the immunosuppressive regimen for these patients will likely differ from those used in recipients of solid organ transplants.[40, 73, 74] Lastly, there is a significant limitation in the ability to track survival of transplanted cells in patients, which is critical for evaluating the therapeutic efficacy and targeted delivery of these cells. As a means to address this, several preclinical studies have demonstrated long-term tracking with single-photon emission computed tomography and magnetic resonance imaging labeling technologies, which may be useful in future trials.[75-77] Overall, limitations notwithstanding, fNPCs have a relatively safe clinical profile, and they continue to be used in clinical trials for numerous indications.

Mesenchymal Stem Cells

Overview

Mesenchymal stem cells are nonhematopoietic, multipotent stem cells that have been the subject of great interest due to their self-renewing capacity. MSCs have been applied in animal models in the CNS, cardiac, hepatic, renal, corneal, and hematologic settings.[78] With regard to the CNS, MSCs have been shown to express characteristics of neuroectodermal lineage after direct administration to the brain and spinal cord in animal models,[52, 79] and can be transdifferentiated to neuronal or glial lineages in vitro.[78] MSCs also have the ability to migrate to lesions following systemic or intrathecal administration in a variety of indications and have a number of secreted proteins (i.e., the MSC secretome) with therapeutic effects.[80]

Mesenchymal stem cells can be obtained from fetal sources, but the current standard in the field is MSCs derived from adult patients for allogeneic or autologous transplantation. For example, bone marrow contains precursor cells that can be isolated and expanded into nonhematopoietic MSCs, termed *bone marrow-derived mesenchymal stem cells* (BMMSCs). A heterogeneous mixture of cells from bone marrow that include BMMSCs are called *bone marrow mononuclear cells* (BMMNCs); these can be applied for therapeutic purposes with or without further purification.[81] In addition to bone marrow, MSCs can also be sourced from the Wharton jelly of the umbilical cord (UC-MSCs), adipose tissue (ASCs), skeletal muscle, peripheral blood, synovial fluid, and amniotic fluid.[82] It should also be noted that, unlike nonhematopoietic MSCs, hematopoietic stem cells (HSCs) in the circulation are a major source of interest for implementation in disorders of the CNS that involve autoimmune dysfunction, such as multiple sclerosis (MS).[83]

Broadly, the most commonly used nonhematopoietic MSCs in preclinical and clinical settings currently include BMMSCs, ASCs, and UC-MSCs. Although these MSCs vary phenotypically, they represent practically available MSCs for preclinical and translational work. One major limitation of MSCs is that no definitive molecular markers define them. Despite this, the Committee of the International Society for Cellular Therapy has proposed criteria for their identification.[84]

Clinical Application and Therapeutic Mechanisms for Mesenchymal Stem Cells

The wide availability of MSCs makes them appealing for use in CNS stem cell therapy. MSCs can be transdifferentiated into endodermal and ectodermal lineages, including neural progenitor cells, neuron-like cells, and other CNS cell fates.[78, 82, 85, 86] MSC therapy can be used for cell replacement therapy, but this application is limited by the lack of functional and phenotypic characterization of terminally differentiated lines. Instead, the most notable paradigm involves direct use of MSCs through intrathecal, intravascular, intraparenchymal, or intranasal administration. This translational paradigm can be attributed to two major characteristics of MSCS: (1) the neuroregulatory MSC secretome and (2) the capacity for MSC migration and homing to a lesion of interest, outlined in Fig. 2.

The MSC secretome releases factors that can elicit neurogenesis, neuroprotection, cell survival, immunomodulation, and limitation of glial scar formation.[80] MSCs are known to facilitate sustained delivery of these neuroregulatory factors in both animal models and in humans, defining the most notable approach for MSC therapy in the CNS.[80, 87] The secretome of MSCs consists of a neuroregulatory milieu that has been demonstrated both in vitro and in vivo to release factors such as BDNF, beta-NGF, IGF-1, HGF, VEGF, TGF-β, FGF-2, and GDNF under different conditions. The paracrine milieu results in several effects (Fig. 3). Although a robust analysis of the secretome of MSCs under various conditions has yet to be carried out, this has not prevented broad implementation. MSC-based therapies have been

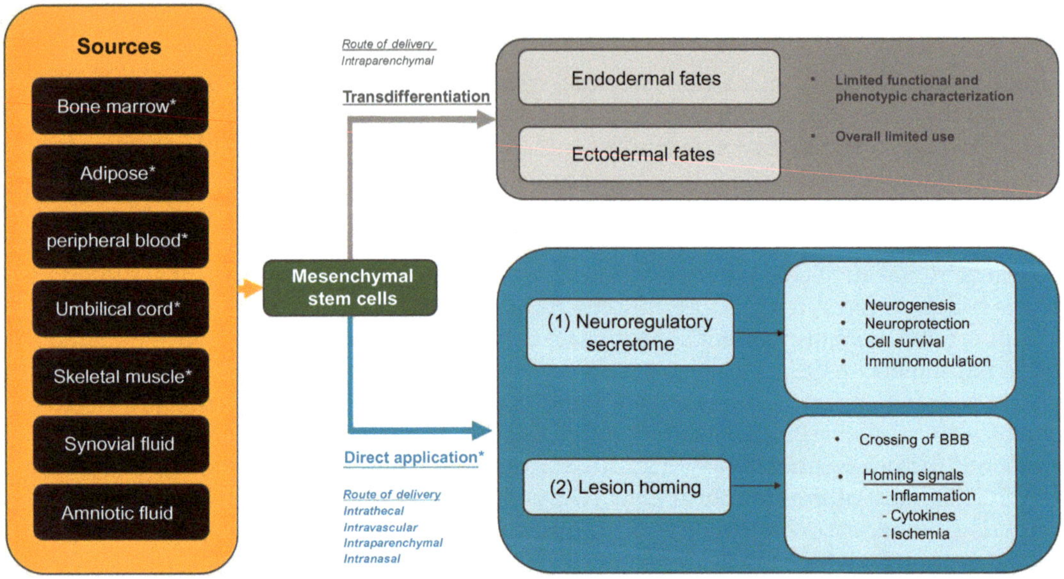

FIG. 2 Sourcing and clinical application of mesenchymal stem cells (MSCs). MSCs have the most widely available allogeneic and autologous sources, and can be implemented either by direct application via transdifferentiation to terminal fates, or by direct application by taking advantage of the (1) neuroregulatory secretome of MSCs, and (2) the native capacity for lesion homing. This represents a vast difference in the delivery routes for MSCs depending on the route of use. **Most commonly used.*

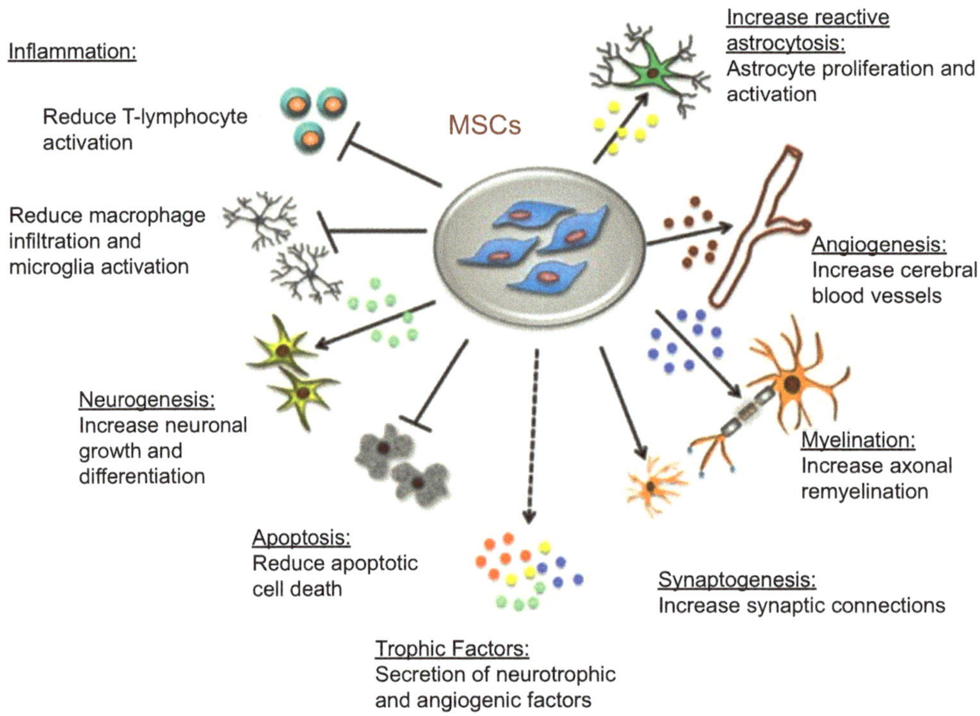

FIG. 3 Effects of the mesenchymal stem cell (MSC) neuroregulatory secretome. The secretion of neuroregulatory factors by MSCs, termed the "MSC secretome," has been demonstrated both in vivo and in vitro. Secreted factors, including BDNF, Beta-NGF, IGF-1, HGF, VEGF, TGF-B, FGF-2, GDNF, and others, are responsible for much of the paracrine effects described in this chapter. *Reproduced with permission from Castillo-Melendez M, Yawno T, Jenkin G, Miller SL. Stem cell therapy to protect and repair the developing brain: a review of mechanisms of action of cord blood and amnion epithelial derived cells.* Front Neurosci. *2013;7:194.*

applied in numerous preclinical studies and clinical trials for PD, stroke, ALS, SCI, TBI, MS, and epilepsy, with promising results.[80, 88-94]

The capacity for MSC migration to a focal lesion of interest or in diffuse disease processes is particularly appealing for intrathecal or systemic administration. Although this capacity for migration is well known, the mechanisms and biodistribution of MSC migration are the subject of debate. Some migratory mechanisms include homing signals from the lesion (e.g., inflammation, cytokines, and ischemia), native capacity for crossing the blood–brain barrier, and endothelial transmigration through expression of cell adhesion molecules.[95, 96] With regard to local migration, one group of researchers who examined direct transplantation of MSCs to the striatum of a healthy primate model reported relatively uniform anteroposterior distribution in the brain.[97]

Use of MSCs in preclinical studies has primarily depended on histopathologic and immunohistochemical techniques. This raises a practical issue for tracking MSC distribution in clinical trials; however, with the advent of novel labeling technologies, this concern is actively being addressed with radiolabeling and magnetic labeling.[95, 98] Despite unclear migratory mechanisms and the paucity of proven cell-tracking paradigms, it should be stated that MSCs do in fact migrate to areas of ischemia and inflammation.[96] Additional preclinical studies and tracking of cell fate in clinical trials should be performed to elucidate these processes and clarify long-term MSC survival, even though MSCs have already been implemented in clinical trials for CNS indications.

Of all types of stem cells, MSCs are the best-tolerated for both autologous and allogeneic applications in humans. The main delivery methods for MSCs include intravascular (i.e., intravenous or intra-arterial) and intraparenchymal administration. For intravascular administration of MSCs, tolerable doses are known and few significant acute and chronic adverse events have been attributed to MSCs in humans.[84] A meta-analysis of 36 studies and 1013 patients with intravenous or intra-arterial administration showed no significant correlation between MSC therapy and acute infusional toxicity, organ system complications, death, or malignancy; in fact, the sole significant finding was transient fever after systemic administration.[99] Despite these promising results, it should be noted that MSCs in animal models are associated with the risk of MSC accumulation in the lungs after intravenous delivery, due to VCAM-1/VLA-4 interactions.[96] This accumulation can lead to embolic phenomena, MSC sequestration, and decreased therapeutic efficacy. Administration of MSCs via the intra-arterial route seems to circumvent this problem.[96, 100] These findings suggest that intravascular administration is a feasible approach, with some concerns for off-target sequestration, which may be avoided with intra-arterial administration. This approach should be considered in future clinical studies of intravascular MSC delivery.

Regarding intrathecal administration, there is preclinical safety data for high-dose allogeneic MSC administration in an equine model, which demonstrated no adverse effects.[101] Furthermore, five individual clinical trials from 2012 to 2017 administered MSC intrathecally and found no adverse effects in patients with ALS, MS, or SCI after follow-up of up to 2 years.[91, 102-105] The summary of these safety data is very promising, but it should be noted that, as with any stem cell therapy, intrathecal administration carries potential risks for infection, ectopic tissue formation, and the theoretical risk of tumorigenesis—which should be monitored in any study implementing MSC therapy, regardless of administration type.

STEM CELL THERAPIES FOR INDICATIONS OF THE CENTRAL NERVOUS SYSTEM

Stem cell therapies have been implemented in numerous clinical trials for indications of the CNS, including TBI, SCI, PD, MS, and ALS. We examine each of these indications in turn, considering the use of stem cells as therapy and reviewing past or ongoing clinical trials stem cell therapies for each indication.

Traumatic Brain Injury

Overview

Traumatic brain injury is the all-encompassing term for sudden injury from a blow to the head that results in cognitive dysfunction.[106] TBI severity is typically classified using the Glasgow Coma Scale (GCS), with mild injury measured as GCS score of 13-15, moderate as GCS score of 9-12, and severe as GCS score of anything less than 9.[107, 108] Currently, it is estimated that the global incidence of mild TBI exceeds roughly 600 per 100,000 people, or 42 million cases annually.[109] In the United States, the average annual incidence of TBI is estimated at 1.7 million cases.[110] This represents a significant burden of disease, where up to 43.3% will develop long-term, TBI-related disability.[111, 112] The most frequent symptoms associated with mild TBI involve headache, dizziness, fatigue, and photosensitivity.[113] The pathophysiology of TBI includes direct tissue damage,[114] immediate loss of cerebral blood flow to affected areas,[114] and disrupted cellular metabolism.[115] Delayed features of TBI include elevated intracranial pressure,[116] cortical spreading depolarization,[117] and inflammation.[118-120]

Effective treatments for TBI are currently limited, and the standard of care is focused on promoting survival and long-term symptom management.[121] However, despite improvements in trauma response systems and surgical techniques, the mortality rate for severe TBI remains relatively high, at approximately 25%.[122] The paucity of effective restorative treatments represents a significant gap for which stem cell therapies may be considered.

Stem Cell Therapies for Traumatic Brain Injury

Preclinical studies have demonstrated that exogenous stem cell transplants in animal models of TBI can reduce cytotoxic neuroinflammation,[123] protect against axonal degeneration,[124, 125] promote angiogenesis,[126, 127] decrease accumulation of neurodegenerative proteins,[128] and improve cognitive recovery.[129, 130] Consequently, several clinical trials examining the use of stem cells for the treatment of acute or subacute TBI have been published.[131-133] Recent clinical trials have used cell types such as BMMNCs, BMMSCs, and ASCs.

Two notable studies by Cox and colleagues used autologous BMMNCs in adults and children with severe TBI. The first trial was a nonrandomized, open-label Phase I clinical trial of 10 children with acute severe TBI.[132] The authors reported neither significant hemodynamic changes after intravenous infusion nor infusion-related toxicity. No treatment-related mortality was reported, and both functional and neuropsychological assessments at follow-up demonstrated improved outcomes in all patients. Strikingly, seven of ten patients demonstrated near-normal functioning. However, the study was underpowered and was not designed to conclusively declare significant differences with stem cell treatment. In a similar Phase I/IIa study of 25 adults, intravenous BMMNCs improved functional outcomes at follow-up without significantly affecting lesion size, in concordance with preclinical studies.[131] Interestingly, radiographic findings demonstrated that BMMNC transplant led to significantly greater white matter preservation, which correlated with better functional outcomes. Though underpowered, these results demonstrate promising results for realistic implementation of cellular therapies for TBI. These studies have also resulted in two additional dose escalation clinical trials.

Several ongoing clinical trials have implemented other cell types for TBI. A Phase I/II trial is implementing intravenous autologous ASCs for concussive TBI, with up to 200 participants planned. Another Phase II trial is examining the contribution of delivery mechanisms on neurological function by using intravenous and intranasal delivery of autologous BMMSCs for several indications, including TBI. Lastly, a double-blinded Phase II, placebo-controlled trial with 52 patients is evaluating the efficacy of modified autologous MSCs in patients with chronic TBI. An overview of these recent clinical trials for stem cell therapies in TBI is outlined in Table 2.

The confluence of these promising published results, although largely underpowered, and ongoing clinical trials represents the increasing use of stem cell therapies in patients with TBI. The knowledge gained from preclinical research has contributed to a steady growth in the translation of stem cell therapy to the clinic yielding promising results. Ultimately, additional clinical investigations that are sufficiently powered, randomized, appropriately designed, and with dose escalation are needed to further elucidate the role of stem cell therapy in the management of TBI.

Traumatic Spinal Cord Injury

Overview

On average, 12,000 new traumatic SCI cases occur annually in the United States. Based on data from the national epidemiologic consortiums, that number is expected to increase to 13,600 in 2020; 14,960 in 2030; and 16,240 in 2040.[134] As the rate of new injuries increases and life expectancy improves, the number of people living with chronic spinal cord-related injuries is also expected to increase. Current estimates of prevalence are approaching 300,000. Symptoms associated with traumatic SCI include long-term paralysis, bladder dysfunction, sexual dysfunction, neuropathic pain, paresthesias, and other symptoms depending on the vertebral level of injury. There are currently no effective therapies for traumatic SCI, and the standard of care now relies on physical and occupational therapy.[135] This represents a lack of restorative treatment options for traumatic SCI, and stem cell therapies may prove useful in this realm.

Stem Cell Therapies for Traumatic Spinal Cord Injury

Stem cell therapies for SCI have undergone promising clinical trials in the United States. Several companies have demonstrated that stem cells can be safely delivered to the spinal cord in patients with SCI, and some have reported preliminary improvements in early testing

TABLE 2 Cell Therapies in Recent Clinical Trials for Traumatic Brain Injury

Trial ID	Year initiated	Status	Cell type	Delivery route	Details
NCT02959294	2016	Phase I/II, recruiting	Autologous ASC	Intravenous	Study will examine the safety and efficacy of parenteral introduction of ASCs in adults and adolescents with concussive TBI, intending to enroll 200 patients
NCT02795052	2016	Phase II, recruiting	Autologous BMMSC	Intravenous Intranasal	Study investigating the effect of intravenous and intranasal administration of BMMSCs on neurological function in patients with a variety of neurological disorders including TBI
NCT02416492	2016	Phase II, active	Autologous MSC	Intraparenchymal	Double-blinded, controlled trial with 52 patients evaluating the efficacy of autologous BMMSCs modified to overexpress neuroprotective trophic factors, termed SB623 cells, on patients with chronic TBI
NCT02525432	2016	Phase IIb, enrolling	Autologous BMMNC	Intravenous	Randomized, double-blinded study with 55 patients evaluating intravenous autologous BMMNC dose escalation on structural and functional outcomes in adults with acute sTBI
NCT01851083	2013	Phase I/II, enrolling	Autologous BMMNC	Intravenous	Multicenter, randomized, blinded, placebo-controlled dose escalation study with 50 patients, evaluating effect of intravenous BMMNC infusion on structural and functional outcomes in children with acute sTBI
NCT01575470	2012	Phase I/II, completed	Autologous BMMNC	Intravenous	Investigating the effect of BMMNC transplant on brain structure preservation and inflammation after TBI. Patients aged 18-55 years received 6×10^6, 9×10^6, or 12×10^6 BMMNCs/kg body weight. Transplant improved white matter integrity in the corpus callosum and corticospinal tract, which correlated with better functional and neuropsychological outcome. A dose-dependent mitigation of the inflammatory response was also seen at 1-month follow-up[131]
NCT02028104	2010	Phase I, completed	Autologous BMMNC	Intrathecal	Assessment of safety and efficacy of intrathecal BMMNCs administration on patients with TBI; patients ranged in age from 6 months to 65 years. Results unpublished
NCT00254722	2006	Phase I, completed	Autologous BMMNC	Intravenous	Study aimed to examine the safety and feasibility of BMMNC transplantation in children 5-14 years of age after isolated TBI. Patients received a single intravenous infusion of 6×10^6 BMMNCs/kg body weight within 36 hours of injury. Follow-up indicated functional improvement, although the study is underpowered. Study has progressed to phase II (NCT01851083). BMMNC transplant did not result in any significant infusion-related toxicity[132]

Abbreviations: ASC, *adipose tissue stem cell;* BMMNC, *bone marrow mononuclear cell;* BMMSC, *bone marrow-derived mesenchymal stem cell;* MSC, *mesenchymal stem cell;* sTBI, *severe traumatic brain injury;* TBI, *traumatic brain injury.*

phases. The mechanism underlying the ability of stem cells to improve outcomes after SCI is still the subject of debate; therefore, preferred cell types, ideal delivery route, and severity of injury vary widely among studies. The types of cells currently under investigation include BMMSCs (both autologous and ex vivo expanded), oligodendrocyte progenitor cells (OPCs) from several sources, hESC-derived cell lines, and fNPCs.

In recent years, several industry-driven clinical trials have been carried out. In Phase I/II clinical trials, Stem Cell Inc. completed manual cervical and thoracic injections of fNPCs. No adverse events were attributed to cell delivery. Despite these findings, patient improvements in the first cohort fell below the threshold for clinical efficacy and the trials were terminated in 2016.[136] Neuralstem Inc. began Phase I testing of fNPCs for chronic SCI in 2014. Four participants received intraparenchymal spinal cord injections. They reportedly tolerated the procedure well and had no adverse events related to the stem cell grafts.[137] Asteria Biotherapeutics completed enrollment for their Phase I/II trial after surpassing their 12-month efficacy target in patients receiving OPCs. Preliminary results suggest tolerability of OPCs, and four of six patients demonstrated motor improvement, with data expected to be announced in 2018 for all 30 patients. Recent clinical trials involving SCI are outlined in Table 3.

Each of these trials sheds light on the feasibility and safety of stem cell delivery to the spinal cord, but questions remain about choice of cell origin and potency, delivery route, and whether some injury types are more amenable to stem cell therapy than others. As previously mentioned, the mechanism underlying spinal cord repair by stem cell treatment is still under investigation. Aside from cell replacement, stem cell therapies can promote healing in SCI by promoting neurogenic signaling and by inhibiting the negative inhibitory processes in the adult spinal cord that stunt chronic tissue repair. Preclinical models have revealed that during the "acute phase" of SCI, endogenous healing processes are initiated, including growth factor release from surviving astrocytes, oligodendrocyte remyelination, and perilesional angiogenesis. However, the adult spinal cord maintains an inhibitory microenvironment that precludes neurogenic signals from persisting beyond the acute injury stage. This results in a second "chronic stage," which is characterized by a cascade of inhibitory processes leading to inflammation, cell death, and tissue scarring.[140]

Future experiments must ask which of these processes is most critical for spinal cord repair. Additionally, matching the correct cell therapy to the proper indication will require an understanding of the contrasting phases of SCI and which therapies may be best suited to each. Certain cells may mimic the permissive acute phase and allow endogenous healing to persist, whereas other types may thrive within the inhibitory milieu of the chronically injured spinal cord, making them a better choice for cell replacement.

Parkinson Disease

Overview

Parkinson disease is a progressive neurodegenerative movement disorder with gradual onset and progression. Most cases are sporadic and appear to be age-related, with onset usually occurring after 50 years of age. The incidence of PD is estimated to be 10-18 per 100,000 people annually.[141] PD classically presents with tremor, rigidity, and bradykinesia.

TABLE 3 Cell Therapies in Recent Clinical Trials for Spinal Cord Injury

Trial ID	Year initiated	Status	Cell type	Delivery route	Details
NCT03225625	2017	Phase II, recruiting	Autologous BMMSC	Intrathecal Intravenous Intranasal	40 patients with SCI enrolled. Study evaluating the use of combination therapy with adjuvant delivery routes and exoskeletal and/or virtual reality augmentation.
NCT02981576	2016	Phase I/II, active, not recruiting	Autologous BMMSCs & ASCs	Intrathecal	14 patients with SCI enrolled, investigating safety and efficacy of BMMSC transplantation.
NCT02917291	2016	Phase I/II, recruiting	Allogeneic ASCs	Intramedullary	Evaluation of safety of single administration of FAB117-HC, allogeneic ASCs expanded and pulsed with H_2O_2. 46 patients enrolled, administration of therapy will be performed during decompression and stabilization of vertebral fractures.
NCT02570932	2015	Phase II, completed	Autologous BMMSCs	Intrathecal	Single-center, nonrandomized, uncontrolled study in patients with chronic SCI. 12 patients enrolled, each received three administrations of autologous BMMSCs that were spaced by 3 months. At 24-month follow-up, no adverse events were reported. Improvements in primary sensitivity, sphincter control. Motor improvement based on neurophysiologic studies.[138]
NCT02302157	2015	Phase I/II, active, not recruiting	AST-OPC1 cells (oligodendrocyte progenitor cells)	Intraparenchymal	Single-center, nonrandomized, uncontrolled study in patients with subacute SCI. 35 patients enrolled. Dose escalation study with 12-month follow-up, evaluating safety and neurologic improvement with AST-OPC1 transplantation.
NCT01772810	2014	Phase I, recruiting	Allogeneic fNPCs	Intraparenchymal	Single-center, open-label, nonrandomized study in patients with chronic SCI. Eight patients enrolled to date. Evaluating safety, graft survival, and efficacy of immunosuppression following allogeneic fNPC transplantation.
NCT01694927	2012	Phase II, unknown	Autologous BMMSCs and ASCs	Intraparenchymal	Evaluation of the safety and efficacy of Autologous BMMSCs and ASCs in patients with SCI. 25 patients enrolled. Related pilot study reported no adverse events and no reports of efficacy at 2 years, suggestion by authors for higher-dose studies.[139]

TABLE 3 Cell Therapies in Recent Clinical Trials for Spinal Cord Injury—cont'd

Trial ID	Year initiated	Status	Cell type	Delivery route	Details
NCT01730183	2012	Phase I/II, complete	BMMSCs	Intrathecal	Multicenter, nonrandomized, open label study to evaluate the safety and efficacy of autologous BMMSC transplantation in patients with SCI. 15 patients enrolled.
NCT01873547	2012	Phase III, completed	UC-MSCs	Intrathecal	Prospective, multicenter, randomized trial. 300 participants with 100 receiving MSCs, 100 with rehabilitation only, and 100 who receive neither intervention. Evaluating safety and efficacy of intrathecal UC-MSC transplantation. Results unpublished.

Abbreviations: ASC, *adipose tissue stem cell;* BMMSC, *bone marrow-derived mesenchymal stem cell;* fNPC, *fetal-derived neural progenitor stem cell;* MSC, *mesenchymal stem cell;* SCI, *spinal cord injury;* UC-MSC, *umbilical cord mesenchymal stem cell.*

Other symptoms include diminished facial expression, stooped posture, or neuropsychiatric disturbances. The pathogenesis of PD is attributed to the loss of dopaminergic neurons in the substantia nigra, resulting in reduced dopamine in the striatum.

Currently, there are no disease-modifying therapies available for PD. Treatment is centered on symptomatic relief and is initiated when symptoms begin to affect daily activities. Dopamine replacement therapy has high initial efficacy in controlling motor symptoms, but its long-term efficacy is questionable. Dopamine receptor agonists are used in younger patients or during later stages of disease.[141, 142] Deep brain stimulation (DBS) has been shown to be an effective treatment in several clinical trials.[141] However, while extremely effective, DBS is not without risks and limitations, and is typically reserved for mid- to late-stage disease, when severe motor symptoms cannot be controlled by medication.[143] With the lack of disease-modifying agents for PD, an unmet need exists for therapies that can slow or even reverse the progression of the disease.

Stem Cell Therapies for Parkinson Disease

Although the first cell therapy studies in PD were performed over 20 years ago, current stem cell studies in PD are still limited in number and early in development.[144] Strategies for PD therapy include cell replacement of dopaminergic neurons, neuromodulation, and neuroprotection to halt disease progression. Today, trials are exploring a variety of cell sources, including fNPCs, allogeneic BMMSCs, and both hiPSCs and hESC-derived neural precursors. Clinical trials for stem cell therapies in PD are summarized in Table 4.

Several recent trials have investigated stem cell therapies for PD. The ongoing open-label, multinational TRANSEURO trial studying fetal ventral mesencephalic neuronal transplant has generated interest, given its scope. The goal of the trial is to optimize manufacturing, trial design, and patient selection for future trials in PD. As of April 2017, 150 patients have been

TABLE 4 Cell Therapies in Recent Trials for Parkinson's Disease

Trial ID	Year initiated	Status	Cell type	Delivery route	Details
NCT03309514	2018	Phase I/II, not yet recruiting	Neural stem cell-derived neurons	Intraparenchymal	To determine the safety and efficacy of intracerebral microinjections of differentiated neurons derived from CNS progenitor cells. Specifics of cell sourcing not published
NCT02611167	2017	Phase I/II, recruiting	Allogeneic MSCs	Intravenous	To determine the safety and efficacy of IV MSCs as a therapy. Allogeneic MSCs will be delivered in escalating doses
NCT03119636	2017	Phase I/II, recruiting	hESC-derived neural progenitor cells	Intraparenchymal	Researchers will enroll 50 patients to evaluate the safety and efficacy of a single injection of NPCs
NCT03128450	2017	Phase II/III, enrolling by invitation	fNPC	Intranasal	Exploring a novel nasal injection delivery system. 12 patients will receive once-weekly dosing for 4 weeks
NCT02452723	2016	Phase I, recruiting	hpNSCs	Intracerebral	Data from the first cohort were presented at a scientific conference in 2017. Early efficacy was seen with improvements in dyskinesia symptoms and responses to levodopa. Given the low number of patients treated at the time of analysis, the data were not yet statistically significant. Cyto Therapeutics, the company running the trial, reported that they are now enrolling patients in the third cohort as of June 2018
NCT01898390	2012	Phase I, active, not recruiting	Allogeneic Fetal ventral-mesencephalic tissue	Intraparenchymal	To assess the safety and efficacy of allogeneic fetal VM tissue transplantation. The study has an observational portion in addition to the transplant cohort. As of April 2017, 150 patients had been recruited and 13 had received transplants, but data are not yet available.[145] The study plans to transplant 20 patients. In 2016, the trial had reported unexpected problems that significantly delayed the trial

Abbreviations: CNS, *central nervous system;* fNPC, *fetal-derived neural progenitor stem cell;* hESC, *human embryonic stem cell;* hpNSC, *human parthenogenetic stem cell-derived neural stem cell;* MSC, *mesenchymal stem cell;* NPC, *neural progenitor stem cell;* VM, *ventral mesencephalic.*

enrolled and 13 had received transplants, but data are not yet available.[145] A Phase I study investigating intraparenchymal injection of human parthenogenetic stem cell-derived neural stem cells (hpNSCs) has early preliminary data. As of November 2017, 6 of 12 patients have received hpNSC doses, and early efficacy was seen in the first cohort with improvements in dyskinesia symptoms and responses to levodopa.[146] However, the study is underpowered. A Phase I/II open-label, nonrandomized trial is currently evaluating the safety and efficacy of a intraparenchymal injection of hESC-derived neuronal precursors in 50 patients with PD. The trial began in April 2017, and is currently enrolling patients.

These select studies highlight the recent use of stem cell therapies in PD. Although some have presented promising preliminary results, the data have yet to provide definitive evidence of the benefit of stem cell therapy in the PD population. Future trials will implement a variety of the aforementioned cell sources and hopefully provide greater clarity.[147]

Multiple Sclerosis

Overview

Multiple sclerosis is an inflammatory autoimmune disorder that destroys myelin and myelin-insulated nerve fibers. MS affects an estimated 2.3 million people worldwide, and it is the leading cause of nontraumatic disability in young adults. Although the etiology of MS is unclear, it is likely the result of the interaction of genetic and environmental factors leading to an immune response against myelin, axons, and neurons.[148]

The clinical manifestations of MS are highly variable and depend on the progression of the disease, lesion location, and tissue destruction. However, patients with MS typically accumulate progressive and eventually permanent disability over time. Common symptoms of MS include paresthesias, impaired balance, double vision, blurred vision, vertigo, and bladder and bowel dysfunction. Relapsing–remitting MS (RRMS) is the most common form of progression and is characterized by acute relapse and remission periods. After the RRMS phase, most patients advance to a phase of progressive disability accumulation known as secondary progressive MS (SPMS), which results in the accumulation of irreversible neurological deficits. Primary progressive MS (PPMS) is characterized by disability progression from the onset of symptoms without relapse or remission.[149]

Treatments for MS focus on recovering from exacerbations, slowing progression, and managing symptoms. The most common treatment for exacerbations is a corticosteroid regimen to reduce nerve inflammation. Physical therapy, muscle relaxants, and other medications are used to relieve symptoms of MS. Disease-modifying therapies, such as beta interferons, are used to slow the progression of the disease; however, none can absolutely stop disease progression or directly promote repair. Patients still experience active inflammation, frequent relapses, and worsening neurological disability. In addition, all therapies require lifelong use and are associated with potential adverse events.[149]

Stem Cell Therapies for Multiple Sclerosis

Current clinical trials are exploring the potential efficacy of stem cell therapies for patients with MS, investigating progression prevention, disability reversal, and safety. Recent clinical trials in this realm have involved autologous HSCs, BMMSCs, and UC-MSCs and have explored strategies for immune modulation, neuroprotection, and CNS repair. One major approach is immunoablation or conditioning, followed by intravenous autologous HSC infusion. Recent clinical trials for stem cell therapies in patients with MS are outlined in Table 5.

Hematopoietic stem cells have been used in several Phase I and Phase II trials, demonstrating high efficacy for remission and stabilizing symptoms, with overall tolerable adverse effects; however, some concern remains regarding toxicity related to pharmacologic conditioning pretransplantation. Evidence suggests that patients with RRMS and less severe disease progression at the start of treatment glean the most benefit from this therapy. A formal multicenter, randomized Phase III trial comparing therapies in a defined patient population is needed to further evaluate this paradigm.[156]

The MSC secretome and the capacity of MSCs for migration and homing makes them an especially attractive option for treatment of MS. So far, small studies have explored the feasibility and safety of transplantation in MS, with preliminary evidence of benefit on clinical and neurophysiological outcomes. An uncontrolled Phase I/II study of SPMS reported tissue

TABLE 5 Cell Therapies in Recent Clinical Trials for Multiple Sclerosis

Trial ID	Year initiated	Status	Cell type	Delivery route	Details
NCT03355365	2018	Phase II, active	Autologous BMMSC-NP	Intrathecal	50 patients enrolled, randomized trial, treatment vs placebo. Evaluating efficacy of bone marrow-derived mesenchymal stem cell-derived neural progenitor cells (MSC-NP)
NCT03269071	2017	Phase I/II, active	fNPCs	Intrathecal	Four patients enrolled, dose-escalation study evaluating safety and feasibility of intrathecal fNPC administration
NCT03282760	2017	Phase I, active	fNPCs	Intraventricular	24 patients with progressive MS enrolled, multicenter, open-label study. Designed to evaluate safety and efficacy of intrathecal administration of fNPC at different doses and time points
NCT01933802	2014	Phase II, completed	MSCs	Intravenous	Nine patients with progressive MS randomized to receive MSCs. At 6-month follow-up, decrease in cumulative number of lesions on MRI. In all, 70% and 50% of patients demonstrated improved muscle strength and bladder function, respectively. No serious adverse events[150]
NCT02034188	2014	Phase I/II, completed	UC-MSCs	Intravenous	20 patients enrolled. No serious adverse events reported. Symptom improvement was most potent 1 month posttransplant. Improvements in bladder, bowel, sexual dysfunction, EDSS scores, motor scores, and quality of life observed. Inactive lesions visible on MRI in 83.3% of subjects at 1 year

TABLE 5 Cell Therapies in Recent Clinical Trials for Multiple Sclerosis—cont'd

Trial ID	Year initiated	Status	Cell type	Delivery route	Details
NCT00813969	2011	Phase I, completed	MSCs	Intravenous	24 patients with RRMS or SPMS patients enrolled in single-arm study. Well-tolerated, no treatment-related severe adverse events or disease activation. No significant benefit on lesion activity[151]
NCT01099930	2010	Phase II, completed	Autologous HSC	Intravenous	Multicenter study. A total of 24 patients underwent immunosuppressive conditioning followed by autologous HSC transplantation. Event-free survival was 69.6% at 3 years posttransplant. One patient died of transplant-related complications[152]
NCT00395200	2008	Phase I/II, completed	MSCs	Intravenous	Patients with SPMS. No serious adverse events. Improvement in visual acuity and optic nerve area, definitive evidence of neuroprotective effect[153]
NCT00288626	2006	Phase II, completed	Autologous HSC	Intravenous	Patients underwent immunosuppressive conditioning followed by autologous HSC transplantation. Event-free survival was 69% at 5 years, and 90% survival MS-related mortality posttransplant[154]
NCT00278655	2003	Phase I/II, completed	Autologous HSC	Intravenous	Multicenter trial. Patients underwent cyclophosphamide and CAMPATH-1H immunosuppressive conditioning followed by autologous HSC transplantation. Event-free survival was 64% at 4 years posttransplant. Several complications attributed to conditioning regimen[155]

Abbreviations: BMMSC-NP, *bone marrow-derived mesenchymal stem cell-derived neural progenitor cells;* EDSS, *expanded disability status scale;* fNPC, *fetal-derived neural progenitor stem cell;* HSC, *hematopoietic stem cell;* MRI, *magnetic resonance imaging;* MSC, *mesenchymal stem cell;* RRMS, *relapsing–remitting multiple sclerosis;* SPMS, *secondary progressive multiple sclerosis;* UC-MSC, *umbilical cord mesenchymal stem cell.*

repair in the afferent visual system, and a small randomized Phase II study in RRMS suggested reduced brain lesions on magnetic resonance imaging. Further clinical trials are necessary, and larger controlled Phase II studies are underway.[156, 157]

Though not in current clinical use, OPCs have been shown to induce myelination after intraparenchymal injection. A Phase I study of OPC transplantation has been proposed to evaluate the feasibility and safety of intracerebral injections of OPCs in patients with MS to augment remyelination. Clinical trials of stem cell therapies for MS are currently largely underpowered, with a lack of randomized and appropriately designed trials. However, these preliminary data represent significant preliminary efficacy. Overall, stem cell therapies for MS have made considerable progress, with multiple clinical trials underway. HSCs have potent efficacy in RRMS patients, and MSC, OPC, fNPC transplantation remain potential options for patients with more progressive, degenerative MS.

Amyotrophic Lateral Sclerosis

Overview

Amyotrophic lateral sclerosis is a progressive neurodegenerative disease that usually results in death within 3 to 5 years of diagnosis.[158, 159] The prevalence of ALS is estimated at 4.3 cases per 100,000.[160] A small percentage (5%-10%) of patients with ALS have familial ALS (fALS), or hereditary ties to the disease.[161, 162] The remaining ALS cases are sporadic ALS (sALS), which have no known familial component. Numerous genetic mutations have been implicated in ALS, but the most prevalent mutations are *C9orf72*, *SOD1*, *TARDBP*, and *FUS*.[163, 164] Multiple mechanisms are purported to cause ALS, including glutamate excitotoxicity, presence of an inflammatory cellular environment in the spinal cord, mitochondrial dysfunction, RNA metabolism, and defective axonal transport.[159, 165, 166] The clinical picture of ALS is a disease of similar clinical features, but a heterogeneous pathophysiology that is a continued subject of debate.

Clinical features of ALS include motor neuron death, resulting in weakness, muscular atrophy, paralysis, and loss of respiratory control. Clinical presentations are heterogeneous, and disease phenotype is usually determined by the initial presentation of the disease (bulbar onset vs spinal onset).[167-169] The heterogeneity of ALS makes it difficult to design practical and efficient parameters of clinical trials, such as inclusion/exclusion criteria, clinical end points, and experimental milestones.

Only two drugs have been approved by the Food and Drug Administration for the treatment of ALS: riluzole[170] and edaravone.[171, 172] Both treatments marginally increase survival and quality of life; however, they are widely used, as they are the only treatment options available for patients with ALS. Experimental gene therapies are under development, but these therapies are associated with limitations that have prevented efficacious implementation in clinical settings.[173]

Stem Cell Therapies for Amyotrophic Lateral Sclerosis

Due to the rapid progression of ALS, limited treatment options, and the contribution of multiple pathogenic molecular processes, stem cell treatment for ALS is a critical area of stem cell research.[174] Cell replacement therapy can theoretically replace dysfunctional or dead motor neurons in the spinal cord, and the use of secreted factors could offer neuroprotection and delay disease progression. This approach intends to address inflammation, neuroprotection, neurogenesis, and excitatory toxicity associated with many of the purported mechanisms of ALS progression.[165, 174-177] To date, several clinical trials have investigated stem cell therapy for patients with ALS, but no studies have demonstrated efficacy. An overview of clinical trials of stem cell therapies is provided in Table 6.

The lack of efficacy in clinical trials highlights obstacles in translation of stem cell therapy for ALS. The largest of these obstacles is disunion between preclinical experiments and clinical trials.[178] Notable trials have demonstrated the safety of stem cell treatment, but none of them have shown any efficacy.[70, 179] Several theories have been suggested to explain why many stem cell treatments are successful in a preclinical setting but fail in clinical trials. Research-grade cells have shown efficacy, but clinical-grade cells (i.e., those derived in good manufacturing practice conditions) of the same type do not yield the same results.[180]

TABLE 6 Cell Therapies in Recent Clinical Trials for Amyotrophic Lateral Sclerosis

Trial ID	Year initiated	Status	Cell type	Delivery route	Details
NCT02478450	2018	Phase I/II, not yet recruiting	fNPC (GRPs)	Intraparenchymal	Patients will receive injections of cells either unilaterally or bilaterally in a dose-escalation study. Multiple cohorts, varying doses, different injection sites (lumbar vs cervical cord)
NCT02943850	2017	Phase I/II, recruiting	fNPCs (hNSCs + GDNF overexpression)	Intraparenchymal	Astrocyte-directed cells expressing GDNF. Two escalating doses will be used. Bilateral injections with control side and treatment side will be evaluated
NCT03280056	2017	Phase III, recruiting	Autologous ASCs (NTF-releasing)	Intrathecal	Follow up to NCT02017912. Unlike NCT02017912, this trial will examine multiple intrathecal doses of NurOwn (MSC-NTF) cells
NCT03268603	2017	Recruiting	Autologous ASCs	Intrathecal	To determine the safety and efficacy of MSCs delivered intrathecally. Cells will be administered every three months for a total of four injections
NCT02017912	2014	Completed	Autologous ASCs (NTF-releasing)	Intrathecal	Therapy is safe, single administration produces transient and clinically meaningful beneficial response based on industry reports. No peer-reviewed results published
NCT01730716	2013	Phase II, complete	fNPCs (hNSCs)	Intraparenchymal	Dose-escalation and safety study in 18 patients. Two participants experienced major complications from treatment. ALS progression was not negatively influenced by treatment. Five groups of three subjects used to observe the effects of dose escalation[71]
NCT01609283	2012	Phase I, active, not recruiting	Autologous ASCs	Intrathecal	The purpose of this study is to determine a safe dose for MSCs injected intrathecally over a 12-month period in patients with ALS. Estimated completion April 2018
NCT01142856	2010	Phase I, completed	Autologous ASCs	Intrathecal	To evaluate the safety of intrathecal injection of MSCs in ALS patients. Patients were originally proposed to be followed for 2 years, but no results available
NCT01348451	2009	Phase I, completed	fNPCs (hNSCs)	Intraparenchymal	Patients received five unilateral or bilateral direct injections in a concentration of 10,000 cells/μL. Direct injection of hNSCs with a stereotaxic frame into the lumbar cord is safe. There is no evidence of toxicity, and the surgery is well tolerated[70]

Abbreviations: ALS, *amyotrophic lateral sclerosis;* ASC, *adipose tissue stem cell;* fNPC, *fetal-derived neural progenitor stem cell;* GDNF, *glial cell line-derived neurotrophic factor;* GRP, *glial restricted progenitor cell;* hNSC, *human neural stem cell;* MSC, *mesenchymal stem cell;* MSC-NTF, *NTF-secreting mesenchymal stromal cells;* NTF, *neurotrophic factor.*

This suggests that several clinical trials were not successful, especially involving MSCs, because cell quality is compromised during mass production.[181] Although the Food and Drug Administration encourages researchers to use the same stem cells throughout the translational pipeline, it is not required. Additionally, stem cell manufacturers do not necessarily have to derive all good manufacturing practice cells from a single fetal source or cell line, which produces inconsistent results.[182] The validity and applicability of preclinical ALS rodent models is controversial; debate is ongoing regarding whether current rodent models are truly representative of ALS in humans and are appropriate for preclinical treatment evaluation.[178, 183]

The aforementioned challenges must be considered when designing clinical trials, especially in regard to inclusion and exclusion criteria, outcomes to assess efficacy, and randomization schemes.[165, 184, 185] The current state of stem cell therapy for ALS has not shown clinical efficacy, but the theoretical possibility remains. Current and future clinical studies must address these challenges before the prospect of an efficacious stem cell therapy is a realistic, broadly applicable strategy.

References

1. Trounson A, McDonald C. Stem cell therapies in clinical trials: progress and challenges. *Cell Stem Cell.* 2015; 17(1):11-22.
2. Huang H, Young W, Chen L, et al. Clinical cell therapy guidelines for neurorestoration (IANR/CANR 2017). *Cell Transplant.* 2018;27(2):310-324.
3. Daley GQ, Hyun I, Apperley JF, et al. Setting global standards for stem cell research and clinical translation: the 2016 ISSCR guidelines. *Stem Cell Rep.* 2016;6(6):787-797.
4. Salgado AJ, Sousa JC, Costa BM, et al. Mesenchymal stem cells secretome as a modulator of the neurogenic niche: basic insights and therapeutic opportunities. *Front Cell Neurosci.* 2015;9:249.
5. Xiao J, Yang R, Biswas S, Qin X, Zhang M, Deng W. Mesenchymal stem cells and induced pluripotent stem cells as therapies for multiple sclerosis. *Int J Mol Sci.* 2015;16(5):9283-9302.
6. Huang L, Zhang L. Neural stem cell therapies and hypoxic-ischemic brain injury. *Prog Neurobiol.* 2018;173:1-17.
7. Goldberg EM, Coulter DA. Seizing the opportunity: stem cells take on epilepsy. *Cell Stem Cell.* 2014;15(5):527-528.
8. Fox IJ, Daley GQ, Goldman SA, Huard J, Kamp TJ, Trucco M. Stem cell therapy. Use of differentiated pluripotent stem cells as replacement therapy for treating disease. *Science.* 2014;345(6199):124739.
9. Thomson JA, Itskovitz-Eldor J, Shapiro SS, et al. Embryonic stem cell lines derived from human blastocysts. *Science.* 1998;282(5391):1145-1147.
10. Revazova ES, Turovets NA, Kochetkova OD, et al. Patient-specific stem cell lines derived from human parthenogenetic blastocysts. *Cloning Stem Cells.* 2007;9(3):432-449. Fall.
11. Tachibana M, Amato P, Sparman M, et al. Human embryonic stem cells derived by somatic cell nuclear transfer. *Cell.* 2013;153(6):1228-1238.
12. Ma H, Morey R, O'Neil RC, et al. Abnormalities in human pluripotent cells due to reprogramming mechanisms. *Nature.* 2014;511(7508):177-183.
13. Dhar D, Hsi-En HJ. Stem cell research policies around the world. *Yale J Biol Med.* 2009;82(3):113-115.
14. King NM, Perrin J. Ethical issues in stem cell research and therapy. *Stem Cell Res Ther.* 2014;5(4):85.
15. Takahashi K, Yamanaka S. Induction of pluripotent stem cells from mouse embryonic and adult fibroblast cultures by defined factors. *Cell.* 2006;126(4):663-676.
16. Fusaki N, Ban H, Nishiyama A, Saeki K, Hasegawa M. Efficient induction of transgene-free human pluripotent stem cells using a vector based on Sendai virus, an RNA virus that does not integrate into the host genome. *Proc Jpn Acad Ser B Phys Biol Sci.* 2009;85(8):348-362.
17. Lukovic D, Moreno Manzano V, Stojkovic M, Bhattacharya SS, Erceg S. Concise review: human pluripotent stem cells in the treatment of spinal cord injury. *Stem Cells.* 2012;30(9):1787-1792.
18. Simonson OE, Domogatskaya A, Volchkov P, Rodin S. The safety of human pluripotent stem cells in clinical treatment. *Ann Med.* 2015;47(5):370-380.

19. Zeng X, Hunsberger JG, Simeonov A, Malik N, Pei Y, Rao M. Concise review: modeling central nervous system diseases using induced pluripotent stem cells. *Stem Cells Transl Med.* 2014;3(12):1418-1428.
20. Hunsberger JG, Efthymiou AG, Malik N, et al. Induced pluripotent stem cell models to enable in vitro models for screening in the central nervous system. *Stem Cells Dev.* 2015;24(16):1852-1864.
21. Funato K, Major T, Lewis PW, Allis CD, Tabar V. Use of human embryonic stem cells to model pediatric gliomas with H3.3K27M histone mutation. *Science.* 2014;346(6216):1529-1533.
22. Badger JL, Cordero-Llana O, Hartfield EM, Wade-Martins R. Parkinson's disease in a dish—using stem cells as a molecular tool. *Neuropharmacology.* 2014;76(Pt A):88-96.
23. Okano H, Yamanaka S. iPS cell technologies: significance and applications to CNS regeneration and disease. *Mol Brain.* 2014;7:22.
24. Grealish S, Diguet E, Kirkeby A, et al. Human ESC-derived dopamine neurons show similar preclinical efficacy and potency to fetal neurons when grafted in a rat model of Parkinson's disease. *Cell Stem Cell.* 2014;15 (5):653-665.
25. Lukovic D, Valdes-Sanchez L, Sanchez-Vera I, et al. Brief report: astrogliosis promotes functional recovery of completely transected spinal cord following transplantation of hESC-derived oligodendrocyte and motoneuron progenitors. *Stem Cells.* 2014;32(2):594-599.
26. Sharp J, Frame J, Siegenthaler M, Nistor G, Keirstead HS. Human embryonic stem cell-derived oligodendrocyte progenitor cell transplants improve recovery after cervical spinal cord injury. *Stem Cells.* 2010; 28(1):152-163.
27. Erceg S, Ronaghi M, Oria M, et al. Transplanted oligodendrocytes and motoneuron progenitors generated from human embryonic stem cells promote locomotor recovery after spinal cord transection. *Stem Cells.* 2010; 28(9):1541-1549.
28. Wang YK, Zhu WW, Wu MH, et al. Human clinical-grade parthenogenetic ESC-derived dopaminergic neurons recover locomotive defects of nonhuman primate models of Parkinson's disease. *Stem Cell Rep.* 2018; 11(1):171-182.
29. Schwartz SD, Regillo CD, Lam BL, et al. Human embryonic stem cell-derived retinal pigment epithelium in patients with age-related macular degeneration and Stargardt's macular dystrophy: follow-up of two open-label phase 1/2 studies. *Lancet.* 2015;385(9967):509-516.
30. Song WK, Park KM, Kim HJ, et al. Treatment of macular degeneration using embryonic stem cell-derived retinal pigment epithelium: preliminary results in Asian patients. *Stem Cell Rep.* 2015;4(5):860-872.
31. Kobayashi Y, Okada Y, Itakura G, et al. Pre-evaluated safe human iPSC-derived neural stem cells promote functional recovery after spinal cord injury in common marmoset without tumorigenicity. *PLoS ONE.* 2012;7(12).
32. Okano H, Nakamura M, Yoshida K, et al. Steps toward safe cell therapy using induced pluripotent stem cells. *Circ Res.* 2013;112(3):523-533.
33. Kikuchi T, Morizane A, Doi D, et al. Human iPS cell-derived dopaminergic neurons function in a primate Parkinson's disease model. *Nature.* 2017;548(7669):592-596.
34. Doi D, Samata B, Katsukawa M, et al. Isolation of human induced pluripotent stem cell-derived dopaminergic progenitors by cell sorting for successful transplantation. *Stem Cell Rep.* 2014;2(3):337-350.
35. Hayashi R, Ishikawa Y, Ito M, et al. Generation of corneal epithelial cells from induced pluripotent stem cells derived from human dermal fibroblast and corneal limbal epithelium. *PLoS ONE.* 2012;7(9).
36. Yoshida S, Yasuda M, Miyashita H, et al. Generation of stratified squamous epithelial progenitor cells from mouse induced pluripotent stem cells. *PLoS ONE.* 2011;6(12).
37. Scheiner ZS, Talib S, Feigal EG. The potential for immunogenicity of autologous induced pluripotent stem cell-derived therapies. *J Biol Chem.* 2014;289(8):4571-4577.
38. Fu X. The immunogenicity of cells derived from induced pluripotent stem cells. *Cell Mol Immunol.* 2014; 11(1):14-16.
39. Tan Y, Ooi S, Wang L. Immunogenicity and tumorigenicity of pluripotent stem cells and their derivatives: genetic and epigenetic perspectives. *Curr Stem Cell Res Ther.* 2014;9(1):63-72.
40. Lamanna JJ, Gutierrez J, Espinosa JR, et al. Peripheral blood detection of systemic graft-specific xeno-antibodies following transplantation of human neural progenitor cells into the porcine spinal cord. *J Clin Neurosci.* 2018;48:173-180.
41. Irion S, Zabierowski SE, Tomishima MJ. Bringing neural cell therapies to the clinic: past and future strategies. *Mol Ther Methods Clin Dev.* 2017;4:72-82.

42. Chambers SM, Qi Y, Mica Y, et al. Combined small-molecule inhibition accelerates developmental timing and converts human pluripotent stem cells into nociceptors. *Nat Biotechnol*. 2012;30(7):715-720.
43. Merkle FT, Maroof A, Wataya T, et al. Generation of neuropeptidergic hypothalamic neurons from human pluripotent stem cells. *Development*. 2015;142(4):633-643.
44. Shi Y, Kirwan P, Livesey FJ. Directed differentiation of human pluripotent stem cells to cerebral cortex neurons and neural networks. *Nat Protoc*. 2012;7(10):1836-1846.
45. Amoroso MW, Croft GF, Williams DJ, et al. Accelerated high-yield generation of limb-innervating motor neurons from human stem cells. *J Neurosci*. 2013;33(2):574-586.
46. Kriks S, Shim JW, Piao J, et al. Dopamine neurons derived from human ES cells efficiently engraft in animal models of Parkinson's disease. *Nature*. 2011;480(7378):547-551.
47. Darbinyan A, Kaminski R, White MK, Darbinian N, Khalili K. Isolation and propagation of primary human and rodent embryonic neural progenitor cells and cortical neurons. *Methods Mol Biol*. 2013;1078:45-54.
48. Kefalopoulou Z, Politis M, Piccini P, et al. Long-term clinical outcome of fetal cell transplantation for Parkinson disease: two case reports. *JAMA Neurol*. 2014;71(1):83-87.
49. Moyer MP, Johnson RA, Zompa EA, Cain L, Morshed T, Hulsebosch CE. Culture, expansion, and transplantation of human fetal neural progenitor cells. *Transplant Proc*. 1997;29(4):2040-2041.
50. Franco PG, Pasquini JM, Silvestroff L. Optimizing culture medium composition to improve oligodendrocyte progenitor cell yields in vitro from subventricular zone-derived neural progenitor cell neurospheres. *PLoS ONE*. 2015;10(4).
51. Liu X, Chu TH, Su H, Guo A, Wu W. Neural progenitor cell apoptosis and differentiation were affected by activated microglia in spinal cord slice culture. *Neurol Sci*. 2014;35(3):415-419.
52. English D, Sharma NK, Sharma K, Anand A. Neural stem cells-trends and advances. *J Cell Biochem*. 2013; 114(4):764-772.
53. Rhim JH, Luo X, Gao D, et al. Cell type-dependent Erk-Akt pathway crosstalk regulates the proliferation of fetal neural progenitor cells. *Sci Rep*. 2016;6:26547.
54. Ruzicka J, Romanyuk N, Hejcl A, et al. Treating spinal cord injury in rats with a combination of human fetal neural stem cells and hydrogels modified with serotonin. *Acta Neurobiol Exp (Wars)*. 2013;73(1):102-115.
55. Wang G, Ao Q, Gong K, Zuo H, Gong Y, Zhang X. Synergistic effect of neural stem cells and olfactory ensheathing cells on repair of adult rat spinal cord injury. *Cell Transplant*. 2010;19(10):1325-1337.
56. Iwanami A, Kaneko S, Nakamura M, et al. Transplantation of human neural stem cells for spinal cord injury in primates. *J Neurosci Res*. 2005;80(2):182-190.
57. Lepore AC, O'Donnell J, Kim AS, et al. Human glial-restricted progenitor transplantation into cervical spinal cord of the SOD1 mouse model of ALS. *PLoS ONE*. 2011;6(10).
58. Hefferan MP, Galik J, Kakinohana O, et al. Human neural stem cell replacement therapy for amyotrophic lateral sclerosis by spinal transplantation. *PLoS ONE*. 2012;7(8).
59. Deng X, Liang Y, Lu H, et al. Co-transplantation of GDNF-overexpressing neural stem cells and fetal dopaminergic neurons mitigates motor symptoms in a rat model of Parkinson's disease. *PLoS ONE*. 2013;8(12).
60. Wang X, Lu Y, Zhang H, et al. Distinct efficacy of pre-differentiated versus intact fetal mesencephalon-derived human neural progenitor cells in alleviating rat model of Parkinson's disease. *Int J Dev Neurosci*. 2004; 22(4):175-183.
61. Sakata H, Niizuma K, Yoshioka H, et al. Minocycline-preconditioned neural stem cells enhance neuroprotection after ischemic stroke in rats. *J Neurosci*. 2012;32(10):3462-3473.
62. Wang Z, Cui C, Li Q, et al. Intracerebral transplantation of foetal neural stem cells improves brain dysfunction induced by intracerebral haemorrhage stroke in mice. *J Cell Mol Med*. 2011;15(12):2624-2633.
63. Mine Y, Tatarishvili J, Oki K, Monni E, Kokaia Z, Lindvall O. Grafted human neural stem cells enhance several steps of endogenous neurogenesis and improve behavioral recovery after middle cerebral artery occlusion in rats. *Neurobiol Dis*. 2013;52:191-203.
64. Skardelly M, Gaber K, Burdack S, et al. Long-term benefit of human fetal neuronal progenitor cell transplantation in a clinically adapted model after traumatic brain injury. *J Neurotrauma*. 2011;28(3):401-414.
65. Gao J, Prough DS, McAdoo DJ, et al. Transplantation of primed human fetal neural stem cells improves cognitive function in rats after traumatic brain injury. *Exp Neurol*. 2006;201(2):281-292.
66. Bachoud-Levi AC, Perrier AL. Regenerative medicine in Huntington's disease: current status on fetal grafts and prospects for the use of pluripotent stem cell. *Rev Neurol (Paris)*. 2014;170(12):749-762.
67. Shin JC, Kim KN, Yoo J, et al. Clinical trial of human fetal brain-derived neural stem/progenitor cell transplantation in patients with traumatic cervical spinal cord injury. *Neural Plast*. 2015;2015:630932.

68. Qiao LY, Huang FJ, Zhao M, et al. A two-year follow-up study of cotransplantation with neural stem/progenitor cells and mesenchymal stromal cells in ischemic stroke patients. *Cell Transplant*. 2014;23(Suppl 1):S65-S72.
69. Kalladka D, Sinden J, Pollock K, et al. Human neural stem cells in patients with chronic ischaemic stroke (PISCES): a phase 1, first-in-man study. *Lancet*. 2016;388(10046):787-796.
70. Glass JD, Boulis NM, Johe K, et al. Lumbar intraspinal injection of neural stem cells in patients with amyotrophic lateral sclerosis: results of a phase I trial in 12 patients. *Stem Cells*. 2012;30(6):1144-1151.
71. Glass JD, Hertzberg VS, Boulis NM, et al. Transplantation of spinal cord-derived neural stem cells for ALS: analysis of phase 1 and 2 trials. *Neurology*. 2016;87(4):392-400.
72. Riley J, Federici T, Polak M, et al. Intraspinal stem cell transplantation in amyotrophic lateral sclerosis: a phase I safety trial, technical note, and lumbar safety outcomes. *Neurosurgery*. 2012;71(2):405-416 [discussion 416].
73. Kulbatski I. Stem/precursor cell-based CNS therapy: the importance of circumventing immune suppression by transplanting autologous cells. *Stem Cell Rev*. 2010;6(3):405-410.
74. Wennersten A, Holmin S, Al Nimer F, Meijer X, Wahlberg LU, Mathiesen T. Sustained survival of xenografted human neural stem/progenitor cells in experimental brain trauma despite discontinuation of immunosuppression. *Exp Neurol*. 2006;199(2):339-347.
75. Lamanna JJ, Gutierrez J, Urquia LN, et al. Ferumoxytol labeling of human neural progenitor cells for diagnostic cellular tracking in the porcine spinal cord with magnetic resonance imaging. *Stem Cells Transl Med*. 2017;6(1):139-150.
76. Lamanna JJ, Urquia LN, Hurtig CV, et al. Magnetic resonance imaging-guided transplantation of neural stem cells into the porcine spinal cord. *Stereotact Funct Neurosurg*. 2017;95(1):60-68.
77. Gleave JA, Valliant JF, Doering LC. 99mTc-based imaging of transplanted neural stem cells and progenitor cells. *J Nucl Med Technol*. 2011;39(2):114-120.
78. Yang YJ, Li XL, Xue Y, et al. Bone marrow cells differentiation into organ cells using stem cell therapy. *Eur Rev Med Pharmacol Sci*. 2016;20(13):2899-2907.
79. Mazzini L, Mareschi K, Ferrero I, et al. Mesenchymal stromal cell transplantation in amyotrophic lateral sclerosis: a long-term safety study. *Cytotherapy*. 2012;14(1):56-60.
80. Teixeira FG, Carvalho MM, Sousa N, Salgado AJ. Mesenchymal stem cells secretome: a new paradigm for central nervous system regeneration? *Cell Mol Life Sci*. 2013;70(20):3871-3882.
81. Cuende N, Rico L, Herrera C. Concise review: bone marrow mononuclear cells for the treatment of ischemic syndromes: medicinal product or cell transplantation? *Stem Cells Transl Med*. 2012;1(5):403-408.
82. Neirinckx V, Coste C, Rogister B, Wislet-Gendebien S. Concise review: adult mesenchymal stem cells, adult neural crest stem cells, and therapy of neurological pathologies: a state of play. *Stem Cells Transl Med*. 2013;2(4):284-296.
83. Karussis D, Petrou P, Vourka-Karussis U, Kassis I. Hematopoietic stem cell transplantation in multiple sclerosis. *Expert Rev Neurother*. 2013;13(5):567-578.
84. Cohen JA. Mesenchymal stem cell transplantation in multiple sclerosis. *J Neurol Sci*. 2013;333(1–2):43-49.
85. Bahat-Stroomza M, Barhum Y, Levy YS, et al. Induction of adult human bone marrow mesenchymal stromal cells into functional astrocyte-like cells: potential for restorative treatment in Parkinson's disease. *J Mol Neurosci*. 2009;39(1–2):199-210.
86. Parivar K, Baharara J, Sheikholeslami A. Neural differentiation of mouse bone marrow-derived mesenchymal stem cells treated with sex steroid hormones and basic fibroblast growth factor. *Cell J*. 2015;17(1):27-36. Spring.
87. Meyerrose T, Olson S, Pontow S, et al. Mesenchymal stem cells for the sustained in vivo delivery of bioactive factors. *Adv Drug Deliv Rev*. 2010;62(12):1167-1174.
88. Canesi M, Giordano R, Lazzari L, et al. Finding a new therapeutic approach for no-option Parkinsonisms: mesenchymal stromal cells for progressive supranuclear palsy. *J Transl Med*. 2016;14(1):127.
89. Petrou P, Gothelf Y, Argov Z, et al. Safety and clinical effects of mesenchymal stem cells secreting neurotrophic factor transplantation in patients with amyotrophic lateral sclerosis: results of phase 1/2 and 2a clinical trials. *JAMA Neurol*. 2016;73(3):337-344.
90. Diez-Tejedor E, Gutierrez-Fernandez M, Martinez-Sanchez P, et al. Reparative therapy for acute ischemic stroke with allogeneic mesenchymal stem cells from adipose tissue: a safety assessment: a phase II randomized, double-blind, placebo-controlled, single-center, pilot clinical trial. *J Stroke Cerebrovasc Dis*. 2014;23(10):2694-2700.
91. Satti HS, Waheed A, Ahmed P, et al. Autologous mesenchymal stromal cell transplantation for spinal cord injury: a phase I pilot study. *Cytotherapy*. 2016;18(4):518-522.

92. Dahbour S, Jamali F, Alhattab D, et al. Mesenchymal stem cells and conditioned media in the treatment of multiple sclerosis patients: clinical, ophthalmological and radiological assessments of safety and efficacy. *CNS Neurosci Ther*. 2017;23(11):866-874.
93. Wang S, Cheng H, Dai G, et al. Umbilical cord mesenchymal stem cell transplantation significantly improves neurological function in patients with sequelae of traumatic brain injury. *Brain Res*. 2013;1532:76-84.
94. Hlebokazov F, Dakukina T, Ihnatsenko S, et al. Treatment of refractory epilepsy patients with autologous mesenchymal stem cells reduces seizure frequency: an open label study. *Adv Med Sci*. 2017;62(2):273-279.
95. Ikegame Y, Yamashita K, Nakashima S, et al. Fate of graft cells: what should be clarified for development of mesenchymal stem cell therapy for ischemic stroke? *Front Cell Neurosci*. 2014;8:322.
96. Leibacher J, Henschler R. Biodistribution, migration and homing of systemically applied mesenchymal stem/stromal cells. *Stem Cell Res Ther*. 2016;7:7.
97. Li J, Zhu H, Chen Y, et al. The distribution of transplanted human mesenchymal stem cells in the CNS of young Macaca fascicularis. *Brain Res*. 2014;1579:1-9.
98. Rosado-de-Castro PH, Schmidt Fda R, Battistella V, et al. Biodistribution of bone marrow mononuclear cells after intra-arterial or intravenous transplantation in subacute stroke patients. *Regen Med*. 2013;8(2):145-155.
99. Lalu MM, McIntyre L, Pugliese C, et al. Safety of cell therapy with mesenchymal stromal cells (SafeCell): a systematic review and meta-analysis of clinical trials. *PLoS ONE*. 2012;7(10).
100. Togel F, Yang Y, Zhang P, Hu Z, Westenfelder C. Bioluminescence imaging to monitor the in vivo distribution of administered mesenchymal stem cells in acute kidney injury. *Am J Physiol Ren Physiol*. 2008;295(1): F315-F321.
101. Barberini DJ, Aleman M, Aristizabal F, et al. Safety and tracking of intrathecal allogeneic mesenchymal stem cell transplantation in healthy and diseased horses. *Stem Cell Res Ther*. 2018;9(1):96.
102. Oh KW, Moon C, Kim HY, et al. Phase I trial of repeated intrathecal autologous bone marrow-derived mesenchymal stromal cells in amyotrophic lateral sclerosis. *Stem Cells Transl Med*. 2015;4(6):590-597.
103. Karussis D, Karageorgiou C, Vaknin-Dembinsky A, et al. Safety and immunological effects of mesenchymal stem cell transplantation in patients with multiple sclerosis and amyotrophic lateral sclerosis. *Arch Neurol*. 2010;67(10):1187-1194.
104. Bonab MM, Sahraian MA, Aghsaie A, et al. Autologous mesenchymal stem cell therapy in progressive multiple sclerosis: an open label study. *Curr Stem Cell Res Ther*. 2012;7(6):407-414.
105. Sykova E, Rychmach P, Drahoradova I, et al. Transplantation of mesenchymal stromal cells in patients with amyotrophic lateral sclerosis: results of phase I/IIa clinical trial. *Cell Transplant*. 2017;26(4):647-658.
106. McKee AC, Daneshvar DH. The neuropathology of traumatic brain injury. *Handb Clin Neurol*. 2015;127:45-66.
107. Saatman KE, Duhaime A-C, Bullock R, Maas AIR, Valadka A, Manley GT. Classification of traumatic brain injury for targeted therapies. *J Neurotrauma*. 2008;25(7):719-738.
108. Narayan RK, Michel ME, Ansell B, et al. Clinical trials in head injury. *J Neurotrauma*. 2002;19(5):503-557.
109. Cassidy JD, Carroll LJ, Peloso PM, et al. Incidence, risk factors and prevention of mild traumatic brain injury: results of the WHO collaborating centre task force on mild traumatic brain injury. *J Rehabil Med*. 2004;(43 Suppl):28-60.
110. Faul M, Xu L, Wald MM, Coronado V, Dellinger AM. Traumatic brain injury in the United States: national estimates of prevalence and incidence, 2002–2006. *Inj Prev*. 2010;16(Suppl 1):A268.
111. Selassie AW, Zaloshnja E, Langlois JA, Miller T, Jones P, Steiner C. Incidence of long-term disability following traumatic brain injury hospitalization, United States, 2003. *J Head Trauma Rehabil*. 2008;23(2):123-131.
112. Taylor CA, Bell JM, Breiding MJ, Xu L. Traumatic brain injury-related emergency department visits, hospitalizations, and deaths—United States, 2007 and 2013. *MMWR Surveill Summ*. 2017;66(9):1-16.
113. McCarthy MT, Kosofsky BE. Clinical features and biomarkers of concussion and mild traumatic brain injury in pediatric patients. *Ann N Y Acad Sci*. 2015;1345(1):89-98.
114. Cunningham AS, Salvador R, Coles JP, et al. Physiological thresholds for irreversible tissue damage in contusional regions following traumatic brain injury. *Brain*. 2005;128(Pt 8):1931-1942.
115. Werner C, Engelhard K. Pathophysiology of traumatic brain injury. *Br J Anaesth*. 2007;99(1):4-9.
116. Marmarou A. Increased intracranial pressure in head injury and influence of blood volume. *J Neurotrauma*. 1992;9(Suppl 1):S327-S332.
117. Hartings JA, Strong AJ, Fabricius M, et al. Spreading depolarizations and late secondary insults after traumatic brain injury. *J Neurotrauma*. 2009;26(11):1857-1866.

118. Corps KN, Roth TL, McGavern DB. Inflammation and neuroprotection in traumatic brain injury. *JAMA Neurol.* 2015;72(3):355-362.
119. Johnson VE, Stewart JE, Begbie FD, Trojanowski JQ, Smith DH, Stewart W. Inflammation and white matter degeneration persist for years after a single traumatic brain injury. *Brain.* 2013;136(Pt 1):28-42.
120. Faden AI, Wu J, Stoica BA, Loane DJ. Progressive inflammation-mediated neurodegeneration after traumatic brain or spinal cord injury. *Br J Pharmacol.* 2016;173(4):681-691.
121. Losiniecki A, Shutter L. Management of traumatic brain injury. *Curr Treat Options Neurol.* 2010;12(2):142-154. 2010/03/01.
122. Langlois JA, Rutland-Brown W, Wald MM. The epidemiology and impact of traumatic brain injury: a brief overview. *J Head Trauma Rehabil.* 2006;21(5):375-378.
123. Zhang R, Liu Y, Yan K, et al. Anti-inflammatory and immunomodulatory mechanisms of mesenchymal stem cell transplantation in experimental traumatic brain injury. *J Neuroinflammation.* 2013;10(1):871.
124. Tajiri N, Acosta SA, Shahaduzzaman M, et al. Intravenous transplants of human adipose-derived stem cell protect the brain from traumatic brain injury-induced neurodegeneration and motor and cognitive impairments: cell graft biodistribution and soluble factors in young and aged rats. *J Neurosci.* 2014;34(1):313-326.
125. Chen Z, Tortella FC, Dave JR, et al. Human amnion-derived multipotent progenitor cell treatment alleviates traumatic brain injury-induced axonal degeneration. *J Neurotrauma.* 2009;26(11):1987-1997.
126. Lee HJ, Kim KS, Park IH, Kim SU. Human neural stem cells over-expressing VEGF provide neuroprotection, angiogenesis and functional recovery in mouse stroke model. *PLoS ONE.* 2007;2(1).
127. Thau-Zuchman O, Shohami E, Alexandrovich AG, Leker RR. Vascular endothelial growth factor increases neurogenesis after traumatic brain injury. *J Cereb Blood Flow Metab.* 2010;30(5):1008-1016.
128. Gao J, Grill RJ, Dunn TJ, et al. Human neural stem cell transplantation-mediated alteration of microglial/macrophage phenotypes after traumatic brain injury. *Cell Transplant.* 2016;25(10):1863-1877.
129. Spurlock MS, Ahmed AI, Rivera KN, et al. Amelioration of penetrating ballistic-like brain injury induced cognitive deficits after neuronal differentiation of transplanted human neural stem cells. *J Neurotrauma.* 2017; 34(11):1981-1995.
130. Haus DL, Lopez-Velazquez L, Gold EM, et al. Transplantation of human neural stem cells restores cognition in an immunodeficient rodent model of traumatic brain injury. *Exp Neurol.* 2016;281:1-16.
131. Cox CS, Hetz RA, Liao GP, et al. Treatment of severe adult traumatic brain injury using bone marrow mononuclear cells. *Stem Cells.* 2017;35(4):1065-1079.
132. Cox Jr CS, Baumgartner JE, Harting MT, et al. Autologous bone marrow mononuclear cell therapy for severe traumatic brain injury in children. *Neurosurgery.* 2011;68(3):588-600.
133. Tian C, Wang X, Wang X, et al. Autologous bone marrow mesenchymal stem cell therapy in the subacute stage of traumatic brain injury by lumbar puncture. *Exp Clin Transplant.* 2013;11(2):176-181.
134. DeVivo MJ. Epidemiology of traumatic spinal cord injury: trends and future implications. *Spinal Cord.* 2012; 50(5):365-372.
135. Daroff RB, Bradley WG, eds. *Bradley's Neurology in Clinical Practice*, 6th ed. Philadelphia, PA: Elsevier/Saunders; 2012.
136. Levi AD, Okonkwo DO, Park P, et al. Emerging safety of intramedullary transplantation of human neural stem cells in chronic cervical and thoracic spinal cord injury. *Neurosurgery.* 2017;82:562-575. https://doi.org/10.1093/neuros/nyx250.
137. Curtis E, Gabel BC, Marsala M, Ciacci JD. A phase i, open-label, single-site, safety study of human spinal cord-derived neural stem cell transplantation for the treatment of chronic spinal cord injury. *Neurosurgery.* 2016; 63(CN_suppl_1):168-169.
138. Vaquero J, Zurita M, Rico MA, et al. An approach to personalized cell therapy in chronic complete paraplegia: the Puerta de Hierro phase I/II clinical trial. *Cytotherapy.* 2016;18(8):1025-1036.
139. Pal R, Venkataramana NK, Bansal A, et al. Ex vivo-expanded autologous bone marrow-derived mesenchymal stromal cells in human spinal cord injury/paraplegia: a pilot clinical study. *Cytotherapy.* 2009;11(7):897-911.
140. Siddiqui AM, Khazaei M, Fehlings MG. Translating mechanisms of neuroprotection, regeneration, and repair to treatment of spinal cord injury. *Prog Brain Res.* 2015;218:15-54.
141. Kalia LV, Lang AE. Parkinson's disease. *Lancet.* 2015;386(9996):896-912.
142. Connolly BS, Lang AE. Pharmacological treatment of Parkinson disease: a review. *JAMA.* 2014; 311(16):1670-1683.

143. Johnson KE. Approach to the patient with Parkinson disease. *Prim Care*. 2015;42(2):205-215.
144. Stoker TB, Barker RA. Cell therapies for Parkinson's disease: how far have we come? *Regen Med*. 2016; 11(8):777-786.
145. TRANSEURO. Result in Brief. Replacement Therapy for Parkinson's Disease. https://cordis.europa.eu/result/rcn/89620_en.html; 2017. Accessed 2018.
146. Globe Newswire. *International Stem Cell Corporation Announces Positive Interim Clinical Results for Parkinson's Disease Clinical Trial*. Globe Newswire; 2017.
147. Barker RA, Parmar M, Studer L, Takahashi J. Human trials of stem cell-derived dopamine neurons for Parkinson's disease: dawn of a new era. *Cell Stem Cell*. 2017;21(5):569-573.
148. Tullman MJ. Overview of the epidemiology, diagnosis, and disease progression associated with multiple sclerosis. *Am J Manag Care*. 2013;19(2 Suppl):S15-S20.
149. Goldenberg MM. Multiple sclerosis review. *P T*. 2012;37(3):175-184.
150. Llufriu S, Sepulveda M, Blanco Y, et al. Randomized placebo-controlled phase II trial of autologous mesenchymal stem cells in multiple sclerosis. *PLoS ONE*. 2014;9(12).
151. Cohen JA, Imrey PB, Planchon SM, et al. Pilot trial of intravenous autologous culture-expanded mesenchymal stem cell transplantation in multiple sclerosis. *Mult Scler*. 2018;24(4):501-511.
152. Atkins HL, Bowman M, Allan D, et al. Immunoablation and autologous haemopoietic stem-cell transplantation for aggressive multiple sclerosis: a multicentre single-group phase 2 trial. *Lancet*. 2016;388(10044):576-585.
153. Connick P, Kolappan M, Crawley C, et al. Autologous mesenchymal stem cells for the treatment of secondary progressive multiple sclerosis: an open-label phase 2a proof-of-concept study. *Lancet Neurol*. 2012;11(2):150-156.
154. Nash RA, Hutton GJ, Racke MK, et al. High-dose immunosuppressive therapy and autologous hematopoietic cell transplantation for relapsing-remitting multiple sclerosis (HALT-MS): a 3-year interim report. *JAMA Neurol*. 2015;72(2):159-169.
155. Burt RK, Balabanov R, Han X, et al. Association of nonmyeloablative hematopoietic stem cell transplantation with neurological disability in patients with relapsing-remitting multiple sclerosis. *JAMA*. 2015;313(3):275-284.
156. Waldman A, Ghezzi A, Bar-Or A, Mikaeloff Y, Tardieu M, Banwell B. Multiple sclerosis in children: an update on clinical diagnosis, therapeutic strategies, and research. *Lancet Neurol*. 2014;13(9):936-948.
157. Dulamea A. Mesenchymal stem cells in multiple sclerosis—translation to clinical trials. *J Med Life*. 2015; 8(1):24-27.
158. Zinman L, Cudkowicz M. Emerging targets and treatments in amyotrophic lateral sclerosis. *Lancet Neurol*. 2011;10(5):481-490.
159. Donsante A, Urquia LN, Boulis NM. In: Tuszynski MH, ed. *Stem Cells for Amyotrophic Lateral Sclerosis*. Boston, MA: Springer; 2016. *Transl Neurosci*.
160. Mehta P, Kaye W, Bryan L, et al. Prevalence of amyotrophic lateral sclerosis—United States, 2012-2013. *MMWR Surveill Summ*. 2016;65(8):1-12.
161. Boylan K. Familial amyotrophic lateral sclerosis. *Neurol Clin*. 2015;33(4):807-830.
162. Kurland LT, Mulder DW. Epidemiologic investigations of amyotrophic lateral sclerosis. 2. Familial aggregations indicative of dominant inheritance. II. *Neurology*. 1955;5(4):249-268.
163. Leblond CS, Kaneb HM, Dion PA, Rouleau GA. Dissection of genetic factors associated with amyotrophic lateral sclerosis. *Exp Neurol*. 2014;262(Pt B):91-101.
164. Zou ZY, Zhou ZR, Che CH, Liu CY, He RL, Huang HP. Genetic epidemiology of amyotrophic lateral sclerosis: a systematic review and meta-analysis. *J Neurol Neurosurg Psychiatry*. 2017;88(7):540-549.
165. Ciervo Y, Ning K, Jun X, Shaw PJ, Mead RJ. Advances, challenges and future directions for stem cell therapy in amyotrophic lateral sclerosis. *Mol Neurodegener*. 2017;12(1):85.
166. Baumer D, Talbot K, Turner MR. Advances in motor neurone disease. *J R Soc Med*. 2014;107(1):14-21.
167. Hardiman O, van den Berg LH, Kiernan MC. Clinical diagnosis and management of amyotrophic lateral sclerosis. *Nat Rev Neurol*. 2011;7(11):639-649.
168. Ravits JM, La Spada AR. ALS motor phenotype heterogeneity, focality, and spread: deconstructing motor neuron degeneration. *Neurology*. 2009;73(10):805-811.
169. Korner S, Kollewe K, Fahlbusch M, et al. Onset and spreading patterns of upper and lower motor neuron symptoms in amyotrophic lateral sclerosis. *Muscle Nerve*. 2011;43(5):636-642.
170. Miller RG, Mitchell JD, Moore DH. Riluzole for amyotrophic lateral sclerosis (ALS)/motor neuron disease (MND). *Cochrane Database Syst Rev*. 2012;(3):CD001447.

171. Kaufman MB. Pharmaceutical approval update. *P T*. 2017;42(10):620-621.
172. Writing G, Edaravone ALSSG. Safety and efficacy of edaravone in well defined patients with amyotrophic lateral sclerosis: a randomised, double-blind, placebo-controlled trial. *Lancet Neurol*. 2017;16(7):505-512.
173. Tora MS, Keifer Jr OP, Lamanna JJ, Boulis NM. The challenges of developing a gene therapy for amyotrophic lateral sclerosis. *Expert Rev Neurother*. 2017;17(4):323-325.
174. Atassi N, Beghi E, Blanquer M, et al. Intraspinal stem cell transplantation for amyotrophic lateral sclerosis: ready for efficacy clinical trials? *Cytotherapy*. 2016;18(12):1471-1475.
175. Caprnda M, Kubatka P, Gazdikova K, et al. Immunomodulatory effects of stem cells: therapeutic option for neurodegenerative disorders. *Biomed Pharmacother*. 2017;91:60-69.
176. Faravelli I, Riboldi G, Nizzardo M, et al. Stem cell transplantation for amyotrophic lateral sclerosis: therapeutic potential and perspectives on clinical translation. *Cell Mol Life Sci*. 2014;71(17):3257-3268.
177. Bonafede R, Mariotti R. ALS pathogenesis and therapeutic approaches: the role of mesenchymal stem cells and extracellular vesicles. *Front Cell Neurosci*. 2017;11:80.
178. Ludolph AC, Bendotti C, Blaugrund E, et al. Guidelines for preclinical animal research in ALS/MND: a consensus meeting. *Amyotroph Lateral Scler*. 2010;11(1–2):38-45.
179. Moura MC, Novaes MR, Zago YS, Eduardo EJ, Casulari LA. Efficacy of stem cell therapy in amyotrophic lateral sclerosis: a systematic review and meta-analysis. *J Clin Med Res*. 2016;8(4):317-324.
180. Anderson AJ, Piltti KM, Hooshmand MJ, Nishi RA, Cummings BJ. Preclinical efficacy failure of human cns-derived stem cells for use in the pathway study of cervical spinal cord injury. *Stem Cell Rep*. 2017;8(2):249-263.
181. Chinnadurai R, Ng S, Velu V, Galipeau J. Challenges in animal modelling of mesenchymal stromal cell therapy for inflammatory bowel disease. *World J Gastroenterol*. 2015;21(16):4779-4787.
182. Anderson AJ, Cummings BJ. Achieving informed consent for cellular therapies: a preclinical translational research perspective on regulations versus a dose of reality. *J Law Med Ethics*. 2016;44(3):394-401.
183. Henderson VC, Kimmelman J, Fergusson D, Grimshaw JM, Hackam DG. Threats to validity in the design and conduct of preclinical efficacy studies: a systematic review of guidelines for in vivo animal experiments. *PLoS Med*. 2013;10(7).
184. Thomsen GM, Gowing G, Svendsen S, Svendsen CN. The past, present and future of stem cell clinical trials for ALS. *Exp Neurol*. 2014;262(Pt B):127-137.
185. Bruijn L, Cudkowicz M. Group ALSCTW. Opportunities for improving therapy development in ALS. *Amyotroph Lateral Scler Frontotemporal Degener*. 2014;15(3–4):169-173.

Further Reading

186. Castillo-Melendez M, Yawno T, Jenkin G, Miller SL. Stem cell therapy to protect and repair the developing brain: a review of mechanisms of action of cord blood and amnion epithelial derived cells. *Front Neurosci*. 2013;7:194

SECTION II

NERVOUS SYSTEM DRUG DELIVERY TECHNIQUES

CHAPTER

9

Intravenous and Intravascular Drug Delivery

James Pan[a], *Yuhao Huang*, *Gerald Grant*
Department of Neurosurgery, Stanford University School of Medicine, Stanford, CA, United States

OVERVIEW OF THE BLOOD–BRAIN BARRIER

The endothelium of the vasculature of the brain and the cells with which it interacts are collectively referred to as the *neurovascular unit* (NVU).[1] The brain vascular endothelium is in constant crosstalk with an orchestra of perivascular components consisting of astrocytes, microglia, neurons, mast cells, pericytes, and circulating immune cells.[2] The blood–brain barrier (BBB) is constantly remodeled through cytokine signaling and interactions with its cellular microenvironment to modify BBB integrity, expression of BBB transporters, and the physical permeability to pathogens and immune cells.[1]

The primary mechanism by which circulating molecules gain access to the brain or cerebrospinal fluid (CSF) is through a transcellular route via vascular endothelial or choroid plexus cells driven by passive transport, active transport, or both. The main passive mechanism is free diffusion.[3] In general, lipophilic molecules between 400 and 600 Da can pass through cells with tight junctions (TJs). Other molecules such as amino acids, glucose, carboxylic acids, and nucleosides pass through transcellularly, facilitated through carrier-mediated transport, receptor-mediated transport (RMT), or absorptive-mediated transport.

The physical integrity of the BBB is maintained by TJs between capillary endothelial cells (ECs), and is an important regulator of brain homeostasis.[4] Lipophilic and small gaseous molecules such as O_2 and CO_2 can readily diffuse through the lipid membranes of the endothelium. Unique features of the BBB endothelium that distinguish it from peripheral ECs include a lower number of endocytic vesicles and a larger number of TJs, which serve to limit transcellular and paracellular

[a]Current address: Department of Neurological Surgery, University of Washington School of Medicine, Seattle, WA, United States

https://doi.org/10.1016/B978-0-12-813997-4.00009-8
© 2019 Elsevier Inc. All rights reserved.

flow, respectively.[5] Specific transporters, such as the adenosine triphosphate-binding cassette (ABC) transporters, are also expressed on the surfaces of ECs and serve to efflux harmful xenobiotic compounds.[6, 7] The vascular endothelium is also associated with a variety of intra- and extracellular enzymes, which metabolize a variety of compounds ranging from peptides to toxic substances.[8]

Tight Junctions

The TJs between brain vascular ECs are also more complex and specialized compared to those found in the periphery. The primary function of TJs is to reduce the permeability of a cellular layer against polar and ionic solutes. A secondary function of TJs is to maintain the polarity of ECs by establishing apical and basal topology. Various transmembrane and paracellular proteins contribute to the biochemical and mechanical characteristics of TJs. Occludin and claudins are two of the most crucial and highly expressed components of the junctional complex (Fig. 1).

Occludin, a 60-65 kDa tetraspan integral membrane protein, was among the first TJ proteins identified and is important for barrier function.[9] Studies have shown that occludin undergoes extensive posttranscriptional and posttranslational modifications,[10] and aberrations in this process can lead to disease states such as autoimmune encephalomyelitis.[11] Occludin also plays an important role in the redox regulation of TJs, as oxidative stress associated with inflammatory conditions promotes TJ disruption.[12] Embryonic stem cells lacking occludin expression could still form TJs, suggesting that occludin is not necessary to barrier formation[13]; however, mouse knockout models of occludin resulted in a complex phenotype involving postnatal growth retardation and brain calcifications.[14]

Claudins comprise a broad family of proteins, with at least 27 members identified.[15] These proteins share the same structural characteristics, as they are integral membrane proteins with four transmembrane domains, two extracellular loops, and two cytoplasmic termini.[16] The C-terminus contains a PDZ-binding motif that interfaces with various scaffolding proteins such as zonula occludens (ZO-1, ZO-2), MAGI-1, PatJ, PALS1, and MUPP1.[17] The dominant claudin protein at the BBB is claudin-5, followed by claudin-3 and claudin-12.[18] Experimental evidence suggests that overexpression of claudin-5 strengthens BBB properties in a rat model,[19] whereas partial claudin-5-knockout mice have compromised BBBs permeable to molecules up to 800 Da in size; a full knockout of claudin-5 results in lethality.[20] Claudins have also been implicated in the remodeling of the BBB in various neoplasms of the brain. A loss of claudin-3 on brain vasculature (demonstrated by decreased immunostaining) has been implicated in the leakiness of vasculature associated with glioblastoma.[21] Downregulation of claudin-1, claudin-5, and occludin in hyperplastic vasculature has also been reported in tumor microvessels, which contribute to the leaky TJs and hyperpermeable ECs seen in tumor microenvironments.[22]

Other components of the junctional complex include junctional adhesion molecules (JAMs) and scaffolding proteins. JAMs are integral membrane proteins that belong to the immunoglobulin superfamily[23] and consist of a single membrane-spanning domain, an extracellular domain with an N terminus, and a short cytoplasmic tail.[24] Like claudins, JAMs also have a PDZ domain on the cytoplasmic tail that interacts with scaffolding proteins including ZO-1, AF-6, ASIP/Par3, and cingulin.[25, 26] Scaffolding proteins, including ZO-1, ZO-2, and ZO-3, are essential in linking the TJ to the intracellular actin cytoskeleton.

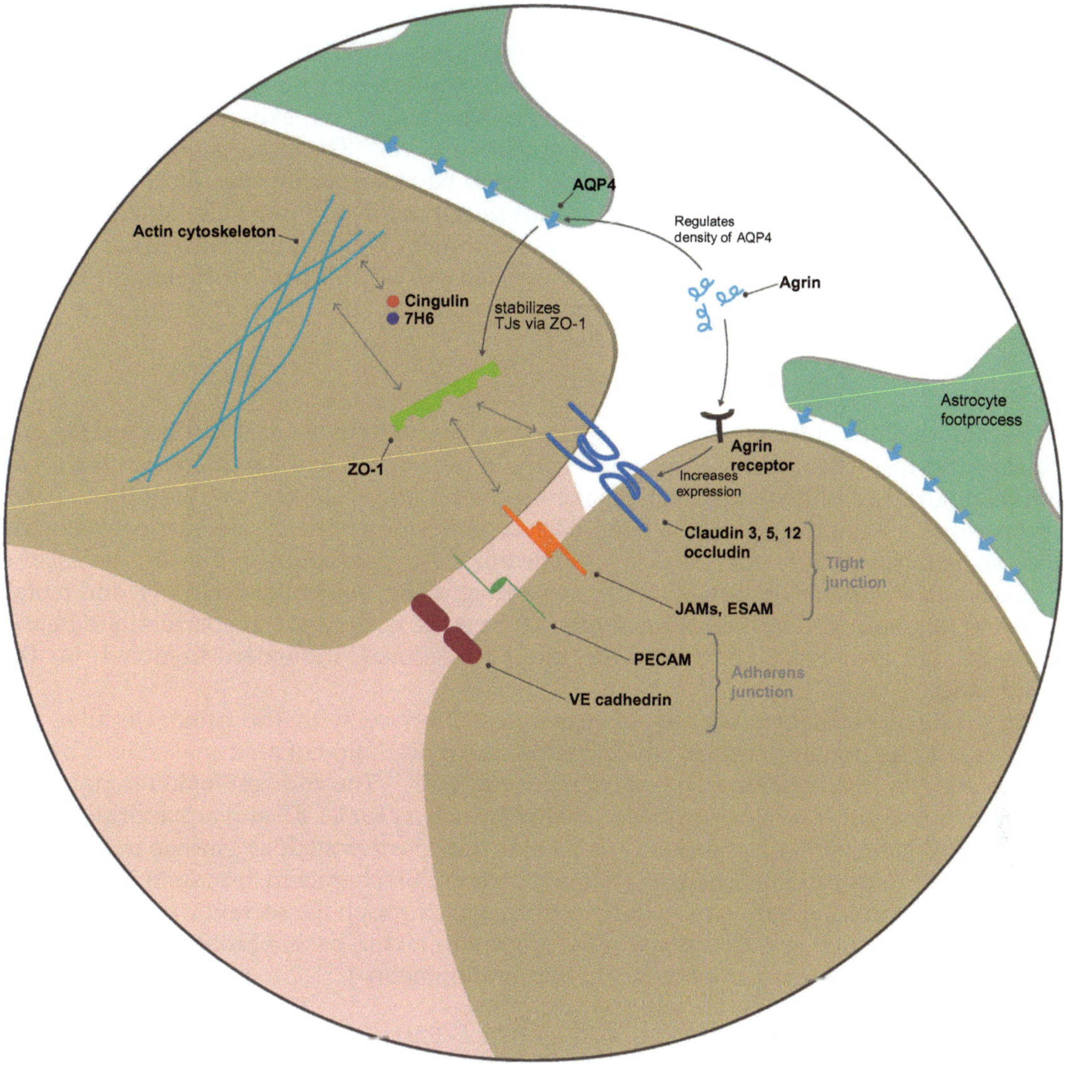

FIG. 1 Schematic of junctional proteins at the blood–brain barrier endothelial junction.

They are members of the membrane-associated guanylate-kinase (MAGUK) protein family and contain a PDZ-binding motif on the C terminus, linking ZO proteins with transmembrane proteins or with PDZ-binding motifs on other proteins (e.g., occludin and JAMs).[23] ZO-1 is involved in endothelial barrier formation, VE-cadherin-mediated cell tension, and actomyosin organization through its association with F-actin.[27]

Cellular Components of The Blood–Brain Barrier

Three main cell types make up the BBB: ECs, astrocytes, and pericytes. We will review each cell type in turn and describe their contributions to function of the BBB.

Endothelial Cells

The specialized brain microvascular ECs regulate the flow of substances into and out of the central nervous system (CNS). This regulation is accomplished by the formation of TJs, which limit the passive diffusion of molecules into the brain, as exhibited by very high transendothelial electrical resistance measurements in vivo.[28] The expression of efflux transporters such as P-glycoprotein (P-gp) contributes to barrier polarity and allows the CNS to exclude small lipophilic molecules that may diffuse across the BBB into ECs. Other properties unique to the BBB endothelium include a lower number of endocytic vesicles (compared to the peripheral endothelium), and a host of intra- and extracellular enzymes that further protect the brain by metabolizing various toxic compounds.[8]

Astrocytes

The endothelium of the cerebrovasculature is surrounded by a basal lamina and the end-feet of astrocytes. Astrocytes are the most common type of cell in the mammalian brain, and with microglia and oligodendrocytes, make up the broader category of glial cells.[29] Astrocytes perform a wide variety of functions, such as promotion of neurovascular coupling, cellular signaling through the release of chemokines, potassium buffering, release of gliotransmitters, release of glutamate by calcium signaling, control of brain pH, and metabolism and uptake of various neurotransmitters.[30] Specific receptors, transporters, and ligands expressed on astrocytes contribute to the bidirectional induction involved in BBB maintenance.[31]

The terminal end-feet processes of astrocytes interface with the brain vascular ECs, pericytes, and neuronal synapses, which enables the modulation of neuronal activity and cerebral blood flow by a calcium-dependent mechanism.[32, 33] The end-feet maintain polarity of astrocytes through expression of surface channels such as Kir4.1 K^+ and aquaporin 4, which regulate BBB ionic concentrations and molecular transporters such as glucose transporter-1 and P-gp.[31, 34] Astrocytes also help to maintain certain BBB characteristics, such as the formation of TJs and the polarization of surface transporters, through the secretion of paracrine signaling molecules such as vascular endothelial growth factor, basic fibroblast growth factor, glial cell line-derived neurotrophic factor, and angiopoietin-1.[35, 36]

Pericytes

Pericytes are cells that occupy the perivascular space and have been shown to be important regulators of the BBB. Experiments show that in a pericyte-deficient mouse model, the permeability of the BBB to water and a range of low-molecular–weight and high-molecular–weight mass tracers increases due to endothelial transcytosis.[37] Pericytes were also shown to be important in regulating BBB-specific gene expression patterns in ECs and inducing polarization of astrocyte end-feet abutting CNS blood vessels.[37] Cell culture experiments in which ECs cocultured with astrocytes and pericytes demonstrated that the presence of pericytes increases the resistance to apoptosis by ECs. These studies support a stabilizing function of pericytes in angiogenesis.[38] Further studies support the notion that pericytes play a role in vascular tone, stability, repair, angiogenesis, and modulation of astrocyte function.[39, 40]

Biochemical Components

The vascular endothelium expresses a multitude of transmembrane transporters that aid in maintaining the integrity and polarity of the BBB (Fig. 2). One such player is the efflux pump P-gp, a 170-kDa transmembrane protein belonging to the ABC transporter family and encoded by the *ABCB1* (or *MDR1*) gene.[41] At the BBB, the expression of P-gp is localized to the apical membrane, which facilitates transport in a unidirectional manner.[42, 43] Physiologically, the expression pattern of P-gp suggests that its normal physiological role is to protect the body from xenobiotic compounds by effluxing cytotoxic molecules into luminal spaces for elimination.[5] Due to the broad substrate specificity of P-gp, especially for chemotherapeutic drugs, the penetration of many CNS pharmacotherapy agents is dramatically decreased (Fig. 2).[44]

Although many P-gp substrates have been characterized by experimental evidence, computational models have predicted additional potential substrates,[45] including the discovery of NSC73306, a thiosemicarbazone derivative that is selectively toxic to cells expressing high levels of *MDR1*.[46] Additionally, many substrates effluxed by P-gp are also shared with similar drug efflux transporters such as *MRP1* (*ABCC2*)[47, 48]and *BCRP* (*ABCG2*), demonstrating that this process is subject to redundant mechanisms.

The importance of P-gp in the maintenance of brain homeostasis is further corroborated by knockout experiments in which *mdr1a*$^{-/-}$ and *mdr1b*$^{-/-}$ mice showed a 100-fold increase in sensitivity to ivermectin, a potent neurotoxin, in addition to dexamethasone, digoxin, and cyclosporin A.[49] Furthermore, in addition to BBB capillary endothelium, P-gp is also expressed in astrocytes, microglia, and neurons.[50-52] With increasing evidence that P-gp plays a central role in maintaining BBB integrity, P-gp is being increasingly targeted either through direct inhibition or modulation of expression and/or protein trafficking.

In addition to the normal physiological role of P-gp, overexpression of P-gp is a common feature of many multidrug-resistant tumors, which highlights the importance of this process in the pathogenesis and treatment of cancer.[53, 54] P-gp has been found to be expressed in glioma,[55-57] and expression levels were correlated with multidrug resistance and tumor grade. These studies showed that low-grade gliomas lack P-gp expression, and glioblastoma multiforme highly expresses P-gp. In relation to the BBB, P-gp expression is disrupted at necrotic core of glioblastoma multiforme, but the BBB and P-gp activity are fully intact at the tumor border.[58] This finding is clinically significant for glioblastoma multiforme after surgical resection, because residual border cells with an intact barrier and potential P-gp overexpression limits drug uptake and often relapses into a larger, more aggressive tumor.[59]

Other transporters in the ABC protein superfamily also play a role in BBB integrity and mechanisms of multidrug resistance. The multidrug resistance protein (MRP) family consists of nine proteins in mammals and is expressed at the BBB as *MRP1 (ABCC1)* and *MRP5 (ABCC5)*.[60, 61] These transporters are ubiquitous on BBB endothelium and are localized to either the abluminal[61] or the luminal surface,[62, 63] unlike P-gp, which is generally found on the luminal membrane. *MRP1* is the best-characterized transporter of the MRP family and has many functional similarities to P-gp, despite having only a 15% amino-acid homology. However, the localization pattern of *MRP1* varies from species to species[63, 64] and depends on the isolation procedure and cell systems used.[65-67]

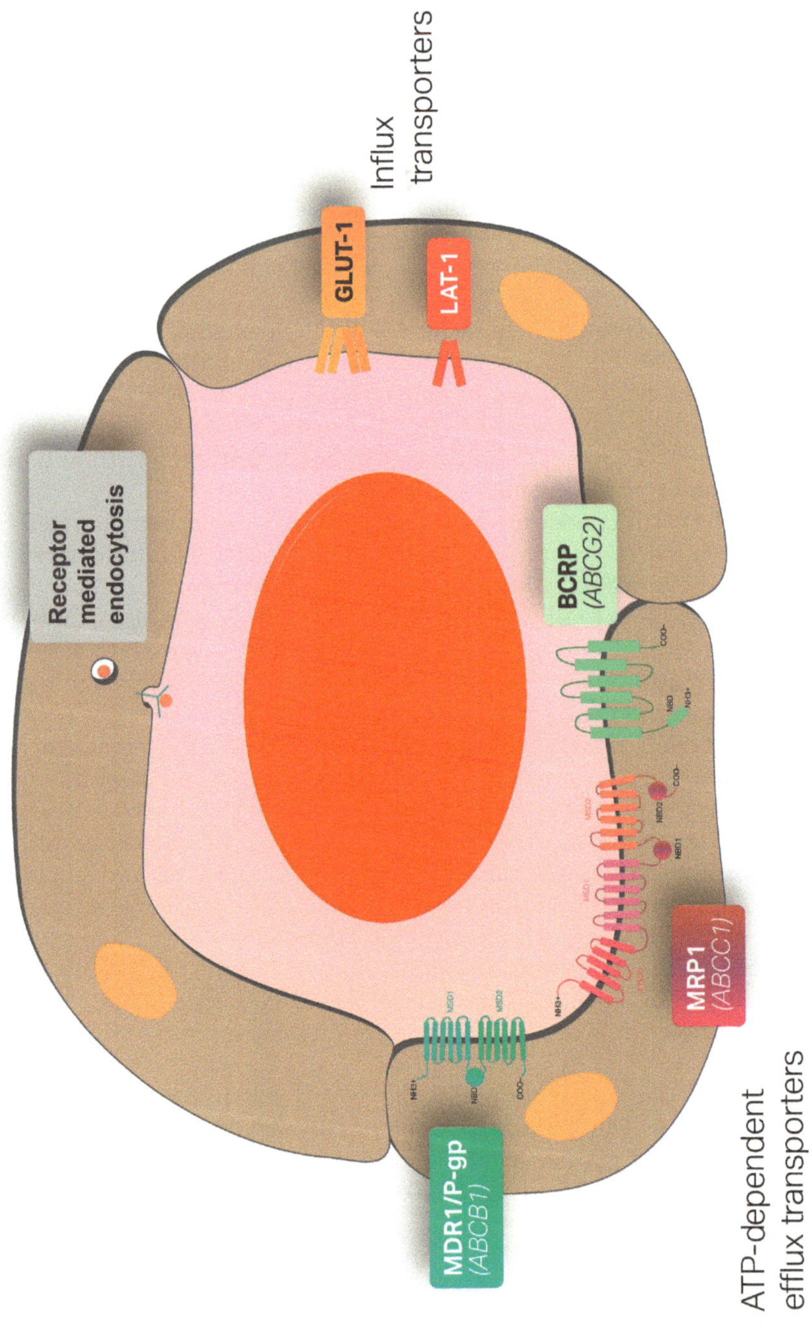

FIG. 2 Multiple mechanisms of efflux and influx into brain vascular endothelial cells exist. This is made possible by polarized facilitated transport systems such as glucose transporter 1 (GLUT1) and large neutral amino acid transporter (LAT1), and adenosine triphosphate-dependent transporters, such as *ABCB1* (P-glycoprotein [P-gp]), *ABCC1* (multidrug resistance-associated protein 1 [MRP1]), and *ABCG2* (breast cancer resistance protein [BCRP]). Receptor-mediated endocytosis is also a major mechanism by which the brain uptakes metabolites from the vasculature.

RATIONAL DRUG DESIGN

Rational drug design, as supposed to drug discovery, is a hypothesis-driven approach that relies on modification of a specific druggable target or existing compound to exert a physiological effect. Chemical modifications of existing drugs to alter properties such as solubility, membrane penetration, substrate for efflux transporters, and other characteristics may also improve their therapeutic functionalities.

Prodrugs

Prodrugs are compounds that must undergo chemical or enzymatic conversion by metabolic processes of the host organism before the compound becomes an active pharmacological agent with therapeutic effects.[68] The release of the active form of the drug can therefore be carefully controlled, and can occur before, during, or after absorption, or at a specific site within the body.[69, 70] The major goal of prodrugs is to overcome various physiochemical, pharmaceutical, biopharmaceutical, and/or pharmacokinetic limitations of the parent drug. Several approaches, described below, involve chemical modification of existing drugs, including "traditional" conjugation of lipophilic esters, chemical delivery systems, exploitation of endogenous transporter systems, and macromolecular delivery mechanisms.[71]

Lipophilic Analogs

Lipid solubility is an important factor in determining whether a drug can passively diffuse across the BBB. Therefore, strategies that chemically modify a drug to increase its lipophilic properties have been explored. This process involves conjugating lipid groups onto the polar ends of drug molecules to assemble lipophilic analogs, which theoretically should have greater cerebrovascular permeability compared to the original drug.[3] For example, the addition of methyl groups to barbiturates[72] and opioids[73] improved their lipophilicity and brain penetration, increasing their hypnotic and analgesic activity. Perhaps the best-known example of this strategy is the diacetylated form of morphine, also known as heroin, which crosses the BBB about 100-fold more efficiently than its unmodified counterpart.[74] Other conjugates reported to increase lipophilicity include dihydropyridine, adamantane, and fatty acyl groups such as *N*-docosahexaenoyl.[75] Modification of enkephalin analogs such as D-penicillamine by a para-chloro halogenation approach[76] or by glycosylation[77] has been reported to increase BBB transport and analgesic activity in vivo. Adding lipophilic groups onto molecules to improve membrane permeability can also be used to exploit native BBB nutrient transport systems, which have a natural affinity for lipophilic compounds. The BBB large neutral amino-acid carrier has been used to deliver L-DOPA for Parkinson disease[78] and D,L-NAM for CNS neoplasms.[79] The delivery of D-MOD—an analog of the antineoplastic agent D-melphalan—across the BBB improved transport across the BBB by exploiting the L-glutamate amino acid system.[80]

This strategy, however, has been limited by trade-offs in drug specificity and poor tissue distribution. Lipidization of drugs greatly enhances the volume of distribution, as these analogs can better bind circulating blood plasma proteins, which present problems in terms of systemic toxicity. Furthermore, lipophilic analogs have altered pharmacokinetics, such as the rate of metabolism by cytochrome P-450 enzymes.[81] Clinical efficacy also remains a challenge despite the demonstration of increased brain-to-plasma concentrations of lipophilic analogs. For example, lipophilic analogs of chlorambucil, an anticancer agent, showed increased penetration into the CNS, but ultimately did not show superior anticancer activity in disease models.[82, 83]

Chemical Drug Delivery Systems

The chemical drug delivery system (CDS) approach refers to a combination strategy of applying prodrug and lipophilic analog designs. The CDS is defined as a prodrug approach that also uses improved lipophilicity and subsequently requires a downstream metabolic conversion step to trap the intermediate in brain tissue.[75, 84] This is distinguished from a pure prodrug approach in that a CDS design requires multiple activation steps to achieve pharmaceutical activity.

A well-studied example of the CDS is linkage of an active drug to a cleavable moiety, 1,4-dihyro-*N*-methylnicotinic acid (1,4-Dihydrotrigonelline), which affords the target molecule significant lipophilic properties.[71] Once the drug distributes into the target organ (i.e., the brain), the lipophilic 1,4-dihydrotrigonelline moiety is oxidized to a cationic intermediate, which accelerates the rate of systemic elimination of this hydrophilic intermediate and captures the ionic drug in tissue of interest. This technique has been investigated in the setting of increasing brain penetration of zidovudine,[85] ganciclovir,[86] benzylpenicillin,[87] and estradiol.[88]

Estradiol-CDS (E2-CDS, Estredox) is the most developmentally advanced example of CDS, having reached Phase I and Phase II trials, for estrogen replacement in postmenopausal women, or for contraceptive purposes.[89] Estrogen is a lipophilic steroid hormone that can readily cross the BBB; however, it is poorly retained and requires frequent or sustained-release administrations. Following administration of E2-CDS, the compound undergoes metabolic transformations such as oxidation to a quaternary pyridinium ester, a much more hydrophilic compound. This intermediate is rapidly eliminated from the periphery;[90] however, it is trapped in the brain parenchyma by the BBB and remains there for an extended time.[91] This approach has been investigated in several preclinical models that have evaluated its effects on luteinizing hormone secretion, copulatory behavior in castrated rats, prostate weight, body weight, and hormone levels.[89]

Carrier-Mediated Transport

Endogenous influx transporters for molecules such as amino acids, glucose, and vitamins expressed on the luminal surfaces of BBB ECs can be used to facilitate drug transport.[92] The goal of this approach is to modify a drug such that it shares structural properties with endogenous substrates, which can be substrates for these transporters. A prodrug approach can be used so that conjugation of a known substrate to a drug will facilitate uptake of the substrate-drug complex, which is subsequently cleaved or modified inside the cell to its pharmacologically active target.

The large neutral amino acid transporter (LAT1) has been a promising target to facilitate uptake of chemically modified drugs. LAT1 is expressed on luminal and abluminal surfaces of BBB ECs, and is responsible for the transport of neutral L-amino acids into the brain.[93] Certain naturally occurring amino acid substrates (e.g., gabapentin and melphalan) are transported by LAT1-mediated transport; therefore, LAT1 was proposed to transport drugs similar in structure to those amino acid substrates. Perhaps the best example illustrating this approach in rational drug design is the development of L-DOPA for use in patients with Parkinson disease, as dopamine cannot cross the BBB.[94] L-DOPA is readily up taken via LAT1 and is decarboxylated into dopamine in brain tissue and in the peripheral circulation.[95] Despite a strong tendency for L-DOPA to be metabolized in the periphery, the remaining L-DOPA in the brain has been enough to exert a therapeutic benefit for patients with Parkinson disease, and has been the standard of care for more than 40 years.[96]

Drugs can also be conjugated with a moiety that is a LAT1 substrate (typically an amino acid) to facilitate transport across the BBB. Studies have suggested that conjugation of L-cysteine to the anticancer agent 6-mercaptopurine was sufficient to help the complex bind to LAT1, evidenced by a decrease in LAT1-mediated brain uptake of [14C]L-leucine.[97] A prodrug design approach to increasing brain uptake of dopamine by conjugation of dopamine to phenylalanine has shown promising results in a rat model.[98] Other studies involving L-tyrosine have found that conjugates of L-tyrosine and phosphoformate inhibited transport to [3H]L-tyrosine in porcine BBB ECs.[99]

Glucose transporter 1 is also expressed on luminal and abluminal surfaces of the BBB endothelium.[100] Glucose transporter 1 is responsible for the cellular uptake of glucose and other hexose sugars and has the highest transport capacity of carrier-mediated transporters at the BBB; therefore, it is an attractive target for prodrug delivery.[71] The conjugated moiety for this approach is glycosylation, and has been used to increase the brain uptake of 7-chlorokynurenic acid (an N-methyl-D-aspartate receptor antagonist) that produces dissociative and anticonvulsant effects, similar to ketamine.[101] L-DOPA and dopamine glycoside analogs have also been tested in morphine-induced locomotion mouse models and reserpine-induced hypolocomotion in rats.[102] Dopamine produgs, which involve using a succinyl linker to conjugate dopamine to glucose at C-3 and dopamine to galactose at C-6, were more active in reversing reserpine-induced hypolocomotion in rats compared to L-DOPA or its prodrugs. The C-6 galactose dopamine conjugate was most effective in reducing morphine-induced locomotion in mice. Although pharmacological efficacy was increased by conjugating dopamine to glycosides, the exact mechanism of action remains unclear.

PHYSICAL DISRUPTION OF THE BLOOD–BRAIN BARRIER

Osmotic Disruption

Intra-arterial (IA) infusion of a hypertonic solution through the carotid artery has been previously reported as a strategy to disrupt the BBB in a reversible manner.[103] Hypertonic agents include arabinose, lactamide, urea, saline, mannitol, and radiographic contrast materials. These agents work by creating an osmotic gradient in the vessel lumen, which results in dehydration of ECs and cerebral vessel vasodilation due to loss of fluids and modulation of the endothelial cytoskeleton from an influx of calcium.[104-106] These physiologic changes open TJs and allow movement of solutes through bulk flow and diffusion. Of the aforementioned agents, mannitol is the most commonly used in both human and animal studies, given its widespread availability.

In early preclinical studies, hypertonic solutions have been used to promote CNS penetration of various substances, including antibodies,[107] plasma proteins,[108] contrast agents,[109] chemotherapeutics,[110] and viruses.[111] In one study, the level of methotrexate in the brain and CSF after osmotic disruption of BBB increased 10- to 100-fold.[112] Administration of Na^+/Ca^{++} exchanger blocker prolonged the effect of hypertonic mannitol in rats up to 1 hour, indicating the importance of calcium regulation in TJ openings.[113] More recent animal studies have begun to define the dynamics of BBB opening during osmotic disruption.[114-117] The time window of BBB opening has been reported to range from 10 minutes to several hours after osmotic disruption, depending on the experimental condition. In an early case series of patients with primary CNS lymphoma, chemotherapy coupled with IA mannitol administration improved survival compared to patients who received cranial radiation alone.[118] Since then, multiple centers have reported the safety and efficacy of IA mannitol infusion to complement chemotherapy delivery;[119-122] however, prospective, multicenter trials are required to ascertain the benefits of osmotic disruption.

Magnetic Resonance Imaging-Guided Focused Ultrasound

When applied transcranially, bursts of focused ultrasound (FUS) have been shown to disrupt the BBB in a noninvasive, reversible manner.[123] The mechanism of such disruption, although unclear at the present, likely involves aspects of cellular shear stress, widening of TJs, and increased transcellular transport.[124-126] An additional component in ultrasound-based disruption of the BBB is the use of ultrasound contrast agent with preformed microbubbles. The microbubbles are injected prior to application of the ultrasound bursts. Microbubbles contract and expand in the presence of ultrasound, lowering the requirement for acoustic power to induce BBB disruption by two orders of magnitude.[123]

The typical preclinical setup for FUS involves the application of multiple bursts of ultrasound at low frequency (i.e., 1 Hz) to the desired cortical region.[127] The animal is anesthetized, with the scalp submerged in a bath of degassed water. Lipid microbubbles and ultrasound contrast agent are injected before BBB disruption. Finally, the effect of FUS is assessed by comparing magnetic resonance imaging (MRI) of the brain before and after FUS.

There are distinct advantages to this approach. It is noninvasive, reproducible, and allows for cortical targeting with high spatiotemporal resolution. Multiple studies have shown that the effects of FUS are transient and do not carry high risk of permanent tissue damage.[128-130] The integrity of the BBB generally recovers in 4-5 hours after FUS.[131] Multiple studies have demonstrated the feasibility of FUS in nonhuman primates.[132-134]

The preclinical applications of FUS have been demonstrated in numerous settings. Delivery of antibodies across the BBB has been shown after FUS;[135] in fact, the ultrasound-facilitated delivery of antibodies against toxic amyloid-beta peptide to the brain reduced Abeta plaque in targeted cortical regions in a mouse model of Alzheimer disease.[136] Similarly, trastuzumab, used in treatment of breast cancer, was demonstrated to cross the BBB under MRI-guided FUS, suggesting that this strategy may be useful in the treatment of intracranial metastasis.[137] FUS has also been shown to increase the intracranial concentration of several widely used chemotherapeutics, including doxorubicin, methotrexate, carmustine, and temozolomide.[138-141] Finally, a nanotechnology platform can be coupled with FUS to improve bioavailability of nanoparticles intracranially. Iron oxide magnetic nanoparticles and gold nanoparticles were shown to have improved uptake following ultrasound-mediated disruption of the BBB.[142] A limitation of FUS is energy attenuation by the skull, which makes targeting the superficial cortex especially difficult. One study has investigated the possibility of an intracranial FUS device, which would completely circumvent the issue of signal attenuation and would lower the FUS energy requirement.[143] Although FUS has shown promise in animal studies, future clinical studies are required to determine the safety and efficacy of this strategy in humans. Of note, high-energy FUS was approved in 2016 by the Food and Drug Administration (FDA) for tissue ablation in patients with essential tremor.[144] In 2017, a trial was initiated to study the use of FUS in patients diagnosed with glioblastoma (NCT03322813). These events highlight the safety of FUS in humans, and suggest great promise for FUS with microbubbles in targeting the BBB. FUS is discussed in greater detail in Chapter 12.

Bradykinin

Intra-arterial infusion of bradykinin has previously been shown to increase BBB permeability.[145] Although the precise mechanism by which bradykinin acts to increase endothelial permeability remains poorly defined, it was found to upregulate potassium adenosine 5′-triphosphate channels and caveolins.[146, 147] Bradykinin analogs have since been developed and have shown promising results in animal studies.[148, 149] However, clinical studies have failed to demonstrate added benefit of bradykinin analog in treatment of intracranial tumors.[150, 151] The main limitation of this strategy is the short BBB opening time after bradykinin administration,[147] which potentially limits the gain in intracranial drug concentration to a maximum of a twofold.[152] Additionally, the bradykinin analog was administered intravenously in the human trial, as opposed to IA administration previously done in animal studies, which may have limited therapeutic efficacy.[150]

Radiation

Increase in BBB permeability has been observed after irradiation to the brain in animal studies.[153-156] For instance, after a single high dose irradiation of rat's brain, extravasation of albumin as a measure of BBB permeability was seen as early as 1 day postradiation.[153] In a clinical study of 14 patients, researchers demonstrated a linear relationship between BBB permeability and the dose of radiation therapy.[157] In a separate study, irradiation of the brain increased the concentration of methotrexate in the CSF threefold, which suggested that chemotherapy should be administered after radiotherapy to optimize drug penetration.[158] These studies indicate that radiotherapy and chemotherapy may have synergistic effects if administered within a certain timeframe, although the ideal timing is still under investigation.

BIOCHEMICAL CIRCUMVENTION OF THE BLOOD–BRAIN BARRIER

Biochemical methods to circumvent the BBB differ from physical methods in that the vasculature remains relatively intact. Biochemical circulation of the BBB, therefore, relies on specific micromolecular processes to achieve increased drug penetration, decreased drug efflux, or both (Fig. 3).

Convection-Enhanced Delivery

Convection-enhanced delivery (CED) is a method of delivering therapeutics directly to the brain parenchyma via a surgically implanted catheter that provides continuous positive-pressure infusion. First described by Bobo et al. in 1994, CED bypasses the BBB, allowing for infusion of both large- and small-molecular-weight compounds, and enables the possibility of real-time magnetic resonance monitoring of infusate distribution in the brain.[159, 160] CED has been used to deliver various antineoplastic agents, including chemotherapeutic compounds, monoclonal antibodies, biologics, and liposomal vectors.[161-166] A recent murine study compared CED and systemic delivery of a liposomal chemotherapeutic agent and demonstrated that a single administration of drug via CED was significantly more effective than multiple systemic administrations.[167] Additionally, CED can be used to deliver multifunctional nanoparticles, which when coupled with a contrast material, lead to not only substantial therapeutic effects in animal glioma models but also facilitate accurate MRI tracking of the nanoparticles.[167, 168] However, to date no Phase III trials have demonstrated clinical efficacy with CED.[169] The major limitations of CED include fixed infusion time from the temporary catheter and poor drug distribution due to inconsistent catheter placement and reflux flow back up the catheter.[170-172] Current advances in catheter design and catheter localization aim to circumvent these limitations and improve the clinical effectiveness of CED, described in greater detail in Chapter 24.

Receptor-Mediated Transport

Receptor-mediated transport is an endogenous mechanism used by ECs to take up certain proteins from the blood into the brain. This process relies on the binding of a ligand to its cognate receptor on the apical surface of ECs, which then leads to endocytosis of the ligand and cellular trafficking to various cellular compartments.[173] For ligands destined for the brain, exocytosis takes place at the basolateral cell membrane, effectively transporting the ligand across the BBB. RMT has been exploited as a drug delivery system by coupling ligands that have receptors on the BBB ECs with the drug of interest. Various ligand-receptor systems have been described, and these systems have been successfully used preclinically to deliver antibodies, viruses, peptides, enzymes, and chemotherapeutics across the BBB.[174-179]

Additionally, RMT is compatible with liposomes and nanoparticles delivery. In either case, the vector (liposomes or nanoparticles) carrying the drug of interest is decorated with ligands, which then target the vector to the BBB.[180, 181] For instance, nanoparticles with covalently bonded apolipoprotein E are targeted to the brain by binding low-density lipoprotein receptors expressed by the BBB ECs.[181] For increased targeting precision, dual-ligand RMT systems

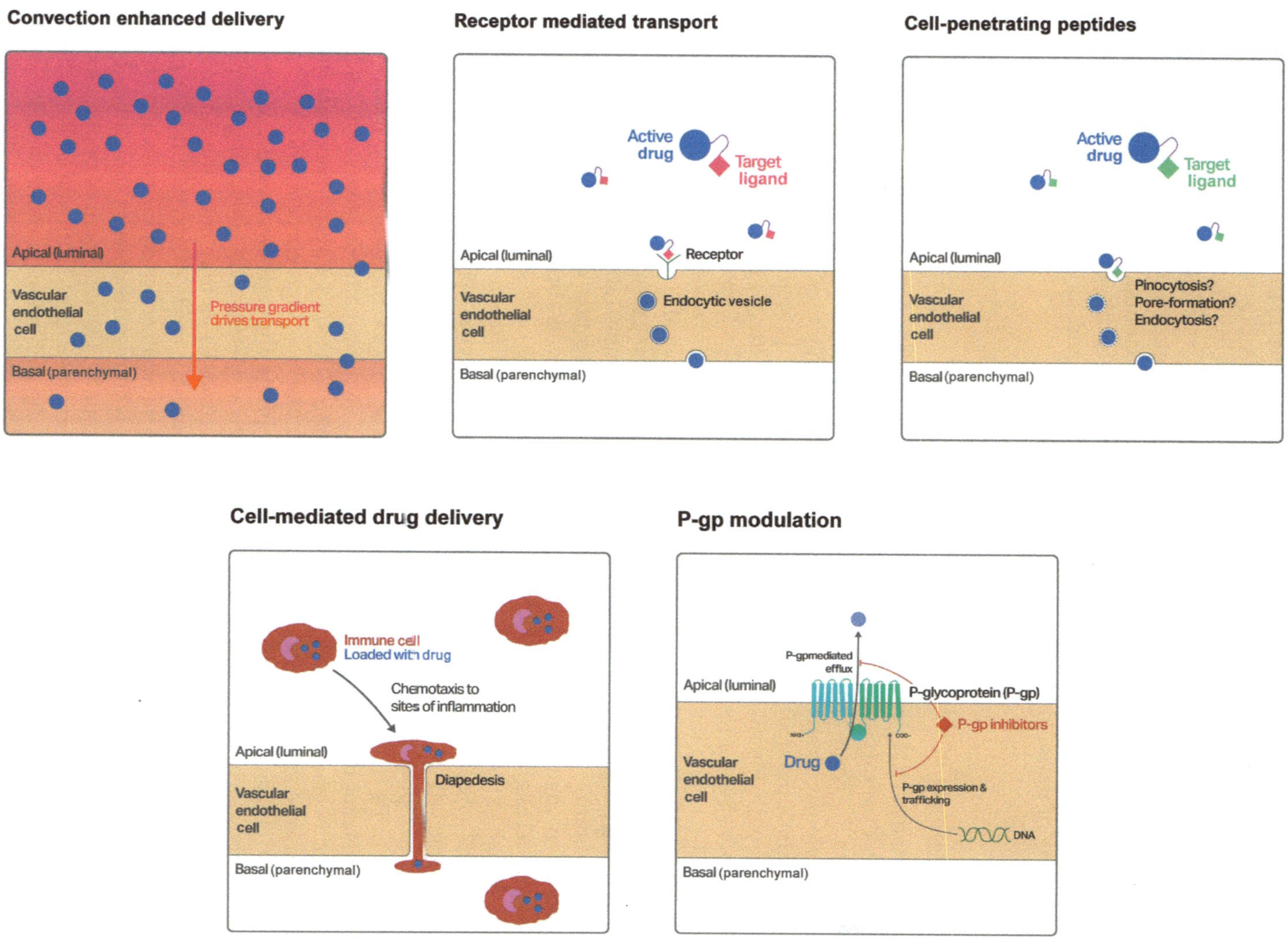

FIG. 3 Summary of various biochemical strategies to circumvent the blood–brain barrier.

have been developed. To target specifically glioma in the brain, Gao et al. conjugated nanoparticles carrying chemotherapeutics with two ligands, one targeting the BBB and the other targeting cancer cells.[182] This led to high intra-tumor concentration of the drug with minimal distribution in other parts of the brain. The major drawback of RMT is that the receptors used in BBB targeting are often expressed in the vasculature of other organs, which results in peripheral uptake.[183] Furthermore, drugs may accumulate in the BBB ECs due to limited transcytosis, preventing transport into the brain parenchyma.[184, 185] Numerous studies are underway to discover new BBB-specific receptors and to improve our mechanistic understanding of transcytosis.

Cell-Penetrating Peptides

Unlike ligands in RMT, which requires specific ligand-receptor interaction at the cell membrane, cell-penetrating peptides (CPPs) can pass through cell membranes independent of surface receptors.[186] These peptides, usually 5-30 amino acids in length, are derived from proteins previously shown to be shuttled across cell membranes.[187-190] A classic example of such protein is the *trans*-activator of transcription (TAT) protein from HIV-1, which was found to be rapidly internalized by cells.[191, 192] Subsequent studies identified a short peptide that was responsible for the membrane translocation of the TAT protein.[189, 193] In one clinical application of this peptide, the antiapoptotic protein Bcl-xL was conjugated to the peptide, creating a fusion protein that was administered intraperitoneally in mice after induction of cerebral ischemia.[194] Bcl-xL expression has previously been associated with neuronal survival following ischemia.[195] Indeed, in this study, the administration of TAT-Bcl-xL fusion protein significantly reduced cerebral infarction after 90 minutes of induced ischemia in a dose-dependent manner. The TAT translocation peptide has also been used to deliver siRNA to mouse model of glioblastoma.[196] In this case, the peptide is conjugated to a dsRNA binding domain, which is used to bind siRNAs. Delivery of this fusion protein carrying siRNA targeting Akt2 and EGFR mRNAs resulted in tumor-specific apoptosis and significantly increased longevity in mice. These results illustrate the ability of CPPs to cross the BBB and deliver therapeutics to brain tissues.

Although CPPs have been studied extensively, the precise mechanism by which CPPs cross the cell membrane is not fully clear. In general, two mechanisms have been described to account for the shuttling properties of CPPs.[197] First, CPPs may directly penetrate the cell membrane in an energy-independent fashion, particularly at high concentrations.[198, 199] In this mode of entry, CPPs localize to the cell membrane through electrostatic interaction and subsequently cross the membrane through several mechanisms, including temporary pore formation and membrane disruption.[200, 201] Second, CPPs may enter the cell through endocytosis, which has been found to be the predominant mode of entry.[202] By the nature of this process, CPPs and their conjugated cargos are internalized as endosomes and are subject to lysosomal degradation. Entrapment of CPPs within endosomes is thought to be the major limitation in delivering CPPs to their target tissue.[203] Development of peptide modifications that promote escape from endosomes have shown increased influx of CPPs to the target tissue.[204-206] The regulation of entry of CPPs at the BBB appears more complex, and a recent study has shown that, although most CPPs penetrate the BBB, there was no dependency on their cell-penetrating properties.[207] A mechanistic understanding of CPP entry at the BBB is required to develop CPPs with specificity to the brain and optimize therapeutic potential.

Cell-Mediated Drug Delivery

Targeted delivery of drugs at sites of CNS pathology has been difficult to optimize with conventional modes of drug transport. Delivery of therapeutics via living cells has emerged as an attractive approach to specifically target certain parts of the body. Immune cells, such as macrophages and neutrophils, migrate toward sites of inflammation through well-characterized processes of margination, diapedesis, and chemotaxis. Specifically, immune cells can readily cross the BBB, especially in settings of trauma or neuroinflammatory disorders.[208] As such, these cells can be loaded with therapeutics for targeted delivery into the brain. Batrakova et al. initially demonstrated that bone-marrow–derived macrophages (BMMs) loaded with a protected cargo of catalase delivered 0.6% of the injected dose to the brain.[209] Subsequent imaging studies showed that BMMs reached the brain by first moving along the vessel wall, then adhering to the ECs, and finally moving across the BBB.[210] To overcome degradation of drugs in phagosomes of BMMs, cellular backpacks were assembled layer by layer and used to carry therapeutics outside the BMMs.[211]

Bone-marrow–derived macrophages have also been used to deliver antineoplastic therapeutics to the brain in mouse models of glioma.[212] In this case, doxorubicin was encapsulated in nanoparticles to protect BMMs from its cytotoxic effects. Greater penetration of nanoparticles was seen when they were combined with cell-mediated delivery compared to when they were simply directly administered. Aside from macrophages, mesenchymal stem cells (MSCs) have also been used in drug delivery.[213-215] In addition to their ability to localize to sites of damage and malignancy, MSCs have regenerative capacity and can potentially restore tissue integrity after treatment. In one study, human MSCs were transduced to secrete tumor necrosis factor-related apoptosis-inducing ligand (S-TRAIL).[216] Injection of human MSCs producing S-TRAIL into mice with an orthotopic glioma implanted resulted in significantly reduced tumor growth and increased survival of the animals. Specifically, an eightfold increase in tumor apoptosis was observed in mice treated with human MSCs compared to control mice. Although the field of cell-mediated delivery is still in its infancy, this approach presents exciting opportunities to precisely deliver therapeutics for treatment of various intracranial pathologies.

Oncolytic Viruses

Genetically engineered viruses designed to target tumor cells, or oncolytic viruses, have been emerging as a promising strategy to target intracranial tumors.[217, 218] These viruses infect the tumor cells while leaving normal cells intact, causing tumor-specific cell lysis. Additionally, the viral genome can be manipulated to carry certain transgenes, the expression of which—in tumor cells—serves to augment the antineoplastic effect. For instance, interleukin-12 is a pro-inflammatory mediator that stimulates T-cell mediated tumor lysis and antiangiogenesis.[219] By genetically arming oncolytic herpes simplex virus (HSV) with murine interleukin-12, the virus significantly increased survival in an immunocompetent mice model of glioblastoma, compared to its unarmed counterpart virus.[220] This extended survival was due to local release of interleukin-12, which mediated T-cell-dependent killing. In another construct of oncolytic HSV, the gene for fusion protein endostatin-angiostatin was inserted into the viral genome, which resulted in potent inhibition of angiogenesis, and increased survival in mice with intracranial glioma compared to the unarmed control.[221]

One limitation of oncolytic viruses is that systemic administration of viruses intravenously may be difficult, especially if an adequate concentration is to be reached at the brain. As such, delivery of oncolytic viruses has recently been coupled to cell-mediated delivery to enable systemic administration.[220, 222] MSCs, pre-infected with oncolytic HSV, have been shown to hone in on sites of brain metastasis and to increase survival in mice bearing melanoma-derived brain tumors.[222] In a separate study, intravenous injection of MSCs carrying oncolytic HSV retargeted for HER2 receptor significantly reduced the number of brain metastases from breast carcinoma in mice.[223] Clinically, the FDA has recently approved oncolytic HSV-carrying granulocyte macrophage colony-stimulating factor (talimogene laherparepvec) for treatment of metastatic melanoma.[224] More clinical trials addressing the role of viral-mediated treatment approaches to intracranial tumors is expected within the next few years.

P-glycoprotein Modulation

Pharmacological interventions often fail in the brain due in part to P-gp-mediated efflux of small molecules out of brain tissue and back into the capillary lumen. Many strategies have been developed to circumvent the BBB, either through inhibition of P-gp or via modulation of its expression, trafficking, or both.

Direct inhibition of P-gp with small molecules and other pharmacological compounds has initially resulted in limited efficacy and safety in a clinical setting. However, recent advances in drug discovery have elucidated promising new molecules with nanomolar specificity and acceptable tolerability. First- and second-generation inhibitors had various limitations, including nonspecific toxicity and deleterious pharmacokinetic interactions with normal chemotherapeutic drugs.[225, 226] Third-generation inhibitors were developed using a rational structure–activity relationship and combinatorial chemistry approach. The most promising drug to result from this process is the anthranilamide derivative tariquidar,[227] which binds P-gp non-competitively at nanomolar concentrations. This drug has been shown to have sufficient activity against P-gp at the BBB in vivo. Using positron emission tomography, Kreisl et al. showed that tariquidar allowed a greater uptake of 11C-N-desmethyl-loperamide, a known P-gp substrate.[228] Patients who received 6 mg/kg of tariquidar intravenously showed a linear correlation of regional influx values with absolute cerebral blood flow, which is consistent with a single high-pass extraction of 11C-dLop. Combined with acceptable tolerability in combination with dose–linear responses, tariquidar shows promise to inhibit P-gp at the human BBB and to allow for effective drug delivery to the CNS.

Alternate approaches to P-gp inhibition in the context of functional expression and protein trafficking suggest that specific signaling pathways can be targeted to selectively reduce P-gp function and improve CNS drug delivery. This is an attractive strategy in contrast to global

P-gp inhibition, in that it offers more precise control of P-gp in a tissue- or tumor-dependent manner while simultaneously preserving basal P-gp expression, thereby maintaining baseline neuroprotection.[229] Cannon et al. used a confocal base activity assay in rat brain capillaries, and showed successful inhibition of P-gp trafficking activity by stimulating the sphingosine-1-phosphate (S1PR) receptor using sphingosine-1-phosphate (S1P) and fingolimod (FTY720, a drug approved for multiple sclerosis), a S1P analog. P-gp trafficking was reduced by a sphingosine kinase-dependent mechanism and in vivo studies in rats showed increased uptake of the P-gp substrates verapamil, loperamide, and paclitaxel upon administration of FTY720. Fine capillary structure (e.g., TJ integrity) was not compromised as measured by 14C–sucrose accumulation. Therefore, FTY720 can be explored as a possible drug to rapidly and reversibly reduce P-gp activity at the BBB.[230]

Pinzón-Daza et al.[231] have elucidated the role of crosstalk between canonical and noncanonical *Wnt* pathways and its relationship to P-gp expression in the human BBB. Inhibition of the RhoA/RhoAK axis by RhoAK inhibitor Y27632 downregulated beta-catenin and has been shown to decrease P-gp expression. In vitro RhoAK inhibition also enhanced delivery of doxorubicin, a P-gp substrate, across a BBB epithelial monolayer against glioblastoma cells.

Other small molecules such as sertraline, an FDA-approved antidepressant, has also been shown to inhibit P-gp acutely at the BBB[232] through an unknown mechanism. In vitro application of sertraline to BBB ECs demonstrated a rapid and potent inhibition of P-gp as measured by cellular calcein concentration. However, in vivo administration of sertraline to mice showed a biphasic accumulation of 3H-digoxin, in that sertraline concentration in BBB ECs rapidly increased 5 minutes after administration and fell to baseline levels 60 minutes later. However, 240 minutes after administration, sertraline concentration was again found to be elevated compared to controls. Combined with its rapid action, tissue specificity for the BBB and testes,[233] and FDA-approved status, sertraline can be an attractive therapeutic option to disrupt the BBB.

Modulation of P-gp has attracted much attention in disrupting the BBB in a noninvasive, specific, and rapid manner. However, despite numerous clinical trials involving P-gp inhibitors, none has been performed in any subjects with primary or metastatic neoplasms of the CNS.[234] Future studies should examine whether coadministration of P-gp inhibitors along with chemotherapy can (1) stop tumor growth and/or reduce tumor size, (2) result in prolonged survival, and (3) result in an outcome that avoids any long-term sequelae.[235] Of course, these studies will need to be carried out in animal models before translation to human trials.

CONCLUSIONS

The BBB plays a crucial role in neurological pathophysiology and is an important entity that can be manipulated for therapeutic purposes. Promising efforts in the treatment of neurological disease involve reversible modulation of the BBB in a noninvasive fashion coupled with precise and targeted delivery of pharmacologically active agents into the brain parenchyma. A multimodal approach will be needed to enhance influx of drugs into the brain and to inhibit the efflux of drugs out of the brain.

References

1. Banks WA. From blood-brain barrier to blood-brain interface: New opportunities for CNS drug delivery. *Nat Rev Drug Discov*. 2016;15(4):275-292.
2. Hawkins BT, Davis TP. The blood-brain barrier/neurovascular unit in health and disease. *Pharmacol Rev*. 2005; 57(2):173-185.
3. Lu C-T, Zhao Y-Z, Wong HL, Cai J, Peng L, Tian X-Q. Current approaches to enhance CNS delivery of drugs across the brain barriers. *Int J Nanomedicine*. 2014;9:2241-2257.
4. Reese TS, Karnovsky MJ. Fine structural localization of a blood-brain barrier to exogenous peroxidase. *J Cell Biol*. 1967;34(1):207-217.
5. Azad TD, Pan J, Connolly ID, Remington A, Wilson CM, Grant GA. Therapeutic strategies to improve drug delivery across the blood-brain barrier. *Neurosurg Focus*. 2015;38(3):E9.
6. Begley DJ, Brightman MW. Structural and functional aspects of the blood-brain barrier. *Prog Drug Res*. 2003;61:39-78.
7. Jones PM, George AM. The ABC transporter structure and mechanism: perspectives on recent research. *Cell Mol Life Sci*. 2004;61(6):682-699.
8. el-Bacha RS, Minn A. Drug metabolizing enzymes in cerebrovascular endothelial cells afford a metabolic protection to the brain. *Cell Mol Biol (Noisy-le-grand)*. 1999;45(1):15-23.
9. Furuse M, Hirase T, Itoh M, Nagafuchi A, Yonemura S, Tsukita S. Occludin: a novel integral membrane protein localizing at tight junctions. *J Cell Biol*. 1993;123(6 Pt 2):1777-1788.
10. Cummins PM. Occludin: one protein, many forms. *Mol Cell Biol*. 2012;32(2):242-250.
11. Morgan L, Shah B, Rivers LE, et al. Inflammation and dephosphorylation of the tight junction protein occludin in an experimental model of multiple sclerosis. *Neuroscience*. 2007;147(3):664-673.
12. Blasig IE, Bellmann C, Cording J, et al. Occludin protein family: oxidative stress and reducing conditions. *Antioxid Redox Signal*. 2011;15(5):1195-1219.
13. Saitou M, Fujimoto K, Doi Y, et al. Occludin-deficient embryonic stem cells can differentiate into polarized epithelial cells bearing tight junctions. *J Cell Biol*. 1998;141(2):397-408.
14. Saitou M, Furuse M, Sasaki H, et al. Complex phenotype of mice lacking occludin, a component of tight junction strands. *Mol Biol Cell*. 2000;11(12):4131-4142.
15. Mineta K, Yamamoto Y, Yamazaki Y, et al. Predicted expansion of the claudin multigene family. *FEBS Lett*. 2011;585(4):606-612.
16. Morita K, Furuse M, Fujimoto K, Tsukita S. Claudin multigene family encoding four-transmembrane domain protein components of tight junction strands. *Proc Natl Acad Sci U S A*. 1999;96(2):511-516.
17. Krause G, Winkler L, Piehl C, Blasig I, Piontek J, Müller SL. Structure and function of extracellular claudin domains. *Ann N Y Acad Sci*. 2009;1165(1):34-43.
18. Daneman R, Zhou L, Agalliu D, Cahoy JD, Kaushal A, Barres BA. The mouse blood-brain barrier transcriptome: a new resource for understanding the development and function of brain endothelial cells. *PLoS ONE*. 2010;5(10).
19. Ohtsuki S, Sato S, Yamaguchi H, Kamoi M, Asashima T, Terasaki T. Exogenous expression of claudin-5 induces barrier properties in cultured rat brain capillary endothelial cells. *J Cell Physiol*. 2007;210(1):81-86.
20. Nitta T, Hata M, Gotoh S, et al. Size-selective loosening of the blood-brain barrier in claudin-5-deficient mice. *J Cell Biol*. 2003;161(3):653-660.
21. Wolburg H, Wolburg-Buchholz K, Kraus J, et al. Localization of claudin-3 in tight junctions of the blood-brain barrier is selectively lost during experimental autoimmune encephalomyelitis and human glioblastoma multiforme. *Acta Neuropathol*. 2003;105(6):586-592.
22. Liebner S, Fischmann A, Rascher G, et al. Claudin-1 and claudin-5 expression and tight junction morphology are altered in blood vessels of human glioblastoma multiforme. *Acta Neuropathol*. 2000;100(3):323-331.
23. Greene C, Campbell M. Tight junction modulation of the blood brain barrier: CNS delivery of small molecules. *Tissue Barriers*. 2016;4(1).
24. Martìn-Padura I, Lostaglio S, Schneemann M, et al. Junctional adhesion molecule, a novel member of the immunoglobulin superfamily that distributes at intercellular junctions and modulates monocyte transmigration. *J Cell Biol*. 1998;142(1):117-127.
25. Bazzoni G, Martinez-Estrada OM, Orsenigo F, Cordenonsi M, Citi S, Dejana E. Interaction of junctional adhesion molecule with the tight junction components ZO-1, cingulin, and occludin. *J Biol Chem*. 2000;275(27):20520-20526.

26. Ebnet K, Schulz CU, Meyer Zu Brickwedde MK, Pendl GG, Vestweber D. Junctional adhesion molecule interacts with the PDZ domain-containing proteins AF-6 and ZO-1. *J Biol Chem*. 2000;275(36):27979-27988.
27. Tornavaca O, Chia M, Dufton N, et al. ZO-1 controls endothelial adherens junctions, cell-cell tension, angiogenesis, and barrier formation. *J Cell Biol*. 2015;208(6):821-838.
28. Butt AM, Jones HC, Abbott NJ. Electrical resistance across the blood-brain barrier in anaesthetized rats: a developmental study. *J Physiol Lond*. 1990;429:47-62.
29. Chen Y, Swanson RA. Astrocytes and brain injury. *J Cereb Blood Flow Metab*. 2003;23(2):137-149.
30. Cabezas R, Avila M, Gonzalez J, et al. Astrocytic modulation of blood brain barrier: perspectives on Parkinson's disease. *Front Cell Neurosci*. 2014;8:211.
31. Abbott NJ, Rönnbäck L, Hansson E. Astrocyte-endothelial interactions at the blood-brain barrier. *Nat Rev Neurosci*. 2006;7(1):41-53.
32. Zonta M, Angulo MC, Gobbo S, et al. Neuron-to-astrocyte signaling is central to the dynamic control of brain microcirculation. *Nat Neurosci*. 2003;6(1):43-50.
33. Maragakis NJ, Rothstein JD. Mechanisms of Disease: astrocytes in neurodegenerative disease. *Nat Clin Pract Neurol*. 2006;2(12):679-689.
34. Nag S. Morphology and properties of brain endothelial cells. *Methods in Molecular Biology*; vol. 686:2011:3-47 [chapter 1].
35. Alvarez JI, Katayama T, Prat A. Glial influence on the blood brain barrier. *Glia*. 2013;61(12):1939-1958.
36. Wong AD, Ye M, Levy AF, Rothstein JD, Bergles DE, Searson PC. The blood-brain barrier: an engineering perspective. *Front Neuroeng*. 2013;6:7.
37. Armulik A, Genové G, Mäe M, et al. Pericytes regulate the blood-brain barrier. *Nature*. 2010;468(7323):557-561.
38. Rascher G, Fischmann A, Kröger S, Duffner F, Grote E-H, Wolburg H. Extracellular matrix and the blood-brain barrier in glioblastoma multiforme: spatial segregation of tenascin and agrin. *Acta Neuropathol*. 2002; 104(1):85-91.
39. Kutcher ME, Herman IM. The pericyte: cellular regulator of microvascular blood flow. *Microvasc Res*. 2009;77 (3):235-246.
40. Tamai I, Tsuji A. Transporter-mediated permeation of drugs across the blood-brain barrier. *J Pharm Sci*. 2000; 89(11):1371-1388.
41. Ambudkar SV, Lelong IH, Zhang J, Cardarelli CO, Gottesman MM, Pastan I. Partial purification and reconstitution of the human multidrug-resistance pump: characterization of the drug-stimulatable ATP hydrolysis. *Proc Natl Acad Sci U S A*. 1992;89(18):8472-8476.
42. Fung KL, Pan J, Ohnuma S, et al. MDR1 synonymous polymorphisms alter transporter specificity and protein stability in a stable epithelial monolayer. *Cancer Res*. 2014;74(2):598-608.
43. Thiebaut F, Tsuruo T, Hamada H, Gottesman MM, Pastan I, Willingham MC. Cellular localization of the multidrug-resistance gene product P-glycoprotein in normal human tissues. *Proc Natl Acad Sci U S A*. 1987; 84(21):7735-7738.
44. Schinkel A. P-Glycoprotein, a gatekeeper in the blood-brain barrier. *Adv Drug Deliv Rev*. 1999;36(2-3):179-194.
45. Szakács G, Annereau J-P, Lababidi S, et al. Predicting drug sensitivity and resistance: profiling ABC transporter genes in cancer cells. *Cancer Cell*. 2004;6(2):129-137.
46. Ludwig JA, Szakács G, Martin SE, et al. Selective toxicity of NSC73306 in MDR1-positive cells as a new strategy to circumvent multidrug resistance in cancer. *Cancer Res*. 2006;66(9):4808-4815.
47. Renes J, de Vries EG, Nienhuis EF, Jansen PL, Müller M. ATP- and glutathione-dependent transport of chemotherapeutic drugs by the multidrug resistance protein MRP1. *Br J Pharmacol*. 1999;126(3):681-688.
48. Doyle LA, Yang W, Abruzzo LV, et al. A multidrug resistance transporter from human MCF-7 breast cancer cells. *Proc Natl Acad Sci U S A*. 1998;95(26):15665-15670.
49. Schinkel AH, Wagenaar E, van Deemter L, Mol CA, Borst P. Absence of the mdr1a P-Glycoprotein in mice affects tissue distribution and pharmacokinetics of dexamethasone, digoxin, and cyclosporin A. *J Clin Invest*. 1995; 96(4):1698-1705.
50. Golden PL, Pardridge WM. P-Glycoprotein on astrocyte foot processes of unfixed isolated human brain capillaries. *Brain Res*. 1999;819(1-2):143-146.
51. Lee G, Dallas S, Hong M, Bendayan R. Drug transporters in the central nervous system: brain barriers and brain parenchyma considerations. *Pharmacol Rev*. 2001;53(4):569-596.
52. Volk HA, Burkhardt K, Potschka H, Chen J, Becker A, Löscher W. Neuronal expression of the drug efflux transporter P-glycoprotein in the rat hippocampus after limbic seizures. *Neuroscience*. 2004;123(3):751-759.
53. Pastan I, Gottesman MM. Multidrug resistance. *Annu Rev Med*. 1991;42(1):277-286.

54. Juliano RL, Ling V. A surface glycoprotein modulating drug permeability in Chinese hamster ovary cell mutants. *Biochim Biophys Acta*. 1976;455(1):152-162.
55. Leweke F, Damian MS, Schindler C, Schachenmayr W. Multidrug resistance in glioblastoma chemosensitivity testing and immunohistochemical demonstration of P-glycoprotein. *Pathol Res Pract*. 1998;194(3):149-155.
56. Henson JW, Cordon-Cardo C, Posner JB. P-glycoprotein expression in brain tumors. *J Neuro-Oncol*. 1992; 14(1):37-43.
57. Matsumoto T, Tani E, Kaba K, Shindo H, Miyaji K. Expression of P-glycoprotein in human glioma cell lines and surgical glioma specimens. *J Neurosurg*. 1991;74(3):460-466.
58. Deeken JF, Löscher W. The blood-brain barrier and cancer: transporters, treatment, and Trojan horses. *Clin Cancer Res*. 2007;13(6):1663-1674.
59. Cheshier SH, Kalani MYS, Lim M, Ailles L, Huhn SL, Weissman IL. A neurosurgeon's guide to stem cells, cancer stem cells, and brain tumor stem cells. *Neurosurgery*. 2009;65(2):237-49-discussion249-50-quizN6.
60. Kruh GD, Guo Y, Hopper-Borge E, Belinsky MG, Chen Z-S. ABCC10, ABCC11, and ABCC12. *Pflugers Arch*. 2007;453(5):675-684.
61. Schinkel AH, Jonker JW. Mammalian drug efflux transporters of the ATP binding cassette (ABC) family: an overview. *Adv Drug Deliv Rev*. 2003;55(1):3-29.
62. Zhang Y, Schuetz JD, Elmquist WF, Miller DW. Plasma membrane localization of multidrug resistance-associated protein homologs in brain capillary endothelial cells. *J Pharmacol Exp Ther*. 2004;311(2):449-455.
63. Soontornmalai A, Vlaming MLH, Fritschy J-M. Differential, strain-specific cellular and subcellular distribution of multidrug transporters in murine choroid plexus and blood-brain barrier. *Neuroscience*. 2006;138(1):159-169.
64. Dallas S, Miller DS, Bendayan R. Multidrug resistance-associated proteins: expression and function in the central nervous system. *Pharmacol Rev*. 2006;58(2):140-161.
65. Sugiyama Y, Kusuhara H, Suzuki H. Kinetic and biochemical analysis of carrier-mediated efflux of drugs through the blood-brain and blood-cerebrospinal fluid barriers: importance in the drug delivery to the brain. *J Control Release*. 1999;62(1-2):179-186.
66. Gutmann H, Fricker G, Drewe J, Toeroek M, Miller DS. Interactions of HIV protease inhibitors with ATP-dependent drug export proteins. *Mol Pharmacol*. 1999;56(2):383-389.
67. Seetharaman S, Barrand MA, Maskell L, Scheper RJ. Multidrug resistance-related transport proteins in isolated human brain microvessels and in cells cultured from these isolates. *J Neurochem*. 1998;70(3):1151-1159.
68. ALBERT A. Chemical aspects of selective toxicity. *Nature*. 1958;182(4633):421-422.
69. Stella VJ, Charman WN, Naringrekar VH. Prodrugs. Do they have advantages in clinical practice? *Drugs*. 1985; 29(5):455-473.
70. Sinkula AA, Yalkowsky SH. Rationale for design of biologically reversible drug derivatives: prodrugs. *J Pharm Sci*. 1975;64(2):181-210.
71. Rautio J, Laine K, Gynther M, Savolainen J. Prodrug approaches for CNS delivery. *AAPS J*. 2008;10(1):92-102.
72. Greig NH. Drug delivery to the brain by blood-brain barrier circumvention and drug modification. In: *Implications of the Blood-Brain Barrier and Its Manipulation*. Boston, MA: Springer; 1989:311-367.
73. Hansen DW, Stapelfeld A, Savage MA, et al. Systemic analgesic activity and delta-opioid selectivity in [2,6-dimethyl-Tyr1,D-Pen2,D-Pen5]enkephalin. *J Med Chem*. 1992;35(4):684-687.
74. Oldendorf WH, Hyman S, Braun L, Oldendorf SZ. Blood-brain barrier: penetration of morphine, codeine, heroin, and methadone after carotid injection. *Science*. 1972;178(4064):984-986.
75. Bodor N, Buchwald P. Drug targeting via retrometabolic approaches. *Pharmacol Ther*. 1997;76(1-3):1-27.
76. Weber SJ, Greene DL, Sharma SD, et al. Distribution and analgesia of [3H][D-Pen2, D-Pen5]enkephalin and two halogenated analogs after intravenous administration. *J Pharmacol Exp Ther*. 1991;259(3):1109-1117.
77. Egleton RD, Mitchell SA, Huber JD, et al. Improved bioavailability to the brain of glycosylated Met-enkephalin analogs. *Brain Res*. 2000;881(1):37-46.
78. Wade LA, Katzman R. Rat brain regional uptake and decarboxylation of L-DOPA following carotid injection. *Am J Phys*. 1975;228(2):352-359.
79. Takada Y, Greig NH, Vistica DT, Rapoport SI, Smith QR. Affinity of antineoplastic amino acid drugs for the large neutral amino acid transporter of the blood-brain barrier. *Cancer Chemother Pharmacol*. 1991;29(2):89-94.
80. Yoshikawa T, Sakaeda nee Kakutani T, Sugawara T, Hirano K, Stella V. A novel chemical delivery system for brain targeting. *Adv Drug Deliv Rev*. 1999;36(2-3):255-275.
81. Bodor N, Buchwald P. Recent advances in the brain targeting of neuropharmaceuticals by chemical delivery systems. *Adv Drug Deliv Rev*. 1999;36(2-3):229-254.

82. Genka S, Deutsch J, Shetty UH, et al. Development of lipophilic anticancer agents for the treatment of brain tumors by the esterification of water-soluble chlorambucil. *Clin Exp Metastasis*. 1993;11(2):131-140.
83. Greig NH, Genka S, Daly EM, Sweeney DJ, Rapoport SI. Physicochemical and pharmacokinetic parameters of seven lipophilic chlorambucil esters designed for brain penetration. *Cancer Chemother Pharmacol*. 1990; 25(5):311-319.
84. Prokai L, Prokai-Tatrai K, Bodor N. Targeting drugs to the brain by redox chemical delivery systems. *Med Res Rev*. 2000;20(5):367-416.
85. Brewster ME, Anderson WR, Helton DO, Bodor N, Pop E. Dose-dependent brain delivery of zidovudine through the use of a zidovudine chemical delivery system. *Pharm Res*. 1995;12(5):796-798.
86. Brewster ME, Raghavan K, Pop E, Bodor N. Enhanced delivery of ganciclovir to the brain through the use of redox targeting. *Antimicrob Agents Chemother*. 1994;38(4):817-823.
87. Wu WM, Pop E, Shek E, Bodor N. Brain-specific chemical delivery systems for beta-lactam antibiotics. In vitro and in vivo studies of some dihydropyridine and dihydroisoquinoline derivatives of benzylpenicillin in rats. *J Med Chem*. 1989;32(8):1782-1788.
88. Estes KS, Brewster ME, Simpkins JW, Bodor N. A novel redox system for CNS-directed delivery of estradiol causes sustained LH suppression in castrate rats. *Life Sci*. 1987;40(13):1327-1334.
89. Buchwald P, Bodor N. Brain-targeting chemical delivery systems and their cyclodextrin-based formulations in light of the contributions of Marcus E. Brewster. *J Pharm Sci*. 2016;105(9):2589-2600.
90. Sarkar DK, Friedman SJ, Yen SS, Frautschy SA. Chronic inhibition of hypothalamic-pituitary-ovarian axis and body weight gain by brain-directed delivery of estradiol-17 beta in female rats. *Neuroendocrinology*. 1989; 50(2):204-210.
91. Brewster ME, Simpkins JW, Bodor N. Brain-targeted delivery of estrogens. *Rev Neurosci*. 1990;2(4):241-285.
92. Pardridge WM, Oldendorf WH. Transport of metabolic substrates through the blood-brain barrier. *J Neurochem*. 1977;28(1):5-12.
93. Boado RJ, Li JY, Nagaya M, Zhang C, Pardridge WM. Selective expression of the large neutral amino acid transporter at the blood-brain barrier. *Proc Natl Acad Sci U S A*. 1999;96(21):12079-12084.
94. Fernández C, Nieto O, Fontenla JA, Rivas E, de Ceballos ML, Fernández-Mayoralas A. Synthesis of glycosyl derivatives as dopamine prodrugs: interaction with glucose carrier GLUT-1. *Org Biomol Chem*. 2003;1(5):767-771.
95. Gomes P, Soares-da-Silva P. L-DOPA transport properties in an immortalised cell line of rat capillary cerebral endothelial cells, RBE 4. *Brain Res*. 1999;829(1-2):143-150.
96. Mena I, Cotzias GC. Protein intake and treatment of Parkinson's disease with levodopa. *N Engl J Med*. 1975; 292(4):181-184.
97. Killian DM, Hermeling S, Chikhale PJ. Targeting the cerebrovascular large neutral amino acid transporter (LAT1) isoform using a novel disulfide-based brain drug delivery system. *Drug Deliv*. 2007;14(1):25-31.
98. Peura L, Malmioja K, Huttunen K, et al. Design, synthesis and brain uptake of LAT1-targeted amino acid prodrugs of dopamine. *Pharm Res*. 2013;30(10):2523-2537.
99. Walker I, Nicholls D, Irwin WJ, Freeman S. Drug delivery via active transport at the blood-brain barrier: affinity of a prodrug of phosphonoformate for the large amino acid transporter. *Int J Pharm*. 1994;104(2):157-167.
100. Farrell CL, Pardridge WM. Blood-brain barrier glucose transporter is asymmetrically distributed on brain capillary endothelial lumenal and ablumenal membranes: an electron microscopic immunogold study. *Proc Natl Acad Sci U S A*. 1991;88(13):5779-5783.
101. Battaglia G, La Russa M, Bruno V, et al. Systemically administered D-glucose conjugates of 7-chlorokynurenic acid are centrally available and exert anticonvulsant activity in rodents. *Brain Res*. 2000;860(1-2):149-156.
102. Bonina F, Puglia C, Rimoli MG, et al. Glycosyl derivatives of dopamine and L-dopa as anti-Parkinson prodrugs: synthesis, pharmacological activity and in vitro stability studies. *J Drug Target*. 2003;11(1):25-36.
103. Rapoport SI, Hori M, Klatzo I. Testing of a hypothesis for osmotic opening of the blood-brain barrier. *Am J Phys*. 1972;223(2):323-331.
104. Rapoport SI, Robinson PJ. Tight-junctional modification as the basis of osmotic opening of the blood-brain barrier. *Ann N Y Acad Sci*. 1986;481:250-267.
105. Rosomoff HL. Distribution of intracranial contents after hypertonic urea. *J Neurosurg*. 1962;19(10):859-864.
106. Rapoport SI. Osmotic opening of the blood-brain barrier: principles, mechanism, and therapeutic applications. *Cell Mol Neurobiol*. 2000;20(2):217-230.
107. Hicks JT, Albrecht P, Rapoport SI. Entry of neutralizing antibody to measles into brain and cerebrospinal fluid of immunized monkeys after osmotic opening of the blood-brain barrier. *Exp Neurol*. 1976;53(3):768-779.

108. Greig NH, Fredericks WR, Holloway HW, Soncrant TT, Rapoport SI. Delivery of human interferon-alpha to brain by transient osmotic blood-brain barrier modification in the rat. *J Pharmacol Exp Ther*. 1988;245(2):581-586.
109. Takahashi M, Tsutsui H, Murayama C, Miyazawa T, Fritz-Zieroth B. Neurotoxicity of gadolinium contrast agents for magnetic resonance imaging in rats with osmotically disrupted blood-brain barrier. *Magn Reson Imaging*. 1996;14(6):619-623.
110. Ohata M, Fredericks WR, Neuwelt EA, Sundaram U, Rapoport SI. [3H]Methotrexate loss from the rat brain following enhanced uptake by osmotic opening of the blood-brain barrier. *Cancer Res*. 1985;45(3):1092-1096.
111. Muldoon LL, Nilaver G, Kroll RA, et al. Comparison of intracerebral inoculation and osmotic blood-brain barrier disruption for delivery of adenovirus, herpesvirus, and iron oxide particles to normal rat brain. *Am J Pathol*. 1995;147(6):1840-1851.
112. Neuwelt EA, Frenkel EP, Rapoport S, Barnett P. Effect of osmotic blood-brain barrier disruption on methotrexate pharmacokinetics in the dog. *Neurosurgery*. 1980;7(1):36-43.
113. Bhattacharjee AK, Nagashima T, Kondoh T, Tamaki N. The effects of the Na(+)/Ca(++) exchange blocker on osmotic blood-brain barrier disruption. *Brain Res*. 2001;900(2):157-162.
114. Bhattacharjee AK, Nagashima T, Kondoh T, Tamaki N. Quantification of early blood-brain barrier disruption by in situ brain perfusion technique. *Brain Res Brain Res Protoc*. 2001;8(2):126-131.
115. Zünkeler B, Carson RE, Olson J, et al. Quantification and pharmacokinetics of blood-brain barrier disruption in humans. *J Neurosurg*. 1996;85(6):1056-1065.
116. Joshi S, Ergin A, Wang M, et al. Inconsistent blood brain barrier disruption by intraarterial mannitol in rabbits: implications for chemotherapy. *J Neuro-Oncol*. 2011;104(1):11-19.
117. Joshi S, Singh-Moon R, Wang M, Bruce JN, Bigio IJ, Mayevsky A. Real-time hemodynamic response and mitochondrial function changes with intracarotid mannitol injection. *Brain Res*. 2014;1549:42-51.
118. Neuwelt EA, Goldman DL, Dahlborg SA, et al. Primary CNS lymphoma treated with osmotic blood-brain barrier disruption: prolonged survival and preservation of cognitive function. *J Clin Oncol*. 1991;9(9):1580-1590.
119. Jahnke K, Kraemer DF, Knight KR, et al. Intraarterial chemotherapy and osmotic blood-brain barrier disruption for patients with embryonal and germ cell tumors of the central nervous system. *Cancer*. 2008;112(3):581-588.
120. Fortin D, Gendron C, Boudrias M, Garant M-P. Enhanced chemotherapy delivery by intraarterial infusion and blood-brain barrier disruption in the treatment of cerebral metastasis. *Cancer*. 2007;109(4):751-760.
121. Fortin D, Desjardins A, Benko A, Niyonsega T, Boudrias M. Enhanced chemotherapy delivery by intraarterial infusion and blood-brain barrier disruption in malignant brain tumors: the Sherbrooke experience. *Cancer*. 2005;103(12):2606-2615.
122. Doolittle ND, Miner ME, Hall WA, et al. Safety and efficacy of a multicenter study using intraarterial chemotherapy in conjunction with osmotic opening of the blood-brain barrier for the treatment of patients with malignant brain tumors. *Cancer*. 2000;88(3):637-647.
123. Hynynen K, McDannold N, Vykhodtseva N, Jolesz FA. Noninvasive MR imaging-guided focal opening of the blood-brain barrier in rabbits. *Radiology*. 2001;220(3):640-646.
124. Sheikov N, McDannold N, Vykhodtseva N, Jolesz F, Hynynen K. Cellular mechanisms of the blood-brain barrier opening induced by ultrasound in presence of microbubbles. *Ultrasound Med Biol*. 2004;30(7):979-989.
125. VanBavel E. Effects of shear stress on endothelial cells: possible relevance for ultrasound applications. *Prog Biophys Mol Biol*. 2007;93(1-3):374-383.
126. van Wamel A, Kooiman K, Emmer M, Cate ten FJ, Versluis M, de Jong N. Ultrasound microbubble induced endothelial cell permeability. *J Control Release*. 2006;116(2):e100-e102.
127. Etame AB, Diaz RJ, Smith CA, Mainprize TG, Hynynen K, Rutka JT. Focused ultrasound disruption of the blood-brain barrier: a new frontier for therapeutic delivery in molecular neurooncology. *Neurosurg Focus*. 2012;32(1):E3.
128. Baseri B, Choi JJ, Tung Y-S, Konofagou EE. Multi-modality safety assessment of blood-brain barrier opening using focused ultrasound and definity microbubbles: a short-term study. *Ultrasound Med Biol*. 2010; 36(9):1445-1459.
129. Mesiwala AH, Farrell L, Wenzel HJ, et al. High-intensity focused ultrasound selectively disrupts the blood-brain barrier in vivo. *Ultrasound Med Biol*. 2002;28(3):389-400.
130. Hynynen K, McDannold N, Vykhodtseva N, Jolesz FA. Non-invasive opening of BBB by focused ultrasound. *Acta Neurochir Suppl*. 2003;86:555-558.
131. Sheikov N, McDannold N, Sharma S, Hynynen K. Effect of focused ultrasound applied with an ultrasound contrast agent on the tight junctional integrity of the brain microvascular endothelium. *Ultrasound Med Biol*. 2008; 34(7):1093-1104.

132. Tung Y-S, Marquet F, Teichert T, Ferrera V, Konofagou EE. Feasibility of noninvasive cavitation-guided blood-brain barrier opening using focused ultrasound and microbubbles in nonhuman primates. *Appl Phys Lett.* 2011;98(16).
133. McDannold N, Arvanitis CD, Vykhodtseva N, Livingstone MS. Temporary disruption of the blood-brain barrier by use of ultrasound and microbubbles: safety and efficacy evaluation in rhesus macaques. *Cancer Res.* 2012; 72(14):3652-3663.
134. Marquet F, Tung Y-S, Teichert T, Ferrera VP, Konofagou EE. Noninvasive, transient and selective blood-brain barrier opening in non-human primates in vivo. *PLoS ONE.* 2011;6(7).
135. Kinoshita M, McDannold N, Jolesz FA, Hynynen K. Targeted delivery of antibodies through the blood-brain barrier by MRI-guided focused ultrasound. *Biochem Biophys Res Commun.* 2006;340(4):1085-1090.
136. Jordão JF, Ayala-Grosso CA, Markham K, et al. Antibodies targeted to the brain with image-guided focused ultrasound reduces amyloid-beta plaque load in the TgCRND8 mouse model of Alzheimer's disease. *PLoS ONE.* 2010;5(5).
137. Kinoshita M, McDannold N, Jolesz FA, Hynynen K. Noninvasive localized delivery of Herceptin to the mouse brain by MRI-guided focused ultrasound-induced blood-brain barrier disruption. *Proc Natl Acad Sci U S A.* 2006;103(31):11719-11723.
138. Wei K-C, Chu P-C, H-YJ W, et al. Focused ultrasound-induced blood-brain barrier opening to enhance temozolomide delivery for glioblastoma treatment: a preclinical study. *PLoS ONE.* 2013;8(3).
139. Liu H-L, Hua M-Y, Chen P-Y, et al. Blood-brain barrier disruption with focused ultrasound enhances delivery of chemotherapeutic drugs for glioblastoma treatment. *Radiology.* 2010;255(2):415-425.
140. Mei J, Cheng Y, Song Y, et al. Experimental study on targeted methotrexate delivery to the rabbit brain via magnetic resonance imaging-guided focused ultrasound. *J Ultrasound Med.* 2009;28(7):871-880.
141. Treat LH, McDannold N, Vykhodtseva N, Zhang Y, Tam K, Hynynen K. Targeted delivery of doxorubicin to the rat brain at therapeutic levels using MRI-guided focused ultrasound. *Int J Cancer.* 2007;121(4):901-907.
142. Diaz RJ, McVeigh PZ, O'Reilly MA, et al. Focused ultrasound delivery of Raman nanoparticles across the blood-brain barrier: potential for targeting experimental brain tumors. *Nanomedicine.* 2014;10(5):1075-1087.
143. Beccaria K, Canney M, Goldwirt L, et al. Opening of the blood-brain barrier with an unfocused ultrasound device in rabbits. *J Neurosurg* 2013 Oct;119(4):887-98.
144. Elias WJ, Huss D, Voss T, et al. A pilot study of focused ultrasound thalamotomy for essential tremor. *N Engl J Med.* 2013;369(7):640-648.
145. Raymond JJ, Robertson DM, Dinsdale HB. Pharmacological modification of bradykinin induced breakdown of the blood-brain barrier. *Can J Neurol Sci.* 1986;13(3):214-220.
146. Zhang H, Gu YT, Xue Y-X. Bradykinin-induced blood-brain tumor barrier permeability increase is mediated by adenosine 5′-triphosphate-sensitive potassium channel. *Brain Res.* 2007;1144:33-41.
147. Liu L-B, Xue Y-X, Liu Y-H. Bradykinin increases the permeability of the blood-tumor barrier by the caveolae-mediated transcellular pathway. *J Neuro-Oncol.* 2010;99(2):187-194.
148. Inamura T, Nomura T, Bartus RT, Black KL. Intracarotid infusion of RMP-7, a bradykinin analog: a method for selective drug delivery to brain tumors. *J Neurosurg.* 1994;81(5):752-758.
149. Elliott PJ, Hayward NJ, Huff MR, Nagle TL, Black KL, Bartus RT. Unlocking the blood-brain barrier: a role for RMP-7 in brain tumor therapy. *Exp Neurol.* 1996;141(2):214-224.
150. Warren K, Jakacki R, Widemann B, et al. Phase II trial of intravenous lobradimil and carboplatin in childhood brain tumors: a report from the Children's Oncology Group. *Cancer Chemother Pharmacol.* 2006;58(3):343-347.
151. Vassart M, Greth A. Hematological and serum chemistry values for Arabian oryx (Oryx leucoryx). *J Wildl Dis.* 1991;27(3):506-508.
152. Emerich DF, Snodgrass P, Dean R, et al. Enhanced delivery of carboplatin into brain tumours with intravenous Cereport (RMP-7): dramatic differences and insight gained from dosing parameters. *Br J Cancer.* 1999;80(7):964-970.
153. Nakata H, Yoshimine T, Murasawa A, et al. Early blood-brain barrier disruption after high-dose single-fraction irradiation in rats. *Acta Neurochir.* 1995;136(1-2):82-6-discussion86-7.
154. Krueck WG, Schmiedl UP, Maravilla KR, Spence AM, Starr FL, Kenney J. MR assessment of radiation-induced blood-brain barrier permeability changes in rat glioma model. *AJNR Am J Neuroradiol.* 1994;15(4):625-632.
155. d'Avella D, Cicciarello R, Angileri FF, Lucerna S, La Torre D, Tomasello F. Radiation-induced blood-brain barrier changes: pathophysiological mechanisms and clinical implications. *Acta Neurochir Suppl.* 1998;71:282-284.

156. d'Avella D, Cicciarello R, Albiero F, et al. Quantitative study of blood-brain barrier permeability changes after experimental whole-brain radiation. *Neurosurgery*. 1992;30(1):30-34.
157. Qin DX, Zheng R, Tang J, Li JX, Hu YH. Influence of radiation on the blood-brain barrier and optimum time of chemotherapy. *Int J Radiat Oncol Biol Phys*. 1990;19(6):1507-1510.
158. Qin D, Ma J, Xiao J, Tang Z. Effect of brain irradiation on blood-CSF barrier permeability of chemotherapeutic agents. *Am J Clin Oncol*. 1997;20(3):263-265.
159. Varenika V, Dickinson P, Bringas J, et al. Detection of infusate leakage in the brain using real-time imaging of convection-enhanced delivery. *J Neurosurg*. 2008;109(5):874-880.
160. Bobo RH, Laske DW, Akbasak A, Morrison PF, Dedrick RL, Oldfield EH. Convection-enhanced delivery of macromolecules in the brain. *Proc Natl Acad Sci U S A*. 1994;91(6):2076-2080.
161. Zhang R, Saito R, Mano Y, et al. Convection-enhanced delivery of SN-38-loaded polymeric micelles (NK012) enables consistent distribution of SN-38 and is effective against rodent intracranial brain tumor models. *Drug Deliv*. 2016;23(8):2780-2786.
162. Voges J, Reszka R, Gossmann A, et al. Imaging-guided convection-enhanced delivery and gene therapy of glioblastoma. *Ann Neurol*. 2003;54(4):479-487.
163. Bogdahn U, Hau P, Stockhammer G, et al. Targeted therapy for high-grade glioma with the TGF-β2 inhibitor trabedersen: results of a randomized and controlled phase IIb study. *Neuro-Oncology*. 2011;13(1):132-142.
164. Patel SJ, Shapiro WR, Laske DW, et al. Safety and feasibility of convection-enhanced delivery of Cotara for the treatment of malignant glioma: initial experience in 51 patients. *Neurosurgery*. 2005;56(6):1243-1252. discussion 1252-3.
165. Vogelbaum MA, Sampson JH, Kunwar S, et al. Convection-enhanced delivery of cintredekin besudotox (interleukin-13-PE38QQR) followed by radiation therapy with and without temozolomide in newly diagnosed malignant gliomas: phase 1 study of final safety results. *Neurosurgery*. 2007;61(5). 1031-7–discussion1037-8.
166. Bruce JN, Fine RL, Canoll P, et al. Regression of recurrent malignant gliomas with convection-enhanced delivery of topotecan. *Neurosurgery*. 2011;69(6):1272-9-discussion1279-80.
167. Chen P-Y, Ozawa T, Drummond DC, et al. Comparing routes of delivery for nanoliposomal irinotecan shows superior anti-tumor activity of local administration in treating intracranial glioblastoma xenografts. *Neuro-Oncology*. 2013;15(2):189-197.
168. Bernal GM, LaRiviere MJ, Mansour N, et al. Convection-enhanced delivery and in vivo imaging of polymeric nanoparticles for the treatment of malignant glioma. *Nanomedicine*. 2014;10(1):149-157.
169. Kunwar S, Chang S, Westphal M, et al. Phase III randomized trial of CED of IL13-PE38QQR vs Gliadel wafers for recurrent glioblastoma. *Neuro-Oncology*. 2010;12(8):871-881.
170. Mandell RM. Fracture of the femoral neck treated with Austin Moore prosthesis: clinical assessment and review of 60 cases. *J Am Osteopath Assoc*. 1972;71(11):953-967.
171. Tanner PG, Holtmannspötter M, Tonn J-C, Goldbrunner R. Effects of drug efflux on convection-enhanced paclitaxel delivery to malignant gliomas: technical note. *Neurosurgery*. 2007;61(4). E880-2-discussionE882.
172. Sampson JH, Archer G, Pedain C, et al. Poor drug distribution as a possible explanation for the results of the PRECISE trial. *J Neurosurg*. 2010;113(2):301-309.
173. Parkar NS, Akpa BS, Nitsche LC, et al. Vesicle formation and endocytosis: function, machinery, mechanisms, and modeling. *Antioxid Redox Signal*. 2009;11(6):1301-1312.
174. Wang D, El-Amouri SS, Dai M, et al. Engineering a lysosomal enzyme with a derivative of receptor-binding domain of apoE enables delivery across the blood-brain barrier. *Proc Natl Acad Sci U S A*. 2013;110(8):2999-3004.
175. Boado RJ, Zhang Y, Zhang Y, Xia C-F, Wang Y, Pardridge WM. Genetic engineering of a lysosomal enzyme fusion protein for targeted delivery across the human blood-brain barrier. *Biotechnol Bioeng*. 2008;99(2):475-484.
176. Boado RJ, Zhou Q-H, Lu JZ, Hui EK-W, Pardridge WM. Pharmacokinetics and brain uptake of a genetically engineered bifunctional fusion antibody targeting the mouse transferrin receptor. *Mol Pharm*. 2010;7(1):237-244.
177. Boado RJ, Zhang Y, Wang Y, Pardridge WM. Engineering and expression of a chimeric transferrin receptor monoclonal antibody for blood-brain barrier delivery in the mouse. *Biotechnol Bioeng*. 2009;102(4):1251-1258.
178. Staquicini FI, Ozawa MG, Moya CA, et al. Systemic combinatorial peptide selection yields a non-canonical iron-mimicry mechanism for targeting tumors in a mouse model of human glioblastoma. *J Clin Invest*. 2011;121(1):161-173.
179. Hajitou A, Trepel M, Lilley CE, et al. A hybrid vector for ligand-directed tumor targeting and molecular imaging. *Cell*. 2006;125(2):385-398.
180. Sharma G, Modgil A, Layek B, et al. Cell penetrating peptide tethered bi-ligand liposomes for delivery to brain in vivo: biodistribution and transfection. *J Control Release*. 2013;167(1):1-10.

181. Zensi A, Begley D, Pontikis C, et al. Albumin nanoparticles targeted with Apo E enter the CNS by transcytosis and are delivered to neurones. *J Control Release.* 2009;137(1):78-86.
182. Gao H, Qian J, Cao S, et al. Precise glioma targeting of and penetration by aptamer and peptide dual-functioned nanoparticles. *Biomaterials.* 2012;33(20):5115-5123.
183. Kuznetsov VD, Iangulova IV. Variation of actinomyces antibioticus 961–producer of oleandomycin. *Antibiotiki.* 1972;17(4):291-295.
184. Gosk S, Vermehren C, Storm G, Moos T. Targeting anti-transferrin receptor antibody (OX26) and OX26-conjugated liposomes to brain capillary endothelial cells using in situ perfusion. *J Cereb Blood Flow Metab.* 2004;24(11):1193-1204.
185. Moos T, Morgan EH. Restricted transport of anti-transferrin receptor antibody (OX26) through the blood-brain barrier in the rat. *J Neurochem.* 2001;79(1):119-129.
186. Raucher D, Ryu JS. Cell-penetrating peptides: strategies for anticancer treatment. *Trends Mol Med.* 2015; 21(9):560-570.
187. Ramsey JD, Flynn NH. Cell-penetrating peptides transport therapeutics into cells. *Pharmacol Ther.* 2015;154:78-86.
188. Futaki S, Suzuki T, Ohashi W, et al. Arginine-rich peptides. An abundant source of membrane-permeable peptides having potential as carriers for intracellular protein delivery. *J Biol Chem.* 2001;276(8):5836-5840.
189. Park J, Ryu J, Kim K-A, et al. Mutational analysis of a human immunodeficiency virus type 1 Tat protein transduction domain which is required for delivery of an exogenous protein into mammalian cells. *J Gen Virol.* 2002;83:1173-1181. Pt 5.
190. Derossi D, Joliot AH, Chassaing G, Prochiantz A. The third helix of the Antennapedia homeodomain translocates through biological membranes. *J Biol Chem.* 1994;269(14):10444-10450.
191. Frankel AD, Pabo CO. Cellular uptake of the tat protein from human immunodeficiency virus. *Cell.* 1988; 55(6):1189-1193.
192. Green M, Loewenstein PM. Autonomous functional domains of chemically synthesized human immunodeficiency virus tat trans-activator protein. *Cell.* 1988;55(6):1179-1188.
193. Vivès E, Brodin P, Lebleu B. A truncated HIV-1 Tat protein basic domain rapidly translocates through the plasma membrane and accumulates in the cell nucleus. *J Biol Chem.* 1997;272(25):16010-16017.
194. Cao G, Pei W, Ge H, et al. In vivo delivery of a Bcl-xL fusion protein containing the TAT protein transduction domain protects against ischemic brain injury and neuronal apoptosis. *J Neurosci.* 2002;22(13):5423-5431.
195. Gillardon F, Lenz C, Waschke KF, et al. Altered expression of Bcl-2, Bcl-X, Bax, and c-Fos colocalizes with DNA fragmentation and ischemic cell damage following middle cerebral artery occlusion in rats. *Brain Res Mol Brain Res.* 1996;40(2):254-260.
196. Michiue H, Eguchi A, Scadeng M, Dowdy SF. Induction of in vivo synthetic lethal RNAi responses to treat glioblastoma. *Cancer Biol Ther.* 2009;8(23):2306-2313.
197. Guidotti G, Brambilla L, Rossi D. Cell-penetrating peptides: from basic research to clinics. *Trends Pharmacol Sci.* 2017;38(4):406-424.
198. Kosuge M, Takeuchi T, Nakase I, Jones AT, Futaki S. Cellular internalization and distribution of arginine-rich peptides as a function of extracellular peptide concentration, serum, and plasma membrane associated proteoglycans. *Bioconjug Chem.* 2008;19(3):656-664.
199. Fretz MM, Penning NA, Al-Taei S, et al. Temperature-, concentration- and cholesterol-dependent translocation of L- and D-octa-arginine across the plasma and nuclear membrane of CD34+ leukaemia cells. *Biochem J.* 2007;403(2):335-342.
200. Thennarasu S, Tan A, Penumatchu R, Shelburne CE, Heyl DL, Ramamoorthy A. Antimicrobial and membrane disrupting activities of a peptide derived from the human cathelicidin antimicrobial peptide LL37. *Biophys J.* 2010;98(2):248-257.
201. Herce HD, Garcia AE. Molecular dynamics simulations suggest a mechanism for translocation of the HIV-1 TAT peptide across lipid membranes. *Proc Natl Acad Sci U S A.* 2007;104(52):20805-20810.
202. Richard JP, Melikov K, Vives E, et al. Cell-penetrating peptides. A reevaluation of the mechanism of cellular uptake. *J Biol Chem.* 2003;278(1):585-590.
203. van den Berg A, Dowdy SF. Protein transduction domain delivery of therapeutic macromolecules. *Curr Opin Biotechnol.* 2011;22(6):888-893.
204. Mäe M, Andaloussi El S, Lundin P, et al. A stearylated CPP for delivery of splice correcting oligonucleotides using a non-covalent co-incubation strategy. *J Control Release.* 2009;134(3):221-227.

205. Lo SL, Wang S. An endosomolytic Tat peptide produced by incorporation of histidine and cysteine residues as a nonviral vector for DNA transfection. *Biomaterials*. 2008;29(15):2408-2414.
206. Wadia JS, Stan RV, Dowdy SF. Transducible TAT-HA fusogenic peptide enhances escape of TAT-fusion proteins after lipid raft macropinocytosis. *Nat Med*. 2004;10(3):310-315.
207. Stalmans S, Bracke N, Wynendaele E, et al. Cell-Penetrating Peptides Selectively Cross the Blood-Brain Barrier In Vivo. Choi J-M. *PLoS ONE*. 2015;10(10).
208. Cucullo L, Marchi N, Hossain M, Janigro D. A dynamic in vitro BBB model for the study of immune cell trafficking into the central nervous system. *J Cereb Blood Flow Metab*. 2011;31(2):767-777.
209. Batrakova EV, Li S, Reynolds AD, et al. A macrophage-nanozyme delivery system for Parkinson's disease. *Bioconjug Chem*. 2007;18(5):1498-1506.
210. Zhao Y, Haney MJ, Mahajan V, et al. Active targeted macrophage-mediated delivery of catalase to affected brain regions in models of Parkinson's disease. *J Nanomed Nanotechnol*. 2011. S4(S4).
211. Klyachko NL, Polak R, Haney MJ, et al. Macrophages with cellular backpacks for targeted drug delivery to the brain. *Biomaterials*. 2017;140:79-87.
212. Pang L, Qin J, Han L, et al. Exploiting macrophages as targeted carrier to guide nanoparticles into glioma. *Oncotarget*. 2016;7(24):37081-37091.
213. Shah K, Hingtgen S, Kasmieh R, et al. Bimodal viral vectors and in vivo imaging reveal the fate of human neural stem cells in experimental glioma model. *J Neurosci*. 2008;28(17):4406-4413.
214. Hingtgen SD, Kasmieh R, van de Water J, Weissleder R, Shah K. A novel molecule integrating therapeutic and diagnostic activities reveals multiple aspects of stem cell-based therapy. *Stem Cells*. 2010;28(4):832-841.
215. Sasportas LS, Kasmieh R, Wakimoto H, et al. Assessment of therapeutic efficacy and fate of engineered human mesenchymal stem cells for cancer therapy. *Proc Natl Acad Sci U S A*. 2009;106(12):4822-4827.
216. Menon LG, Kelly K, Yang HW, Kim S-K, Black PM, Carroll RS. Human bone marrow-derived mesenchymal stromal cells expressing S-TRAIL as a cellular delivery vehicle for human glioma therapy. *Stem Cells*. 2009; 27(9):2320-2330.
217. Liu T-C, Galanis E, Kirn D. Clinical trial results with oncolytic virotherapy: a century of promise, a decade of progress. *Nat Clin Pract Oncol*. 2007;4(2):101-117.
218. Selznick LA, Shamji MF, Fecci P, Gromeier M, Friedman AH, Sampson J. Molecular strategies for the treatment of malignant glioma–genes, viruses, and vaccines. *Neurosurg Rev*. 2008;31(2):141-155. discussion 155.
219. Del Vecchio M, Bajetta E, Canova S, et al. Interleukin-12: biological properties and clinical application. *Clin Cancer Res*. 2007;13(16):4677-4685.
220. Cheema TA, Wakimoto H, Fecci PE, et al. Multifaceted oncolytic virus therapy for glioblastoma in an immunocompetent cancer stem cell model. *Proc Natl Acad Sci U S A*. 2013;110(29):12006-12011.
221. Zhang G, Jin G, Nie X, et al. Enhanced antitumor efficacy of an oncolytic herpes simplex virus expressing an endostatin-angiostatin fusion gene in human glioblastoma stem cell xenografts. *PLoS ONE*. 2014;9(4).
222. Du W, Seah I, Bougazzoul O, et al. Stem cell-released oncolytic herpes simplex virus has therapeutic efficacy in brain metastatic melanomas. *Proc Natl Acad Sci U S A*. 2017;114(30):E6157-E6165.
223. Leoni V, Gatta V, Palladini A, et al. Systemic delivery of HER2-retargeted oncolytic-HSV by mesenchymal stromal cells protects from lung and brain metastases. *Oncotarget*. 2015;6(33):34774-34787.
224. Andtbacka RHI, Kaufman HL, Collichio F, et al. Talimogene Laherparepvec improves durable response rate in patients with advanced melanoma. *J Clin Oncol*. 2015;33(25):2780-2788.
225. Thomas H, Coley HM. Overcoming multidrug resistance in cancer: an update on the clinical strategy of inhibiting p-glycoprotein. *Cancer Control*. 2003;10(2):159-165.
226. Lampidis TJ, Krishan A, Planas L, Tapiero H. Reversal of intrinsic resistance to adriamycin in normal cells by verapamil. *Cancer Drug Deliv*. 1986;3(4):251-259.
227. Roe M, Folkes A, Ashworth P, et al. Reversal of P-glycoprotein mediated multidrug resistance by novel anthranilamide derivatives. *Bioorg Med Chem Lett*. 1999;9(4):595-600.
228. Kreisl WC, Liow J-S, Kimura N, et al. P-glycoprotein function at the blood-brain barrier in humans can be quantified with the substrate radiotracer 11C-N-desmethyl-loperamide. *J Nucl Med*. 2010;51(4):559-566.
229. Ronaldson PT, Davis TP. Targeted drug delivery to treat pain and cerebral hypoxia. *Pharmacol Rev*. 2013; 65(1):291-314.
230. Cannon RE, Peart JC, Hawkins BT, Campos CR, Miller DS. Targeting blood-brain barrier sphingolipid signaling reduces basal P-glycoprotein activity and improves drug delivery to the brain. *Proc Natl Acad Sci U S A*. 2012; 109(39):15930-15935.

231. Pinzón-Daza ML, Salaroglio IC, Kopecka J, et al. The cross-talk between canonical and non-canonical Wnt-dependent pathways regulates P-glycoprotein expression in human blood-brain barrier cells. *J Cereb Blood Flow Metab*. 2014;34(8):1258-1269.
232. Kapoor A, Iqbal M, Petropoulos S, Ho HL, Gibb W, Matthews SG. Effects of sertraline and fluoxetine on p-glycoprotein at barrier sites: in vivo and in vitro approaches. *PLoS ONE*. 2013;8(2).
233. Bhuiyan M, Petropoulos S, Gibb W, Matthews SG. Sertraline alters multidrug resistance phosphoglycoprotein activity in the mouse placenta and fetal blood-brain barrier. *Reprod Sci*. 2012;19(4):407-415.
234. Szakács G, Paterson JK, Ludwig JA, Booth-Genthe C, Gottesman MM. Targeting multidrug resistance in cancer. *Nat Rev Drug Discov*. 2006;5(3):219-234.
235. Agarwal S, Hartz AMS, Elmquist WF, Bauer B. Breast cancer resistance protein and P-glycoprotein in brain cancer: two gatekeepers team up. *Curr Pharm Des*. 2011;17(26):2793-2802.

CHAPTER

10

Blood–Brain Barrier Disruption

John M. McGregor, Nancy D. Doolittle†, Emily Youngers†, Susan D. Bell*, Edward A. Neuwelt‡*

*Department of Neurological Surgery, The Ohio State University Wexner Medical Center, Columbus, OH, United States †Department of Neurology, Oregon Health & Science University, Portland, OR, United States ‡Departments of Neurology and Neurosurgery, Oregon Health & Science University, Portland, OR, United States

INTRODUCTION

Whether primary or metastatic, malignant brain tumors remain difficult to manage and treat in neuro-oncology. One of the overarching challenges of treatment is the successful delivery of therapeutic agents. Many are available to treat and affect the systemic disease burden, but the blood–brain barrier (BBB) limits the successful passage of many molecules into the central nervous system (CNS) at therapeutic levels. The BBB is a functional unit of regulation of the influx of molecules from the bloodstream to the CNS. Current analysis identifies the neurovascular unit as the anatomic construct that forms the BBB. This unit consists of endothelial cells and their interconnecting tight junctions, surrounding pericytes, astrocyte and neuron processes, and the surrounding basement membrane and extracellular matrix.[1-3] This anatomic and physiologic structure exists to limit the passage of molecules into the brain from the bloodstream on the basis of size, electric charge, and lipid solubility. The degree of barrier permeability has been shown to be altered in various disease states, including brain tumors. Depending on the contribution of the various components, this complex neurovascular unit allows for barrier blockade to be variably incomplete, and thus inconsistent in its integrity. The overall effect, then, is limited ability of agents introduced via the bloodstream to attain the necessary concentrations to provide the clinical efficacy they can achieve in the systemic environment.[4-7]

Osmotic disruption of the BBB is a way to overcome the challenges of drug delivery imposed by the BBB. The procedure consists of the intra-arterial administration of hyperosmolar

Nervous System Drug Delivery
https://doi.org/10.1016/B978-0-12-813997-4.00010-4

© 2019 Elsevier Inc. All rights reserved.

mannitol, in the bilateral carotid and vertebral arteries. The hyperosmolar infusion has been shown to acutely shrink the endothelial cells at the capillary level in the CNS enough to temporarily separate their tight junctions. In addition, cellular cytoskeleton changes and cellular influx of calcium and nitrous oxide are associated with the opening of the BBB. During the time the barrier is open, many more substances are able to cross, including molecules of high molecular weight, proteins, antibodies, viral vectors, and immunoconjugates.[1, 8] This method of enhanced drug delivery to the CNS has been shown to benefit patients with chemosensitive tumors, such as primary CNS lymphoma.

The timing of the temporary and reversible opening and closing of the BBB in osmotic BBB disruption allows the opportunity to treat patients under the paradigm known as the *two-compartment model*. Conceptually, a patient has dynamic drug distribution depending on the timing of drug delivery with respect to the timing of the delivery of mannitol. Any circulating agent will remain in the serum compartment until the barrier is open. Additional agents may then be delivered, and may pass the BBB to enter the CNS. The barrier then closes, and the two compartments are again separated. This is a useful concept in mitigating potential adverse effects of chemotherapy. Platinum agents and chemoprotectant thiols, for example, can be administered to the patient at separate times, providing therapeutic benefits while minimizing toxicities (Fig. 1).[9]

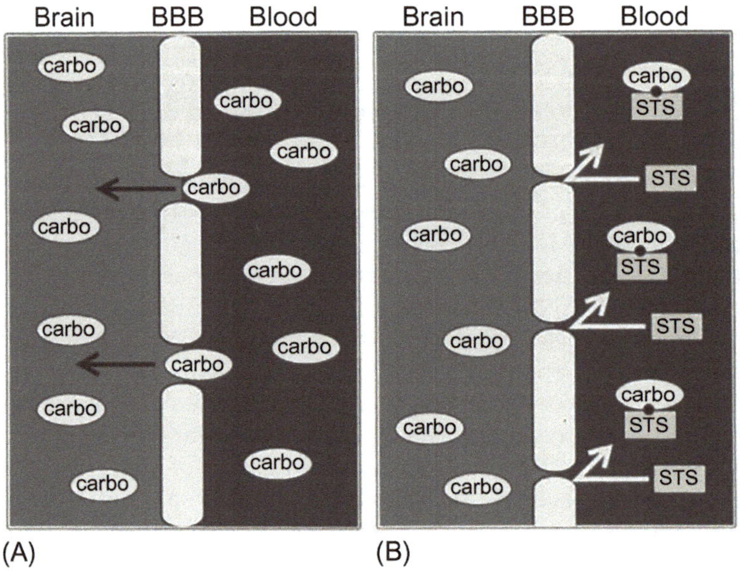

FIG. 1 The two-compartment model. Carboplatin (carbo) is administered intra-arterially immediately after osmotic disruption of the blood–brain barrier (BBB) with hypertonic mannitol. (A) Delivery of carboplatin across the BBB. Sodium thiosulfate (STS) is administered intravenously 4 hours after osmotic disruption (or 4 and 8 hours after disruption), after BBB permeability has returned to baseline levels. (B) Exclusion of STS from the brain. *Reproduced with permission from McGregor JM, Bell SD, Doolittle ND, Murillo TP, Neuwelt EA. Blood–brain barrier disruption chemotherapy. In: Newton HB, ed.* Handbook of Brain Tumor Chemotherapy, Molecular Therapeutics, and Immunotherapy. *2nd ed. London: Academic Press, Elsevier; 2018:145–153 [chapter 10].*

DELIVERING AGENTS ACROSS THE BBB: FACTORS THAT IMPACT CHEMOTHERAPY DELIVERY TO BRAIN TUMORS, FROM PRECLINICAL STUDIES TO CLINICAL TRIALS

Current understanding of the clinical opportunities and clinical effects of drug delivery across a modified BBB is based on extensive preclinical studies that showed how greater volumes of drug delivery could be achieved. Intra-arterial administration of methotrexate affords a substantial increase in concentrations of methotrexate in the brain and cerebrospinal fluid. Osmotic BBB disruption allows for administration of agents throughout the CNS, not just at areas of relative tumor disruption. There is variability in delivery based on the relative integrity of the BBB, which is a useful concept in trying to treat tumors known to be widespread at presentation (even if not yet otherwise imaged). Osmotic BBB disruption allows for substances otherwise blocked from crossing the BBB. Their permeability is maximal at 15 minutes following proper intra-arterial mannitol dose administration and returns to normal about 2 hours later.[3, 10-13]

The effects of ionizing radiation on drug delivery, the BBB, and tumor response to therapy have provided interesting but sometimes conflicting results. Studies have shown that CNS irradiation is associated with a decrease in the delivery of chemotherapeutic agents and increased CNS toxicities. In a series of studies by Remson et al., the effects of 2000 cGy single-fraction radiation, either before or concurrent with BBB disruption, on drug delivery across the BBB were studied. Decreased drug delivery was observed in irradiated animals compared to nonirradiated controls. Seizures were more frequent after BBB disruption in animals receiving cranial radiation compared to nonirradiated controls. Furthermore, animal mortality was higher in the previously irradiated mice compare to nonirradiated controls.[14-16] Clinical studies by Cao et al. have suggested that low initial dose cranial radiation, in the setting of fractionated radiation therapy, may provide an initial opening of the BBB that allows some pharmaceuticals to cross. This opening closes as the radiation dose accumulates.[17]

The potential of radiation to influence the BBB and the efficacy of radiation with targeted therapeutics has been explored in clinical and laboratory settings and results are mixed. Rodents irradiated before or concurrently with chemotherapy with BR96-DOX, an antitumor mAb-doxorubicin immunoconjugate, decreased survival compared to rodents that had the antibody administered before irradiation.[15] In another study, patients with primary CNS lymphoma treated with BBB disruption chemotherapy with no previous CNS radiation had higher median survival compared with patients who had undergone previous radiation therapy.[18] However, additional clinical evidence in the era of immunotherapy has indicated that there may be an enhanced immune response as a result of radiation. Focused stereotactic radiation in particular may have a beneficial tumoricidal inflammatory effect.[19] CD68-positive macrophages and CD3-positive T lymphocytes have been identified on immunohistochemistry studies of path specimens following stereotactic radiation to metastatic tumors, suggesting a possible change in the BBB allowing for their markedly increased presence.[20]

The BBB is a protector of the CNS. All efforts to circumvent it can have consequences when agents traditionally excluded from the compartment are allowed access. Safety and toxicity profiles for chemotherapy agents have been tested extensively in the preclinical studies before Phase 1 clinical trials. These studies have uncovered important limitations to the use of several agents over the years, including doxorubicin, cisplatin, and 5-FU, and the associated neurotoxicity identified when these agents were administered intra-arterially after osmotic BBB disruption.[21, 22] Other examples include the enhanced neurotoxicity identified when etoposide phosphate was administered with other agents (i.e., carboplatin, melphalan, or methotrexate) after BBB disruption. This was mitigated if the agent was given before BBB disruption, and the toxicity was enhanced if propofol was used as the anesthetic.[23] These types of studies are requisite prior to the administration of novel agents across the BBB.

Chemotherapeutic agents currently available for use after osmotic BBB disruption include methotrexate, carboplatin, melphalan, cyclophosphamide, etoposide, and etoposide phosphate. Their use as individual agents or in combination depends on tumor histology or the particular IRB-approved protocol under investigation. Acceptable safety and toxicity profiles have been shown when the following agents are used after osmotic BBB disruption, within the proper methods and dose parameters as indicated here:

- Methotrexate (intra-arterial, 2500 mg/day for 2 consecutive days),
- Carboplatin (intra-arterial, 200 mg/m^2/day for 2 consecutive days),
- Melphalan (intra-arterial, 4 mg/m^2/day for 2 consecutive days),
- Cyclophosphamide (intravenous, 500 mg/m^2/day for 2 consecutive days when given with methotrexate; 330 mg/m^2/day for 2 consecutive days when given with carboplatin), and
- Etoposide and etoposide phosphate (intravenous, 150 mg/m^2/day for 2 consecutive days when given with methotrexate; 200 mg/m^2/day for 2 consecutive days when given with carboplatin).[18, 24-28]

THE CLINICAL METHOD OF OSMOTIC BLOOD–BRAIN BARRIER OPENING

In a clinical setting, osmotic BBB disruption is performed by a multidisciplinary team. Patients are evaluated, treated, and monitored throughout their hospital stay and beyond by the team approach, which ensures maximal attention to detail as well as close monitoring for safety and efficacy. Long before treatment, this multidisciplinary team is educated on proper patient selection, preparation, acceptable medications, appropriate anesthetics, requisite vascular imaging needed, intra-arterial drug delivery technique, postprocedure imaging requirements, postprocedure recovery, timing of additional medications, and follow-up after discharge. The personnel involved include a neuro-oncologist, neurosurgeon, nurse coordinator, interventional neuroradiologist/neurosurgeon, neuroanesthesiologist, pharmacist, neuropsychologist, neuroophthalmologist, audiologist, physical therapist, and social worker.

The treatment itself consists of monthly admissions. During each admission, mannitol and chemotherapy are delivered intra-arterially with the patient under general anesthesia on 2 consecutive days, rotating CNS arterial distributions for each admission, over 1 year (12 admissions total). At baseline, patients undergo a neuropsychological evaluation, a central venous access port placement, and an initial electrocardiogram. Also at baseline and again at monthly admissions, patients undergo contrast magnetic resonance imaging (MRI) of the brain, chest radiograph, complete blood count, chemistry panel, urinalysis, neurologic examination, and Karnofsky performance scale evaluation. Patients are evaluated for adequate neurologic, hematologic, pulmonary, cardiac, renal, and hepatic function to tolerate a general anesthetic and an intra-arterial cerebral angiogram. Ophthalmologic assessment is indicated for certain tumor types and if symptoms warrant. Monthly audiograms are indicated for patients undergoing treatment with carboplatin in conjunction with osmotic BBB disruption.

Monthly MRIs of the brain are assessed to ensure that certain patients meet proper safety criteria prior to BBB disruption. There is a risk of transient increase in cerebral volume and mass effect during and immediately after BBB disruption in patients with rapidly progressive disease. Therefore, every effort is made to recommend BBB disruption for patients whose tumor burden is not excessive.[26]

Patients may be admitted the day before the procedure, depending on whether premedications are indicated. They are to receive BBB disruption chemotherapy on Days 1 and 2, observed overnight, and released on Day 3. Care is taken to ensure that no unexpected medications are in the patient's system prior to BBB disruption. On the day of the procedure, patients are brought to the cerebral angiography suite, where they are placed under general anesthesia. Alternating femoral arteries are cannulated on successive days as needed. The intracranial artery selected alternates with each treatment, either the right carotid, left carotid, or a vertebral artery. The catheter is placed at the C1-C2 level for carotid infusions, and the C6-C7 level for vertebral infusions. Contrast injection is performed to detect the ideal rate of infusion of mannitol that matches cerebral flow. Catheter position is confirmed under fluoroscopy, and mannitol is given at the identified rate, between 3 and 12 cc/second, via a power injector, for 30 seconds.

On completion of mannitol administration, the intra-arterial chemotherapeutic agents are then injected, each over 10 minutes. Intravenous agents meant to be available on the opening of the BBB are infused in advance of disruption, either at the time of induction of general anesthesia, or in some cases well in advance of the procedure, depending on the agent. Nonionic computed tomography (CT) contrast agent is injected intravenously immediately following the mannitol administration. On completion of the chemotherapy infusion, the patient directly undergoes a head CT to document the degree of disruption obtained. This determination is made based on the degree of contrast agent seen in the disrupted arterial distribution compared to the nontreated brain. The grading is based on the scale suggested in Roman-Goldstein et al. (Fig. 2).[9, 29, 30]

As indicated above, during each disruption session one intracranial arterial distribution (i.e., right carotid, left carotid, and vertebrobasilar) is accessed on Day 1, and a different artery is accessed on Day 2. The following month, one arterial distribution previously disrupted and the one not treated are chosen. The patient is treated repeatedly until 12 courses of therapy are completed, typically after 12 monthly admissions. For multifocal disease, such as seen in patients diagnosed with glioblastoma multiforme or primary CNS lymphoma, the disruptions are rotated among the various distributions over the course of the 24 sessions, with each artery receiving eight cannulations and disruptions. This achieves the goal of global therapeutic distribution throughout the CNS.

After the disruption procedure, patients are monitored for vital signs, fluid intake and output, and neurologic status. Adequate urine output is maintained with fluid boluses and diuretics as needed. For protocols that include methotrexate, the urine pH is monitored and sodium bicarbonate is added to intravenous fluids being administered, with the amount titrated to keep urine pH greater than 6.5. Leucovorin rescue is also performed in patients

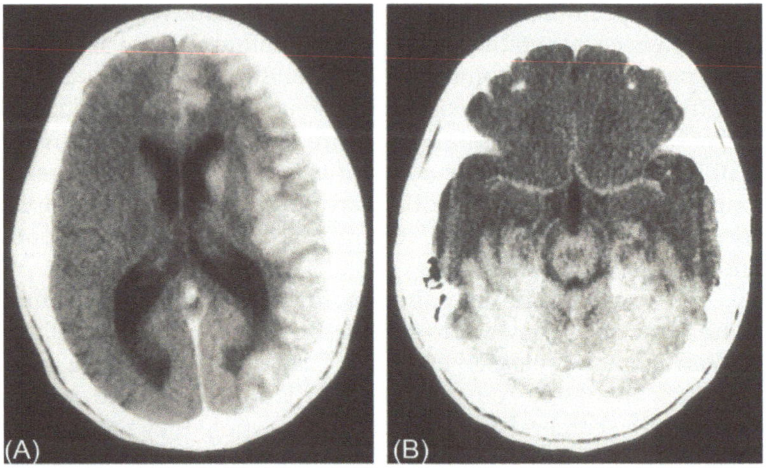

FIG. 2 Computed tomograms showing blood–brain barrier (BBB) disruption of (A) the left cerebral hemisphere after mannitol infusion in the left internal carotid artery, and (B) BBB disruption of the posterior fossa region after mannitol infusion in the vertebral artery. *Reproduced with permission from McGregor JM, Bell SD, Doolittle ND, Murillo TP, Neuwelt EA. Blood–brain barrier disruption chemotherapy. In: Newton HB, ed.* Handbook of Brain Tumor Chemotherapy, Molecular Therapeutics, and Immunotherapy. *2nd ed. London: Academic Press, Elsevier; 2018:145–153 [chapter 10].*

receiving methotrexate. They receive 80 mg leucovorin intravenously 36 hours after the first infusion of methotrexate, then 50 mg intravenously or orally every 6 hours, for a total of 20 doses. Granulocyte-colony stimulating factor is administered subcutaneously 48 hours after the Day 2 disruption. A total of 5 μg/kg of filgrastim are administered subcutaneously each day, until the white blood cell count is greater than 5000/mcl. Alternatively, pegfilgrastim, 6 μg subcutaneously in a single dose, may be used. Monitoring after discharge includes twice-weekly complete blood counts while receiving granulocyte-colony stimulating factor, and once a week thereafter.

Transient neurologic deficits may be seen immediately following BBB disruption and during recovery. Among patients who achieve good or excellent disruption, 10% will have neurologic findings such as decreased level of consciousness, possibly associated with a temporary aphasia, or motor weakness of the extremities. This may last up to 48 hours before complete resolution to baseline. Patients may be treated successfully with scheduled intravenous dexamethasone until resolution, usually by 48 hours. A subintimal tear in the cerebral artery being cannulated is the most common arterial injury to occur during the BBB disruption procedure. These are most often noted at fluoroscopy and are asymptomatic. If observed, the infusions of mannitol and chemotherapy are not performed in that vessel. The artery is reassessed at the next angiogram, 4 weeks later, before consideration of resuming infusions in that distribution.

Patients undergoing BBB disruption chemotherapy for brain tumors are at risk for deep venous thrombosis. These patients undergo routine venous Doppler monitoring and are placed on anticoagulation therapy as needed, depending on their risk assessment. Evidence of ischemic injury in the CNS may be seen on routine studies in up to 5% of patients. These are predominately asymptomatic and are noted as an incidental finding on brain MRI obtained routinely just prior to each monthly procedure.[18, 24-26] Symptomatic stroke with changes on MRI and neurologic deficit are rare, and are usually noted within the first 24 hours following a disruption. Symptoms may consist of hemiparesis or speech and language disturbances. They are the result of embolic events happening at the time of arterial cannulation and infusion of drug. Patients with neurologic deficits may be symptomatic beyond 48 hours; however, they are usually at their baseline status within a month of the procedure.

Several patients in our series had evidence of cervical cord deficit and abnormal MRI signals were noted in the high cervical cord.[31] These results were seen in patients receiving carboplatin chemotherapy. This toxicity is treated most successfully with dexamethasone therapy and inpatient monitoring. High-frequency hearing loss has been noted in patients receiving carboplatin chemotherapy intra-arterially via BBB disruption.[32] We have identified substantial reduction in incidence with the introduction of a delayed infusion of a high-dose thiol chemo-protectant, sodium thiosulfate.[33, 34] In addition to the above, patients undergoing intra-arterial chemotherapy have noted more typical adverse effects related to the drugs themselves, including nausea and fatigue. Myelodepression is noted in some patients. These adverse effects have been substantially limited by adherence to the strict guidelines developed in conjunction with the multicenter BBB Consortium.

The BBB Consortium comprises institutions participating in the application of the technique of BBB disruption as described above. The consortium members have included The Ohio State University, University of Oklahoma Health Science Center, University of Minnesota, Cleveland Clinic Foundation, University of Kentucky, Hadassah-Hebrew University Medical Center, Center Hospital Universitaire de Sherbrooke, and Oregon Health & Science University (the coordinating center). The consortium has developed guidelines for the application of BBB disruption that include proper anesthesia, techniques for placement of a transfemoral arterial catheter, a grading system for measuring the degree of BBB disruption on postdisruption head CT scans, radiographic assessment of tumor response, guidelines for the technique and rate of mannitol and chemotherapy infusion, and guidelines for the care of patients before, during, and after the procedures. These guidelines have been developed over the course of decades of use of the technique in the treatment of neuro-oncologic patients. The consortium continues to hold annual meetings, partially funded by an NIH R13 grant. The meeting provides an opportunity for participating centers to discuss patient care, guideline development, recent advances in BBB science, and future directions. The proceedings are summarized by the participants and are submitted for publication.[35-37]

CLINICAL RESULTS WITH BLOOD–BRAIN BARRIER DISRUPTION

One of the most successful effects of osmotic BBB disruption chemotherapy has been in the care and treatment of patients with primary CNS lymphoma. This is a highly chemosensitive tumor that responds to methotrexate-based chemotherapy. Dose intensity is important for therapeutic benefit, and is achieved with osmotic BBB disruption. Initial improvements in primary CNS lymphoma survival were first achieved with a combination of whole brain radiation therapy (WBRT) and chemotherapy; however, WBRT was soon associated with increased risk of neurotoxicity and cognitive decline, particularly in elderly patients.[38] In an effort to assess the potential for cognitive neurotoxicity, trials of BBB disruption have included an extensive battery of neuropsychological testing and quality of life analyses for each patient before enrolment in a study protocol. Long-term follow-up of neuropsychological function has been obtained for patients treated in this manner. The increased dose intensity achieved with barrier disruption has provided an opportunity to enhance clinical outcomes without the addition of cognitive decline secondary to WBRT.

Our initial report described a series of 74 patients diagnosed with primary CNS lymphoma who were treated with intra-arterial methotrexate and osmotic BBB disruption. They had undergone no prior radiation therapy. In this group, 86% of patients had a complete response at 1 year. The estimated 5-year survival rate was 42%, and there was no evidence of cognitive decline.[25] Subsequent analysis demonstrated that longer survival was associated with increased total dose intensity as measured by the number of successful BBB disruptions performed and the cumulative quality of the disruption scores.[7] Additional analysis of cognitive function was performed in a subset of 16 patients whose primary CNS lymphoma was treated with BBB disruption. This group had complete response noted 1 year after treatment. They demonstrated cognitive improvements from baseline. MRI findings in this group of patients showed stable, improved, or absent T2 signal abnormalities in the areas of their previously enhancing tumor burden. These findings indicated an association between cognitive improvement and results of MRI.[39] The addition of methotrexate injected intravitreally in patients with ocular involvement has also been shown to achieve clinical remission without significant morbidity.[40]

We have previously reported our multicenter consortium results.[41] A total of 149 patients with newly diagnosed primary CNS lymphoma were treated with BBB disruption in conjunction with intra-arterial methotrexate-based chemotherapy without the use of whole-brain radiation therapy. The protocol consisted of intra-arterial administration to two separate circulations (i.e., carotid and vertebrobasilar) on successive days, repeated monthly over the course of 1 year. A complete response of 56% and a partial response rate of 24% were noted. The overall response rate was 82%. The median overall survival was 3.1 years. There was a 25% estimated survival at 8.5 years. The median progression-free survival was 1.8 years. The 5- and 7-year progression-free survivals were 31% and 25%, respectively. Based on Karnofsky performance scale scores, risk groups were stratified. The low-risk group included patients younger than 60 years and with Karnofsky performance scale scores of 70 or higher. Median overall survival in that group was 14 years, with a survival plateau noted at 8 years.[41]

Subsequent long-term follow-up of patients with primary CNS lymphoma treated with osmotic BBB disruption without WBRT has been reported with regard to quality of life and cognitive outcome. The patients who were evaluated remained in complete response. One of the stated efforts of BBB disruption chemotherapy is to avoid the neurocognitive decline seen in patients who had undergone WBRT. Long-term results in cognitive capabilities in this patient population were evaluated by Doolittle et al.[42] Neurocognitive examinations were performed in patients who had experienced complete response. Examinations were performed between 2 and 26 years after treatment; median time to evaluation was 12 years. Comparisons were made to pretreatment cognitive evaluations. The results identified significant improvements in attention/executive function and improvements in verbal memory as compared to pretreatment evaluations.[42]

In comparison, long-term cognitive outcomes have been assessed for other patients with primary CNS lymphoma who were treated with different forms of therapy. In this analysis,

80 patients treated with four different regimens were evaluated with neurocognitive testing pretreatment and at a median of 5.5 years after treatment. Patients who had WBRT as part of their methotrexate-based treatment regimen had significantly decreased attention/executive function, motor skills, and neuropsychological composite scores compared to pretreatment baseline. MRI findings of increased T2 signal abnormalities were associated with worse quality of life outcomes and neurocognitive scores. T2 signal abnormalities were found more frequently in patients treated with WBRT.[43]

Even as new targeted therapeutics are developed, limitations to delivery at the NVU remain significant. Rituximab, a monoclonal antibody that reacts specifically with the B-cell antigen CD20, induces apoptosis and acts synergistically with platinum drugs.[44] It has been studied as an additional drug in conjunction with the methotrexate regimen and BBB disruption to enhance therapeutic response in primary CNS lymphoma. Dose delivery across the BBB is accomplished by infusing the drug 12 hours in advance of disruption on Day 1 of the monthly 2-day disruption sessions. Clinical analysis shows an improved complete response of 75% in a series of 24 patients. This compares favorably to the findings of the above-mentioned studies. The median progression-free survival was 2.1 years, and the median overall survival was 5.1 years. The data suggest improved tumor treatment with the addition of rituximab to the BBB disruption regimen.[45]

Patients with relapsed primary CNS lymphoma after initial methotrexate-based BBB disruption have shown positive responses to BBB disruption using carboplatin-based regimens with the addition of rituximab.[27] The results have been encouraging and the adverse events were not increased. We have reported on three patients who have subsequently undergone two complete yearlong treatments for initial and subsequently recurrent primary CNS lymphoma, with a total number of disruptions of 152 for all three patients. The regimens were tolerated well, and no increases in adverse events were noted. These results suggest that patients can undergo repeat BBB disruption regimens as treatment needs change and as additional therapeutic agents develop.[46]

Another chemosensitive tumor is the anaplastic oligodendroglioma with allelic loss of heterozygosity on chromosome 1p, 19q, or both. There tumors have shown favorable responses to alkylating agents. We have reported previously on a small group of 13 patients with recurrent progressing disease after temozolomide therapy. They were treated with a regimen of intra-arterial melphalan, intra-arterial carboplatin, and intravenous etoposide phosphate in conjunction with BBB disruption. Their results have also been encouraging and the safety has been reasonable. Preliminary results showed a complete or partial response in five patients, stable disease in five patients, and disease progression in three patients.[28]

Osmotic BBB disruption with a methotrexate and carboplatin intra-arterial regimen has been shown to be effective for patients diagnosed with germ cell tumor. In a study by Jahnke et al., 54 patients with various germ cell tumors (29 primitive neuro ectodermal tumors, 13 germ cell tumors, and 12 medulloblastoma) underwent BBB disruption therapy. The median overall survival for all patients was 2.8 years (2.5 years for patients with primitive neuro ectodermal tumors, 5.4 for patients with germ cell tumors, and 1.7 years for patients with medulloblastoma). Out of 54, 16 were deemed "long-term survivors," living between 4 and 18 years after treatment.[47]

The use of chemotherapeutic agents for the treatment of metastatic brain tumors presents an area of study that is still evolving, and much work remains to be done. The BBB presents variable limitations to effective drug delivery.[3] Preclinical and clinical studies are needed to identify safety and efficacy of identified molecules to be introduced in a dose-intense manner across the NVU. Temozolomide has shown some benefit in metastatic disease in addition to its utility in primary brain tumors. Muldoon et al. noted initial preclinical evidence for improved therapeutic benefit following intra-arterial administration of temozolomide in a rat model, but initial clinical introduction of the drug suggested a high potential for neurotoxicity, due perhaps more to the formulation than to the drug itself.[48]

A small series of five patients with gynecological cancer (four with ovarian cancer, one with endometrial cancer), all metastatic to the brain, were treated with intra-arterial carboplatin with or without BBB disruption, etoposide or etoposide phosphate (intravenous or intra-arterial), and intravenous cyclophosphamide. Figs. 3A and B[9] show a dramatic response to intra-arterial carboplatin, intravenous cyclophosphamide, and intravenous etoposide phosphate with osmotic BBB disruption in one patient with ovarian cancer who chose not to proceed with either surgery or radiation therapy. The patient subsequently agreed to have radiosurgery to the residual enhancement and was in remission for 41 months, with a Karnofsky performance scale score of 90. Table 1 outlines the mean overall survival from the time of primary diagnosis in this group of patients with brain metastases from gynecological cancers treated with BBB disruption and intra-arterial chemotherapy. Mean overall survival from initial diagnosis was 47 months (range, 25-65 months) and mean overall survival from diagnosis of brain metastasis is 26 months. The Karnofsky performance scale scores remained steady during treatment. These data indicate a potential for further study of dose-intense chemotherapy delivery using BBB disruption in patients diagnosed with metastatic disease.

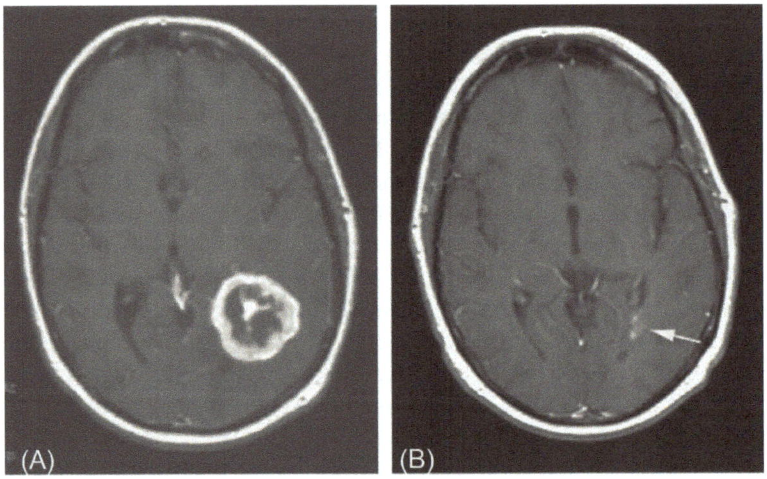

FIG. 3 (A) Magnetic resonance image (MRI) with gadolinium shows a large deep temporal-occipital brain metastasis in a patient with a history of ovarian carcinoma. The patient underwent five rounds of intra-arterial carboplatin, intravenous cyclophosphamide, and intravenous etoposide phosphate with blood–brain barrier (BBB) disruption. (B) MRI with gadolinium shows a dramatic response after BBB disruption chemotherapy. Radiosurgery was then administered to the small residual enhancing lesion *(arrow). Reproduced with permission from McGregor JM, Bell SD, Doolittle ND, Murillo TP, Neuwelt EA. Blood–brain barrier disruption chemotherapy. In: Newton HB, ed.* Handbook of Brain Tumor Chemotherapy, Molecular Therapeutics, and Immunotherapy. *2nd ed. London: Academic Press, Elsevier; 2018:145–153 [chapter 10].*

TABLE 1 Mean Overall Survival of 5 Patients With Gynecological Cancers (5 Ovarian Cancer, 1 Endometrial Cancer) Treated With Blood–Brain Barrier Disruption and Intraarterial Chemotherapy

Patient no.	Location of brain metastases (*n*)	Additional treatment for brain metastases	Survival (mo.) from diagnosis of gynecological cancer	Survival (mo.) from diagnosis of brain metastases
1	Multiple deep and superficial supratentorial (4)	P: Whole brain radiation	46	32
2	Left temporal-occipital (1)	A: Stereotactic radiation	65	41
3	Right cerebellar hemisphere (1)	A: Whole brain radiation	36	10
4	Left frontal (1), left parietal (1), right temporal (1)	None	25	9
5	Left cerebellar hemisphere (1)	A: Stereotactic radiation	63	41

Abbreviations: P, *treatment prior to intra-arterial chemotherapy;* A, *treatment after chemotherapy.*

FUTURE DIRECTIONS FOR BLOOD–BRAIN BARRIER DISRUPTION

Recurrent primary CNS lymphoma remains a difficult problem to solve. Although high-dose intravenous methotrexate regimens have shown promise in obtaining complete response in patients with upfront disease, over 50% of patients relapse within 2 years of diagnosis. Of patients who relapse, 81% relapse in a location distant from the initial presentation, suggesting that sanctuary disease may be present behind a more intact NVU during the course of initial therapy.[49] These findings confirm that BBB integrity is variable across CNS regions in the disease state. The variability of the BBB is noted in primary and secondary tumors, as well as in settings such as trauma, inflammation, and degenerative conditions. Better methods of upfront therapy that encourage therapeutic concentrations throughout the CNS may be required to improve initial response rates and overall survival to even targeted therapeutics.

Recent advances in treatment for recurrent primary CNS lymphoma show promise with combination immunotherapy, combining a checkpoint inhibitor with an anti-CD20 antibody. Ambady et al. reported six patients with recurrent primary CNS lymphoma or isolated CNS recurrence of systemic lymphoma who were treated using rituximab in combination with either nivolumab or pembrolizumab. Three patients had a complete response, and two of three still had complete response at 7 months.[50] High-intensity treatment including high-dose methotrexate chemotherapy, BBB disruption, and autologous stem cell transplant produced a 76% complete response, 2-year progression-free survival of 61%, a 2-year overall survival rate of 57%, and a 5-year overall survival rate of 47%.[51]

Although high-dose chemotherapy is necessary for therapeutic response in the CNS, systemic toxicities can be problematic. Future developments in chemotherapy delivery will include the timely addition of chemoprotectants to mitigate systemic toxicities. These have been evaluated both in ototoxicity and in renal toxicity. Cisplatin chemotherapy is an effective agent for many tumors, particularly in children. The most common cancers in pediatric patients treated with cisplatin include medulloblastoma, osteosarcoma, hepatoblastoma, neuroblastoma, and germ cell tumors. Cisplatin, however, is associated with significant risk for considerable permanent hearing loss, which can have a particularly profound effect on a developing child.[52, 53] Sodium thiosulfate has demonstrated an otoprotective effect, and when delivered 6 hours after administration of cisplatin, the drug provides maximum otoprotection without evidence of tumor protection. This has also been shown for children with hepatoblastoma.[54]

In the field of neuro-oncology, imaging techniques are a fundamental source of information for both the state of the disease, its burden on the CNS, and the effect of any therapeutic intervention. Clinicians continue to struggle with how to acquire best imaging information that will assess for both progression of disease and for disease burden. Contrast agents help visualize abnormalities in the CNS, but like their therapeutic cousins, also are subject to the limitations established by the NVU and the BBB. Accurate and nuanced imaging of the CNS to identify details about extent of disease, degree of infiltration, tumor volume, and extent of treatment effects in both primary and metastatic tumors remains challenging in neuro-oncology. Recent developments in iron-containing nanoparticles have shown some benefit in neuro-oncological imaging. Ultrasmall iron oxide particles consist of superparamagnetic iron oxide surrounded by a carbohydrate shell. In this form, the nanoparticle has been shown to offer improvement in the imaging of tumors and in tumor response to treatment.[55-57] The agent appears to have a delayed entrance across the BBB. As a result, using dynamic susceptibility contrast perfusion MRI techniques, initial imaging can give an improved estimation of cerebral blood volume.[58, 59] With the initial infusion, these agents tend not to have the degree of leakage noted with gadolinium, and therefore may act as an improved blood pool agent.[60, 61] Blood volume analysis is enhanced and an estimation of the perfusion characteristics of a tumor are improved. Tumor vascularity has been predictive of primary tumor grade, and has helped differentiate between tumor progression and treatment effects, or pseudoprogression.[62] With subsequent delayed imaging, nanoparticles are found to cross the BBB and to enter the CNS in the extracellular space. The preferred method of transit for nanoparticles seems to be related to tumor-activated macrophages, which sequester the nanoparticles and act as carriers for the iron across the BBB. After 24-48 hours, the enhancement seen on MRI is likely more indicative of the actual extent of tumor in the brain, and over the course of a patient's treatments, may be able to identify and quantify the degree of immunological response to therapy, especially immunotherapy.[63]

CONCLUSIONS

Delivery of therapeutic molecules into the CNS is hampered by the BBB; however, various methods have been developed to allow passage of agents into the CNS. Translational studies in BBB research attempt to evaluate toxicity and efficacy of global CNS delivery of chemotherapeutics. Osmotic BBB disruption is a safe and effective means of delivering agents to the CNS when safety guidelines are carefully followed. Studies conducted by participating BBB Consortium institutions have paved the way for safe delivery of agents to the CNS with minimal local and systemic toxicities. Advances in the two-compartment model of pharmacokinetics have created options for increasing therapeutic benefit and mitigating systemic toxicities for CNS-acting agents. The concept has proved beneficial in situations in which drugs can be used in combination and with proper timing to act as chemoprotectants. Further development of BBB techniques will allow for improved monitoring of therapeutic responses, as suggested by iron nanoparticles.

Acknowledgments

The studies reported in this chapter were supported by a Veterans Administration Merit Review Grant; National Institutes of Health R01 Grants NS 44687, NS 34608, and NS 33618 from the National Institute of Neurological Disorders and Stroke (NINDS); R13 CA 86959 from the National Cancer Institute, NINDS, and the National Institute of Deafness and Other Communication Disorders; and PHS Grant 5 M01 RR000334.

References

1. Neuwelt EA. Mechanisms of disease: the blood-brain barrier. *Neurosurgery*. 2004;54:131-142.
2. Doolittle ND, Muldoon LL, Culp AY, Neuwelt EA. Delivery of chemotherapeutics across the blood-brain barrier: challenges and advances. *Adv Pharmacol*. 2014;71:203-243. https://doi.org/10.1016/bs.apha.2014.06.002. Review.
3. Puhalla S, Elmquist W, Freyer D, et al. Unsanctifying the sanctuary: challenges and opportunities with brain metastases. *Neuro-Oncology*. 2015;17(5):639-651. https://doi.org/10.1093/neuonc/nov023. Review.
4. Groothuis DR, Vriesendorp FJ, Kupfer B, et al. Quantitative measurements of capillary transport in human brain tumors by computed tomography. *Ann Neurol*. 1991;30:581-588.
5. Levin VA, Clancy TP, Ausman JI, Rall DP. Uptake and distribution of 3H-methotrexate by the murine ependymoblastoma. *J Natl Cancer Inst*. 1972;48:875-883.
6. Groothuis DR. The blood-brain and blood-tumor barriers: a review of strategies for increasing drug delivery. *Neuro Oncol*. 2000;2:45-59.
7. Kraemer DF, Fortin D, Doolittle ND, Neuwelt EA. Association of total dose intensity of chemotherapy in primary central nervous system lymphoma (human non-acquired immunodeficiency syndrome) and survival. *Neurosurgery*. 2001;48:1033-1041.
8. Neuwelt EA, Barnett PA, Hellstrom KE, et al. Delivery of melanoma-associated specific immunoglobulin monoclonal antibody and Fab fragments to normal brain utilizing osmotic blood-brain barrier disruption. *Cancer Res*. 1988;48:4725-4729.
9. McGregor JM, Bell SD, Doolittle ND, Murillo TP, Neuwelt EA. Blood–brain barrier disruption chemotherapy. In: *Handbook of Brain Tumor Chemotherapy, Molecular Therapeutics, and Immunotherapy*. 2nd ed. London, UK: Academic Press, Elsevier; 2018:145-153. [chapter 10].
10. Neuwelt EA, Frenkel EP, Rapoport S, Barnett P. Effect of osmotic blood-bRain barrier disruption on methotrexate pharmacokinetics in the dog. *Neurosurgery*. 1980;7:36-43.
11. Neuwelt EA, ed. vol.2. *Implications of the Blood-Brain Barrier and Its Manipulation*. New York: Plenum Publishing Corporation; 1989.
12. Kroll RA, Neuwelt EA. Outwitting the blood-brain barrier for therapeutic purposes: osmotic opening and other means. *Neurosurgery*. 1998;42:1083-1100.
13. Percy DB, Ribot EJ, Chen Y, et al. In vivo characterization of changing blood-tumor barrier permeability in a mouse model of breast cancer metastasis: a complementary magnetic resonance imaging approach. *Investig Radiol*. 2011;46(11):718-725.
14. Remson LG, McCormick CI, Sexton G, Pearse HD, Garcia R, Neuwelt EA. Decreased delivery and acute toxicity of cranial irradiation and chemotherapy given with osmotic blood-brain barrier disruption in a rodent model: the issue of sequence. *Clin Cancer Res*. 1995;1:731-739.
15. Remsen LG, Marquez C, Garcia R, Thrun LA, Neuwelt EA. Efficacy after sequencing of brain radiotherapy and enhanced antibody targeted chemotherapy delivery in a rodent human lung cancer brain xenograft model. *Int J Radiat Oncol Biol Phys*. 2001;51:1045-1049.

16. Remsen LG, McCormick CI, Sexton G, et al. Long-term toxicity and neuropathology associated with the sequencing of cranial irradiation and enhanced chemotherapy delivery. *Neurosurgery*. 1997;40:1034-1042.
17. Cao Y, Tsien CI, Shen Z, et al. Use of magnetic resonance imaging to assess blood-brain/blood-glioma barrier opening during conformal radiotherapy. *J Clin Oncol*. 2005;23(18):4127-4136.
18. Dahlborg SA, Henner WD, Crossen JR, et al. Non-AIDS primary CNS lymphoma: first example of a durable response in a primary brain tumor using enhanced chemotherapy delivery without cognitive loss and without radiotherapy. *Cancer J Sci Am*. 1996;2:166-174.
19. Finkelstein SE, Timmerman R, WH MB, et al. The confluence of stereotactic ablative radiotherapy and tumor immunology. *Clin Dev Immunol*. 2011;2011:439752.
20. Szeifert GT, Kondziolka D, Levivier M, et al. Histopathology of brain metastases after radiosurgery. *Prog Neurol Surg*. 2012;25:30-38.
21. Neuwelt EA, Pagel M, Barnett P, Glasberg M, Frenkel EP. Pharmacology and toxicity of intracarotid adriamycin administration following osmotic blood-brain barrier modification. *Cancer Res*. 1981;41:4466-4470.
22. Neuwelt EA, Barnett PA, Glasberg M, Frenkel EP. Pharmacology and neurotoxicity of cis-Diamminedichloroplatinum, bleomycin, 5-fluorouracil, and cyclophosphamide administration following osmotic blood-brain barrier modification. *Cancer Res*. 1983;43:5278-5285.
23. Fortin D, McCormick CI, Remsen LG, Nixon R, Neuwelt EA. Unexpected neurotoxicity of etoposide phosphate administered in combination with other chemotherapeutic agents after blood-brain barrier modification to enhance delivery, using propofol for general anesthesia, in a rat model. *Neurosurgery*. 2000;47:199-207.
24. Dahlborg SA, Petrillo A, Crossen JR, et al. The potential for complete and durable response in nonglial primary brain tumors in children and young adults with enhanced chemotherapy delivery. *Cancer J Sci Am*. 1998;4:110-124.
25. McAllister LD, Doolittle ND, Guastadisegni PE, et al. Cognitive outcomes and long-term follow-up results after enhanced chemotherapy delivery for primary central nervous system lymphoma. *Neurosurgery*. 2000;46:51-61.
26. Doolittle ND, Miner ME, Hall WA, et al. Safety and efficacy of a multicenter study using intraarterial chemotherapy in conjunction with osmotic opening of the blood-brain barrier for the treatment of patients with malignant brain tumors. *Cancer*. 2000;88:637-647.
27. Tyson RM, Siegal T, Doolittle ND, Lacy C, Kraemer DF, Neuwelt EA. Current status and future of relapsed primary central nervous system lymphoma (PCNSL). *Leuk Lymphoma*. 2003;44:627-633.
28. Guillaume DJ, Doolittle ND, Gahramanov S, Hedrick NA, Delashaw JB, Neuwelt EA. Intra-arterial chemotherapy with osmotic blood-brain barrier disruption for aggressive oligodendroglial tumors: results of a phase I study. *Neurosurgery*. 2010;66(1):48-58. discussion 58.
29. Doolittle ND, Petrillo A, Bell S, Cummings P, Eriksen S. Blood-brain barrier disruption for the treatment of malignant brain tumors: the national program. *J Neurosci Nurs*. 1998;30:81-90.
30. Roman-Goldstein S, Clunie DA, Stevens J, et al. Osmotic blood-brain barrier disruption: CT and radionuclide imaging. *AJNR Am J Neuroradiol*. 1994;15:581-590.
31. Fortin D, McAllister LD, Nesbit G, et al. Unusual cervical spinal cord toxicity associated with intra-arterial carboplatin, intra-arterial or intravenous etoposide phosphate, and intravenous cyclophosphamide in conjunction with osmotic blood-brain barrier disruption in the vertebral artery. *AJNR Am J Neuroradiol*. 1999;20:1794-1802.
32. Williams PC, Henner DW, Roman-Goldstein S, et al. Toxicity and efficacy of carboplatin and etoposide in conjunction with disruption of the blood-brain barrier in the treatment of intracranial neoplasms. *Neurosurgery*. 1995;37:17-28.
33. Neuwelt EA, Brummett RE, Doolittle ND, et al. First evidence of otoprotection against carboplatin-induced hearing loss with a two-compartment system in patients with central nervous system malignancy using sodium thiosulfate. *J Pharmacol Exp Ther*. 1998;286:77-84.
34. Doolittle ND, Muldoon LL, Brummett RE, et al. Delayed sodium thiosulfate as an otoprotectant against carboplatin-induced hearing loss in patients with malignant brain tumors. *Clin Cancer Res*. 2001;7:493-500.
35. Doolittle ND, Anderson CP, Bleyer WA, et al. for the Blood-Brain Barrier Disruption Consortium. Importance of dose intensity in neuro-oncology clinical trials: summary report of the sixth annual meeting of the Blood-Brain Barrier Disruption Consortium. *Neuro Oncol*. 2001;3:46-54.
36. Doolittle ND, Abrey LE, Ferrari N, et al. Targeted delivery in primary and metastatic brain tumors: summary report of the seventh annual meeting of the Blood-Brain Barrier Disruption Consortium. *Clin Cancer Res*. 2002;8:1702-1709.

37. Doolittle ND, Abrey LE, Bleyer WA, et al. New frontiers in translational research in neuro-oncology and the blood-brain barrier: report of the tenth annual Blood-Brain Barrier Disruption Consortium meeting. *Clin Cancer Res*. 2005;11(2 Pt 1):421-428.
38. Abrey LE, DeAngelis LM, Yahalom J. Long-term survival in primary CNS lymphoma. *J Clin Oncol*. 1998;16:859-863.
39. Neuwelt EA, Guastadisegni PE, Varallyay P, Doolittle ND. Imaging changes and cognitive outcome in primary CNS lymphoma after enhanced chemotherapy delivery. *AJNR Am J Neuroradiol*. 2005;26(2):258-265.
40. Smith JR, Rosenbaum JT, Wilson DJ, et al. Role of intravitreal methotrexate in the management of primary central nervous system lymphoma with ocular involvement. *Ophthalmology*. 2002;109:1709-1716.
41. Angelov L, Doolittle ND, Kraemer DF, et al. Blood-brain barrier disruption and intra-arterial methotrexate-based therapy for newly diagnosed primary CNS lymphoma: a multi-institutional experience. *J Clin Oncol*. 2009; 27(21):3503-3509.
42. Doolittle ND, Dósa E, Fu R, et al. Preservation of cognitive function in primary CNS lymphoma survivors a median of 12 years after enhanced chemotherapy delivery. *J Clin Oncol*. 2013;31(31):4026-4027.
43. Doolittle ND, Korfel A, Lubow MA, et al. Long-term cognitive function, neuroimaging, and quality of life in primary CNS lymphoma. *Neurology*. 2013;81(1):84-92.
44. Alas S, Ng CP, Bonavida B. Rituximab modifies the cisplatin-mitochondrial signaling pathway, resulting in apoptosis in cisplatin-resistant non-Hodgkins lymphoma. *Clin Cancer Res*. 2002;8:836-845.
45. Doolittle ND, Fu R, Muldoon LL, Tyson RM, Lacy C, Neuwelt EA. Rituximab in combination with methotrexate-based chemotherapy with blood–brain barrier disruption in newly diagnosed primary CNS lymphoma. *Hematol Oncol*. 2013;31(S1):179.
46. McGregor J, Bourekas E, Bell S. Repeat blood-brain barrier disruption in patients with recurrent primary central nervous system lymphoma. *Neuro Oncol*. 2008;10(5):759-915. ST-17.
47. Jahnke K, Kraemer DF, Knight KR, et al. Intraarterial chemotherapy and osmotic blood–brain barrier disruption for patients with embryonal and germ cell tumors of the central nervous system. *Cancer*. 2008; 112(3):581-588.
48. Muldoon LL, Pagel MA, Netto JP, Neuwelt EA. Intra-arterial administration improves temozolomide delivery and efficacy in a model of intracerebral metastasis, but has unexpected brain toxicity. *J Neuro-Oncol*. 2016; 126(3):447-454. https://doi.org/10.1007/s11060-015-2000-1.
49. Ambady P, Fu R, Netto JP, et al. Patterns of relapse in primary central nervous system lymphoma: inferences regarding the role of the neuro-vascular unit and monoclonal antibodies in treating occult CNS disease. *Fluids Barriers CNS*. 2017;14(1):16. https://doi.org/10.1186/s12987-017-0064-3.
50. Ambady P, Szidonya L, Firkins J, et al. Combination immunotherapy as a non-chemotherapy alternative for refractory or recurrent CNS lymphoma. *Leuk Lymphoma*. 2018;1-4. https://doi.org/10.1080/10428194.2018.1480771.
51. Kuitunen H, Tokola S, Siniluoto T, et al. Promising treatment results with blood brain barrier disruption (BBB DISRUPTION) based immunochemotherapy combined with autologous stem cell transplantation (ASCT) in patients with primary central nervous system lymphoma (PCNSL). *J Neuro-Oncol*. 2017;131(2):293-300. https://doi.org/10.1007/s11060-016-2293-8.
52. Bass JK, Knight KR, Yock TI, Chang KW, Cipkala D, Grewal SS. Evaluation and management of hearing loss in survivors of childhood and adolescent cancers: a report from the Children's Oncology Group. *Pediatr Blood Cancer*. 2016;63(7):1152-1162.
53. Brooks B, Knight K. Ototoxicity monitoring in children treated with platinum chemotherapy. *Int J Audiol*. 2018; 57:S34-S40. suppl. 4.
54. Brock PR, Maibach R, Childs M, et al. Sodium thiosulfate for protection from cisplatin-induced hearing loss. *N Engl J Med*. 2018;378(25):2376-2385. https://doi.org/10.1056/NEJMoa1801109.
55. Varallyay P, Nesbit G, Muldoon LL, et al. Comparison of two super paramagnetic viral-sized iron oxide particles ferumoxides and ferumoxtran-10 with a gadolinium chelate in imaging intracranial tumors. *AJNR Am J Neuroradiol*. 2002;23:510-519.
56. Neuwelt EA, Varallyay P, Bago A, Muldoon LL, Nesbit G, Nixon R. Imaging of iron oxide nanoparticles by MR and light microscopy in patients with malignant brain tumours. *Neuropathol Appl Neurobiol*. 2004;30(1):70. https://doi.org/10.1111/j.1365-2990.2004.00557.x.
57. Toth GB, Varallyay CG, Horvath A, et al. Current and potential imaging applications of ferumoxytol for magnetic resonance imaging. *Kidney Int*. 2017;92(1):47-66.

58. Gahramanov S, Muldoon LL, Varallyay CG, et al. Pseudoprogression of glioblastoma after chemo- and radiation therapy: diagnosis by using dynamic susceptibility-weighted contrast-enhanced perfusion MR imaging with ferumoxytol versus gadoteridol and correlation with survival. *Radiology*. 2013;266(3):842-852.
59. Varallyay CG, Nesbit E, Horvath A, et al. Cerebral blood volume mapping with ferumoxytol in dynamic susceptibility contrast perfusion MRI: Comparison to standard of care. *J Magn Reson Imaging*. 2018;48(2):441-448. https://doi.org/10.1002/jmri.25943.
60. Weinstein JS, Varallyay CG, Dosa E, et al. Superparamagnetic iron oxide nanoparticles: diagnostic magnetic resonance imaging and potential therapeutic applications in neurooncology and central nervous system inflammatory pathologies, a review. *J Cereb Blood Flow Metab*. 2010 Jan;30(1):15-35.
61. Horvath A, Varallyay CG, Schwartz D, et al. Quantitative comparison of delayed ferumoxytol T1 enhancement with immediate gadoteridol enhancement in high grade gliomas. *Magn Reson Med*. 2018;80(1):224-230.
62. Barajas JRF, Hamilton BE, Schwartz D, et al. Combined iron oxide nanoparticle ferumoxytol and gadolinium contrast enhanced MRI define glioblastoma pseudoprogression. *Neuro Oncol*. 2018;noy160.
63. McConnell HL, Schwartz DL, Richardson BE, Woltjer RL, Muldoon LL, Neuwelt EA. Ferumoxytol nanoparticle uptake in brain during acute neuroinflammation is cell-specific. *Nanomedicine*. 2016;12(6):1535-1542.

CHAPTER

11

Ultrasonic Methods

Ying Meng[*,†], Karim Mithani*[*], Laura Vecchio[†], Isabelle Aubert[†,‡], Nir Lipsman[*,†]*

[*]Division of Neurosurgery, Sunnybrook Health Sciences Centre, University of Toronto, Toronto, Canada [†]Sunnybrook Research Institute, Sunnybrook Health Sciences Centre, University of Toronto, Toronto, Canada [‡]Department of Laboratory Medicine and Pathobiology, University of Toronto, Toronto, Canada

INTRODUCTION

The blood–brain barrier (BBB) is notorious for obstructing the delivery of effective therapy for many different neurological diseases. The BBB is a selective boundary between the central nervous system (CNS) and the systemic circulation, and it plays a critical role in maintaining the appropriate milieu within the CNS and in protecting cells from hazardous toxins. In addition to protecting the brain, the BBB also impedes drug delivery, which restricts the use of many compounds proven to be effective in preclinical studies. For example, studies show that less than 1% of antiamyloid antibodies administered into the systemic circulation are measured in the CNS.[1-3]

Various methods of bypassing the BBB have been investigated, such as direct convection-enhanced delivery, biomimetics (e.g., Trojan horses), carriers (e.g., nanoparticles), as well as chemical and physical approaches. For instance, there exists evidence of BBB breakdown after a radiotherapy dose of 20 Gy,[4, 5] which helps sensitize the tissue to certain chemotherapies. Unfortunately, healthy brain tissue is subjected to the potentially harmful effects of radiation toxicity. Chemical opening of the BBB (e.g., with mannitol, carriers, and biomimetics) is a method that lacks spatial and temporal specificity, and is additionally associated with adverse systemic effects. Although convection-enhanced delivery allows precise spatial and temporal control, an open neurosurgical procedure is required to implant the catheter. Optimization of therapeutic delivery to the CNS requires a means of noninvasively and

Nervous System Drug Delivery
https://doi.org/10.1016/B978-0-12-813997-4.00011-6

© 2019 Elsevier Inc. All rights reserved.

repeatedly bypassing the BBB with a high degree of spatial and temporal specificity and minimal adverse effects. Over the past two decades, focused ultrasound (FUS) has emerged as a potential solution to this longstanding problem. In this chapter, we review the mechanisms of FUS-mediated BBB opening and the clinical applications of this method, which are currently under investigation.

FOCUSED ULTRASOUND

Basic Physical Principles

Ultrasound technology involves sound waves produced at frequencies greater than the upper limit of normal human hearing, which is approximately 20 kHz. Ultrasound was discovered in the late 1800s as the result of the piezoelectric effect, which describes a reciprocal conversion between mechanical and electrical energy in certain materials. Transducers that generate ultrasound were improved during World War I to enhance underwater submarine navigation. Since these transducers were introduced to the medical field, the applications of ultrasound have been predominantly diagnostic; however, it was apparent early on that ultrasound could heat biological tissue. In fact, the first ultrasonic device (Sonostat by Siemens) was advertised as a treatment option for patients diagnosed with inflammatory musculoskeletal conditions.

The propagation of sound waves through a material depends on the density and stiffness of that material; waves travel fastest in solids and slowest in gases. Modes of ultrasound waves used in devices include "continuous" and "pulsed;" the latter of which is more common in medical applications. Continuous-wave ultrasound is characterized by frequency, wavelength, and propagation speed in the medium (Fig. 1A). Pulsed ultrasound describes bursts of ultrasound waves, interspersed with intervals of no signal at all. It involves pulse repetition frequency and period, pulse duration, and duty cycle (Fig. 1B). The duty cycle communicates the proportion of the pulse repetition period that contains a signal; in other words, pulse duration over pulse repetition period.

The effect of ultrasound on biological tissue is altered by the intensity and duration of the exposure. Whereas ultrasonic power or energy is measured in watts (W), the intensity is the rate of energy passing through a unit area, measured in W/m^2 (Fig. 1C). Therefore, the spatial average intensity is the total power divided by the cross-sectional area. The intensity of pulsed-wave ultrasound is dependent on time, where the greatest intensity during the pulse defines the temporal peak. Pulse average is calculated as the average intensity during the pulse duration, whereas the temporal average is calculated during the interval without signal. Consequently, the spatial peak temporal average (SPTA) intensity indicates the intensity measured at the center of the beam, averaged over the entire pulse repetition period, and is clinically important as an index of safety. Intensity also varies as ultrasound propagates through the medium as well as at the boundaries between two mediums, where the ultrasound waves may be focused, diffracted, absorbed, scattered, reflected, or refracted to varying degrees.

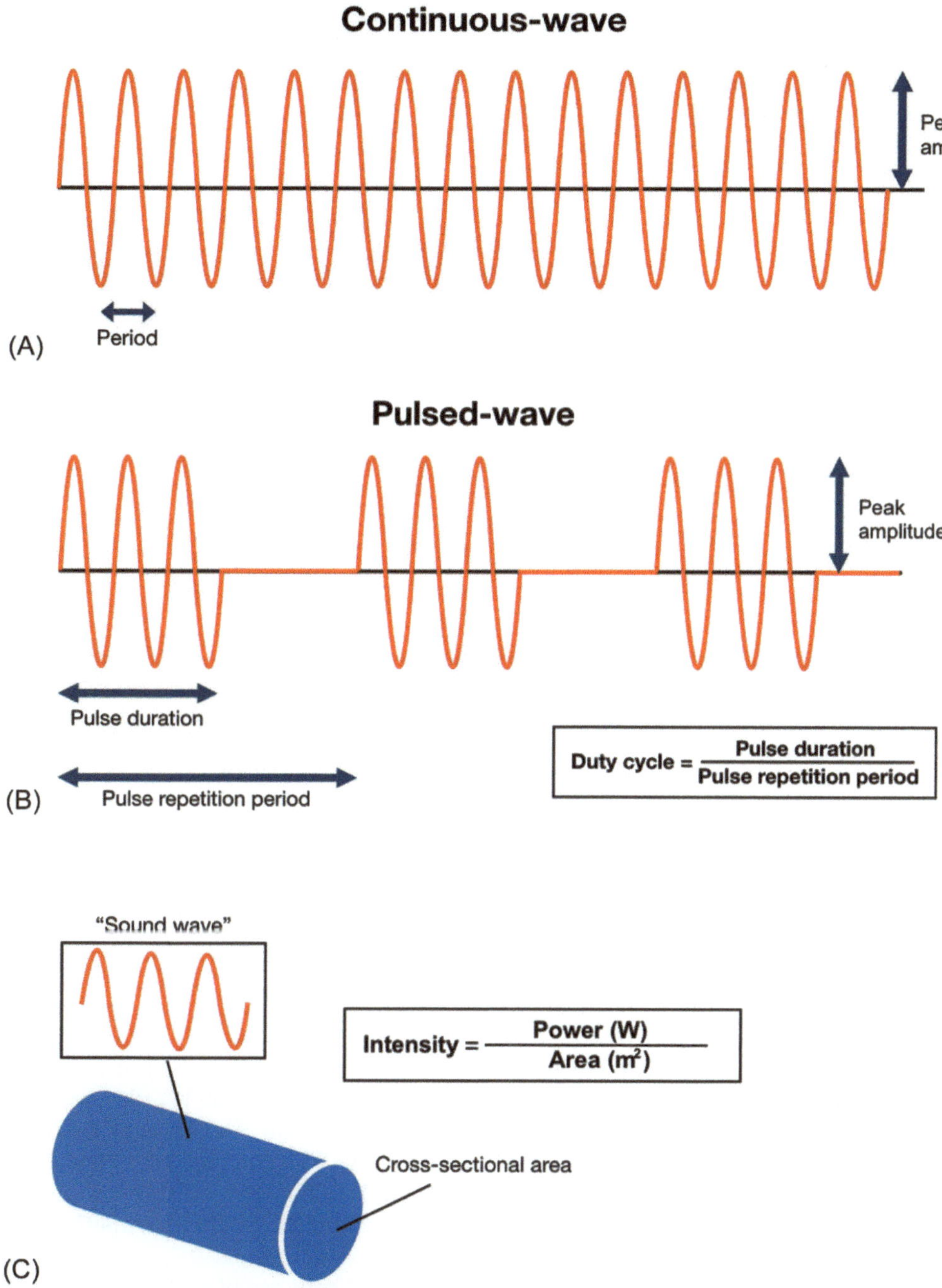

FIG. 1 Schematic diagrams of ultrasound parameters that characterize (A) continuous and (B) pulsed waves. (C) Intensity is defined as the ultrasound power over the cross-sectional area of the ultrasound beam. Intensity is useful in measuring and predicting the effect of ultrasound on biological tissue.

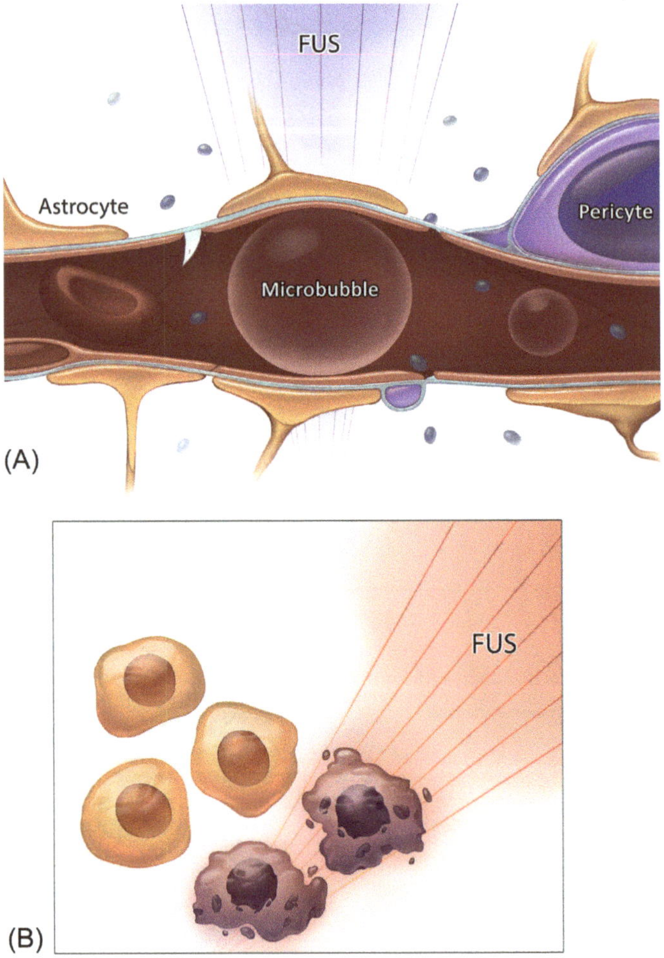

FIG. 2 Biological effects of ultrasound in the central nervous system. (A) Interaction between low-intensity focused ultrasound and endogenous or exogenously introduced gas in the vessels results in mechanical stress on the vessel walls, thereby inducing opening of the blood–brain barrier. (B) High-intensity focused ultrasound delivers energy to biological tissue, resulting in heat and coagulative necrosis. *FUS, focused ultrasound.*

Biological Effects

The bioeffects of ultrasound can be broadly classified as thermal (Fig. 2A) or mechanical (Fig. 2B). At a frequency of 650 kHz, with acoustic power as low as 300 W, ultrasound can generate sufficient heat at the target zone to create a thermal lesion. Elevation of tissue temperature to 55-60°C will induce coagulation necrosis (Fig. 2A). This thermal effect is the result of time-dependent energy deposition. Furthermore, at sublesional temperatures, sonications can result in transient lesional effect as well as other effects, such as chemosensitization of the tissue.[6, 7]

Ultrasound sonications can also induce the formation and collapse of gas bubbles inherent in blood vessels, in a mechanical process known as *acoustic cavitation*. The oscillation of bubbles in the ultrasound field results in expansion and contraction, enabling disruption of adjacent BBB structures, such as tight junctions (Fig. 2B). In early investigations, the biological sequelae were highly variable, ranging from transient BBB opening to hemorrhage and tissue injury.[8] Although the ability to disrupt the BBB by ultrasound has important clinical applications, the translation of ultrasound at these particular parameters has been problematic. In an effort to leverage this phenomenon for reversible BBB opening more safely and predictably, exogenous, preformed microbubbles were introduced into the circulation prior to sonication.[9] These microbubbles concentrate the acoustic energy within blood vessels, and serve as nuclei for stable cavitation with a 100-fold reduction in the required energy. Combining this procedure with magnetic resonance imaging (MRI) guidance and real-time acoustic feedback further improved its overall safety profile, resulting in the eventual clinical translation of the device.

BLOOD–BRAIN BARRIER DISRUPTION

The BBB is a selective, semipermeable membrane composed of tight junctions between adjacent epithelial cells (Fig. 3). The tight junctions are formed by transmembrane proteins such as occludins, claudins, and junctional adhesion molecules that interact with cytoplasmic scaffolding proteins such as zonula occluden (ZO) and other membrane-associated proteins that are important in cell signaling. The BBB is further fortified by pericytes, astrocyte end-feet processes, and neurons, and prevents the passage of hydrophilic proteins greater than 400 Da. In fact, pharmacokinetic studies reveal that less than 1% of large therapeutics, such as antiamyloid antibodies, reach the CNS.[1-3] In the field of oncology, numerous antibodies developed for various bodily tumors are ineffective for brain tumors due to their inability to cross the BBB. Specifically, antibodies that are larger in size (>150 kDa) or those that have

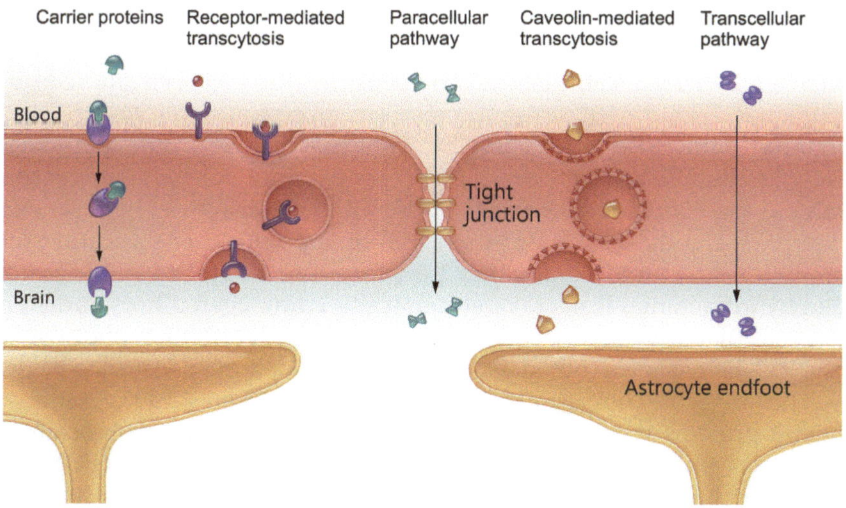

FIG. 3 Schematic diagrams of various transport pathways across the blood–brain barrier. Blood–brain barrier opening, when induced by focused ultrasound, may be facilitated by increased paracellular transport and caveolin-mediated transcytosis, in addition to a decrease in P-glycoproteins.

a longer circulatory half-life (>21 days) are less likely to permeate the BBB. In addition, solid tumors exhibit high interstitial fluid pressures and correspondingly reduced blood flow, further restricting immunoglobulin penetration.[10] Moreover, uncontrolled cellular proliferation in tumors reduces microvascular density, shortens the width of interstitial pathways, and increases the tortuosity of vessels, all of which further impede antibody delivery through the BBB.[11]

Multridrug resistance protein 1 (MDR1) or P-glycoproteins (P-gps) present on the endothelial cells of the BBB are also responsible for decreased permeability of the barrier to drugs. These are energy-dependent efflux pumps that remove various molecules from cells. They are especially relevant in neuro-oncology therapeutics, as certain cancer cells express large amounts of P-gp. Accordingly, MDR modulation by various compounds such as verapamil and cyclosporine is a significant area of research in overcoming drug resistance. Nevertheless, the BBB remains a considerable obstacle to the delivery of countless neuro-therapeutics; mechanical disruption of the barrier with spatiotemporal specificity is therefore critically important for treating neurological disease.

Focused ultrasound in combination with intravenously injected microbubbles permits noninvasive, focal, and transient disruption of the BBB by controlled cavitation.[9] FUS has several advantages over other means of BBB disruption. It is highly focal, with spatial specificity in intracranial targeting down to the submillimeter. Furthermore, FUS-mediated BBB opening can be performed noninvasively and under MRI guidance, resulting in safe, temporary, and reproducible disruption. Third, compared to other intracranial delivery systems such as convection-enhanced therapy, FUS is associated with more uniform drug distribution. Finally, the characteristics of BBB disruption can be readily modified by various ultrasound parameters, such as sonication power and duration.

Mechanism of BBB Disruption

The mechanism by which stable cavitation opens the BBB has been studied using various imaging techniques, including electron microscopy, two-photon microscopy, and dynamic contrast-enhanced MRI. BBB disruption by FUS may result from an increase in both paracellular and transcellular transport (Fig. 3). Using immunoelectron microscopy, Sheikov et al.[12] quantified the expression and distribution of the microvessel tight junction proteins occludin, claudin-1, claudin-5, and ZO-1 at serial time points after sonication with intravenous microbubbles. At 1 and 2 hours after the procedure, increased BBB permeability was demonstrated by intracranial leakage of systemically administered horseradish peroxidase, which has a molecular weight of 40 kDa. A concurrent redistribution and loss of occludin, claudin-5, and ZO-1 signals was noted. At 6 hours, BBB function appeared to be restored such that there was no longer any horseradish peroxidase leakage, and the localization and density of tight junction proteins were restored. Using electron microscopy, the same group also found more vesicles and vacuoles after sonications to facilitate increased BBB transcytosis.[13]

Various tight junction proteins and multidrug transporters including MDRs such as P-gp can contribute to the treatment resistance of brain tumors, epilepsy, neurodegenerative diseases (e.g., amyotrophic lateral sclerosis), and psychiatric diseases.[14-17] In addition to physically disrupting the BBB, microbubble-enhanced FUS may also induce functional changes in the barrier. Several studies have shown suppression of multidrug transporters in BBB cells after FUS treatment.[18-22] One proposed mechanism for this involves temporary, local down-regulation of protein expression after FUS, as was demonstrated recently for P-gp.[23, 24] Low P-gp expression is seen in arterioles of rats where shear stress is high, suggesting that this effect may be a result of stress caused by the microbubbles and mediated by mechanosensitive receptors. Furthermore, FUS can also disrupt tight junction proteins as detected by electron microscopy, including occludin and claudin-5.[12, 13] Connexin 36, a gap junction protein, was also downregulated after FUS treatment with microbubbles.[25]

Upregulation of the structural protein caveolin-1[25a] has recently been identified as a potential contributor to FUS-mediated BBB disruption. This scaffold protein is an integral component of caveolae, which are invaginations of the plasma membrane on most vertebrate cell types. These caveolae are believed to play a role in signal transduction and endocytosis, among many other functions. Therefore, changes in membrane proteins may be another way FUS modulates BBB function.

The mechanical and functional changes mediated by microbubble-enhanced FUS on tight junction proteins and multidrug transporters (e.g., P-gp) can potentially reduce treatment resistance in oncologic, neurodegenerative, and intractable psychiatric patients. Further research is required to elucidate all the mechanisms by which this occurs, so as to maximize the clinical utility and safety of FUS technology.

Safety

Microbubble-enhanced, FUS-mediated BBB opening is both noninvasive and reversible. Low-frequency ultrasound parameters enable BBB opening but avoid tissue damage, with the disrupted region typically closing within 6 to 12 hours.[9, 26-28] Notably, the duration of opening may vary based on the size of the agent being delivered.[27]

Some have suggested that the time to BBB closure depends on the volume of BBB disruption, which in turn is related to the pressure and pulse length of the applied ultrasound.[29, 30] However, by increasing the volume of BBB disruption using multiple overlapping foci rather than by increasing acoustic pressure, O'Reilly et al.[31] found that the time to closure was independent of the volume targeted by FUS. BBB closure as measured by serial gadolinium scans showed complete closure in 9 out of 10 hemispheres by 6 hours, with the remaining location showing significantly decreased enhancement at 6 hours and none at 24 hours. This finding is critical in approximating the safety of FUS in clinical applications, which generally require relatively large volumes of BBB opening. In fact, whole-hemisphere opening of the BBB has been successfully performed using MRI-guided FUS in aged beagles, which also naturally accumulate amyloid deposits, resulting in cognitive decline. The FUS parameters and targeted volumes in this study—employing a frequency of 0.28 MHz, with pulse repetition frequency of 1 Hz for 2 minutes total and 10 ms bursts weekly over 4 weeks—are clinically relevant and readily translatable.[32] The investigators found no significant or lasting tissue damage associated with FUS-mediated whole-hemisphere BBB disruption, based on neuroimaging and histological analyses.

Recent research has shown evidence of acute sterile inflammation after microbubble-enhanced FUS.[33] Although a subsequent study by McMahon and Hynynen also found evidence of acute inflammation (due to NFκB signaling pathway activation after FUS), this occurred only at high microbubble doses.[34] The results indicated that FUS-induced BBB disruption is possible at microbubble doses that do not trigger an inflammatory response. This conclusion was supported by studies from a range of laboratories, which use lower microbubble doses than those that induce acute inflammation.[35] Although microvascular transcriptome in the acute phase after FUS did suggest a transient inflammatory response with upregulated expression of pro-inflammatory cytokines (e.g., Ccl2, Ccl7, Cxcl1, Cxcl11, IL1b, IL6, Tnf, Sele, Sod3, etc.), these returned to baseline by 24 hours.[36] The same group also reported downregulated expression of several BBB transporter genes and upregulation of angiogenesis-related genes 6 hours after sonication. Furthermore, biweekly FUS-induced striatum opening over 6 months in mice did not result in any motor impairments (on rotarod performance tests), behavioral changes, or morphologic alterations, suggesting long-term safety of the procedure.[37] Nonhuman primate studies have also reported safe and effective BBB disruption with microbubble-enhanced FUS, at a variety of ultrasound parameters, microbubble doses, and brain targets.[38-40] The safety of an implantable ultrasound device has been explored in nonhuman primates, revealing no CNS damage in neuroimaging, electroencephalogram, somatosensory evoked potential, behavioral, and histological analyses following 4 months of biweekly microbubble-enhanced sonications.[41]

An important concern with intracranial FUS procedures is the extravasation of plasma proteins into the CNS parenchyma. Albumin and its bound substances are of particular importance, as they constitute approximately 50% of plasma proteins[42] and may be neurotoxic to neurons. One group found that extravasated albumin following BBB disruption is found mostly in the periphery of blood vessels and, to a lesser extent, in the sonicated brain parenchyma.[25, 43] In a subsequent study, that group demonstrated that cellular uptake of this albumin begins as early as 30 minutes after sonication, increases constantly over 24 hours, and is mediated predominantly by activated microglia, astrocytes, and endothelial cells.[44] This suggests that neuroprotective mechanisms may mitigate or prevent potential damage to CNS cells induced by FUS-mediated BBB opening.

Although the molecular, cellular, and inflammatory effects of microbubble-enhanced FUS are still under investigation, the procedure has been repeatedly tested in preclinical models and in phase I clinical trials, with no significant safety concerns. Further rigorous investigations into FUS-mediated BBB disruption are required to more comprehensively ascertain its bioeffects, and to ultimately optimize its safety in clinical applications.

Factors Affecting Opening and Closure of the BBB

Factors known to affect the size and duration of BBB opening include: physical parameters of the administered ultrasound, microbubble properties, size of the agent being delivered, and vascular/tissue properties of the target.

Opening of the BBB is positively correlated with both acoustic pressure and microbubble diameter.[45, 46] The transfer, or "permeability," constant across the BBB (as a function of time) has been assessed with fluorescent dye and 2-photon fluorescent microscopy.[47] A linear regression analysis suggested that this permeability constant is positively correlated with the applied acoustic pressure. At 0.6 MPa, which is a commonly used acoustic pressure for BBB opening in animals, transfer constants of 1, 10, and 70 kDa agents were found to be

0.039/min, 0.019/min, and 0.007/min, respectively. Permeability was also inversely related to vessel diameter and the molecular size of the agent being delivered. Furthermore, the researchers identified two distinct "leakage types" based on the onset of BBB disruption, defined as initiation of leakage at either <10 min (i.e., fast) or >10 min (i.e., slow) after sonication. Fast leakage exhibited a significantly higher permeability constant than slow leakage, and was speculated to be a result of transjunctional transport across the barrier. In contrast, slow leakage was thought to be a result of transcytosis, which is known to have a noticeable lag (see "Mechanism of BBB Disruption" section). Importantly, the onset of BBB disruption, leakage kinetics, and enhanced permeability are all affected by the acoustic pressure of the applied ultrasound, suggesting that delivery of therapeutics across the barrier can be tailored by modifying FUS parameters.[47]

Vascular density and size appear to play an important role in FUS-mediated BBB opening. Studies show that acoustic pressure required for BBB opening varies with vessel diameter—this is potentially due to the higher number of microbubbles present in larger capillaries. Nevertheless, the independent effects of vascular smooth muscle, structure, and tight junctions on opening pressure (with other variables held constant) have yet to be investigated. Notably, a significant difference in opening pressure between gray and white matter was detected in nonhuman primates, with the probability of successful BBB opening in grey matter 3 times higher than in white matter.[48]

Ultrasound frequencies between approximately 500 kHz and 1.18 MHz, and pressures between 0.3 and 0.54 MPa, have been used to successfully open the BBB.[49-51] Higher frequency waves have shorter wavelengths, enabling finer spatial resolution of the target. Pulse duration also varies, as do the type and dose of microbubbles used (e.g., SonoVue [Bracco], DEFINITY [Lantheus], and USphere Prime [Trust Bio Sonics]). Wu et al.[52] found these microbubbles to be comparable in their BBB-opening effects at identical concentrations, and Song et al.[53] concluded that gas volume dose was positively correlated to the extent of BBB opening, independent of microbubble size.

To facilitate clinical translation, multiple indicators of the likelihood of BBB opening with microbubble-enhanced FUS have been identified and investigated. Mechanical index (defined as peak negative acoustic pressure over the square root of the frequency) and cavitation index (defined as peak negative acoustic pressure over frequency) are both correlated with the scale of BBB opening, as assessed by dynamic-contrast-enhanced MRI.[49] These indicators can allow clinicians to gauge the capacity for therapeutic delivery into the CNS with microbubble-enhanced FUS.

An algorithm to estimate power for efficient BBB opening was developed from a 2017 preclinical parametric study.[50] Based on microbubble emissions during sonication, starting from 0.6 to 1 MPa and increased by 0.016 MPa, 50% of the power at which ultraharmonic emissions were detected was found to be the optimal power setting. Ultraharmonic emissions have been shown to precede wideband emissions that indicate inertial cavitation, which increases the risk of tissue damage.

THERAPEUTIC DELIVERY

Introducing microbubbles exogenously not only decreases the ultrasound energy required to open the BBB but it also provides another parameter for tailoring the technology to the treatment, and offers a conduit for delivery of compounds specifically to the area of BBB opening.[51] The enhanced delivery of various biological agents such as growth factors, antibodies to viruses for gene therapy, and stem cells for cell therapy have been investigated in small animal to nonhuman primate models.

Neuro-oncology

Despite major advances in oncological care and novel targeted therapies for other organs, patients with tumors of the CNS (e.g., primary brain tumors and brain metastasis) have not enjoyed similar medical breakthroughs. The BBB and efflux transporters, such as P-gp, preclude a vast majority of cancer therapeutics from affecting both normal brain tissue and target tumoral cells. Furthermore, the CNS is largely an immunoprivileged organ, due to the functions of the BBB. Traditionally, the BBB has been considered to be disrupted in malignant tumors, such as glioblastoma multiforme (GBM). However, advanced imaging with dynamic contrast-enhanced MRI has shown greater heterogeneity than previously thought. The BBB is also disrupted by radiation therapy, and clinicians take advantage of this increased permeability with concurrent chemotherapy to improve drug delivery. For the most part, this mode of BBB disruption cannot be achieved in healthy tissue, and the temporal resolution is difficult to control or predict. Overall, the improvement of cancer therapy delivery to tumoral cells in the CNS through minimally invasive BBB disruption can revolutionize neuro-oncology and will hopefully bring a cure for brain tumors closer within reach.

A number of studies in preclinical animal models have demonstrated enhanced delivery of various cancer therapeutics, including methotrexate,[28] doxorubicin, temozolomide, and carmustine, as well as liposomal or brain-penetrating nanoparticle encapsulation with the aid of FUS. In addition, immune targeted therapies such as antibodies (e.g., trastuzumab), IL-12, and NK-92 for innate immune response to tumor cells[54] have also been delivered by FUS with overall statistically significant augmentation of the clinical benefit over medical therapy alone. Apart from BBB disruption, the observed decrease in MDR proteins after FUS may further improve the retention of chemotherapy in the parenchyma once it has crossed the BBB.[24] Alternatively, FUS may be used to cause mild hyperthermia around 42°C to induce increased uptake of chemotherapy or induced unloading of carrier content at the targeted region.[6, 7]

In this section, we review the benefits and challenges of augmenting chemotherapeutic options with BBB disruption through FUS in disease models of breast cancer and GBM specifically. However, the successful development of FUS has exciting implications for many other brain tumors, including all intracranial metastases with drugs already developed for systemic disease, and highly inaccessible tumors such as diffuse intrinsic pontine glioma in children.

Breast Cancer Brain Metastasis

Breast cancer is the most common cancer and the second leading cause of cancer death in women.[55] Surface receptors on breast cancer cells—namely, human epidermal growth factor receptor 2 (HER2), estrogen receptor, and progesterone receptor, help guide oncology treatment paradigms and provide prognostic information. For instance, HER2-positive breast cancers are poorly differentiated and are associated with aggressive proliferation, invasion, and metastasis. Trastuzumab is a humanized monoclonal antibody that binds to the extracellular domain of the HER2 transmembrane protein, leading to signal disruption and reduced cellular proliferation. Multiple phase III clinical trials have demonstrated that adjuvant trastuzumab therapy confers significant clinical benefit to both primary and metastatic HER2-positive breast cancers.[56, 57]

The number of patients presenting with intracranial HER2-positive breast metastasis has increased due to the improved therapeutic options (e.g., trastuzumab) available for the systemic tumor burden. Trastuzumab is approximately 150 kDa in size, so its efficacy in treating brain metastasis has been limited due to the impermeability of the BBB to antibodies. Trastuzumab was one of the first drugs tested in FUS BBB studies in animals. In a wild-type mouse model, the concentration of trastuzumab in unsonicated tissue was below the range of detection (780 ng/g), but the average concentration in sonicated tissue was 3257 ng/g.[58] In a rat xenograft model with HER2/neu-positive human breast cancer cells, weekly FUS enhanced delivery of trastuzumab for 6 weeks, and led to complete resolution of the tumor in 4 of 10 animals.[59] Another study, which used a more translational xenograft model of MDA-MB-361 cells derived from intracranial breast metastasis, found that 6 weeks of repeated FUS and trastuzumab treatments significantly suppressed tumor growth during treatment.[60] Unfortunately, neither of these two studies found a survival difference between FUS plus trastuzumab compared to trastuzumab alone. However, neither study measured trastuzumab concentration. It is worth noting that a single set of ultrasound parameters may be insufficient to create a uniform BBB opening, given heterogeneous vascularity and tissue density in tumor tissue. Future studies may address these limitations. The delivery of additional therapeutics, such as immune cell therapy, has also shown promising early results.[54]

Glioblastoma Multiforme

Although GBM is the most common and aggressive primary CNS tumor, medical and surgical options are limited for patients diagnosed with this tumor. The prognosis for patients with GBM has not improved appreciably despite rigorous biological and clinical research, standing at a median survival of approximately 12 months. The current standard of treatment is maximal surgical resection, followed by adjuvant radiation therapy and temozolomide. Temozolomide has good bioavailability within the CNS. However, due to the diffuse and infiltrative nature of GBM, a substantial volume of peritumoral area that appears normal on imaging may in fact contain microscopic tumor cells. This microscopic infiltration of GBM cells into surrounding brain leads to high risk of recurrence. We argue that administration of FUS to open the BBB in the peritumoral area would help treat these infiltrative cells, and will significantly affect patients' treatment. To date, a range of chemotherapies have been reinvestigated in xenograft animal models in conjunction with FUS: doxorubicin,[61] liposomal doxorubicin,[62] carmustine,[63] cisplatin,[64] and liposomal paclitaxel.[65]

The efficacy of brain–tumor barrier (BTB) disruption by FUS is different than that of the BBB disruption, presumably due to different vascular and tissue properties. Evidence suggests that the BTB responds differently to the same ultrasound parameters, and the efficiency of chemotherapy delivery to tumor tissue compared to healthy tissue is less homogeneous and less predictable.[61, 63] In xenograft rat models with C6 glioma cells, the concentration of carmustine after FUS was significantly higher in healthy tissue than in tumor tissue. However, in 2017, Park and colleagues[61] observed that doxorubicin could cross the BBB more easily than the BTB after FUS. To achieve a more uniform barrier opening, a possible solution is to utilize the ramp test (described by Huang et al.[50]) at each location. In using this method, the optimal power for BBB disruption is a percentage of power at which broadband emissions are detected by acoustic feedback.[50] A closed-loop cavitation controlling paradigm, another approach designed to achieve a more uniform BBB opening, was found feasible and superior to prior designs in the delivery of liposomal doxorubicin to a rat F98 glioma model.[66]

Carriers (e.g., liposomes, brain-penetrating particles, and microbubbles) are another area of active research in optimizing drug delivery in conjunction with FUS. The advantages of carriers include simplified pharmacokinetics when drugs are encapsulated in microbubbles, reduced systemic absorption, and improved brain penetration once past the BBB.[64] The delivery of liposomal paclitaxel and doxorubicin, as well as brain-penetrating nanoparticles loaded with cisplatin have been shown to be enhanced by FUS in rodent xenograft model of GBM.[62, 64, 65] Overall, these preclinical studies demonstrate the feasibility of delivering targeted therapy with adjuvant FUS to intracranial lesions with potential efficacy. Importantly, they found a repeated dosing regimen with FUS to be safe.[54] Since FUS-induced BBB disruption is transient and reversible, it is reasonable to assume that multiple or repeated use is feasible in a clinical setting. This opens the possibility of increasing the bioavailability and efficacy of systemic chemotherapies, which have already been proven to be effective in other regions of the body, for treatment of intracranial metastatic disease.

Diffuse Intrinsic Pontine Glioma

Diffuse intrinsic pontine glioma is one of the most lethal types of pediatric brain tumor, typically fatal within 2 years of diagnosis. No definitive treatment exists for diffuse intrinsic pontine glioma,[67] and it is surgically inaccessible and resistant to systemic therapies, as the BTB is relatively intact.[68] Yet, given its compactness, it may be more amenable to FUS-aided delivery of therapeutics. Clinical trials using convection-enhanced delivery to facilitate delivery of antibodies (e.g., bevacizumab) and chemotherapy (e.g., gemcitabine) are currently underway.[69] However, convection-enhanced delivery is associated with two significant limitations. First, there is restricted diffusion of the drug past the injection tip, due to the size of the drug and reflux pressure. Second, the flow or accumulation of the drug in low-pressure regions results in neurotoxicity, observed in a substantial number of patients in early-phase trials for treatment of malignant tumors.[70, 71] These issues may be mitigated by the use of FUS and an ideal adjuvant therapeutic modality.

Opening of the BBB using FUS can enhance the delivery of a wide range of cancer therapeutics, all shown largely to have further positive impact on its anticancer effects. The addition of carriers or conjugation to microbubbles injected for BBB opening can also improve this effect and simplify treatment protocols. Which therapy should be tested first likely depends on two factors: tumor biology and the treatment that is expected to have the greatest clinical efficacy (assuming adequate bioavailability). Furthermore, because BBB permeability and the pressure conditions within and around the tumor are heterogeneous, additional optimization of ultrasound parameters is likely to be necessary for superior clinical outcomes.

Neurodegenerative Disease

Neurodegenerative diseases impose major negative effects on the health and wellbeing of individuals, and also pose a significant socioeconomic burden on society. Almost all neurodegenerative diseases lack disease-modifying therapy. This is due in part to our incomplete understanding of their complex pathophysiology, but is also partly the result of the impermeability of the BBB to therapeutics that have passed the preclinical phase of testing. Drugs in development, such as antibody therapy, gene therapy, and cell therapy, are all associated with problems with delivery and insufficient bioavailability in the CNS without intolerable systemic toxicity. For a more spatially targeted treatment, previous researchers have predominantly used direct injection via convection-enhanced delivery. However, studies have shown limited diffusion, and diffusion is exponentially decreased with distance away from the catheter tip.[72] In this section, we review the preclinical work that has delved into the application of FUS-mediated BBB opening to animal models of Alzheimer disease (AD) and Parkinson disease (PD).

Alzheimer Disease

The most common cause of dementia, AD has an expected global prevalence of 65.7 million by 2030.[73] As the population ages, this dramatic growth in disease prevalence will result in an estimated 1 trillion dollars in healthcare costs during 2018 in the United States alone.[74] The amyloid cascade hypothesis is the prevalent theory for the pathophysiology of AD, which refers to an accumulation of extracellular amyloid-beta (Aβ) plaques and intracellular tau neurofibrillary tangles. Current treatment paradigms center around acetylcholinesterase inhibitors and low-affinity N-methyl-D-aspartate receptor antagonists, which are the only medications presently approved by the Food and Drug Administration. Although these medications have shown benefit, the neurological and cognitive gains have been modest. There is an urgent need for more effective treatment regimens for AD, for which passive immunization has been actively investigated.

In the TgCRND8 mouse model of AD, FUS-mediated BBB opening in the cortex (FUS delivered at 0.3 MPa for 2 min) enhanced the delivery of antiamyloid antibody, BAM-10.[75] The BAM-10 antibody is directed against the N-terminus of Aβ, having an effect on dissociating Aβ aggregates. At a low peripheral intravenous dose of BAM-10 (40 μg), the antibody was detected only in the sonicated hemisphere, with a subsequent 12% reduction in number of plaques and 23% reduction in surface area of plaques in the sonicated hemisphere 4 days after treatment. This level of effect typically required a 10-fold higher concentration of systemic passive immunization after repeated weekly treatments. Antiamyloid antibodies have also been administered with FUS, repeated over three treatments, in rabbits.[76]

Neurogenesis, a fundamental process for neuroplasticity and memory function, is altered in rodent models of AD.[77] FUS-mediated opening of the BBB, even in the absence of exogenous therapeutics, has been shown to increase neurogenesis in nontransgenic mice.[78] In mouse models of AD, FUS-induced BBB permeability resulted in amyloid reduction, improved spatial memory performance in APP23 and TgCRND8 mice, and increased neuroplasticity of immature neurons in the dentate gyrus.[79-81] BBB opening in the hippocampus increased the number of immature neurons by approximately fourfold compared to the unsonicated side, and it induced greater dendritic complexity.[79] Increased levels of endogenous antibodies in the sonicated regions were also detected in TgCRND8 mice.[80] Activated microglia emerged as early as 4 hours after treatment and was found to persist for at least 4 days, with evidence that they contributed to Aβ clearance.[80] Microglial activation was shown to resolve by 15 days post-FUS. Astrocytic activation was also present and potentially contributed to reduction in amyloid pathology at 4 days postsonication.[80]

Subsequent BBB disruption studies in aged beagles, a naturalistic in vivo model of amyloid deposition, also demonstrated that brain-wide coverage with ultrasound was safe and feasible.[32] Therefore, clinical improvement may result from stimulation of neurogenesis and/or mechanisms, inducing Aβ clearance. Enhanced delivery of an antitau agent was also demonstrated after BBB opening in a mouse model with intracellular neurofibrillary tangles.[82] RN2N, an antitau single chain variable fragment (scFv) specific for the 2N isoform of tau, was selected for increased tissue penetration and potentially decreased inflammatory response. Ultrasound plus RN2N resulted in significantly greater reduction in certain isoforms than RN2N alone, in addition to greater improvements in spatial memory.[83]

Parkinson Disease

Parkinson disease is a chronic neurodegenerative disorder resulting in both motor deficits (e.g., tremor, bradykinesia, rigidity) and nonmotor deficits (e.g., anosmia, sleep disturbances, depression). Treatment of symptomatic PD consists of medical and surgical intervention, with no clinically proven, disease-modifying therapy to date. Pharmaceutic therapies for PD include MAO-B inhibitors, dopamine agonists, and dopamine replacement therapy. Deep brain stimulation of the subthalamic nucleus or the globus pallidus interna is effective at reducing motor features.

Over the past two decades, several clinical trials have investigated direct intracranial injection of neuroprotective factors (e.g., GDNF) or viral constructs expressing these factors (e.g., AAV2-Neurturin) in patients diagnosed with PD. Although early studies were promising, these therapeutics ultimately failed to demonstrate efficacy in late-phase trials. Effective delivery has been a key limitation, as drugs must diffuse far enough past the cannula site against high intraparenchymal pressure. An autopsy study for patients who received multiple intraparenchymal injections of AAV2-Neurturin found sparse viral expression and only a modest increase in tyrosine hydroxylase.[84] Therefore, optimizing delivery of therapeutics in the brain is an important goal.

Preclinical models have demonstrated that FUS-mediated BBB opening enhanced intraparenchymal concentrations of brain-derived neurotrophic factor,[85] GDNF,[86] and neurturin,[82] as well as gene expression of viral vectors AAV2-GFP up to 7 times.[87, 88] Further upregulation of downstream signaling pathways was also observed.[82, 86, 89] In rat models of PD, FUS-mediated delivery of GDNF restored dopamine levels without evidence of local or systemic toxicity,[90] with FUS animals performing better in rotarod performance tests compared with those who underwent systemic administration[89] and direct intraparenchymal infusion.[91]. MRgFUS delivery of short hairpin RNA-expressing vectors targeting alpha-synuclein was also shown to reduce alpha-synuclein pathology in key affected areas in an A30P transgenic mouse model of PD.[92] Finally, neural stem cells have also been noninvasively delivered to the striatum and hippocampus using FUS, with demonstration of subsequent differentiation and survival.[93] Overall, these results are encouraging and may revitalize the therapeutic potential of neuroprotective factors in PD.

Other Emerging Indications

The potential disease applications for ultrasound-aided therapeutic delivery are diverse. Hsu et al.[94] tested the feasibility of FUS-delivered enzyme-replacement therapy in a mouse model of mucopolysaccharidosis type I. Mucopolysaccharidosis type I is a lysosomal storage disorder; its most severe form is referred to as *Hurler Syndrome*. Patients diagnosed with mucopolysaccharidosis type I lack enzyme alpha-L-iduronidase, resulting in accumulation of cytotoxic products, followed by multiple organ failure and significantly shortened life expectancy. A phase III study of systemic administration of the enzyme in affected children was successful only in reducing nonneurological symptoms and pain, but a significant percentage of patients also experienced an immune reaction. The enzyme does not pass through the BBB, but intraparenchymal concentration can increase almost eightfold by FUS.

In addition to delivering therapeutics, noninvasive delivery of BBB-impenetrable neurotoxin and inhibitory compounds has also been studied. For instance, tissue close to the skull is currently outside the treatment envelope for high-intensity FUS thermoablation due to significant overheating from the absorptive properties of the bone. Low-intensity ultrasound-mediated delivery of quinolinic acid, which does not normally cross the BBB, is an alternative to thermoablation and has been tested in the rodent hippocampus.[95]

Finally, the delivery of substances that affect neuronal excitability, such as GABA,[96] propofol,[97] and viral construct for optogenetic transfection,[98] are potential uses of FUS for neuromodulation. Drugs that naturally cross the BBB with systemic administration can be encapsulated in nanoparticles and unloaded by ultrasound specifically at the targeted site. This circumvents the potential unwanted widespread effects of these drugs on the CNS.

FUTURE PROSPECTS

Focused ultrasound is predicted to have a transformative effect on medical therapy. The number of searchable FUS studies in the US National Library of Medicine has increased from approximately 200 in 2000 to 1250 in 2017, and promising results of preclinical FUS studies have driven the initiation of clinical trials. Today, two major clinical ultrasound devices for BBB disruption are available for investigational uses: ExAblate Neuro (INSIGHTEC), a transcranial MR-guided FUS device, and SonoCloud (CarThera), an implanted device that uses unfocused ultrasound. The low-intensity ExAblate Neuro is designed as a phased-array helmet with over 1000 ultrasound transducers. To undergo the procedure, the patient must have their hair completely shaven, then be fitted in a MR-compatible stereotactic frame. A complete shave is necessary to minimize interruption of ultrasound waves by air bubbles.[99] In combination with stereotaxy and MRI, the ExAblate Neuro system allows accurate image-guided targeting of the brain down to the submillimeter scale.

Current clinical procedures are limited by the spatial constraints of the target. Although an MR-guided FUS system can theoretically treat the entire brain in patients diagnosed with diffuse neurological disorders, such a procedure would likely be too time-consuming for most patients. Not only are microbubbles quickly expelled from the body, requiring readministration before every sonication, but the phased-array helmet must be mechanically moved so that its geometric center encompasses the target. Furthermore, the procedure requires a complete head shave and placement of a stereotaxic frame, which is unpleasant for some patients, and again prohibitive of prolonged, repeated treatments. Another challenge to be addressed is how to achieve large, uniform BBB disruption in the context of heterogeneity in brain tissue. To this end, we must achieve a greater understanding of how the optimal ultrasound parameters change depending on tissue composition, vessel density, size, and morphology. Furthermore, a better understanding of the biological effects of BBB disruption, including neurogenesis and inflammation, continues to be an important focus of ongoing research.

Continued research efforts in device and protocol design will increase the safety, efficiency, and efficacy of the procedure. With safe, noninvasive, targeted delivery of therapeutics, effective treatment for previously inaccessible CNS diseases will be made possible.

References

1. Banks WA, Farr SA, Morley JE. Entry of blood-borne cytokines into the central nervous system: effects on cognitive processes. *Neuroimmunomodulation*. 2002;10(6):319-327.
2. Bard F, Cannon C, Barbour R, et al. Peripherally administered antibodies against amyloid beta-peptide enter the central nervous system and reduce pathology in a mouse model of Alzheimer disease. *Nat Med*. 2000;6(8):916-919.
3. DeMattos RB, Bales KR, Cummins DJ, Dodart JC, Paul SM, Holtzman DM. Peripheral anti-A beta antibody alters CNS and plasma A beta clearance and decreases brain A beta burden in a mouse model of Alzheimer's disease. *Proc Natl Acad Sci U S A*. 2001;98(15):8850-8855.
4. Lim WH, Choi SH, Yoo RE, et al. Does radiation therapy increase gadolinium accumulation in the brain? Quantitative analysis of T1 shortening using R1 relaxometry in glioblastoma multiforme patients. *PLoS ONE*. 2018;13(2).
5. Steen RG, Spence D, Wu S, Xiong X, Kun LE, Merchant TE. Effect of therapeutic ionizing radiation on the human brain. *Ann Neurol*. 2001;50(6):787-795.
6. Centelles MN, Wright M, Gedroyc W, Thanou M. Focused ultrasound induced hyperthermia accelerates and increases the uptake of anti-HER-2 antibodies in a xenograft model. *Pharmacol Res*. 2016;114:144-151.
7. Wu SK, Chiang CF, Hsu YH, Liou HC, Fu WM, Lin WL. Pulsed-wave low-dose ultrasound hyperthermia selectively enhances nanodrug delivery and improves antitumor efficacy for brain metastasis of breast cancer. *Ultrason Sonochem*. 2017;36:198-205.
8. Vykhodtseva NI, Hynynen K, Damianou C. Histologic effects of high intensity pulsed ultrasound exposure with subharmonic emission in rabbit brain in vivo. *Ultrasound Med Biol*. 1995;21(7):969-979.
9. Hynynen K, McDannold N, Vykhodtseva N, Jolesz FA. Noninvasive MR imaging-guided focal opening of the blood-brain barrier in rabbits. *Radiology*. 2001;220(3):640-646.
10. Boucher Y, Baxter LT, Jain RK. Interstitial pressure gradients in tissue-isolated and subcutaneous tumors: implications for therapy. *Cancer Res*. 1990;50(15):4478-4484.
11. Dewhirst MW, Secomb TW. Transport of drugs from blood vessels to tumour tissue. *Nat Rev Cancer*. 2017;17(12):738-750.
12. Sheikov N, McDannold N, Sharma S, Hynynen K. Effect of focused ultrasound applied with an ultrasound contrast agent on the tight junctional integrity of the brain microvascular endothelium. *Ultrasound Med Biol*. 2008;34(7):1093-1104.
13. Sheikov N, McDannold N, Jolesz F, Zhang YZ, Tam K, Hynynen K. Brain arterioles show more active vesicular transport of blood-borne tracer molecules than capillaries and venules after focused ultrasound-evoked opening of the blood-brain barrier. *Ultrasound Med Biol*. 2006;32(9):1399-1409.
14. Adkins CE, Mittapalli RK, Manda VK, et al. P-glycoprotein mediated efflux limits substrate and drug uptake in a preclinical brain metastases of breast cancer model. *Front Pharmacol*. 2013;4:136.
15. Cordon-Cardo C, O'Brien JP, Boccia J, Casals D, Bertino JR, Melamed MR. Expression of the multidrug resistance gene product (P-glycoprotein) in human normal and tumor tissues. *J Histochem Cytochem*. 1990;38(9):1277-1287.
16. Loscher W. Critical review of current animal models of seizures and epilepsy used in the discovery and development of new antiepileptic drugs. *Seizure*. 2011;20(5):359-368.

17. Milane A, Fernandez C, Dupuis L, et al. P-glycoprotein expression and function are increased in an animal model of amyotrophic lateral sclerosis. *Neurosci Lett*. 2010;472(3):166-170.
18. He Y, Bi Y, Hua Y, et al. Ultrasound microbubble-mediated delivery of the siRNAs targeting MDR1 reduces drug resistance of yolk sac carcinoma L2 cells. *J Exp Clin Cancer Res*. 2011;30:104.
19. Wan CP, Jackson JK, Pirmoradi FN, Chiao M, Burt HM. Increased accumulation and retention of micellar paclitaxel in drug-sensitive and P-glycoprotein-expressing cell lines following ultrasound exposure. *Ultrasound Med Biol*. 2012;38(5):736-744.
20. Wu F, Shao ZY, Zhai BJ, Zhao CL, Shen DM. Ultrasound reverses multidrug resistance in human cancer cells by altering gene expression of ABC transporter proteins and Bax protein. *Ultrasound Med Biol*. 2011;37(1):151-159.
21. Zhai BJ, Shao ZY, Wu F, Wang ZB. Reversal of multidrug resistance of human hepatocarcinoma HepG2/Adm cells by high intensity focused ultrasound. *Ai Zheng*. 2003;22(12):1284-1288.
22. Zhang Z, Xu K, Bi Y, et al. Low intensity ultrasound promotes the sensitivity of rat brain glioma to Doxorubicin by down-regulating the expressions of p-glucoprotein and multidrug resistance protein 1 in vitro and in vivo. *PLoS ONE*. 2013;8(8):e70685.
23. Aryal M, Fischer K, Gentile C, Gitto S, Zhang YZ, McDannold N. Effects on P-glycoprotein expression after blood-brain barrier disruption using focused ultrasound and microbubbles. *PLoS ONE*. 2017;12(1).
24. Cho H, Lee HY, Han M, et al. Localized down-regulation of p-glycoprotein by focused ultrasound and microbubbles induced blood-brain barrier disruption in rat brain. *Sci Rep*. 2016;6.
25. Alonso A, Reinz E, Jenne JW, et al. Reorganization of gap junctions after focused ultrasound blood-brain barrier opening in the rat brain. *J Cereb Blood Flow Metab*. 2010;30(7):1394-1402.
25a. Deng J, Huang Q, Wang F, et al. The role of caveolin-1 in blood–brain barrier disruption induced by focused ultrasound combined with microbubbles. *J Mol Neurosci*. 2012;46(3):677-687.
26. Hynynen K, McDannold N, Sheikov NA, Jolesz FA, Vykhodtseva N. Local and reversible blood-brain barrier disruption by noninvasive focused ultrasound at frequencies suitable for trans-skull sonications. *NeuroImage*. 2005;24(1):12-20.
27. Marty B, Larrat B, Van Landeghem M, et al. Dynamic study of blood-brain barrier closure after its disruption using ultrasound: a quantitative analysis. *J Cereb Blood Flow Metab*. 2012;32(10):1948-1958.
28. Mei J, Cheng Y, Song Y, et al. Experimental study on targeted methotrexate delivery to the rabbit brain via magnetic resonance imaging-guided focused ultrasound. *J Ultrasound Med*. 2009;28(7):871-880.
29. Samiotaki G, Konofagou EE. Dependence of the reversibility of focused- ultrasound-induced blood-brain barrier opening on pressure and pulse length in vivo. *IEEE Trans Ultrason Ferroelectr Freq Control*. 2013;60(11):2257-2265.
30. Samiotaki G, Vlachos F, Tung YS, Konofagou EE. A quantitative pressure and microbubble-size dependence study of focused ultrasound-induced blood-brain barrier opening reversibility in vivo using MRI. *Magn Reson Med*. 2012;67(3):769-777.
31. O'Reilly MA, Hough O, Hynynen K. Blood-brain barrier closure time after controlled ultrasound-induced opening is independent of opening volume. *J Ultrasound Med*. 2017;36(3):475-483.
32. O'Reilly MA, Jones RM, Barrett E, Schwab A, Head E, Hynynen K. Investigation of the safety of focused ultrasound-induced blood-brain barrier opening in a natural canine model of aging. *Theranostics*. 2017;7(14):3573-3584.
33. Kovacs ZI, Kim S, Jikaria N, et al. Disrupting the blood–brain barrier by focused ultrasound induces sterile inflammation. *Proc Natl Acad Sci U S A*. 2017;114(1):E75-E84.
34. McMahon D, Hynynen K. Acute inflammatory response following increased blood-brain barrier permeability induced by focused ultrasound is dependent on microbubble dose. *Theranostics*. 2017;7(16):3989-4000.
35. Silburt J, Lipsman N, Aubert I. Disrupting the blood–brain barrier with focused ultrasound: perspectives on inflammation and regeneration. *Proc Natl Acad Sci*. 2017;114(33):E6735-E6736.
36. McMahon D, Bendayan R, Hynynen K. Acute effects of focused ultrasound-induced increases in blood-brain barrier permeability on rat microvascular transcriptome. *Sci Rep*. 2017;7.
37. Olumolade OO, Wang S, Samiotaki G, Konofagou EE. Longitudinal motor and behavioral assessment of blood-brain barrier opening with transcranial focused ultrasound. *Ultrasound Med Biol*. 2016;42(9):2270-2282.
38. Karakatsani MEM, Samiotaki GM, Downs ME, Ferrera VP, Konofagou EE. Targeting effects on the volume of the focused ultrasound-induced blood-brain barrier opening in nonhuman primates in vivo. *IEEE Trans Ultrason Ferroelectr Freq Control*. 2017;64(5):798-810.

39. Marquet F, Tung YS, Teichert T, Ferrera VP, Konofagou EE. Noninvasive, transient and selective blood-brain barrier opening in non-human primates in vivo. *PLoS ONE*. 2011;6(7).
40. McDannold N, Arvanitis CD, Vykhodtseva N, Livingstone MS. Temporary disruption of the blood-brain barrier by use of ultrasound and microbubbles: safety and efficacy evaluation in rhesus macaques. *Cancer Res*. 2012;72(14):3652-3663.
41. Horodyckid C, Canney M, Vignot A, et al. Safe long-term repeated disruption of the blood-brain barrier using an implantable ultrasound device: a multiparametric study in a primate model. *J Neurosurg*. 2017;126(4):1351-1361.
42. Nadal A, Fuentes E, McNaughton PA. Glial cell responses to lipids bound to albumin in serum and plasma. *Prog Brain Res*. 2001;132:367-374.
43. Alonso A, Reinz E, Fatar M, Jenne J, Hennerici MG, Meairs S. Neurons but not glial cells overexpress ubiquitin in the rat brain following focused ultrasound-induced opening of the blood-brain barrier. *Neuroscience*. 2010;169(1):116-124.
44. Alonso A, Reinz E, Fatar M, Hennerici MG, Meairs S. Clearance of albumin following ultrasound-induced blood-brain barrier opening is mediated by glial but not neuronal cells. *Brain Res*. 2011;1411:9-16.
45. Vlachos F, Tung YS, Konofagou E. Permeability dependence study of the focused ultrasound-induced blood-brain barrier opening at distinct pressures and microbubble diameters using DCE-MRI. *Magn Reson Med*. 2011;66(3):821-830.
46. Vlachos F, Tung YS, Konofagou EE. Permeability assessment of the focused ultrasound-induced blood-brain barrier opening using dynamic contrast-enhanced MRI. *Phys Med Biol*. 2010;55(18):5451-5466.
47. Nhan T, Burgess A, Cho EE, Stefanovic B, Lilge L, Hynynen K. Drug delivery to the brain by focused ultrasound induced blood-brain barrier disruption: quantitative evaluation of enhanced permeability of cerebral vasculature using two-photon microscopy. *J Control Release*. 2013;172(1):274-280.
48. Wu SY, Sanchez CS, Samiotaki G, Buch A, Ferrera VP, Konofagou EE. Characterizing focused-ultrasound mediated drug delivery to the heterogeneous primate brain in vivo with acoustic monitoring. *Sci Rep*. 2016;6.
49. Chu PC, Chai WY, Tsai CH, Kang ST, Yeh CK, Liu HL. Focused ultrasound-induced blood-brain barrier opening: association with mechanical index and cavitation index analyzed by dynamic contrast-enhanced magnetic-resonance imaging. *Sci Rep*. 2016;6.
50. Huang Y, Alkins R, Schwartz ML, Hynynen K. Opening the blood-brain barrier with mr imaging-guided focused ultrasound: preclinical testing on a trans-human skull porcine model. *Radiology*. 2017;282(1):123-130.
51. Hernot S, Klibanov AL. Microbubbles in ultrasound-triggered drug and gene delivery. *Adv Drug Deliv Rev*. 2008;60(10):1153-1166.
52. Wu SK, Chu PC, Chai WY, et al. Characterization of Different Microbubbles in Assisting Focused Ultrasound-Induced Blood-Brain Barrier Opening. *Sci Rep*. 2017;7.
53. Song KH, Fan AC, Hinkle JJ, Newman J, Borden MA, Harvey BK. Microbubble gas volume: A unifying dose parameter in blood-brain barrier opening by focused ultrasound. *Theranostics*. 2017;7(1):144-152.
54. Alkins R, Burgess A, Kerbel R, Wels WS, Hynynen K. Early treatment of HER2-amplified brain tumors with targeted NK-92 cells and focused ultrasound improves survival. *Neuro-Oncology*. 2016;18(7):974-981.
55. DeSantis CE, Ma J, Goding Sauer A, Newman LA, Jemal A. Breast cancer statistics, 2017, racial disparity in mortality by state. *CA Cancer J Clin*. 2017;67(6):439-448.
56. Escriva-de-Romani S, Arumi M, Bellet M, Saura C. HER2-positive breast cancer: Current and new therapeutic strategies. *Breast*. 2018;39:80-88.
57. Piccart-Gebhart MJ, Procter M, Leyland-Jones B, et al. Trastuzumab after adjuvant chemotherapy in HER2-positive breast cancer. *N Engl J Med*. 2005;353(16):1659-1672.
58. Kinoshita M, McDannold N, Jolesz FA, Hynynen K. Noninvasive localized delivery of Herceptin to the mouse brain by MRI-guided focused ultrasound-induced blood-brain barrier disruption. *Proc Natl Acad Sci U S A*. 2006;103(31):11719-11723.
59. Park EJ, Zhang YZ, Vykhodtseva N, McDannold N. Ultrasound-mediated blood-brain/blood-tumor barrier disruption improves outcomes with trastuzumab in a breast cancer brain metastasis model. *J Control Release*. 2012;163(3):277-284.
60. Kobus T, Zervantonakis IK, Zhang Y, McDannold NJ. Growth inhibition in a brain metastasis model by antibody delivery using focused ultrasound-mediated blood-brain barrier disruption. *J Control Release*. 2016;238:281-288.
61. Park J, Aryal M, Vykhodtseva N, Zhang YZ, McDannold N. Evaluation of permeability, doxorubicin delivery, and drug retention in a rat brain tumor model after ultrasound-induced blood-tumor barrier disruption. *J Control Release*. 2017;250:77-85.

62. Aryal M, Vykhodtseva N, Zhang YZ, Park J, McDannold N. Multiple treatments with liposomal doxorubicin and ultrasound-induced disruption of blood-tumor and blood-brain barriers improve outcomes in a rat glioma model. *J Control Release*. 2013;169(1-2):103-111.
63. Liu HL, Hua MY, Chen PY, et al. Blood-brain barrier disruption with focused ultrasound enhances delivery of chemotherapeutic drugs for glioblastoma treatment. *Radiology*. 2010;255(2):415-425.
64. Timbie KF, Afzal U, Date A, et al. MR image-guided delivery of cisplatin-loaded brain-penetrating nanoparticles to invasive glioma with focused ultrasound. *J Control Release*. 2017;263:120-131.
65. Shen Y, Pi Z, Yan F, et al. Enhanced delivery of paclitaxel liposomes using focused ultrasound with microbubbles for treating nude mice bearing intracranial glioblastoma xenografts. *Int J Nanomedicine*. 2017;12:5613-5629.
66. Sun T, Zhang Y, Power C, et al. Closed-loop control of targeted ultrasound drug delivery across the blood-brain/tumor barriers in a rat glioma model. *Proc Natl Acad Sci U S A*. 2017;114(48):E10281-E10290.
67. Kaye EC, Baker JN, Broniscer A. Management of diffuse intrinsic pontine glioma in children: current and future strategies for improving prognosis. *CNS Oncol*. 2014;3(6):421-431.
68. Jansen MH, Veldhuijzen van Zanten SE, Sanchez Aliaga E, et al. Survival prediction model of children with diffuse intrinsic pontine glioma based on clinical and radiological criteria. *Neuro-Oncology*. 2015;17(1):160-166.
69. Zhou Z, Singh R, Souweidane MM. Convection-enhanced delivery for diffuse intrinsic pontine glioma treatment. *Curr Neuropharmacol*. 2017;15(1):116-128.
70. Weaver M, Laske DW. Transferrin receptor ligand-targeted toxin conjugate (Tf-CRM107) for therapy of malignant gliomas. *J Neuro-Oncol*. 2003;65(1):3-13.
71. Weber F, Asher A, Bucholz R, et al. Safety, tolerability, and tumor response of IL4-Pseudomonas exotoxin (NBI-3001) in patients with recurrent malignant glioma. *J Neuro-Oncol*. 2003;64(1-2):125-137.
72. Salvatore MF, Ai Y, Fischer B, et al. Point source concentration of GDNF may explain failure of phase II clinical trial. *Exp Neurol*. 2006;202(2):497-505.
73. Gulland A. Number of people with dementia will reach 65.7 million by 2030, says report. *BMJ*. 2012;344:e2604.
74. Wimo A, Guerchet M, Ali GC, et al. The worldwide costs of dementia 2015 and comparisons with 2010. *Alzheimers Dement*. 2017;13(1):1-7.
75. Jordao JF, Ayala-Grosso CA, Markham K, et al. Antibodies targeted to the brain with image-guided focused ultrasound reduces amyloid-beta plaque load in the TgCRND8 mouse model of Alzheimer's disease. *PLoS ONE*. 2010;5(5).
76. Alecou T, Giannakou M, Damianou C. Amyloid beta plaque reduction with antibodies crossing the blood-brain barrier, which was opened in 3 sessions of focused ultrasound in a rabbit model. *J Ultrasound Med*. 2017;36(11):2257-2270.
77. Hollands C, Bartolotti N, Lazarov O. Alzheimer's disease and hippocampal adult neurogenesis; exploring shared mechanisms. *Front Neurosci*. 2016;10:178.
78. Scarcelli T, Jordao JF, O'Reilly MA, Ellens N, Hynynen K, Aubert I. Stimulation of hippocampal neurogenesis by transcranial focused ultrasound and microbubbles in adult mice. *Brain Stimul*. 2014;7(2):304-307.
79. Burgess A, Dubey S, Yeung S, et al. Alzheimer disease in a mouse model: MR imaging-guided focused ultrasound targeted to the hippocampus opens the blood-brain barrier and improves pathologic abnormalities and behavior. *Radiology*. 2014;273(3):736-745.
80. Jordao JF, Thevenot E, Markham-Coultes K, et al. Amyloid-beta plaque reduction, endogenous antibody delivery and glial activation by brain-targeted, transcranial focused ultrasound. *Exp Neurol*. 2013;248:16-29.
81. Leinenga G, Gotz J. Scanning ultrasound removes amyloid-beta and restores memory in an Alzheimer's disease mouse model. *Sci Transl Med*. 2015;7(278):278ra233.
82. Samiotaki G, Acosta C, Wang S, Konofagou EE. Enhanced delivery and bioactivity of the neurturin neurotrophic factor through focused ultrasound-mediated blood–brain barrier opening in vivo. *J Cereb Blood Flow Metab*. 2015;35(4):611-622.
83. Nisbet RM, Van der Jeugd A, Leinenga G, Evans HT, Janowicz PW, Gotz J. Combined effects of scanning ultrasound and a tau-specific single chain antibody in a tau transgenic mouse model. *Brain*. 2017;140(5):1220-1230.
84. Bartus RT, Kordower JH, Johnson Jr. EM, et al. Post-mortem assessment of the short and long-term effects of the trophic factor neurturin in patients with alpha-synucleinopathies. *Neurobiol Dis*. 2015;78:162-171.
85. Chen H, Yang GZ, Getachew H, Acosta C, Sierra Sanchez C, Konofagou EE. Focused ultrasound-enhanced intranasal brain delivery of brain-derived neurotrophic factor. *Sci Rep*. 2016;6:28599.
86. Baseri B, Choi JJ, Deffieux T, et al. Activation of signaling pathways following localized delivery of systemically administered neurotrophic factors across the blood-brain barrier using focused ultrasound and microbubbles. *Phys Med Biol*. 2012;57(7):N65-N81.

87. Hsu PH, Wei KC, Huang CY, et al. Noninvasive and targeted gene delivery into the brain using microbubble-facilitated focused ultrasound. *PLoS ONE*. 2013;8(2).
88. Mead BP, Mastorakos P, Suk JS, Klibanov AL, Hanes J, Price RJ. Targeted gene transfer to the brain via the delivery of brain-penetrating DNA nanoparticles with focused ultrasound. *J Control Release*. 2016;223:109-117.
89. Lin CY, Hsieh HY, Chen CM, et al. Non-invasive, neuron-specific gene therapy by focused ultrasound-induced blood-brain barrier opening in Parkinson's disease mouse model. *J Control Release*. 2016;235:72-81.
90. Mead BP, Kim N, Miller GW, et al. Novel focused ultrasound gene therapy approach noninvasively restores dopaminergic neuron function in a rat Parkinson's disease model. *Nano Lett*. 2017;17(6):3533-3542.
91. Fan CH, Ting CY, Lin CY, et al. Noninvasive, targeted, and non-viral ultrasound-mediated GDNF-plasmid delivery for treatment of Parkinson's disease. *Sci Rep*. 2016;6:19579.
92. Xhima K, Nabbouh F, Hynynen K, Aubert I, Tandon A. Non-invasive delivery of an α-synuclein gene silencing vector with MR-guided focused ultrasound. *Mov Disord*. 2018;33:1567-1579.
93. Burgess A, Ayala-Grosso CA, Ganguly M, Jordao JF, Aubert I, Hynynen K. Targeted delivery of neural stem cells to the brain using MRI-guided focused ultrasound to disrupt the blood-brain barrier. *PLoS ONE*. 2011;6(11).
94. Hsu YH, Liu RS, Lin WL, Yuh YS, Lin SP, Wong TT. Transcranial pulsed ultrasound facilitates brain uptake of laronidase in enzyme replacement therapy for Mucopolysaccharidosis type I disease. *Orphanet J Rare Dis*. 2017;12 (1):109.
95. Zhang Y, Tan H, Bertram EH, et al. Non-invasive, focal disconnection of brain circuitry using magnetic resonance-guided low-intensity focused ultrasound to deliver a neurotoxin. *Ultrasound Med Biol*. 2016;42(9):2261-2269.
96. McDannold N, Zhang Y, Power C, Arvanitis CD, Vykhodtseva N, Livingstone M. Targeted, noninvasive blockade of cortical neuronal activity. *Sci Rep*. 2015;5:16253.
97. Airan RD, Foss CA, Ellens NP, et al. MR-Guided delivery of hydrophilic molecular imaging agents across the blood-brain barrier through focused ultrasound. *Mol Imaging Biol*. 2017;19(1):24-30.
98. Wang S, Kugelman T, Buch A, et al. Non-invasive focused ultrasound-facilitated gene delivery for optogenetics. *Sci Rep*. 2017;7:39955.
99. Meng Y, Huang Y, Solomon B, et al. MRI-guided Focused Ultrasound Thalamotomy for Patients with Medically-refractory Essential Tremor. *J Vis Exp*. 2017;130.

CHAPTER

12

Nanoparticles for Brain Tumor Delivery

Tista Roy Chaudhuri, Robert M. Straubinger*,†,‡*

*Department of Pharmaceutical Sciences, University at Buffalo, State University of New York, Buffalo, NY, United States †Department of Molecular and Cellular Biophysics and Biochemistry, Roswell Park Comprehensive Cancer Center, Buffalo, NY, United States ‡Department of Pharmacology and Therapeutics, Roswell Park Comprehensive Cancer Center, Buffalo, NY, United States

NANOPARTICLES FOR TREATMENT OF BRAIN TUMORS

Primary brain and central nervous system tumors are the leading cause of cancer-related deaths among children and adolescents, with glioblastoma multiforme (GBM) being the most common, and having an overall 5-year survival rate of 5%.[1, 2] Despite advances in diagnostic and surgical techniques, the location of brain tumors often hinders or precludes surgery, and hypoxic conditions make radiotherapy a challenge. Additionally, malignant forms of non-small cell lung cancer, breast cancer, and melanoma have a propensity to metastasize to the brain, which typically portends a poor prognosis.[3] Current treatment options include surgery (if feasible), followed by radiotherapy and temozolomide (TMZ) as the standard of care.[4] A combination of vincristine and the alkylating agents lomustine and procarbazine may be administered to patients with 1p19q chromosomal deletions,[5] and a controlled-release formulation of BCNU wafers may be implanted in the resection cavity after surgery to treat local metastases or extensions of tumor that remain after surgery.[6] Nevertheless, tumors develop drug resistance quickly, and frequently reappear within centimeters of the resected primary tumor.[7]

Historically, brain tumor therapy has been hindered by poor penetration of most therapeutic agents into the tumor interstitium. Network models and large-scale genome analysis identified targetable derangements of MEK1 phosphorylation, and alterations in the EGFR, TP53, RB1, and PI3K/PTEN signaling pathways in most patients with GBM.[8, 9] Although these pathways are typically altered in other forms of cancer as well, small molecule inhibitors targeting aberrant GBM pathways have largely failed clinically[10] because of robust drug efflux systems and an almost impenetrable blood–brain barrier (BBB) that shields tumor cells

Nervous System Drug Delivery
https://doi.org/10.1016/B978-0-12-813997-4.00012-8

© 2019 Elsevier Inc. All rights reserved.

from chemotherapeutic exposure.[11] Similarly, brain metastases remain refractory to treatment even when the patient responds to systemic therapy with reduction in primary tumor size.[12] For example, the incidence of brain metastases increases in patients with breast cancer who receive trastuzumab, because the BBB prevents trastuzumab entry even when the primary tumor responds to treatment, thus providing a sanctuary for metastases.[13] Therapy development has also been hampered by the lack of translationally relevant preclinical models that reflect the vascular barrier properties of clinical tumors. Preclinical brain tumor models often have a compromised BBB, which allows more facile passage of most therapeutic agents into the tumor; this may partially account for clinical failure of therapies that show preclinical promise. Patient-derived xenograft (PDX) models can recapitulate the heterogeneous distribution of regions of intact and compromised BBB within patient tumors, providing a clinically relevant platform for testing novel therapeutics engineered to cross the BBB.[11, 14]

Nanoparticulate carriers provide the opportunity to deposit large doses of encapsulated agents in tumors because their size (typically 80-120 nm) provides a large drug cargo capacity, and regions of compromised tumor vasculature allow extravasation of both therapeutic and contrast-enhancing macromolecular particulates[15] with minimal systemic toxicity compared to the corresponding free drug. Another advantage conferred by the uniquely "tunable" structure and properties of nanoparticles is their versatility; they allow encapsulation and sustained release of a wide range of chemical compounds (including extremely hydrophobic drugs),[16] provide a stable platform for nucleic acid delivery, and may be used to encapsulate and deliver multiple agents simultaneously.[17] Nanotechnology therefore represents a promising approach for delivering both diagnostic and therapeutic agents to brain tumors.[18, 19] The objective of this review is to discuss the current state of nanotherapeutics designed to penetrate the BBB and to explore the pharmacological effects of nanotherapeutics. We also discuss aspects of brain tumor physiology that exert considerable impact upon the performance of nanoparticle carriers, particularly in terms of nanoparticle transport, and the pivotal role of physiological differences between clinical tumors and preclinical models in the development and clinical translation of macromolecular therapeutics.

BRAIN TUMOR PHYSIOLOGY AND MICROENVIRONMENT

Brain tumors are composed of tumor cells, tumor-suppressive neural precursor cells, and nonneoplastic parenchyma cells, which consist of vascular endothelial cells, astrocytes, glial cells, T cells, and other peripheral immune cells.[20] The cellular components are enmeshed in an extracellular matrix (ECM) that provides mechanical support and modulates critical events in tumor progression such as angiogenesis[21] and tumor cell invasion.[22] Paracrine interactions between glioma and parenchyma cells maintain the BBB and sustain an immunosuppressed tumor microenvironment (TME). Astrocytes are integral parts of the BBB that are attached to endothelial cells. In addition to providing mechanical support through integrin linkages, paracrine interactions between the end-foot processes of astrocytes and the vascular endothelial cells via FGF-2, TGF-β, Ang-1, and thrombospondin-1 signaling help promote tight junction connections and induce the expression of the P-glycoprotein (Pgp) transporters in endothelial cells that expel therapeutic agents from the tumor, thereby supporting the barrier function of the BBB.[23]

The angiogenic switch is an important event in brain tumor progression that compromises the BBB in the tumor core and promotes invasion.[24] As primary tumors progress, glioma cells activate astrocytes to secrete neurotropic factors that remodel the ECM and enhance vascular permeability, thereby indirectly aiding in BBB disruption in the tumor core.[22] Other nonneoplastic parenchymal cells, such as glial cells (resident macrophages of the brain), also interact with tumor cells via paracrine factors to maintain an immunosuppressed TME, and contributing to BBB maintenance. Thus, the molecular interplay among various components of the brain TME creates a complex and dynamic mechanical and physiological barrier to drug delivery. This interplay could be targeted pharmacologically to promote barrier compromise, vascular permeability, and drug delivery.

Preclinical murine models based on tumor cell lines frequently have been used to develop experimental therapeutics, but they neither replicate the cellular heterogeneity of clinical tumors nor recapitulate the permeability barriers and TME that exist clinically. Although magnetic resonance imaging (MRI) studies demonstrate approximately 20-fold higher vascular permeability to small-molecule agents in patient tumors compared to normal brain,[25] brain tumors display peripheral invasive regions with an intact BBB that is highly resistant to the penetration of therapeutic agents. Therefore, many existing therapies offer moderate efficacy in primary tumors but fail against invasive and metastatic niches.

In contrast, most preclinical tumor models consist entirely of regions having a compromised BBB[26] and have higher rates of blood flow than patient tumors,[27] which leads to discrepancies between preclinical and clinical outcomes of investigational therapeutic agents and combination regimens. For example, tyrosine kinase inhibitors, such as inhibitors of EGFR and VEGF/PDGFR, induce "vascular normalization" preclinically, pruning the chaotic and permeable tumor vasculature into an organized network and alleviating tumor interstitial fluid pressure (IFP). However, tyrosine kinase inhibitors are dual substrates of Pgp (ABCB1), multidrug resistance-associated protein 1 MRP1 (ABCC1), and breast cancer resistance protein (BCRP/ABCG2) efflux transporters. As a result, they fail to penetrate the metastatic regions and show dismal efficacy in the clinic.[28] PDX models of clinical tumors in immunocompromised mice reflect the genetic and cellular constitution of clinical tumors, and also recapitulate a number of vascular and barrier features[14, 29]; orthotopic implants of PDX tumors in immunocompromised mice do recapitulate the clinical efficacy of agents such as PARP inhibitors, and demonstrate poor drug penetration as the cause of clinical failure of the agent.[11] However, these preclinical models may exhibit reduced sensitivity toward therapeutic agents compared to clinical tumors because of a compromised immune system, which aids clinically in mediating cytotoxicity during chemo- and radiation therapy.[30] Preclinical evaluations of genetically engineered mouse models of spontaneously generated brain tumors mimicking the physiology and de novo developmental process of clinical tumors can fill this gap and provide a platform for the genetic manipulations that aid mechanistic studies.[31, 32] Thus, genetically engineered mouse models and PDX models that better represent the unique physiology of clinical brain tumors provide clinically translatable platforms for testing of novel nanoparticle formulations and their delivery to brain tumors.

PHYSIOLOGICAL BARRIERS TO NANOPARTICLE DELIVERY

Brain tumors are typically well vascularized. However, heterogeneity in microvessel distribution, intracranial location, and the presence of an almost impermeable BBB collude to escalate delivery challenges in brain tumors compared to other solid tumors. The intracranial location of brain tumors leads to high intratumor pressure, resulting in constricted blood vessels and higher flow resistance[33] and posing a significant challenge for delivery of blood-borne agents. The tumor penetration of systemically administered anticancer agents is hindered by several physiological barriers, including:

1. the chaotic flow pattern of tumor blood circulation, which leads to heterogeneous drug distribution;
2. endothelial junctional complexes lining microvessel walls, which modulate vascular permeability and form the BBB;
3. elevated IFP and solid tissue stress that results from high cellular density, anisotropic deposition of ECM, and non-functional lymphatic drainage, all of which hinder diffusion through the tumor interstitial space[34-36]; and
4. the robust physiological barrier formed by Pgp, MRP1, and BCRP drug efflux systems, which recognize and transport molecularly targeted agents, such as tyrosine kinase inhibitors.[11, 31]

Nanoparticulate drug carrier systems have many advantages over traditional low-molecular–weight chemotherapeutics, such as greater tumor accumulation of the encapsulated drug because of high cargo capacity, and excellent capacity for hydrophobic drug solubilization.[16, 37] However, nanoparticles encounter different resistive forces within the tumor compared to small-molecule cytotoxic agents, because their comparatively large size limits diffusion. For example, antiangiogenic agents increase tumor perfusion of small-molecule drugs in the tumor core by remodeling microvasculature into an organized network and relief of IFP, thereby directing blood flow toward the tumor core and promoting more homogenous drug distribution.[38, 39] However, macromolecular agents and particulates typically rely on the aberrant permeability of tumor microvessels for deposition; therefore, antivascular agents often fail to promote nanoparticulate delivery.[40] In fact, antiangiogenic therapy may reestablish the BBB, which impedes the delivery of subsequently administered macromolecular agents.[41]

In terms of nanoparticle interactions with physiological barriers, several phenomena affect nanoparticle deposition in brain tumors. First, nanoparticles encounter greater resistance to flow within tumor blood vessels than do small-molecule agents because of elevated viscosity associated with erratic tumor blood flow.[35, 38] Second, nanoparticle size is an important determinant of tumor deposition. Nanoparticle extravasation occurs through gaps in the tumor wall that measure up to 500 nm in diameter. The enhanced permeability and retention (EPR) phenomenon promotes macromolecular deposition, trapping the nanoparticles within the ECM and establishing a sustained tumor drug depot that persists after circulating nanoparticles have been cleared from the blood.[42-45] Typically, particles larger than 500 nm are excluded from the tumor.[44] In most solid tumors, the presence of leaky vasculature and lack of lymphatic drainage favor the retention of higher-molecular–weight agents in tumors via the EPR mechanism. In brain tumors, however, the lack of sufficient functional lymphatic drainage results in elevated IFP, which can induce an outward convective force that directs incoming nanoparticles toward the tumor periphery and away from the tumor core.[36] This results in localized accumulation in only a fraction of the peripheral tumor volume having a compromised BBB. Third, upon extravasation, the size of nanoparticles and the dense ECM reduces their diffusional transport through the interstitial space, in contrast to the more

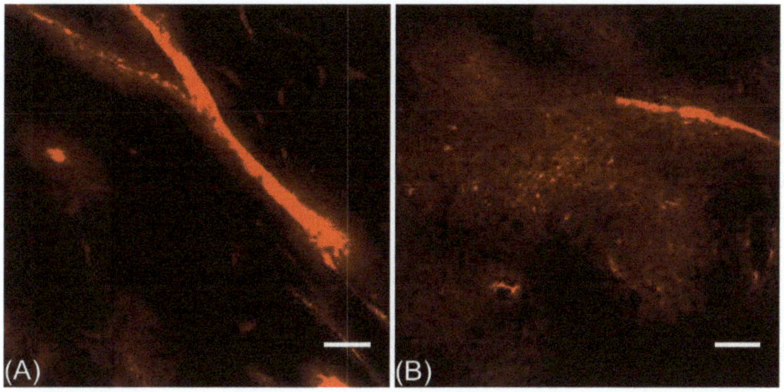

FIG. 1 Tumor distribution of 80 nm fluorescent nanoparticles in intracranial brain tumors. Rats bearing intracranial 9L brain tumors received intravenous injections of sterically stabilized liposomes labeled with a dialkyl carbocyanine membrane label and killed 24 hours later, at the time of peak liposome deposition in the tumor. (A) Perivascular deposition of fluorescent liposomes in regions of intact blood–brain barrier (BBB). (B) Extravasation and diffusion of the fluorescent nanoparticles in regions of compromised BBB. Bars: 50 μm. Additional details can be found in Roy Chaudhuri et al.[40]

rapid diffusional transport of low-molecular–weight agents.[46] Therefore, macromolecules and nanoparticulates cannot diffuse farther than a few micrometers from the afferent microvessels. Fluorescently labeled liposomes of approximately 80 nm display a focal pattern of distribution in regions of the tumor with an intact BBB, and are confined to the perivascular space (Fig. 1A). In contrast, regions with a disrupted BBB show more diffuse distribution of extravasated liposomes, which can move away from the afferent vasculature (Fig. 1B).

Nanoparticulate uptake into brain tumors is also mediated by transcytosis pathways [44, 47] that can be enhanced by particle surface functionalization that promotes receptor-mediated transport. When the BBB is intact, nanoparticles are primarily transported by the transcytosis mechanism, which involves the binding of nanoparticles to the luminal surface of brain endothelial cells, either nonspecifically or by binding to specific receptors that then undergo vesicular transport into the tumor interstitium.[43] However, if the affinity of binding between nanoparticles and endothelial receptors is too high, the nanoparticles remain strongly bound to the BBB even after endosomal uptake.[48] This represents another functional, physiological barrier to delivery.

PASSIVE ACCUMULATION OF NANOPARTICLES BY THE EPR EFFECT

Despite advances in the design of nanoparticle vehicles, passive accumulation of large molecular chemotherapeutic agents by the EPR mechanism in brain tumors remains the primary means of their deposition, and continues to pose challenges. Nanoparticle extravasation from microvessels depends upon their size, charge, and surface properties.[43-45, 47] The most commonly investigated nanoparticles can be broadly classified based on their material composition; they include phospholipid-based carriers, poly(lactide-coglycolide) (PLGA), polybutylcyanoacrylate, poly(isohexyl cyanoacrylate), poly(amine-co-ester) polymeric carriers, and more recently, chitosan nanoparticles. All of these differ in size, surface charge, characteristics, stability, and drug-release rates. Among them, sterically stabilized liposomal (SSL) formulations currently dominate the clinical landscape, with FDA-approved products containing doxorubicin (DXR) or irinotecan.[49]

Strategies for improving brain penetration of nanoparticles include reducing their size and coating their surface with various polymers. Methods for controlling size are typically specific to the nanoparticle composition. Whereas phospholipid-based particles can be size-controlled during preparation by extrusion[50] or sonication,[51] polymeric and inorganic nanoparticle sizes are controlled by the conditions of synthesis.[52-54] The structure of PLGA nanoparticles offers size control,[53] and reduction of their diameter to 70 nm resulted in a 7-fold increase in brain penetration compared to PLGA particles with a mean diameter of 150 nm.[55] However, PLGA nanoparticles have a relatively limited cargo capacity, and often demonstrate an early burst release of the encapsulated drug, limiting tumor exposure and hindering their clinical development.

Methods of surface coating are similarly controlled during synthesis or by postprocessing.[56] Surface coating of cytotoxic DXR-loaded liposomes with hydrophilic polymers such as polyethylene glycol (PEG)[56] results in SSL formulations with greatly increased circulation time, as they have reduced opsonization by serum proteins and clearance by the reticuloendothelial system. Extended circulation time results in preferential uptake of SSL in tumors via the EPR effect and establishes a tumor drug depot that increases cumulative exposure of the tumor to the encapsulated DXR.[57, 58] Surface coating of polymeric nanoparticles with dense, hyperbranched PEG was reported to mediate even greater tumor and cellular uptake, tissue retention, and therapeutic efficacy of cytotoxic PLGA nanoparticles than traditional linear PEG chains.[59, 60] In another study, surface coating of polybutylcyanoacrylate nanoparticles with polysorbate-80 promoted transcytosis across the BBB by surface adsorption of apolipoproteins[61]; plasma proteins apoE and apoB anchor onto the polysorbate coating and bind to low-density lipoprotein receptors on the luminal side of endothelial cells to mediate transport across the BBB.[43] Surface properties of nanoparticles can be manipulated during synthesis by compositional variation, such as in the amphiphilic design of monomers used for synthesis of polymeric nanoparticle that improved BBB penetration; micellar formulations of Pluronic block copolymers were reported to interact with cellular membranes and spontaneously cross the BBB.[62, 63]

Controlling both size and surface coating can yield improvements in tumor delivery. Additional factors, such as high drug-encapsulation capacity and slow drug release rate are essential parameters for sustained drug exposure, and thereby for therapeutic efficacy and clinical success. A PEGylated SSL formulation encapsulating a large cargo of irinotecan (1:1 molar drug:lipid), which was approved for metastatic pancreatic cancer,[49] was reported to mediate a steep decline in the progression of brain metastases in a preclinical triple-negative breast cancer model having a relatively intact BBB.[64] Another FDA-approved SSL encapsulating DXR (SSL-DXR), approved clinically as Doxil/Caelix, lowers DXR-mediated cardiotoxicity significantly, and promoted a 10-fold increase in DXR deposition compared to free DXR in preclinical models of intracranial GBM with a somewhat compromised BBB. However, this significant increase in tumor drug deposition mediated only a modest increase in median survival (i.e., 20%-30%).[15, 65] It was observed that only a small fraction (roughly 0.04%-0.2%) of the injected dose underwent tumor deposition, which is attributable to the more realistic blood-tumor barrier properties of this intracranial tumor model. These reports underscore the challenges of systemic administration of nanoparticles, which may potentially be overcome by local routes of administration, such as convection-enhanced delivery (CED) and intra-arterial (IA) delivery.

Convection-enhanced delivery involves application of a pressure gradient at the injection site, which overrides systemic barriers including the BBB to increase the volume of drug distribution and enhance brain penetration. Therefore, several attempts have been made to deliver small-molecule drugs, liposomes, conjugated nanoparticles, gene therapies, and viral particles by CED in clinical and preclinical settings, which have been reviewed in detail.[66, 67] Nanoparticle penetration into brain tumors upon CED application can also be enhanced by size reduction, steric shielding, and surface charge manipulation.[68] Therefore, small-sized (i.e., 70-nm) PEGylated PLGA nanoparticles were best suited for combination with CED,[60, 69] and paclitaxel (PTX) and dithiazanine iodide (a highly toxic antihelmintic agent) loaded in PLGA nanoparticles demonstrated superior survival benefit in preclinical intracranial models.[55, 60]

Intra-arterial administration also bypasses the systemic circulation and deposits high a concentration of the drug in the brain. Whereas small-molecule chemotherapeutics evoke neurotoxicity in patients more frequently with IA delivery than with intravenous delivery,[70] nanoparticulate carriers with high affinity toward brain tumors can be used to reduce toxicity. Because nanoparticles are subject to greater hydrodynamic forces in arteries than are small molecules, the shear stress on the arterial wall and binding affinity of nanoparticles to vascular endothelium are critical factors to be considered for nanoparticle delivery via IA administration. Cationic liposomes and lipid micelles that bind to cells with high affinity show enhanced tumor uptake and retention compared with liposomes having neutral or negative surface charge in intracranial glioma models, when administered via the IA route.[71, 72] Clinically feasible manipulations of blood flow and pressure during IA administration provides additional opportunities for increased local exposure and delivery.[73-75] To date, the clinical success of IA administration is limited. IA administration of bevacizumab did not prolong survival significantly compared to systemic administration in patients.[76] Engineering nanoparticles for high-efficiency and specific tumor uptake requires additional understanding of both particle design and how design interacts with the unique characteristics of IA delivery for clinical deployment.

siRNA Delivery

RNA interference (RNAi) therapeutics represent an exciting area of research. These therapeutics can target "undruggable" proteins and provide individualization of cancer therapy; however, these macromolecular agents are difficult to deliver. Oligonucleotides are recognized by the reticuloendothelial and immune systems, are prone to enzymatic degradation in the plasma and the tumor interstitium, and may cause systemic inflammation.[77] These challenges can be overcome by nanoparticle encapsulation. Attempts have been made to encapsulate siRNA in various nanoparticulate systems, such as silica, metal (e.g., iron oxide and gold nanoparticles), liposomes, and polymers, all of which offer unique delivery advantages.[77] Surface derivatization of such nanoparticles provides an opportunity for specific targeting and for endocytic uptake. In cases in which the cellular uptake of RNAi nanoparticles is mediated primarily via endocytosis, pH-responsive release of siRNA to the cytoplasm has been investigated to avoid lysosomal degradation. Delivery by pH-mediated triggers can be achieved by using polymeric nanoparticles such as PLGA, whose unit-by-unit construction allows several active agents to be encapsulated in layers. Such multifunctional nanoparticulate systems can be released from lysosomes via acid-cleavable links in an outer layer before degradation, and can also perform other functions.[78]

Attempts have been made to use RNAi against numerous potential targets in GBM, including stem cell transcription factors, EGFR, Bcl2, c-Met, and multiple micro RNAs.[79] Nanoparticulate vehicles that encapsulate nucleic acids and show promise in brain tumors include spherical nucleic acid complexes, which consist of a gold nanoparticle core around which linear strands of DNA are densely conjugated. These structures evoke minimal immune response because of their tightly packed structure, wherein the DNA is relatively unexposed, resists nucleases, permits high cellular uptake because of polyvalent interactions,[80] and allows BBB penetration and tumor accumulation via both the EPR phenomenon and scavenger receptor-mediated endocytosis.[17] Spherical nucleic acid complexes have progressed to the clinic and represent the first trial of RNAi-based therapeutics in patients with brain tumors.[81] NU-0129 consists of a spherical nucleic acid complex targeting the mRNA encoding BCL2L12 (antiapoptotic Bcl-2 related proline-rich protein) in GBM, which would exert a pro-apoptotic effect in tumor cells. Other vehicles for delivering RNAi include cationic liposomes that complex the siRNA with the lipid bilayer, as well as self-assembling cationic polymers, but their distribution in the brain parenchyma is yet to be determined.[82-84] Despite many promising advances, siRNA nanotherapeutics remain in their infancy, and the clinical application of RNAi-based nanotherapeutics has been hindered by limited biological activity, challenges intrinsic to the requirement for intracellular delivery of macromolecules, and a high incidence of adverse events.[85] Further advances in reducing the rate of nucleic acid release and increasing BBB penetration, tumor deposition, and intracellular delivery are necessary to overcome these hurdles.

Antivascular Effects of Drug-containing Nanoparticles

Nanoparticles encapsulate active agents in high concentrations, which can alter tumor deposition and overall pharmacokinetic properties of the drug, as well as its pharmacodynamics. Sustained-release formulations of liposome-encapsulated chemotherapeutic agents have been reported to exert antiangiogenic effects[40, 86] because of altered intra-tumor distribution and extended duration of drug exposure, reminiscent of that induced by metronomic chemotherapy regimens.[87] The high hydrodynamic radius of liposomes resists interstitial diffusion and localizes them within micrometers of afferent microvessels (Fig. 1A). A persistent drug depot in close proximity to microvasculature may induce selective cytotoxicity in the vascular endothelial and supporting cells, leading to vascular collapse. The temporal effects of SSL-DXR upon a tumor are complex, consisting of an initial vascular normalization response with restricted perfusion and permeability within days, followed by a period of increased tumor permeability and perfusion within a week.[40] In a highly aggressive intracranial model of rat glioma derived from 9L spheroid cultures, SSL-DXR pretreatment 5 days before TMZ resulted in significantly reduced tumor volume compared to simultaneous administration of SSL-DXR and TMZ (Fig. 2). Additionally, successive doses of SSL-DXR were found to exert vascular effects, leading to compromised vascular permeability, intra-tumor micro-hemorrhage (Fig. 3),[15, 88, 89] and an increase in total DXR deposition that was not merely the additive result of multiple administrations.[15] The findings were consistent with a mechanism in which each dose resulted in high, sustained local drug concentrations that depleted the tumor vascular endothelium and progressively compromised the blood-tumor barrier, permitting greater deposition of successive doses.

Additional reports demonstrate the effects of liposome-based formulations to increase tumor deposition of either subsequent liposome doses, or of other chemotherapy drugs. In rat brain tumors, nanoliposomal formulations of PTX demonstrated greater efficacy with sequential administrations and were more effective at lower doses applied more frequently.[90] A formulation of nanoliposomal irinotecan similarly increased tumor delivery of

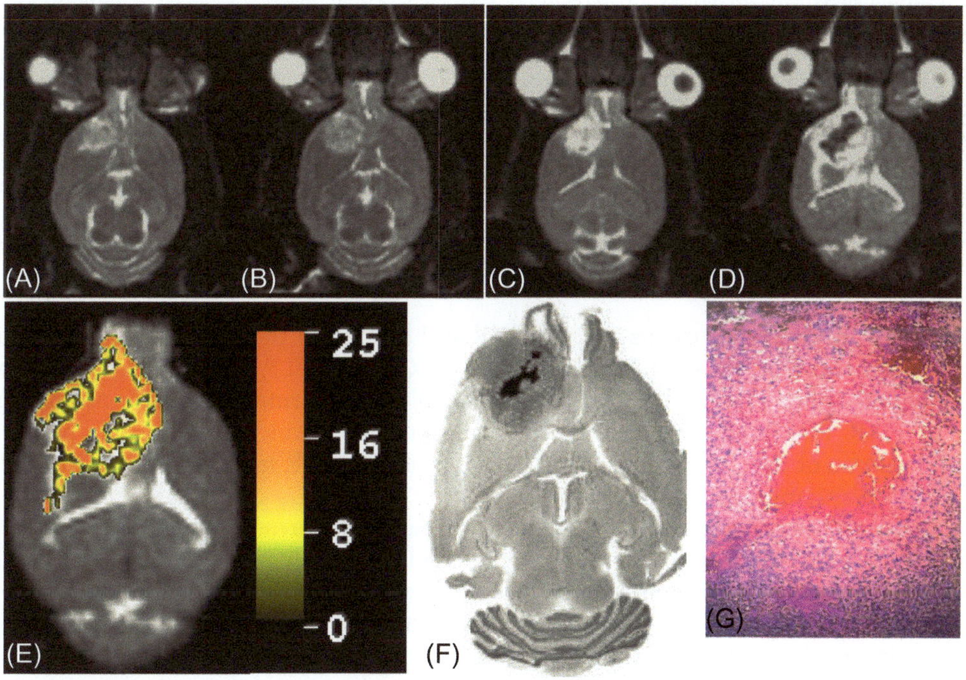

FIG. 2 Antivascular effect of sterically stabilized liposomes encapsulating doxorubicin (SSL-DXR). Rats bearing intracranial 9L tumors received weekly doses of SSL-DXR (5.67 mg/kg/week) or saline, starting on day 7 postimplantation, and were imaged using high-resolution T2-weighted magnetic resonance imaging (MRI) and blood-oxygen-level-dependent functional MRI (fMRI) protocols 24 hours before and 48 hours after receiving the second dose. (A, B) Representative coronal image slices from a control animal, obtained 24 hours before an intravenous saline dose (A), and 48 hours after saline administration (B). (C, D) Representative images from an SSL-DXR-treated animal, taken 24 hours before the second SSL-DXR dose (C), and 48 hours after, showing induction of a hypointense region in the tumor core (D). (E) fMRI map of the corresponding coronal slice in the SSL-DXR-treated animal shows increased blood flow into the tumor core. The fMRI maps were prepared by quantifying blood oxygenation from T2-weighted images acquired while the animals were breathing room air or carbogen (7% CO_2, 93% O_2). The color gradient is proportional to the percent increase in MR signal intensity as a result of increased oxygenation of hemoglobin during carbogen breathing. (F, G) Hematoxylin and eosin stained section of the region corresponding to the hypointense tumor region (F) in an SSL-DXR-treated animal demonstrating the extravasation of erythrocytes (G). *Reproduced with permission from Zhou R, Mazurchuk RV, Straubinger RM. Antivasculature effects of doxorubicin-containing liposomes in an intracranial rat brain tumor model.* Cancer Res. *2002;62(9):2561-2566.*

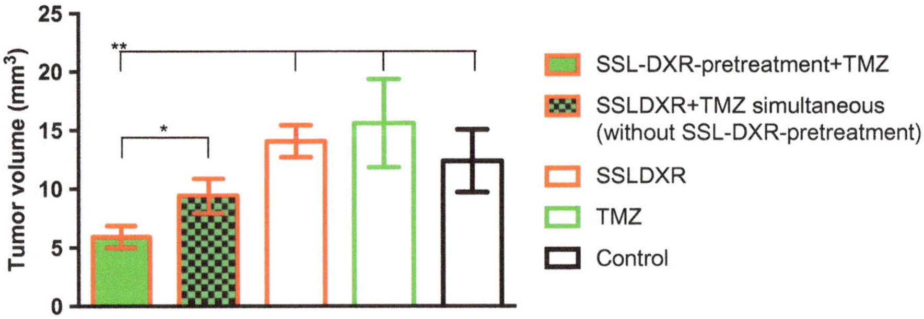

FIG. 3 Effect of sterically stabilized liposomes encapsulating doxorubicin (SSL-DXR) priming combined with subsequently administered temozolomide (TMZ) on tumor volume of orthotopically implanted 9L tumors in rats. Rats ($n=6$/treatment group) were implanted with aggressive intracranial tumors derived from 9L cells grown as tumor spheroids in stem cell media (Dulbecco modified Eagle medium/Ham's F-12 medium containing 20 ng/mL bFGF, 20 ng/mL epidermal growth factor, 50 ng/mL heparin, and 1x B-27 nutrient supplement). Animals were pretreated with intravenous SSL-DXR (5.67 mg/kg) or saline (control) 7 days after tumor implantation. TMZ was administered per oral (5 mg/kg for 5 days) starting either on the day of SSL-DXR treatment or 5 days after SSL-DXR. Tumor volume was measured using high-resolution T2-weighted MRI on day 18 and showed that animals that received SSL-DXR pretreatment 5 days prior to TMZ had significantly lower tumor volumes compared to those in which treatment with both agents was initiated at the same time. Animals receiving the combination of SSL-DXR and TMZ also displayed significantly lower tumor volumes compared to untreated controls and animals receiving single agents ($^*P<.05$; $^{**}P<.01$).

subsequently administered drugs such as DXR and 5-FU, and the mechanism was the result of downregulation of pro-angiogenic factors VEGF and IL8, and upregulation of the antiangiogenic factor TIMP-1.[91] Other examples include a thermosensitive liposomal formulation of DXR that mediated rapid, transient vascular shutdown upon hyperthermia-triggered release of the drug,[92] and a formulation of SSL-DXR targeted to $\alpha_v\beta_3$ integrin that demonstrated antiangiogenic and antimetastatic properties associated with transient vascular shutdown.[93] The complex temporal window of reduced permeability after treatment with cytotoxic nanoparticles, followed potentially by a window of increased tumor perfusion and permeability, suggests that the timing of repeated administrations or subsequently administered chemotherapy drugs can be of critical importance in determining the therapeutic efficacy of regimens that include cytotoxic nanoparticles. Overall, findings suggest that altered pharmacokinetics conferred by nanoparticulate formulations, and the pattern of tumor deposition (which differs from that of free drug) can result in unique mechanisms of pharmacological effects. Detailed investigation into the pharmacodynamic responses mediated by novel nanoparticulate carriers is therefore necessary for their successful clinical translation, and to utilize fully the tumor response mechanisms that result from the particulate nature of macromolecular delivery vehicles.

VASCULAR PERMEABILITY MODULATION TO ENHANCE EPR-MEDIATED DELIVERY

Although the challenges of delivering nanoparticulates are formidable, the structural components of the BBB or the paracrine interactions that preserve BBB integrity are potential targets for transient compromise of the BBB and EPR-mediated delivery. Tumor "priming" to increase vascular permeability and to permit macromolecular entry into the tumor interstitium can be mediated by cytotoxic nanoparticles, as described earlier, or by other molecularly targeted, pharmacologically active agents that target cell–cell tight junction proteins or cell–ECM adhesion proteins. Paracrine signaling that maintains BBB integrity presents additional targets to achieve similar results.

Molecular Modulation of Barrier Properties

Targeting the structural components of the BBB with brain-penetrating peptides was intended to pinpoint peptide-bound macromolecules or nanoparticles specific to tumors.[94] Using multiple rounds of phage display and lead refinement to identify tumor-selective targeting ligands, peptides with a motif of arginylglycylaspartic acid (RGD) that bind to integrins of vascular basement membranes were identified. Further development of this strategy by refinement of peptide targeting to the $\alpha_v\beta_3$ integrin with a CRGDKGPDC motif and the addition of a second functionality that binds the neuropilin 1/2 receptor (thereby activating transcytosis) resulted in a 9-amino acid iRGD targeting peptide with an attached CendR translocation motif. The concept, supported by experimental data, was that upon cleavage from RGD-bound-integrin by unidentified proteases in the TME, the CendR motif would undergo endocytosis and then oriented exocytosis, carrying along any other moieties attached to this homing/transporting peptide and improving tissue penetration of the attached cargo.[95] Further investigations demonstrated that covalent attachment of the bifunctional homing/transport iRGD peptide to nab-paclitaxel, a 130-nm albumin nanoparticle, resulted in enhanced, targeted delivery to solid tumors, with deeper tumor penetration and efficacy.[96]

Studies in a preclinical model of glioblastoma also showed increased deposition and therapeutic efficacy of paclitaxel-loaded-nanoparticles coadministered with the iRGD peptide.[97] Perplexingly, intravenous coadministration of free iRGD peptide and untargeted nab-paclitaxel resulted in even greater tumor nanoparticle deposition, compared to iRGD-targeted nanoparticles.[98] Subsequent work to replicate experiments demonstrating targeting and delivery mediated by the iRGD strategy[96] was unsuccessful.[99] Data in documents filed with the FDA demonstrated that nab-paclitaxel dissociates rapidly upon dilution, possibly into noncovalent albumin/paclitaxel monomers.[100] Thus, although evidence has shown that increased magnitude and depth tumor penetration can be mediated by cotransport of nanoparticles in endosomes, mediated by covalently attached iRGD and its CendR functionalities, alternative mechanisms to those hypothesized may be operant. More detailed understanding of the temporal profile of compromised tumor vascular permeability promoted by iRGD may provide insights that could be used for successful clinical translation of the iRGD-priming strategy.

Other strategies to overcome the BBB delivery barrier include the use of Pluronic block copolymer unimers that readily cross the BBB because of their amphiphilic nature. Such particles may also cause endoplasmic reticulum stress and deplete intracellular adenosine triphosphate in endothelial cells, ultimately inhibiting drug transporters such as Pgp and promoting CNS accumulation of a subsequently administered agent.[63] Clearer definition of the delivery mechanisms involved is essential for wider deployment of this strategy.

Paracrine interactions that maintain the integrity of the BBB also represent a potential strategy for compromise of drug delivery barrier properties. Adenosine receptor signaling mediates cross-talk between interstitial cells and brain cells. It modulates BBB permeability and maintains the immune microenvironment by preventing lymphocyte infiltration in the brain.[101] Activation of the adenosine receptor A_{2A} on endothelial cells with the A_{2A} agonist regadenoson temporarily compromised BBB integrity by reducing the expression of tight junction proteins and inhibiting Pgp,[102] permitting a seven- to ninefold increase in nanoparticle extravasation and drug accumulation in the CNS.[103] This approach is not specific to tumors, however, and would result in increased penetration of nanoparticles throughout the CNS. To overcome this hurdle, one strategy developed "autocatalytic" nanoparticles, encapsulating various modulators of BBB integrity, including adenosine receptor agonist regadenoson, for a two-step tumor priming function.[104] Poly(amine-co-ester) nanoparticles containing the BBB modulators were derivatized with chlorotoxin, a short Cl^- channel-blocking peptide that binds to matrix metalloprotease 2 (MMP2), which is overexpressed in brain tumors compared to the normal brain. The surface-displayed chlorotoxin ligand mediated a twofold increase in brain tumor deposition compared to nontargeted nanoparticles, which would undergo deposition via transcytosis, endocytosis, or EPR-mediated delivery. Sustained, controlled release of the encapsulated regadenoson exerted a tumor-priming effect that was largely selective for tumor regions, and allowed entry and accumulation of the same nanoparticles administered subsequently, amplifying the tumor priming effect drastically. Adenosine administered via the IA route, combined with contralateral carotid artery occlusion to increase blood flow to the tumor, creates transient cerebral hypoperfusion (TCH), which can achieve an even higher arterial concentration of therapeutic agents compared to IA. Cationic liposomes delivered using this technique showed fourfold higher tumor deposition compared to IA alone, and 100-fold higher tumor accumulation than liposomes administered intravenously.[73, 74]

Tumor priming strategies have yet to be deployed clinically. A major hurdle is identifying patients who respond to the priming agent, and the time of peak priming response. Although animal studies have demonstrated that regadenoson increases tumor dextran deposition,[103] contrast-enhanced computed tomography and single-photon emission computed tomography did not detect changes in the deposition of small-molecule contrast agents commonly used in clinics.[105] The need to develop a means to detect tumor priming clinically is evident. Ideally, the onset of increased permeability would be quantified noninvasively, using modalities such as contrast-enhanced MRI or positron emission tomography. Further complicating the matter, enhancing EPR by inducing BBB compromise is likely more complicated in brain tumors than in other cancers because of the physiological role of the BBB in protecting the CNS and the requirement that permeability enhancement is restricted to regions with tumors. In principle, particle and immune cell infiltration in the normal brain could be harmful. Therefore, further investigation is necessary to identify and target pharmacological modulators of the BBB to tumors specifically, in order to increase both vascular permeability and nanoparticle deposition selectively in tumors.

Mechanical Modulation of Barrier Properties

Physiological stressors such as osmotic imbalance and hyperthermia cause transient opening of the BBB by modulation of tight junction proteins, suggesting another strategy for allowing macromolecular or nanoparticulate therapeutic agents to pass through endothelial gaps. Osmotic shock in the brain by intracarotid administration of hypertonic solutions for 30 min shrinks endothelial cells via cytoskeletal rearrangement and diminished expression of the tight junction proteins occludin, claudin-5, and ZO-1,[106] opening gaps of up to 40 nm in diameter.[107] Although this strategy has been investigated clinically, intracarotid administration of hypertonic mannitol mediates a 10-fold enhanced permeability to small molecules, but not to macromolecules.[108] Additionally, the vascular barrier compromise is propagated throughout the brain, and is not selective for the tumor. As a result, the therapeutic efficacy has been disappointing.[109]

Microparticles themselves can mediate transiently increased BBB permeability. Sonication of gas-filled microbubbles with focused ultrasound (FUS) creates local pressure disturbances and cavitation within microvessels, which opens the BBB reversibly and transiently in selected regions of the brain.[110] FUS induces the formation of numerous intracellular vesicles, vacuoles, and channels in the endothelial cells and adjoining pericytes, which promotes transcellular transport.[111] Additionally, endothelial gaps are created by the disintegration of the tight junctional complexes via suppressed expression of occludin, claudin-5, and ZO-1 proteins, which promotes paracellular macromolecule transport.[112] BBB disruption by FUS enhances macromolecular deposition[113]; FUS has been shown to mediate up to 10-fold increase in anti–Her2 monoclonal antibody deposition in the brain.[114] Similarly, FUS has been shown to result in a significant increase in tumor uptake and retention of cationic liposomes in rat glioma.[115] FUS has also been applied to release an encapsulated drug from microbubbles in tumors, but did not mediate a significant therapeutic benefit.[116] Additionally, this may lead to rapid renal elimination of the drug from circulation, which would limit tumor drug exposure and potentially increase systemic toxicity by exposing the encapsulated drug to other organs.

Another ultrasound-based approach for enhancing macromolecular permeation is high-intensity focal ultrasound (HIFU), which mediates a transient temperature increase or mild hyperthermia, depending upon the ultrasound parameters (i.e., intensity and frequency) and the energy absorption coefficient of the tissue. Mild hyperthermia in the fever range (42°C) halts nucleic acid synthesis and inhibits repair enzymes, inducing apoptosis in tumor cells.[117] The rise in temperature enhances tumor perfusion and permeability by altering ECM and cytoskeletal structural proteins and by inducing a direct cytotoxic effect on tumor cells. This effect decreases cellular density and relieves IFP, thereby decompressing tumor vasculature and improving perfusion.[118, 119] Hyperthermia induced by HIFU has been shown to promote nanoparticle delivery: a 2-fold increase in tumor deposition of liposomal DXR was observed after application of HIFU, which mediated tumor stasis in an intracranial model of metastatic breast cancer.[120] Nevertheless, substantial technological challenges remain, and concerns of tissue damage must be addressed before HIFU can be clinically implemented. Despite being a relatively new technology, FUS represents a tumor-specific priming strategy to enhance drug delivery, and may find clinical utility in the treatment of brain metastases with small nanoparticles or antibody-based therapeutics for metastatic breast cancer.

Regional tumor hyperthermia may also be induced by systemic or intratumor administration of magnetic nanoparticles and application of an alternating magnetic field to generate heat in regions of nanoparticle accumulation.[121] Nanoparticle formulations of iron oxide core encapsulated within a lipidic or polymeric shell and having a size range of 10–100 nm undergo passive, EPR-driven accumulation in the tumor. The alternating magnetic field then produces heat only in regions of nanoparticle deposition, which partially overcomes the disadvantages of peritumoral thermal damage associated with other tissue-heating sources, such as microwave and electrically induced hyperthermia.[121, 122] Additionally, the intrinsic ferromagnetic properties of these superparamagnetic nanoparticles enable their visualization by MRI. Thermotherapy has been reported to promote interstitial hyperthermia and cellular ablation,[123] and increases radiotherapy efficacy by improving tissue oxygenation.[124] However, a considerable tumor depot of magnetic nanoparticles is necessary to mediate these pharmacological effects, and clinical success has been modest. Nonetheless, application of an alternating magnetic field mediated a 10-fold increase in systemically administered magnetic nanoparticle deposition in brain tumors,[125] and improved liposomal penetration in other cancers.[126]

Other approaches to increase tumor permeability in glioma are reported in the literature and have been recently reviewed.[127] With advances in surgical techniques, novel approaches such as retro-CED, which involves removal of fluid from the interstitial space in order to create a positive flow of therapeutic agents from systemic circulation to the interstitium, may become feasible to overcome the BBB.[128] Nevertheless, there is still an urgent need to identify physiological and molecular mediators of nanoparticle transport across the BBB and in the tumor interstitium to promote tumor accumulation of both targeted and nontargeted nanoparticles. Strategies for inducing a transient window of increased vascular permeability may make many experimental nanotherapeutics clinically feasible. However, variation in patient responses and the time-sensitive nature of the approach complicate clinical deployment. Nonetheless, further investigation of temporal tumor permeability responses, and a robust imaging platform to detect and quantify increases in vascular permeability, may aid in the clinical translation of tumor permeation strategies.

TARGETED NANOPARTICLES

Target-specific uptake mediated by antibody- or peptide-conjugated nanoparticles reduces systemic toxicity and facilitates receptor-mediated transport across an intact BBB via endocytic or transcytotic pathways in endothelial cells. In brain tumors, potential targets include agents that bind to factors associated with the BBB or glioma cells (or both), and ECM/stromal elements. The rate of nanoparticle transport across the endothelium and the BBB depends on affinity of the antibody to target, nanoparticulate retention within endothelial cells, molecular driving forces, and rate of efflux.[47] Recent advances in nanoparticle design enable protein/peptide derivatization via linkers (e.g., activated PEG) and covalent linkages (e.g., sulfhydryl-reactive groups). Endothelial cells have been targeted with tumor-specific peptides such as the brain-penetrating iRGD peptide and glutathione. Additionally, receptors that have been explored extensively to achieve transcytotic transport of nanoparticles across the BBB, as well as for endocytic uptake by glioma cells, include the transferrin, insulin, low-density lipoprotein,[129] and chlorotoxin receptor families.

Glutathione, an endogenous tripeptide with antioxidant properties, readily crosses the BBB by unknown mechanisms, and has been used to transport SSL-DXR into brain tumors. Covalent linkage of the PEG on SSL-DXR with glutathione enhances BBB transit and tumor retention of DXR fourfold, which slowed the tumor growth rate significantly compared to untargeted SSL-DXR.[130] The FDA-approved status of SSL-DXR expedited a clinical trial for testing the efficacy of glutathione-SSL-DXR clinically, and the initial results were promising.[131] Other agents show enhanced penetration of both endothelial and glioma cells via multiple mechanisms; chlorotoxin has been conjugated with a fluorescent dye and has been used clinically to detect residual tumor in the surgical resection site, and is under preclinical investigation for targeting polymeric nanoparticles for BBB compromise (described earlier), as well as targeting iron oxide nanoparticles as a contrast agent for MRI.[132]

The transferrin (Tf) receptor is densely expressed on cells surrounding the invading edge of tumors, and provides a strategy for exploiting the receptor-mediated endocytic pathway to promote BBB penetration of diverse nanoparticulate formulations such as gold, iron oxide, liposomes, albumin, and PLGA nanoparticles that encapsulate therapeutic and diagnostic cargos.[133] However, anti-Tf antibodies that bind to the receptor with high affinity, such as OX26, result in nanoparticle accumulation on the luminal side of endothelium, whereas lowering the avidity of the nanoparticle-bound antibody for the Tf receptor results in activation of a transcytotic process. This results in greater BBB penetration because the nanoparticles are more efficiently released from the receptor into the interstitium.[48] This finding has also led to the development of acid-cleavable antibody–nanoparticle linkers that are released by the acidic environment within endocytic vesicles, which showed increased efficiency in crossing the BBB and penetrating the tumor, and yielded greater therapeutic efficacy.[134] Because of their superior brain-penetrating ability, diagnostic and p53-silencing siRNA formulations of Tf-nanoparticles are in various stages of clinical development,[85] and preclinically demonstrate potent p53 mediated-chemosensitization in glioma.[135]

The large surface area of nanoparticulate carriers allows derivatization with multiple proteins that may be used for dual targeting of receptors either on the same cell or on adjacently placed cells.[136] However, the need for oriented conjugation of antibodies on the nanoparticle surface is apparent, and the strategies to achieve this are advancing rapidly.[137] Overall, despite the emergence of numerous combinations of nanoparticle systems and targets, few have reached clinical evaluation, and further refinement is necessary for clinical realization of this strategy.

QUANTITATIVE TOOLS FOR DEVELOPMENT OF NANOPARTICLE DELIVERY STRATEGIES

Quantitative systems pharmacological modeling that captures both the pharmacokinetic and pharmacodynamic behavior of nanoparticulate delivery vehicles is a powerful emerging tool for developing testable hypotheses that can lead to the optimization of treatment strategies for nanoparticulate carriers.[138] Models of glioblastoma disease processes, including tumor growth and invasion, and of nanoparticle delivery to brain tumors have been the subject of recent investigation.[139] Computational models have been developed to explore quantitatively the effect of nanoparticle physical characteristics upon pharmacokinetics,[140] and to provide insights into the effects of tumor priming mediated by drug-containing nanoparticles on nanoparticle pharmacokinetics in glioblastoma models.[15] Modeling and simulation have been used to explore the conditions that may control whether nanoparticle-encapsulated drugs are more or less efficacious than the conventional free drug, and the reasons that efficacy observed in preclinical rodent models may not translate clinically to humans.[141]

In tumor priming strategies, linked pharmacokinetic/pharmacodynamic models have been developed to provide experimentally testable predictions as to the optimal inter-dose interval between the priming treatment and the subsequent administration of nanoparticle vehicles, which is essential for successful clinical use.[142] Models have also been used to generate hypotheses relating the stability and drug release characteristics of nanoparticulate formulations to their toxicity[143] and to explore the effects of different dosing intervals on nanoparticulate drug formulations in glioblastoma models.[90] Suitable mathematical models can be used to quantify the role of numerous and complex processes controlling drug delivery to brain tumors, and to assist in the multidimensional optimization that would be necessary for successful development of nanoparticulate systems. Going forward, they represent important tools to accelerate the development of therapeutic nanoparticle formulations.

CONCLUSIONS

Nanoparticles alter the pharmacokinetics and biodistribution of conventional chemotherapeutic agents to alleviate systemic toxicity and promote drug accumulation in regions of tumor vascular permeability. Brain tumors display regions in which BBB integrity is compromised to varying degrees. Although numerous formulations and strategies have shown preclinical promise, a very small proportion of nanoparticle carriers display high brain tumor penetration, retention, and the requisite low clinical toxicity. Despite breakthroughs in material sciences and rapid emergence of a variety of nanoparticle conjugates, the physical and molecular mediators of nanoparticulate transport into and within tumor tissue have received relatively less attention, to the detriment of efforts to achieve substantial improvement in brain tumor therapy. The BBB remains a formidable obstacle. Advances in molecular biology have increased our understanding of the various structural components of the BBB, but further investigation of the paracrine interactions that maintain and modulate BBB integrity is needed, as manipulation of the BBB holds the potential to improve entry of nanoparticles into brain tumors. Additionally, the transport of particulates through the ECM is poorly understood. To date, intra-tumor transport has been assumed to be diffusion-mediated, which is inefficient in conveying nanoparticles the distances into tissues that are required for effective brain tumor therapy. Recent findings have identified primordial channels in the interstitium[144] that may aid in convective transport of blood-borne particulates, and manipulation of tumor-stromal paracrine interactions could also mediate modification of the ECM to improve intra-tumor transport.

Because tumor vasculature serves as both the mode of and the hindrance to tumor drug delivery, the vasculature represents a promising target for overcoming the barriers to drug delivery. Additionally, the alteration of pharmacokinetics that encapsulation in nanoparticulates confers upon therapeutic agents imparts novel pharmacological properties, including those resulting from sustained rather than transient drug exposure, such as the BBB-compromising antivascular effects observed following extravasation of long-circulating liposomes containing cytotoxic agents. Clinical trials of nanoparticulates typically are conducted in such a way as to leave their particulate properties unexploited, and those properties continue to hold considerable potential in combination chemotherapeutic regimens that could enhance antitumor efficacy. Modulation of vascular permeability has been implemented clinically. However, with greater understanding of the complex molecular interactions that form and maintain the integrity of the BBB, hypothesis-driven design of strategies to modulate tumor vascular permeability on-demand, and advances in the design of nanoparticles for improved tumor selectivity and local drug exposure, nanoparticulates may provide the means to improve the outcomes of patients with highly fatal brain cancers.

Acknowledgments

We extend our apologies to many authors whose work could not be cited directly owing to space limitations. Support for this work was provided by NIH/NCI grants R01CA107570 and R01CA198096 to RMS. Comprehensive Cancer Center support grant NIH/NCI P30CA016056 to Roswell Park Comprehensive Cancer Center provided shared resources used in the work.

References

1. Ostrom QT, Gittleman H, Liao P, et al. CBTRUS Statistical Report: Primary brain and other central nervous system tumors diagnosed in the United States in 2010-2014. *Neuro Oncol*. 2017;19(suppl 5):v1-v88.
2. Siegel RL, Miller KD, Jemal A. Cancer Statistics, 2017. *CA Cancer J Clin*. 2017;67(1):7-30.
3. Aminoff MJ, Josephson SA. *Aminoff's Neurology and General Medicine*. 5th ed. Amsterdam: Elsevier/Academic Press; 2014. ClinicalKey, *https://www.clinicalkey.com/dura/browse/bookChapter/3-s2.0-C20120030311*.
4. Stupp R, Mason WP, van den Bent MJ, et al. Radiotherapy plus concomitant and adjuvant temozolomide for glioblastoma. *N Engl J Med*. 2005;352(10):987-996.
5. Lassman AB. Procarbazine, lomustine and vincristine or temozolomide: which is the better regimen? *CNS Oncol*. 2015;4(5):341-346.
6. Brem SS, Bierman PJ, Black P, et al. Central nervous system cancers. *J Natl Compr Cancer Netw*. 2008;6(5):456-504.
7. Hochberg FH, Pruitt A. Assumptions in the radiotherapy of glioblastoma. *Neurology*. 1980;30(9):907-911.
8. Parsons DW, Jones S, Zhang X, et al. An integrated genomic analysis of human glioblastoma multiforme. *Science*. 2008;321(5897):1807-1812.
9. Tuncbag N, Milani P, Pokorny JL, et al. Network modeling identifies patient-specific pathways in glioblastoma. *Sci Rep*. 2016;6:28668.
10. De Witt Hamer PC. Small molecule kinase inhibitors in glioblastoma: a systematic review of clinical studies. *Neuro Oncol*. 2010;12(3):304-316.
11. Parrish KE, Cen L, Murray J, et al. Efficacy of parp inhibitor rucaparib in orthotopic glioblastoma xenografts is limited by ineffective drug penetration into the central nervous system. *Mol Cancer Ther*. 2015;14(12):2735-2743.
12. Trunzer K, Pavlick AC, Schuchter L, et al. Pharmacodynamic effects and mechanisms of resistance to vemurafenib in patients with metastatic melanoma. *J Clin Oncol*. 2013;31(14):1767-1774.
13. Bendell JC, Domchek SM, Burstein HJ, et al. Central nervous system metastases in women who receive trastuzumab-based therapy for metastatic breast carcinoma. *Cancer*. 2003;97(12):2972-2977.
14. Giannini C, Sarkaria JN, Saito A, et al. Patient tumor EGFR and PDGFRA gene amplifications retained in an invasive intracranial xenograft model of glioblastoma multiforme. *Neuro Oncol*. 2005;7(2):164-176.
15. Arnold RD, Mager DE, Slack JE, Straubinger RM. Effect of repetitive administration of doxorubicin-containing liposomes on plasma pharmacokinetics and drug biodistribution in a rat brain tumor model. *Clin Cancer Res*. 2005;11(24):8856-8865.
16. Sharma A, Straubinger RM, Ojima I, Bernacki RJ. Antitumor efficacy of taxane liposomes on a human ovarian tumor xenograft in nude athymic mice. *J Pharm Sci*. 1995;84(12):1400-1404.
17. Jensen SA, Day ES, Ko CH, et al. Spherical nucleic acid nanoparticle conjugates as an RNAi-based therapy for glioblastoma. *Sci Transl Med*. 2013;5(209):209ra152.
18. Patel T, Zhou J, Piepmeier JM, Saltzman WM. Polymeric nanoparticles for drug delivery to the central nervous system. *Adv Drug Deliv Rev*. 2012;64(7):701-705.
19. Zhou J, Atsina KB, Himes BT, Strohbehn GW, Saltzman WM. Novel delivery strategies for glioblastoma. *Cancer J*. 2012;18(1):89-99.

20. Charles NA, Holland EC, Gilbertson R, Glass R, Kettenmann H. The brain tumor microenvironment. *Glia*. 2012;60(3):502-514.
21. Jallo GI, Friedlander DR, Kelly PJ, Wisoff JH, Grumet M, Zagzag D. Tenascin-C expression in the cyst wall and fluid of human brain tumors correlates with angiogenesis. *Neurosurgery*. 1997;41(5):1052-1059.
22. Lee J, Lund-Smith C, Borboa A, Gonzalez AM, Baird A, Eliceiri BP. Glioma-induced remodeling of the neurovascular unit. *Brain Res*. 2009;1288:125-134.
23. Kim JH, Kim JH, Park JA, et al. Blood-neural barrier: intercellular communication at glio-vascular interface. *J Biochem Mol Biol*. 2006;39(4):339-345.
24. Folkman J. Tumor angiogenesis: therapeutic implications. *N Engl J Med*. 1971;285(21):1182-1186.
25. Jain RK, di Tomaso E, Duda DG, Loeffler JS, Sorensen AG, Batchelor TT. Angiogenesis in brain tumours. *Nat Rev Neurosci*. 2007;8(8):610-622.
26. Leten C, Struys T, Dresselaers T, Himmelreich U. In vivo and ex vivo assessment of the blood brain barrier integrity in different glioblastoma animal models. *J Neuro-Oncol*. 2014;119(2):297-306.
27. Wolf RL, Wang J, Wang S, et al. Grading of CNS neoplasms using continuous arterial spin labeled perfusion MR imaging at 3 Tesla. *J Magn Reson Imaging*. 2005;22(4):475-482.
28. Agarwal S, Sane R, Gallardo JL, Ohlfest JR, Elmquist WF. Distribution of gefitinib to the brain is limited by P-glycoprotein (ABCB1) and breast cancer resistance protein (ABCG2)-mediated active efflux. *J Pharmacol Exp Ther*. 2010;334(1):147-155.
29. William D, Mullins CS, Schneider B, et al. Optimized creation of glioblastoma patient derived xenografts for use in preclinical studies. *J Transl Med*. 2017;15(1):27.
30. Formenti SC, Demaria S. Systemic effects of local radiotherapy. *Lancet Oncol*. 2009;10(7):718-726.
31. Agarwal S, Mittapalli RK, Zellmer DM, et al. Active efflux of Dasatinib from the brain limits efficacy against murine glioblastoma: broad implications for the clinical use of molecularly targeted agents. *Mol Cancer Ther*. 2012; 11(10):2183-2192.
32. Liu C, Sage JC, Miller MR, et al. Mosaic analysis with double markers reveals tumor cell of origin in glioma. *Cell*. 2011;146(2):209-221.
33. Boucher Y, Salehi H, Witwer B, Harsh GR, Jain RK. Interstitial fluid pressure in intracranial tumours in patients and in rodents. *Br J Cancer*. 1997;75(6):829-836.
34. Jain RK. Vascular and interstitial barriers to delivery of therapeutic agents in tumors. *Cancer Metastasis Rev*. 1990;9(3):253-266.
35. Jain RK. Physiological barriers to delivery of monoclonal antibodies and other macromolecules in tumors. *Cancer Res*. 1990;50(3 Suppl):814s-819s.
36. Jain RK. Haemodynamic and transport barriers to the treatment of solid tumours. *Int J Radiat Biol*. 1991;60(1-2):85-100.
37. Sharma A, Straubinger RM. Novel taxol formulations: preparation and characterization of taxol-containing liposomes. *Pharm Res*. 1994;11(6):889-896.
38. Sun C, Jain RK, Munn LL. Non-uniform plasma leakage affects local hematocrit and blood flow: Implications for inflammation and tumor perfusion. *Ann Biomed Eng*. 2007.
39. Jain RK, Tong RT, Munn LL. Effect of vascular normalization by antiangiogenic therapy on interstitial hypertension, peritumor edema, and lymphatic metastasis: Insights from a mathematical model. *Cancer Res*. 2007; 67(6):2729-2735.
40. Roy Chaudhuri T, Arnold RD, Yang J, et al. Mechanisms of tumor vascular priming by a nanoparticulate doxorubicin formulation. *Pharm Res*. 2012;29(12):3312-3324.
41. Claes A, Wesseling P, Jeuken J, Maass C, Heerschap A, Leenders WP. Antiangiogenic compounds interfere with chemotherapy of brain tumors due to vessel normalization. *Mol Cancer Ther*. 2008;7(1):71-78.
42. Maeda H, Sawa T, Konno T. Mechanism of tumor-targeted delivery of macromolecular drugs, including the EPR effect in solid tumor and clinical overview of the prototype polymeric drug SMANCS. *J Control Release*. 2001; 74(1-3):47-61.
43. Kreuter J. Drug delivery to the central nervous system by polymeric nanoparticles: what do we know? *Adv Drug Deliv Rev*. 2014;71:2-14.
44. Yuan F, Dellian M, Fukumura D, et al. Vascular permeability in a human tumor xenograft: molecular size dependence and cutoff size. *Cancer Res*. 1995;55(17):3752-3756.
45. Yuan F, Leunig M, Huang SK, Berk DA, Papahadjopoulos D, Jain RK. Microvascular permeability and interstitial penetration of sterically stabilized (stealth) liposomes in a human tumor xenograft. *Cancer Res*. 1994; 54(13):3352-3356.

46. Jain RK. Interstitial transport in tumors: Barriers and strategies for improvement. In: *American Association of Cancer Research. 96th Annual Meeting*; 2005:108-113.
47. Dreher MR, Liu W, Michelich CR, Dewhirst MW, Yuan F, Chilkoti A. Tumor vascular permeability, accumulation, and penetration of macromolecular drug carriers. *J Natl Cancer Inst*. 2006;98(5):335-344.
48. Wiley DT, Webster P, Gale A, Davis ME. Transcytosis and brain uptake of transferrin-containing nanoparticles by tuning avidity to transferrin receptor. *Proc Natl Acad Sci U S A*. 2013;110(21):8662-8667.
49. Ko AH, Tempero MA, Shan YS, et al. A multinational phase 2 study of nanoliposomal irinotecan sucrosofate (PEP02, MM-398) for patients with gemcitabine-refractory metastatic pancreatic cancer. *Br J Cancer*. 2013;109 (4):920-925.
50. Olson F, Hunt CA, Szoka FC, Vail WJ, Papahadjopoulos D. Preparation of liposomes of defined size distribution by extrusion through polycarbonate membranes. *Biochim Biophys Acta*. 1979;557(1):9-23.
51. Deamer D, Bangham AD. Large volume liposomes by an ether vaporization method. *Biochim Biophys Acta*. 1976;443(3):629-634.
52. Meledandri CJ, Stolarczyk JK, Brougham DF. Hierarchical gold-decorated magnetic nanoparticle clusters with controlled size. *ACS Nano*. 2011;5(3):1747-1755.
53. Danhier F, Ansorena E, Silva JM, Coco R, Le Breton A, Preat V. PLGA-based nanoparticles: an overview of biomedical applications. *J Control Release*. 2012;161(2):505-522.
54. Maximova K, Aristov A, Sentis M, Kabashin AV. Size-controllable synthesis of bare gold nanoparticles by femtosecond laser fragmentation in water. *Nanotechnology*. 2015;26(6).
55. Zhou J, Patel TR, Sirianni RW, et al. Highly penetrative, drug-loaded nanocarriers improve treatment of glioblastoma. *Proc Natl Acad Sci U S A*. 2013;110(29):11751-11756.
56. Gabizon A, Papahadjopoulos D. Liposome formulations with prolonged circulation time in blood and enhanced uptake by tumors. *Proc Natl Acad Sci U S A*. 1988;85(18):6949-6953.
57. Lasic DD. Doxorubicin in sterically stabilized liposomes. *Nature*. 1996;380(6574):561-562.
58. Ceh B, Lasic DD. A rigorous theory of remote loading of drugs into liposomes: transmembrane potential and induced ph-gradient loading and leakage of liposomes. *J Colloid Interface Sci*. 1997;185(1):9-18.
59. Deng Y, Saucier-Sawyer JK, Hoimes CJ, et al. The effect of hyperbranched polyglycerol coatings on drug delivery using degradable polymer nanoparticles. *Biomaterials*. 2014;35(24):6595-6602.
60. Nance E, Zhang C, Shih TY, Xu Q, Schuster BS, Hanes J. Brain-penetrating nanoparticles improve paclitaxel efficacy in malignant glioma following local administration. *ACS Nano*. 2014;8(10):10655-10664.
61. Tian XH, Lin XN, Wei F, et al. Enhanced brain targeting of temozolomide in polysorbate-80 coated polybutylcyanoacrylate nanoparticles. *Int J Nanomedicine*. 2011;6:445-452.
62. Batrakova EV, Kabanov AV. Pluronic block copolymers: evolution of drug delivery concept from inert nanocarriers to biological response modifiers. *J Control Release*. 2008;130(2):98-106.
63. Kabanov AV, Batrakova EV, Miller DW. Pluronic block copolymers as modulators of drug efflux transporter activity in the blood-brain barrier. *Adv Drug Deliv Rev*. 2003;55(1):151-164.
64. Mohammad AS, Griffith JI, Adkins CE, et al. Liposomal irinotecan accumulates in metastatic lesions, crosses the blood-tumor barrier (btb), and prolongs survival in an experimental model of brain metastases of triple negative breast cancer. *Pharm Res*. 2018;35(2):31.
65. Sharma US, Sharma A, Chau RI, Straubinger RM. Liposome-mediated therapy of intracranial brain tumors in a rat model. *Pharm Res*. 1997;14(8):992-998.
66. Vogelbaum MA, Aghi MK. Convection-enhanced delivery for the treatment of glioblastoma. *Neuro Oncol*. 2015;17(Suppl 2):ii3-ii8.
67. Jahangiri A, Chin AT, Flanigan PM, Chen R, Bankiewicz K, Aghi MK. Convection-enhanced delivery in glioblastoma: a review of preclinical and clinical studies. *J Neurosurg*. 2017;126(1):191-200.
68. MacKay JA, Deen DF, Szoka Jr. FC. Distribution in brain of liposomes after convection enhanced delivery; modulation by particle charge, particle diameter, and presence of steric coating. *Brain Res*. 2005;1035(2):139-153.
69. Seo YE, Bu T, Saltzman WM. Nanomaterials for convection-enhanced delivery of agents to treat brain tumors. *Curr Opin Biomed Eng*. 2017;4:1-12.
70. Newton HB. Intra-arterial chemotherapy of primary brain tumors. *Curr Treat Options Oncol*. 2005;6 (6):519-530.
71. Joshi S, Cooke JR, Chan DK, et al. Liposome size and charge optimization for intraarterial delivery to gliomas. *Drug Deliv Transl Res*. 2016;6(3):225-233.
72. Nguyen J, Cooke JRN, Ellis JA, et al. Cationizable lipid micelles as vehicles for intraarterial glioma treatment. *J Neuro-Oncol*. 2016;128(1):21-28.

73. Joshi S, Singh-Moon RP, Ellis JA, et al. Cerebral hypoperfusion-assisted intra-arterial deposition of liposomes in normal and glioma-bearing rats. *Neurosurgery*. 2015;76(1):92-100.
74. Joshi S, Singh-Moon RP, Wang M, et al. Transient cerebral hypoperfusion assisted intraarterial cationic liposome delivery to brain tissue. *J Neuro-Oncol*. 2014;118(1):73-82.
75. Nguyen J, Hossain SS, Cooke JRN, et al. Flow arrest intra-arterial delivery of small TAT-decorated and neutral micelles to gliomas. *J Neuro-Oncol*. 2017;133(1):77-85.
76. Burkhardt JK, Riina H, Shin BJ, et al. Intra-arterial delivery of bevacizumab after blood-brain barrier disruption for the treatment of recurrent glioblastoma: progression-free survival and overall survival. *World Neurosurg*. 2012;77(1):130-134.
77. Ozcan G, Ozpolat B, Coleman RL, Sood AK, Lopez-Berestein G. Preclinical and clinical development of siRNA-based therapeutics. *Adv Drug Deliv Rev*. 2015;87:108-119.
78. Zhou J, Patel TR, Fu M, Bertram JP, Saltzman WM. Octa-functional PLGA nanoparticles for targeted and efficient siRNA delivery to tumors. *Biomaterials*. 2012;33(2):583-591.
79. Lozada-Delgado EL, Grafals-Ruiz N, Vivas-Mejia PE. RNA interference for glioblastoma therapy: Innovation ladder from the bench to clinical trials. *Life Sci*. 2017;188:26-36.
80. Cutler JI, Auyeung E, Mirkin CA. Spherical nucleic acids. *J Am Chem Soc*. 2012;134(3):1376-1391.
81. A Phase 0 First-In-Human Study Using NU-0129: A Spherical Nucleic Acid (SNA) Gold Nanoparticle Targeting BCL2L12 in Recurrent Glioblastoma Multiforme or Gliosarcoma Patients. https://clinicaltrials.gov/ct2/show/NCT03020017.
82 Yu D, Khan OF, Suva ML, et al. Multiplexed RNAi therapy against brain tumor-initiating cells via lipopolymeric nanoparticle infusion delays glioblastoma progression. *Proc Natl Acad Sci U S A*. 2017;114(30):E6147-E6156.
83. Jin J, Bae KH, Yang H, et al. In vivo specific delivery of c-Met siRNA to glioblastoma using cationic solid lipid nanoparticles. *Bioconjug Chem*. 2011;22(12):2568-2572.
84. Guerrero-Cazares H, Tzeng SY, Young NP, Abutaleb AO, Quinones-Hinojosa A, Green JJ. Biodegradable polymeric nanoparticles show high efficacy and specificity at DNA delivery to human glioblastoma in vitro and in vivo. *ACS Nano*. 2014;8(5):5141-5153.
85. Pirollo KF, Nemunaitis J, Leung PK, Nunan R, Adams J, Chang EH. Safety and efficacy in advanced solid tumors of a targeted nanocomplex carrying the p53 gene used in combination with docetaxel: A phase 1b study. *Mol Ther*. 2016;24(9):1697-1706.
86. Verreault M, Strutt D, Masin D, et al. Vascular normalization in orthotopic glioblastoma following intravenous treatment with lipid-based nanoparticulate formulations of irinotecan (Irinophore C), doxorubicin (Caelyx(R)) or vincristine. *BMC Cancer*. 2011;11:124.
87. Lau DH, Xue L, Young LJ, Burke PA, Cheung AT. Paclitaxel (Taxol): an inhibitor of angiogenesis in a highly vascularized transgenic breast cancer. *Cancer Biother Radiopharm*. 1999;14(1):31-36.
88. Straubinger RM, Arnold RD, Zhou R, Mazurchuk R, Slack JE. Antivascular and antitumor activities of liposome-associated drugs. *Anticancer Res*. 2004;24(2A):397-404.
89. Zhou R, Mazurchuk RV, Straubinger RM. Antivasculature effects of doxorubicin-containing liposomes in an intracranial rat brain tumor model. *Cancer Res*. 2002;62(9):2561-2566.
90. Zhou R, Mazurchuk RV, Tamburlin JH, Harrold JM, Mager DE, Straubinger RM. Differential pharmacodynamic effects of paclitaxel formulations in an intracranial rat brain tumor model. *J Pharmacol Exp Ther*. 2010;332(2):479-488.
91. Baker JH, Lam J, Kyle AH, et al. Irinophore C, a novel nanoformulation of irinotecan, alters tumor vascular function and enhances the distribution of 5-fluorouracil and doxorubicin. *Clin Cancer Res*. 2008;14(22):7260-7271.
92. Chen Q, Tong S, Dewhirst MW, Yuan F. Targeting tumor microvessels using doxorubicin encapsulated in a novel thermosensitive liposome. *Mol Cancer Ther*. 2004;3(10):1311-1317.
93. Murphy EA, Majeti BK, Barnes LA, et al. Nanoparticle-mediated drug delivery to tumor vasculature suppresses metastasis. *Proc Natl Acad Sci U S A*. 2008;105(27):9343-9348.
94. Ruoslahti E. Tumor penetrating peptides for improved drug delivery. *Adv Drug Deliv Rev*. 2017;110-111:3-12.
95. Teesalu T, Sugahara KN, Kotamraju VR, Ruoslahti E. C-end rule peptides mediate neuropilin-1-dependent cell, vascular, and tissue penetration. *Proc Natl Acad Sci U S A*. 2009;106(38):16157-16162.
96. Sugahara KN, Teesalu T, Karmali PP, et al. Tissue-penetrating delivery of compounds and nanoparticles into tumors. *Cancer Cell*. 2009;16(6):510-520.
97. Gu G, Gao X, Hu Q, et al. The influence of the penetrating peptide iRGD on the effect of paclitaxel-loaded MT1-AF7p-conjugated nanoparticles on glioma cells. *Biomaterials*. 2013;34(21):5138-5148.

98. Sugahara KN, Teesalu T, Karmali PP, et al. Coadministration of a tumor-penetrating peptide enhances the efficacy of cancer drugs. *Science*. 2010;328(5981):1031-1035.
99. Mantis C, Kandela I, Aird F. Reproducibility Project: Cancer B. Replication Study: Coadministration of a tumor-penetrating peptide enhances the efficacy of cancer drugs. *Elife*. 2017;6.
100. Abraxis Bioscience. Nanoparticle Albumin bound (nab) technology: A nanotechnology platform for biologically interactive drug delivery and targeting. FDA Oncologic Drugs Advisory Committee Meeting September 7, 2006. Accessed 7 August 2011.
101. Mills JH, Thompson LF, Mueller C, et al. CD73 is required for efficient entry of lymphocytes into the central nervous system during experimental autoimmune encephalomyelitis. *Proc Natl Acad Sci U S A*. 2008; 105(27):9325-9330.
102. Kim DG, Bynoe MS. A2A adenosine receptor modulates drug efflux transporter P-glycoprotein at the blood-brain barrier. *J Clin Invest*. 2016;126(5):1717-1733.
103. Carman AJ, Mills JH, Krenz A, Kim DG, Bynoe MS. Adenosine receptor signaling modulates permeability of the blood-brain barrier. *J Neurosci*. 2011;31(37):13272-13280.
104. Han L, Kong DK, Zheng MQ, et al. Increased nanoparticle delivery to brain tumors by autocatalytic priming for improved treatment and imaging. *ACS Nano*. 2016;10(4):4209-4218.
105. Jackson S, George RT, Lodge MA, et al. The effect of regadenoson on the integrity of the human blood-brain barrier, a pilot study. *J Neuro-Oncol*. 2017;132(3):513-519.
106. Dobrogowska DH, Vorbrodt AW. Immunogold localization of tight junctional proteins in normal and osmotically-affected rat blood-brain barrier. *J Mol Histol*. 2004;35(5):529-539.
107. Rapoport SI. Osmotic opening of the blood-brain barrier: principles, mechanism, and therapeutic applications. *Cell Mol Neurobiol*. 2000;20(2):217-230.
108. Kroll RA, Neuwelt EA. Outwitting the blood-brain barrier for therapeutic purposes: osmotic opening and other means. *Neurosurgery*. 1998;42(5):1083-1099. discussion 1099-1100.
109. Boockvar JA, Tsiouris AJ, Hofstetter CP, et al. Safety and maximum tolerated dose of superselective intraarterial cerebral infusion of bevacizumab after osmotic blood-brain barrier disruption for recurrent malignant glioma. Clinical article. *J Neurosurg*. 2011;114(3):624-632.
110. Deng CX. Targeted drug delivery across the blood-brain barrier using ultrasound technique. *Ther Deliv*. 2010; 1(6):819-848.
111. Sheikov N, McDannold N, Vykhodtseva N, Jolesz F, Hynynen K. Cellular mechanisms of the blood-brain barrier opening induced by ultrasound in presence of microbubbles. *Ultrasound Med Biol*. 2004;30(7):979-989.
112. Sheikov N, McDannold N, Sharma S, Hynynen K. Effect of focused ultrasound applied with an ultrasound contrast agent on the tight junctional integrity of the brain microvascular endothelium. *Ultrasound Med Biol*. 2008; 34(7):1093-1104.
113. Choi JJ, Wang S, Tung YS, Morrison 3rd B, Konofagou EE. Molecules of various pharmacologically-relevant sizes can cross the ultrasound-induced blood-brain barrier opening in vivo. *Ultrasound Med Biol*. 2010;36(1):58-67.
114. Kinoshita M, McDannold N, Jolesz FA, Hynynen K. Noninvasive localized delivery of Herceptin to the mouse brain by MRI-guided focused ultrasound-induced blood-brain barrier disruption. *Proc Natl Acad Sci U S A*. 2006;103(31):11719-11723.
115. Joshi S, Singh-Moon R, Wang M, et al. Cationic surface charge enhances early regional deposition of liposomes after intracarotid injection. *J Neuro-Oncol*. 2014;120(3):489-497.
116. Fan CH, Ting CY, Chang YC, Wei KC, Liu HL, Yeh CK. Drug-loaded bubbles with matched focused ultrasound excitation for concurrent blood-brain barrier opening and brain-tumor drug delivery. *Acta Biomater*. 2015;15:89-101.
117. Wong RS, Kapp LN, Krishnaswamy G, Dewey WC. Critical steps for induction of chromosomal aberrations in CHO cells heated in S phase. *Radiat Res*. 1993;133(1):52-59.
118. Dewhirst MW, Vujaskovic Z, Jones E, Thrall D. Re-setting the biologic rationale for thermal therapy. *Int J Hyperth*. 2005;21(8):779-790.
119. Xu Y, Choi J, Hylander B, et al. Fever-range whole body hyperthermia increases the number of perfused tumor blood vessels and therapeutic efficacy of liposomally encapsulated doxorubicin. *Int J Hyperth*. 2007;23(6):513-527.
120. Wu SK, Chiang CF, Hsu YH, et al. Short-time focused ultrasound hyperthermia enhances liposomal doxorubicin delivery and antitumor efficacy for brain metastasis of breast cancer. *Int J Nanomedicine*. 2014;9:4485-4494.
121. Banerjee R, Katsenovich Y, Lagos L, McIintosh M, Zhang X, Li CZ. Nanomedicine: magnetic nanoparticles and their biomedical applications. *Curr Med Chem*. 2010;17(27):3120-3141.

122. Silva AC, Oliveira TR, Mamani JB, et al. Application of hyperthermia induced by superparamagnetic iron oxide nanoparticles in glioma treatment. *Int J Nanomedicine*. 2011;6:591-603.
123. Jordan A, Scholz R, Maier-Hauff K, et al. The effect of thermotherapy using magnetic nanoparticles on rat malignant glioma. *J Neuro-Oncol*. 2006;78(1):7-14.
124. Maier-Hauff K, Rothe R, Scholz R, et al. Intracranial thermotherapy using magnetic nanoparticles combined with external beam radiotherapy: results of a feasibility study on patients with glioblastoma multiforme. *J Neuro-Oncol*. 2007;81(1):53-60.
125. Chertok B, David AE, Yang VC. Brain tumor targeting of magnetic nanoparticles for potential drug delivery: effect of administration route and magnetic field topography. *J Control Release*. 2011;155(3):393-399.
126. Kong G, Braun RD, Dewhirst MW. Hyperthermia enables tumor-specific nanoparticle delivery: effect of particle size. *Cancer Res*. 2000;60(16):4440-4445.
127. Zhang F, Xu CL, Liu CM. Drug delivery strategies to enhance the permeability of the blood-brain barrier for treatment of glioma. *Drug Des Devel Ther*. 2015;9:2089-2100.
128. Motion JP, Huynh GH, Szoka Jr FC, Siegel RA. Convection and retro-convection enhanced delivery: some theoretical considerations related to drug targeting. *Pharm Res*. 2011;28(3):472-479.
129. Stenehjem DD, Hartz AM, Bauer B, Anderson GW. Novel and emerging strategies in drug delivery for overcoming the blood-brain barrier. *Future Med Chem*. 2009;1(9):1623-1641.
130. Gaillard PJ, Appeldoorn CC, Dorland R, et al. Pharmacokinetics, brain delivery, and efficacy in brain tumor-bearing mice of glutathione pegylated liposomal doxorubicin (2B3-101). *PLoS ONE*. 2014;9(1).
131. Pieter J, Gaillard BMK, Aftimos P, et al. Abstract CT216: Phase I dose escalating study of 2B3-101, glutathione PEGylated liposomal doxorubicin, in patients with solid tumors and brain metastases or recurrent malignant glioma. In: *Paper presented at: Annual meeting of American Association of Cancer Research*; 2014.
132. Veiseh O, Kievit FM, Fang C, et al. Chlorotoxin bound magnetic nanovector tailored for cancer cell targeting, imaging, and siRNA delivery. *Biomaterials*. 2010;31(31):8032-8042.
133. Tortorella S, Karagiannis TC. The significance of transferrin receptors in oncology: the development of functional nano-based drug delivery systems. *Curr Drug Deliv*. 2014;11(4):427-443.
134. Clark AJ, Davis ME. Increased brain uptake of targeted nanoparticles by adding an acid-cleavable linkage between transferrin and the nanoparticle core. *Proc Natl Acad Sci U S A*. 2015;112(40):12486-12491.
135. Kim SS, Rait A, Kim E, et al. A nanoparticle carrying the p53 gene targets tumors including cancer stem cells, sensitizes glioblastoma to chemotherapy and improves survival. *ACS Nano*. 2014;8(6):5494-5514.
136. Gao H. Perspectives on dual targeting delivery systems for brain tumors. *J NeuroImmune Pharmacol*. 2017; 12(1):6-16.
137. Greene MK, Richards DA, Nogueira JCF, et al. Forming next-generation antibody-nanoparticle conjugates through the oriented installation of non-engineered antibody fragments. *Chem Sci*. 2018;9(1):79-87.
138. Ait-Oudhia S, Mager DE, Straubinger RM. Application of pharmacokinetic and pharmacodynamic analysis to the development of liposomal formulations for oncology. *Pharmaceutics*. 2014;6(1):137-174.
139. Ozdemir-Kaynak E, Qutub AA, Yesil-Celiktas O. Advances in glioblastoma multiforme treatment: new models for nanoparticle therapy. *Front Physiol*. 2018;9:170.
140. Mager DE, Mody V, Xu C, et al. Physiologically based pharmacokinetic model for composite nanodevices: effect of charge and size on in vivo disposition. *Pharm Res*. 2012;29(9):2534-2542.
141. Hendriks BS, Reynolds JG, Klinz SG, et al. Multiscale kinetic modeling of liposomal Doxorubicin delivery quantifies the role of tumor and drug-specific parameters in local delivery to tumors. *CPT Pharmacometrics Syst Pharmacol*. 2012;1.
142. Ait-Oudhia S, Straubinger RM, Mager DE. Systems pharmacological analysis of paclitaxel-mediated tumor priming that enhances nanocarrier deposition and efficacy. *J Pharmacol Exp Ther*. 2013;344(1):103-112.
143. Ait-Oudhia S, Straubinger RM, Mager DE. Meta-analysis of nanoparticulate paclitaxel delivery system pharmacokinetics and model prediction of associated neutropenia. *Pharm Res*. 2012;29(10):2833-2844.
144. Benias PC, Wells RG, Sackey-Aboagye B, et al. Structure and distribution of an unrecognized interstitium in human tissues. *Sci Rep*. 2018;8(1):4947.

CHAPTER

13

Solute Transport in the Cerebrospinal Fluid: Physiology and Practical Implications

Mikhail Papisov, Vasily Belov

Massachusetts General Hospital, Harvard Medical School, and Shriners Hospitals for Children, Boston, MA, United States

PHYSIOLOGICAL MECHANISMS OF DRUG DELIVERY TO THE CENTRAL NERVOUS SYSTEM VIA CEREBROSPINAL FLUID AND PERIVASCULAR CHANNELS

Outside of pain and spasticity management fields, the subarachnoid (i.e., intrathecal) route for delivery of therapeutics to the central nervous system (CNS) has not yet been systematically explored. Preclinical and clinical studies that test this route are scarce, but are attracting interest in light of the increasing need to deliver macromolecular therapeutics to CNS targets—specifically, biopharmaceuticals, which cannot enter the CNS through the blood–brain barrier from the systemic circulation. In animal models, the intrathecal drug delivery route has shown promising results (i.e., detectable biological effects) for large molecules, such as enzyme-replacement therapeutics, antibodies, nerve growth factor, Sonic Hedgehog, siRNA, and dynorphins.[1-16]

The ability of large molecules to penetrate the brain parenchyma along perivascular channels from cerebrospinal fluid (CSF) was established decades ago.[17] However, the intrathecal modality of therapy administration to the CNS is still in its infancy, due to several uncertainties surrounding the processes that transport solutes in the CSF. These ambiguities relate to both the efficacy and safety of intrathecal therapy administration, and the main roadblock to preclinical and clinical study plan development is the lack of rational concepts for method optimization. Complete mechanistic descriptions of each step of solute transport from the

https://doi.org/10.1016/B978-0-12-813997-4.00013-X
© 2019 Elsevier Inc. All rights reserved.

administration site to the target (and all concurrent processes) are needed to identify the optimal administration site, administration volume, injection/infusion rate, and other factors.

The current clinical paradigm of leptomeningeal (i.e., intrathecal) drug administration is rooted in the work of J. Corning (1885) and his pioneering studies of spinal anesthesia with cocaine[18] and his subsequent research in anesthesiology. Corning made great strides with his investigation of the risk of respiratory distress when an anesthetic is administered at a higher location than the lower thoracic region.[19] Significant developments were based on the realization that hydrophobic anesthetics introduced in the leptomeningeal space induced rapid segmental anesthesia, whereas hydrophilic anesthetics induced a more gradual onset of anesthesia, with actions extending from the administration site.[20, 21]

Published observations generally fit into a mechanistic scheme in which the drug administered into the spinal CSF acted locally, contained by its association with the lipids of the arachnoid and, presumably, by a hypothetical caudad CSF flow. The cephalad (or upward) spread of an anesthetic agent from an injection site close to the head was considered risky because of the proximity to the medullary vasomotor centers.[22] Coincidentally, due to the anatomy of the skull and of the vertebrae encasing the compartment, lumbar administration is the most clinically feasible and least invasive way to introduce a drug to the leptomeningeal space. As a result, intrathecal anesthetic delivery was focused on slow infusion in the lumbar region. Implantable, slow-infusion devices have been developed; these devices are capable of chronic (over a period of years), uninterrupted delivery of pain or spasticity management therapies, with the action focused mostly on the distal segments of the spinal cord.[23, 24] Solute transport in the cerebral, cervical, and proximal spinal CSF was not actively studied, due in part to the perceived dependence of CSF drug transport on CSF "circulation"[25] and other expected limitations.[26]

The "flow" of the CSF from the ventricles (where most of it is produced by the choroid plexus)[27, 28] toward the drainage sites in the arachnoid granulations and hypothetical perineural drainage sites in the spine (Fig. 1), was considered the primary barrier preventing drug entrance to the CNS. One recent paper describing the results of early clinical studies states: "...[Intrathecally administered brain derived neurotrophic factor] can be delivered cranially against CSF flow."[29] Some experimental data seemed to confirm the existence of such flow.[30] However, such considerations do not take into account the data on the pulsatile movement of the CSF induced by the cardiac waves that propagate along the blood vessels in the subarachnoid (leptomeningeal) space. In fact, the pulsatile movement of CSF has a much more significant effect on solute spread in the liquid phase than the CSF flux from the ventricles to the drainage sites.

Finally, the leptomeningeal therapy transport route was often considered inefficient a priori because of the prohibitively slow diffusion of the therapeutic entities into the CNS parenchyma from the CSF, which, considering the known rapid clearance of the (low-molecular–weight) experimental therapeutics from the CSF to the systemic circulation, gave the impression that "...drug injection into the CSF is a slow intravenous infusion."[31] However, solute clearance from the CSF depends on the hydrodynamic size of the solute, enabling macromolecules to stay in the CSF long enough to be translocated into the parenchyma along the perivascular liquid conduits (Virchow-Robin spaces) around large vessels, and thinner spaces along the smaller branching arteries, first described by Wagner et al.[17] According to our most recent data, the perivascular transport system covers the entire CNS, providing a potentially robust terminal transport route for intrathecal therapy delivery.

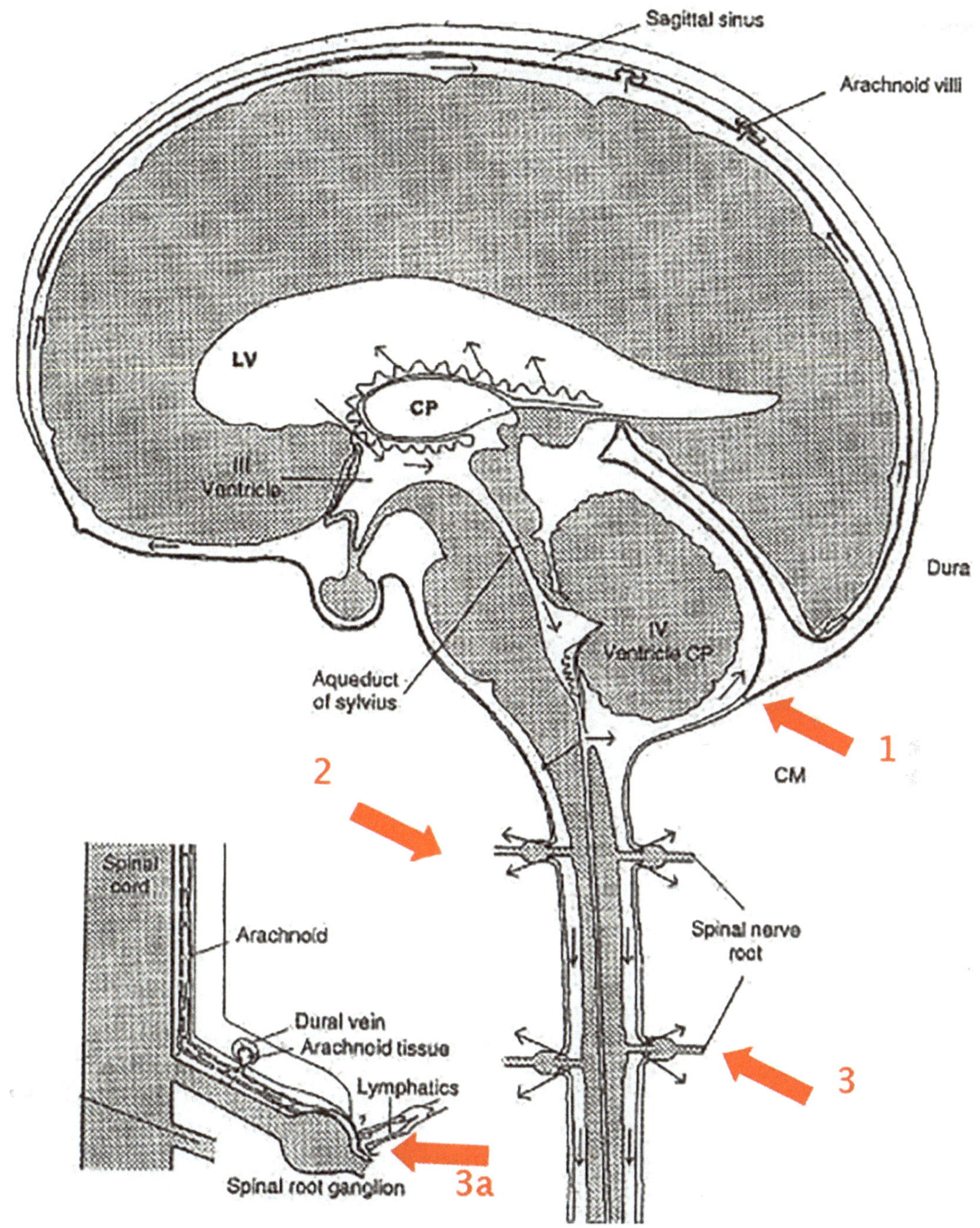

FIG. 1 Illustration depicting the circulation and drainage of cerebrospinal fluid (CSF). The *red arrows* demonstrate examples of CSF flows, the significance of which can be misinterpreted in terms of drug delivery. *Reproduced with permission from Segal MB. The anatomy and physiology of intracranial fluids. In: Reid A, Marshbanks R, Ernst A, eds.* Intracranial and Inner Ear Physiology and Pathophysiology. *London: Whurr Publishers Ltd., Wiley; 1998.*

In this chapter, we provide key data on the characteristics and dynamics of the observed components and stages of solute transport in the leptomeningeal space that, as we expect, will be essential for further research and instrumental in designing preclinical and clinical studies of macromolecular and supramolecular therapeutics. These data include experimental results obtained in hypothesis-driven experiments, as well as excerpts from multiple preclinical studies of prospective therapeutics in rats and nonhuman primates conducted in our laboratory.

METHODS

Investigating physiological transport in a liquid-filled space such as the CSF requires minimally invasive methods that preserve the integrity of the boundaries and components of the space. Some of the known controversies in the interpretation of leptomeningeal transport are due to the intrinsic difficulties of performing this type of investigation; they appear to originate from the lack of real-time data and overreliance on experimental results obtained in studies affecting the physiological processes. (There is also lack of consistent comparative data on the interspecies differences in the physiology of cerebrospinal solute transport.)

In order to minimize the risk of affecting any elements of the leptomeningeal transport system and to prevent artifacts, we used animals with preinstalled and healed intrathecal ports whenever possible. This allowed us to ensure that transdural CSF leakages would not occur (confirmed on imaging). We also investigated the impact of intrathecal liquid injection and infusion on CSF pressure to minimize the risk of pressure-induced distortions in the physiological data. All animal studies were carried out in accordance with institutionally approved protocols.

The studies were carried out on two different scales. First, at a whole-body or whole-organ scale, by quantitative real-time positron emission tomography (PET) with radiolabeled macromolecular tracers. Second, at microscale, by fluorescence microscopy with macromolecular tracers labeled (or co-labeled) with fluorophores.

Research Animals

All studies were carried out in accordance with institutionally approved protocols. Intrathecal administration of macromolecules and tracers in nonhuman primates (*Macaca fascicularis* and *Macaca mulatta*) was carried out either through the atlanto-occipital membrane into the cisterna magna, or through a preinstalled subcutaneous injection port (P.A.S. PORT Elite, Smiths Medical) connected to polyurethane kink-resistant catheters (Smiths Medical). Preinstalled ports entered the subarachnoid space at the L4/L5 spinal segment and advanced to the L1/T12 area (Fig. 2). In two animals, catheters led to the cerebral ventricles.

In rats, most injections were carried out through the atlanto-occipital membrane into the cisterna magna, following the technique developed by Farris and Griffith.[32] One group of animals had catheters placed into the cisterna magna by the vendor (Charles River Laboratories). Animals intended for CSF pressure monitoring were equipped with a temporary catheter as follows. Animals were anesthetized by inhalation of the isoflurane/air mixture

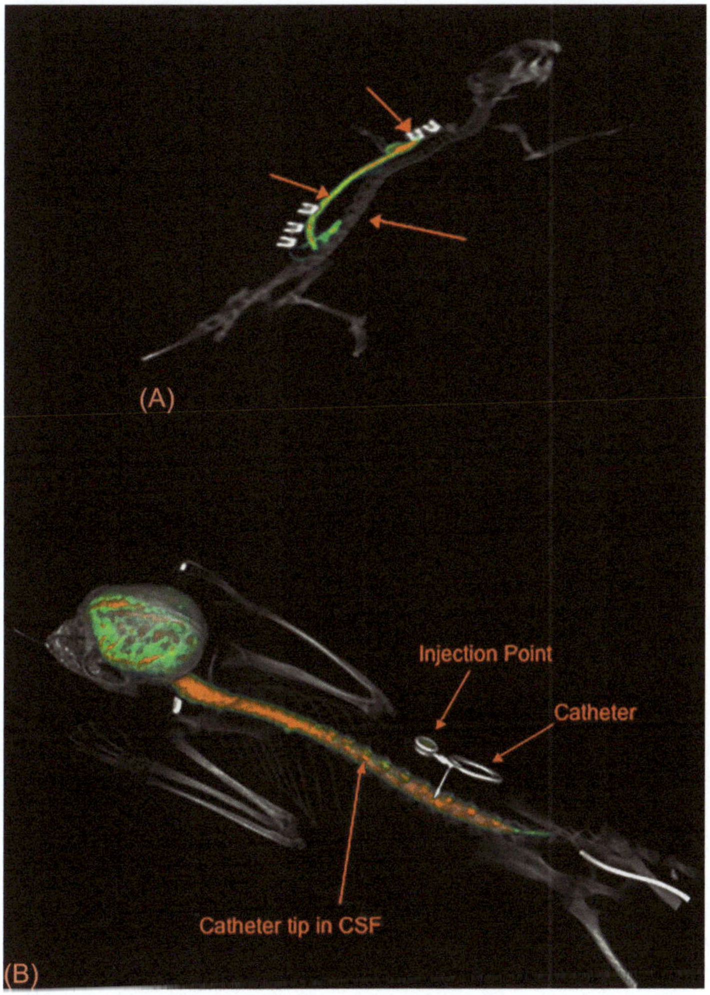

FIG. 2 Positron emission tomography and computed tomography images of the lumbar porting system used to inject (A) Sprague Dawley rats and (B) nonhuman primates (shown: cynomolgus macaque).

(3% for induction, 2% for maintenance) given at a 300 mL/minutes flow rate using a SomnoSuite small animal anesthesia platform (Kent Scientific). The animal was then mounted on a stand set at a 50° angle to horizontal with the head tilted down, thereby positioning the occipital bone in the horizontal plane. The back of the neck and the base of the skull were shaved and disinfected with 70% ethyl alcohol. The location of the cisterna magna was identified by palpation of a 3 mm^2 rhomboid depressed area (the atlanto-occipital joint) at the intersection of the line connecting the ear bases with the midsagittal plane. A 24G 0.75″ intravenous catheter (Angiocath, Becton Dickinson) was then slowly inserted to a depth of 7 mm. The needle guide was immediately removed after the appearance of clear CSF in the outer catheter. The Luer adapter of the catheter was promptly capped, and the catheter was affixed to the skin with cyanoacrylate surgical gel. Localization of the tip of the radiopaque catheter in the cisterna magna was confirmed at the end of the study by computed tomography (CT).

After catheterization, a Tuohy-Borst adapter (Cook Medical) equipped with a side arm (initially closed) and prefilled with saline was connected to the intravenous catheter via a Luer adapter. A micro-tip pressure transducer was then slowly advanced into the cisterna magna through the entire length of the Tuohy-Borst adapter and the intravenous catheter. The adaptor was tightened up, and baseline intracranial pressure referenced to the atmospheric pressure (preset to 0 mm Hg) was recorded for 5 to 15 minutes. Animals then underwent either PET or fluorescence microscopy with in vivo labeling of perivascular cells, both of which are described in greater detail below.

Positron Emission Tomography

To observe solute transport in real-time at the organism and organ scale, we used PET/CT with long-living positron emitting labels, ^{124}I and ^{89}Zr. PET is a well-established imaging modality that, unlike other imaging modalities, offers quantitative data reconstruction with correction for gamma photon attenuation (i.e., absorption) in the tissues and other corrections.[33] The quantitative nature of PET relies upon detection of the pair of photons emitted by an annihilating positron along the same line but in opposite directions. Detection of the pair (rather than of a single photon, as in other radionuclide-based methods), enables acquisition of a three-dimensional (3D) map of label concentration in the tissues; a set of such maps can be compiled into a "movie" or used to produce same-animal kinetic curves from any region of the image.

For attenuation correction, PET requires (in addition to the main dataset [emission image]) a second set of data known as the transmission image, which can be obtained using a source of either gamma photons or an X-ray source external to the animal. Absorption of the radiation of the source is measured by a respective detector, and the data are used to correct the emission image for attenuation. When an X-ray CT is used for acquisition of the transmission image, the dataset also contains a 3D X-ray image that provides a sound anatomical reference for interpretation of PET data.

For labeling of macromolecular tracers with Iodine-124, the solutions of macromolecules, 10-25 mg/mL in 0.25 M sodium phosphate buffer, pH = 7.5 (0.1 mL), were incubated in Pierce Iodogen iodinating tubes (ThermoFisher Scientific) with 0.1-3.0 mCi (0.01-0.3 mL) of [^{124}I]NaI solution containing 10^{-3}M NaOH received from IBA Molecular (Sofie) for 40 minutes. The product was desalted on Sephadex G-25 (PD-10 column, GE Healthcare Life Sciences), equilibrated with sterile saline, and analyzed by HPLC (BioRad BioSil 125 column) with UV and gamma detection. Radiochemical yield was 60%-80%, depending on the macromolecule. Radioactivity was measured on an Atomlab 100 dose calibrator (Biodex Medical Systems).

For labeling with Zirconium-89, macromolecules were first conjugated with 1-(4-isothiocyanatophenyl) deferoxamine (Macrocyclics) in 0.1 M sodium carbonate buffer, pH = 8.5, at 10 and 0.1 mg/mL, respectively. After an overnight incubation at ambient temperature, the reaction mixture was spun at 9000 g for 10 minutes, and the supernatant was desalted and analyzed as described earlier. [^{89}Zr]Zr oxalate in 1 M oxalic acid from IBA Molecular was neutralized with 2 M Na_2CO_3 and then reacted with the deferoxamine-macromolecule conjugate in 0.2 M HEPES buffer, pH = 7.0, at ambient temperature for 25 minutes. The products were desalted on PD-10 columns equilibrated with 0.9% NaCl (radiochemical yield: 70%-90%).

The radiolabeled macromolecules were administered to rats and nonhuman primates at doses from 0.1 to 3.0 mg per kg; label content per dose 0.05-0.5 mCi and 0.2-1 mCi, respectively. Imaging studies were carried out using either microPET P4 scanner (rats with I2S, hGalC, ARSA, HNS and nonhuman primates with hGalC) or a custom PET/CT imaging system consisting of a microPET Focus 220 PET scanner and a CereTom NL 3000 CT (all other studies). In the latter method, imagers were aligned and equipped with a custom imaging bed extending through both imagers along the alignment axis, ensuring reliable PET/CT image registration.[34]

Imaging on microPET (Concorde Microsystems/Siemens) was carried out with transmission image acquisition using a point source containing 185 Mbq of ^{57}Co (Eckert & Ziegler Isotope Products). Imaging on the PET/CT system was carried out with CT image acquisition on CereTom NL 3000 CT scanner (Neurologica). The transmission and CT images were used for anatomical reference as well as for attenuation and scatter correction of PET images.

Both the microPET P4 and Focus 220 work in 3D mode and have axial fields of view of 7.6 cm and transaxial fields of view of 19 cm. The scanner's detection systems enabled 2.5 mm (P4) and 2.1 mm (Focus 220) spatial resolution for ^{124}I, and 1.7 mm for ^{89}Zr (Focus 220). The energy windows were set to 350-650 keV for all studies, and the coincidence timing windows were set to 6 ns.[33] CereTom NL 3000 is a six-slice tomograph with high contrast resolution of 0.8 mm. The image acquisition settings were tube voltage, 100 kV; tube current, 5 mA; resolution, 6 s/projection. For both imagers, axial slice thickness was set to 1.25 mm. Fiducial markers (Eckert & Ziegler) were used for PET/CT image co-registration, which was carried out manually using ASIProVM software (Siemens/CTI Concorde Microsystems).

PET data were acquired using microPET software (Siemens Medical Solutions, Inc.). CT image acquisition was carried out using Neurologica software. Imaging procedures and data quality control are described in more detail in our previous publications.[33, 35]

Fluorescence Microscopy With in Vivo Labeling of Perivascular Cells

Fluorescent macromolecules were administered to rats and nonhuman primates at doses ranging from 0.5 to 1 mg/kg. Animals were euthanized 24 hours after the administration. Macromolecules were labeled with activated fluorescent dyes (Alexa series, Texas Red, and Fluorescein isothiocyanate) from Thermo/Molecular probes. The labeling procedures were carried out generally in accordance with the manufacturer's recommendations, with optimization of the protein:dye ratio for maximal fluorescence intensity per protein mass. Fluorescence was measured on a photon-counting spectrometer (PTI Quantamaster, Photon Technology International/Horiba Scientific) for a 10-μL aliquot of the test solution in 3 mL of 0.15 M NaCl buffered with 50 mM sodium phosphate buffer solution, pH = 7.4.

Tissue samples were collected 24 hours after administration of the fluorescent macromolecules, when PET data showed that the material had cleared from the CSF. Tissues were frozen unfixed on an aluminum plate cooled to −78°C and incubated at −21°C overnight for temperature equilibration. Cryosections of the CNS, 10-100 μm, were prepared on a Leica CM3050 S cryostat, placed on microscopy slides, dried in ambient air, and covered with OCT mounting medium.

The slides were imaged on a Nikon Eclipse epifluorescence TE300 microscope equipped with long working distance objectives (CFI Plan Fluor ELWD 20X and 60X, Plan Fluor 4X, and Plan Achromat 1X). Narrow-band filter cubes (high throughput filter sets # 49000, 49009, 49011, Chroma Technology) and Nikon DS-Qi2 photon counting camera were used for image acquisition. The background of the camera in the absence of light was 395 ± 5 photons $\cdot s^{-1}$ per well. Fluorescence background in all three channels was registered daily using a covered microscopy slide with a layer of OCT. The three fluorescence channels were found to be well isolated from each other, with no detectable FITC cross talk to the blue and red channels, no detectable Alexa 350 cross talk to the green and red channels, no detectable Texas Red cross talk to the green channel, and $0.16 \pm 0.21\%$ Texas Red cross talk to the blue channel.

The images of the CNS cryosections were acquired in all three fluorescence channels, in phase contrast (X20 and X60) and in bright field with crossed polarizers (X4). The images were processed using Nikon Elements AR 4.30.02 software. For each studied CNS cryosection, a series of 4× or 20× frames completely covering the slice was acquired and merged in one image. 20× and 60× images were then acquired from multiple regions.

Sections obtained from the areas corresponding to plates # 3, 25, 51, 87, 115, and 161 of the Paxinos atlas[36] were used for counting the perivascular channel entering the parenchyma in rats. In monkeys, the channel counting was performed in sections made through the middle of the brain, approximately corresponding to slice 3 in Fig. 3 published in Calabrese et al.[37] Identification of sulci was in accordance with Von Bonin and Bailey.[38] Images were used for analysis without further processing.

Images intended for deconvolution were acquired as 3D stacks (2 μm per subslice) using Nikon Plan Apochromate λ 60× Oil objective and processed using Nikon EDF or 3D deconvolution algorithms. For presentation, images were exported to monochrome TIFF format. Monochrome TIFF images were pasted into the respective layers of RGB templates to obtain mono- or multicolored images. In all images and individual RGB channels, the LUTs are linear and cover the full range of the data.

Monitoring CSF Pressure

Cerebrospinal spinal fluid pressure in rats was measured directly in the cisterna magna pool of the CSF using a 1F piezoresistive diffused semiconductor pressure sensor mounted on the tip of a 20-cm-long 0.8 F polyimide line (Millar, Inc.). The line was inserted through a temporary catheter installed through the atlanto-occipital membrane. In monkeys, port configuration did not allow sensor insertion into the leptomeningeal space. The pressure was measured in the catheter line connected to the subcutaneous intrathecal port.

The sensor signal was processed using a FE221 Bridge Amplifier (ADInstruments) coupled with a PowerLab 4/35 data acquisition platform (ADInstruments). Control of both modules, signal acquisition, and the subsequent CSF pressure waveform analysis were performed using LabChart 8 software (ADInstruments). Sampling rate was set to 1000 measurements per second. Prior to all CSF pressure measurements, the pressure transducer was soaked in sterile saline at ambient temperature for at least 30 minutes. After that, atmospheric pressure was set to 0 mm Hg and the stability of the sensor was ensured by recording the baseline for at least 10 minutes.

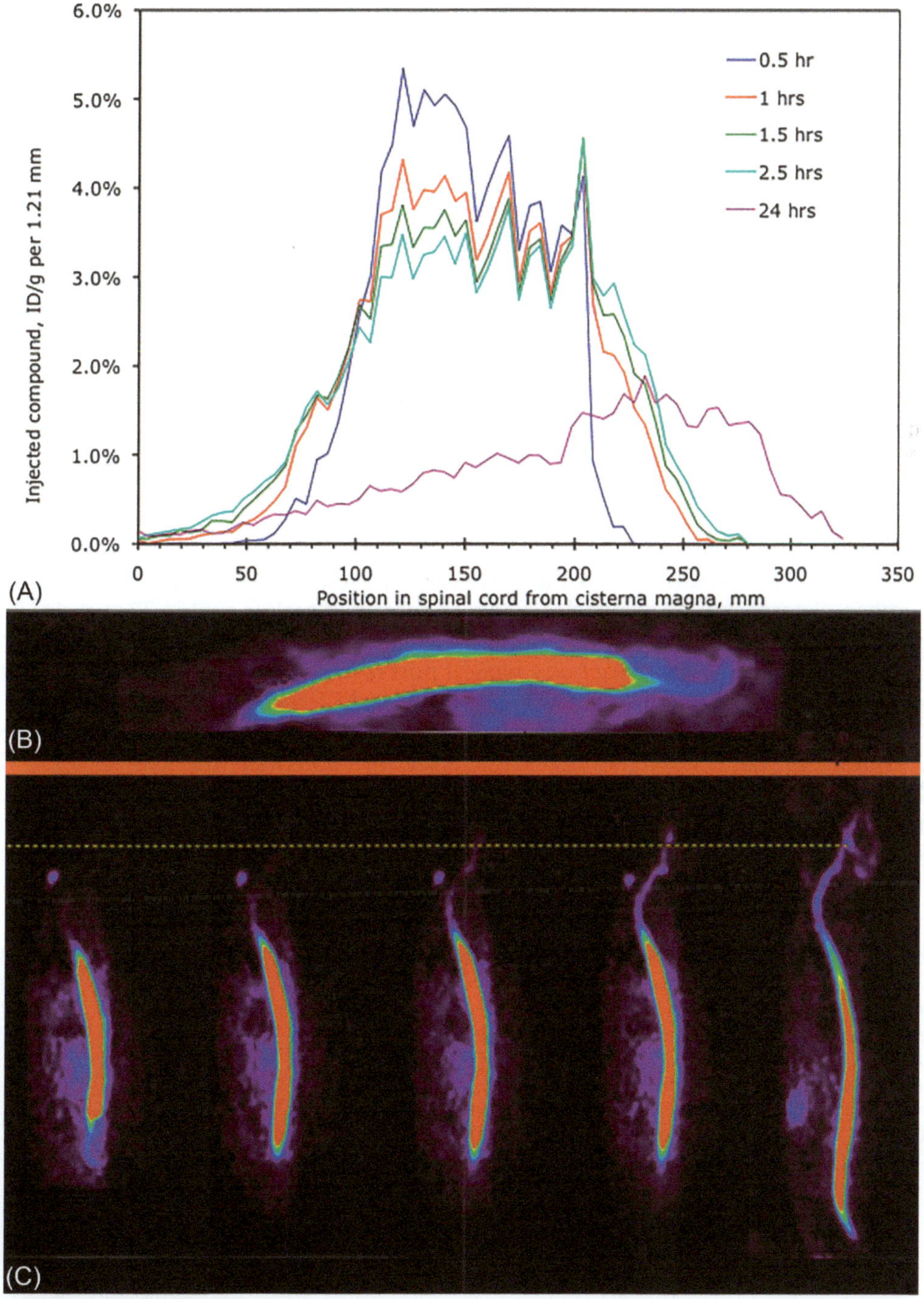

FIG. 3 Spread of ^{124}I-labeled iduronate sulfatase in the spinal cerebrospinal fluid, small-volume administration at L1-L2. (A) Protein content per 1.2-mm image slice along the spine. (B) Segment of the spine (positron emission tomogram) aligned with the axis of the graph shown in (A). (C) Time sequence of positron emission tomography images at 0.5, 1.0, 1.5, 2.5, and 24 hours.

KEY RESULTS AND DISCUSSION

The Initial Translocation of the Intrathecal Bolus

Early in our nonhuman primate studies, we found that the immediate disposition of a solute administered in an intrathecal bolus depends on both the administration site and the administered volume. The established facts are as follows.

First, tracer administered in a small volume bolus (<0.5 mL/kg body weight) into the distal spinal CSF occupies a segment of the spinal CSF pool; the solute slowly spreads in both directions. The solute front translocation rate is 2-6 mm/hour (Fig. 3), which is at least three orders of magnitude faster than diffusional spread would be. However, there is no evidence of directional movement of the solute-labeled segment of the CSF. Overall, this solute transport pattern suggests absence of directional CSF flows in the spine (below T2-T3), and pulsation-facilitated "pseudo-diffusion" (i.e., Taylor dispersion[39, 40]) along the spinal CSF pool.

Second, tracer administered to the cerebral or cervical CSF pool (e.g., ventricle, cisterna magna) in any volume spreads over the entire cerebro-cervical CSF pool within 30 minutes, but not to the spinal CSF (Fig. 4B). In rats, unlike in monkeys, administration of a large-volume bolus into cisternal CSF (in the cisterna magna) resulted in immediate solute translocation to the spinal CSF, including distal segments.

Third, tracer administered in a large-volume lumbar intrathecal bolus (>0.5 mL/kg body weight) rapidly translocates toward the basal cisterns, regardless of the administration site (e.g., lumbar, cisterna magna, ventricle). Then, as described earlier, within 30 minutes the solute spreads over the entire cerebro-cervical CSF (Fig. 4A and 5). Administration of the tracer in 0.5 mL followed by a 0.5 mL/kg body weight port flush with saline consistently results in the delivery of >90% of the dose to the cerebro-cervical CSF (Fig. 6).

The safety of a large-volume intrathecal bolus was positively established for both humans and nonhuman primates; the data suggest that bolus injection of up to at least 33% and 42% of total CSF volume is well tolerated by humans and monkeys, respectively.[41] In our studies, rats tolerated equally large bolus injections. The physiological processes and anatomical elements responsible for acceptance of large additional volumes of liquid by a compartment encased in solid walls (e.g., bones, vertebral discs) are not fully understood.

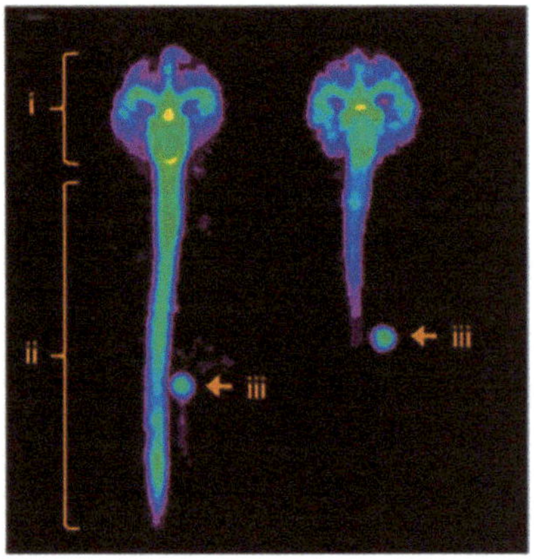

FIG. 4 Disposition of ^{124}I-labeled iduronate sulfatase in the cerebrospinal fluid (CSF) 30 minutes after administration. Left: Lumbar large-volume bolus at L1-L2. Right: Intraventricular small-volume bolus. (i) cerebral CSF; (ii) spinal CSF, and (iii), subcutaneous injection port. *Reproduced with permission from Calias P, Papisov M, Pan J, et al. CNS penetration of intrathecal-lumbar idursulfase in the monkey, dog, and mouse: implications for neurological outcomes of lysosomal storage disorder.* PLoS One. *2012;7(1):e30341.*

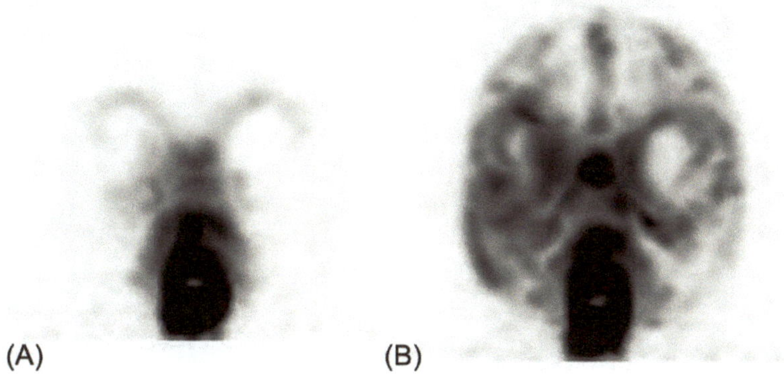

FIG. 5 ^{89}Zr-labeled α-N-acetylglucosaminidase, (A) immediately and (B) 30 minutes after a high-volume lumbar intrathecal bolus administered over 2 minutes. Projection positron emission tomogram (head, coronal), inverted color scale.

Acceptance of Large Additional Volume and Hydrostatic Compliance

We hypothesized that the hydrostatically compliant (i.e., physically compressible or expandable under pressure) anatomical elements may include large cerebral veins, cervical membranes, elements of the inner ear, and nerve roots. Compression of blood vessels could be considered a risk factor due to the potential attenuation or disruption of the blood supply to the CNS (Figs. 5 and 6).

To determine whether a high-volume intrathecal bolus compresses the cerebral veins or other blood vessels, we labeled the blood pool of nonhuman primates with [^{89}Zr] albumin and then used PET to determine the amount of blood contained by the major venous sinuses, and the total amount of blood in the head before and after intrathecal administration of nonradioactive saline (1 mL + 0.5 mL/kg body weight) in *Macaca mulatta* and *Macaca fascicularis*. Transient vasculature compression (10%-15% of the venous blood volume for 1-3 minutes) was detected in the regions drawn closely around the superior and inferior sagittal and transverse venous sinuses. However, in the regions drawn around the entire brain or other subcompartments of the CNS, no blood volume changes exceeding normal baseline fluctuations were detected.[42] The pressure in the intrathecal catheter was elevated during the bolus delivery up to 81-141 mm Hg (largely due to the hydrostatic resistance of the catheter), and returned to 30-50 mm Hg when the bolus delivery was stopped. Subsequently, the pressure rapidly ($t_{1/2} = 55 \pm 20$ s) returned to normal values (8-10 mm Hg). CSF pressure relaxation after a high-volume intrathecal bolus was even faster in rats (Fig. 7).

Therefore, blood vessels are not compressed significantly enough during the high-volume intrathecal bolus delivery to be responsible for the hydrostatic compliance of the compartment. The rapid pressure relaxation after the bolus suggest that a concerted action of other anatomical elements and possibly increased drainage of CSF at the elevated pressure during the bolus delivery may be responsible for both acceptance of the extra fluid to the compartment and translocation of the administered solute toward the head.

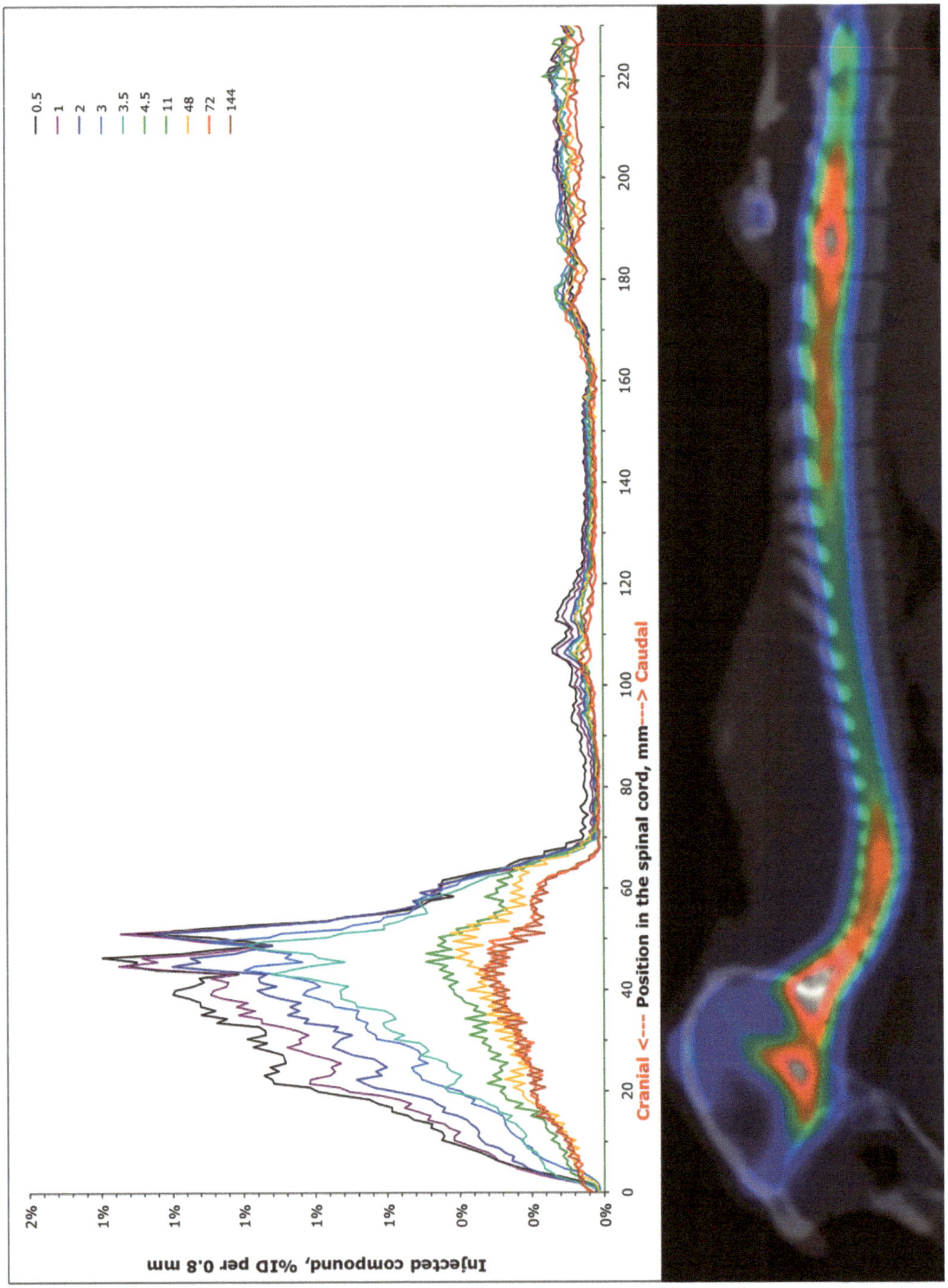

FIG. 6 Concentration profile of ^{89}Zr-labeled α-N-acetylglucosaminidase in the cerebrospinal fluid after a high-volume bolus administration at L1-L2. Top: Protein content per 0.8-mm image slice. Bottom: Positron emission tomogram aligned with the axis of the graph derived from the positron emission tomography data (top image).

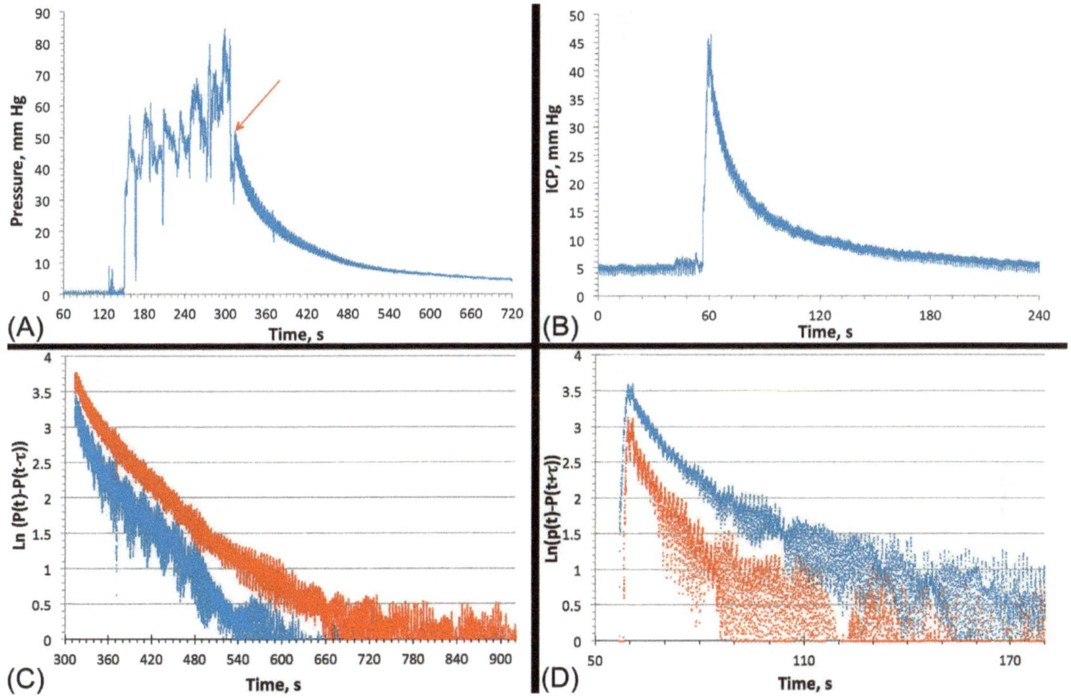

FIG. 7 Cerebrospinal fluid pressure during and after intrathecal bolus administration. A 5-kg monkey, 5 mL at 1.5 mL/minute, (A) pressure graph and (C) Guggenheim linearization.[43] A 0.3-kg rat, 0.1 mL at 1.9 mL/minute, (B) pressure graph and (D) Guggenheim linearization. Frame widths (τ) for Guggenheim linearizations are 5.5 and 8.0 minutes (monkey) and 60 and 10 seconds (rat). The effective initial pressure relaxation half-lives in these animals are 1 ± 0.05 minutes (monkey) and 9.8 ± 0.3 seconds (rat).

We have found that the first sign of an excessively fast bolus delivery is significant, progressive reduction of the respiration rate, followed by elevation of the heart rate. The results of pressure monitoring in the intrathecal line in monkeys (with correction for the hydrodynamic resistance of the catheter/port) and in the cisternal CSF in rats during the routine manual and pump-controlled intrathecal bolus delivery suggest that a short-term increase in CSF pressure (up to 45 mm Hg) is safe in these animals. Changes in the respiratory and heart rates appeared when this threshold was exceeded (and in some individually sensitive animals at lower pressures). However, both respiratory and heart rates returned to baseline within about a minute of delivery cessation. This correlates with the rapid pressure relaxation after completion (or interruption) of the bolus delivery (Fig. 7).

Overall, the direction of the action of the hydrostatically compliant elements (and, possibly, CSF drainage) enables selective initial delivery of the dose to the cisternal CSF pool or to any segment of the spinal CSF pool. The subsequent spread of the solute in the CSF is controlled by hydrodynamic factors. We note that intraventricular administration of solutes immediately (i.e., during the bolus delivery) resulted in solute translocation to the cisternal CSF, and the initial solute disposition map for intracerebroventricular and large-volume lumbar intrathecal bolus in monkeys differed only in the distal spine, which is not covered by the intraventricular administration route (Fig. 4).

Solute Spread in the CSF: Hydrodynamic Factors

The real-time PET imaging data on solute translocations within the CSF-filled compartment from the initial disposition segment revealed several findings. Solute transport in the compartment is not governed or affected by CSF flux from the choroid plexus toward arachnoid granulations or other drainage sites. Furthermore, solute spread in the CSF occurs too quickly to be explained by diffusion, by at least three orders of magnitude in the spinal pool and even faster in the cerebro-cervical pool. In addition, solute spread is nondirectional. In the spinal compartment, the solute spreads from the initial segment of disposition in both directions, cranially and caudally, at an equal rate (front movement rate: 1-6 mm/hour, Fig. 3). In the cerebro-cervical CSF, solute spreads into all CSF-filled spaces at a rate that enables complete equilibration of concentrations within 20 minutes (Figs. 5 and 8). This characteristic of transport can be explained only by the mixing action of the pulsation wave induced in the CSF by pulsation of arteries and perhaps of the entire brain.

Noninvasive flow measurement with phase-contrast magnetic resonance imaging enabled in vivo evaluation of the basic parameters of CSF movement.[44-47] Computational fluid dynamics models have been used to complement magnetic resonance imaging measurements to explain the complex fluid-flow patterns in CSF-filled spaces.[45, 48, 49] Generally, CSF flow waveform reflects that of the arterial cardiac cycle and has its own systole (spinal column

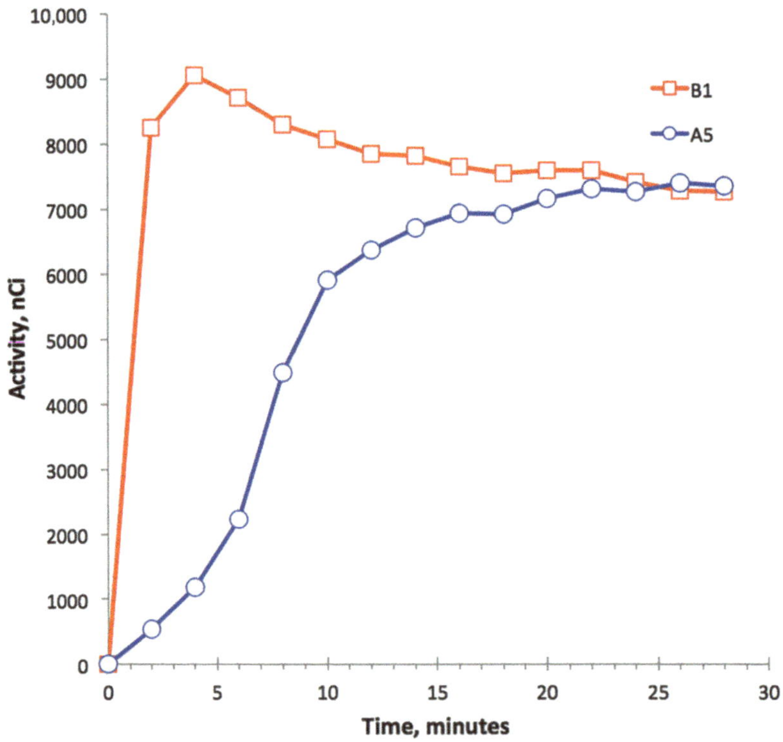

FIG. 8 The kinetics of radioactivity concentration in the basal cistern (B1), to which the solute is delivered immediately from the lumbar injection site, and over the central fissure (A5). Note that the curves merge by the 25th minute.

inflow) and diastole (outflow). Peak pulsatile flow rates measured in humans at the cervical level by different authors range from 120 to 360 mL/min.[45, 48, 50] Variations in the velocity are significant along the spinal canal, ranging from 0.5 to 8 cm/s, with the maximum at about 20 cm below the skull.[45] Computational fluid dynamics studies show that inertial effects should dominate the flow field, causing significant turbulence, especially in the cervical leptomeningeal space.[45] The pulsatile movements of CSF in the cervical and ventricular areas are sometimes interpreted as "pumps," driving a directional flow of CSF.[51] The real-time PET imaging data suggest a completely nondirectional mechanism of the Taylor dispersion type.[39, 40]

Solute Clearance from the CSF: Clearance Routes and Solute Molecule/Particle Size Dependence

Cerebrospinal fluid drainage is a process that removes the dissolved drug from the CSF, which reduces the efficacy of drug delivery to the entrances of the perivascular channels that carry macromolecules into the parenchyma. CSF is produced by the choroid plexus in the brain ventricles. The ventricles communicate with the basal channels and through them with other cranial subcompartments.[27, 28] Excess CSF drains to the systemic circulation and in some species, such as rodents, the excess CSF drains partially to the lymphatic system.

The rate of CSF exchange is about four times every 24 hours, with some interspecies variation.[52] Drainage routes of CSF are insufficiently studied due to significant methodological difficulties. While the principal draining site is believed to be in the venous sinuses connected to the CSF through specialized structures (i.e., arachnoid granulations),[53] the exact anatomical location and distribution of drainage sites varies by species. In humans and nonhuman primates, arachnoid granulations were found in the superior sagittal sinus, in lateral sinuses, and at the nerve roots. In rabbits, drainage sites were found only at the cranial base[54]; in rats, the drainage site is reportedly in the superior sagittal sinus, although arachnoid granulations are less well defined in rats compared with other animals.[55] In the preparations of arachnoid granulations, multiple pores are apparently formed by merged giant vacuoles (Fig. 9). Whether these channels are permanent or dynamic remains unknown. In rats, the

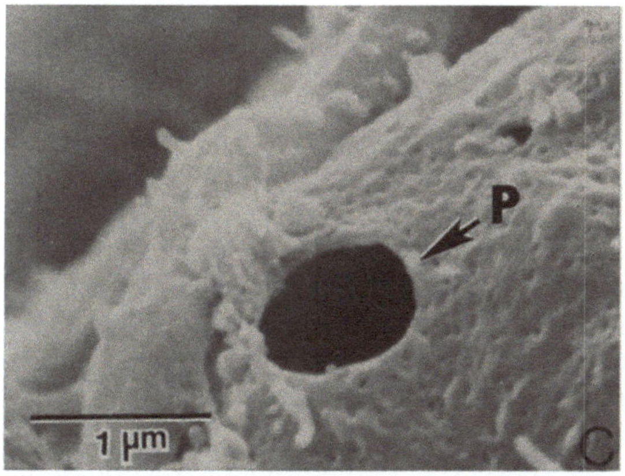

FIG. 9 Leptomeningeal pore (P). Scanning electron micrograph. *Reproduced with permission from Levine JE, Povlishock JT, Becker DP. The morphological correlates of primate cerebrospinal fluid absorption.* Brain Res. *1982;241 (1):31-41.*

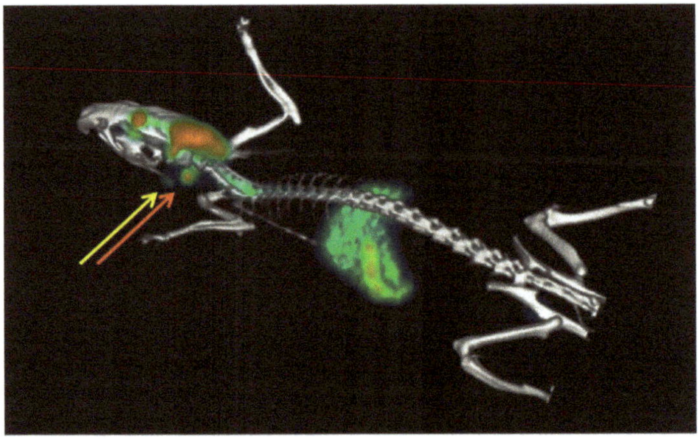

FIG. 10 Deep *(red arrows)* and superficial cervical lymph nodes *(yellow arrows)* 3 hours after cisterna magna administration of ^{124}I-labeled iduronate sulfatase.

administered solute (i.e., peroxidase) was also found in numerous transendothelial vesicles, suggesting a dual-filtration mechanism.[56]

In addition to direct drainage, paraneural CSF drainage into the interstitium with subsequent lymphatic drainage has been suggested, in rabbits, for example.[57] The estimates of the CSF fraction drained through lymphatics vary widely, with some estimating up to 50% of total CSF reabsorption.[58] Data from a recent study[59] suggest that, in mice, the lymphatic system rather than the venous route appears to be the major exit pathway for both macromolecular and small molecule tracers from CSF (Figs. 9 and 10).

We analyzed the PET and CT scans of 43 rats injected intrathecally (into the cisterna magna) with radiolabeled highly glycosylated proteins, galactosylceramidase (hGalC), Idursulfase (I2S), Heparan-N-sulfatase (HNS), or arylsulfatase A (ARSA), through both manual injection (31 animals) and through a surgically installed catheter (12 animals). In each rat, only cervical lymph nodes were visible 4 hours after injection. No lymph nodes accumulating the radiotracer were visible in any other area of the body, including the para-aortic area along the spine (Fig. 10). Based on PET imaging data, obtained with highly glycosylated proteins that accumulate in the lymph nodes through which CSF is drained, we estimate that at least 2.4% of CSF in rats drains through deep cervical lymph nodes.[60] No other nodes consistently accumulated the tracer, except for sentinel nodes draining sites in which the dura had been damaged (e.g., from needle injection).

We analyzed the PET scans of 17 nonhuman primates that had received intrathecal administration of the same radiotracers used in rats. The tracer was not detected to be accumulating (nor appearing transiently) in any of the nonhuman primates' deep cervical lymph nodes. 34 studies were conducted in these 17 animals. Throughout the body at least one node was visible 5 hours after tracer administration in 4/17 animals, and 12 to 24 hours after administration, at least one node was visible in 15/17 animals. Bilateral tracer accumulation in the superficial cervical nodes was detectable in 6 of 17 animals, unilateral accumulation in 4 animals, and no accumulation was noted in any cervical nodes in 7 animals. Bilateral accumulation in para-aortic lymph nodes in the thoracic area (T1-T4) was detectable in 6 of 17 animals (2 were likely from port leaks); unilateral accumulation in the thoracic area was seen in 4 animals; and no accumulation was observed in 7 animals. The total lymphatic fraction of the administered dose (all lymph nodes) ranged from 0% to 0.5%. Thus, in nonhuman primates, there is no consistent lymphatic drainage from the CSF in any specific region of the body.

The large mesothelial pores are evidently not the only mechanism of solute clearance from the CSF. In our preliminary studies in rats on the dependence of macromolecule clearance from the CSF from its hydrodynamic size, we confirmed that the functional size of the pores is at least 1.2 μm. Radiolabeled amorphous silica particles of this diameter were cleared from the CSF and accumulated in the liver, suggesting direct drainage of the CSF to the blood through large pores. We have also found size dependence of the macromolecule clearance from the CSF in the 2-20 nm region (probed with [^{124}I] tyrosine dextrans of varying sizes), which suggests the presence of another set of either true pores or transcytosis processes functionally indistinguishable from pores (not unlike "large endothelial pores"[61]). The kinetics of clearance of all dextrans and particles were found to be nearly monoexponential for the first 4 hours after the injection, with the following half-lives in CSF: 1.5 hours (3.7 nm dextran), 3 hours (7.1 nm dextran), 8 hours (187 nm particle), and 10 hours (1260 nm particle). The clearance half-life of a small molecule (left rotating deoxyfluorinated glucose) was 15 minutes.

Notably, the process of clearance is concurrent with the process of solute entrance into the CNS (see the next section), which is relatively slow. Therefore, fast clearance of smaller macromolecules reduces the exposure of the entrances of perivascular channels, and their uptake in the CNS will be not as effective as of larger molecules.

Perivascular Entrance and Transport of Macromolecules into the CNS

Solute translocation into the brain and cerebellum from the CSF was observed by PET/CT during the first 3 to 5 hours after administration, with subsequent biphasic clearance. The routes of translocation were further investigated with fluorophores labeling multiple cell types.

Arteries and veins of the CNS, unlike the blood vessels in other tissues, run inside liquid-filled "tubes" (i.e., Virchow-Robin spaces[62, 63]). Despite a large amount of anatomical study with light and electron microscopy, the exact relationship of the perivascular and leptomeningeal spaces remains unclear and may vary among mammalian species.[52] While in some reports the continuation of the leptomeningeal sheath into the perivascular space is clearly seen,[64, 65] in others it is less apparent.[66-68] It has been suggested that the perivascular space is connected with the lymphatic system.[69] Some studies suggest that there is a "subpial space" that extends into the perivascular spaces.[68, 70] Functional studies, however, clearly demonstrate the continuity of the leptomeningeal and perivascular spaces. Protein penetration from the CSF to the perivascular space was detected as early as 4 minutes after administration,[71] with spread to the "subpial space" and with indications of back-and-forth flow.[70]

Although macromolecular fluorescent probes binding cell surfaces can be used for staining perivascular channels (Fig. 11), their emission is relatively low and not all channels can be reliably detected. We used a highly mannose-6-phosphate glycosylated protein expressed in chicken eggs, α-N-acetylglucosaminidase (rhNAGLU[72]), to label highly endocytosing perivascular cells (also known as *perivascular macrophages*) with a large amount of fluorophore conjugated with the protein (roughly five Fluorescein or Alexa 350 dye molecules per protein molecule). The studies were conducted in rats and rhesus monkeys. The fluorescent protein was administered into the cisterna magna at 1 mg/kg. The animals were euthanized 24 hours

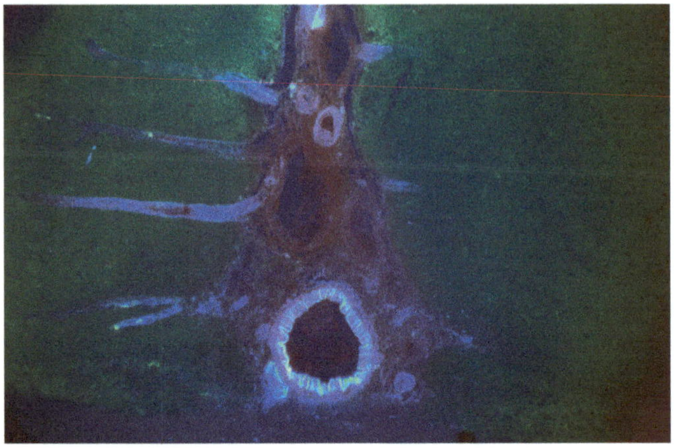

FIG. 11 Branching of perivascular channels from the central fissure. Fluorescence microphotograph, 20× (field: 0.9×0.6 mm). Rat, staining via intrathecal administration of 1 mg/kg wheat germ agglutinin labeled with Alexa 350.

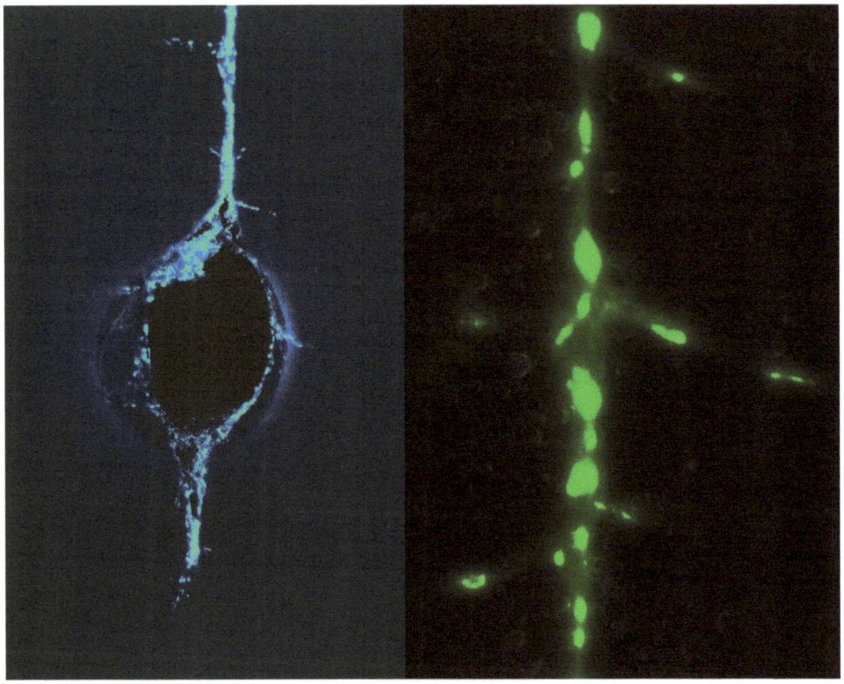

FIG. 12 Distribution of perivascular channels stained via intrathecal administration of fluorescent conjugates of N-acetyl-alpha-glucosaminidase (NAGLU). Fluorescence microphotographs. Left: monkey, 4× (field: 4.4×2.9 mm), NAGLU-Alexa 350, 1 mg/kg. Right: rat, 20× (field: 0.9×0.6 mm), NAGLU-FITC, 1 mg/kg.

after protein administration, when no free protein remained in the liquid phase (as determined by PET/CT in a separate experiment). In a small group of animals, the blood vessels were counterstained by perfusion with Texas Red. Unstained, unfixed cryosections were studied to determine the number of channels entering the CNS surface and the depth of their extension into the CNS (Figs. 11 and 12).

Prominent labeling of the perivascular channels entering the CNS from the outer and inner boundaries was observed throughout the CNS (i.e., brain, cerebellum, spinal cord) with indications of probe exit to the parenchyma (Fig. 12). The rate and depth of the entrance exclude diffusion-driven mechanisms. The coverage of the parenchyma by large and small transporting channels of various morphologies was found to be very significant. We have found that in both rats and monkeys, the density of perivascular entrances ranges from about 40 to 180 mm^{-2} on the external and internal (e.g., fissures, sulci) surfaces of the CNS.

While the data (including our own[73]) demonstrate the presence of large molecules administered to CSF in the brain parenchyma, the mechanism(s) of their exit from the perivascular space has not been studied in detail. Hypothetically, they can be transcytosed into the parenchyma by the cells (of the epithelial type) lining perivascular space. According to Hager,[74] only large vessels have perivascular space layered with such cells; the perivascular spaces of smaller vessels have only glial sheaths. The structure of the latter has not been studied in detail; they may be either penetrable for macromolecules and particles or transport them actively.

There may be several potential transport mechanisms in the parenchymal interstitium, including diffusion, directional transfer with interstitial fluid, and axonal transport. Anterograde axonal transport of gene vectors, detected by expression of intrathecally administered gene vectors in dorsal root ganglia, was demonstrated by Wang et al.[75]

The State of Perivascular Channels in Disease

Cancer, inflammation, and many other conditions are known to affect the structure and function of tissues potentially involved in the leptomeningeal solute transport. In many pathologies, perivascular spaces are dilated, which may result in an enhanced solute delivery to the CNS parenchyma. Brain trauma results in a pathologically dilated perivascular space.[76-78] A dilated perivascular space has also been associated with migraine,[79] mucopolysaccharidosis,[80] cryptococcosis,[81] advanced age,[82] hypertension, dementia, incidental white matter lesions,[83] and other conditions. The mechanism of dilation is believed to be associated with inflammatory changes, but this has not been established.[77] Although some cranial cancers reportedly do not invade the perivascular space,[84] most meningeal and brain cancers do invade or occlude the perivascular space,[85-88] which should limit or halt solute transport. Infection-related factors can induce expression of inflammation-associated receptors in the meninges,[89, 90] and can potentially alter the functionality of the leptomeningeal pores, which can also attenuate solute transport. The influence of these factors on the cerebrospinal and perivascular drug transport requires investigation.

Overall Physiological Pharmacokinetic Scheme

Development of novel therapeutics benefits from the availability of analytical mechanistic pharmacokinetic models. Unlike computational methods (for example, those based on computational fluid dynamics), analytical mechanistic pharmacokinetic models enable data scaling from research animals to humans, and help optimize therapeutic entities and procedures based on organism-specific constants.

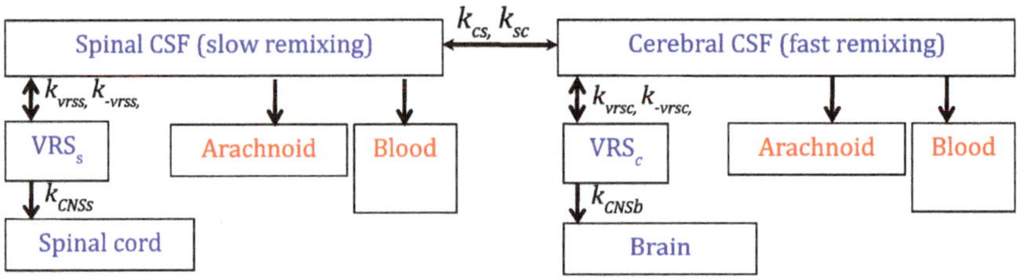

FIG. 13 Major solute transport processes in/from the cerebrospinal fluid (CSF; simplified scheme). Transport in the CSF and perivascular/Virchow-Robin spaces (VRS) is biomechanically driven. Solute exchange between the spinal and cerebral CSF pools is slow (see Fig. 3). From each CSF subcompartment, the solute can be transferred to the VRS, the arachnoid, or directly to the blood. The processes are hypothetically solute size-dependent, and their rates are proportional to the solute concentration (first order). For estimation of the maximal dose deliverable to the parenchyma, the transfer of macromolecules from the liquid compartments to the central nervous can be considered irreversible. The scheme does not take into account the solute transport along the spinal CSF pool.

Based on the physiology of the cerebrospinal/perivascular solute transport, a simplified model of leptomeningeal/perivascular solute transport may consist of three consecutive major processes: active leptomeningeal spread, active periarterial transport, and (active or passive) entrance into the brain interstitium from the perivascular spaces. Concurrent clearance processes divert the solute from the CNS: drug absorption from the CSF by the arachnoid, and direct or transcytosis-mediated drainage of CSF to the blood (Fig. 13).

The kinetic constants of the processes (k) likely depend upon the species, properties of the solute (e.g., hydrodynamic size, presence of cell-binding motifs), and tissue state, which may be affected by the pathology. The kinetic parameters of most of these processes can be measured noninvasively in both humans and animals, using quantitative imaging.

CONCLUSIONS

The perivascular system of the CNS provides a potentially robust therapy delivery route for delivering macromolecular and supramolecular pharmaceuticals to parenchymal targets throughout the CNS, but this method is not suitable for delivery of small molecules to the CNS. Perivascular entrances take up and transport into the CNS pharmaceuticals administered to the CSF. Selection of the injection site and optimization of the volume at which the therapeutic is administered enable delivery of selective therapies to the cerebro-cervical CSF or to any segment of the spinal CSF, and a large-volume lumbar bolus is safe and suitable for selective delivery. The safety of a large-volume lumbar bolus can also be ensured through monitoring of CSF pressure and respiration and heart rates. The observed transport timeframes suggest a possibility for optimizing intrathecal schedules for existing drugs, as well as strong potential for developing highly effective, targeted novel intrathecal therapies in the near future. The observed transport phenomena can explain most, if not all, known but insufficiently understood effects (or absence of effects) of intrathecally administered drugs.

Acknowledgments

This work was supported by NIH Grants R21NS090049, R01NS092838, and R21CA191979. Tracer precursors and partial funding were provided by Shire (ARSA, HNS, I2S, hGalC), Synageva (now Alexion) (rhNAGLU), and Neurophage (phage M13).

We thank our students, research assistants, and veterinarians who conducted the studies and provided logistical support.

References

1. Bambakidis NC, Petrullis M, Kui X, et al. Improvement of neurological recovery and stimulation of neural progenitor cell proliferation by intrathecal administration of Sonic hedgehog. *J Neurosurg*. 2012;116(5):1114-1120.
2. Cahill CM, Dray A, Coderre TJ. Intrathecal nerve growth factor restores opioid effectiveness in an animal model of neuropathic pain. *Neuropharmacology*. 2003;45(4):543-552.
3. Crawley AC, Marshall N, Beard H, et al. Enzyme replacement reduces neuropathology in MPS IIIA dogs. *Neurobiol Dis*. 2011;43(2):422-434.
4. Dickson P, McEntee M, Vogler C, et al. Intrathecal enzyme replacement therapy: successful treatment of brain disease via the cerebrospinal fluid. *Mol Genet Metab*. 2007;91(1):61-68.
5. Dickson PI, Ellinwood NM, Brown JR, et al. Specific antibody titer alters the effectiveness of intrathecal enzyme replacement therapy in canine mucopolysaccharidosis I. *Mol Genet Metab*. 2012;106(1):68-72.
6. Hemsley KM, Hopwood JJ. Delivery of recombinant proteins via the cerebrospinal fluid as a therapy option for neurodegenerative lysosomal storage diseases. *Int J Clin Pharmacol Ther*. 2009;47(Suppl 1):S118-S123.
7. Im YB, Jee MK, Choi JI, Cho HT, Kwon OH, Kang SK. Molecular targeting of NOX4 for neuropathic pain after traumatic injury of the spinal cord. *Cell Death Dis*. 2012;3:e426.
8. Ishigaki A, Aoki M, Nagai M, et al. Intrathecal delivery of hepatocyte growth factor from amyotrophic lateral sclerosis onset suppresses disease progression in rat amyotrophic lateral sclerosis model. *J Neuropathol Exp Neurol*. 2007;66(11):1037-1044.
9. Kakkis E, McEntee M, Vogler C, et al. Intrathecal enzyme replacement therapy reduces lysosomal storage in the brain and meninges of the canine model of MPS I. *Mol Genet Metab*. 2004;83(1-2):163-174.
10. Meuli-Simmen C, Liu Y, Yeo TT, et al. Gene expression along the cerebral-spinal axis after regional gene delivery. *Hum Gene Ther*. 1999;10(16):2689-2700.
11. Milligan ED, Sloane EM, Langer SJ, et al. Repeated intrathecal injections of plasmid DNA encoding interleukin-10 produce prolonged reversal of neuropathic pain. *Pain*. 2006;126(1-3):294-308.
12. Okuda Y, Sakoda S, Fujimura H, Nagata S, Yanagihara T, Bernard CC. Intrathecal administration of neutralizing antibody against Fas ligand suppresses the progression of experimental autoimmune encephalomyelitis. *Biochem Biophys Res Commun*. 2000;275(1):164-168.
13. Shi L, Tang GP, Gao SJ, et al. Repeated intrathecal administration of plasmid DNA complexed with polyethylene glycol-grafted polyethylenimine led to prolonged transgene expression in the spinal cord. *Gene Ther*. 2003;10(14):1179-1188.
14. Tsai SY, Markus TM, Andrews EM, et al. Intrathecal treatment with anti-Nogo-A antibody improves functional recovery in adult rats after stroke. *Exp Brain Res*. 2007;182(2):261-266.
15. Vanderah TW, Laughlin T, Lashbrook JM, et al. Single intrathecal injections of dynorphin A or des-Tyr-dynorphins produce long-lasting allodynia in rats: blockade by MK-801 but not naloxone. *Pain*. 1996;68(2-3):275-281.
16. Xu S, Wang L, El-Banna M, Sohar I, Sleat DE, Lobel P. Large-volume intrathecal enzyme delivery increases survival of a mouse model of late infantile neuronal ceroid lipofuscinosis. *Mol Ther*. 2011;19(10):1842-1848.
17. Wagner HJ, Pilgrim C, Brandl J. Penetration and removal of horseradish peroxidase injected into the cerebrospinal fluid: role of cerebral perivascular spaces, endothelium and microglia. *Acta Neuropathol*. 1974;27(4):299-315.
18. Corning JL. Spinal anaesthesia and local medication of the cord. *New York Med J*. 1885;42:483-485.
19. Koster H, Kasman LP. Spinal anesthesia for the head, neck, and thorax: its relation to respiratory paralysis. *Surg Gynecol Obstet*. 1929;49:617-630.
20. Cousins MJ, Mather LE. Intrathecal and epidural administration of opioids. *Anesthesiology*. 1984;61(3):276-310.

21. Yaksh TL, Reddy SV. Studies in the primate on the analgetic effects associated with intrathecal actions of opiates, alpha-adrenergic agonists and baclofen. *Anesthesiology*. 1981;54(6):451-467.
22. Mackey DC. The history of spinal drug delivery: the evolution of lumbar puncture and spinal narcosis. In: Yaksh TL, ed. *Spinal Drug Delivery*. New York: Elsevier; 1999:1-41.
23. Maythaler JM. Intrathecally delivered medications for spasticity and dystonia. In: Yaksh TL, ed. *Spinal Drug Delivery*. New York: Elsevier; 1999:513-529.
24. Penn RD, Kroin JS. Continuous intrathecal baclofen for severe spasticity. *Lancet*. 1985;2(8447):125-127.
25. Di Chiro G. Observations on the circulation of the cerebrospinal fluid. *Acta Radiol Diagn (Stockh)*. 1966;5:988-1002.
26. Soderquist RG, Mahoney MJ. Central nervous system delivery of large molecules: challenges and new frontiers for intrathecally administered therapeutics. *Expert Opin Drug Deliv*. 2010;7(3):285-293.
27. Dandy WE, Blackfan KD. An experimental, clinical and pathological study: Part 1—Experimental studies. *Am J Dis Child*. 1914;VIII(6):406-482.
28. Ingraham FD, Matson DD, et al. Studies in the treatment of experimental hydrocephalus. *J Neuropathol Exp Neurol*. 1948;7(2):123-143.
29. Ochs G, Penn RD, York M, et al. A phase I/II trial of recombinant methionyl human brain derived neurotrophic factor administered by intrathecal infusion to patients with amyotrophic lateral sclerosis. *Amyotroph Lateral Scler Other Motor Neuron Disord*. 2000;1(3):201-206.
30. Chiro GD, Hammock MK, Bleyer WA. Spinal descent of cerebrospinal fluid in man. *Neurology*. 1976;26(1):1-8.
31. Pardridge WM. Drug transport in brain via the cerebrospinal fluid. *Fluids Barriers CNS*. 2011;8(1):7.
32. Farris EJ, Griffith JQ. *The Rat in Laboratory Investigation*. Philadelphia: J. B. Lippincott; 1949.
33. Belov VV, Bonab AA, Fischman AJ, Heartlein M, Calias P, Papisov MI. Iodine-124 as a label for pharmacological PET imaging. *Mol Pharm*. 2011;8(3):736-747.
34. Carney JPJ, Flynn JL, Cole KS, et al. Preclinical PET/CT system for imaging non-human primates. In: *IEEE Medical Imaging Conference Vol Abstract Number M06-67. Orlando, FL*; 2009.
35. Papisov MI, Belov V, Fischman AJ, et al. Delivery of proteins to CNS as seen and measured by positron emission tomography. *Drug Deliv Transl Res*. 2012;2(3):201-209.
36. Paxinos G, Watson C. *The Rat Brain in Stereotaxic Coordinates*. 7th ed. London, UK: Elsevier Inc; 2013.
37. Calabrese E, Badea A, Coe CL, et al. A diffusion tensor MRI atlas of the postmortem rhesus macaque brain. *NeuroImage*. 2015;117:408-416.
38. Von Bonin G, Bailey P. *The neocortex of* Macaca mulatta. Champaign, IL: University of Illinois Press; 1947.
39. Hettiarachchi HD, Hsu Y, Harris Jr. TJ, Penn R, Linninger AA. The effect of pulsatile flow on intrathecal drug delivery in the spinal canal. *Ann Biomed Eng*. 2011;39(10):2592-2602.
40. Taylor GI. Dispersion of soluble matter in solvent flowing slowly through a tube. *Proc Roy Soc A*. 1953;219:186-203.
41. Rieselbach RE, Di Chiro G, Freireich EJ, Rall DP. Subarachnoid distribution of drugs after lumbar injection. *N Engl J Med*. 1962;267:1273-1278.
42. Pass J, Belov V, Appleton J, Giffenig P, Papisov M. *The effect of large volume intrathecal bolus on the cerebral blood vessels. Paper presented at: SNMMI Annual Meeting, Philadelphia*; 2018.
43. Guggenheim EA. XLVI. On the determination of the velocity constant of a unimolecular reaction. *Lond Edinb Dubl Philos Mag J Sci*. 1926;2(9):538-543.
44. Bhadelia RA, Bogdan AR, Kaplan RF, Wolpert SM. Cerebrospinal fluid pulsation amplitude and its quantitative relationship to cerebral blood flow pulsations: a phase-contrast MR flow imaging study. *Neuroradiology*. 1997; 39(4):258-264.
45. Enzmann DR, Pelc NJ. Normal flow patterns of intracranial and spinal cerebrospinal fluid defined with phase-contrast cine MR imaging. *Radiology*. 1991;178(2):467-474.
46. Henry-Feugeas MC, Idy-Peretti I, Blanchet B, Hassine D, Zannoli G, Schouman-Claeys E. Temporal and spatial assessment of normal cerebrospinal fluid dynamics with MR imaging. *Magn Reson Imaging*. 1993; 11(8):1107-1118.
47. Yoshida K, Takahashi H, Saijo M, et al. Phase-contrast MR studies of CSF flow rate in the cerebral aqueduct and cervical subarachnoid space with correlation-based segmentation. *Magn Reson Med Sci*. 2009;8(3):91-100.
48. Linge SO, Haughton V, Lovgren AE, Mardal KA, Langtangen HP. CSF flow dynamics at the craniovertebral junction studied with an idealized model of the subarachnoid space and computational flow analysis. *AJNR Am J Neuroradiol*. 2010;31(1):185-192.

49. Sweetman B, Linninger AA. Cerebrospinal fluid flow dynamics in the central nervous system. *Ann Biomed Eng.* 2011;39(1):484-496.
50. Linninger AA, Xenos M, Sweetman B, Ponkshe S, Guo X, Penn R. A mathematical model of blood, cerebrospinal fluid and brain dynamics. *J Math Biol.* 2009;59(6):729-759.
51. Du Boulay G, O'Connell J, Currie J, Bostick T, Verity P. Further investigations on pulsatile movements in the cerebrospinal fluid pathways. *Acta Radiol Diagn (Stockh).* 1972;13:496-523.
52. Davson H, Segal MB. *Physiology of the CSF and Blood-Brain Barriers.* Boca Raton, FL: CRC Press; 1995.
53. Weed LH. The absorption of cerebrospinal fluid into the venous system. *Am J Anat.* 1923;31:191-221.
54. McComb JG, Davson H, Hollingsworth JR. Attempted separation of blood-brain and blood-cerebrospinal fluid barriers in the rabbit. *Exp Eye Res.* 1977;25(Suppl):333-343.
55. Butler A, Mann J, Bass N. Identification of the major site for cerebrospinal fluid efflux in the albino rat. *Anat Rec.* 1975;181:323.
56. Butler A, Mann J, Bass N. Vesicular transport of peridoxase across the endothelium of the rat arachnoid villus. *Anat Rec.* 1977;187(543).
57. Yamada S, DePasquale M, Patlak CS, Cserr HF. Albumin outflow into deep cervical lymph from different regions of rabbit brain. *Am J Phys.* 1991;261(4 Pt 2):H1197-H1204.
58. Boulton M, Flessner M, Armstrong D, Mohamed R, Hay J, Johnston M. Contribution of extracranial lymphatics and arachnoid villi to the clearance of a CSF tracer in the rat. *Am J Phys.* 1999;276(3 Pt 2):R818-R823.
59. Ma Q, Ineichen BV, Detmar M, Proulx ST. Outflow of cerebrospinal fluid is predominantly through lymphatic vessels and is reduced in aged mice. *Nat Commun.* 2017;8(1):1434.
60. Gillooly C, Belov V, Belova E, Levine D, Appleton J, Papisov M. *Cerebrospinal Fluid Drainage to the Lymphatic System in Rats and Nonhuman Primates. Paper Presented at: SNMMI Annual Meeting, Denver, CO;* 2017.
61. Michel CC. Transport of macromolecules through microvascular walls. *Cardiovasc Res.* 1996;32(4):644-653.
62. Robin C. Recherches sur quelques particularités de la structure des capillaires de l'encephale. *J Physiol Homme Anim.* 1859;2:537-548.
63. Virchow R. Ueber die Erweiterung kleinerer Gefaesse. *Archiv Pathol Anat Physiol Klin Med.* 1851;3:427-462.
64. Millen JW, Woollam DH. On the nature of the pia mater. *Brain.* 1961;84:514-520.
65. Woollam DHM, Millen JW. The perivascular spaces of the mammalian central nervous system. *Biol Rev.* 1954;29:251-283.
66. Hutchings M, Weller RO. Anatomical relationships of the pia mater to cerebral blood vessels in man. *J Neurosurg.* 1986;65(3):316-325.
67. Pollock H, Hutchings M, Weller RO, Zhang ET. Perivascular spaces in the basal ganglia of the human brain: their relationship to lacunes. *J Anat.* 1997;191(Pt 3):337-346.
68. Zhang ET, Inman CB, Weller RO. Interrelationships of the pia mater and the perivascular (Virchow-Robin) spaces in the human cerebrum. *J Anat.* 1990;170:111-123.
69. Schley D, Carare-Nnadi R, Please CP, Perry VH, Weller RO. Mechanisms to explain the reverse perivascular transport of solutes out of the brain. *J Theor Biol.* 2006;238(4):962-974.
70. Ichimura T, Fraser PA, Cserr HF. Distribution of extracellular tracers in perivascular spaces of the rat brain. *Brain Res.* 1991;545(1-2):103-113.
71. Rennels ML, Gregory TF, Blaumanis OR, Fujimoto K, Grady PA. Evidence for a 'paravascular' fluid circulation in the mammalian central nervous system, provided by the rapid distribution of tracer protein throughout the brain from the subarachnoid space. *Brain Res.* 1985;326(1):47-63.
72. Quinn A, Leavitt MC, Zhinan X, Rutkowski JV. *Inventors. Recombinant Human NaGlu Protein and Uses Thereof,* 2013.
73. Calias P, Papisov M, Pan J, et al. CNS penetration of intrathecal-lumbar idursulfase in the monkey, dog and mouse: implications for neurological outcomes of lysosomal storage disorder. *PLoS ONE.* 2012;7(1).
74. Hager H. Elektronenmikroskopische Untersuchungen über die Feinstruktur der Blutgefässe und perivascuären Räume im Säugetiergehirn. *Acta Neuropathol.* 1961;1:9-33.
75. Wang X, Wang C, Zeng J, et al. Gene transfer to dorsal root ganglia by intrathecal injection: effects on regeneration of peripheral nerves. *Mol Ther.* 2005;12(2):314-320.
76. Esiri MM, Gay D. Immunological and neuropathological significance of the Virchow-Robin space. *J Neurol Sci.* 1990;100(1-2):3-8.
77. Inglese M, Bomsztyk E, Gonen O, Mannon LJ, Grossman RI, Rusinek H. Dilated perivascular spaces: hallmarks of mild traumatic brain injury. *AJNR Am J Neuroradiol.* 2005;26(4):719-724.

78. Inglese M, Grossman RI, Diller L, et al. Clinical significance of dilated Virchow-Robin spaces in mild traumatic brain injury. *Brain Inj*. 2006;20(1):15-21.
79. Schick S, Gahleitner A, Wober-Bingol C, et al. Virchow-Robin spaces in childhood migraine. *Neuroradiology*. 1999;41(4):283-287.
80. Murata R, Nakajima S, Tanaka A, et al. MR imaging of the brain in patients with mucopolysaccharidosis. *AJNR Am J Neuroradiol*. 1989;10(6):1165-1170.
81. Miszkiel KA, Hall-Craggs MA, Miller RF, et al. The spectrum of MRI findings in CNS cryptococcosis in AIDS. *Clin Radiol*. 1996;51(12):842-850.
82. Maclullich AM, Wardlaw JM, Ferguson KJ, Starr JM, Seckl JR, Deary IJ. Enlarged perivascular spaces are associated with cognitive function in healthy elderly men. *J Neurol Neurosurg Psychiatry*. 2004;75(11):1519-1523.
83. Heier LA, Bauer CJ, Schwartz L, Zimmerman RD, Morgello S, Deck MD. Large Virchow-Robin spaces: MR-clinical correlation. *AJNR Am J Neuroradiol*. 1989;10(5):929-936.
84. Hassin GB. Histopathology of carcinoma of the cerebral meninges. *Arch Neurol Psych*. 1919;1:705-716.
85. Herman C, Kupsky WJ, Rogers L, Duman R, Moore P. Leptomeningeal dissemination of malignant glioma simulating cerebral vasculitis. Case report with angiographic and pathological studies. *Stroke*. 1995;26(12):2366-2370.
86. Kooistra KL, Rodriguez M, Powis G, et al. Development of experimental models for meningeal neoplasia using intrathecal injection of 9L gliosarcoma and Walker 256 carcinosarcoma in the rat. *Cancer Res*. 1986;46(1):317-323.
87. Mammoser AG, Groves MD. Biology and therapy of neoplastic meningitis. *Curr Oncol Rep*. 2010;12(1):41-49.
88. Mirfakhraee M, Crofford MJ, Guinto Jr FC, Nauta HJ, Weedn VW. Virchow-Robin space: a path of spread in neurosarcoidosis. *Radiology*. 1986;158(3):715-720.
89. Deckert-Schluter M, Schluter D, Hof H, Wiestler OD, Lassmann H. Differential expression of ICAM-1, VCAM-1 and their ligands LFA-1, Mac-1, CD43, VLA-4, and MHC class II antigens in murine Toxoplasma encephalitis: a light microscopic and ultrastructural immunohistochemical study. *J Neuropathol Exp Neurol*. 1994;53(5):457-468.
90. Endo H, Sasaki K, Tonosaki A, Kayama T. Three-dimensional and ultrastructural ICAM-1 distribution in the choroid plexus, arachnoid membrane and dural sinus of inflammatory rats induced by LPS injection in the lateral ventricles. *Brain Res*. 1998;793(1-2):297-301.

Further Reading

91. Segal MB. The anatomy and physiology of intracranial fluids. In: Reid A, Marshbanks R, Ernst A, eds. *Intracranial and Inner Ear Physiology and Pathophysiology*. London: Whurr Publishers Ltd., Wiley; 1998.
92. Levine JE, Povlishock JT, Becker DP. The morphological correlates of primate cerebrospinal fluid absorption. *Brain Res*. 1982;241(1):31-41.

CHAPTER

14

Drug-Impregnated Polymer Delivery

Riccardo Serra, Joshua Casaos, Betty Tyler, Henry Brem
Department of Neurosurgery and Hunterian Neurosurgical Research Laboratory, Johns Hopkins University School of Medicine, Baltimore, MD, United States

TUMORS OF THE CENTRAL NERVOUS SYSTEM

Tumors of the central nervous system (CNS) are among the most aggressive cancers in humans. Of astrocytic tumors specifically, glioblastoma multiforme (GBM), a grade IV glial tumor, is the most common and lethal brain malignancy. Increasing the number of treatment options for patients diagnosed with GBM has been a focus of research over the past few decades. One of the most daunting obstacles that impedes chemotherapeutic access to a brain tumor is the blood–brain barrier (BBB),[1] a mechanical barrier that blocks therapeutics from diffusing within the CNS and, therefore, throughout the bulk of the tumor. Although lipophilic molecules generally have good penetration through the BBB,[2] other more hydrophilic compounds are often unable to reach significant intracerebral concentrations.[3] The need for a reliable system capable of delivering an efficacious concentration of chemotherapeutics across the BBB led to the development of the biodegradable polymeric wafer, Gliadel (Eisai Inc. for Arbor Pharmaceuticals). The Gliadel wafer is impregnated with a chemotherapeutic agent and is directly implanted at the time of tumor surgical resection.

NATURAL BARRIERS TO THE CENTRAL NERVOUS SYSTEM

The Neurovascular Unit

The neurovascular unit (NVU) is a specialized anatomical barrier[4] made of endothelial cells that are linked by tight junctions and supported by various other cell types (i.e., astrocytes,[5] pericytes,[6] and neurons[7]). Its main function is to hamper the diffusion of highly hydrophilic macro- and micromolecules, creating a selective membrane that allows the passage of lipophilic compounds with a molecular weight (MW) less than 400 Da (Figs. 1 and 2).[8] Drugs are generally unable to permeate this barrier and increased doses of therapeutics are

Nervous System Drug Delivery
https://doi.org/10.1016/B978-0-12-813997-4.00014-1

© 2019 Elsevier Inc. All rights reserved.

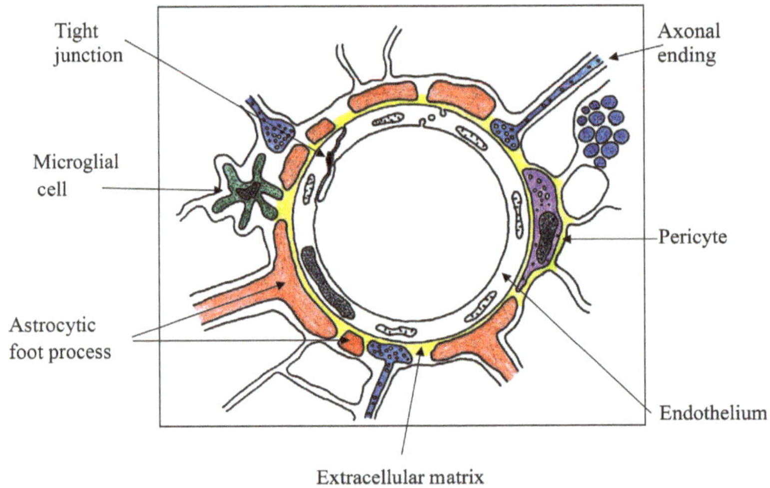

FIG. 1 Diagrammatic representation of the neurovascular unit and cell association forming the blood-brain barrier (BBB). Endothelial cells and their basement membrane form the internal layer of the BBB; astrocytic foot processes, microglia, and axons contribute to barrier function and central nervous system homeostasis. *Reproduced with permission from Begley DJ. Delivery of therapeutic agents to the central nervous system: the problems and the possibilities. Pharmacol Ther. 2004;104 (1):29-45.*

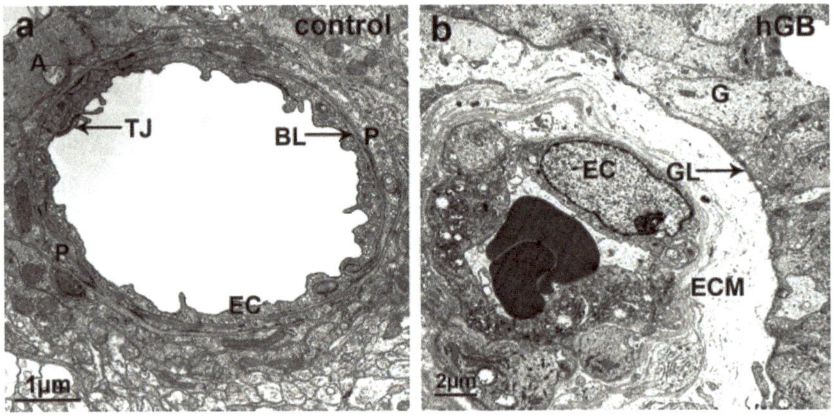

FIG. 2 Electron microscopy of (A) a normal brain and (B) a human glioblastoma. In astrocytic tumors, blood vessels and capillaries are less integrated into the neuropil and are separated by a large extracellular space and matrix. A, astrocyte; BL, basal lamina; EC, endothelial cell; G, glioma cell; GL, glial-limiting membrane; P, pericyte; TJ, tight junction. *Reproduced with permission from Noell S, Wolburg-Buchholz K, Mack AF, Wolburg H, Fallier-Becker P. The blood-brain barrier in brain tumours. Manag CNS Tumors: InTech; 2011.*

necessary for efficient and effective diffusion through the NVU. Recently, due to computerized high-throughput tools and specific software, a database of small molecules capable of crossing the BBB has been developed, which has provided a foundation for future drug and treatment development.[9]

The BBB is highly efficient at shielding the brain from toxins, drugs, and other compounds, but is sometimes subjected to damage that, depending on the severity, can increase its permeability and allow the nonselective passage of larger and more hydrophilic molecules. Infections, tumors and, more generally, inflammatory processes, are capable of disrupting the physical integrity of the BBB. This disruption can allow various compounds to gain access to the brain parenchyma.[10, 11]

The Extracellular Space

The extracellular space (ECS) is an additional compartment that lies between the BBB and the cell membranes of neurons, astrocytes, and other cell types that constitute the brain parenchyma.[12, 13] This functional barrier[14] helps to maintain optimal osmotic and oncotic concentrations, moderating the diffusion of drugs while diluting toxins and other molecules. Once thought to be irrelevant in the process of limiting drug and molecule diffusion, the ECS has recently been shown to include complex interactions that influence the pharmacokinetics and pharmacodynamics of drugs and other compounds.

Solutes are capable of moving across the ECS in two ways.[15] First, they may move across the ECS via diffusion, a nonspecific and non-oriented random movement that allows molecules to follow their own concentration gradient. Second, they may cross the ECS via bulk-flow movement, which is a pressure-driven phenomenon that occurs where a hydrostatic pressure gradient is in place.[16] These mechanisms decrease the concentration of therapeutics and reduce the ability of a therapeutic compound to target a specific cell type. The mechanisms set a threshold for molecules and their movement across the NVU, thereby protecting the brain parenchyma and creating an environment in which the distribution and retention of drugs and other high-molecular–weight substances are regulated via endocytosis and receptor-mediated transcytosis.[16]

Another central contributor that has been investigated for decades[17] is the cerebral lymphatic system, an elusive pathway of paravenous spaces that drains fluids and solutes outside the ECS, maintaining constant pressure and osmolarity while allowing for the migration of T cells and other cell types.[18, 19] The discovery of this complementary system not only adds another confounding aspect to neuroinflammatory and neurodegenerative disease pathogenesis; it also has an effect on drug diffusion in the CNS.

The Blood–Brain Tumor Barrier

Having addressed these protective elements in place under normal circumstances, this microenvironment is drastically altered under abnormal conditions—in particular, in cases of tumor development and growth within the brain parenchyma. An important aspect to consider is the profound morphological and functional modification of the BBB and the genesis of the blood–brain tumor barrier (BBTB), a unique structure that closely resembles the unit from which it takes its name, but with some additional aspects that are specific to the tumor interface. The physiologic functioning and anatomical appearance of the BBB can be subverted by a variety of diseases and conditions, such as stroke, infection, inflammatory processes, and brain tumors.[20-22] The degree of damage and its location are extremely irregular and, in the vast majority of cases, the type and grade of the underlying malignancy have a central role in their determination. Moreover, histology of high-grade gliomas shows diverse characteristics and aggressiveness in relation to the BBB.

The most lethal and anaplastic of these tumors, GBM, is capable of resulting in near-uniform damage to the blood–tumor interface, increasing the leakiness of its vasculature and dismantling the physiologic and microscopic anatomy of its surroundings.[23] These findings have been confirmed by magnetic resonance imaging and positron emission tomography, where highly permeable vessels lead to abnormal contrast-enhanced images in the tumor area.[23] Electron microscopy has shown the detachment of astrocytic pedicles from the basal membrane that lines parenchymal capillaries.[24] However, the degree of damage is not sufficient to allow a perceptible diffusion of therapeutics across the BBTB or BBB, mainly because of two factors. The BBB is not uniformly impaired across the tumor and along the tumor–brain interface,[25] which results in uneven drug distribution within the tumor parenchyma. Even more importantly, anaplastic astrocytoma and GBM tumors are so aggressive that a consistent fraction of their cells is capable of infiltrating the normal tissues that surround the main bulk of the tumor. Occasionally, a small number of cells migrate as far as the contralateral hemisphere,[26] where the BBB is completely intact and able to protect the brain, and by extension the tumor, from therapeutic molecules.[27] In these cases, the malignant tumor recurs far from the original site, and its aggressiveness is also typically more pronounced.

DELIVERY OF NEUROONCOLOGY THERAPEUTICS

This brief review of the elements and qualities of the BBB and BBTB should help elucidate the protective characteristics imposed by the brain's anatomy and physiology to therapeutics for effective treatments for malignant brain tumors. The complexity of the obstacles faced by neurooncology therapeutics is daunting, in that a molecule must cross the endothelial lining, diffuse into the ECS, and reach its cellular target—all while maintaining a therapeutic concentration within the CNS that will achieve a cytotoxic effect on malignant cells.

Different techniques and approaches have been developed and tested to achieve significant concentrations of therapeutics across the BBB. Unfortunately, only a small fraction of these approaches has reached clinical application. Among these, drug-impregnated polymers, thermosensitive hydrogels, and nanoparticles and micelles have led to significant progress in the field and have developed a new generation of treatment options for highly aggressive primary and metastatic brain tumors. We now review the various approaches used for the delivery of pharmaceuticals into the CNS, exploring the development of Gliadel and continuing with the most recent generation of nanoparticles and micelles, a new family of carriers capable of targeting malignant cells with extreme specificity and efficacy, while simultaneously minimizing systemic toxicity.

Polymer Development

Biodegradable polymers have been used for many clinical applications and surgical devices, in part because of their safety and malleability. These polymers can degrade both in vitro and in vivo, and release metabolites that are nontoxic and can be eliminated from the body. Degradation can occur by passive hydrolysis or by active enzymatic activity; however, passive hydrolysis is preferable to active enzymatic activity because of its lower site-to-site and patient-to-patient variability. Polymer hydrolysis can be one of two types: surface erosion[28] or bulk degradation. Most polymers undergo bulk erosion because of their innate hydrophilicity and water-lability, including hydrophilic polyesters such as poly(lactic-*co*-glycolic) acid (PLGA), poly(lactic acid) (PLA), and their copolymers. Water permeates the polymer and brings it rapidly to complete dissolution, affecting not only its release kinetics, but also the stability and activity of the loaded drug.

Alternatively, polymers that degrade by surface erosion have considerable advantages when used for the local delivery of drugs: surface erosion occurs at a fairly constant rate, which can have a significant effect on release kinetics and pharmacodynamics. Molecular stability and the ability to maintain zero-order release kinetics are two vital aspects, which can allow this type of polymer to release its payload in vivo over a predetermined span of time. The rate of release of chemotherapeutics can be nearly constant from the time of implantation to total dissolution of the matrix, resulting in more gradual and uniform kinetics than that of polymers which dissolve through bulk erosion.[29]

The Polyanhydride Wafer

Polyanhydrides have been investigated for in vivo use due to their low toxicity and slow erosion, which is mainly dictated by their hydrolytic stability.[30] A small subset of polyanhydrides was developed for in vitro and in vivo use: highly hydrophobic polyanhydrides withstand hydrolysis and rapid erosion of their surface much more efficiently than other molecules, and therefore guarantee a longer release of their payload.[31] Leong et al.[29] considered the highly hydrophobic poly(bis[p-carboxyphenoxy]propane) anhydride (p[CPP]) a good candidate for in vivo use, given its degradation rate. These researchers subsequently added different percentages of sebacic acid (SA), a hydrophilic monomer, to the p(CPP) matrix. This modification resulted in a copolymer with a faster rate of erosion that could be finely tuned simply by adding or subtracting SA. Three other variables that can modify erosion kinetics are the area of the polymer form, the thickness of the formulation, and the pH of its surroundings.[29] Leong et al.[29] demonstrated that polyanhydrides had a faster degradation at basic pH compared to environments with a pH of 7.4 or an acidic pH. Additionally, a number of studies confirmed the kinetics and the mechanisms of polymer erosion and absorption in vivo, demonstrating that there was an initial lag phase with a rapid decrease of polymer molecular weight. The subsequent rates of absorption of water and loss of mass showed both surface- and bulk-eroding systems.[32, 33] Wu et al.[33] showed that neither the encapsulation of carmustine nor the presence of tumor or hemostatic material had any effect on polymer degradation, and the p(CPP):SA ratio (i.e., 20:80) had consistent in vivo degradation time of 6–8 weeks in a rat model. The polyanhydride preserved hydrophilic drugs, shielding them from the surrounding environment and prolonging their half-lives. p(CPP):SA polymers also have another significant characteristic: they are metabolized by surface erosion, a mechanism that allows wafers to release a predetermined amount of drug each day, with near-zero-order kinetics.[34] The polymeric matrix is highly moldable, so it can be shaped into multiple forms. This malleability makes it the perfect candidate for intracranial implantation, considering the irregular surface of postexcision cavities. p(CPP):SA is also completely degradable, nontoxic, and barely traceable in the blood weeks after implantation.[33, 35]

These results prompted researchers to further investigate these polymers as intracerebral implants and to identify potential adverse effects and toxicities, including allergic responses, mutagenesis, increased risk of thrombosis and infection, inflammation, irritation, and toxicity in vivo. Inflammation was tested by applying samples of p(CPP):SA (20:80) and (50:50) to the corneas of live rabbits. These tests showed that no inflammatory response was elicited locally or systemically.[36] The polymer also elicited no adverse reaction or significant degree of inflammation when implanted subcutaneously in rats.[37] Systemic signs of toxicity were investigated by Laurencin et al.[30] in animals implanted with subcutaneous p(CPP):SA (20:80) polymers at two different doses. Rats were tested for organ and systemic parameters of toxicity, with blood samples drawn during the study and formalin-fixed slides examined postmortem. No signs of macroscopic, microscopic, or molecular toxicity were observed, except for minor local fibrosis and foreign body reaction at the subcutaneous implantation site.[30] Biocompatibility of the p(CPP):SA polymer in rodent and primate brains was also assessed, and it was concluded that neither local toxicities nor carcinogenesis were among the adverse effects attributable to intracerebral polymeric implantation.[36, 38, 39]

CARMUSTINE

Nitrosoureas are a class of cytotoxic agents characterized by strong activity against GBM cells, a pronounced lipophilicity that allows them to cross the BBB, and a short half-life determined by their sensitivity to hydrolysis. Carmustine (bis-chloroethyl-nitrosourea, or BCNU) is the compound in this class with the highest capacity for penetration of the CNS, conferring a significant advantage over other drugs in the same family. Lomustine (CCNU) and semustine (methyl-CCNU),[40] two compounds with additional aliphatic groups added to them, were specifically designed and modified for better BBB penetration, but clinical trials have not shown a consistent advantage in terms of improved clinical outcomes over the better-established carmustine.[41-47] Lipophilicity is important because 90% of patients diagnosed with GBM have recurrence within 2 cm of the original site of resection, and a highly lipophilic drug is capable of diffusing more efficiently and for longer distances within the CNS parenchyma compared with a hydrophilic compound. Carmustine, which was already being given to patients through systemic administration, was therefore the ideal candidate to deliver the drug payload from the intracranial biodegradable wafer.

Preclinical Carmustine Studies

The development of carmustine delivered by p(CPP):SA wafers was initially focused on kinetics and safety studies. Intracranial and systemic biodistribution of carmustine:p(CPP):SA were assessed by means of ^{14}C radiolabeling and autoradiography in rats and rabbits. While SA was mainly excreted through the respiratory system and urine, a remarkably high fraction of p(CPP) molecules was found at the implantation site a week after polymer implantation. Carmustine, on the other hand, was subject to widespread biodistribution and hydrolysis inside the brain parenchyma, and only a small percentage of its total radioactivity was imputable to the intact drug.[48-50]

Efficacy was evaluated by comparing intracranial placement of 10-mg polymer wafers impregnated with 20% carmustine with intraperitoneal administration of carmustine.[51] Among rats with 9L gliosarcoma flank tumors, locally delivered carmustine wafers resulted in a significant delay of tumor growth of 16.3 days, compared to the growth delay of 9.3 and 11.2 days observed with carmustine administered systemically (systemic subcutaneous contralateral controlled release BCNU polymer or systemic intraperitoneal BCNU injection, respectively). In a survival study of animals with intracranial 9L gliosarcoma, a 5.4- to 7.3-fold increase in survival was observed in the groups that received carmustine wafer compared to controls. These interstitial treatment groups also had long-term survivors. Intraperitoneal injections of carmustine only slightly improved overall animal survival compared to controls and there were no long-term survivors.[51]

Local drug concentration and distribution of tritiated carmustine after intracranial implantation of carmustine:p(CPP):SA (20:80) polymer was studied in rabbits.[52] In this study, sustained intracranial release of carmustine was achieved during the first days of treatment and the volume of brain exposed to carmustine gradually decreased between 14 and 21 days.[52] Another study that used 20% carmustine-loaded wafers in rats found tumoricidal concentrations of this drug 4.7 mm from the resection margin 24 hours postimplantation.[53] In another intracranial efficacy experiment, carmustine-impregnated p(CPP):SA wafers were compared with intracranial injections of an equivalent dose of carmustine.[54] Rats that received the carmustine wafers showed a 271% increase in survival, whereas the survival of rats treated with direct intracranial injection of carmustine solution was only 36% longer than that of the controls.[54]

Sipos et al.[55] demonstrated that different proportions of p(CPP) and SA can influence release kinetics in vivo and prolong polymer erosion. p(CPP):SA 50:50 wafers were capable of resisting degradation much longer than p(CPP):SA 20:80 wafers (18 days vs. 7 days). A range of carmustine concentrations was also employed, and wafers loaded with 0%, 4%, 8%, 12%, 20%, and 32% carmustine were administered locally to rats with intracranial 9L gliosarcoma. The 20% and 32% formulations showed the most remarkable results in terms of increased survival, and the 20% carmustine wafer delivered the most balanced concentration in terms of toxicity and antitumoral activity.[55] This formulation was further tested for biodistribution and toxicity in nonhuman primates, and no signs of local toxicity were found around the implantation site, confirming the safety of this locally delivered chemotherapy.[56]

Clinical Applications of Carmustine

The evidence acquired and analyzed from many preclinical studies paved the way for the clinical testing of carmustine wafers in patients with recurrent and newly diagnosed GBM, translating a novel concept into practice and broadening the range for further developments and clinical applications.

Interstitial Chemotherapy for Recurrent GBM

After promising results from preclinical studies, two clinical trials were carried out in patients diagnosed with recurrent glioblastoma. An initial Phase I–II trial included 21 patients with recurrent GBM.[57] Wafers were placed inside the tumor-resection cavity and patients were carefully evaluated for neurological deterioration and signs of local and systemic toxicities. Escalating percent loadings of BCNU:p(CPP):SA wafer were studied, including doses of 1.93%, 3.86%, and 6.3% BCNU. The treatment was well tolerated at all three doses, and the wafer released BCNU for approximately 3 weeks. Patients treated with the BCNU polymeric wafers showed an overall mean survival from implant surgery of 48 weeks, with a total mean survival of 94 weeks from their primary tumor excision. From this study, the 3.8% BCNU dosage was chosen for further testing.[57]

A multicenter, prospective, randomized, double-blind, placebo-controlled Phase III trial was then conducted with patients presenting with recurrent GBM after primary resection.[58] Interstitial chemotherapy of 3.85% BCNU wafers with a maximum patient dose of 62 mg was administered after tumor resection. In 27 medical centers, 222 patients were randomly assigned to receive the wafers with or without BCNU. The median survival of the 110 patients who received the carmustine wafers was 31 weeks compared to 23 weeks in the control group of 112 patients ($P=0.006$). The 6-month survival was 60% in the treatment group and 47% in the placebo group (BCNU wafers caused a 50% increase in survival at 6 months when compared to controls; $P=0.02$). Adverse events were evaluated by means of Karnofsky performance score. Postmortem analysis of brain samples showed no signs of local or systemic toxicity, except for mild necrosis and inflammatory reaction around the implantation site.[58] BCNU-impregnated wafers were therefore determined to be a safe and effective alternative to systemic chemotherapy for recurrent GBM.

Interstitial Chemotherapy for Newly Diagnosed GBM

The data obtained from the usage of carmustine wafers in patients with recurrent GBM led to testing in patients with newly diagnosed GBM, due to the evidence that cancers that have not undergone previous chemotherapeutic treatment are usually more responsive to chemotherapeutics than recurrent cancers.[59] Gliadel was assessed after primary excision of a high-grade glioma, with a maximum number of 8 wafers implanted inside the residual cavity, in close contact with its margins. Patients then underwent radiotherapy, which is a standard treatment for high-grade glioma.[60-62] Two Phase III randomized, multicenter, double-blind, placebo-controlled clinical trials were designed to evaluate the effects of the wafers on survival and to evaluate the presence of toxicities or adverse events related to local chemotherapy.

Westphal et al.[60] designed the first trial, in which 240 patients presenting with high-grade glioma were randomized to receive either Gliadel wafers or empty wafers. Up to eight discs were inserted inside the cavity after primary surgery, followed by external beam radiotherapy. Overall survival, the primary endpoint, was 13.8 months in the carmustine wafer group and 11.6 months in the empty wafer group ($P=0.017$), with a hazard ratio of 0.73 ($P=0.018$). Long-term follow-up revealed that the difference between the two groups became more significant over time, and 4 years after the initial treatment, the survival benefit reached a fivefold increase among patients who received the carmustine wafers.[61] Similar results were reported by Valtonen et al.[62] in their cohort of 32 patients with Grade III and IV gliomas: The carmustine wafer group achieved a median survival of 58.1 weeks after implantation; median survival in the control group, on the other hand, was 39.9 weeks ($P < 0.012$).

Based on these results, carmustine wafers became an essential part of the neurosurgeon's armamentarium (Fig. 3). The Food and Drug Administration approved the use of carmustine wafers for patients with recurrent GBM as an adjunct to surgery in 1996, and approved their use for patients with newly diagnosed high-grade glioma as an adjunct to surgery and radiation in 2003, making it the first new treatment for patients with GBM in 23 years. Based on the Sipos et al.[55] study that demonstrated safety of escalating dosages of carmustine intracranially delivered by polymer in rats, Olivi et al.[63] conducted a Phase I study to determine the maximum tolerated dose of carmustine delivered by polymer in humans. They found that the maximum tolerated dose of carmustine from polymer was 20%. This higher dosage contained five times more carmustine than commercially available carmustine wafers, and resulted in minimal systemic carmustine exposure.[63]

In an effort to determine whether patients receiving carmustine wafers had increased rates of infection, a retrospective study was conducted to examine patients who had undergone surgical resection of an intracranial GBM followed by carmustine wafer implantation.[64] Of the 401 patients in the study, 21 (5%) developed an infection at a median time of 40 days after surgery. This incidence of infection was not higher in patients who had received the wafer. In multivariate analyses, the factors significantly associated with an increased risk of infection were prior surgery, diabetes mellitus, and longer hospital stay.[64] Over the past decade, a series of meta-analyses and literature reviews have shed light on the role and efficacy of the carmustine wafer, highlighting a number of advantages over other therapeutics and confirming its role in complementing systemic chemotherapies and radiotherapy.[65, 66]

The Stupp protocol, established in 2005, consists of radiotherapy and concomitant chemotherapy with oral temozolomide (TMZ).[67] This therapeutic combination has since been used consistently in patients diagnosed with GBM.[67] Considering the role that TMZ has had since the Stupp protocol was established, carmustine wafers have been used as either a strategic alternative for patients with disease progression after TMZ-based chemotherapy or as an

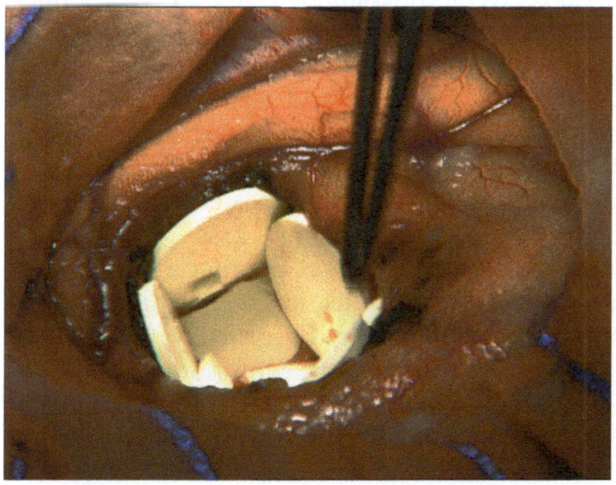

FIG. 3 Intracavitary placement of Gliadel wafers (Eisai Inc. for Arbor Pharmaceuticals) after human brain tumor excision. *Reproduced with permission from Moses MA, Brem H, Langer R. Advancing the field of drug delivery: taking aim at cancer. Cancer Cell. 2003;4 (5):337-341.*

additional therapy capable of enhancing the effect of TMZ plus radiotherapy.[68, 69] This combination therapeutic approach has led to significant increases in median and overall survival for this patient population.[69] Additionally, due to the combination of alkylating agents, patients with high levels of MGMT or with a non-methylated promoter can overcome innate mechanisms of resistance and experience prolonged survival.[70]

Another patient population that may benefit from the use of carmustine wafers is the geriatric population. Increasing evidence has supported the adoption of interstitial chemotherapy in patients who were originally excluded from randomized controlled trials because of their advanced age. Careful selection of candidates on the basis of tumor location, parenchymal invasion, focality, and overall health and mental status can lead to a significant benefit in increased survival and quality of life among elderly patients who receive Gliadel.[66] A deeper understanding of outcomes and potential adverse events in this population is therefore needed. Although case-control studies can give a hint of the benefit provided by the interstitial administration of BCNU, a randomized controlled trial of patients older than 65 years is highly recommended in order to determine the benefit to this patient population.[71]

Interstitial Chemotherapy for Brain Metastases

Metastatic lesions are still the most common malignancy of the CNS, with an incidence of 8.3–11.1 cases of brain metastasis per 100,000 individuals.[72] These malignancies account for a significant burden in terms of quality of life and treatment-related side effects. Among the most common origins of these lesions are, in decreasing relative frequency, lung and breast cancer, melanoma, renal cell carcinoma, colon cancer, and genitourinary malignancies. Approximately 25% of these secondary lesions are asymptomatic[72] and are diagnosed only postmortem. Treatment and prognosis of brain metastases vary significantly among patients. The Radiation Therapy Oncology Group (RTOG)[73] subdivided patients with brain metastases into three categories with different survivals and symptoms. Patient-related variables used for the analysis included age, Karnofsky performance score, neurologic function, and neurologic signs and symptoms. A series of criteria related to the tumor were primary pathology, status of primary lesion, positive or negative evidence of extracranial metastases, number of metastases in the brain (1 or more), and interval between the diagnosis of the primary mass and the development of brain metastases. In the RTOG Study 79-16, other valuable prognostic factors included mass effect and midline shift (present or absent), necrotic core, location, volume of sentinel lesion, prior brain surgery, total radiation dose, and response to radiation therapy, as well as continuous variables (i.e., age, radiation dose, and tumor size).[73]

Based on the RTOG classification, patients can be subdivided into three main classes with different survival and prognostic factors. A common trait shared by these categories is the poor overall patient survival, usually ranging from 7.1 months of Class 1, to 4.2 and 2.3 months for patients in Classes 2 and 3, respectively. A novel set of criteria was created by the Graded Prognostic Assessment group,[74] and their classification takes into account a series of additional prognostic factors, such as the histology of the primary malignancy. The Graded Prognostic Assessment system can be considered as an evolution of the RTOG subgroups, and facilitates accurate prediction of the behavior of each tumor and its corresponding clinical outcome.

In terms of therapeutic options, brain metastases have been treated by the following three major approaches: surgery, chemotherapy, and whole brain radiotherapy (WBRT)/stereotactic radiosurgery.[75] Surgery is the preferred choice for large lesions with mass effect, but WBRT has been considered the mainstay of treatment for intracerebral metastatic lesions. Remarkable advancements in neurosurgical and radiotherapeutic techniques and increasingly efficacious systemic therapies over the past decade have improved local control and decreased mortality of brain metastases, which has decreased the use of WBRT.[75] Historically systemic chemotherapy, given its poor BBB penetration and limited activity against metastatic lesions, has been considered a second-line treatment. Recently, mainly due to the advent of modern targeted therapy and immunotherapy, the incidence and aggressiveness of brain lesions from these metastasis-prone malignancies have been greatly reduced.[76]

Ewend et al.[77] modeled brain metastatic lesions by implanting B16-F10 melanoma cells, Lewis lung carcinoma cells, CT26 colon carcinoma cells, or renal carcinoma cells (RenCA) intracranially in athymic mice. They tested radiation alone or radiation plus locally implanted p(CPP):SA wafers loaded with carmustine, carboplatin, or sodium camptothecin. Overall survival was significantly increased in all groups that received radiation and locally delivered carmustine. Carmustine wafers alone proved to be effective against intracranial melanoma and intracranial renal cell carcinoma. The use of a carboplatin-loaded wafer alone prolonged survival when used against the melanoma and colon adenocarcinoma and, in combination with radiation, was effective against colon and renal cell carcinomas. The camptothecin-loaded wafer showed an effect against melanoma cells when given in combination with radiotherapy, but it did not prolong survival in the other cell lines, with or without the addition of radiation therapy. Local delivery of these agents, and in particular carmustine, was therefore evaluated as an intriguing option, capable of restricting tumor growth while acting as a radiosensitizing complementary therapy, which helped reduce the burden of WBRT-related adverse effects.

Gliadel and other drug-impregnated wafers, despite being originally developed for local treatment of primary malignant gliomas, could therefore have a role in decreasing the volume of secondary brain lesions and enhancing the cytotoxicity of standard radiotherapy. In 2007, Ewend et al.[78] published their result of a Phase I clinical trial. Patients with single, unilateral, resectable cerebral metastases were treated with surgical resection, interstitial chemotherapy, and postoperative fractionated external beam radiotherapy.[78] A total of 25 patients with brain metastases from melanoma, lung, colon, or breast cancer had carmustine wafers placed in the tumor resection cavity in the brain, and the local recurrence rate of tumor was 0%. These data provide evidence for further clinical trials, which should increase understanding of these wafers—a safe therapy capable of extending local control of a metastatic CNS lesion.[79-85]

Future Directions with the Polymeric Wafer

Intracranial drug delivery with the p(CPP):SA 20:80 polymer was intended to enhance local drug concentration and decrease the systemic adverse effects caused by chemotherapeutic agents. By loading different compounds with numerous mechanisms of action into this polymer matrix, the wafers (initially conceived for delivering carmustine) may be able to effectively and efficiently deliver various monotherapies as well as combination drug therapies. Minocycline, a tetracycline antibiotic and an angiogenesis inhibitor, was formulated to be released from the polymeric matrix and tested against VX2 carcinoma in the cornea of rodents.[86] Local release of this drug ensured a higher local concentration and achieved a significant reduction in tumor growth and neovascularization, which resulted in further investigations of antiangiogenic agents. These experiments led to intracranial implantation of the minocycline polymer wafer, which significantly increased median survival in a rodent glioma model.[87]

TMZ, a second-generation imidazotetrazine alkylating agent used in the Stupp protocol, has prolonged patient survival through oral administration.[64] Local delivery of TMZ from a polymeric wafer achieved significantly longer survival in a rodent model of 9L gliosarcoma compared to controls and animals receiving orally delivered TMZ.[88] In 2009, Recinos et al.[89] administered a wafer loaded with both TMZ and carmustine and found significantly prolonged overall survival in rats with intracranial 9L gliosarcoma. The intracranial delivery of these drugs increased the number of long-term survivors among experimental animals, without leading to local or systemic toxicities. Most recently, acriflavine, an antiseptic and HIF-1 inhibitor, was intracranially delivered from the p(CPP):SA polymer to rats with gliosarcoma. Control animals had a median survival of 12 days; median survival for the locally delivered acriflavine group was never reached, with treated animals living years after tumor implantation.[90] Extraordinary outcomes of 83% and 100% of tumor-bearing rats surviving nearly when treated with 50% and 25% acriflavine wafers, respectively, provides encouraging preclinical data for the future treatment of GBM.

POLYMERIC DRUG DELIVERY: PRECLINICAL MODELS AND FUTURE APPLICATIONS

The clinical translation of polymeric drug delivery was followed by efforts to effectively deliver drugs across the BBB and inside the CNS. Various intracranial delivery concepts have been in preclinical development, including the use of moldable hydrogels that could fill the tumor resection cavity[91] and polymeric nanoparticles that selectively target malignant cells and are retained inside the tumor mass.[92] These biomedical applications are at the forefront of medical innovation, and may contribute to addressing the challenging therapeutic limitations of aggressive intracranial malignancies.

Hydrogels

Hydrogels are three-dimensional polymeric networks that can retain large amounts of water within their structures. Due to their hydrophilicity, hydrogels can maintain intact molecular bonds and a cross-linked structure. These polymers differ from rigid polymeric discs in that they tend to swell when in contact with water,[93] which allows them to closely adapt to the internal margins of resection cavities. Hydrogels have emerged as candidates for local controlled delivery and release of therapeutics to malignant brain tumors.[94] The gels can be impregnated with cytotoxic drugs and then slowly injected inside the tumor cavity after resection. In situ gelation introduces stable and nonreversible bonds, allowing them to adhere firmly to the brain parenchyma. Gelation can be triggered by a variety of stimuli, including change in temperature or pH, or exposure to visible or ultraviolet light.[95] Hydrogels can be administered locally and are designed to gradually release the embedded drug.

Therapeutic compounds can either be loaded directly into the hydrogel matrix or incorporated in nanoparticles that are dispersed in the hydrogel and slowly released over the course of days or weeks.

PLGA-Based Hydrogels

PLA and PLGA are two of the most commonly used polymers for hydrogel synthesis. Their breakdown results in monomers or lactic or glycolic acid, which are nontoxic and hydrosoluble compounds. PLGA is approved by both the Food and Drug Administration and the European Medicine Agency for use in parenteral devices, and can be included in copolymers with other molecules without significant adverse reaction. Copolymers made with PGLA:plasticizers (40:60) are already used in preclinical studies, and when loaded with 30% TMZ, they showed a significant decrease in tumor burden in a C6 rodent glioma model.[96]

ReGel (MacroMed Inc.) is a thermosensitive copolymer of PLGA and polyethylene glycol (PEG) that behaves as a fluid at temperatures between 2°C and 15°C. It solidifies around 37°C, becoming insoluble and releasing its contents at a controlled fixed rate over 6 weeks.[97] OncoGel, which was ReGel loaded with paclitaxel, was applied intracranially as an adjuvant treatment to radiation therapy alone and in combination with oral TMZ in the 9L gliosarcoma rodent model.[98] OncoGel showed a significant increase in survival among animals implanted concomitantly with 9L gliosarcoma compared to controls and ReGel ($P < 0.0001$) and when the formulation was implanted 5 days after tumor inoculation (compared to controls; $P = 0.02$). A beneficial effect was also reported when OncoGel was combined with radiation therapy, resulting in an overall survival that was significantly longer than that of controls and animals treated with radiation alone.[98]

PLGA–PEG microparticles have also been used to create a free-floating powder that, after mixing with saline, results in a paste that can be molded inside the surgical resection cavity and solidifies at body temperature. In vitro experiments tested the release kinetics of etoposide, methotrexate, and Trichostatin-A, and showed gradual release of these drugs. Researchers have also conducted in vivo safety, biocompatibility, and biodegradability studies in rodents (Fig. 4).[99, 100]

FIG. 4 Electron microscopy scan of PEG/PLGA microparticles kept at 37°C for 30 days. Intact particles and pores between single nanoparticles are clearly visible. *Reproduced with permission from Rahman CV, Smith SJ, Morgan PS, et al. Adjuvant chemotherapy for brain tumors delivered via a novel intracavity moldable polymer matrix. PLoS ONE. 2013;8 (10):e77435 [open access].*

Photosensitive Hydrogels

Visible or ultraviolet light can induce a polymerization reaction of monomers and form three-dimensional structures that can be used for therapeutic devices and photosensitive hydrogels.[101] A polymer made with PEG dimethacrylate (PEG-DMA) and water was mixed in a ratio of 75:25 to embed TMZ. This was then used for the treatment of subcutaneous human xenograft U87 gliomas in athymic mice, and led to a significant decrease in tumor weight and a higher fraction of apoptotic cells compared to untreated controls and animals treated with intravenous TMZ. Photosensitive matrices require a photo initiator (in this case, researchers used a 0.5% concentration of Lucirin-TPO [UL LLC], a molecule sensitive to wavelengths of ~400 nm for 15 s) to sensitize the mass and to initiate the chemical reaction that led to the final cross-linked linear polymer.[102] This hydrogel photopolymerized almost instantaneously (in <2 min), producing a viscous modulus (≈10 kPa). TMZ was released in two phases: an initial linear burst release of around 45% of its payload, and a logarithmic diffusion of a fraction close to 20% of its initial weight during the period between day 2 and day 7 postimplantation.

Polymeric Nanoparticle-Based Systems

Polymeric micelles and nanoparticles are core-shell-type particles created through self-assembly of block copolymers. Drugs and compounds can be incorporated inside the solid core and, due to their innate stability, are ideal carriers for chemotherapeutic agents that target GBM.[103] A key feature of these particles is their size; nanoparticles are by definition just 1–1000 nm in diameter, with the vast majority of those used for BBB penetration ranging between 10 and 100 nm. Their small size allows them to cross the BBB, since endothelial pores have a diameter of approximately 100 nm. They can also permeate the BBTB, given its intrinsic leakiness and the structural disarray typical of tumor vasculature.[104]

The electric charge of these nanostructures strongly modifies their ability to permeate the endothelial lining—anionic nanoparticles can cross the BBB more easily than cationic or neutral particles can. Because strongly anionic and cationic nanoparticles are toxic to the brain parenchyma, nanoparticles used in vivo tend to be neutral or only mildly anionic.[105] Tumor targeting is usually achieved through a nonspecific mechanism. Nanoparticles are retained in the malignant tissue because of the enhanced permeability and retention *effect* typical of highly vascular tumors.[106] In situ retention, increased local permeability, and slow washout of nanoparticles lead to increased accumulation of the active agent inside the tumor core, which in turn results in higher concentrations in tumors than in healthy tissue.

Another strategy to improve the homing and tumor-targeting ability of nanoparticles is to conjugate antibodies, nucleic acids, and peptides to their surface, thereby increasing local uptake and reducing adverse effects and toxicity. This surface conjugation also reduces nanoparticle degradation and dilution and increases tumor targeting. The antibody or peptide bound to the nanoparticle surface can then alter and modify the cellular pathways that are hyperactivated in cancers.[107]

Nanoparticles are subdivided into biodegradable and nonbiodegradable nanoparticles. Biodegradable nanoparticles have an intrinsic advantage over nonbiodegradable nanoparticles because adverse effects and toxicities can be minimized through active metabolism and rapid excretion of the polymeric matrix. These particles are therefore able to deliver highly hydrophobic compounds and drugs that do not cross the BBB to the CNS by virtue of the solid core and physical shielding provided by the nanoparticle membrane.[108]

Various materials and formulations have been used over the years to synthesize nanoparticles. Polymeric matrices have played a central role, mainly because of their low toxicity, rapid bioavailability, and almost inert metabolites. Furthermore, the composition and molecular properties of nanoparticles can be drastically modified based on their purpose. Synthetic polymers (e.g., e-poly caprolactone, polyacrylate, and polyacrylamide) and natural materials (e.g., gelatin, albumin, gelatin, and polysaccharide)[109] are all broadly exploited for nanoparticle synthesis. Materials including PLA, PLGA, and poly(glycolic acid) (PGA) have played a pivotal role in biomedical engineering due to their unique biocompatibility and hydrolysis-mediated metabolism.[110] Many of these have been approved by the Food and Drug Administration for use in humans for numerous applications, and therefore should be explored for their delivery of active compounds, proteins, and nucleic acids for possible therapeutic intervention in GBM.

Synthetic Polymers

Synthetic polymers are among the most diffuse and safe materials for nanocarrier synthesis due to their innate biocompatibility and toxin-free metabolism. These molecules have become a cornerstone of modern nanoparticle-mediated therapies. A number of polymer-based nanocarriers have been developed and used in preclinical and clinical trials; PLGA, PLA, PGA, and PLGA–PEG are among the most common. For instance, PEG–PGA nanoparticles were used with cisplatin in a Phase I trial,[111] and are now undergoing a Phase III clinical trial in patients diagnosed with pancreatic cancer. This strategy significantly reduces cisplatin-related adverse events and should maximize the therapeutic effect of the drug. Docetaxel was loaded into a PLGA–PEG nanocarrier at concentrations 10 times higher than those achieved via customary routes of administration. This formula was tested with a murine model of prostate cancer with positive results in terms of drug distribution, intratumoral retention, and overall animal survival.[112] Wohlfart et al.[113] was able to boost doxycycline delivery across the BBB by packaging this compound inside poloxamer 188-coated PLGA nanoparticles. Additionally, surfactants are commonly used as a coating to help nanoparticles and other carriers decrease surface tension and cross the BBB in greater quantities.[113]

Natural Polymers

Abraxane (Celgene), a paclitaxel-loaded albumin nanoparticle, uses a formulation of natural polymers and has been tested clinically for non-small cell lung cancer,[114] prostate cancer,[115] pancreatic cancer,[116] and breast cancer.[117] Paclitaxel is a poorly water-soluble molecule that exerts its cytotoxic effect by inhibiting microtubule assembly. To overcome its extreme hydrophobicity, paclitaxel is commonly mixed with polyethoxylated castor oil, a compound known for triggering severe allergic reactions. Albumin nanoparticles have supplied a double benefit; they diminish the adverse effects of castor oil and reduce the toxicity of paclitaxel in the CNS.[118] Albumin mediates transcytosis via the gp60 receptor on the endothelium and allows nanoparticles to deliver their load directly into malignant cells.[119]

Glycans such as cyclodextrins have been tested in Phase I and II trials as polymeric nanocarriers for drugs like camptothecin in patients with solid cancers,[120] showing a significant decrease in side effects and stable disease in three of five patients with pancreatic cancer.[120] Chitosan nanoparticles have also been loaded with gemcitabine and administered orally, showing a pattern of rapid release in the first 8 hours, and a more gradual and continuous drug release thereafter. The fraction of gemcitabine transported across the intestinal wall was also increased three- to fivefold with chitosan-based nanoparticles.[121] This multitude of local delivery concepts demonstrates the complex problems that must be addressed and overcome for effective chemotherapeutic delivery across the BBB and the BBTB for patients with GBM.

Microchips

An innovation in the field of local delivery of therapeutics has been achieved through microfabrication technology. This technology has resulted in active micro-meter-scale pumps, valves, and flow channels to deliver liquid therapeutics. A solid-state silicon microchip was developed that can provide controlled release of one or more substances in a predetermined pulsatile manner; the chip can house over 1000 independent reservoirs in 17 mm of surface area.[122] This microelectromechanical system (MEMS) was shown to release carmustine in a rodent flank model and demonstrated spatial and temporal release profiles of intact bioavailable compound and clearance through the urine.[123] The miniaturized MEMS device has also delivered carmustine and TMZ in both subcutaneous and intracranial glioma models in rodents, and has reliably delivered these compounds to decrease tumor growth.[124, 125] This implantable "pharmacy on a chip" was then used in humans to deliver human parathyroid hormone fragment for the treatment of osteoporosis.[126] The pharmacokinetics, safety, tolerability, and bioequivalence studies showed no toxic or adverse events due to the device or drug. Bone marker evaluations also detected increased bone formation. Advantages to the MEMS system include customization of drug delivery and injection-like pharmacokinetic profiles without the need for frequent needle injections.

Another implantable, programmable device known as the *passive chip* contains PLA with poly(D,L-lactic-*co*-glycolic acid) membranes of varying molecular masses. These devices can deliver a predetermined amount of drug, achieving a multipulse release based on the degradation of the membranes.[127] Carmustine stability in the passive chip was superior to that of the carmustine wafer; after 48 hours in aqueous solution, the chip contained nearly twice as much intact carmustine as the wafer (70% and 38%, respectively).[128] The passive chip, or microcapsule, was also effective in delivering carmustine and TMZ to treat flank and intracranial tumor models in rodents.[128, 129] Most recently, microcapsules were used to deliver TMZ and doxorubicin to a metastatic breast adenocarcinoma in the brain.[130] The two chemotherapeutic agents displayed different distribution profiles, but resulted in safe and efficacious results. Both the microcapsule and the MEMS are implantable, programmable options that can reliably deliver various therapeutic compounds for a host of pathogenic conditions.

ADDITIONAL TECHNIQUES

Various techniques to enhance drug penetration through the parenchyma and to permeabilize the BBB are being used. One such technique includes the use of focused ultrasound systems that can penetrate these significant obstacles and increase dispersion through selected regions of the brain. Nance et al.[131] used an animal model to show that, with intravascular microbubbles and magnetic-resonance- guided focused ultrasound, the BBB could be reversibly disrupted and nanoparticles diffused farther and circulated for longer periods in the brain. This technique has multiple CNS applications in addition to neuro-oncology, including Parkinson disease, epilepsy,[132] and Alzheimer disease.[133]

CONCLUSIONS

Polymeric local drug delivery, an innovative strategy ideated in the 1980s, has significantly evolved over the past decades. A variety of different applications, including local delivery of chemotherapeutics for GBM therapy and nanoparticle-based approaches to help treat malignant and nonmalignant diseases have been developed, expanding the armamentarium that can be used to treat brain tumors. Intracranial delivery of carmustine ushered in an era in which a therapeutic drug can be locally delivered to the brain, which was previously protected by the BBB and effectively shielded from therapeutics. The development and optimization of Gliadel encompassed many mathematical and physical models of kinetics, diffusion, cytotoxicity, and efficacy, including numerous preclinical biocompatibility, safety, and efficacy studies. Now, more than 25 years since its first use in patients with recurrent GBM, a significant number of different polymeric formulations based on the local delivery rationale are under investigation for the treatment of intracranial neoplasms. Nanoparticles, micelles, hydrogels, microchips, and polymeric wafers are at the forefront of modern research, with their ability to overcome the BBB and their capacity to deliver high-dose therapeutics with neither local nor systemic toxicities. These novel therapeutic strategies should therefore be maximized to improve the quality of life and survival for patients with these cancers.

References

1. Reese T, Karnovsky MJ. Fine structural localization of a blood-brain barrier to exogenous peroxidase. *J Cell Biol.* 1967;34(1):207-217.
2. Greig NH. Optimizing drug delivery to brain tumors. *Cancer Treat Rev.* 1987;14(1):1-28.
3. Abbott NJ, Romero IA. Transporting therapeutics across the blood-brain barrier. *Mol Med Today.* 1996;2 (3):106-113.
4. Daneman R, Prat A. The blood–brain barrier. *Cold Spring Harb Perspect Biol.* 2015;7(1):a020412.
5. Yao Y, Chen Z-L, Norris EH, Strickland S. Astrocytic laminin regulates pericyte differentiation and maintains blood brain barrier integrity. *Nat Commun.* 2014;5:3413.
6. Armulik A, Genové G, Mäe M, et al. Pericytes regulate the blood–brain barrier. *Nature.* 2010;468(7323):557.
7. Serlin Y, Shelef I, Knyazer B, Friedman A. Anatomy and physiology of the blood–brain barrier. *Semin Cell Dev Biol.* 2015;38:2-6.
8. Lecrux C, Hamel E. The neurovascular unit in brain function and disease. *Acta Physiol (Oxf).* 2011;203(1):47-59.
9. Saraiva C, Praça C, Ferreira R, Santos T, Ferreira L, Bernardino L. Nanoparticle-mediated brain drug delivery: overcoming blood–brain barrier to treat neurodegenerative diseases. *J Control Release.* 2016;235:34-47.

10. Wang Z, Sun H, Sebastian Yakisich J. Overcoming the blood-brain barrier for chemotherapy: limitations, challenges and rising problems. *Anticancer Agents Med Chem*. 2014;14(8):1085-1093.
11. Drapeau A, Fortin D. Chemotherapy delivery strategies to the central nervous system: neither optional nor superfluous. *Curr Cancer Drug Targets*. 2015;15(9):752-768.
12. Baish JW, Stylianopoulos T, Lanning RM, et al. Scaling rules for diffusive drug delivery in tumor and normal tissues. *Proc Nat Acad Sci*. 2011;108(5):1799-1803.
13. Wong C, Stylianopoulos T, Cui J, et al. Multistage nanoparticle delivery system for deep penetration into tumor tissue. *Proc Nat Acad Sci*. 2011;108(6):2426-2431.
14. Lei Y, Han H, Yuan F, Javeed A, Zhao Y. The brain interstitial system: anatomy, modeling, in vivo measurement, and applications. *Prog Neurobiol*. 2017;157:230-246.
15. Marcoli M, Agnati LF, Benedetti F, et al. On the role of the extracellular space on the holistic behavior of the brain. *Rev Neurosci*. 2015;26(5):489-506.
16. Pardridge WM. CSF, blood-brain barrier, and brain drug delivery. *Expert Opin Drug Deliv*. 2016;13(7):963-975.
17. Iliff JJ, Goldman SA, Nedergaard M. Implications of the discovery of brain lymphatic pathways. *Lancet Neurol*. 2015;14(10):977-979.
18. Louveau A, Smirnov I, Keyes TJ, et al. Structural and functional features of central nervous system lymphatic vessels. *Nature*. 2015;523(7560):337.
19. Iliff JJ, Wang M, Liao Y, et al. A paravascular pathway facilitates CSF flow through the brain parenchyma and the clearance of interstitial solutes, including amyloid β. *Sci Transl Med*. 2012;4(147):147ra111.
20. Abbott N, Revest P, Romero I. Astrocyte-endothelial interaction: physiology and pathology. *Neuropathol Appl Neurobiol*. 1992;18(5):424-433.
21. Obermeier B, Daneman R, Ransohoff RM. Development, maintenance and disruption of the blood-brain barrier. *Nat Med*. 2013;19(12):1584.
22. de Vries HE, Kooij G, Frenkel D, Georgopoulos S, Monsonego A, Janigro D. Inflammatory events at blood–brain barrier in neuroinflammatory and neurodegenerative disorders: implications for clinical disease. *Epilepsia*. 2012;53(s6):45-52.
23. Dhermain FG, Hau P, Lanfermann H, Jacobs AH, van den Bent MJ. Advanced MRI and PET imaging for assessment of treatment response in patients with gliomas. *Lancet Neurol*. 2010;9(9):906-920.
24. Watkins S, Robel S, Kimbrough IF, Robert SM, Ellis-Davies G, Sontheimer H. Disruption of astrocyte–vascular coupling and the blood–brain barrier by invading glioma cells. *Nat Commun*. 2014;5:4196.
25. Hardee ME, Zagzag D. Mechanisms of glioma-associated neovascularization. *Am J Pathol*. 2012;181(4):1126-1141.
26. Demuth T, Berens ME. Molecular mechanisms of glioma cell migration and invasion. *J Neurooncol*. 2004;70 (2):217-228.
27. Plate KH, Scholz A, Dumont DJ. Tumor angiogenesis and anti-angiogenic therapy in malignant gliomas revisited. *Acta Neuropathol*. 2012;124(6):763-775.
28. Göpferich A, Teßmar J. Polyanhydride degradation and erosion. *Adv Drug Deliv Rev*. 2002;54(7):911-931.
29. Leong K, Brott B, Langer R. Bioerodible polyanhydrides as drug-carrier matrices. I: Characterization, degradation, and release characteristics. *J Biomed Mater Res A*. 1985;19(8):941-955.
30. Laurencin C, Domb A, Morris C, et al. Poly (anhydride) administration in high doses in vivo: studies of biocompatibility and toxicology. *J Biomed Mater Res A*. 1990;24(11):1463-1481.
31. Kumar N, Langer RS, Domb AJ. Polyanhydrides: an overview. *Adv Drug Deliv Rev*. 2002;54(7):889-910.
32. Akbari H, D'Emanuele A, Attwood D. Effect of geometry on the erosion characteristics of polyanhydride matrices. *Int J Pharm*. 1998;160(1):83-89.
33. Wu MP, Tamada JA, Brem H, Langer R. In vivo versus in vitro degradation of controlled release polymers for intracranial surgical therapy. *J Biomed Mater Res A*. 1994;28(3):387-395.
34. Liechty WB, Kryscio DR, Slaughter BV, Peppas NA. Polymers for drug delivery systems. *Annu Rev Chem Biomol Eng*. 2010;1:149-173.
35. Akbari H, D'emanuele A, Attwood D. Effect of fabrication technique on the erosion characteristics of polyanhydride matrices. *Pharm Dev Technol*. 1998;3(2):251-259.
36. Brem H, Kader A, Epstein JI, et al. Biocompatibility of a biodegradable, controlled-release polymer in the rabbit brain. *Sel Cancer Ther*. 1989;5(2):55-65.
37. Leong K, D'Amore P, Marletta M, Langer R. Bioerodible polyanhydrides as drug-carrier matrices. II. Biocompatibility and chemical reactivity. *J Biomed Mater Res A*. 1986;20(1):51-64.
38. Tamargo RJ, Epstein JI, Reinhard CS, Chasin M, Brem H. Brain biocompatibility of a biodegradable, controlled-release polymer in rats. *J Biomed Mater Res A*. 1989;23(2):253-266.

39. Brem H, Domb A, Lenartz D, Dureza C, Olivi A, Epstein JI. Brain biocompatibility of a biodegradable controlled release polymer consisting of anhydride copolymer of fatty acid dimer and sebacic acid. *J Control Release*. 1992;19 (1–3):325-329.
40. Frost P, DeVita VT. Pigmentation due to a new antitumor agent: effects of topical application of BCNU [1,3-bis (2-chloroethyl)-1-nitrosourea]. *Arch Dermatol*. 1966;94(3):265-268.
41. Cianfriglia F, Pompili A, Riccio A, Grassi A. CCNU-chemotherapy of hemispheric supratentorial glioblastoma multiforme. *Cancer*. 1980;45(6):1289-1299.
42. Garrett M, Hughes H, Freedman L. A comparison of radiotherapy alone with radiotherapy and CCNU in cerebral glioma. *Clin Oncol*. 1978;4(1):71-76.
43. Eyre HJ, Quagliana JM, Eltringham JR, et al. Randomized comparisons of radiotherapy and CCNU versus radiotherapy, CCNU plus procarbazine for the treatment of malignant gliomas following surgery. *J Neurooncol*. 1983;1(3):171-177.
44. Payne DG, Simpson WJ, Keen C, Platts ME. Malignant astrocytoma. Hyperfractionated and standard radiotherapy with chemotherapy in a randomized prospective clinical trial. *Cancer*. 1982;50(11):2301-2306.
45. Seiler R, Zimmermann A, Markwalder H. Adjuvant chemotherapy with VM 26 and CCNU after operation and radiotherapy of high-grade supratentorial astrocytomas. *Surg Neurol*. 1980;13(1):65-68.
46. Solero C, Monfardini S, Brambilla C, et al. Controlled study with BCNU vs. CCNU as adjuvant chemotherapy following surgery plus radiotherapy for glioblastoma multiforme. *Cancer Clin Trials*. 1979;2(1):43-48.
47. Walker MD, Alexander Jr E, Hunt WE, et al. Evaluation of BCNU and/or radiotherapy in the treatment of anaplastic gliomas: a cooperative clinical trial. *J Neurosurg*. 1978;49(3):333-343.
48. Domb AJ, Nudelman R. In vivo and in vitro elimination of aliphatic polyanhydrides. *Biomaterials*. 1995;16 (4):319-323.
49. Domb AJ, Rock M, Schwartz J, et al. Metabolic disposition and elimination studies of a radiolabelled biodegradable polymeric implant in the rat brain. *Biomaterials*. 1994;15(9):681-688.
50. Domb A, Rock M, Perkin C, Yipchuck G, Broxup B, Villemure J. Excretion of a radiolabelled anticancer biodegradable polymeric implant from the rabbit brain. *Biomaterials*. 1995;16(14):1069-1072.
51. Tamargo RJ, Myseros JS, Epstein JI, Yang MB, Chasin M, Brem H. Interstitial chemotherapy of the 9L gliosarcoma: controlled release polymers for drug delivery in the brain. *Cancer Res*. 1993;53(2):329-333.
52. Grossman SA, Reinhard C, Colvin OM, et al. The intracerebral distribution of BCNU delivered by surgically implanted biodegradable polymers. *J Neurosurg*. 1992;76(4):640-647.
53. Fung LK, Shin M, Tyler B, Brem H, Saltzman WM. Chemotherapeutic drugs released from polymers: distribution of 1,3-bis (2-chloroethyl)-l-nitrosourea in the rat brain. *Pharma Res*. 1996;13(5):671-682.
54. Buahin KG, Brem H. Interstitial chemotherapy of experimental brain tumors: comparison of intratumoral injection versus polymeric controlled release. *J Neurooncol*. 1995;26(2):103-110.
55. Sipos EP, Tyler B, Piantadosi S, Burger PC, Brem H. Optimizing interstitial delivery of BCNU from controlled release polymers for the treatment of brain tumors. *Cancer Chemother Pharmacol*. 1997;39(5):383-389.
56. Fung LK, Ewend MG, Sills A, et al. Pharmacokinetics of interstitial delivery of carmustine, 4-hydroperoxycyclophosphamide, and paclitaxel from a biodegradable polymer implant in the monkey brain. *Cancer Res*. 1998;58(4):672-684.
57. Brem H, Mahaley Jr MS, Vick NA, et al. Interstitial chemotherapy with drug polymer implants for the treatment of recurrent gliomas. *J Neurosurg*. 1991;74(3):441-446.
58. Brem H, Piantadosi S, Burger PC, et al. Placebo-controlled trial of safety and efficacy of intraoperative controlled delivery by biodegradable polymers of chemotherapy for recurrent gliomas. *Lancet*. 1995;345 (8956):1008-1012.
59. Szakács G, Paterson JK, Ludwig JA, Booth-Genthe C, Gottesman MM. Targeting multidrug resistance in cancer. *Nat Rev Drug Discov*. 2006;5(3):219.
60. Westphal M, Hilt DC, Bortey E, et al. A phase 3 trial of local chemotherapy with biodegradable carmustine (BCNU) wafers (Gliadel wafers) in patients with primary malignant glioma. *Neurooncology*. 2003;5(2):79-88.
61. Westphal M, Ram Z, Riddle V, Hilt D, Bortey E. Group ECotGS. Gliadel® wafer in initial surgery for malignant glioma: long-term follow-up of a multicenter controlled trial. *Acta Neurochir*. 2006;148(3):269-275.
62. Valtonen S, Ul T, Toivanen P, et al. Interstitial chemotherapy with carmustine-loaded polymers for high-grade gliomas: a randomized double-blind study. *Neurosurgery*. 1997;41(1):44-49.
63. Olivi A, Grossman SA, Tatter S, et al. Dose escalation of carmustine in surgically implanted polymers in patients with recurrent malignant glioma: a new approaches to brain tumor therapy CNS consortium trial. *J Clin Oncol*. 2003;21(9):1845-1849.

64. Chaichana KL, Kone L, Bettegowda C, et al. Risk of surgical site infection in 401 consecutive patients with glioblastoma with and without carmustine wafer implantation. *Neurol Res*. 2015;37(8):717-726.
65. W-k X, Shao C, Qi Z-y, Yang C, Wang Z. The role of Gliadel wafers in the treatment of newly diagnosed GBM: a meta-analysis. *Drug Des Devel Ther*. 2015;9:3341.
66. Chowdhary SA, Ryken T, Newton HB. Survival outcomes and safety of carmustine wafers in the treatment of high-grade gliomas: a meta-analysis. *J Neurooncol*. 2015;122(2):367-382.
67. Stupp R, Mason WP, Van Den Bent MJ, et al. Radiotherapy plus concomitant and adjuvant temozolomide for glioblastoma. *N Engl J Med*. 2005;352(10):987-996.
68. Miglierini P, Bouchekoua M, Rousseau B, Hieu PD, Malhaire J-P, Pradier O. Impact of the per-operatory application of GLIADEL wafers (BCNU, carmustine) in combination with temozolomide and radiotherapy in patients with glioblastoma multiforme: efficacy and toxicity. *Clin Neurol Neurosurg*. 2012;114(9):1222-1225.
69. McGirt MJ, Than KD, Weingart JD, et al. Gliadel (BCNU) wafer plus concomitant temozolomide therapy after primary resection of glioblastoma multiforme. *J Neurosurg*. 2009;110(3):583-588.
70. Lechapt-Zalcman E, Levallet G, Dugue AE, et al. O6-methylguanine-DNA methyltransferase (MGMT) promoter methylation and low MGMT-encoded protein expression as prognostic markers in glioblastoma patients treated with biodegradable carmustine wafer implants after initial surgery followed by radiotherapy with concomitant and adjuvant temozolomide. *Cancer*. 2012;118(18):4545-4554.
71. Chaichana KL, Zaidi H, Pendleton C, et al. The efficacy of carmustine wafers for older patients with glioblastoma multiforme: prolonging survival. *Neurol Res*. 2011;33(7):759-764.
72. Nayak L, Lee EQ, Wen PY. Epidemiology of brain metastases. *Curr Oncol Rep*. 2012;14(1):48-54.
73. Gaspar L, Scott C, Rotman M, et al. Recursive partitioning analysis (RPA) of prognostic factors in three radiation therapy oncology group (RTOG) brain metastases trials. *Int J Radiat Oncol Biol Phys*. 1997;37(4):745-751.
74. Sperduto PW, Kased N, Roberge D, et al. Summary report on the graded prognostic assessment: an accurate and facile diagnosis-specific tool to estimate survival for patients with brain metastases. *J Clin Oncol*. 2012;30(4):419.
75. Thiagarajan A, Yamada Y. Radiobiology and radiotherapy of brain metastases. *Clin Exp Metastasis*. 2017;34 (6–7):411-419.
76. Ahluwalia MS, Winkler F. Targeted and immunotherapeutic approaches in brain metastases. Paper Presented at: American Society of Clinical Oncology Educational Book. American Society of Clinical Oncology. Meeting 2015.
77. Ewend MG, Williams JA, Tabassi K, et al. Local delivery of chemotherapy and concurrent external beam radiotherapy prolongs survival in metastatic brain tumor models. *Cancer Res*. 1996;56(22):5217-5223.
78. Ewend MG, Brem S, Gilbert M, et al. Treatment of single brain metastasis with resection, intracavity carmustine polymer wafers, and radiation therapy is safe and provides excellent local control. *Clin Cancer Res*. 2007;13 (12):3637-3641.
79. Ewend MG, Sampath P, Williams JA, Tyler BM, Brem H. Local delivery of chemotherapy prolongs survival in experimental brain metastases from breast carcinoma. *Neurosurgery*. 1998;43(5):1185-1192.
80. Mu F, Lucas JT, Watts JM, et al. Tumor resection with carmustine wafer placement as salvage therapy after local failure of radiosurgery for brain metastasis. *J Clin Neurosci*. 2015;22(3):561-565.
81. Ene CI, Nerva JD, Morton RP, et al. Safety and efficacy of carmustine (BCNU) wafers for metastatic brain tumors. *Surg Neurol Int*. 2016;7(Suppl 11):S295.
82. Brem S, Meyers CA, Palmer G, Booth-Jones M, Jain S, Ewend MG. Preservation of neurocognitive function and local control of 1 to 3 brain metastases treated with surgery and carmustine wafers. *Cancer*. 2013;119 (21):3830-3838.
83. Abel TJ, Ryken T, Lesniak MS, Gabikian P. Gliadel for brain metastasis. *Surg Neurol Int*. 2013;4(Suppl 4):S289.
84. Kerklaan BM, van Tellingen O, Huitema A, et al. Strategies to target drugs to gliomas and CNS metastases of solid tumors. *J Neurol*. 2016;263(3):428-440.
85. Zagar TM, Van Swearingen A, Kaidar-Person O, Ewend MG, Anders CK. Multidisciplinary management of breast cancer brain metastases. *Oncology*. 2016;30(10).
86. Tamargo RJ, Bok RA, Brem H. Angiogenesis inhibition by minocycline. *Cancer Res*. 1991;51(2):672-675.
87. Weingart JD, Sipos EP, Brem H. The role of minocycline in the treatment of intracranial 9L glioma. *J Neurosurg*. 1995;82(4):635-640.
88. Brem S, Tyler B, Li K, et al. Local delivery of temozolomide by biodegradable polymers is superior to oral administration in a rodent glioma model. *Cancer Chemother Pharmacol*. 2007;60(5):643-650.

89. Recinos VR, Tyler BM, Bekelis K, et al. Combination of intracranial temozolomide with intracranial carmustine improves survival when compared with either treatment alone in a rodent glioma model. *Neurosurgery*. 2010;66 (3):530-537.
90. Mangraviti A, Raghavan T, Volpin F, et al. HIF-1α-targeting Acriflavine provides long term survival and radiological tumor response in brain Cancer therapy. *Sci Rep*. 2017;7(1).
91. Kamata H, Li X, Ui C, Sakai T. Design of hydrogels for biomedical applications. *Adv Healthc Mat*. 2015;4 (16):2360-2374.
92. Paolillo M, Boselli C, Schinelli S. Glioblastoma under siege: an overview of current therapeutic strategies. *Brain Sci*. 2018;8(1):15.
93. Peppas N, Bures P, Leobandung W, Ichikawa H. Hydrogels in pharmaceutical formulations. *Eur J Pharm Biopharm*. 2000;50(1):27-46.
94. Lin C-C, Metters AT. Hydrogels in controlled release formulations: network design and mathematical modeling. *Adv Drug Deliv Rev*. 2006;58(12 – 13):1379-1408.
95. Van Tomme SR, Storm G, Hennink WE. In situ gelling hydrogels for pharmaceutical and biomedical applications. *Int J Pharm*. 2008;355(1–2):1-18.
96. Akbar U, Jones T, Winestone J, et al. Delivery of temozolomide to the tumor bed via biodegradable gel matrices in a novel model of intracranial glioma with resection. *J Neurooncol*. 2009;94(2):203-212.
97. Vellimana AK, Recinos VR, Hwang L, et al. Combination of paclitaxel thermal gel depot with temozolomide and radiotherapy significantly prolongs survival in an experimental rodent glioma model. *J Neurooncol*. 2013;111 (3):229-236.
98. Tyler B, Fowers KD, Li KW, et al. A thermal gel depot for local delivery of paclitaxel to treat experimental brain tumors in rats. *J Neurosurg*. 2010;113(2):210-217.
99. Rahman CV, Smith SJ, Morgan PS, et al. Adjuvant chemotherapy for brain tumors delivered via a novel intracavity moldable polymer matrix. *PLoS ONE*. 2013;8(10).
100. Smith SJ, Rahman CV, Clarke P, et al. Surgical delivery of drug releasing poly (lactic-co-glycolic acid)/poly (ethylene glycol) paste with in vivo effects against glioblastoma. *Ann R Coll Surg Engl*. 2014;96(7):495-501.
101. Nguyen KT, West JL. Photopolymerizable hydrogels for tissue engineering applications. *Biomaterials*. 2002;23 (22):4307-4314.
102. Fourniols T, Randolph LD, Staub A, et al. Temozolomide-loaded photopolymerizable PEG-DMA-based hydrogel for the treatment of glioblastoma. *J Control Release*. 2015;210:95-104.
103. Miura Y, Takenaka T, Toh K, et al. Cyclic RGD-linked polymeric micelles for targeted delivery of platinum anticancer drugs to glioblastoma through the blood–brain tumor barrier. *ACS Nano*. 2013;7(10): 8583-8592.
104. Zhou Y, Peng Z, Seven ES, Leblanc RM. Crossing the blood-brain barrier with nanoparticles. *J Control Release*. 2017.
105. Brand W, Noorlander CW, Giannakou C, et al. Nanomedicinal products: a survey on specific toxicity and side effects. *Int J Nanomedicine*. 2017;12:6107.
106. Danhier F. To exploit the tumor microenvironment: since the EPR effect fails in the clinic, what is the future of nanomedicine? *J Control Release*. 2016;244:108-121.
107. Nam J, Won N, Bang J, et al. Surface engineering of inorganic nanoparticles for imaging and therapy. *Adv Drug Deliv Rev*. 2013;65(5):622-648.
108. Kumari A, Yadav SK, Yadav SC. Biodegradable polymeric nanoparticles based drug delivery systems. *Colloids Surf B Biointerfaces*. 2010;75(1):1-18.
109. Wilczewska AZ, Niemirowicz K, Markiewicz KH, Car H. Nanoparticles as drug delivery systems. *Pharmacol Rep*. 2012;64(5):1020-1037.
110. Couvreur P, Vauthier C. Nanotechnology: intelligent design to treat complex disease. *Pharm Res*. 2006;23 (7):1417-1450.
111. Plummer R, Wilson R, Calvert H, et al. A Phase I clinical study of cisplatin-incorporated polymeric micelles (NC-6004) in patients with solid tumours. *Br J Cancer*. 2011;104(4):593.
112. Farokhzad OC, Cheng J, Teply BA, et al. Targeted nanoparticle-aptamer bioconjugates for cancer chemotherapy in vivo. *Proc Nat Acad Sci*. 2006;103(16):6315-6320.
113. Wohlfart S, Khalansky AS, Gelperina S, et al. Efficient chemotherapy of rat glioblastoma using doxorubicin-loaded PLGA nanoparticles with different stabilizers. *PLoS ONE*. 2011;6(5).
114. Paik PK, James LP, Riely GJ, et al. A phase 2 study of weekly albumin-bound paclitaxel (Abraxane®) given as a two-hour infusion. *Cancer Chemother Pharmacol*. 2011;68(5):1331-1337.

115. Shepard D, Dreicer R, Garcia J, et al. Phase II trial of neoadjuvant nab-paclitaxel in high risk patients with prostate cancer undergoing radical prostatectomy. *J Urol*. 2009;181(4):1672-1677.
116. Von Hoff DD, Ervin T, Arena FP, et al. Increased survival in pancreatic cancer with nab-paclitaxel plus gemcitabine. *N Engl J Med*. 2013;369(18):1691-1703.
117. Rugo HS, Barry WT, Moreno-Aspitia A, et al. Randomized phase III trial of paclitaxel once per week compared with nanoparticle albumin-bound nab-paclitaxel once per week or ixabepilone with bevacizumab as first-line chemotherapy for locally recurrent or metastatic breast cancer: CALGB 40502/NCCTG N063H (Alliance). *J Clin Oncol*. 2015;33(21):2361.
118. Damascelli B, Cantu G, Mattavelli F, et al. Intraarterial chemotherapy with polyoxyethylated castor oil free paclitaxel, incorporated in albumin nanoparticles (ABI-007). *Cancer*. 2001;92(10):2592-2602.
119. Predescu D, Vogel SM, Malik AB. Functional and morphological studies of protein transcytosis in continuous endothelia. *Am J Physiol Lung Cell Mol Physiol*. 2004;287(5):L895-L901.
120. Yen Y, Synold T, Schluep T, Hwang J, Oliver J, Davis M. First-in-human phase I trial of a cyclodextrin-containing polymer-camptothecin nanoparticle in patients with solid tumors. *J Clin Oncol*. 2007;25(18_suppl):14078.
121. Derakhshandeh K, Fathi S. Role of chitosan nanoparticles in the oral absorption of gemcitabine. *Int J Pharm*. 2012;437(1–2):172-177.
122. Santini Jr JT, Cima MJ, Langer R. A controlled-release microchip. *Nature*. 1999;397(6717):335.
123. Li Y, Shawgo RS, Tyler B, et al. In vivo release from a drug delivery MEMS device. *J Control Release*. 2004;100(2):211-219.
124. Li Y, Duc HLH, Tyler B, et al. In vivo delivery of BCNU from a MEMS device to a tumor model. *J Control Release*. 2005;106(1–2):138-145.
125. Masi BC, Tyler BM, Bow H, et al. Intracranial MEMS based temozolomide delivery in a 9L rat gliosarcoma model. *Biomaterials*. 2012;33(23):5768-5775.
126. Farra R, Sheppard NF, McCabe L, et al. First-in-human testing of a wirelessly controlled drug delivery microchip. *Sci Transl Med*. 2012;4(122):122ra121.
127. Grayson ACR, Choi IS, Tyler BM, et al. Multi-pulse drug delivery from a resorbable polymeric microchip device. *Nat Mater*. 2003;2(11):767.
128. Kim GY, Tyler BM, Tupper MM, et al. Resorbable polymer microchips releasing BCNU inhibit tumor growth in the rat 9L flank model. *J Control Release*. 2007;123(2):172-178.
129. Scott AW, Tyler BM, Masi BC, et al. Intracranial microcapsule drug delivery device for the treatment of an experimental gliosarcoma model. *Biomaterials*. 2011;32(10):2532-2539.
130. Upadhyay UM, Tyler B, Patta Y, et al. Intracranial microcapsule chemotherapy delivery for the localized treatment of rodent metastatic breast adenocarcinoma in the brain. *Proc Natl Acad Sci U S A*. 2014;111(45):16071-16076.
131. Nance E, Timbie K, Miller GW, et al. Non-invasive delivery of stealth, brain-penetrating nanoparticles across the blood – brain barrier using MRI-guided focused ultrasound. *J Control Release*. 2014;189:123-132.
132. Quadri SA, Waqas M, Khan I, et al. High-intensity focused ultrasound: past, present, and future in neurosurgery. *Neurosurg Focus*. 2018;44(2):E16.
133. Burgess A, Nhan T, Moffatt C, Klibanov A, Hynynen K. Analysis of focused ultrasound-induced blood–brain barrier permeability in a mouse model of Alzheimer's disease using two-photon microscopy. *J Control Release*. 2014;192:243-248.

Further Reading

134. Begley DJ. Delivery of therapeutic agents to the central nervous system: the problems and the possibilities. *Pharmacol Ther*. 2004;104(1):29-45.
135. Noell S, Wolburg-Buchholz K, Mack AF, Wolburg H, Fallier-Becker P. The blood-brain barrier in brain tumours. In: *Management of CNS Tumors*. InTech; 2011.
136. Moses MA, Brem H, Langer R. Advancing the field of drug delivery: taking aim at cancer. *Cancer Cell*. 2003;4(5):337-341.

CHAPTER

15

Immunomodulatory Methods

Cynthia Kassab, Brittany Parker Kerrigan, Hillary Caruso, Sanaalarab Al Enazy, Amy B. Heimberger

Department of Neurosurgery, The University of Texas MD Anderson Cancer Center, Houston, TX, United States

IMMUNOTHERAPY FOR MALIGNANCIES OF THE CENTRAL NERVOUS SYSTEM

The first pioneer to use immunotherapy in the setting of malignancy was the surgeon William Bradley Coley[1]; however, this therapeutic strategy has only gained widespread acceptance in the past decade. The application of immunotherapy to malignancies of the central nervous system (CNS) has lagged, partly due to presumed lack of immune responses within the CNS, and partly because of unclear characterization of different immune responses among various organs and types of malignancies. The immune system is a multifaceted network that defends the host against internal and external immune challenges. The different immune facets communicate via mediators such as cytokines, chemokines, and growth factors to ensure a well-coordinated response. Immunomodulation involves an induced manipulation of the immune system to either suppress or enhance its response to specific biological conditions. Immunosuppressive methods have changed the course of numerous autoimmune diseases over the years. However, the focus has now shifted to a more constructive and restorative modulation of immune responses, especially oncology, rendering immunotherapy the fourth pillar in cancer treatment (the first three pillars being surgical resection, radiation therapy, and chemotherapy).

There has been a long-standing presumption that the CNS, like the eye, was "immunoprivileged,"[2] meaning that it possessed its own unique defenses and did not rely on the systemic immune system for surveillance. This notion emanated in part from our "anatomo-histological ignorance" of the CNS. The CNS was thought to be immunologically isolated due to tight junctions of the blood–brain barrier (BBB) inhibiting circulating immunological cells from entering the brain, and a lack of lymphatic circuits draining the CNS, thereby precluding antigen presentation in draining lymph nodes.[3] These erroneous

Nervous System Drug Delivery
https://doi.org/10.1016/B978-0-12-813997-4.00015-3

© 2019 Elsevier Inc. All rights reserved.

assumptions impeded enthusiasm for application of immunotherapy to the diseases of the CNS. However, this is not meant to imply that there are not differences in the types of immune responses to tumors within the CNS relative to other organ sites.

The final dismantling of the presumption of immunoprivilege of the CNS largely stems from intracranial brain imaging with multiphoton microscopy. Using this type of imaging in combination with flow cytometric analysis of meninges, Louveau et al. demonstrated that functional lymphatic vessels line the dural sinuses that carry both fluid and immune cells from the cerebrospinal fluid (CSF) directly to deep cervical lymph nodes.[4] Similarly, Cserr et al. described the afferent pathways, by which antigens exit the brain to be presented to the effector cells in the lymph nodes through the arachnoid villi and granulations to the venous sinuses.[5] Additional routes have also been described involving the perivascular network in the Virchow-Robin space[6] and via the subarachnoid space through the pores of the cribriform plate to the nasal mucosa, where it connects the nasal lymphatics to the deep cervical and retropharyngeal lymph nodes. In the cervical lymph nodes, the effector cells are activated on encountering antigen-presenting cells (APCs) and circulate back to the brain by three extravasation efferent conduits: (1) through the leptomeningeal microcapillaries into the subarachnoid space,[7] (2) through the choroid plexus,[8,9] and (3) across the postcapillary venules of the BBB[9] (Fig. 1). In patients with inflammatory conditions, the BBB is disrupted and immune cells can cross it effortlessly.[10-12]

The brain has a unique immune population of microglia, known as resident macrophages of the brain. Microglia can constitute up to 40% of the tumor mass of a glioblastoma, and can predominate in the mesenchymal glioblastoma subtype relative to the proneural or classical subtypes.[13, 14] The immune functional activities of microglia can be heterogeneous in the tumor microenvironment, but may assume a phenotype that does not support antitumor immune effector function overall.[15, 16] More specifically, although microglia express the major histocompatibility complex (MHC), they are inefficient at presenting tumor antigens, and they may lack expression of T cell costimulatory molecules.[17] A variety of CNS immune cells, such as macrophages and bone-marrow derived dendritic cells, express the MHC[18] and are able to act as APCs,[19] but the degree to which they mediate this activity in the tumor microenvironment within the CNS is unknown. These cells can travel from the CNS to the lymph nodes, where they initiate a systemic response.[5] T cells are not physiologically indigenous inhabitants of the brain, and their entry requires a high affinity for brain markers and integrin expression.[20] Kivisakk et al. showed that T-cell trafficking occurs via P-selectin in the choroid plexus and the meninges and that T cells in the noninflamed CNS are characterized by the presence of CD45RO+/CD27+/CCR7/L-selectins, which are markers for previously activated memory cells that preserve the ability of these cells to home back to the lymphoid organs.[9] Cumulatively, based on these immune cell trafficking studies, immunotherapy could be logically applied to CNS malignancies.

In this chapter, we provide an overview of immunotherapeutic strategies, including immune checkpoint inhibitors, active immunotherapy/vaccination, effector cell therapy, and oncolytic viruses, and provide a critical assessment of those that hold the most promise. Notably, although most immunotherapy clinical trials have focused on glioblastoma, current data are now emerging that other types of CNS malignancies and disorders may be amenable to immunotherapeutic strategies.

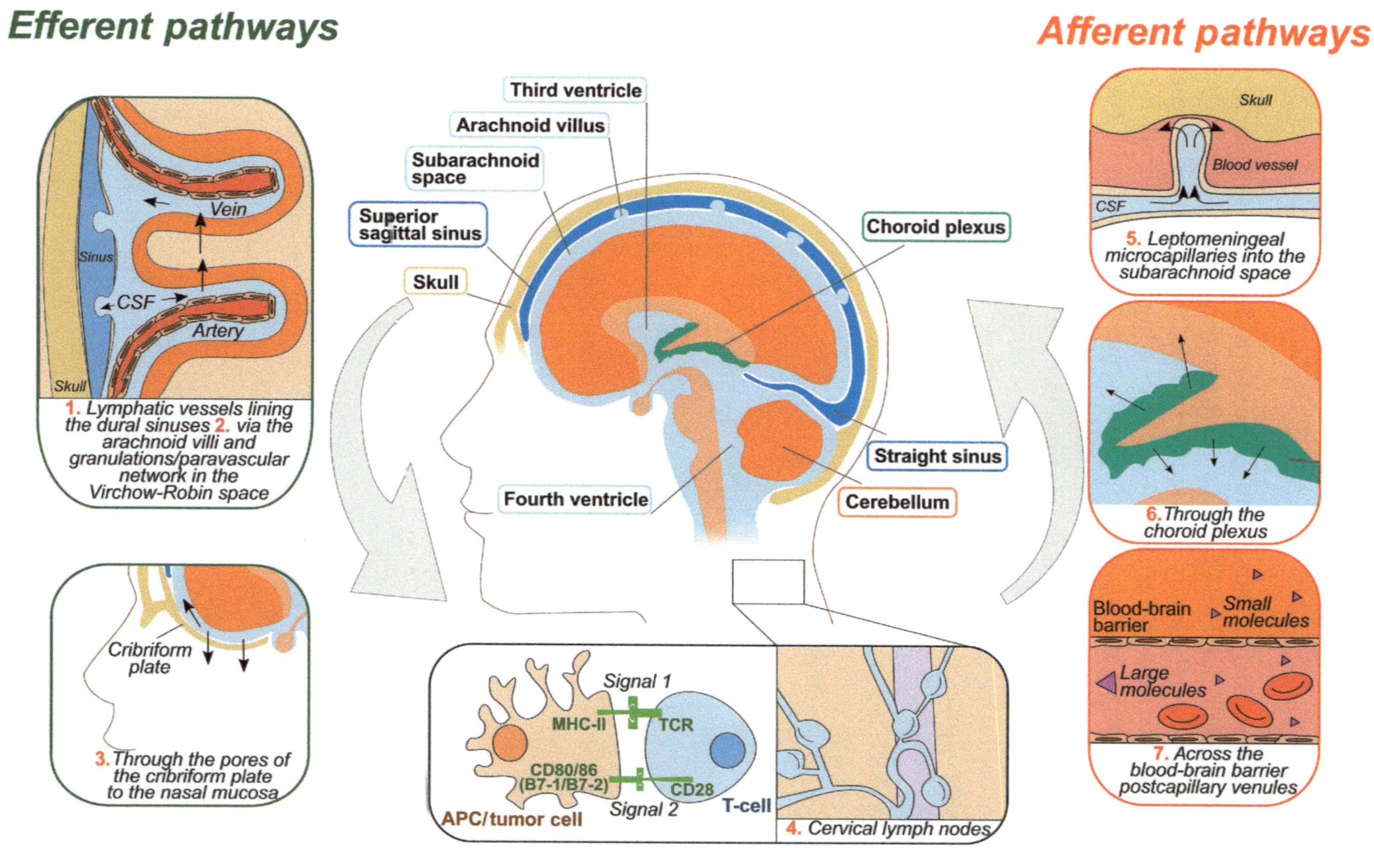

FIG. 1 Afferent (right) and efferent (left) immune cells circulation in the central nervous system. Antigens exit the brain to be presented to the effector cells in the deep cervical and retropharyngeal lymph nodes via (1) functional lymphatic vessels that line the dural sinuses, (2) through the granulations of the arachnoid villi to the venous sinuses and the perivascular network in the Virchow-Robin space, and (3) via the subarachnoid space through the pores of the cribriform plate to the nasal mucosa, where it joins the nasal lymphatics. (4) In the cervical lymph nodes, the effector cells are activated on encountering the antigen-presenting cell. Activated immune cells then circulate back to the brain via (5) the leptomeningeal microcapillaries into the subarachnoid space, (6) through the choroid plexus, and (7) across the blood–brain barrier postcapillary venules.

IMMUNE BIOLOGICS: ANTIBODIES FOR THE TREATMENT OF DISEASES OF THE CENTRAL NERVOUS SYSTEM

Production and History

Monoclonal antibodies (mAbs) are lab-produced proteins designed to recognize specific antigens expressed on target cells (Fig. 2). mAbs can be highly specific and sensitive, with minimal cross-reactivity and can be easily conjugated to other molecules for a variety of purposes. However, as homogeneous molecules, mAbs have a high susceptibility for degradation and a short shelf life if not stored properly. Furthermore, mAbs can be costly due to the complexity of their production and dosing requirements. mAb therapy was the first type of immunotherapy to gain approval from the Federal Drug Administration (FDA) roughly 30 years ago. The first-approved mAB was muromonab (against CD3), which was developed as a prophylactic measure to prevent kidney allograft rejection.[21] As of March 2017, more than 60 antibodies have been approved for treatment of a wide variety of diseases affecting different organ systems. Malignancies and autoimmune disorders are the primary indications for mAb therapy.

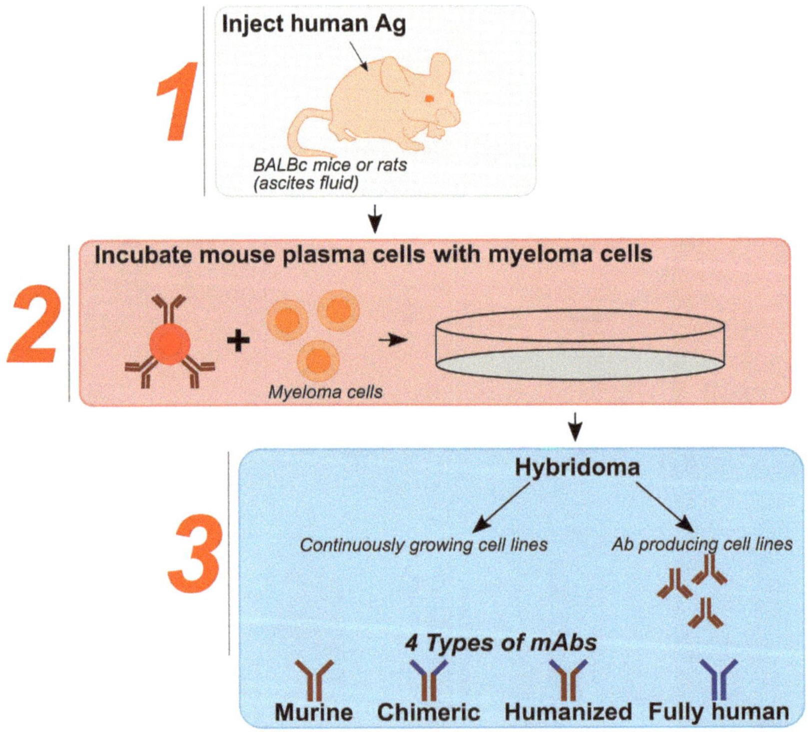

FIG. 2 Production of monoclonal antibodies: (1) to create monoclonal antibodies, scientists inoculate mice with the human antigen and harvest the plasma cells that produce these antibodies, (2) the antibody-producing cells are then mixed with rapidly dividing myeloma B cell lines to produce a hybridoma, and (3) hybridomas divide and produce 2 types of cells: a continuously growing cell line and an antibody-producing cell line. Antibodies are typically designated as murine, chimeric, humanized, or fully humanized.

Mechanism of Action and Applications

mAbs can have four mechanisms of action pertaining to the pathophysiology of the disease and the purpose of their application: (1) neutralization, (2) antibody-dependent cell-mediated cytotoxicity (ADCC), (3) complement-dependent cytotoxicity (CDC), and (4) drug conjugation with an associated toxin. As macromolecules, mAbs typically do not cross the BBB and enter the brain (brain:blood ratio is 1:1000 for immunoglobulin G [IgG]),[22] and mAbs have been modified to protect them from degradation and to optimize their concentration in the brain. Yet, depending on the antigenic target, some mAbs can cross into the brain, especially when the BBB is disrupted, such as with anti-CD20,[23] or if they are targeting the tumor vasculature, such as with bevacizumab.[24] mAbs are typically administered intravenously to avoid degradation by the acidity of the stomach and by the first passage in the enterohepatic cycle.[25]

NAKED MONOCLONAL ANTIBODIES

Monoclonal Antibodies in Neuro-Oncology

The most extensive use of mAbs has been in the field of oncotherapy, and mAbs have demonstrated marked therapeutic success in the management of hematologic and solid tumors. Examples include trastuzumab (anti-HER2 humanized monoclonal) for breast cancer, alemtuzumab for chronic lymphoid leukemia, and dinutuximab for neuroblastoma. In 2009, the FDA approved the mAb bevacizumab for patients with recurrent glioblastoma based on finding an increase in their progression-free survival time (but not overall survival).[26] Bevacizumab is an anti-vascular endothelial growth factor (anti-VEGF) humanized mAb that inhibits the binding of the angiogenic cytokine VEGF to its receptors, thereby blocking signal transduction and blood vessel growth to the tumor. Because bevacizumab targets the vasculature of the tumor and requires no access to the tumor cells, BBB penetration is not necessarily a limitation for this agent to hit its target. The antiangiogenic effect also considerably reduces peritumoral cerebral edema; as such, bevacizumab is considered in some applications as a substitute for steroids.[27] However, therapeutic resistance to anti-VEGF therapies develops early and is mediated by the treatment itself. Furthermore, inhibiting angiogenesis in the tumor leads to hypoxia, which in turn induces upregulation of the phosphorylated signal transducer and activator of transcription 3 (p-STAT3), a transcription factor implicated in tumor proliferation, invasion, and immunosuppression.[28] In addition, inhibition of VEGF results in a shift to other proangiogenic pathways using new growth factors such as the metalloproteinases[29] and neuropilin-1.[30] Other mechanisms at play in mediating therapeutic resistance include the participation of myeloid cells[31]—further potentiating tumor-mediated immunosuppression.

Clinical Implementation Limitations

Monoclonal antibody therapeutics have experienced exponential growth in the market over the past 2 decades, with more than 60 mAbs approved by the FDA by 2017. However, they are associated with a variety of intrinsic limitations. Hepatic enzymes limit the capacity for systemic administration of mAbs,[32] and the price tag on these biologics is still high, partly due to the complexity of the production, and partly due to the multiple injections required (because of their short half-life).[33] Overcoming the selectivity of the BBB and maintaining a sufficient concentration in the CNS remains a challenge.[34] In addition, even the most specificity-refined mAbs are not completely devoid of toxicity mediated by their cross-reaction, immunogenicity, and interaction with other drugs.[35, 36] Finally, bevacizumab is the only FDA-approved monoclonal antibody in treatment of glioblastoma[37]; however, it only functions passively on blood vessels and lacks vigor in killing tumor cells.

Immune Checkpoint Inhibitors

Immune checkpoints are modulatory molecules expressed on immune and tumor cells that switch off the immune response, especially the activation of the cytotoxic and helper T cells that play key roles in long-lasting adaptive immunity. Targeting the immune checkpoints with mAbs is a therapeutic approach that prevents immune downregulation, thereby halting tumor recognition and promoting eradication of tumors. Notably, this is one of many immunosuppressive mechanisms that operate in the tumor microenvironment, such as immunosuppression by cytokines TGF-β,[38] IL-10,[39] or PGE2[40]; downregulation of the MHC[41]; induction of T-cell apopotis via Fas expression[42]; induction of regulatory T cells (Tregs)[43]; and upregulation of the STAT3 pathway,[44] among many others (Table 1).

The immune checkpoints are a family of molecules being targeted, including cytotoxic T lymphocyte-associated protein 4 (CTLA-4), programmed cell death 1 (PD-1), PD-L1, TIM-3, LAG-3, among others (Fig. 3). The most extensively studied and therapeutically targeted include CTLA-4, PD-1, and PD-L1. CTLA-4 is a T cell surface marker, expressed on CD4+ and CD8+ cells whose main function is to switch off the immune cells whenever a self-antigen is presented, resulting in immune tolerance. As such, blocking CTLA-4 removes

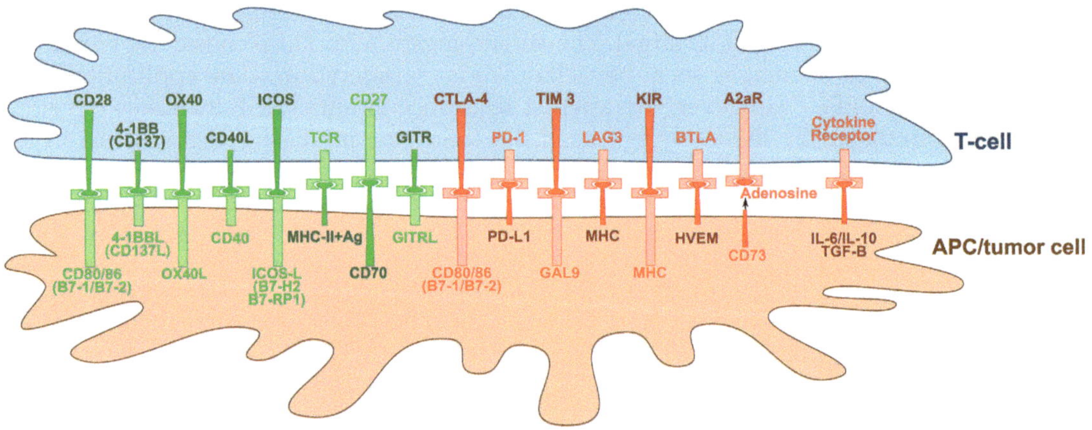

FIG. 3 Summary of immunomodulatory receptors and ligands. The T-cell response is regulated on the interaction with the antigen-presenting cell or the tumor cell by either stimulatory (green) or inhibitory (red) signals. The CD40–CD40L interaction has a reciprocal costimulatory activity on T cells and antigen-presenting cells. CD73 stimulates the production of adenosine, which inhibits the A2aR.

TABLE 1 Summary of Glioma-Mediated Mechanisms of Immune Suppression

Tumor cell mechanism	Effect on immune system
Immune checkpoints	Inhibition of IL-2 and T-cell expansion + apoptosis of CTLs
Downregulation of MHC-I	Inactive cytotoxic T cells
Immunosuppressive elements	
CSF-1	Polarization of macrophages to M2 phenotype
VEGF	Inhibit DC maturation + promote tumor growth
TGF-β	Increase in M2 macrophages + inhibition of CTLs
IL-10	Increase in M2 macrophages + inhibition of DCs and CTLs
Arginase 1	PMN degranulation
PGE2	Downregulation of MHC-II + induction of Tregs
Regeneration and tolerance factors	Inhibit NK cells
Induction of STAT3 pathway	Impaired phagocytosis and upregulation of Tregs
Upregulation of CD70, CD95 (Fas ligand)	Fas-induced apoptosis of T lymphocytes
Upregulation of lectin-like transcript 1	NK inhibition

Abbreviations: CD, *cluster of differentiation;* CSF, *colony-stimulating factor;* CTL, *cytotoxic T lymphocytes;* DC, *dendritic cells;* IL, *interleukin;* MHC, *major histocompatibility complex;* NK, *natural killer;* PGE, *prostaglandin E;* PMN, *polymorphonuclear neutrophil;* STAT, *signal transducer and activator of transcription;* TGF, *transforming growth factor;* Tregs, *regulatory T cells;* VEGF, *vascular endothelial growth factor.*

this immune tolerance, intensifies T-cell activation, and inhibits the immune supportive effect of the Tregs.[45] Ipilimumab, a humanized IgG1 mAb that targets CTLA-4, became the first FDA-approved immune checkpoint inhibitor in 2011 and is now one of the cornerstones in the treatment of melanoma patients with[46] or without brain metastasis.[47] Ipilimumab is currently being evaluated in patients with glioblastoma in various combinations with other agents (NCT02311920; NCT02017717). Unlike CTLA-4, PD-1 is not only expressed on T cells but also on B cells, NK cells, and on myeloid cell lineage-derived monocytes and macrophages.[48] PD-1 can interact with a variety of ligands such as PD-L1 or PD-L2 on the APCs and/or tumor cells, thereby causing immune dysfunction.[49] The anti-PD-1 mAbs (pembrolizumab, pidilizumab, and nivolumab) and anti-PD-L1 antibodies (durvalumab, avelumab, and atezolizumab) work to inhibit the interaction of the ligand with its receptor, thereby reinstating the cytotoxic T cell effector function.[50]

Although monotherapeutic responses are observed using these agents, the greatest clinical efficacy is seen when they are used in combination with other immune checkpoint inhibitors.[51, 52] Tumor-mediated immune cell apoptosis, exhaustion, and anergy pathways were proposed as justifications for the resistance to immune checkpoint inhibitor therapy. Anergy occurs when the cell enters an unresponsive state secondary to lack of costimulation. Immune cell exhaustion has been extensively studied in the tumor microenvironment and was found to occur when the repeatedly stimulated immune cell loses its function and resists reactivation. With immune exhaustion, other members of the immune checkpoint family can be upregulated, and the synergistic effect observed with using multiple mAbs to target the immune checkpoint family may result from the redundancy of family member expression. However, there may be a point when, despite the application of multiple immune checkpoint inhibitors, the recovery of T cell effector function is a lost cause. Whether different types of CNS tumors have different degrees of T-cell exhaustion remains unknown. Another combinatorial strategy worth serious consideration is the use of immune checkpoints with irradiation.[53, 54] Enhanced therapeutic results may result from release of more antigens, due to cytolysis of the tumor cells or enhanced penetration of the immune cells into the tumor compartment.

Despite these therapeutics having marked clinical activity in many types of malignancies, clinical trial results have been disappointing for most patients with glioblastoma, although some responses have been noted. These responses were mostly associated with specific genetic alterations, such as a somatic loss of the DNA mismatch repair acquired after temozolomide treatment,[55] or a mutation in the exonuclease domain of the polymerase epsilon (POLE) gene.[56, 57] Similarly, constitutional homogeneous germline mutation of DNA mismatch repair genes (PMS2, MLH1, MSH2, and MSH6) was correlated with an increase in survival in two pediatric patients treated with the immune checkpoint inhibitor nivolumab.[58] However, DNA mismatch repair, POLE mutations, or bi-allelic mismatch repair defect are rarely found in most gliomas.[59] Additionally, the mutational load in CNS gliomas is relatively low,[59] unlike in melanoma and squamous cell carcinoma of the lung.[60] Gliomas also have a relative paucity of T cells in the tumors and are mostly infiltrated with macrophages and other myeloid cell lineages.[61] The fact that brain metastases from lung cancer and melanoma showed significant responses to immune checkpoint inhibitors[62] provided support for their use in glioblastoma; however, the number of PD-1-expressing T cells and the degree of PD-L1 expression in brain metastases are also substantially higher than in glioblastoma (Ferguson, *Int J Cancer*, in revision). Cumulatively, the expression of these various biomarkers indicates that only certain glioma subpopulations may respond to immune checkpoint inhibitors.

Indoleamine 2,3-Dioxygenase Pathway Approaches

Indoleamine 2,3-dioxygenase (IDO) is an inducible, rate-limiting enzyme in tryptophan catabolism. It promotes tryptophan depletion and accumulation of kynurenine, which is correlated with upregulation of Tregs and inhibition of effector T cells[63] within the tumor microenvironment. IDO expression is also associated with a poor prognosis.[64] IDO levels are low in the brain in normal physiological conditions, but rise quickly with inflammatory signals, regardless of the disorder. Targeting IDO pathways is appealing because more than 90% of malignant gliomas express IDO.[65] Studies of extracranial solid tumors have shown that inhibiting IDO as a monotherapy is insufficient; these tumors typically require a combinatorial strategy.[66] Subsequently, IDO inhibitors were tested preclinically together with immune checkpoint inhibitors such as anti-PD-L1 and anti-CTLA-4.[67] IDO inhibitors are now entering clinical trials for treatment of several tumor types, including glioblastoma (NCT02327078). PF-06840003 is an oral IDO1 inhibitor currently in a Phase I trial for oligodendroglioma, astrocytoma, and higher-grade gliomas (NCT02764151). Indoximod is also an IDO inhibitor that interferes with IDO-mediated tryptophan signaling. A trial that combined the IDO1 inhibitor indoximod with chemotherapy and radiation therapy showed promising preliminary results, and it is currently recruiting patients for treatment of glioblastoma and other malignant brain tumors (NCT02052648) in a Phase II study. Indoximod is also being studied in combination with chemoradiation in a first-in-children, Phase I study for pediatric brain tumors (NCT02502708).

Clinical Implementation Limitations

Immune checkpoint inhibitor monotherapy for treatment of patients with glioblastoma has not had the same positive impact on survival as it has for patients with other malignancies,[68] probably secondary to relative lower tumor mutation load, T-cell tumor infiltration, and PD-L1 expression. Adverse effects, although manageable, cannot be overlooked, especially given that outcomes have not been impressive.[69] Furthermore, it is thought that the steep price of immune checkpoint inhibitor therapy is unjustified, based on the outcomes of this treatment in patients with glioblastoma multiforme (GBM).[70] In addition to these factors limiting clinical implementation, IDO inhibitors have failed against several tumor types as a monotherapy. Their use is theoretically appealing, but they are newly discovered molecules still under investigation and development; therefore, drawing overly optimistic conclusions regarding their use in patients with brain tumors is premature.

Monoclonal Antibodies for Patients with Multiple Sclerosis

Multiple sclerosis (MS) is an autoimmune demyelinating disorder that mostly affects women in their thirties. In patients with MS, high IgG levels and oligoclonal bands of immunoglobulins are detected in the CSF, and white matter plaque can be demonstrated on magnetic resonance images. MS is mediated by immune cell infiltration and BBB breakdown. B cells have a pathogenic role in MS, as they produce antibodies that target the myelin sheath in a nonspecific manner, which to date is not completely understood.[71] The use of monoclonal therapeutics in patients with MS may be particularly relevant. Multiple immunomodulatory antibodies have been approved for use in relapsing MS, including daclizumab (anti-CD25),[72] natalizumab (anti-VLA-4),[73] and alemtuzumab (anti-CD52).[74] Ocrelizumab (anti-CD20) has demonstrated an advantage against MS over other mAbs because it was the first to be approved for the primary progressive form of MS. Ocrelizumab acts by depleting B cells via ADCC and CDC, and can cross the BBB in measurable amounts[75] compared with the other mAbs that inhibit the migration of T cells across the BBB. Long-term safety of ocrelizumab is still being assessed by postmarketing surveillance programs. Another anti-CD20 antibody, ofatumumab, is being assessed in clinical trials, which are demonstrating an acceptable tolerance profile and promising results.[76] Ofatumumab can easily be administered subcutaneously, which may be preferable to patients (relative to intravenous administration).

Monoclonal Antibodies for Patients with Alzheimer Disease

Alzheimer disease (AD) is the most common cause of dementia in elderly patients. It involves a progressive cortical atrophy pathologically associated with the accumulation of plaques of amyloid and neurofibrillary tangles of tau proteins. In contrast to MS and glioblastoma, the use of mAbs in patients with AD is theoretically a greater challenge, given the intact BBB in these patients. mAbs targeting amyloid proteins, tau proteins, and amyloidogenic enzymes have been developed and used in conjunction with opening the BBB osmotically, biochemically, ultrasonically, or by irradiation; however, they have not resulted in sufficiently compelling efficacy to prompt FDA approval. So far, 20 recombinant mAbs that target aggregated forms of amyloid beta have been studied in clinical trials.[77, 78] These antibodies were manufactured to either mediate the removal of the toxic proteins locally through microglia-mediated phagocytosis of the aggregated β-amyloid substances (e.g., aducanumab)[79] or to help with the efflux of the toxins out of the brain for peripheral removal and degradation (e.g., solanezumab). However, solanezumab failed in a Phase III clinical trial.[80] Another recognized hallmark and potential cause of AD is the accumulated neurofibrillary tangles of abnormally phosphorylated tau protein.[81] Thus, mAbs targeting tau kinases have been synthesized to inhibit atypical tau protein aggregation.[82, 83] To date, four Phase I and Phase II trials have been conducted to study anti-tau mAbs in early and mild to moderate forms of the disease.[77]

Monoclonal Antibodies for Patients with Headaches

Targeting the Trigeminovascular Calcitonin Gene-Related Pain Mediated Pathway

Migraine is one of the most incapacitating maladies in the world and has an incompletely deciphered pathophysiology. One migraine disorder is believed to be mediated through the trigeminovascular calcitonin gene-related peptide (CGRP) pain-mediated pathway. The CGRP is a 37-amino-acid pain mediator, which is also a potent vasodilator.[84] Thus, four anti-CGRP mAbs (i.e., eptinezumab, erenumab, fremanezumab, and galcanezumab-gnlm) are now being tested and have shown significantly positive results in Phase II and III clinical trials, with good tolerability and minor adverse effects.[85] Two of these drugs, fremanezumab and galcanezumab-gnlm, are also being tested for alleviation of cluster headaches (FNCT02945046, NCT02964338, NCT02397473, and NCT02797951).

CONJUGATED MONOCLONAL ANTIBODIES: BISPECIFIC ANTIBODIES, ANTIBODIES–DRUG CONJUGATES, RECOMBINANT IMMUNOTOXIN THERAPY, RADIOIMMUNOTHERAPY, IMMUNOLIPOSOMES

Monoclonal antibodies can be modified to improve specificity and targeting, to deliver a toxic payload, or to enhance their delivery. Specificity can be promoted by using a dual-targeting or bispecific antibody that recognizes two antigen binding sites. Conjugating a mAb to an active toxin (immunotoxin or a chemotherapeutic drug), forming an antibody-drug conjugate (ADC), can enhance the function of the antibody. Furthermore, mAb delivery can be enhanced with nanoparticles, particularly liposomes, thereby generating immunoliposomes. All of these approaches have been exploited for treatment of a variety of CNS diseases.

Bispecific Antibodies

Sufficient delivery of antibodies into the CNS has been a confounding factor for clinical efficacy, which is why more recent efforts have focused on the creation of bispecific antibodies to shuttle these agents across the BBB. These dual-valent molecules are made of one antibody that binds to an antigen on the luminal side of the BBB, allowing a receptor-mediated transport of the other antibody, which has a specificity for the target antigen.[86] The first generations of bispecific antibodies were made of antigen-binding fragments linked at their hinge regions. However, the cost–benefit ratio was unappealing. New generations of bispecific antibodies are produced solely from the variable fragments, which are advantageous in that they are easy to produce, do not require purification, and achieve better tumor penetration due to their small size.[87]

Bispecific T-cell engagers are a form of tandem single-chain variable fragment (scFv) that merits special attention in oncology. They are made of two scFvs, where one recognizes the tumor-associated antigen and the other attaches to the CD3ε receptor on the T cell. Their efficacy depends on the specificity conferred toward the tumor antigen side and not toward the T-cell receptor side; which reduces the nonspecific cross-linking and recruits more T cells that are specific to the tumor-associated antigen–Abs complex. These divalent antibodies can stimulate the anergic T cells without antigen presentation or any other costimulatory method.[88] In brain tumors, bispecific T-cell engagers that target the epidermal growth factor variant III (EGFRvIII) showed long-lasting survival in animal models.[89] Several studies are now weighing the efficacy of directing bispecific antibodies against two antigens within the tumor, such as targeting epidermal growth factors EGFRvIII and HER2neu at the same time using armed, activated T cells.[90] Likewise, bispecific antibodies directed against the apoptosis mediators angiopoietin-2 and translocator protein showed an increase in survival via suppression of tumor cell proliferation, reduction in vascular permeability, and enhanced T-cell infiltration in mouse models of glioblastoma in which mice were previously treated with bevacizumab.[91] Certainly, human studies are required to ascertain the efficacy of this strategy.

Bispecific antibodies are also being evaluated in the context of neurodegenerative diseases. Targeting β-secretase 1 (known as BACE-1 for β-site APP cleaving enzyme 1) is a mechanism of preventing amyloid deposition in patients with AD. BACE-1 is the primary enzyme involved in the cleavage of the amyloid precursor protein in the amyloidogenic process cascade. Because achieving an effective concentration of the antibody targeting BACE-1 in the brain remained a challenge,[92] a bispecific antibody was created that allowed for endothelial receptor-mediated transcytosis across the BBB.[93] An alternative strategy using a BACE-1-specific peptide using receptor-mediated transcytosis across the BBB endothelial cells has been devised,[94] but neither of these strategies has been tested in clinical trials.

Clinical Implementation Limitations

Toxicity from persistent T-cell activation was observed with the bispecific antibody 4G7xH22 for hematologic neoplasms, which led to the termination of a clinical trial (NCT00014560). This raised many questions about the immunogenicity and purity of bispecific antibodies for use in multitarget recognition.[95]

Modifying the composition of bispecific antibodies affects their pharmacology (i.e., a small configuration change allows them to penetrate the brain and gain access to the tumor). However, they are unstable and are rapidly immunologically cleared, which can compromise efficacy in the biological setting, requiring further engineering.[96] An additional limitation of the use of bispecific antibodies is that, although bispecific T-cell engagers should be administered systemically by continuous infusion[97] to allow tight control of the concentration, this administration is associated with infusion complications (e.g., infection, extravasation, allergic reactions, and technical problems). Finally, as with other immunomodulatory methods, bioengineering is costly, especially for recombinant antibodies.

Antibody-Drug Conjugates

In the case of ADCs, the mAb is the targeting moiety.[98] Once the antibody binds to its antigens on the cell, the biological conjugate is internalized by receptor-mediated endocytosis and executes its cytotoxic effect on the cell. The mechanisms of cytotoxicity mediated by the conjugated chemotherapeutic mainly involve either intercalation between the DNA nucleotides, inhibition of the polymerization of the microtubules (thus blocking the mitotic cell cycle), or the induction of apoptosis by interruption of the signal transduction pathways.

A prototypical example of an ADC tested in glioblastoma is depatuxizumab mafodotin (ABT-414), which is an anti-EGFR mAb linked to monomethyl auristatin F (MMAF, mafodotin) by the noncleavable maleimidocaproyl linker. MMAF is a powerful chemotherapeutic drug that binds to the β-subunit of tubulin dimers, thereby inhibiting microtubule polymerization and thus stopping the cell cycle at the M phase. Three trials are currently recruiting patients to study the efficacy of this ADC in the treatment of high-grade glioma in adults: NCT02590263 (Phases I and II), NCT02573324 (Phase III), and NCT02343406, which is studying children diagnosed with glioblastoma (Phase II). Previous trials have demonstrated that ABT-414 had clinical activity in GBM, but produced adverse ocular effects,[99,100] required modification of the dose limitation, and triggered refinement of the conjugated cytotoxic agent. ABBV-221 consists of the anti-EGFR ABT-806 that targets the EGFR deletion variant and is conjugated to monomethyl auristatin E (MMAE).[101] The improved cellular uptake, greater potency to EGFR-amplified tumor, and reduced adverse corneal effects of the new formulation are now being tested in Phase I clinical trials (NCT02365662).[102]

Clinical Implementation Limitations

Manufacturing an ADC can be complicated and pricey. Creation of an ideal, stable, specific antibody-linker-conjugate combination requires tremendous expertise and optimization. All three components can produce undesirable nonspecific binding and eventual adverse effects. Furthermore, ADCs attach to tumor antigens that may not be expressed by all the cells; therefore, nonexpressing cells can continue to grow.

Recombinant Immunotoxin Therapy

Recombinant immunotoxin therapy (RIT) is another way to conjugate mAbs to biologically active molecules—usually toxins produced by bacteria, animals, or plants. The most commonly used toxins are the *Pseudomonas* exotoxin A (PE) and the *Diphtheria* toxin (DT), because they can be easily generated from eukaryotic cells and are less toxic than those that originate from plants.[103] One of the earliest *Pseudomonas* immunotoxin studies was published by Chaudhary et al.[104] in the late 1980s; these researchers created a truncated pseudomonal exotoxin that showed potency against T-cell leukemia.[105] In glioma treatment, monoclonal immune toxins have been created to target the wildtype and mutant variants of EGFR. D2C7-IT is currently being evaluated in a Phase I/II clinical trial (NCT02303678), where it is administered via convection-enhanced delivery (CED) to patients with recurrent glioblastoma (NCT02303678).[106] CED uses a microcatheter to infuse an agent directly into the brain parenchyma by creating a pressure gradient that theoretically ensures a homogeneous distribution. From a scientific standpoint, CED offers solutions to overcome the BBB obstacle and to minimize systemic toxicity. CED has been widely used for a variety of RITs[107] that target IL-4,[108] TGF-α,[109] and IL13R.[110] More than 90% of glioblastomas express IL-13 receptors compared with the normal brain.[110] Preclinical studies using a *Pseudomonas*-bound IL-13 toxin (IL13-PE83) showed striking results in preclinical studies. It was well tolerated and was associated with significant improvement in survival in the early phase clinical trials.[111] However, this strategy was not FDA approved when Phase III international studies revealed that median survival time was not significantly enhanced.[112] Poor drug distribution was one of several explanations for the lack of therapeutic effect observed,[113] and investigators have been working on improving CED technology to increase intratumoral drug distribution.

Clinical Implementation Limitations

Immunotoxins can be enzymatically active against the endothelium, which can cause a vascular leak syndrome with variable severity.[114, 115] Several modifications were attempted to overcome this limitation, but they sabotaged drug efficacy. Administration of dexamethasone and a fluids bolus before immunotoxin infusion can reduce these symptoms,[116] but this exposes the patient to the adverse effects associated with steroids, especially when multiple treatments are required. Immunosuppressants administered before the immunotoxins can negate the therapeutic activity, and hepatotoxicity can occur with systemic administration.[117, 118]

Direct intratumoral injection can reduce immunogenicity,[108] but there are risks associated with the craniotomy and/or CED, and these approaches limit treatment to a single dose. Finally, immunotoxins are relatively large molecules that cannot penetrate solid tumors easily.[119]

Radioimmunotherapy

For patients treated with radioimmunotherapy, the mAb is coupled to a radiotracer or to a radioactive particle. These can be delivered systemically by intravenous infusion or into a surgical cavity postoperatively. Once the mAb reaches the tumor cell, the antibodies attach to the surface antigen and deliver a high dose of radiation presumably sufficient to destroy the neoplastic cells. This strategy is FDA-approved in patients with non-Hodgkin lymphoma using the anti-CD20 mAbs (ibritumomab tiuxetan and tositumomab). The former, ibritumomab coupled with a tiuxetan chelating agent and labeled by a β-emitter yttrium-90 (^{90}Y), is still on the market; however, tositumomab coupled with ^{131}I, is no longer available.[120]

In the field of neuro-oncology, radioimmunotherapy strategies have been tested in newly diagnosed and recurrent malignant gliomas,[121] in medulloblastoma,[122] and in secondary metastasis of solid cancers to the brain.[123] The most extensively tested radioimmunotherapy in the CNS was a chimeric anti-tenascin mAb 81C6, which is a human IgG2 constant fragment linked to a variable murine anti-tenascin 81C6 labeled with ^{131}I. Early-phase studies in patients with high-grade gliomas who received intratumoral injections using a Rickham reservoir were encouraging, with minimal neurologic side effects[123] and satisfactory median survival times compared with brachytherapy and stereotactic radiosurgery.[124, 125] Based on these results, a study was conducted on glioblastoma and anaplastic astrocytoma using patient-specific dosing of I^{131}-labeled anti-tenascin 81C6 to obtain a 44-Gy radiation dose to the 2-cm tumor resection margin, followed by temozolomide treatment and conventional radiation therapy. Tailoring the I^{131}-81C6 dose to each patient produced a modest survival advantage, but grade III toxicities were observed, which were reversible.[126]

Clinical Implementation Limitations

An Ommaya reservoir is placed in the surgical cavity to allow for multiple administrations, but the cavity needs to be appropriately sealed and typically requires verification with radiographic flow studies, which increase the cost of this treatment strategy. Furthermore, the reservoir does not solve the problem of distant tumor recurrence because radioimmunotherapy is limited to simple diffusion from the surgical cavity and the field of the radiation-emiting particle. Surgically created resection cavities may also be limited secondary to anatomical constraints, such as the location of eloquent cortex and the ventricles.

Immunoliposomes

Nanoparticles hold great promise for biomedical application. Since doxorubicin liposomal, a chemotherapeutic agent formulated with liposomal nanoparticles, was approved by the FDA in 1995, a multitude of biocompatible and biodegradable nanoparticles have been used in clinical trials or have gained regulatory approval for treatment of a variety of diseases.[127-130] Among all nanoparticle varieties, liposomes are the best studied, and due to their success, approximately 40 types of liposomal nanoparticles are being investigated in clinical trials or have been approved by the FDA.[130, 131] Liposomes used for drug delivery are vesicles (50–1000 nm) composed of a lipid membrane with an aqueous core; this unique construction allows for encapsulation of hydrophilic and/or hydrophobic molecules in the core or lipid membrane, respectively. Liposomal nanoparticles can encapsulate a large quantity of either micro- or macromolecules, such as drugs,[132-134] proteins,[135] peptides,[136] or nucleic acids.[137-139] Encapsulating molecules within liposomes can protect therapeutic agents from rapid inactivation, degradation, and clearance, thereby enhancing potency, efficacy, and reducing adverse effects.[128, 140, 141]

Liposomes can be classified into 1 of 3 groups: conventional, stealth, or ligand-targeted liposomes (Fig. 4). Conventional liposomes are those with no surface modifications, and they were the first generation of liposomes developed. However, conventional liposomes undergo rapid clearance due to opsonization and uptake by macrophages and the reticuloendothelial system.[142] To overcome rapid reticuloendothelial system clearance and to enhance liposomal

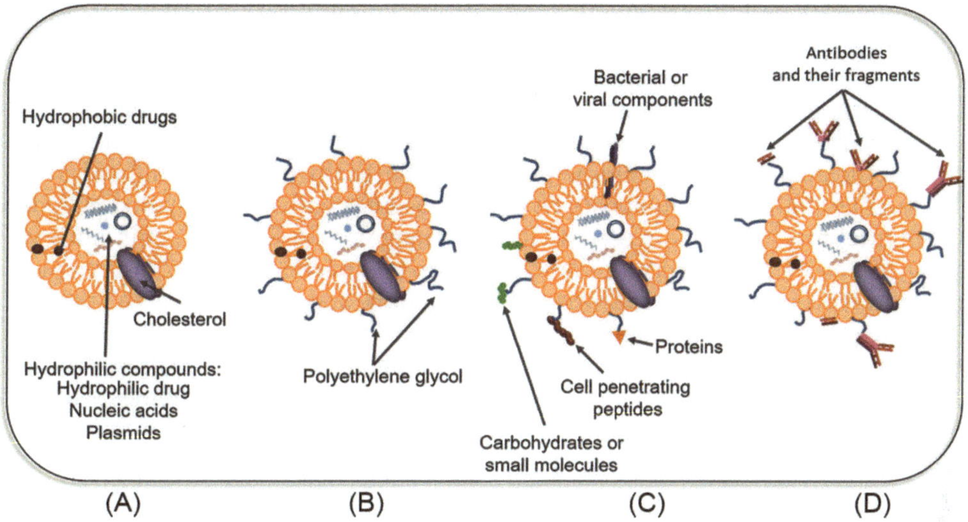

FIG. 4 Schematic representation of immunoliposomes as a drug delivery system. (A) Conventional liposomes with hydrophilic and hydrophobic therapeutic components; (B) stealth liposomes incorporating a hydrophilic polyethylene glycol that shields liposomes from rapid recognition by the reticuloendothelial system; (C) ligand-targeting liposomes, which are used for selective targeting; and (D) immunoliposomes that combine the characteristics of stealth liposomes and ligand-targeted liposomes where antibodies or biologically active fragments of antibodies are used for specific targeting. These antibodies are covalently attached to the liposome phospholipid membrane or grafted to the hydrophilic polyethylene glycol.

circulation time and stability, stealth liposomes were introduced. Stealth liposomes have surface modification with a hydrophilic polymer such as polyethylene glycol, which provides sufficient steric hindrance for eluding recognition by the reticuloendothelial system, thereby prolonging the stealth liposomal half-life compared with the half-life of conventional liposomes.[143, 144] Currently, the most widely used approach uses ligand-targeted liposomes, which utilize targeting moieties (e.g., mAbs, fragments of antibodies, proteins, peptides, and/or small molecules) that are covalently or noncovalently attached to the surface of liposomes.[138] Most targeting ligands used to modify the liposome surface are mAbs and fragments of antibodies (either antigen-binding or scFV) and can be subclassified as immunoliposomes (Fig. 4). Immunoliposomes enjoy selective uptake by cells that overexpress the ligand of interest, thus enhancing liposomal accumulation at the targeted site and minimizing off target adverse effects.[145]

Within this context, several studies have shown that immunoliposomes incorporating antibodies to target brain capillary endothelial cells augment the bioavailability of liposomes inside the brain. Liposomes coupled with antibodies targeting transferrin receptors have been shown to enhance the delivery of polyphenolic xanthon α-M (a neuroprotective agent),[146] neuronal growth factor,[147] and epigallocatechin-3-gallate[148] by providing effective drug delivery through transferrin receptor-mediated transcytosis for the treatment of AD,[146] Parkinson disease,[149] and glioma.[150] In a transgenic AD mouse model, stealth liposomes modified with anti-amyloid-β antibody demonstrated superior therapeutic efficacy relative to free antibodies without a liposome.[151]

Several studies have begun to explore dual targeting of immunoliposomes; for example, as proof of concept, immunoliposomes targeting low-density lipoprotein receptors (LDLs) and glucose transporter-1 effectively crossed the BBB.[152] Similarly, lactoferrin immunoliposomes showed significantly higher accumulation in glioma brain tumors in animal model systems.[153] Recent literature has demonstrated that liposomes with a combination of ligands that target both receptors highly expressed at the BBB along with potent cell-penetrating peptides (CPPs) can increase liposome accumulation in brain compartments.[154, 155] More specifically, LDL-related protein-1 receptor is expressed on the BBB, and angiopep-2 is a specific ligand for LDL-related protein-1. Liposomes dually targeting angiopep-2 and TAT, a CPP, led to significant enhanced permeability of doxorubicin in gliomas relative to free drug and nontarget liposomes.[154] Another preclinical study of dually targeted immunoliposomes with potential therapeutic value for treating glioma was temozolomide encapsulated in immunoliposomes, linked with anti-CD33 and the angiopep-2 peptide.[156]

Clinical Implementation Limitations

Repeated administration of stealth liposomes has been associated with accelerated blood clearance between the first and next dose. Although the exact mechanism of this process remains unknown, it may be due to the interaction of the immune system with the liposome components. Furthermore, the addition of surface modifications to liposomes and incorporation of multiple components in one immunoliposome increases physicochemical variability, which complicates the assessment and evaluation of biocompatibility, pharmacokinetics, pharmacodynamics, and the toxicological profile of the nanoformulations.[127, 145, 157] Pharmaceutical development of immunoliposomes is limited due to quality assurance, which involves scalability, reliability, reproducibility, chemical instability, and the long-term stability of immuoliposomes and their cargo; and due to the balance between cost and benefits, when compared with an approved existing therapeutic counterpart with similar efficacy and with similar associated adverse effects.[158, 159]

CYTOKINES FOR THE TREATMENT OF DISEASES OF THE CENTRAL NERVOUS SYSTEM

Definition and Function

Cytokines are glycoproteins that synchronize immune system responses, and their concentrations vary during the course of a disease. Cytokines can have either pro- or anti-inflammatory effects, and these effects can be contextual. Cytokines are classified according to their cell of origin or their mechanism of action.[160]

Cytokines Across the BBB

Some cytokines are produced in the brain sanctuary and others can originate from the periphery. Some, such as epidermal growth factor (EGF), tumor necrosis factor-α (TNF), and interleukins (IL)-1 and 6, can cross the BBB; others, such as transforming growth factor (TGF)-α and β1, accumulate in the vicinity of the barrier and are degraded in the blood circulation.[161]

Applications of Cytokines in Neurological Disease: Stroke, Alzheimer Disease, Multiple Sclerosis, Tumors

Strategies to modulate immune responses using cytokines have been evaluated in a wide variety of diseases in the CNS, including stroke, MS, AD, and malignancies.[160] In patients with stroke, cytokine trials were aimed at increasing perfusion in the ischemic penumbra, and subsequently, the diminishing clinical deficit. Several studies have shown a role for IL-1α and IL-1β in the pathophysiology of neuronal injury after an ischemic accident in the brain.[162, 163] As such, an antagonist to the IL-1 receptor (anakinra), which can bypass the BBB and reach sufficient concentrations in the brain to perform neuroprotective functions,[164] has been used to safely treat patients with subarachnoid hemorrhage.[165] Anakinra may also be beneficial in patients with stroke to reduce inflammatory IL-6 levels, which have been implicated in poststroke tissue remodeling.[166]

In patients diagnosed with AD, amyloid deposition is accompanied by an increase in several pro-inflammatory cytokines, such as TNF-α and IL-6, which may contribute to neuron loss.[167] A randomized Phase II trial of subcutaneously administered etanercept, an anti-TNF-α agent, demonstrated acceptable tolerability by patients but did not bring about significant cognitive improvements.[168] However, intrathecal administration of etanercept in a pilot study did show some improvement in cognition, suggesting that insufficient levels may have been a shortcoming of the prior Phase II studies.[169] Other immunomodulatory approaches to reduce IL-6 and TNF-α include a semisynthetic tetracycline by-product, minocycline, which crosses the BBB[170] and reduces tau proteins and amyloid deposition in xenograft models.[171]

The inflammatory response in patients diagnosed with MS is associated with interferon (IFN)-γ, IL-12, IL-17, and IL-23. Targeting these components, particularly at an early stage of the disease to stop the demyelination process, is the mainstay of MS management. In patients with acute flare-ups, intravenous methylprednisolone is used to suppress the inflammatory response when there is no response to oral steroids. Intervention through disease-modifying drugs intends to modify the immune system, either by treating it as a whole or by selectively targeting the molecules believed to contribute to the inflammatory process. IFN-β, glatiramer acetate, teriflunomide, and dimethyl fumarate are immunomodulators that inhibit inflammatory cytokines such as IL-23 and IL-17[172] and inhibit the activated T and B cells that escape Treg-arbitrated control.[173] Other approved drugs act indirectly, performing more general immunosuppressive mechanisms. For example, cyclophosphamide and mitoxantrone deplete lymphocytes and decrease their availability in the brain. Natalizumab selectively blocks heterodimeric transmembrane integrin, thereby inhibiting T-cell trafficking.[174] Fingolimod also inhibits T-cell trafficking by sequestration in the lymph nodes. Finally, alemtuzumab depletes CD52+ T and B cells.[173] Treatment can be challenging because inflammatory responses can be heterogeneous in nature and different at the various stages of the disease.

Cytokine modulation has also been used in patients with malignancies of the CNS.[175] The rationale behind this modulation is to enhance the antitumor immune response, which can overcome the immunosuppressive properties of the tumors. One of the first agents to be tested clinically was IL-2, which stimulates proliferation and activation of the adaptive immune response. In preclinical models of glioma, IL-2 demonstrated antitumor activity when injected into the tumor.[176] However, the clinical use of IL-2 monotherapy has been limited due to a narrow therapeutic index. IL-2 can exert toxicity by binding to the CD25 receptor subunit, which triggers vascular leak syndrome,[177] pulmonary edema,[178] and the expansion of immunosuppressive Tregs.[179] IL-2 has also been used in combination with adoptive cell therapy, such as lymphokine-activated killer (LAK) cells and tumor-infiltrating lymphocytes (TILs) for renal cell carcinoma[180] and malignant melanoma.[181] However, toxic adverse effects from this treatment were observed in glioma clinical trials.[182] A novel bioengineered IL-2 devoid of the CD25 function has been designed, but it has not yet been tested for safety and efficacy in tumor treatment.[183] Another method for CNS tumor treatment is to combine IL-2 with a suicide gene. When this mix is injected directly into the tumor, this strategy showed an acceptable safety margin and a desirable response in half of the treated population.[184]

An alternative strategy to inducing a T-cell effector response with immunostimulatory IL-2 is to target tumor-elaborated immunosuppressive TGF-β. In brain tumors, TGF-β has a paradoxical role—it plays a tumor-supportive role in high-grade glioma, but inhibits tumor growth in low-grade gliomas.[185] Three isoforms have been isolated from mammalian cells, TGF-β1, TGF-β2, and TGF-β3, with isoform 1 predominating in normal tissues and isoform 2 elaborated by malignant gliomas.[186] As a tumorigenic factor, TGF-β2 inhibits antigen presentation and activation of T cells. It also mediates angiogenesis by stimulating the secretion of VEGF and tumor invasiveness via the metalloproteinases.[185, 187] Recently, TGF-β2 was implicated in the pathways of autophagy involved in tumor invasion.[188] Trabedersen, a synthetic antisense oligodeoxynucleotide, was designed to inhibit TGF-β production at the translational level and showed promising results in early-phase clinical trials; however, it failed to demonstrate responses in Phase II studies.[189] An anti-TGF-β mAb, fresolimumab, was also tested in glioma patients, but despite showing good penetration into the tumor, it also failed to show a beneficial effect.[30] Several Phase I and II trials are in the recruiting phases to study the effects of TGF-β kinase inhibitors (NCT01682187, NCT01220271, and NCT01582269); however, it should be noted that TGF-β is one of many agents that a tumor uses to suppress the immune system.

IL-4 can also exert a proliferative and stimulatory effect on a variety of immune cells.[190] Chimeric IL-4 conjugated to a *Pseudomonas* immunotoxin and administered systemically demonstrated regression of human glioblastoma xenografts[191] and was well tolerated in primates.[192] Based on these results, several Phase I dose escalation studies were conducted to study the safety and efficacy of its direct intratumoral injection. Tumor regression after injection was evident on magnetic resonance imaging, but tumor recurrence was noted within 4 weeks, and no sign of a significant increase in survival was observed.[108, 193, 194] Phase II studies have been initiated in which chimeric IL-4 conjugated to a *Pseudomonas aeruginosa* immunotoxin A has been injected into the resection cavity (NCT02858895) or after surgery (NCT00014677), but no results are yet available. Furthermore, although this drug is well tolerated, it has not shown tumor responses in other solid malignancies.[195] IL-4 was also tested in combination with autologous glioma cells or dendritic cells for treatment of patients with

GBM or anaplastic astrocytoma,[196] but in some instances, the tumor recurred before vaccine production could be completed.[197]

The field of cytokine immunotherapy for gliomas has been revived by the development of intratumoral IL-12. IL-12 is a 4-helix heterodimeric cytokine that is secreted primarily by macrophages and dendritic cells and stimulates innate immunity through NK and T cytotoxicity.[198] Furthermore, IL-12 has a chemokine-inducible protein 10 (CXCL10)-mediated antiangiogenic effect.[199] Promising results in preclinical studies triggered human studies at several centers, but these were confounded by toxicity, and resulted in multi-organ failure and diffuse gastrointestinal hemorrhage.[200, 201] Direct intratumoral injection was investigated as a method to mitigate the systemic toxicity of intravenous injection.[202] Promising results were shown in preclinical models of glioma using this strategy[203, 204] and in combination with a CTLA-4 inhibitor.[205] The latter combinatorial strategy showed an increase in effector T cells and a decrease in the Tregs in the tumor microenvironment.[205] A clinically viable strategy for intratumoral IL-12 was devised using viral gene therapy constructs such as adenovirus,[206, 207] herpes simplex virus,[208] canarypox virus,[209, 210] and semliki virus.[211] Among those, the IL-12-armed herpes simplex virus[208] showed enhanced tumor killing in brain tumors and has been evaluated in Phase I clinical trials for recurrent and progressive GBM, anaplastic astrocytoma, and gliosarcoma (NCT02062827).[212] Most recently, the triple combination of the oncolytic herpes simplex virus armed with IL-12 along with two immune checkpoint inhibitors (anti-CTLA-4 and anti-PD-L1) showed a boost in the immune system within the tumor and a complete cure in mice[213]; however, this approach needs further safety and tolerance investigations before translation to human studies. Likewise, IL-12 coupling to an adenovirus showed a favorable modulation of the immune response against several tumors such as melanoma,[214] gastrointestinal neoplasms,[215] and brain tumors,[216] and is currently in Phase I trials for patients diagnosed with GBM (NCT03330197, NCT02026271). The use of oncolytic viruses as immunomodulatory agents is discussed in detail later in this chapter.

Clinical Implementation Limitations

Cytokines are ubiquitous molecules that circulate in the bloodstream and lack specificity, which is the biggest drawback in their application. When administered systemically, they are susceptible to rapid biodegradation and they therefore require prolonged, repetitive administration. High doses of cytokines can be associated with significant toxicity,[217] raising apprehension about their viability as a therapeutic strategy.

CELL-BASED THERAPIES FOR THE TREATMENT OF DISEASES OF THE CENTRAL NERVOUS SYSTEM

Cell-based therapy mainly aims to replace the defective cells underlying pathology and to improve the immune defenses against the causative agent. Proficiencies in ex vivo cell engineering are rendering this method more applicable to the treatment of CNS diseases, despite significant challenges. Ongoing clinical trials using cell therapy are being tested for a variety of neurological diseases such as stroke, spinal cord injury, and MS; however, malignant gliomas are the prevailing target. Different cell types used in this therapeutic strategy include neural stem cells, mesenchymal stem cells, and immune cells such as NK cells, T cells, and dendritic cells.

For this strategy to work against a tumor in the CNS, the T cell should have tumor specificity.[218] Moreover, cell trafficking to the tumor should be sufficient to mount a cytotoxic response that cannot be overcome by the immunosuppressive tumor microenvironment. An alternative strategy is to exploit NK cells that can target tumor cells independently of the MHC and antigen expression. NK cells have been tested in an unmodified format[219] and after genetic modification to secrete immunomodulatory cytokines,[220] but they have not yet achieved much clinical success.[221] The most recent approach is to manufacture NK cells that express chimeric antigen receptors (CARs),[222] which are being evaluated in the treatment of HIV[223, 224] and in various cancers such as multiple myeloma,[225] breast cancer brain metastasis,[226] and glioblastoma.[227, 228] They hold promise but need further exploration. Cellular immunotherapy for malignant gliomas is divided in two main categories: adoptive cell therapy and active immunotherapy.

Adoptive Cell Therapy

Immune tolerance poses a major challenge in the fight against cancer because the T-cell repertoire that recognizes the self-antigens is depleted. To overcome this depletion, scientists have developed adoptive cell therapy, which involves harvesting autologous T cells, priming them against tumor antigen(s) and then reinjecting them into the host circulation. The first human experiment with adoptive cell therapy was a prophylactic measure against cytomegalovirus (CMV) in a high-risk transplant patient. T cells were expanded numerically in vitro against a CMV antigen and reinfused into the patient, thereby inducing significant protection against the virus.[229] Adoptive T-cell immunotherapy involves harvesting this immune cell population from either the peripheral blood or the tumor environment, stimulating these cells or genetically retargeting them to attack glioma cells, and then reinfusing them into the patient. LAK cells were used in the first adoptive cell transfer performed in the setting of malignancy. In this technique, $CD8^+$ T cells are isolated from the patient's peripheral blood, treated in vitro with IL-2 to stimulate antitumor function, and then reinfused without genetic alterations. The mechanism of action of LAKs is not specific.[230] Intracavitary injection of LAKs and IL-2 improved long-term survival time in patients with recurrent high-grade glioma.[231] Consequently, a Phase II trial was conducted in which LAKs were directly injected into the GBM tumor; however, survival was compromised by corticosteroid treatment, which significantly reduced the number of immune cells before leukapheresis.[230] Using LAKs in glioma treatment is limited not only by steroids but also by the high level of CNS toxicity associated with the concomitant administration of IL-2.

The next unmodified form of adoptive T-cell immunotherapy was the use of TILs, which are collected from the resected tumor or adjacent lymph node, expanded with IL-2, exposed to inactive tumor cells to ensure specificity, and then injected back into the host. TILs showed impressive results against malignant metastatic melanoma[232] compared with IL-2 and LAKs, but the infused cells did not persist long.[233, 234] To increase the lifespan of the TILs and to potentiate their effect, patients were preconditioned with chemotherapy and radiation therapy for lymphoablative purposes.[235, 236] The rationale was that the preexisting immune clones and cytokines are incapacitating the new cells. Regardless of the mechanism, this strategy demonstrated favorable outcomes in melanoma patients.[237] However, although TILs directly injected into glioma along with recombinant IL-2 were well tolerated in patients with glioma, these patients did not have favorable outcomes.[238] Furthermore, because gliomas are not highly populated by T cells, the yield of TILs was costly, time consuming, and ultimately, insufficient.

Tumor-infiltrating lymphocytes can track efficiently to the brain and their killing is specific; however, these cells are nonetheless restricted to the expression of MHC class I molecules, which is downregulated in tumor cells. One way to overcome the MHC dependency is by using CARs. CARs are artificial receptors composed of an extracellular single chain variable fragment of a mAb, an extracellular stalk, a transmembrane domain, and an intracellular signaling element derived from the zeta-subunit of the T-cell receptor (CD3ζ). T cells genetically modified with CARs hold a great advantage in immunotherapy over other agents because they can bypass MHC-mediated activation, circulate into the tumor core,[239,240] and amplify the immune response by recruiting other immune system effectors.[241] Moreover, their antigen recognition is not limited to proteins—they can recognize glycolipids and carbohydrate epitopes as well.[242] To optimize the function of the signaling domain of the CAR T cells, second and third generations have been developed with 1 or 2 additional cytosolic costimulatory domains, respectively, the most common consisting of CD28, 4-1BB, or OX40 attached to the CD3ζ endodomain.[243] This alteration potentiates the cytotoxic effect, and enhances the proliferation and the persistence of the T cells.[244]

Chimeric antigen receptors targeting EGFRvIII have been developed, as EGFRvIII is tumor-specific and therefore there is minimal potential for off-tissue toxicity. EGFRvIII results from an in-frame deletion in the extracellular domain of the original EGFR. It is the most common mutation in glioblastoma, with 30% of GBM tumors expressing the variant receptor on their cell surface.[245] EGFRvIII CAR T cells demonstrated tumor regression in immunodeficient animal model systems[246] and significant cure rates in fully immunocompetent animal models.[247] Because EGFRvIII is not uniformly expressed by all cells within the glioblastoma, there is a significant risk of antigen escape by targeting this antigen. A pivotal clinical trial in patients with recurrent GBM showed that intravenously injected EGFRvIII CAR T cells were capable of trafficking to the CNS tumor; however, there also was trafficking of Tregs, upregulation of IDO1 and PD-L1, and downregulation of the targeted antigen EGFRvIII, all of which contributed to eventual tumor progression.[248] One way to overcome antigen escape may be to induce epitope spreading by immunization with the prevailing antigen, thus priming the CAR T cells to bind to the antigen-negative cells within the tumor.[249] Induction of eptiope spreading in a murine model of GBM treated with EGFRvIII CAR T cells resulted in clearance of $EGFRvIII^{neg}$ tumor; however, this has not yet been demonstrated in clinical trials.[247]

A more ubiquitous and uniformly expressed antigen, IL-13 receptor 2 alpha (IL13R2α), is found in more than 80% of glioblastomas.[250-252] Cytolytic CD8+ T cells have been genetically engineered to express an IL-13R2α-specific CAR generated by fusing recombinant IL-13 to the intracellular zeta unit, hence known as "IL-13-zetakine." It has a tyrosine-to-glutamic acid substitution at position 13 (E13Y) in IL-13 that preferentially binds IL-13Rα2 and reduces binding to the IL-13α1 or the IL-4Rα, both of which are expressed on white blood cells and share similar subunits that can bind IL-13.[253] IL-13Rα CAR T cells produced prolonged survival in murine model systems of glioma when injected directly into the tumor.[254] A Phase I clinical trial demonstrated a favorable toxicity profile for direct intracranial infusion of CD8 + CAR T cells targeting IL13Rα2 engineered from a GBM patient's peripheral blood mononuclear cells; however, there was evidence of antigen escape.[240] To further enhance the specificity and persistence of CARs within the tumor, a second-generation CAR was engineered by adding a second mutation (lysine to arginine on position 109) IL13.E13K. R109K in the IL-13 and a CD28 to the endodomain.[255] Autologous T cells transfected with the IL13Rα2-specific CAR are now being used in a Phase I clinical trial for refractory and recurrent malignant glioma (NCT02208362). However, these IL3Rα2 "zetakine" CARs lack exclusive specificity to IL13Rα2 and demonstrated significant residual binding to IL-13Rα1.[256] To overcome this, IL13Rα2 CARs have been generated from an scFv specific for IL13Rα2. A third generation of IL-13Rα CAR T cells using scFv with a different combination of intracellular domains (CD3z, 4-1BB, CD28, and OX40) showed more efficient cytotoxicity in vitro and in vivo than the previous generations, and appear to have specificity for IL13Rα2, with no recognition of IL13Rα1 or IL4R in in vitro assays.[257]

A key limitation of CARs for treatment of CNS malignancies such as GBM is antigen escape, which has been observed in both EGFRvIII and IL-13Rα2 CAR clinical trials.[248, 258] This was previously observed in both preclinical glioma models of a peptide vaccine targeting EGFRvIII[259] and in subsequent Phase II clinical trials.[260, 261] Various other antigens overexpressed in GBM, but are not tumor-restricted to GBM have been proposed, but there are risks of off target toxicity—an example of which is the death of a patient treated with ERBB2-CAR T cells.[262] The erythropoietin-producing hepatocellular carcinoma A2 (EphA2) receptor belongs to the ephrin tyrosine kinase receptor family and is an attractive target in GBM. More than 60% of GBMs express the EphA2 receptor,[263, 264] and its presence indicates a high level of invasiveness and poor prognosis.[265] A second generation, EphA2-specific CAR coupled to a CD28-ζ intracellular domain showed good affinity for glioma-initiating cells and EphA2+ glioma cells, with significant killing of tumor cells in vitro and in vivo. It also prolonged the survival of mice when injected locally.[266] Although it is a potential target, to date no clinical trials have been initiated in glioma patients with the EphA2 receptor. EGFR is homogeneously expressed in approximately 60% of patients with GBM, making it a potential target for CAR T cell therapy. However, its widespread normal tissue expression also increases the risk for off-tissue toxicity. An EGFR-specific CAR has been generated from the clinical antibody nimotuzumab, which demonstrates a favorable toxicity profile in clinical trials. This EGFR-specific CAR shows antigen density-dependent activation, thus sparing normal tissue with low density of EGFR expression from toxicity.[267] In addition, strategies to express EGFR-specific CAR as an RNA species have shown transient redirection of T cell activity to antigen.[267] Other strategies to restrict CAR activity from normal tissue toxicity are being actively pursued in various other tumor models.[268] Several key limitations must be overcome in order for CAR immunotherapy to be suitable for use in patients with malignancies of the CNS; those are discussed below.

For dendritic cell immunotherapy, monocytes are isolated by leukapheresis from the peripheral blood of the patient; the cells are differentiated, loaded with tumor antigens, and then infused back into the patient. Dendritic cells are systemically infused, which makes it a more convenient treatment strategy for patients requiring multiple dosage regimens, and the naïve autologous T cells are activated in vivo. The antigen can be a previously identified synthetic tumor-specific peptide, a tumor lysate, or mRNA. Trials using dendritic cells with tumor lysate have shown acceptable tolerability, with some evidence of a signal of response in patients with glioblastoma.[269] Furthermore, dendritic cells loaded with acid-eluted peptides were shown to induce intracranial T-cell responses.[270] Additionally, dendritic cells loaded with the EGFRvIII peptide induced significant T-lymphocyte responses and were nontoxic in adults with newly diagnosed GBM.[271] Notably, as these trials moved into later stage, efficacy-assessment clinical trials and increases in progression-free or overall survival time were too insufficient to justify a more labor-intensive cellular therapy.[272, 273] More recently, a novel strategy has been investigated in which the dendritic cells target CMV pp65.[274] In a small cohort of patients with GBM, dendritic cells loaded with pp65 mRNA were preconditioned with a tetanus toxoid to improve the lymph node homing of the dendritic cells, which resulted in the enhanced function of the CMV-specific CD8+ T cells along with an increase in overall survival time.[275] A simplified cellular strategy of using autologous CMV-specific T cells is currently being evaluated in patients with recurrent GBM (NCT02661282).

Clinical Implementation Limitations

In order for CAR immunotherapy to become a viable treatment for patients with malignancies of the CNS, several advances must take place: (1) the production time must be significantly diminished, because high-grade gliomas progress rapidly; (2) the actual production must be streamlined so that CAR immunotherapy is more universally available; (3) the cost must be reduced by performing systemic infusion; (4) the in vivo persistence and trafficking throughout the tumor must be enhanced; and (5) a tumor-specific target that is homogeneously expressed needs to be identified. Dendritic cell expansion, meanwhile, is a complex, technically challenging treatment strategy.[276]

ACTIVE IMMUNOTHERAPY FOR THE TREATMENT OF DISEASES OF THE CENTRAL NERVOUS SYSTEM

Vaccination induces immunological memory by exposing the body to a nonvirulent, live attenuated organism or to an antigen that has been typically used to prevent infectious disease. Over the past several decades, immune biology and bioengineering laboratories have been designing DNA, RNA, and polypeptide vaccines that alter immune system function for oncologic therapeutic and preventive measures.[277, 278] These have been particularly successful in the prevention of human papillomavirus- and hepatitis B-induced malignancies, such as cervical and hepatocellular carcinomas, respectively.[279] Therapeutic vaccines prime the immune system to recognize the tumor as a nonself-entity and attack it. APCs phagocytose, process, and present the tumor antigens.[218] However, this strategy can be mitigated by cancer-induced T-cell immunosuppression. In the context of CNS disease, glatiramer acetate/copolymer 1 is the only successful peptide vaccine agent, approved for treatment of MS.[280] It consists of four amino acids found in myelin basic protein that induce the expansion of Tregs. Other peptide vaccine strategies have not been as successful in patients with MS.[281] Likewise, a peptide vaccine for patients with AD demonstrated positive results in preclinical models,[282] but failed in clinical trials when the vaccine did not mount the expected T-cell response,[283] and cortical microhemorrhages were detected.[284]

A wide variety of vaccine approaches have been studied for treatment of glioblastoma, including autologous formalin-fixed tumor cells, peptides, and heat shock proteins.[285-288]

Among peptide vaccines, those directed against EGFRvIII have received the most attention clinically. EGFRvIII contains a deletion in the extracellular domain that results in a unique junction with a novel glycine residue that renders this a tumor-specific antigen.[289] Peptides derived from this fusion junction, when chemically conjugated to carrier proteins (e.g., keyhole limpet hemocyanin), have been shown to induce EGFRvIII-specific immune responses[259, 290] in preclinical models. Although there were marked increases in progression-free survival and overall survival times during Phase II clinical trials,[260, 261] in March 2016, Phase III results demonstrated that EGFRvIII-positive patients who received the vaccine had no improvement in overall survival relative to standard-of-care therapy. Notably, on recurrence, many tumors were noted to be EGFRvIII-negative, indicating either instability of the expression longitudinally or a result of antigen-negative selection—an observation similar to that made when using the CAR approaches targeting EGFRvIII. Other antigens have been tested in patients newly diagnosed with GBM, including a vaccine that includes 11 tumor-associated peptides, but the progression-free survival time was not particularly compelling.[291] A Phase I clinical trial is now underway for a peptide vaccine targeting the R132H (arginine to histidine substitution at the 132 position)-specific point mutation in the IDH1 gene in patients with grade III and IV glioma (NCT02454634).

Clinical Implementation Limitations

One limitation of peptide vaccines is that they are usually HLA-restricted.[292] Furthermore, all vaccines are associated with an unpredictable risk of toxicity and immunogenicity. Personalized therapies require a companion CLIA-compliant biomarker strategy for enrollment.

Virotherapy

Viruses require actively dividing cells to survive, which makes the cancer cell a perfect milieu for that purpose. Taking advantage of that tropism, viruses are viewed as emerging vehicles for inducing an immunological response within the tumor. Virotherapy has previously been viewed from the mechanistic perspective as either operating by a gene therapy approach or by direct oncolytic activity. The latter involves a virus designed as a tumor-selective strain that preserves its infective and lytic capacity against the tumor cells but that spares the surrounding normal cells. The cytotoxic effect is exerted at different levels, depending on the type of virus: (1) stimulation of extracellular lytic mechanisms (e.g., cytokines and proteases); (2) intracellular mechanisms (nuclear and cytoplasmic); or (3) via cell membrane targeting and signal transduction. In gene therapy, the virus is devoid of its replication and virulence factors and is used as a vector to deliver a gene. More recently, these viruses have been modified to enhance their ability to generate local immune responses in the tumor microenvironment. More than 20 different viruses have been tested in various cancers,[293] neurodegenerative diseases,[294, 295] and syndromic metabolic diseases that affect the brain.[296, 297] Among these, a genetically engineered herpes virus (talimogene laherparepvec, or T-VEC) has received FDA approval for use against malignant melanoma.[298, 299] Likewise, the adenovirus (dl1520/ONYX-015) has received approval in China for use in patients with head and neck cancer.[300] Many others, like herpes simplex type-1,[301] adenoviruses,[302] polioviruses,[303] and Newcastle disease viruses[304] are currently being tested in clinical trials. The FDA provided expedited approval for the oncolytic poliovirus PVS-RIPO for the treatment of patients with recurrent glioblastoma after three patients survived for 36 months during the Phase I trial (NCT01491893).

The Delta-24 adenovirus has also been extensively tested in patients with GBM. It acts on the dysfunctional retinoblastoma tumor-suppressor gene pathway and inhibits the progression of the cell cycle into the synthesis phase required for cell proliferation.[305] The first generation of Delta-24 needed improvement or combination with other agents for a more specific binding.[306] The virus subsequently underwent multiple reiterations to enhance its infectivity and cytotoxicity. One of the successful variants is a replication-competent adenovirus with retargeted infectivity via EGFR (Delta-24-RIVER), which has a particular tropism against the variant type of EFGR found on glioma cells. It showed high infectivity and efficacy against EGFR mutant cells and, to a lesser extent, against the EGFR wild-type cells.[307] Delta-24-RGD, known as DNX-2401, is an improved version of Delta-24 that was developed with the insertion of Arg-Gly-Asp in the fiber knob of the virus to strengthen its binding to the glioma-expressed integrins.[308] It increased survival time significantly in glioma-implanted mice[309] and showed complete responses in a Phase I trial treating patients with recurrent GBM and showed no drug-related adverse effects.[310] To further enhance immunogenicity, Delta-24-RGDOX was designed with an OX40 ligand expression cassette to boost antigen-presentation and T-cell activation.[311] Delta-24-RGD (DNX-2401) combined with pembrolizumab has currently reached a Phase II clinical trial and is still recruiting patients at 13 centers (NCT02798406). Oncolytic viruses are now being combined with other immunomodulatory methods to overcome the heterogeneity of glioblastoma and to target tumor-mediated immunosuppression. Measles virus combined with anti-PD-1 improved survival in syngeneic preclinical models,[312] and Delta-24-RGD (DNX-2401) combined with pembrolizumab is currently in Phase II clinical trials at 13 centers (NCT02798406).

Clinical Implementation Limitations

Virotherapy typically requires site-specific delivery, which is especially challenging in patients who require multiple doses and interventions. Although systemic administration can be considered, the amount delivered to the CNS tumor is relatively low with this method. Finally, generation of a clinical-grade product is expensive and requires a great deal of labor.

CONCLUSIONS

Immunotherapy holds great promise even for incurable diseases in the CNS. However, early preclinical results have not correlated with activity in human patients—partly due to the prematurity and cost of newly discovered immunomodulatory methods. Some of the most successful drugs against glioblastoma cells in vitro and against glioma xenografts do not translate well to these tumors in human subjects, but for the drugs that do, delivery to the brain remains a challenge, and direct intracranial injection is not always a satisfactory option for the patient. Thus far, no specific immunomodulatory method has emerged to address the multiple confounding clinical trial limitations we have outlined. The brain is not static, and it undergoes continuous immune reorganization in response to new therapies, which in turn produce therapeutic resistance. The price tag, the adverse effects, and the challenges of drug delivery to the brain may be overcome with a long-term commitment to finding the ideal drug that causes minimal adverse effects and offers a durable response. Combination therapy may ameliorate some of these limitations and reduce the emergence of resistance, but further investigation is required. Finally, one needs to consider the dominant immune cell populations within a glioma and the role of these immune cells in tumor clearance. Modulating the innate immune system is an emerging therapeutic direction that may be particularly relevant to glioma.

References

1. The treatment of malignant tumors by inoculations of erysipelas. *J Am Med Assoc*. 1893;(22):615-616.
2. Medawar PB. Immunity to homologous grafted skin; the fate of skin homografts transplanted to the brain, to subcutaneous tissue, and to the anterior chamber of the eye. *Br J Exp Pathol*. 1948;29(1):58-69.
3. Lampson LA. Molecular bases of the immune response to neural antigens. *Trends Neurosci*. 1987;10(5):211-216.
4. Louveau A, Smirnov I, Keyes TJ, et al. Structural and functional features of central nervous system lymphatic vessels. *Nature*. 2015;523(7560):337-341.
5. Cserr HF, Knopf PM. Cervical lymphatics, the blood-brain barrier and the immunoreactivity of the brain: a new view. *Immunol Today*. 1992;13(12):507-512.
6. Sun BL, Wang LH, Yang T, et al. Lymphatic drainage system of the brain: a novel target for intervention of neurological diseases. *Prog Neurobiol*. 2017;163-164:118-143.
7. Bartholomäus I, Kawakami N, Odoardi F, et al. Effector T cell interactions with meningeal vascular structures in nascent autoimmune CNS lesions. *Nature*. 2009;462:94.
8. Reboldi A, Coisne C, Baumjohann D, et al. C-C chemokine receptor 6-regulated entry of TH-17 cells into the CNS through the choroid plexus is required for the initiation of EAE. *Nat Immunol*. 2009;10:514.
9. Kivisakk P, Mahad DJ, Callahan MK, et al. Human cerebrospinal fluid central memory CD4+ T cells: evidence for trafficking through choroid plexus and meninges via P-selectin. *Proc Natl Acad Sci U S A*. 2003;100(14):8389-8394.
10. Kooij G, van Horssen J, de Lange EC, et al. T lymphocytes impair P-glycoprotein function during neuroinflammation. *J Autoimmun*. 2010;34(4):416-425.
11. Kooij G, Mizee MR, van Horssen J, et al. Adenosine triphosphate-binding cassette transporters mediate chemokine (C-C motif) ligand 2 secretion from reactive astrocytes: relevance to multiple sclerosis pathogenesis. *Brain*. 2011;134(Pt 2):555-570.
12. Avison MJ, Nath A, Greene-Avison R, et al. Inflammatory changes and breakdown of microvascular integrity in early human immunodeficiency virus dementia. *J Neurovirol*. 2004;10(4):223-232.
13. Gabrusiewicz K, Rodriguez B, Wei J, et al. Glioblastoma-infiltrated innate immune cells resemble M0 macrophage phenotype. *JCI Insight*. 2016;1(2).
14. Bhat KPL, Balasubramaniyan V, Vaillant B, et al. Mesenchymal differentiation mediated by NF-kappaB promotes radiation resistance in glioblastoma. *Cancer Cell*. 2013;24(3):331-346.
15. Doucette T, Rao G, Rao A, et al. Immune heterogeneity of glioblastoma subtypes: extrapolation from the cancer genome atlas. *Cancer Immunol Res*. 2013;1(2):112-122.
16. Hussain SF, Yang D, Suki D, Aldape K, Grimm E, Heimberger AB. The role of human glioma-infiltrating microglia/macrophages in mediating antitumor immune responses. *Neuro-Oncology*. 2006;8(3):261-279.
17. Schartner JM, Hagar AR, Van Handel M, Zhang L, Nadkarni N, Badie B. Impaired capacity for upregulation of MHC class II in tumor-associated microglia. *Glia*. 2005;51(4):279-285.
18. Serot JM, Foliguet B, Bene MC, Faure GC. Ultrastructural and immunohistological evidence for dendritic-like cells within human choroid plexus epithelium. *Neuroreport*. 1997;8(8):1995-1998.
19. Kanegawa N, Collste K, Forsberg A, et al. In vivo evidence of a functional association between immune cells in blood and brain in healthy human subjects. *Brain Behav Immun*. 2016;54(Suppl C):149-157.
20. Abadier M, Haghayegh Jahromi N, Cardoso Alves L, et al. Cell surface levels of endothelial ICAM-1 influence the transcellular or paracellular T-cell diapedesis across the blood-brain barrier. *Eur J Immunol*. 2015;45(4):1043-1058.
21. Emmons C, Hunsicker LG. Muromonab-CD3 (Orthoclone OKT3): the first monoclonal antibody approved for therapeutic use. *Iowa Med*. 1987;77(2):78-82.

22. Reiber H, Felgenhauer K. Protein transfer at the blood cerebrospinal fluid barrier and the quantitation of the humoral immune response within the central nervous system. *Clin Chim Acta*. 1987;163(3):319-328.
23. Batchelor TT, Grossman SA, Mikkelsen T, Ye X, Desideri S, Lesser GJ. Rituximab monotherapy for patients with recurrent primary CNS lymphoma. *Neurology*. 2011;76(10):929-930.
24. Field KM, Jordan JT, Wen PY, Rosenthal MA, Reardon DA. Bevacizumab and glioblastoma: scientific review, newly reported updates, and ongoing controversies. *Cancer*. 2015;121(7):997-1007.
25. Keizer RJ, Huitema ADR, Schellens JHM, Beijnen JH. Clinical pharmacokinetics of therapeutic monoclonal antibodies. *Clin Pharmacokinet*. 2010;49(8):493-507.
26. Cohen MH, Shen YL, Keegan P. Pazdur R. FDA drug approval summary: bevacizumab (Avastin) as treatment of recurrent glioblastoma multiforme. *Oncologist*. 2009;14(11):1131-1138.
27. Dietrich J, Rao K, Pastorino S, Kesari S. Corticosteroids in brain cancer patients: benefits and pitfalls. *Expert Rev Clin Pharmacol*. 2011;4(2):233-242.
28. de Groot J, Liang J, Kong LY, et al. Modulating antiangiogenic resistance by inhibiting the signal transducer and activator of transcription 3 pathway in glioblastoma. *Oncotarget*. 2012;3(9):1036-1048.
29. Lucio-Eterovic AK, Piao Y, de Groot JF. Mediators of glioblastoma resistance and invasion during antivascular endothelial growth factor therapy. *Clin Cancer Res*. 2009;15(14):4589-4599.
30. Kwiatkowski SC, Guerrero PA, Hirota S, et al. Neuropilin-1 modulates TGFbeta signaling to drive glioblastoma growth and recurrence after anti-angiogenic therapy. *PLoS ONE*. 2017;12(9).
31. Shojaei F, Wu X, Malik AK, et al. Tumor refractoriness to anti-VEGF treatment is mediated by CD11b+Gr1+ myeloid cells. *Nat Biotechnol*. 2007;25(8):911-920.
32. Wang W, Wang E, Balthasar J. Monoclonal antibody pharmacokinetics and pharmacodynamics. *Clin Pharmacol Ther*. 2008;84(5):548-558.
33. Park T, Griggs SK, Suh DC. Cost effectiveness of monoclonal antibody therapy for rare diseases: a systematic review. *BioDrugs*. 2015;29(4):259-274.
34. Rubenstein JL, Combs D, Rosenberg J, et al. Rituximab therapy for CNS lymphomas: targeting the leptomeningeal compartment. *Blood*. 2003;101(2):466-468.
35. Viardot A, Goebeler M-E, Hess G, et al. Phase 2 study of the bispecific T-cell engager (BiTE) antibody blinatumomab in relapsed/refractory diffuse large B-cell lymphoma. *Blood*. 2016;127(11):1410-1416.
36. Brennan FR, Morton LD, Spindeldreher S, et al. Safety and immunotoxicity assessment of immunomodulatory monoclonal antibodies. *MAbs*. 2010;2(3):233-255.
37. FDA approves first biosimilar to treat cancer. *Cancer Discov*. 2017;7(11):1206.
38. Joseph JV, Balasubramaniyan V, Walenkamp A. Kruyt FA TGF-beta as a therapeutic target in high grade gliomas—promises and challenges. *Biochem Pharmacol*. 2013;85(4):478-485.
39. Hao C, Parney IF, Roa WH, Turner J, Petruk KC, Ramsay DA. Cytokine and cytokine receptor mRNA expression in human glioblastomas: evidence of Th1, Th2 and Th3 cytokine dysregulation. *Acta Neuropathol*. 2002;103 (2):171-178.
40. Cook PJ, Thomas R, Kingsley PJ, et al. Cox-2-derived PGE2 induces Id1-dependent radiation resistance and self-renewal in experimental glioblastoma. *Neuro-Oncology*. 2016;18(10):1379-1389.
41. Taniguchi Y, Ono K, Yoshida S, Tanaka R. Antigen-presenting capability of glial cells under glioma-harboring conditions and the effect of glioma-derived factors on antigen presentation. *J Neuroimmunol*. 2000;111 (1-2):177-185.
42. Badie B, Schartner J, Prabakaran S, Paul J, Vorpahl J. Expression of Fas ligand by microglia: possible role in glioma immune evasion. *J Neuroimmunol*. 2001;120(1):19-24.
43. Humphries W, Wei J, Sampson JH, Heimberger AB. The role of tregs in glioma-mediated immunosuppression: potential target for intervention. *Neurosurg Clin N Am*. 2010;21(1):125-137.
44. Yu H, Kortylewski M, Pardoll D. Crosstalk between cancer and immune cells: role of STAT3 in the tumour microenvironment. *Nat Rev Immunol*. 2007;7(1):41-51.
45. Peggs KS, Quezada SA, Chambers CA, Korman AJ, Allison JP. Blockade of CTLA-4 on both effector and regulatory T cell compartments contributes to the antitumor activity of anti-CTLA-4 antibodies. *J Exp Med*. 2009;206(8):1717-1725.
46. Hodi FS, O'Day SJ, McDermott DF, et al. Improved survival with ipilimumab in patients with metastatic melanoma. *N Engl J Med*. 2010;363(8):711-723.
47. O'Day SJ, Maio M, Chiarion-Sileni V, et al. Efficacy and safety of ipilimumab monotherapy in patients with pretreated advanced melanoma: a multicenter single-arm phase II study. *Ann Oncol*. 2010;21(8):1712-1717.

48. Gordon SR, Maute RL, Dulken BW, et al. PD-1 expression by tumour-associated macrophages inhibits phagocytosis and tumour immunity. *Nature*. 2017;545(7655):495-499.
49. Dejaegher J, Verschuere T, Vercalsteren E, et al. Characterization of PD-1 upregulation on tumor-infiltrating lymphocytes in human and murine gliomas and preclinical therapeutic blockade. *Int J Cancer*. 2017;141(9):1891-1900.
50. Reardon DA, Gokhale PC, Klein SR, et al. Glioblastoma eradication following immune checkpoint Blockade in an orthotopic, immunocompetent model. *Cancer Immunol Res*. 2016;4(2):124-135.
51. Callahan MK, Postow MA, Wolchok JD. CTLA-4 and PD-1 pathway blockade: combinations in the clinic. *Front Oncol*. 2014;4:385.
52. Larkin J, Chiarion-Sileni V, Gonzalez R, et al. Combined nivolumab and ipilimumab or monotherapy in untreated melanoma. *N Engl J Med*. 2015;373(1):23-34.
53. Dovedi SJ, Adlard AL, Lipowska-Bhalla G, et al. Acquired resistance to fractionated radiotherapy can be overcome by concurrent PD-L1 blockade. *Cancer Res*. 2014;74(19):5458-5468.
54. Derer A, Spiljar M, Baumler M, et al. Chemoradiation increases PD-L1 expression in certain melanoma and glioblastoma cells. *Front Immunol*. 2016;7:610.
55. Cahill DP, Levine KK, Betensky RA, et al. Loss of the mismatch repair protein MSH6 in human glioblastomas is associated with tumor progression during temozolomide treatment. *Clin Cancer Res*. 2007;13(7):2038-2045.
56. Johanns TM, Miller CA, Dorward IG, et al. Immunogenomics of hypermutated glioblastoma: a patient with germline POLE deficiency treated with checkpoint blockade immunotherapy. *Cancer Discov*. 2016; 6(11):1230-1236.
57. Erson-Omay EZ, Caglayan AO, Schultz N, et al. Somatic POLE mutations cause an ultramutated giant cell high-grade glioma subtype with better prognosis. *Neuro-Oncology*. 2015;17(10):1356-1364.
58. Bouffet E, Larouche V, Campbell BB, et al. Immune checkpoint inhibition for hypermutant glioblastoma multiforme resulting from germline biallelic mismatch repair deficiency. *J Clin Oncol*. 2016;34(19):2206-2211.
59. Hodges TR, Ott M, Xiu J, et al. Mutational burden, immune checkpoint expression, and mismatch repair in glioma: implications for immune checkpoint immunotherapy. *Neuro-Oncology*. 2017;19(8):1047-1057.
60. Alexandrov LB, Nik-Zainal S, Wedge DC, et al. Signatures of mutational processes in human cancer. *Nature*. 2013;500:415.
61. Wang Q, Hu B, Hu X, et al. Tumor evolution of glioma-intrinsic gene expression subtypes associates with immunological changes in the microenvironment. *Cancer Cell*. 2018;33(1):152.
62. Goldberg SB, Gettinger SN, Mahajan A, et al. Pembrolizumab for patients with melanoma or non-small-cell lung cancer and untreated brain metastases: early analysis of a non-randomised, open-label, phase 2 trial. *Lancet Oncol*. 2016;17(7):976-983.
63. Hwu P, Du MX, Lapointe R, Do M, Taylor MW, Young HA. Indoleamine 2,3-dioxygenase production by human dendritic cells results in the inhibition of T cell proliferation. *J Immunol*. 2000;164(7):3596-3599.
64. Wainwright DA, Balyasnikova IV, Chang AL, et al. IDO expression in brain tumors increases the recruitment of regulatory T cells and negatively impacts survival. *Clin Cancer Res*. 2012;18(22):6110-6121.
65. Mitsuka K, Kawataki T, Satoh E, Asahara T, Horikoshi T, Kinouchi H. Expression of indoleamine 2,3-dioxygenase and correlation with pathological malignancy in gliomas. *Neurosurgery*. 2013;72(6):1031-1038 discussion 1038-1039.
66. Hou DY, Muller AJ, Sharma MD, et al. Inhibition of indoleamine 2,3-dioxygenase in dendritic cells by stereoisomers of 1-methyl-tryptophan correlates with antitumor responses. *Cancer Res*. 2007;67(2):792-801.
67. Wainwright DA, Chang AL, Dey M, et al. Durable therapeutic efficacy utilizing combinatorial blockade against IDO, CTLA-4, and PD-L1 in mice with brain tumors. *Clin Cancer Res*. 2014;20(20):5290-5301.
68. Reardon DA, Omuro A, Brandes AA, et al. OS10.3 randomized phase 3 study evaluating the efficacy and safety of nivolumab vs bevacizumab in patients with recurrent glioblastoma: checkmate 143. *Neuro-Oncology*. 2017; 19(Suppl 3):iii21.
69. Kourie HR, Klastersky J. Immune checkpoint inhibitors side effects and management. *Immunotherapy*. 2016; 8(7):799-807.
70. Andrews A. Treating with checkpoint inhibitors-figure $1 million per patient. *Am Health Drug Benefits*. 2015; 8(Spec Issue):9.
71. Breij EC, Brink BP, Veerhuis R, et al. Homogeneity of active demyelinating lesions in established multiple sclerosis. *Ann Neurol*. 2008;63(1):16-25.
72. Kim AP, Baker DE. Daclizumab. *Hosp Pharm*. 2016;51(11):928-939.

73. Ramos-Cejudo J, Oreja-Guevara C, Stark Aroeira L, Rodriguez de Antonio L, Chamorro B, Diez-Tejedor E. Treatment with natalizumab in relapsing–remitting multiple sclerosis patients induces changes in inflammatory mechanism. *J Clin Immunol*. 2011;31(4):623-631.
74. Coles AJ, Twyman CL, Arnold DL, et al. Alemtuzumab for patients with relapsing multiple sclerosis after disease-modifying therapy: a randomised controlled phase 3 trial. *Lancet*. 2012;380(9856):1829-1839.
75. Amanda MS, Kathryn MH. Ocrelizumab: a new B-cell therapy for relapsing remitting and primary progressive multiple sclerosis. *Ann Pharmacother*. 2017;52:473-483. https://doi.org/10.1177/1060028017747635.
76. Bar-Or A, Grove R, Austin D, et al. The MIRROR study: a randomized, double-blind, placebo-controlled, parallel-group, dose-ranging study to investigate the safety and MRI efficacy of subcutaneous ofatumumab in subjects with relapsing-remitting multiple sclerosis (RRMS) (S23.006). *Neurology*. 2014;82(10 Suppl).
77. Cummings J, Lee G, Mortsdorf T, Ritter A, Zhong K. Alzheimer's disease drug development pipeline: 2017. *Alzheimers Dement (NY)*. 2017;3(3):367-384.
78. van Dyck CH. Anti-amyloid-β monoclonal antibodies for Alzheimer's disease: pitfalls and promise. *Biol Psychiatry*. 2018;83(4):311-319.
79. Ferrero J, Williams L, Stella H, et al. First-in-human, double-blind, placebo-controlled, single-dose escalation study of aducanumab (BIIB037) in mild-to-moderate Alzheimer's disease. *Alzheimers Dement*. 2016;2(3):169-176.
80. Doody RS, Thomas RG, Farlow M, et al. Phase 3 trials of solanezumab for mild-to-moderate Alzheimer's disease. *N Engl J Med*. 2014;370(4):311-321.
81. Martin L, Latypova X, Wilson CM, et al. Tau protein kinases: involvement in Alzheimer's disease. *Ageing Res Rev*. 2013;12(1):289-309.
82. Brici D, Gotz J, Nisbet RM. A novel antibody targeting tau phosphorylated at serine 235 detects neurofibrillary tangles. *J Alzheimers Dis*. 2018;61(3):899-905.
83. Shahpasand K, Sepehri Shamloo A, Nabavi SM, Ping Lu K, Zhen Zhou X. Tau immunotherapy: hopes and hindrances. *Hum Vaccin Immunother*. 2017;14:1-8.
84. Lassen LH, Haderslev PA, Jacobsen VB, Iversen HK, Sperling B, Olesen J. CGRP may play a causative role in migraine. *Cephalalgia*. 2002;22(1):54-61.
85. Sabrina K, Astrid O, Messoud A. CGRP, a target for preventive therapy in migraine and cluster headache: systematic review of clinical data. *Cephalalgia*. 2017. https://doi.org/10.1177/0333102417741297.
86. Pardridge WM. CSF, blood-brain barrier, and brain drug delivery. *Expert Opin Drug Deliv*. 2016;13(7):963-975.
87. Suryadevara CM, Gedeon PC, Sanchez-Perez L, et al. Are BiTEs the "missing link" in cancer therapy? *OncoImmunology*. 2015;4(6):e1008339.
88. Choi BD, Cai M, Bigner DD, Mehta AI, Kuan CT, Sampson JH. Bispecific antibodies engage T cells for antitumor immunotherapy. *Expert Opin Biol Ther*. 2011;11(7):843-853.
89. Choi BD, Kuan C-T, Cai M, et al. Systemic administration of a bispecific antibody targeting EGFRvIII successfully treats intracerebral glioma. *Proc Natl Acad Sci*. 2013;110(1):270-275.
90. Zitron IM, Thakur A, Norkina O, Barger GR, Lum LG, Mittal S. Targeting and killing of glioblastoma with activated T cells armed with bispecific antibodies. *BMC Cancer*. 2013;13:83.
91. Li J, Zhang Z, Lv L, Qiao H, Chen X, Zou C. A bispecific antibody (ScBsAbAgn-2/TSPO) target for Ang-2 and TSPO resulted in therapeutic effects against glioblastomas. *Biochem Biophys Res Commun*. 2016;472(2):384-391.
92. Atwal JK, Chen Y, Chiu C, et al. A therapeutic antibody targeting BACE1 inhibits amyloid-beta production in vivo. *Sci Transl Med*. 2011;3(84):84ra43.
93. Ruderisch N, Schlatter D, Kuglstatter A, et al. Potent and selective BACE-1 peptide inhibitors lower brain abeta levels mediated by brain shuttle transport. *EBioMedicine*. 2017;24:76-92.
94. Pardridge WM. Re-engineering therapeutic antibodies for Alzheimer's disease as blood-brain barrier penetrating bi-specific antibodies. *Expert Opin Biol Ther*. 2016;16(12):1455-1468.
95. Teachey DT, Rheingold SR, Maude SL, et al. Cytokine release syndrome after blinatumomab treatment related to abnormal macrophage activation and ameliorated with cytokine-directed therapy. *Blood*. 2013;121(26):5154-5157.
96. Kontermann RE. Recombinant bispecific antibodies for cancer therapy. *Acta Pharmacol Sin*. 2005;26(1):1-9.
97. Nagorsen D, Kufer P, Baeuerle PA, Bargou R. Blinatumomab: a historical perspective. *Pharmacol Ther*. 2012;136(3):334-342.
98. Nagayama A, Ellisen LW, Chabner B, Bardia A. Antibody–drug conjugates for the treatment of solid tumors: clinical experience and latest developments. *Target Oncol*. 2017;12(6):719-739.

99. Gan HK, Reardon DA, Lassman AB, et al. Safety, pharmacokinetics and antitumor response of depatuxizumab mafodotin as monotherapy or in combination with temozolomide in patients with glioblastoma. *Neuro-Oncology*. 2017;20:838-847.
100. van den Bent M, Gan HK, Lassman AB, et al. Efficacy of depatuxizumab mafodotin (ABT-414) monotherapy in patients with EGFR-amplified, recurrent glioblastoma: results from a multi-center, international study. *Cancer Chemother Pharmacol*. 2017;80(6):1209-1217.
101. Reilly EB, Phillips AC, Buchanan FG, et al. Characterization of ABT-806, a humanized tumor-specific anti-EGFR monoclonal antibody. *Mol Cancer Ther*. 2015;14(5):1141-1151.
102. Phillips AC, Boghaert ER, Vaidya KS, et al. Characterization of ABBV-221, a tumor-selective EGFR targeting antibody drug conjugate. *Mol Cancer Ther*. 2018;17:795-805.
103. Zhu S, Liu Y, Wang PC, Gu X, Shan L. Recombinant immunotoxin therapy of glioblastoma: smart design, key findings, and specific challenges. *Biomed Res Int*. 2017;2017.
104. Chaudhary VK, Queen C, Junghans RP, Waldmann TA, FitzGerald DJ, Pastan I. A recombinant immunotoxin consisting of two antibody variable domains fused to *Pseudomonas* exotoxin. *Nature*. 1989;339(6223):394-397.
105. Kreitman RJ, Chaudhary VK, Waldmann T, Willingham MC, FitzGerald DJ, Pastan I. The recombinant immunotoxin anti-Tac(Fv)-*Pseudomonas* exotoxin 40 is cytotoxic toward peripheral blood malignant cells from patients with adult T-cell leukemia. *Proc Natl Acad Sci*. 1990;87(21):8291-8295.
106. Chandramohan V, Pegram CN, Piao H, et al. Production and quality control assessment of a GLP-grade immunotoxin, D2C7-(scdsFv)-PE38KDEL, for a phase I/II clinical trial. *Appl Microbiol Biotechnol*. 2017; 101(7):2747-2766.
107. Patel SJ, Shapiro WR, Laske DW, et al. Safety and feasibility of convection-enhanced delivery of Cotara for the treatment of malignant glioma: initial experience in 51 patients. *Neurosurgery*. 2005;56(6):1243-1252. discussion 1243-1252.
108. Weber F, Asher A, Bucholz R, et al. Safety, tolerability, and tumor response of IL4-*Pseudomonas* exotoxin (NBI-3001) in patients with recurrent malignant glioma. *J Neuro-Oncol*. 2003;64(1-2):125-137.
109. Sampson JH, Akabani G, Archer GE, et al. Intracerebral infusion of an EGFR-targeted toxin in recurrent malignant brain tumors. *Neuro Oncol*. 2008;10(3):320-329.
110. Debinski W, Gibo DM, Hulet SW, Connor JR, Gillespie GY. Receptor for interleukin 13 is a marker and therapeutic target for human high-grade gliomas. *Clin Cancer Res*. 1999;5(5):985-990.
111. Kunwar S, Prados MD, Chang SM, et al. Direct intracerebral delivery of cintredekin besudotox (IL13-PE38QQR) in recurrent malignant glioma: a report by the Cintredekin Besudotox Intraparenchymal Study Group. *J Clin Oncol*. 2007;25(7):837-844.
112. Kunwar S, Chang S, Westphal M, et al. Phase III randomized trial of CED of IL13-PE38QQR vs Gliadel wafers for recurrent glioblastoma. *Neuro-Oncology*. 2010;12(8):871-881.
113. Sampson JH, Archer G, Pedain C, et al. Poor drug distribution as a possible explanation for the results of the PRECISE trial. *J Neurosurg*. 2010;113(2):301-309.
114. Kuan CT, Pai LH, Pastan I. Immunotoxins containing *Pseudomonas* exotoxin that target LeY damage human endothelial cells in an antibody-specific mode: relevance to vascular leak syndrome. *Clin Cancer Res*. 1995;1 (12):1589-1594.
115. Baluna R, Rizo J, Gordon BE, Ghetie V, Vitetta ES. Evidence for a structural motif in toxins and interleukin-2 that may be responsible for binding to endothelial cells and initiating vascular leak syndrome. *Proc Natl Acad Sci U S A*. 1999;96(7):3957-3962.
116. Foss FM, Bacha P, Osann KE, Demierre MF, Bell T, Kuzel T. Biological correlates of acute hypersensitivity events with DAB(389)IL-2 (denileukin diftitox, ONTAK) in cutaneous T-cell lymphoma: decreased frequency and severity with steroid premedication. *Clin Lymphoma*. 2001;1(4):298-302.
117. Rosenblum MG, Shawver LK, Marks JW, Brink J, Cheung L, Langton-Webster B. Recombinant immunotoxins directed against the c-erb-2/HER2/neu oncogene product: in vitro cytotoxicity, pharmacokinetics, and in vivo efficacy studies in xenograft models. *Clin Cancer Res*. 1999;5(4):865-874.
118. Selvaggi K, Saria EA, Schwartz R, et al. Phase I/II study of murine monoclonal antibody-ricin A chain (XOMAZYME-Mel) immunoconjugate plus cyclosporine A in patients with metastatic melanoma. *J Immunother Emphasis Tumor Immunol*. 1993;13(3):201-207.
119. Allahyari H, Heidari S, Ghamgosha M, Saffarian P, Amani J. Immunotoxin: a new tool for cancer therapy. *Tumor Biol*. 2017;39(2).

120. Kawashima H. Radioimmunotherapy: a specific treatment protocol for cancer by cytotoxic radioisotopes conjugated to antibodies. *Sci World J*. 2014;2014:10.
121. Paganelli G, Bartolomei M, Grana C, Ferrari M, Rocca P, Chinol M. Radioimmunotherapy of brain tumor. *Neurol Res*. 2006;28(5):518-522.
122. Kramer K, Pandit-Taskar N, Humm JL, et al. A phase II study of radioimmunotherapy with intraventricular (131) I-3F8 for medulloblastoma. *Pediatr Blood Cancer*. 2018;65(1).
123. Cokgor I, Akabani G, Kuan CT, et al. Phase I trial results of iodine-131-labeled antitenascin monoclonal antibody 81C6 treatment of patients with newly diagnosed malignant gliomas. *J Clin Oncol*. 2000;18(22):3862-3872.
124. Reardon DA, Akabani G, Coleman RE, et al. Phase II trial of murine (131)I-labeled antitenascin monoclonal antibody 81C6 administered into surgically created resection cavities of patients with newly diagnosed malignant gliomas. *J Clin Oncol*. 2002;20(5):1389-1397.
125. Reardon DA, Akabani G, Coleman RE, et al. Salvage radioimmunotherapy with murine iodine-131-labeled antitenascin monoclonal antibody 81C6 for patients with recurrent primary and metastatic malignant brain tumors: phase II study results. *J Clin Oncol*. 2006;24(1):115-122.
126. Reardon DA, Zalutsky MR, Akabani G, et al. A pilot study: 131I-antitenascin monoclonal antibody 81c6 to deliver a 44-Gy resection cavity boost. *Neuro-Oncology*. 2008;10(2):182-189.
127. Allen TM, Cullis PR. Liposomal drug delivery systems: from concept to clinical applications. *Adv Drug Deliv Rev*. 2013;65(1):36-48.
128. Hawkins MJ, Soon-Shiong P, Desai N. Protein nanoparticles as drug carriers in clinical medicine. *Adv Drug Deliv Rev*. 2008;60(8):876-885.
129. Kim J, Wilson DR, Zamboni CG, Green JJ. Targeted polymeric nanoparticles for cancer gene therapy. *J Drug Target*. 2015;23(7-8):627-641.
130. van der Meel R, Vehmeijer LJ, Kok RJ, Storm G, van Gaal EV. Ligand-targeted particulate nanomedicines undergoing clinical evaluation: current status. *Adv Drug Deliv Rev*. 2013;65(10):1284-1298.
131. Bobo D, Robinson KJ, Islam J, Thurecht KJ, Corrie SR. Nanoparticle-based medicines: a review of FDA-approved materials and clinical trials to date. *Pharm Res*. 2016;33(10):2373-2387.
132. Anselmo AC, Mitragotri S. Nanoparticles in the clinic. *Bioeng Transl Med*. 2016;1(1):10-29.
133. Laginha KM, Verwoert S, Charrois GJ, Allen TM. Determination of doxorubicin levels in whole tumor and tumor nuclei in murine breast cancer tumors. *Clin Cancer Res*. 2005;11(19 Pt 1):6944-6949.
134. Straubinger RM, Arnold RD, Zhou R, Mazurchuk R, Slack JE. Antivascular and antitumor activities of liposome-associated drugs. *Anticancer Res*. 2004;24(2A):397-404.
135. Johnston D, Reynolds SR, Bystryn JC. Interleukin-2/liposomes potentiate immune responses to a soluble protein cancer vaccine in mice. *Cancer Immunol Immunother*. 2006;55(4):412-419.
136. Accardo A, Morelli G. Review peptide-targeted liposomes for selective drug delivery: advantages and problematic issues. *Biopolymers*. 2015;104(5):462-479.
137. Chen CW, Lu DW, Yeh MK, Shiau CY, Chiang CH. Novel RGD-lipid conjugate-modified liposomes for enhancing siRNA delivery in human retinal pigment epithelial cells. *Int J Nanomedicine*. 2011;6:2567-2580.
138. Sapra P, Allen TM. Ligand-targeted liposomal anticancer drugs. *Prog Lipid Res*. 2003;42(5):439-462.
139. Yaghi NK, Wei J, Hashimoto Y, et al. Immune modulatory nanoparticle therapeutics for intracerebral glioma. *Neuro-Oncology*. 2016;19:372-382.
140. Glaser T, Han I, Wu L, Zeng X. Targeted nanotechnology in glioblastoma multiforme. *Front Pharmacol*. 2017;8:166.
141. Wolfram J, Zhu M, Yang Y, et al. Safety of nanoparticles in medicine. *Curr Drug Targets*. 2015;16(14):1671-1681.
142. Poste G, Papahadjopoulos D, Vail WJ. Lipid vesicles as carriers for introducing biologically active materials into cells. *Methods Cell Biol*. 1976;14:33-71.
143. Gabizon AA, Barenholz Y, Bialer M. Prolongation of the circulation time of doxorubicin encapsulated in liposomes containing a polyethylene glycol-derivatized phospholipid: pharmacokinetic studies in rodents and dogs. *Pharm Res*. 1993;10(5):703-708.
144. Hua S, Wu SY. The use of lipid-based nanocarriers for targeted pain therapies. *Front Pharmacol*. 2013;4:143.
145. Eloy JO, Petrilli R, Topan JF, et al. Co-loaded paclitaxel/rapamycin liposomes: development, characterization and in vitro and in vivo evaluation for breast cancer therapy. *Colloids Surf B: Biointerfaces*. 2016;141:74-82.

146. Wang Y, Xia Z, Xu JR, et al. Alpha-mangostin, a polyphenolic xanthone derivative from mangosteen, attenuates beta-amyloid oligomers-induced neurotoxicity by inhibiting amyloid aggregation. *Neuropharmacology*. 2012; 62(2):871-881.
147. Kuo YC, Chou PR. Neuroprotection against degeneration of sk-N-mc cells using neuron growth factor-encapsulated liposomes with surface cereport and transferrin. *J Pharm Sci*. 2014;103(8):2484-2497.
148. Ying X, Wang Y, Xu H, et al. The construction of the multifunctional targeting ursolic acids liposomes and its apoptosis effects to C6 glioma stem cells. *Oncotarget*. 2017;8(38):64129-64142.
149. Kang YS, Jung HJ, Oh JS, Song DY. Use of PEGylated immunoliposomes to deliver dopamine across the blood-brain barrier in a rat model of Parkinson's disease. *CNS Neurosci Ther*. 2016;22(10):817-823.
150. Mamot C, Drummond DC, Noble CO, et al. Epidermal growth factor receptor-targeted immunoliposomes significantly enhance the efficacy of multiple anticancer drugs in vivo. *Cancer Res Dec*. 2005;65(24):11631-11638.
151. Tanifum EA, Dasgupta I, Srivastava M, et al. Intravenous delivery of targeted liposomes to amyloid-beta pathology in APP/PSEN1 transgenic mice. *PLoS One*. 2012;7(10).
152. Bu G. Apolipoprotein E and its receptors in Alzheimer's disease: pathways, pathogenesis and therapy. *Nat Rev Neurosci*. 2009;10(5):333-344.
153. Chen H, Qin Y, Zhang Q, et al. Lactoferrin modified doxorubicin-loaded procationic liposomes for the treatment of gliomas. *Eur J Pharm Sci*. 2011;44(1-2):164-173.
154. Han W, Yin G, Pu X, Chen X, Liao X, Huang Z. Glioma targeted delivery strategy of doxorubicin-loaded liposomes by dual-ligand modification. *J Biomater Sci Polym Ed*. 2017;28(15):1695-1712.
155. Li Y, Zheng X, Gong M, Zhang J. Delivery of a peptide-drug conjugate targeting the blood brain barrier improved the efficacy of paclitaxel against glioma. *Oncotarget*. 2016;7(48):79401-79407.
156. Kim JS, Shin DH, Kim JS. Dual-targeting immunoliposomes using angiopep-2 and CD133 antibody for glioblastoma stem cells. *J Control Release*. 2018;269:245-257.
157. Narang AS, Chang RK, Hussain MA. Pharmaceutical development and regulatory considerations for nanoparticles and nanoparticulate drug delivery systems. *J Pharm Sci*. 2013;102(11):3867-3882.
158. Narang S, Kim D, Aithala S, et al. Tumor image-derived texture features are associated with CD3 T-cell infiltration status in glioblastoma. *Oncotarget*. 2017;8(60):101244-101254.
159. Sercombe L, Veerati T, Moheimani F, Wu SY, Sood AK, Hua S. Advances and challenges of liposome assisted drug delivery. *Front Pharmacol*. 2015;6:286.
160. Azodi S, Jacobson S. Cytokine therapies in neurological disease. *Neurotherapeutics*. 2016;13(3):555-561.
161. Pan W, Stone KP, Hsuchou H, Manda VK, Zhang Y, Kastin AJ. Cytokine signaling modulates blood-brain barrier function. *Curr Pharm Des*. 2011;17(33):3729-3740.
162. Hutchinson PJ, O'Connell MT, Rothwell NJ, et al. Inflammation in human brain injury: intracerebral concentrations of IL-1alpha, IL-1beta, and their endogenous inhibitor IL-1ra. *J Neurotrauma*. 2007;24(10):1545-1557.
163. Rothwell NJ, Luheshi GN. Interleukin 1 in the brain: biology, pathology and therapeutic target. *Trends Neurosci*. 2000;23(12):618-625.
164. Ogungbenro K, Hulme S, Rothwell N, Hopkins S, Tyrrell P, Galea J. Study design and population pharmacokinetic analysis of a phase II dose-ranging study of interleukin-1 receptor antagonist. *J Pharmacokinet Pharmacodyn*. 2016;43(1):1-12.
165. Singh N, Hopkins SJ, Hulme S, et al. The effect of intravenous interleukin-1 receptor antagonist on inflammatory mediators in cerebrospinal fluid after subarachnoid haemorrhage: a phase II randomised controlled trial. *J Neuroinflammation*. 2014;11:1.
166. Erta M, Quintana A, Hidalgo J. Interleukin-6, a major cytokine in the central nervous system. *Int J Biol Sci*. 2012; 8(9):1254-1266.
167. Wang WY, Tan MS, Yu JT, Tan L. Role of pro-inflammatory cytokines released from microglia in Alzheimer's disease. *Ann Transl Med*. 2015;3(10):136.
168. Butchart J, Brook L, Hopkins V, et al. Etanercept in Alzheimer disease: a randomized, placebo-controlled, double-blind, phase 2 trial. *Neurology*. 2015;84(21):2161-2168.
169. Tobinick E, Gross H, Weinberger A, Cohen H. TNF-alpha modulation for treatment of Alzheimer's disease: a 6-month pilot study. *MedGenMed*. 2006;8(2):25.
170. Familian A, Eikelenboom P, Veerhuis R. Minocycline does not affect amyloid β phagocytosis by human microglial cells. *Neurosci Lett*. 2007;416(1):87-91.
171. Corbett A, Williams G, Ballard C. Drug repositioning: an opportunity to develop novel treatments for Alzheimer's disease. *Pharmaceuticals (Basel)*. 2013;6(10):1304-1321.

172. Kurtuncu M, Tuzun E, Turkoglu R, et al. Effect of short-term interferon-beta treatment on cytokines in multiple sclerosis: significant modulation of IL-17 and IL-23. *Cytokine*. 2012;59(2):400-402.
173. Grigoriadis N, van Pesch V. A basic overview of multiple sclerosis immunopathology. *Eur J Neurol*. 2015; 22(Suppl 2):3-13.
174. Engelhardt B, Kappos L. Natalizumab: targeting alpha4-integrins in multiple sclerosis. *Neurodegener Dis*. 2008; 5(1):16-22.
175. Goldstein D, Laszlo J. The role of interferon in cancer therapy: a current perspective. *CA Cancer J Clin*. 1988; 38(5):258-277.
176. Jacobs SK, Wilson DJ, Kornblith PL, Grimm EA. Interleukin-2 or autologous lymphokine-activated killer cell treatment of malignant glioma: phase I trial. *Cancer Res*. 1986;46(4 Pt 2):2101-2104.
177. Melencio L, McKallip RJ, Guan H, et al. Role of CD4(+)CD25(+) T regulatory cells in IL-2-induced vascular leak. *Int Immunol*. 2006;18(10):1461-1471.
178. Krieg C, Létourneau S, Pantaleo G, Boyman O. Improved IL-2 immunotherapy by selective stimulation of IL-2 receptors on lymphocytes and endothelial cells. *Proc Natl Acad Sci*. 2010;107(26):11906-11911.
179. Hallett WHD, Ames E, Álvarez M, et al. Combination therapy using IL-2 and Anti-CD25 results in augmented natural killer cell-mediated antitumor responses. *Biol Blood Marrow Transplant*. 2008;14(10):1088-1099.
180. Fyfe G, Fisher RI, Rosenberg SA, Sznol M, Parkinson DR, Louie AC. Results of treatment of 255 patients with metastatic renal cell carcinoma who received high-dose recombinant interleukin-2 therapy. *J Clin Oncol*. 1995; 13(3):688-696.
181. Rosenberg SA, Yang JC, White DE, Steinberg SM. Durability of complete responses in patients with metastatic cancer treated with high-dose interleukin-2: identification of the antigens mediating response. *Ann Surg*. 1998;228(3):307-319.
182. Holladay FP, Heitz-Turner T, Bayer WL, Wood GW. Autologous tumor cell vaccination combined with adoptive cellular immunotherapy in patients with grade III/IV astrocytoma. *J Neurooncol*. 1996;27(2):179-189.
183. Levin AM, Bates DL, Ring AM, et al. Exploiting a natural conformational switch to engineer an interleukin-2 'superkine'. *Nature*. 2012;484(7395):529-533.
184. Colombo F, Barzon L, Franchin E, et al. Combined HSV-TK/IL-2 gene therapy in patients with recurrent glioblastoma multiforme: biological and clinical results. *Cancer Gene Ther*. 2005;12(10):835-848.
185. Christofides A, Kosmopoulos M, Piperi C. Pathophysiological mechanisms regulated by cytokines in gliomas. *Cytokine*. 2015;71(2):377-384.
186. Ohno M, Natsume A, Wakabayashi T. Cytokine therapy. *Adv Exp Med Biol*. 2012;746:86-94.
187. Zhu VF, Yang J, Lebrun DG, Li M. Understanding the role of cytokines in glioblastoma multiforme pathogenesis. *Cancer Lett*. 2012;316(2):139-150.
188. Zhang C, Zhang X, Xu R, et al. TGF-beta2 initiates autophagy via Smad and non-Smad pathway to promote glioma cells' invasion. *J Exp Clin Cancer Res*. 2017;36(1):162.
189. Bogdahn U, Hau P, Stockhammer G, et al. Targeted therapy for high-grade glioma with the TGF-beta2 inhibitor trabedersen: results of a randomized and controlled phase IIb study. *Neuro-Oncology*. 2011;13(1):132-142.
190. Paul WE. Interleukin 4: signalling mechanisms and control of T cell differentiation. *CIBA Found Symp*. 1997,204:208-216. discussion 209-216.
191. Husain SR, Behari N, Kreitman RJ, Pastan I, Puri RK. Complete regression of established human glioblastoma tumor xenograft by interleukin-4 toxin therapy. *Cancer Res*. 1998;58(16):3649-3653.
192. Puri RK, Hoon DS, Leland P, et al. Preclinical development of a recombinant toxin containing circularly permuted interleukin 4 and truncated *Pseudomonas* exotoxin for therapy of malignant astrocytoma. *Cancer Res*. 1996;56(24):5631-5637.
193. Rand RW, Kreitman RJ, Patronas N, Varricchio F, Pastan I, Puri RK. Intratumoral administration of recombinant circularly permuted interleukin-4-*Pseudomonas* exotoxin in patients with high-grade glioma. *Clin Cancer Res*. 2000;6(6):2157-2165.
194. Weber FW, Floeth F, Asher A, et al. Local convection enhanced delivery of IL4-*Pseudomonas* exotoxin (NBI-3001) for treatment of patients with recurrent malignant glioma. *Acta Neurochir Suppl*. 2003;88:93-103. Vienna.
195. Garland L, Gitlitz B, Ebbinghaus S, et al. Phase I trial of intravenous IL-4 *Pseudomonas* exotoxin protein (NBI-3001) in patients with advanced solid tumors that express the IL-4 receptor. *J Immunother*. 2005;28(4):376-381.
196. Okada H, Lieberman FS, Edington HD, et al. Autologous glioma cell vaccine admixed with interleukin-4 gene transfected fibroblasts in the treatment of recurrent glioblastoma: preliminary observations in a patient with a favorable response to therapy. *J Neurooncol*. 2003;64(1-2):13-20.

197. Rainov NG, Gorbatyuk K, Heidecke V. Clinical trials with intracerebral convection-enhanced delivery of targeted toxins in malignant glioma. *Rev Recent Clin Trials*. 2008;3(1):2-9.
198. Perussia B, Chan SH, D'Andrea A, et al. Natural killer (NK) cell stimulatory factor or IL-12 has differential effects on the proliferation of TCR-alpha beta+, TCR-gamma delta+ T lymphocytes, and NK cells. *J Immunol*. 1992;149 (11):3495-3502.
199. Luster AD, Greenberg SM, Leder P. The IP-10 chemokine binds to a specific cell surface heparan sulfate site shared with platelet factor 4 and inhibits endothelial cell proliferation. *J Exp Med*. 1995;182(1):219-231.
200. Leonard JP, Sherman ML, Fisher GL, et al. Effects of single-dose interleukin-12 exposure on interleukin-12-associated toxicity and interferon-gamma production. *Blood*. 1997;90(7):2541-2548.
201. Atkins MB, Robertson MJ, Gordon M, et al. Phase I evaluation of intravenous recombinant human interleukin 12 in patients with advanced malignancies. *Clin Cancer Res*. 1997;3(3):409-417.
202. Sabel MS, Arora A, Su G, et al. Generation of a tumor-specific systemic response after intratumoral injection of IL-12 and IL-18-loaded polylactic acid microspheres. *J Immunother*. 2007;30(8):808-816.
203. Kikuchi T, Joki T, Saitoh S, et al. Anti-tumor activity of interleukin-2-producing tumor cells and recombinant interleukin 12 against mouse glioma cells located in the central nervous system. *Int J Cancer*. 1999;80(3):425-430.
204. DiMeco F, Rhines LD, Hanes J, et al. Paracrine delivery of IL-12 against intracranial 9L gliosarcoma in rats. *J Neurosurg*. 2000;92(3):419-427.
205. Vom Berg J, Vrohlings M, Haller S, et al. Intratumoral IL-12 combined with CTLA-4 blockade elicits T cell-mediated glioma rejection. *J Exp Med*. 2013;210(13):2803-2811.
206. Addison CL, Bramson JL, Hitt MM, Muller WJ, Gauldie J, Graham FL. Intratumoral coinjection of adenoviral vectors expressing IL-2 and IL-12 results in enhanced frequency of regression of injected and untreated distal tumors. *Gene Ther*. 1998;5(10):1400-1409.
207. Bramson JL, Hitt M, Addison CL, Muller WJ, Gauldie J, Graham FL. Direct intratumoral injection of an adenovirus expressing interleukin-12 induces regression and long-lasting immunity that is associated with highly localized expression of interleukin-12. *Hum Gene Ther*. 1996;7(16):1995-2002.
208. Parker JN, Gillespie GY, Love CE, Randall S, Whitley RJ, Markert JM. Engineered herpes simplex virus expressing IL-12 in the treatment of experimental murine brain tumors. *Proc Natl Acad Sci U S A*. 2000;97 (5):2208-2213.
209. Triozzi PL, Allen KO, Carlisle RR, Craig M, LoBuglio AF, Conry RM. Phase I study of the intratumoral administration of recombinant canarypox viruses expressing B7.1 and interleukin 12 in patients with metastatic melanoma. *Clin Cancer Res*. 2005;11(11):4168-4175.
210. Triozzi PL, Strong TV, Bucy RP, et al. Intratumoral administration of a recombinant canarypox virus expressing interleukin 12 in patients with metastatic melanoma. *Hum Gene Ther*. 2005;16(1):91-100.
211. Quetglas JI, Dubrot J, Bezunartea J, et al. Immunotherapeutic synergy between anti-CD137 mAb and intratumoral administration of a cytopathic Semliki Forest virus encoding IL-12. *Mol Ther*. 2012;20(9):1664-1675.
212. Patel DM, Foreman PM, Nabors LB, Riley KO, Gillespie GY, Markert JM. Design of a phase I clinical trial to evaluate M032, a genetically engineered HSV-1 expressing IL-12, in Patients with recurrent/progressive glioblastoma multiforme, anaplastic astrocytoma, or gliosarcoma. *Hum Gene Ther Clin Dev*. 2016;27(2):69-78.
213. Saha D, Martuza RL, Rabkin SD. Macrophage polarization contributes to glioblastoma eradication by combination immunovirotherapy and immune checkpoint blockade. *Cancer Cell*. 2017;32(2). 253-267.e5.
214. Linette GP, Hamid O, Whitman ED, et al. A phase I open-label study of Ad-RTS-hIL-12, an adenoviral vector engineered to express hIL-12 under the control of an oral activator ligand, in subjects with unresectable stage III/IV melanoma. *J Clin Oncol*. 2013;31(15 Suppl):3022.
215. Sangro B, Mazzolini G, Ruiz J, et al. Phase I trial of intratumoral injection of an adenovirus encoding interleukin-12 for advanced digestive tumors. *J Clin Oncol*. 2004;22(8):1389-1397.
216. Mineharu Y, Muhammad AK, Yagiz K, et al. Gene therapy-mediated reprogramming tumor infiltrating T cells using IL-2 and inhibiting NF-kappaB signaling improves the efficacy of immunotherapy in a brain cancer model. *Neurotherapeutics*. 2012;9(4):827-843.
217. Leonard JP, Sherman ML, Fisher GL, et al. Effects of single-dose interleukin-12 exposure on interleukin-12–associated toxicity and interferon-γ production. *Blood*. 1997;90(7):2541-2548.
218. Buonaguro L, Petrizzo A, Tornesello ML, Buonaguro FM. Translating tumor antigens into cancer vaccines. *Clin Vaccine Immunol*. 2011;18(1):23-34.
219. Kanold J, Paillard C, Tchirkov A, et al. NK cell immunotherapy for high-risk neuroblastoma relapse after haploidentical HSCT. *Pediatr Blood Cancer*. 2012;59(4):739-742.

220. Vivier E, Ugolini S, Blaise D, Chabannon C, Brossay L. Targeting natural killer cells and natural killer T cells in cancer. *Nat Rev Immunol*. 2012;12(4):239-252.
221. Rubnitz JE, Inaba H, Ribeiro RC, et al. NKAML: a pilot study to determine the safety and feasibility of haploidentical natural killer cell transplantation in childhood acute myeloid leukemia. *J Clin Oncol*. 2010; 28(6):955-959.
222. Shimasaki N, Campana D. Natural killer cell reprogramming with chimeric immune receptors. *Methods Mol Biol*. 2013;969:203-220.
223. Tran AC, Zhang D, Byrn R, Roberts MR. Chimeric zeta-receptors direct human natural killer (NK) effector function to permit killing of NK-resistant tumor cells and HIV-infected T lymphocytes. *J Immunol*. 1995;155 (2):1000-1009.
224. Zhen A, Kamata M, Rezek V, et al. HIV-specific immunity derived from chimeric antigen receptor-engineered stem cells. *Mol Ther*. 2015;23(8):1358-1367.
225. Jiang H, Zhang W, Shang P, et al. Transfection of chimeric anti-CD138 gene enhances natural killer cell activation and killing of multiple myeloma cells. *Mol Oncol*. 2014;8(2):297-310.
226. Chen X, Han J, Chu J, et al. A combinational therapy of EGFR-CAR NK cells and oncolytic herpes simplex virus 1 for breast cancer brain metastases. *Oncotarget*. 2016;7(19):27764-27777.
227. Zhang C, Burger MC, Jennewein L, et al. ErbB2/HER2-specific NK cells for targeted therapy of glioblastoma. *J Natl Cancer Inst*. 2016;108(5).
228. Genssler S, Burger MC, Zhang C, et al. Dual targeting of glioblastoma with chimeric antigen receptor-engineered natural killer cells overcomes heterogeneity of target antigen expression and enhances antitumor activity and survival. *Oncoimmunology*. 2016;5(4):e1119354.
229. Greenberg PD, Reusser P, Goodrich JM, Riddell SR. Development of a treatment regimen for human cytomegalovirus (CMV) infection in bone marrow transplantation recipients by adoptive transfer of donor-derived CMV-specific T cell clones expanded in vitro. *Ann N Y Acad Sci*. 1991;636:184-195.
230. Dillman RO, Duma CM, Ellis RA, et al. Intralesional lymphokine-activated killer cells as adjuvant therapy for primary glioblastoma. *J Immunother*. 2009;32(9):914-919.
231. Hayes RL, Koslow M, Hiesiger EM, et al. Improved long term survival after intracavitary interleukin-2 and lymphokine-activated killer cells for adults with recurrent malignant glioma. *Cancer*. 1995;76(5):840-852.
232. Khammari A, Knol AC, Nguyen JM, et al. Adoptive TIL transfer in the adjuvant setting for melanoma: long-term patient survival. *J Immunol Res*. 2014;2014:186212.
233. Rosenberg SA, Yannelli JR, Yang JC, et al. Treatment of patients with metastatic melanoma with autologous tumor-infiltrating lymphocytes and interleukin 2. *J Natl Cancer Inst*. 1994;86(15):1159-1166.
234. Rosenberg SA, Packard BS, Aebersold PM, et al. Use of tumor-infiltrating lymphocytes and interleukin-2 in the immunotherapy of patients with metastatic melanoma. A preliminary report. *N Engl J Med*. 1988;319 (25):1676-1680.
235. Dudley ME, Gross CA, Langhan MM, et al. CD8+ enriched "young" tumor infiltrating lymphocytes can mediate regression of metastatic melanoma. *Clin Cancer Res*. 2010;16(24):6122-6131.
236. Dudley ME, Wunderlich JR, Yang JC, et al. Adoptive cell transfer therapy following non-myeloablative but lymphodepleting chemotherapy for the treatment of patients with refractory metastatic melanoma. *J Clin Oncol*. 2005;23(10):2346-2357.
237. Hong JJ, Rosenberg SA, Dudley ME, et al. Successful treatment of melanoma brain metastases with adoptive cell therapy. *Clin Cancer Res*. 2010;16(19):4892-4898.
238. Quattrocchi KB, Miller CH, Cush S, et al. Pilot study of local autologous tumor infiltrating lymphocytes for the treatment of recurrent malignant gliomas. *J Neuro-Oncol*. 1999;45(2):141-157.
239. Yaghoubi SS, Jensen MC, Satyamurthy N, et al. Noninvasive detection of therapeutic cytolytic T cells with 18F-FHBG PET in a patient with glioma. *Nat Clin Pract Oncol*. 2009;6(1):53-58.
240. Brown CE, Badie B, Barish ME, et al. Bioactivity and safety of IL13Ralpha2-redirected chimeric antigen receptor CD8+ T cells in patients with recurrent glioblastoma. *Clin Cancer Res*. 2015;21(18):4062-4072.
241. Milone MC, Fish JD, Carpenito C, et al. Chimeric receptors containing CD137 signal transduction domains mediate enhanced survival of T cells and increased antileukemic efficacy in vivo. *Mol Ther*. 2009;17(8):1453-1464.
242. Kumaresan PR, Manuri PR, Albert ND, et al. Bioengineering T cells to target carbohydrate to treat opportunistic fungal infection. *Proc Natl Acad Sci*. 2014;111(29):10660-10665.
243. Hombach AA, Abken H. Costimulation by chimeric antigen receptors revisited the T cell antitumor response benefits from combined CD28-OX40 signalling. *Int J Cancer*. 2011;129(12):2935-2944.

244. Kalaitsidou M, Kueberuwa G, Schutt A, Gilham DE. CAR T-cell therapy: toxicity and the relevance of preclinical models. *Immunotherapy*. 2015;7(5):487-497.
245. Heimberger AB, Hlatky R, Suki D, et al. Prognostic effect of epidermal growth factor receptor and EGFRvIII in glioblastoma multiforme patients. *Clin Cancer Res*. 2005;11(4):1462-1466.
246. Johnson LA, Scholler J, Ohkuri T, et al. Rational development and characterization of humanized anti-EGFR variant III chimeric antigen receptor T cells for glioblastoma. *Sci Transl Med*. 2015;7(275):275ra222.
247. Sampson JH, Choi BD, Sanchez-Perez L, et al. EGFRvIII mCAR-modified T-cell therapy cures mice with established intracerebral glioma and generates host immunity against tumor-antigen loss. *Clin Cancer Res*. 2014;20(4):972-984.
248. O'Rourke DM, Nasrallah MP, Desai A, et al. A single dose of peripherally infused EGFRvIII-directed CAR T cells mediates antigen loss and induces adaptive resistance in patients with recurrent glioblastoma. *Sci Transl Med*. 2017;9(399).
249. Pilon SA, Kelly C, Wei WZ. Broadening of epitope recognition during immune rejection of ErbB-2-positive tumor prevents growth of ErbB-2-negative tumor. *J Immunol*. 2003;170(3):1202-1208.
250. Jarboe JS, Johnson KR, Choi Y, Lonser RR, Park JK. Expression of interleukin-13 receptor alpha2 in glioblastoma multiforme: implications for targeted therapies. *Cancer Res*. 2007;67(17):7983-7986.
251. Mintz A, Gibo DM, Slagle-Webb B, Christensen ND, Debinski W. IL-13Ralpha2 is a glioma-restricted receptor for interleukin-13. *Neoplasia*. 2002;4(5):388-399.
252. Sengupta S, Thaci B, Crawford AC, Sampath P. Interleukin-13 receptor alpha 2-targeted glioblastoma immunotherapy. *Biomed Res Int*. 2014;2014:952128.
253. Thaci B, Brown CE, Binello E, Werbaneth K, Sampath P, Sengupta S. Significance of interleukin-13 receptor alpha 2-targeted glioblastoma therapy. *Neuro-Oncology*. 2014;16(10):1304-1312.
254. Kahlon KS, Brown C, Cooper LJ, Raubitschek A, Forman SJ, Jensen MC. Specific recognition and killing of glioblastoma multiforme by interleukin 13-zetakine redirected cytolytic T cells. *Cancer Res*. 2004;64(24):9160-9166.
255. Kong S, Sengupta S, Tyler B, et al. Suppression of human glioma xenografts with second-generation IL13R-specific chimeric antigen receptor-modified T cells. *Clin Cancer Res*. 2012;18(21):5949-5960.
256. Krebs S, Chow KK, Yi Z, et al. T cells redirected to interleukin-13Ralpha2 with interleukin-13 mutein–chimeric antigen receptors have anti-glioma activity but also recognize interleukin-13Ralpha1. *Cytotherapy*. 2014;16(8):1121-1131.
257. Krenciute G, Krebs S, Torres D, et al. Characterization and functional analysis of scFv-based chimeric antigen receptors to redirect T CELLS to IL13Ralpha2-positive glioma. *Mol Ther*. 2016;24(2):354-363.
258. Brown CE, Alizadeh D, Starr R, et al. Regression of glioblastoma after chimeric antigen receptor T-Cell therapy. *N Engl J Med*. 2016;375(26):2561-2569.
259. Heimberger AB, Crotty LE, Archer GE, et al. Epidermal growth factor receptor VIII peptide vaccination is efficacious against established intracerebral tumors. *Clin Cancer Res*. 2003;9(11):4247-4254.
260. Sampson JH, Heimberger AB, Archer GE, et al. Immunologic escape after prolonged progression-free survival with epidermal growth factor receptor variant III peptide vaccination in patients with newly diagnosed glioblastoma. *J Clin Oncol*. 2010;28(31):4722-4729.
261. Schuster J, Lai RK, Recht LD, et al. A phase II, multicenter trial of rindopepimut (CDX-110) in newly diagnosed glioblastoma: the ACT III study. *Neuro-Oncology*. 2015;17(6):854-861.
262. Morgan RA, Yang JC, Kitano M, Dudley ME, Laurencot CM, Rosenberg SA. Case report of a serious adverse event following the administration of T cells transduced with a chimeric antigen receptor recognizing ERBB2. *Mol Ther*. 2010;18(4):843-851.
263. Wykosky J, Gibo DM, Stanton C, Debinski W. EphA2 as a novel molecular marker and target in glioblastoma multiforme. *Mol Cancer Res*. 2005;3(10):541-551.
264. Wykosky J, Gibo DM, Stanton C, Debinski W. Interleukin-13 receptor alpha 2, EphA2, and Fos-related antigen 1 as molecular denominators of high-grade astrocytomas and specific targets for combinatorial therapy. *Clin Cancer Res*. 2008;14(1):199-208.
265. Wang LF, Fokas E, Bieker M, et al. Increased expression of EphA2 correlates with adverse outcome in primary and recurrent glioblastoma multiforme patients. *Oncol Rep*. 2008;19(1):151-156.
266. Chow KK, Naik S, Kakarla S, et al. T cells redirected to EphA2 for the immunotherapy of glioblastoma. *Mol Ther*. 2013;21(3):629-637.
267. Caruso HG, Hurton LV, Najjar A, et al. Tuning sensitivity of CAR to EGFR density limits recognition of normal tissue while maintaining potent antitumor activity. *Cancer Res*. 2015;75(17):3505-3518.

268. Caruso H, Heimberger A, Cooper LJN. Steering CAR T cells to distinguish friend from foe. *OncoImmunology*. 2017. https://doi.org/10.1080/2162402X.2016.1271857.
269. Yu JS, Liu G, Ying H, Yong WH, Black KL, Wheeler CJ. Vaccination with tumor lysate-pulsed dendritic cells elicits antigen-specific, cytotoxic T-cells in patients with malignant glioma. *Cancer Res*. 2004;64(14): 4973-4979.
270. Liau LM, Prins RM, Kiertscher SM, et al. Dendritic cell vaccination in glioblastoma patients induces systemic and intracranial T-cell responses modulated by the local central nervous system tumor microenvironment. *Clin Cancer Res*. 2005;11(15):5515-5525.
271. Sampson JH, Archer GE, Mitchell DA, et al. An epidermal growth factor receptor variant III-targeted vaccine is safe and immunogenic in patients with glioblastoma multiforme. *Mol Cancer Ther*. 2009;8(10):2773-2779.
272. De Vleeschouwer S, Fieuws S, Rutkowski S, et al. Postoperative adjuvant dendritic cell-based immunotherapy in patients with relapsed glioblastoma multiforme. *Clin Cancer Res*. 2008;14(10):3098-3104.
273. Cho DY, Yang WK, Lee HC, et al. Adjuvant immunotherapy with whole-cell lysate dendritic cells vaccine for glioblastoma multiforme: a phase II clinical trial. *World Neurosurg*. 2012;77(5-6):736-744.
274. Mitchell DA, Batich KA, Gunn MD, et al. Tetanus toxoid and CCL3 improve dendritic cell vaccines in mice and glioblastoma patients. *Nature*. 2015;519(7543):366-369.
275. Reap EA, Suryadevara CM, Batich KA, et al. Dendritic cells enhance polyfunctionality of adoptively transferred T cells that target cytomegalovirus in glioblastoma. *Cancer Res*. 2018;78(1):256-264.
276. Cintolo JA, Datta J, Mathew SJ, Czerniecki BJ. Dendritic cell-based vaccines: barriers and opportunities. *Future Oncol*. 2012;8(10):1273-1299.
277. Melief CJ, van Hall T, Arens R, Ossendorp F, van der Burg SH. Therapeutic cancer vaccines. *J Clin Invest*. 2015; 125(9):3401-3412.
278. McNamara MA, Nair SK, Holl EK. RNA-based vaccines in cancer immunotherapy. *J Immunol Res*. 2015;2015.
279. Liu JK. Anti-cancer vaccines—a one-hit wonder? *Yale J Biol Med*. 2014;87(4):481-489.
280. Goodin DS, Frohman EM, Garmany GP, et al. Disease modifying therapies in multiple sclerosis: subcommittee of the American Academy of Neurology and the MS Council for Clinical Practice Guidelines. *Neurology*. 2002; 58(2):169-178.
281. Bielekova B, Goodwin B, Richert N, et al. Encephalitogenic potential of the myelin basic protein peptide (amino acids 83-99) in multiple sclerosis: results of a phase II clinical trial with an altered peptide ligand. *Nat Med*. 2000; 6(10):1167-1175.
282. Schenk D, Barbour R, Dunn W, et al. Immunization with amyloid-beta attenuates Alzheimer-disease-like pathology in the PDAPP mouse. *Nature*. 1999;400(6740):173-177.
283. Orgogozo JM, Gilman S, Dartigues JF, et al. Subacute meningoencephalitis in a subset of patients with AD after Abeta42 immunization. *Neurology*. 2003;61(1):46-54.
284. Boche D, Zotova E, Weller RO, et al. Consequence of Aβ immunization on the vasculature of human Alzheimer's disease brain. *Brain*. 2008;131(12):3299-3310.
285. Crane CA, Han SJ, Ahn B, et al. Individual patient-specific immunity against high-grade glioma after vaccination with autologous tumor derived peptides bound to the 96 KD chaperone protein. *Clin Cancer Res*. 2013; 19(1):205-214.
286. Bloch O, Crane CA, Fuks Y, et al. Heat-shock protein peptide complex-96 vaccination for recurrent glioblastoma: a phase II, single-arm trial. *Neuro-Oncology*. 2014;16(2):274-279.
287. Bloch O, Lim M, Sughrue ME, et al. Autologous heat shock protein peptide vaccination for newly diagnosed glioblastoma: impact of peripheral PD-L1 expression on response to therapy. *Clin Cancer Res*. 2017; 23(14):3575-3584.
288. Ishikawa E, Tsuboi K, Yamamoto T, et al. Clinical trial of autologous formalin-fixed tumor vaccine for glioblastoma multiforme patients. *Cancer Sci*. 2007;98(8):1226-1233.
289. Bigner SH, Humphrey PA, Wong AJ, et al. Characterization of the epidermal growth factor receptor in human glioma cell lines and xenografts. *Cancer Res*. 1990;50(24):8017-8022.
290. Humphrey PA, Wong AJ, Vogelstein B, et al. Anti-synthetic peptide antibody reacting at the fusion junction of deletion-mutant epidermal growth factor receptors in human glioblastoma. *Proc Natl Acad Sci U S A*. 1990; 87(11):4207-4211.
291. Rampling R, Peoples S, Mulholland PJ, et al. A cancer research UK first time in human phase I trial of IMA950 (novel multipeptide therapeutic vaccine) in patients with newly diagnosed glioblastoma. *Clin Cancer Res*. 2016;22(19):4776-4785.

292. Slingluff Jr CL. The present and future of peptide vaccines for cancer: single or multiple, long or short, alone or in combination? *Cancer J*. 2011;17(5):343-350.
293. Kaufman HL, Kohlhapp FJ, Zloza A. Oncolytic viruses: a new class of immunotherapy drugs. *Nat Rev Drug Discov*. 2015;14(9):642-662.
294. Kou J, Kim H, Pattanayak A, et al. Anti-amyloid-beta single-chain antibody brain delivery via AAV reduces amyloid load but may increase cerebral hemorrhages in an Alzheimer's disease mouse model. *J Alzheimers Dis*. 2011;27(1):23-38.
295. Palfi S, Gurruchaga JM, Ralph GS, et al. Long-term safety and tolerability of ProSavin, a lentiviral vector-based gene therapy for Parkinson's disease: a dose escalation, open-label, phase 1/2 trial. *Lancet*. 2014;383 (9923):1138-1146.
296. Passini MA, Watson DJ, Vite CH, Landsburg DJ, Feigenbaum AL, Wolfe JH. Intraventricular brain injection of adeno-associated virus type 1 (AAV1) in neonatal mice results in complementary patterns of neuronal transduction to AAV2 and total long-term correction of storage lesions in the brains of beta-glucuronidase-deficient mice. *J Virol*. 2003;77(12):7034-7040.
297. Baek RC, Broekman ML, Leroy SG, et al. AAV-mediated gene delivery in adult GM1-gangliosidosis mice corrects lysosomal storage in CNS and improves survival. *PLoS ONE*. 2010;5(10).
298. Andtbacka RH, Kaufman HL, Collichio F, et al. Talimogene laherparepvec improves durable response rate in patients with advanced melanoma. *J Clin Oncol*. 2015;33(25):2780-2788.
299. Greig SL. Talimogene laherparepvec: first global approval. *Drugs*. 2016;76(1):147-154.
300. Garber K. China approves world's first oncolytic virus therapy for cancer treatment. *J Natl Cancer Inst*. 2006; 98(5):298-300.
301. Markert JM, Razdan SN, Kuo HC, et al. A phase 1 trial of oncolytic HSV-1, G207, given in combination with radiation for recurrent GBM demonstrates safety and radiographic responses. *Mol Ther*. 2014;22(5):1048-1055.
302. Muhammad AK, Puntel M, Candolfi M, et al. Study of the efficacy, biodistribution, and safety profile of therapeutic gutless adenovirus vectors as a prelude to a phase I clinical trial for glioblastoma. *Clin Pharmacol Ther*. 2010;88(2):204-213.
303. Dobrikova EY, Broadt T, Poiley-Nelson J, et al. Recombinant oncolytic poliovirus eliminates glioma in vivo without genetic adaptation to a pathogenic phenotype. *Mol Ther*. 2008;16(11):1865-1872.
304. Koks CA, Garg AD, Ehrhardt M, et al. Newcastle disease virotherapy induces long-term survival and tumor-specific immune memory in orthotopic glioma through the induction of immunogenic cell death. *Int J Cancer*. 2015;136(5):E313-E325.
305. Fueyo J, Gomez-Manzano C, Alemany R, et al. A mutant oncolytic adenovirus targeting the Rb pathway produces anti-glioma effect in vivo. *Oncogene*. 2000;19(1):2-12.
306. Tsamis KI, Alexiou GA, Vartholomatos E, Kyritsis AP. Combination treatment for glioblastoma cells with tumor necrosis factor-related apoptosis-inducing ligand and oncolytic adenovirus Delta-24. *Cancer Investig*. 2013;31 (9):630-638.
307. Piao Y, Jiang H, Alemany R, et al. Oncolytic adenovirus retargeted to Delta-EGFR induces selective antiglioma activity. *Cancer Gene Ther*. 2009;16(3):256-265.
308. Fueyo J, Alemany R, Gomez-Manzano C, et al. Preclinical characterization of the antiglioma activity of a tropism-enhanced adenovirus targeted to the retinoblastoma pathway. *J Natl Cancer Inst*. 2003;95(9):652-660.
309. Jiang H, Clise-Dwyer K, Ruisaard KE, et al. Delta-24-RGD oncolytic adenovirus elicits anti-glioma immunity in an immunocompetent mouse model. *PLoS One*. 2014;9(5).
310. Lang FF, Conrad C, Gomez-Manzano C, et al. Phase I study of DNX-2401 (Delta-24-RGD) oncolytic adenovirus: replication and immunotherapeutic effects in recurrent malignant glioma. *J Clin Oncol*. 2018;36(14):1419-1427. JCO.2017.2075.8219.
311. Jiang H, Rivera-Molina Y, Gomez-Manzano C, et al. Oncolytic adenovirus and tumor-targeting immune modulatory therapy improve autologous cancer vaccination. *Cancer Res*. 2017;77(14):3894-3907.
312. Hardcastle J, Mills L, Malo CS, et al. Immunovirotherapy with measles virus strains in combination with anti-PD-1 antibody blockade enhances antitumor activity in glioblastoma treatment. *Neuro-Oncology*. 2017;19 (4):493-502.

CHAPTER

16

Convection-Enhanced Drug Delivery in the Central Nervous System

Jerusha Naidoo[*], *Massimo Fiandaca*[†], *Russell R. Lonser*[‡], *Krystof Bankiewicz*[*]

[*]Interventional Neuro Center, Department of Neurological Surgery, University of California–San Francisco, San Francisco, CA, United States [†]Brain Neurotherapy Bio, Inc., Oakland, CA, United States [‡]Department of Neurological Surgery, The Ohio State University Wexner Medical Center, Columbus, OH, United States

INTRODUCTION

Based on emerging biologic information, potent putative therapeutics have been developed to address neurologic disorders that are currently untreatable. These therapeutic compounds have shown success in the reversal or amelioration in vitro and/or in animal models of neurologic disease. Unfortunately, these findings have yet to be translated into successful clinical paradigms. The most critical obstacle to the conversion of these therapeutic agents into efficacious clinical therapies is the inability to deliver these agents across the central nervous system (CNS) blood–brain barrier (BBB) in a homogeneous, targeted, and traceable manner using methods and/or featuring systemic delivery, intrathecal/intraventricular distribution, and polymer implantation.[1-4]

Convection-enhanced delivery (CED) of putative therapeutics overcomes many limitations associated with other CNS drug delivery strategies, and is being more frequently used in neurological disease research as well as in therapeutic clinical trials for neurologic disorders. CED involves distribution of an infusate, delivered via an intraparenchymal cannula, through the interstitial spaces of the CNS parenchyma. CED distributes the infusate via a small hydrostatic pressure that results in bulk or convective flow,[5] enabling homogeneous distribution of small- and large-molecular–weight therapeutics to targeted regions of the nervous system across the BBB. Co-infusion of surrogate imaging tracers allows for accurate simultaneous, noninvasive monitoring of the convective distribution using magnetic

Nervous System Drug Delivery
https://doi.org/10.1016/B978-0-12-813997-4.00016-5

© 2019 Elsevier Inc. All rights reserved.

resonance imaging (MRI). This chapter reviews the biophysical principles of convective flow, convective delivery methodology, and clinical applications in the CNS.

BIOMECHANICS AND PROPERTIES OF CED IN THE CNS

General Properties of CED

Edward Oldfield first described CED in the early 1990s as a method to directly deliver putative therapeutic agents across the BBB.[4] The discovery was based on the observation that small- and large-molecular–weight molecules move freely and rapidly through the interstitial spaces of the CNS in regions of peritumoral edema associated with intracranial metastatic deposits. Later, it was discovered that the permeable vasculature of metastatic deposits created a source of increased hydrostatic pressure derived from the systemic circulation (systemic blood pressure) that drove a plasma ultrafiltrate through the CNS parenchyma, including small and large molecules.[6]

Similarly, CED utilizes an applied small hydrostatic pressure gradient to the infusate solution, derived from a syringe pump, and allows them to distribution of the convective fluid within the parenchyma via specially designed cannulae. Because convective delivery relies on a small hydrostatic pressure gradient to drive infusate through the interstitial or extracellular spaces of the nervous system, infusate distribution is based on bulk or convective flow properties. The unique properties of convective delivery allow for safe, targeted, and homogeneous distribution of therapeutic agents (of small and large molecular weights) over clinically relevant volumes (small or large), with low morbidity and mortality, while bypassing the BBB. These properties can be tailored and exploited to treat a wide variety of specific neurological disorders.

CED Bypasses the BBB

Therapeutic agents that do not often cross the BBB may be optimal candidates for delivery by CED. Convective delivery of infusate bypasses the BBB because the putative therapeutics are infused directly into the CNS parenchyma via surgically placed cannula. Agents that do not cross the BBB that are distributed by CED will remain sequestered on the abluminal side of the BBB for prolonged periods until they are cleared from the interstitial spaces or are metabolized.[2, 5, 7-9] As a result, ideal agents for CED include large-molecular–weight (i.e., >300 Da) and/or hydrophilic compounds, with limited access across the BBB. Small lipophilic compounds, on the other hand, are readily cleared into the systemic circulation from the CNS tissues and have restricted distribution volumes when delivered via direct convection.

Targeted Delivery

Deep brain structures (e.g., nuclear and white matter), as well as other CNS loci, can be selectively perfused with small- and large-molecular–weight compounds from a cannula tip that has been stereotactically placed within the target site.[10-12] Furthermore, animal studies have demonstrated retrograde and anterograde transport of putative therapeutics, including proteins and products of transgene expression delivered by viral capsids, from perfused regions to other distant nonperfused sites via interconnecting pathways in a predictable manner.[11, 13-16] This allows for the delivery of therapeutics to potentially widespread anatomic regions of the CNS via the precise manipulation of infusate delivery to specifically targeted regions and their corresponding connecting structures.

Homogeneous Distribution

Convection-enhanced delivery enables homogenous distribution of therapeutic compounds within CNS parenchyma. In contrast, diffusion-driven distribution results in restricted tissue penetration (1-2 mm) from the CNS surface exposed to drug. Associated with the limited tissue penetration is a steep drop-off in drug concentration over that short distance (a 100- to 1000-fold decrease).[1, 6, 17] Analysis of radiolabeled compounds delivered by convection in animal models immediately after infusion revealed a homogeneous "square-shaped" distribution pattern in gray or white matter. This is consistent with high uniform infusate concentrations throughout the perfused region of the CNS undergoing CED (Fig. 1).[5, 18, 19]

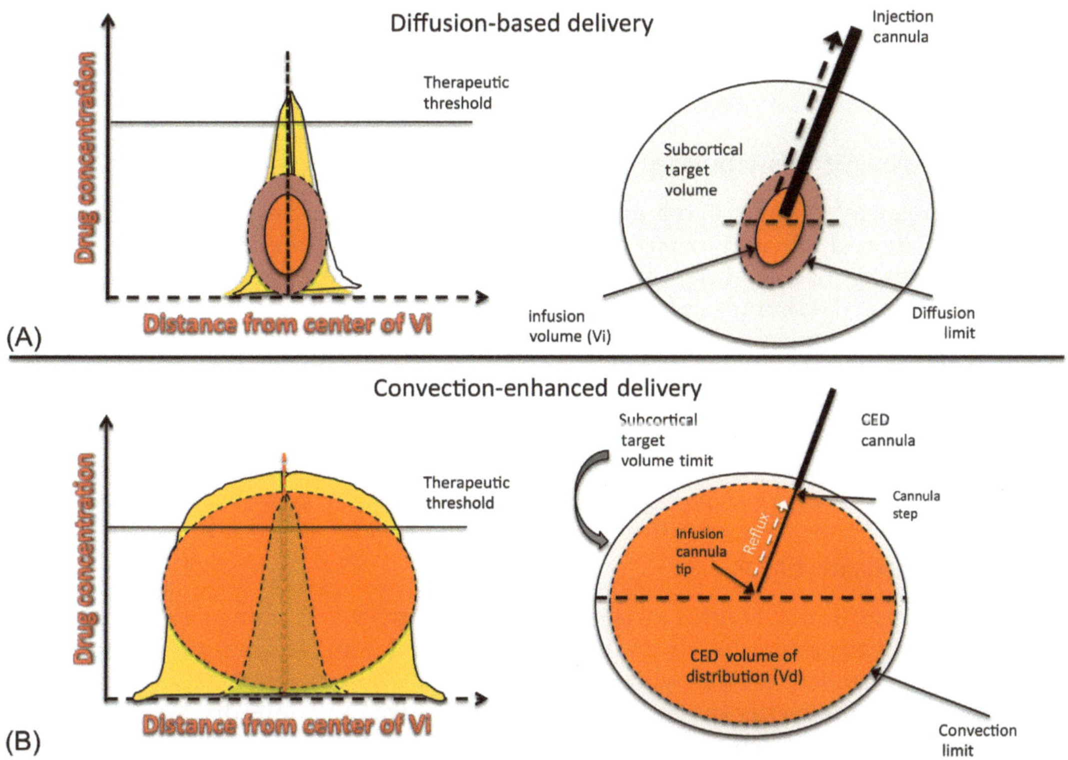

FIG. 1 Schematic representation of infusate distribution with diffusion versus convection-enhanced delivery. (A) Diffusion-based drug delivery results in therapeutic threshold being reached only within a narrow region close to the cannula tip. (B) Convection-enhanced delivery results in homogenous drug concentration over regions both proximal and distal to the cannula tip placement.

Reproducible Distribution

Convection-enhanced delivery of therapeutic infusate in the CNS is reliable in its distribution. The tissue volume of distribution (Vd) using convective delivery of an infused volume is inversely proportional to the available extracellular space within the specific region of perfused tissue. Specifically, the smaller amount of interstitial space, the greater the Vd with CED. Similarly, infusion into a region of expanded extracellular space (e.g., areas of vasogenic edema) leads to a reduction in the ratio of Vd to volume of infusion (Vi).[20, 21] Vd to Vi ratios in nondiseased gray or white matter of the brain, spinal cord, and peripheral nerves range from 4:1 to 7:1.[11, 18, 22-24] Convective delivery into the tightly compacted extracellular spaces of the brainstem is associated with larger Vd to Vi ratios (ranging from 6:1 to 10:1).[8, 19, 25] This convective property can be exploited to rapidly distribute drug infusate to large and clinically effective volumes of CNS tissue.

Although a linear relationship exists between Vd and Vi, the slope (Vd to Vi ratio) can be altered by physicochemical properties of the infusate, including chemical and metabolic stability, molecular size, and lipophilicity. Postinfusion T2-weighted analysis of convected distribution of adeno-associated viral (AAV) vectors, which are large in size (approximately 24 nm), demonstrated that AAV2 vectors have a Vd to Vi ratio of approximately three in gray matter. Intraoperative MRI-guided infusion of the contrast reagent gadoteridol co-infused into the nonhuman primate (NHP) putamen with AAV vector confirmed that delivery of 450 μL of infusate results in Vd of approximately 1500 mm^3.

Clinically Relevant Distribution

Clinically relevant volumes of therapeutic compounds (small or large) can be infused with CED. The duration of infusion time is correlated with the Vd required for effective delivery. Complete perfusion of deep brain nuclear targets can be accomplished via a single acute short-term infusion (i.e., minutes to hours).[6, 12, 14, 26] Perfusion of large CNS volumes (cerebral hemisphere) require longer periods of time and may necessitate placement of an indwelling cannula for more extended delivery intervals. [14, 27, 28] Nevertheless, based on the bulk flow properties of CED, large regions of nervous system structures can be reliably perfused in relatively short time periods, which permits the therapeutic treatment of large anatomic regions of the CNS, based on the specific neurological disorder.

Patterns of Infusate Flow

Convective distribution of infusate is isotropic in gray matter (creating a spherical or teardrop distribution pattern at the tip of the infusion cannula) and anisotropic (preferentially occurring along parallel tracts) in white matter.[18, 24, 29] These features of convective distribution in gray and white matter can be exploited for treatment of various diseases. Preferential flow along white matter tracts can be used to fill large regions of white and cortical gray matter from a single-point source infusion.[28, 30, 31] For example, a single-infusion cannula can be placed in white matter of the corona radiata to perfuse large portions of the white matter and immediately subjacent cortical gray matter.[28] Similarly, convective infusion of the parallel lateral corticospinal tracts can result in perfusion of several segments of the spinal cord.[18]

Effect of Anatomic Regions

The effectiveness of CED is adversely affected after the infusion-leading edges encounter pial and ependymal surfaces. Primate and clinical data indicate that, once the leading front of a convective infusion contacts a pial or an ependymal surface, the Vd to Vi volume ratios decline and flatten (to 1 or less), as infusate preferentially leaks into the cerebrospinal fluid space of the cistern or ventricle.[32] The latter is typically confirmed with co-infused contrast agents and real-time monitoring with MRI.

CONVECTIVE DELIVERY METHODOLOGY

Acute Delivery in the Brain

Baseline T1-weighted MR images are acquired for trajectory planning. After the target structure has been identified and the trajectory planned, a stereotactic guidance system is mounted on the skull, enabling access to the target site via the desired trajectory. Cannulae are attached to an infusion pump, and the system is primed by setting the initial infusion rate at 1-3 μL/min and run until fluid is visualized exiting from the cannula tip. A burr or twist drill hole is placed, and the dura is pierced with a lancet at the desired position. A depth-stop is secured on the cannula at the marked location, limiting the insertion distance verified prior to insertion. The cannula is then manually inserted through the targeting guide into the brain, to the desired initial depth while maintaining the low initial flow rate, creating a pressure gradient to drive bulk flow and prevent tip occlusion by tissue. After confirmation of depth and flow via imaging, final adjustments are made and infusion rates may be increased from 1 to 5 μL/min and up, gradually to a maximum rate of 30 μL/min. After turning off the delivery pumps at the end of the infusion session, the cannula remains in place for another 5 min to allow the built-up tissue pressure to dissipate. The cannula is then gradually removed along with the trajectory and the wound site is closed.

Infusion Rates

Defining an optimal infusion rate for CED has historically been problematic because of the development of reflux along the cannula tract. The optimal infusion rate is that which allows the largest volume of distribution with the smallest infusion volume in the shortest period of time without the development of reflux. Reflux can be induced regardless of cannula design if the infusion rate is too high. Initial studies estimated that flows >0.5-1.0 μL/min were enough to initiate significant backflow.[2] Low infusion rates cannot, however, achieve convective flow and are associated with unacceptably long durations of treatment, which exposes patients to greater risks of anesthesia and infection, as well as to an extended period of physical discomfort and emotional stress. We have found that the reflux-resistant cannula enables infusion

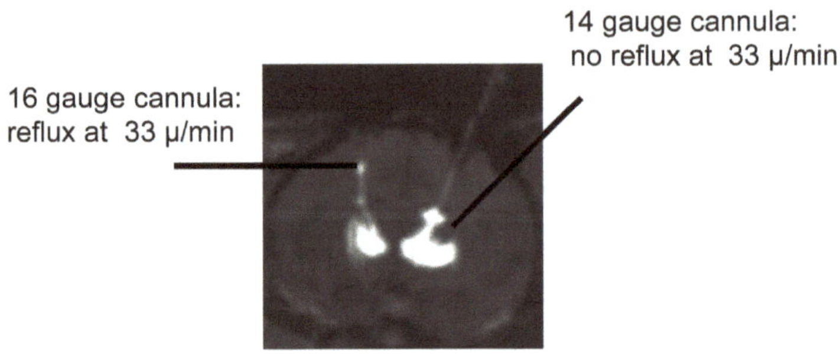

FIG. 2 Reflux of infusate in the nonhuman primate brain. Despite the step design of the infusion cannula, reflux can be induced by high-flow infusions. In the nonhuman primate brain, reflux is seen at 33 μL/min with a small-bore cannula (i.e., 16 gauge) but not with a large-bore cannula (i.e., 14 gauge).

rates to increase up to 30.0 μL/min with the absence of significant reflux, thereby greatly reducing total infusion times in the clinic. Reflux can be induced by high flow rate infusions (>30.0 μL/min), despite the use of a step design cannula. Fig. 2 shows an AAV vector infusion in the NHP brain, with reflux observed at 33 μL/min with a small-bore cannula (16 gauge) but not with a large-bore cannula (14 gauge). A possible explanation is that pressure flow is reduced at the tip of the 16 gauge cannula, permitting the infusate to flow up the path of least resistance.

Process for Chronic Delivery

We have also developed in-dwelling cannulae to allow for prolonged infusion times and intermittent infusions and to enable patient mobility between sites, such as between the surgical suite and the MRI suite, without increasing the risk of cannula dislocation or infection. In-dwelling cannulae are implanted in a separate surgery before the infusion session; however, the design requirements for in-dwelling cannulae are distinct from those required for acute infusions.

In-dwelling cannulae must be subcutaneously secured to prevent infection and to avoid parenchymal injury related to motion. This is typically carried out by tunneling the infusion line under the skin and connecting it to an external pump, or to a pump implanted into the patient's abdomen.[33, 34] In-dwelling cannulae must be rigid to prevent bending during stereotactic placement and to ensure accurate placement, but has to be flexible once placed in the brain to prevent shearing during normal brain pulsation and movement. Cannulae used for single "acute" infusions beginning immediately after cannula insertion, in contrast, do not need to be flexible because they can be left protruding directly from the skull while the patient is in a sterile environment.

IMAGING OF CONVECTIVE DELIVERY

Benefits of Intraoperative Imaging

Intraoperative MRI (iMRI) of CED drug delivery in the brain is crucial for a number of reasons. iMRI enables confirmation of adequate and accurate drug distribution within the target structure, ensuring that the targeted structure is sufficiently perfused. This is a key concept because the true efficacy of a drug can only be accurately determined if it was effectively delivered. iMRI greatly enhances the safety of these procedures by avoiding delivery of drug to off-target sites. Critically, iMRI also allows infusion parameters to be altered during the procedure, as needed. Disease can be associated with parenchymal edema or atrophy, and adjustments are often required in these settings. Patients may also require accommodations for individual differences in anatomical boundaries. The flexibility provided by iMRI enhances overall reproducibility, reliability, and safety of CED infusions.

Imaging Tracers

The most effective method to visualize CED in real time is achievable with imaging tracers that are co-infused with the drug (as opposed to being conjugated to the drug itself). The contrast reagents can be detected via CT (iopamidol and iopamic acid being the most common—for reviews, please see Lonser et al. 2015 and 2017),[4, 35] MR imaging, or a combination of both. Gadolinium and gadoteridol are commercially available small-molecular–weight imaging tracers that have been extensively characterized for their utility for MRI-based imaging of CED infusions in the brain by our group and by others in rodents, NHPs, and humans.[36-40]

Our initial work with gadolinium-loaded liposomes[41-43] has progressed to the co-infusion of free gadoteridol with protein[44] and AAV vectors.[40, 45-48] We have also followed a similar strategy for co-infusion of anticancer drugs with gadoteridol to treat brainstem lesions (NCT00734682). In our experience, regardless of the encoded transgene, co-infusion of AAV2 and gadoteridol with MRI monitoring shows an excellent correlation with the transgene expression as assessed in the NHP brain by immunohistochemistry (Fig. 3).[40]

Dose-ranging studies have revealed that a 5 mM concentration is most effective for tracking molecules ranging from 400 to 70,000 Da,[19, 49] whereas nanometer-sized molecules such

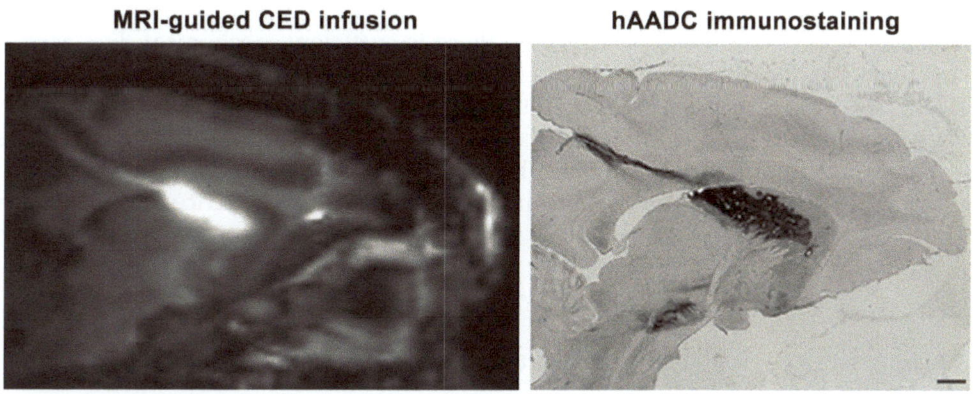

FIG. 3 Left: Magnetic resonance image showing co-infusion of AAV2-AADC co-infused with gadolinium into nonhuman primate caudate nucleus. Right: AADC immunostaining (AADC gene expression) in the caudate nucleus that closely approximates distribution of the gadolinium. *Modified with permission from Bankiewicz KS, et al. AAV viral vector delivery to the brain by shape-conforming MR-guided infusions. J Control Release. 2016.*

as AAV vector are best visualized with concentrations ranging between 1 and 2 mM.[40] Gadolinium co-infusion with AAV2 vector for real-time imaging of CED has already been used in more than 30 patients treated at the National Institutes of Health and at University of California–San Francisco. This strategy has been shown to be safe within the human brain parenchyma, providing clear, well-defined, high-resolution imaging in real time with conventional MR imaging. Other MR imaging tracers have been explored for tracing high-molecular-weight compounds in animal models (e.g., gadolinium-tagged liposomes).[50, 51] Although these compounds appear safe and effective for clear, high-resolution imaging in real time, they have not yet been tested for CED in clinical trials and are not yet commercially available.

Intraoperative MR Imaging

Infusate distribution can be monitored during dosing by acquiring consecutive spoiled gradient echo MR images. The MRI protocol begins with a high-resolution T1-weighted coronal scan to visualize the anatomy and plan the procedure. The scan time depends on the number of slices required to cover the entirety of the infusion, but typically ranges from 9 min and 44 s to 11 min and 53 s. For imaging in NHP, TR/TE/flip angle, field of view, and slice thickness imaging parameters are adjusted to culminate in voxel sizes that range from 1.2×10^{-4} cm^3 to 1.2×10^{-3} cm^3 during the procedure. Both T1- and T2-weighted scans are captured at the end of the procedure to document postinfusion coverage.

REFLUX-RESISTANT INFUSION CANNULAE

Mechanics and Common Causes of Reflux

Backflow or reflux along the edges of the cannula tract can greatly reduce the volume of infusate distribution within the intended target structure and lead to off-target coverage, which may be toxic. Reflux can occur when there is sufficient mechanical displacement of tissue to create a space between the cannula and the tissue through which infusate preferentially flows. As such, care should be taken to reduce rotation of the cannula during insertion and when adjusting during dosing. Reflux prevents the generation of continued pressure within the extracellular space that drives bulk flow from the cannula tip. After reflux develops, the Vd plateaus while the Vi continues to increase—the difference represents infusate lost as reflux. Common causes of backflow include poor cannula insertion technique, cannula design, and large-diameter cannulae.

Step Design Reflux-Resistant Cannulae

Initial animal experiments performed by Morrison and colleagues demonstrated that reflux can be decreased and CED can be improved by decreasing the diameter of the needle from 27 gauge to 32 gauge. Even the 32-gauge needle, one of the smallest metal needles commercially available, must be used at a flow rate of 0.5 μL/min to avoid reflux.[52] To overcome this issue, new "step design" cannulae have been developed to be reflux-resistant and permit CED at a higher flow rate.[53]

The first generation of step cannula, used in gene therapy trials at University of California–San Francisco, was composed of 1.2 m of fused silica surrounded by a 24.6-cm surgical steel cannula to provide rigidity. The fused silica extended 10 mm beyond the tip of the steel cannula. More recently, we have designed an MRI-compatible, reflux-resistant cannula consisting of a 27-gauge (0.2-mm) needle with an internal silica tubing (outer diameter 168 μm and inner diameter of 102 μm) that is glued in. The difference in diameters creates a step that extends beyond the tip of the needle ~10 mm in humans and 5 mm or less in NHPs.[53, 54]

Shape-Conforming of Infusate Using Controlled Backflow

As described earlier, step design cannulae effectively prevent reflux and allow fast infusions. The step design also provides an additional advantage when performing CED in the brain. The current design incorporates two steps: the first is located 3 mm from the cannula tip (composed of fused silica), and the second is located 10, 15, or 18 mm from the cannula tip depending on the design. This second step is composed of ceramic and makes up the outer portion of the cannula. As the infusate exits the cannula tip, backflow is created to the first step (3 mm). For brain structures with an elongated shape, such as the putamen and the caudate, the cannula may be advanced farther into the structure during infusion to create additional backflow, resulting in an elliptical infusate distribution that more closely resembles the shape of the target structure. Combining this method with aligning the cannula trajectory along the long axis of the structure enables elliptical shape fitting from a single infusion. When transfrontal, occipital, and parietal trajectories were compared for putaminal infusions in the NHP brain, we found that an occipital approach was significantly more effective. A parietal approach, on the other hand, resulted in significantly more coverage of the caudate (Fig. 4).[55] In current clinical trials that involve CED into the putamen of patients diagnosed with Parkinson disease, multiple trajectories are used, with entry points close to the coronal

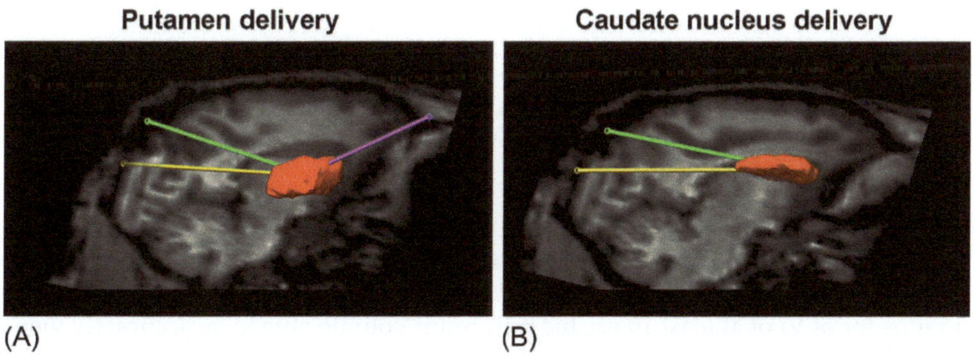

FIG. 4 Gadolinium co-infusion closely correlates with adeno-associated virus (AAV) vector distribution in the nonhuman primate (NHP) brain. (A) Example of AAV-AADC/gadolinium infusion into the NHP caudate. (B) AADC immunostaining (AADC gene expression) closely correlates with the distribution of gadolinium. *Modified with permission from Bankiewicz KS, et al. AAV viral vector delivery to the brain by shape-conforming MR-guided infusions.* J Control Release. *2016. doi:https://doi.org/10.1016/j.jconrel.2016.02.034 (web archive link).*

suture (coronal approach). An occipital trajectory could facilitate similar coverage from a single cannula insertion, reducing the incidence of hemorrhage.

Optimized Cannula Placement

Despite rigorous validation of drug targets, the successful implementation of CED requires optimized cannula placement to ensure maximal coverage of the target structure. This point was highlighted via retrospective analysis of the CED arm of the PRECISE trial, performed by our laboratory. The PRECISE trial was a randomized controlled Phase III trial that used CED for the intracerebral infusion of an anticancer drug for recurrent glioblastoma. The survival of patients in this study was compared to survival of patients treated with local implantation of carmustine-impregnated wafers, and no significant difference was observed between study groups. We found that only 49.8% of cannulae met all positioning criteria, and estimated coverage of the target structure was low, which had a significant effect on progression-free survival.[56] The importance of precise cannula placement was also evident from the AAV2-GAD trial conducted by Kaplitt and colleagues, in which patients with Parkinson disease who had suboptimal cannula placement showed no significant clinical improvement.[57]

High Reproducibility with Optimized Cannula Placement

Convection-enhanced delivery infusions into the brain are remarkably reproducible when care is taken to optimize cannula tip position. We have defined infusion parameters, referred to as "red," "blue," and "green" zones, that describe cannula placements resulting in poor, suboptimal, and optimal volumes of distribution, respectively, for the putamen,[58] thalamus, and brainstem[59] of NHPs. Retrospective analysis was performed of the coverage of MR images with gadoteridol as a contrast reagent, which were obtained during CED infusions into the NHP putamen (25 cases), thalamus (14 cases), and brainstem (8 cases). Green zone cannula placements were defined as those that generated ~99% containment of infused gadoteridol within the target structure, 85%-87% in the blue zone, and 49% in the red zone (Fig. 5).

INFUSION VOLUME

Vd to Vi Ratio in AAV Clinical Trials

Extensive NHP studies and current clinical AAV trials have led us to optimize the infusion volume of AAV2 vector delivered by real time CED into the putamen. Vd can be measured throughout the infusion by monitoring the distribution of gadoteridol, permitting calculation Vd/Vi ratios for of Vi of at least 10 μL, the minimum volume able to be accurately visualized by MRI. MR images acquired during the infusion procedure are correlated with Vi in each series using BrainLab software. This involves defining the pixel threshold value for gadoteridol signal, then the software calculates the signal above this value establishing the volume of distribution at a given time point, which can be reconstructed as a three-dimensional image.

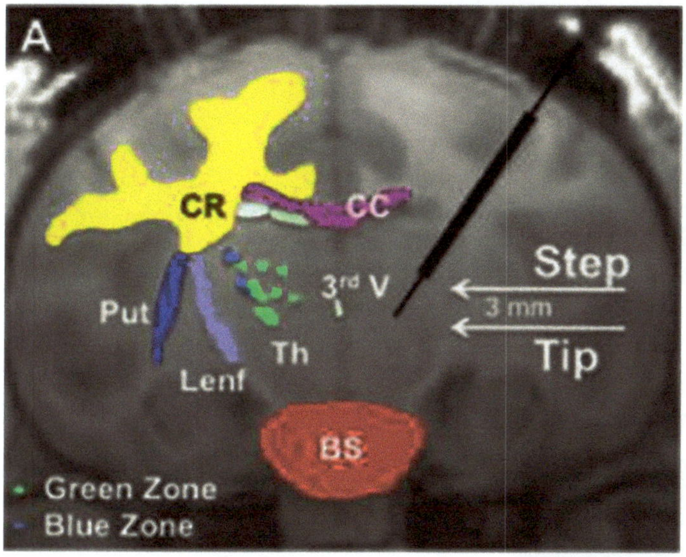

FIG. 5 Schematic representation of the step cannula placement in the thalamus. Both the step and tip portion of the cannula placement in the green and blue zones for each case are shown. *CC*, corpus callosum; *CR*, corona radiata; *Lenf*, lenticular fasciculus; *Put*, putamen; *3rd V*, third ventricle. *Modified with permission from Yin D, et al. Cannula placement for effective convection-enhanced delivery in the nonhuman primate thalamus and brainstem: implications for clinical delivery of therapeutics.* J Neurosurg. *2010;113:240-248.*

The human putamen has volume of approximately 4000 mm^3. In the first AAV2-NTN trial, just 40 μL was delivered per putamen, via simple needle injection into eight sites, resulting in NTN expression localized only to the injection sites.[60] CED was subsequently used in the first AAV2-hAADC trial, where 100 μL was delivered into the putamen at two sites with a reflux resistant cannula covering 300 mm^3 of the putamen.[61] The second AAV2-NTN trial involved infusion of 120 μL split into three sites in the putamen. If a reflux-resistant cannula had been used, the infusion would have resulted in coverage of approximately 450 mm^3 of the putamen, representing just 12% of the putamen. In ongoing clinical trials of AAV2-hAADC and AAV2-GDNF at University of California–San Francisco and the National Institutes of Health, 450 μL of infusate is being delivered per putamen, which we have confirmed via 3D reconstruction of gadoteridol coverage translated to a Vd of approximately 1500 mm^3, consistent with our predictions from T2-weighted imaging. This represents about 40%-50% coverage of the putamen, or 80% of the postcommissural putamen.[62]

GUIDANCE SYSTEMS

Shift from Frame-Based to Frameless Systems

The past decade has seen a shift from the traditional, often bulky frame-based systems for stereotactic surgery toward frameless systems. Presently there are two commercially available MRI compatible frameless options for preclinical and clinical gene therapy procedures: the Nexframe system (Medtronic Inc.) and the ClearPoint system (Fig. 6; MRI Interventions Inc.). These devices have proven useful for targeting certain brain structures, often through

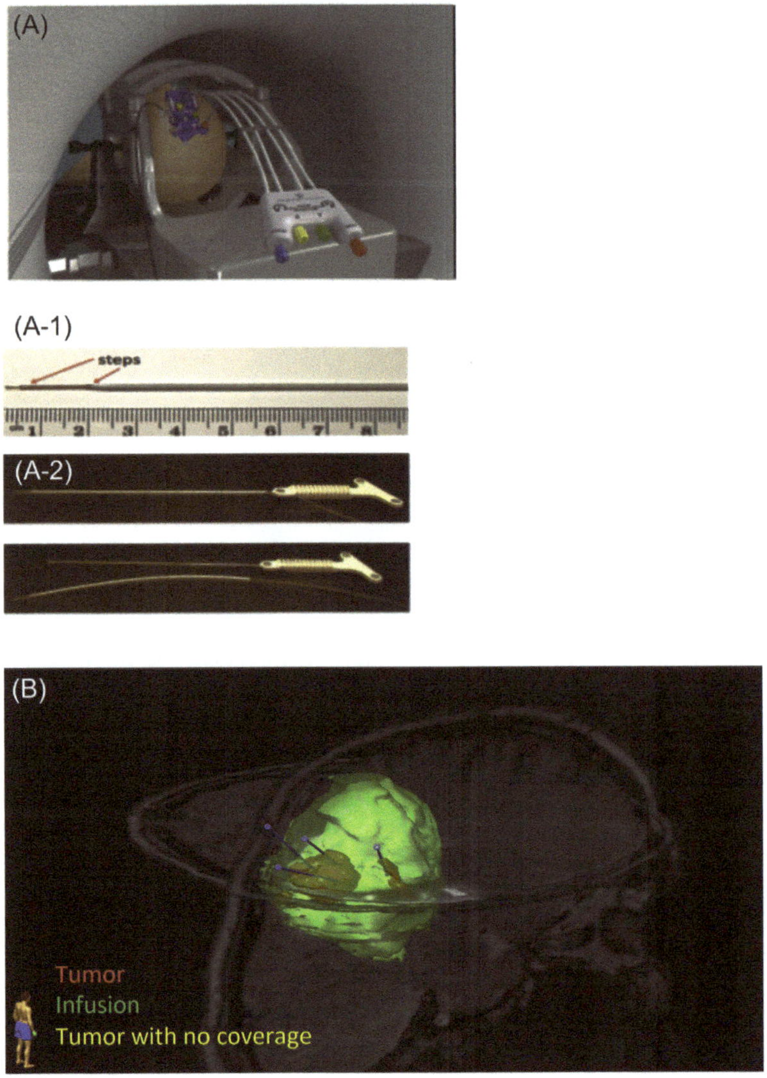

FIG. 6 Magnetic resonance imaging (MRI) guidance system for convection-enhanced delivery (CED) in the brain. (A) ClearPoint MR guiding system used for CED in the brain. This system allows for neuronavigation in real time while the patient is in the MR scanner. (A-1, A-2) MR-compatible drug administration cannulae developed at University of California–San Francisco and marketed by BrainLAB and MRI Interventions. (B) Example of MR-guided CED-based delivery of a cancer drug (MDNA55—IL-4 fused to the catalytic domain of *Pseudomonas* exotoxin A plus gadolinium) in a patient diagnosed with recurring glioblastoma multiforme. Four catheters (blue) trajectory planning and infusate distribution was performed using BrainLab iPlan Flow software. 95% coverage was calculated based on the fraction of the target volumes covered by the determined gadolinium distribution at end of infusion. Both tumor and peritumoral regions were covered by the infusion of 60 mL of the drug based on MR detection of the gadolinium in the infusate (shown as green on 3D postinfusion reconstruction).

the use of specialized MRI software, but their large profile imposes limitations on efficient targeting of deeper brain structures.

Ball-Joint Guide Array

We have developed a novel frameless guidance system: the skull-mounted ball-joint guide array (BJGA). The BJGA is smaller than the Nexframe and ClearPoint systems, features a wider range of motion, and is compatible with most MRI platforms and software. The implantable BJGA (Fig. 7) is composed of a threaded base, outfitted with three holes for securing the device to the skull; a three-hole ball array with locking side screw that can fit cannulae up to 16 gauge; a knurled threaded collar that releases/locks the array to the threaded base; and a three-prong fiducial. Each fiducial contains three parallel tubes filled with 2 mM gadolinium-based solution that appears hyperintense on T1-weighted images. The BJGA is economical, lightweight, has a similar degree of angulation from the vertical, a full 360° rotation over the base, and can be used with any on-board MRI software. Rigorous precision analysis in the NHP revealed that the BJGA can be used to reliably target subcortical brain structures with comparable accuracy to the Nexframe and Clearpoint systems, with the added advantages of small profile allowing for greater flexibility in executing MRI-guided experiments within the NHP brain.

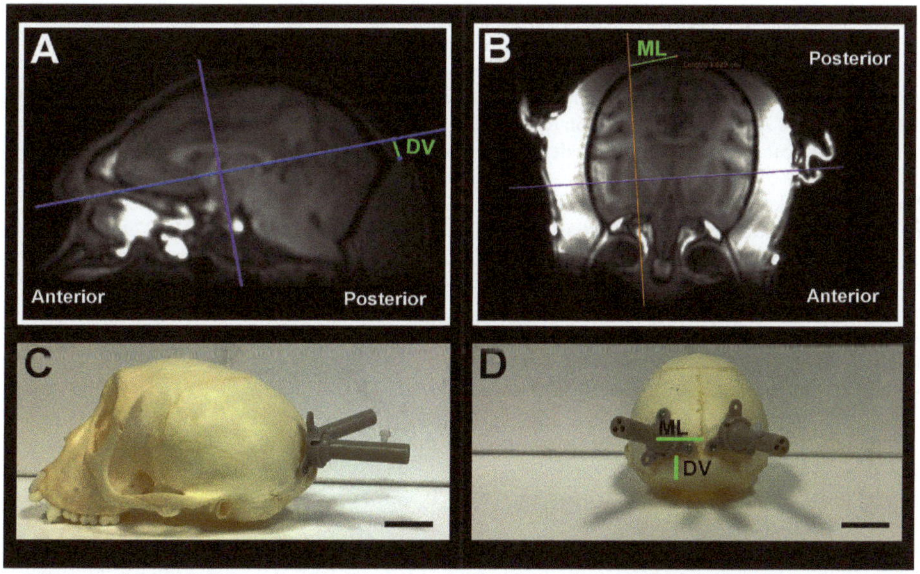

FIG. 7 Schematic representation of the ball joint guide array (BJGA). (A, B) Magnetic resonance imaging planning for the dorsoventral and mediolateral placement of the BJGA for targeting the nonhuman primate putamen. (C, D) Longitudinal and apical views of the ball array and the skull-mounted base showing the three-hole ball array with adjustable side screw. The threaded skull-mounted base is secured onto the skull by three titanium fasteners. The knurled threaded locking collar interlocks with the base and secures the device in the desired position.

CONCLUSIONS

The biophysical principles of CED have been confirmed via real-time imaging under a variety of pathological conditions, across a wide spectrum of target structures in the brain. The ability to place the cannula accurately within the target structure, use of surrogate imaging tracers, real-time imaging, and use of reflux-resistant cannulae will be important for clinical trial development looking forward.

References

1. Blasberg RG, Patlak C, Fenstermacher JD. Intrathecal chemotherapy: brain tissue profiles after ventriculocisternal perfusion. *J Pharmacol Exp Ther*. 1975;195:73-83.
2. Chen P-Y, et al. Comparing routes of delivery for nanoliposomal irinotecan shows superior anti-tumor activity of local administration in treating intracranial glioblastoma xenografts. *Neuro Oncol*. 2013;15:189-197.
3. Langer R. New methods of drug delivery. *Science*. 1990;249:1527-1533.
4. Lonser RR, Sarntinoranont M, Morrison PF, Oldfield EH. Convection-enhanced delivery to the central nervous system. *J Neurosurg*. 2015;122:697-706.
5. Bobo RH, et al. Convection-enhanced delivery of macromolecules in the brain. *Proc Natl Acad Sci USA*. 1994;91:2076-2080.
6. Asthagiri AR, et al. Prospective evaluation of radiosurgery for hemangioblastomas in von Hippel-Lindau disease. *Neuro Oncol*. 2010;12:80-86.
7. Corem-Salkmon E, et al. Convection-enhanced delivery of methotrexate-loaded maghemite nanoparticles. *Int J Nanomedicine*. 2011;6:1595-1602.
8. Lonser RR, et al. Successful and safe perfusion of the primate brainstem: in vivo magnetic resonance imaging of macromolecular distribution during infusion. *J Neurosurg*. 2002;97:905-913.
9. Morrison PF, Laske DW, Bobo H, Oldfield EH, Dedrick RL. High-flow microinfusion: tissue penetration and pharmacodynamics. *Am J Physiol*. 1994;266:R292-R305.
10. Heiss JD, et al. Local distribution and toxicity of prolonged hippocampal infusion of muscimol. *J Neurosurg*. 2005;103:1035-1045.
11. Lieberman DM, Laske DW, Morrison PF, Bankiewicz KS, Oldfield EH. Convection-enhanced distribution of large molecules in gray matter during interstitial drug infusion. *J Neurosurg*. 1995;82:1021-1029.
12. Lonser RR, Corthésy ME, Morrison PF, Gogate N, Oldfield EH. Convection-enhanced selective excitotoxic ablation of the neurons of the globus pallidus internus for treatment of parkinsonism in nonhuman primates. *J Neurosurg*. 1999;91:294-302.
13. Ciesielska A, Mittermeyer G, Hadaczek P, Kells AP. Anterograde axonal transport of AAV2-GDNF in rat basal ganglia. *Mol Ther*. 2011.
14. Naidoo J, et al. Extensive transduction and enhanced spread of a modified AAV2 capsid in the non-human primate CNS. *Mol Ther*. 2018;26:2418-2430.
15. Kells AP, et al. Efficient gene therapy-based method for the delivery of therapeutics to primate cortex. *Proc Natl Acad Sci USA*. 2009;106:2407-2411.
16. Ksendzovsky A, et al. Convection-enhanced delivery of M13 bacteriophage to the brain. *J Neurosurg*. 2012;117:197-203.
17. Strasser JF, Fung LK, Eller S, Grossman SA, Saltzman WM. Distribution of 1,3-bis(2-chloroethyl)-1-nitrosourea and tracers in the rabbit brain after interstitial delivery by biodegradable polymer implants. *J Pharmacol Exp Ther*. 1995;275:1647-1655.
18. Lonser RR, Gogate N, Morrison PF, Wood JD, Oldfield EH. Direct convective delivery of macromolecules to the spinal cord. *J Neurosurg*. 1998;89:616-622.
19. Murad GJA, et al. Image-guided convection-enhanced delivery of gemcitabine to the brainstem. *J Neurosurg*. 2007;106:351-356.
20. Lonser RR, et al. Real-time image-guided direct convective perfusion of intrinsic brainstem lesions. Technical note, *J Neurosurg*. 2007;107:190-197.

21. Wood JD, Lonser RR, Gogate N, Morrison PF, Oldfield EH. Convective delivery of macromolecules into the naive and traumatized spinal cords of rats. *J Neurosurg*. 1999;90:115-120.
22. Asthagiri AR, Walbridge S, Heiss JD, Lonser RR. Effect of concentration on the accuracy of convective imaging distribution of a gadolinium-based surrogate tracer. *J Neurosurg*. 2011;115:467-473.
23. Croteau D, et al. Real-time in vivo imaging of the convective distribution of a low-molecular-weight tracer. *J Neurosurg*. 2005;102:90-97.
24. Lonser RR, Weil RJ, Morrison PF, Governale LS, Oldfield EH. Direct convective delivery of macromolecules to peripheral nerves. *J Neurosurg*. 1998;89:610-615.
25. Sandberg DI, Edgar MA, Souweidane MM. Convection-enhanced delivery into the rat brainstem. *J Neurosurg*. 2002;96:885-891.
26. Lieberman DM, Corthésy ME, Cummins A, Oldfield EH. Reversal of experimental parkinsonism by using selective chemical ablation of the medial globus pallidus. *J Neurosurg*. 1999;90:928-934.
27. Laske DW, et al. Chronic interstitial infusion of protein to primate brain: determination of drug distribution and clearance with single-photon emission computerized tomography imaging. *J Neurosurg*. 1997;87:586-594.
28. Lonser RR, et al. Convection perfusion of glucocerebrosidase for neuronopathic Gaucher's disease. *Ann Neurol*. 2005;57:542-548.
29. Chen MY, Lonser RR, Morrison PF, Governale LS, Oldfield EH. Variables affecting convection-enhanced delivery to the striatum: a systematic examination of rate of infusion, cannula size, infusate concentration, and tissue-cannula sealing time. *J Neurosurg*. 1999;90:315-320.
30. Kunwar S, et al. Safety of intraparenchymal convection-enhanced delivery of cintredekin besudotox in early-phase studies. *Neurosurg Focus*. 2006;20:E15.
31. Laske DW, Youle RJ, Oldfield EH. Tumor regression with regional distribution of the targeted toxin TF-CRM107 in patients with malignant brain tumors. *Nat Med*. 1997;3:1362-1368.
32. Jagannathan J, Walbridge S, Butman JA, Oldfield EH, Lonser RR. Effect of ependymal and pial surfaces on convection-enhanced delivery. *J Neurosurg*. 2008;109:547-552.
33. Barua NU, et al. Intermittent convection-enhanced delivery to the brain through a novel transcutaneous bone-anchored port. *J Neurosci Methods*. 2013;214:223-232.
34. Bienemann A, et al. The development of an implantable catheter system for chronic or intermittent convection-enhanced delivery. *J Neurosci Methods*. 2012;203:284-291.
35. Lonser RR. Imaging of convective drug delivery in the nervous system. *Neurosurg Clin N Am*. 2017;28:615-622.
36. Forsayeth JR, et al. A dose-ranging study of AAV-hAADC therapy in Parkinsonian monkeys. *Mol Ther*. 2006;14:571-577.
37. Richardson RM, et al. Interventional MRI-guided putaminal delivery of AAV2-GDNF for a planned clinical trial in Parkinson's disease. *Mol Ther*. 2011;19:1048-1057.
38. San Sebastian W, et al. Safety and tolerability of MRI-guided infusion of AAV2-hAADC into the mid-brain of non-human primate. *Mol Ther Methods Clin Dev*. 2014;3.
39. Christine CW, et al. Safety and tolerability of putaminal AADC gene therapy for Parkinson disease. *Neurology*. 2009;73:1662-1669.
40. Su X, et al. Real-time MR imaging with Gadoteridol predicts distribution of transgenes after convection-enhanced delivery of AAV2 vectors. *Mol Ther*. 2010;18:1490-1495.
41. Krauze MT, et al. Effects of the perivascular space on convection-enhanced delivery of liposomes in primate putamen. *Exp Neurol*. 2005;196:104-111.
42. Krauze MT, Forsayeth J, Yin D, Bankiewicz KS. Convection-enhanced delivery of liposomes to primate brain. *Meth Enzymol*. 2009;465:349-362.
43. Krauze MT, Forsayeth J, Park JW, Bankiewicz KS. Real-time imaging and quantification of brain delivery of liposomes. *Pharm Res*. 2006;23:2493-2504.
44. Gimenez F, et al. Image-guided convection-enhanced delivery of GDNF protein into monkey putamen. *Neuroimage*. 2011;54(Suppl 1):S189-S195.
45. Fiandaca MS, et al. Real-time MR imaging of adeno-associated viral vector delivery to the primate brain. *Neuroimage*. 2009;47(Suppl 2):T27-T35.
46. Samaranch L, et al. MR-guided parenchymal delivery of adeno-associated viral vector serotype 5 in non-human primate brain. *Gene Ther*. 2017;24:253-261.
47. Salegio EA, et al. Axonal transport of adeno-associated viral vectors is serotype-dependent. *Gene Ther*. 2013;20:348-352.

48. San Sebastian W, et al. Adeno-associated virus type 6 is retrogradely transported in the non-human primate brain. *Gene Ther*. 2013;20:1178-1183.
49. Murad GJA, et al. Real-time, image-guided, convection-enhanced delivery of interleukin 13 bound to pseudomonas exotoxin. *Clin Cancer Res*. 2006;12:3145-3151.
50. Saito R, et al. Distribution of liposomes into brain and rat brain tumor models by convection-enhanced delivery monitored with magnetic resonance imaging. *Cancer Res*. 2004;64:2572-2579.
51. Krauze MT, et al. Safety of real-time convection-enhanced delivery of liposomes to primate brain: a long-term retrospective. *Exp Neurol*. 2008;210:638-644.
52. Morrison PF, Chen MY, Chadwick RS, Lonser RR, Oldfield EH. Focal delivery during direct infusion to brain: role of flow rate, catheter diameter, and tissue mechanics. *Am J Physiol*. 1999;277:R1218-R1229.
53. Krauze MT, et al. Reflux-free cannula for convection-enhanced high-speed delivery of therapeutic agents. *J Neurosurg*. 2005;103:923-929.
54. Sanftner LM, et al. AAV2-mediated gene delivery to monkey putamen: evaluation of an infusion device and delivery parameters. *Exp Neurol*. 2005;194:476-483.
55. Bankiewicz KS, et al. AAV viral vector delivery to the brain by shape-conforming MR-guided infusions. *J Control Release*. 2016; https://doi.org/10.1016/j.jconrel.2016.02.034.
56. Sampson JH, et al. Poor drug distribution as a possible explanation for the results of the PRECISE trial. *J Neurosurg*. 2010;113:301-309.
57. LeWitt PA, et al. AAV2-GAD gene therapy for advanced Parkinson's disease: a double-blind, sham-surgery controlled, randomised trial. *Lancet Neurol*. 2011;10:309-319.
58. Yin D, et al. Optimal region of the putamen for image-guided convection-enhanced delivery of therapeutics in human and non-human primates. *NeuroImage*. 2011;54(Suppl 1):S196-S203.
59. Yin D, et al. Cannula placement for effective convection-enhanced delivery in the nonhuman primate thalamus and brainstem: implications for clinical delivery of therapeutics. *J Neurosurg*. 2010;113:240-248.
60. Marks WJ, et al. Gene delivery of AAV2-neurturin for Parkinson's disease: a double-blind, randomised, controlled trial. *Lancet Neurol*. 2010;9:1164-1172.
61. Valles F, et al. Qualitative imaging of adeno-associated virus serotype 2-human aromatic L-amino acid decarboxylase gene therapy in a phase I study for the treatment of Parkinson disease. *Neurosurgery*. 2010;67:1377-1385.
62. Yin D, et al. Striatal volume differences between non-human and human primates. *J Neurosci Methods*. 2009;176:200-205.

CHAPTER

17

Stem Cell Transplantation for Neurological Disease: Technical Considerations and Delivery Devices

Ethan A. Winkler, Daniel A. Lim*,†,‡*

*Department of Neurological Surgery, University of California San Francisco, San Francisco, CA, United States †Eli and Edythe Broad Center of Regeneration Medicine and Stem Cell Research, University of California San Francisco, San Francisco, CA, United States ‡Veterans Affairs Medical Center, San Francisco, CA, United States

STEM CELLS IN NEUROLOGIC DISEASE

Continued interest in the evolution of our understanding of cell-based therapeutics have led to therapeutic successes in multiple preclinical models of human neurologic diseases, including Parkinson disease (PD),[1] Huntington disease,[2] multiple sclerosis,[3] and stroke.[4] However, results have been mixed in human clinical trials.[5] Although many investigators have focused on identifying, isolating, and modifying appropriate neural cell populations for transplantation, relatively few resources have been invested in the development of new devices and techniques to facilitate optimal delivery of cell-based therapy to the human brain.[6] Improper cell delivery has been speculated to be a major contributor to the variable outcomes observed in double-blinded, randomized controlled trials for cell transplantation in patients with PD, and flawed delivery may have confounded many other trials.[7-9] It is difficult to envision a successful human cell transplantation trial unless surgical strategies are developed to ensure accurate and reliable delivery of cell-based therapeutics.

The brain poses a number of unique challenges for the delivery of cell therapeutics, including the blood–brain barrier (BBB), dense vasculature, and complex regional anatomy that requires meticulous targeting. For decades, preclinical and clinical human trials have relied on a single straight cannula with a syringe. While this may be effective for small animals such as rodents, this method involves considerable limitations and often multiple injections when

https://doi.org/10.1016/B978-0-12-813997-4.00017-7

© 2019 Elsevier Inc. All rights reserved.

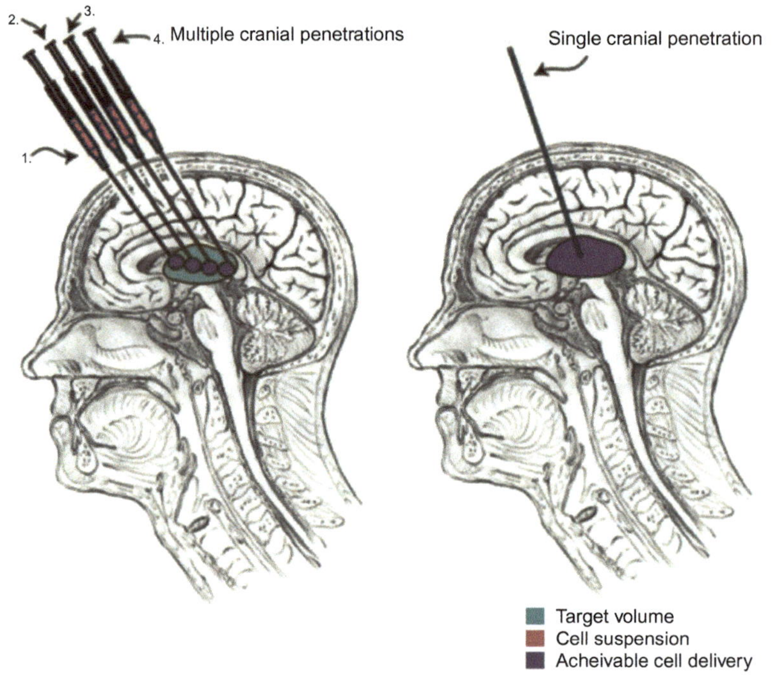

FIG. 1 A need for instrumentation that minimizes the number of current transcortical penetrations. *Left panel*: Conventional straight-cannula delivery platforms require multiple transcortical cell injections *(Purple)* to achieve adequate cell distribution over a target volume *(green)*. Each individual transcortical puncture is associated with additive potential neurologic morbidity. *Right panel*: A more idealized delivery system would distribute cells throughout the target volume with single transcortical puncture.

scaled up to target larger, more complex neuroanatomical structures in humans (Fig. 1). In this chapter, we describe some of the unique challenges of delivering stem cell-based therapeutics to the brain and discuss limitations of conventional syringes coupled with straight cannulas. We also describe recently designed radial trajectory cannulas that ameliorate some of the neurosurgical challenges of cell delivery. Finally, we identify areas for future development to maximize patient safety and optimize the efficacy of cell-based neurotherapeutics.

SELECTION OF THE OPTIMAL STRATEGY FOR STEM CELL DELIVERY

A fundamental requirement for cell-based therapeutics to be effective is that the transplanted cells must be delivered in sufficient numbers to the intended target brain region. For example, this may involve targeting cells to a cortical infarct zone of an ischemic stroke or to the subcortical regions vulnerable to neurodegenerative diseases such as PD or Huntington disease. Transport from the systemic circulation into the brain is tightly regulated by the BBB,[10] which, due to the tightly connected endothelial lining of the cerebrovasculature, largely prevents the transport of cells from the peripheral vasculature into the brain parenchyma. Although some have reported success in delivering cells to the rodent brain via systemic vascular injections, these studies have had mixed results.[11] Efforts are presently underway to selectively disrupt the BBB in specified regions, either through infusion of osmotic agents such as mannitol, or via focused ultrasound.[12,13] Until these techniques are fully developed, it is unlikely that systemic intravascular or more selective intra-arterial delivery of stem cell-based therapeutics will be of clinical utility when treating diseases of the brain.

Accordingly, great effort has been focused on refining delivery strategies that bypass the BBB, including intracerebroventricular, intrathecal, and direct intraparenchymal delivery. Intracerebroventricular and intrathecal delivery involve injection of a therapeutic into the lateral ventricle or lumbar cisterns, respectively. The injected therapy may then be disseminated within the central nervous system through bulk flow of the cerebrospinal fluid and diffusion into the brain interstitial fluid. This approach is currently used in clinical practice to deliver soluble, noncell-based therapies in human patients—mainly for local delivery of chemotherapeutics in disseminated cancers or antibiotics in infections—through either repeated injection or a surgically implanted reservoir.[14] Efforts are underway to adapt such an approach for delivery of cell-based therapeutics, and promising preliminary results have been published in rodent models of neurologic disease, including multiple sclerosis and ischemic stroke.[3,15,16] However, several obstacles exist in translating this into effective therapy for humans. Most notably, cells would be required to migrate over distances that, in many cases, could exceed several centimeters. Without effective chemotrophic targeting (an underdeveloped technology), this is unlikely to result in effective engraftment in disease-specific brain regions.

To ensure accurate targeting, many investigators have adopted strategies to directly deliver stem cell-based therapies within the brain parenchyma. This approach is not without its own limitations, however. Direct injection involves an invasive surgical procedure in which the brain parenchyma must be surgically traversed. This comes at the cost of potential of neurologic morbidity through inadvertent vascular injury and hemorrhage, direct injury to eloquent brain regions, or more subtle tissue trauma and activation of potentially deleterious neuroinflammatory cascades.[6,17,18] Despite these risks, direct stereotactic injection into carefully selected brain regions remains the most widely used delivery strategy when adapting stem cell-based technologies for treatment of neurologic disease, and will be the focus of the remaining chapter.

Technical Considerations for Intraparenchymal Stem Cell Delivery

Unlike delivery of soluble, noncell-based therapies, injection of biologically viable stem cells requires careful attention to a number of considerations, including cell viability, cell reflux, and accuracy of targeting. Proper experimental design and selection of the optimal delivery platform are essential for therapeutic success, and multiple parameters must be systematically optimized for each indication and may differ based on cell lines and disease states.

Cell Viability

Efficiency of cell transplantation is often relatively low, with as little as 5%-10% of the injected cells surviving.[16] Impaired viability may result from cellular damage incurred either as a result of mechanical forces during delivery or later, due to suboptimal biological conditions in the host microenvironment. During flow through a delivery cannula, cells experience mechanical forces that may deform or disrupt cell membranes, including a pressure drop across the cell at transition points of diameter change, shearing forces from differential fluid flow, and stretching forces from extensional flow. Cells and fluid in the middle of a cannula travel faster than those along the outer boundaries, giving rise to shear force along the cell membrane. With standard needle sizes, such as 16- to 22-gauge diameters, shear forces have negligible impact on cell viability, but may exert deleterious effects on viability with smaller inner diameter cannulas.[19]

Extensional force occurs with abrupt changes in caliber of the delivery conduit (e.g., the transition point between syringe and cannula), which result in a dramatic increase in linear velocity. Studies have demonstrated that extensional flow and these transition points decrease cell viability and should be minimized with optimal cannula design.[20] Other studies have identified that cell density of the suspended cells, duration of injection, preparation of the cell suspension, and selection of carrier media all influence cell viability.[21-23] Factors affecting viability also differ and may depend on the stem cell line selected for delivery[22]; therefore, strategies must be individually optimized based on the cell population. Host factors, such as inflammation from catheter entry or access to blood supply and circulating nutrients, often limit the scalability of any single injection and may lead to a precipitous drop in viable cells.[22]

Avoiding Reflux of Delivered Cells

In addition to cell death, unintended reflux along the injection catheter tract further compromises delivery of stem cells to their intended targets. The parenchyma of the brain and parenchyma of the spinal cord have low-elastic modulus properties. During infusion, the infusate creates a potential space, and the buildup of positive pressure within this potential space leads to intermittent dispersion of cells as shown by real-time magnetic resonance imaging (MRI) during infusion of prelabeled neural stem cells (NSCs).[24] However, if the volume of the potential space is overcome too quickly, reflux of the infused cells occurs down the path of least resistance and often occurs up the tract made by the cannula itself, resulting in fewer cells deposited at the intended target (Fig. 2). Depending on the design of the cannula, up to 75% of injected cells may be distributed away from the target, resulting in unpredictable cell dosing.[25] To help limit reflux, cannulas with smaller diameters and a tapered design in which the diameter decreases distally toward the tip in a series of steps, slow injection rates, and injection of small molecules or soluble therapeutics with convection-enhanced delivery, have been used and may be useful with cell-based therapeutics.[26,27]

Accuracy of Delivery: Frame-based or Frameless Stereotactic Targeting

Accurate and reliable placement of a cannula or catheter in a target of interest is often the first step in ensuring that cells are delivered to the right location. Neurosurgical device delivery, such as implantation of deep brain stimulation (DBS) electrodes, is often accomplished with submillimeter accuracy with either frame-based stereotaxy or frameless, fiducial-based navigation in tandem with patient-mounted aiming devices.[28] In both approaches, anatomic targets are selected based on preoperative imaging studies and trajectories are planned with

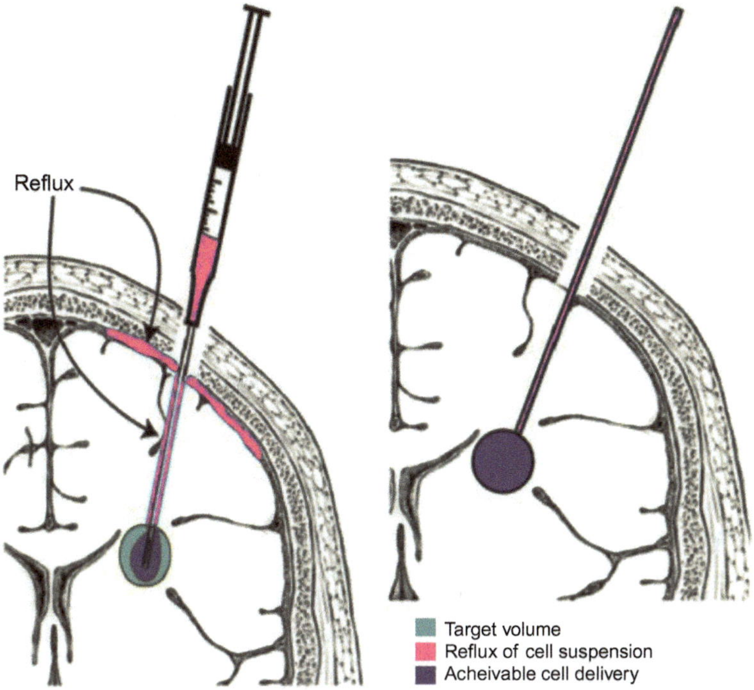

FIG. 2 Problems associated with reflux of cell suspension. *Left panel*: Injection of cell suspension *(pink)* into a desired target region *(green)* with a straight cannula may result in reflux of cells along the cannula tract *(arrows)*. Reflux reduces the achievable volume of cells that may be delivered *(purple)* to the target of interest. *Right panel*: The ideal cannula system or delivery device eliminates reflux and ensures that cells are delivered to the entire target region *(purple)*.

three-dimensional (3D) coordinates to localize the target. Conventional frame-based stereotaxy has been suggested to have higher accuracy and precision for hitting small brain targets than frameless techniques.[29] However, this notion has been disputed by others,[30] and either technique may be carefully adapted to permit accurate placement of a rigid cannula for cell delivery.

One limitation of these conventional approaches is that their trajectory planning relies on preoperative imaging, which is not always precise. Inaccuracies with registration, mechanical imprecisions of the device itself, or brain shift as a result of intraoperative loss of cerebrospinal fluid can all introduce error and result in suboptimal cannula placement. These limitations have led to the development of interventional MRI (iMRI)-based procedures. With iMRI, the procedure is performed in the MRI magnet. Real-time imaging then allows for target selection and optimization of trajectory planning. Growing clinical experience with iMRI with DBS has shown that this approach may avoid some of the errors commonly associated with targeting.[28,31,32] Viewing cannula placement in real time also has the added benefits of allowing for immediate repositioning of a misplaced cannula and the early detection of potential complications such as hemorrhage.[33] Many existing cell-delivery platforms have now been rendered MRI-compatible,[34,35] and prelabeling cells for delivery with ferromagnetic compounds, such as iron oxide nanoparticles, allows the neurosurgeon to monitor cell transplantation in real time and to continuously optimize injection parameters.[24,36] iMRI may therefore help control or identify some of the variabilities and inaccuracies of earlier trials.

Available Stem Cell Delivery Platforms

The Straight Cannula–syringe Delivery System

A stereotactically implanted straight cannula is the most common method for cell transplantation into the brain across both preclinical studies and human trials. The cannula can be inserted into the target region with either a frame-based or frameless, fiducial-based neuronavigational system. The cannula then acts as a guide conduit through which a needle or smaller delivery catheter is passed into the region of interest and connected via a Luer lock to a manual or automated syringe system. Neural cells suspended in carrier media are then delivered from the distal tip or more proximal side ports in the radial axis of the cannula without angulation. The concentration of cellular suspension, the rate of infusion, and the location of the distal tip of the cannula can all be altered to adjust the distribution of transplanted cells, optimizing delivery to small targets of interest.[22-24,36]

Over the past several decades, multiple innovations have further refined cannula design to improve viability, accuracy, and reliability of cell delivery. Reduction in cannula diameters from 2.5 mm to <100 μm decreases tissue trauma and inflammation, and improves graft survival by making the host microenvironment more favorable to cell engraftment.[18,37] Another useful innovation is the automated syringe system, which improves consistency in volumetric cell delivery.[38] Optimization of graft viability has led to a decrease in time from cell preparation to implantation, as there can be a 10%-15% decrease in viability with each hour that elapses between preparation and delivery.[21] Careful scrutiny has also gone into selecting the right carrier media—phosphate-buffered saline, artificial cerebrospinal fluid, Dulbecco's modified Eagle media (Millipore Sigma), HypoThermosol (BioLife Solutions), and Pluronic (Millipore Sigma) are all used—with respect to maintaining cells in suspension and preserving viability.[21,23] The straight cannula has been more recently adapted for use with iMRI to monitor and enhance infusions of prelabeled cells in real time.[24,36]

LIMITATIONS IN SCALABILITY

Despite these advances, the single cannula–syringe delivery system still faces considerable limitations in its ability to improve delivery in larger neural structures with complex 3D anatomy. For small rodents, a single injection in the target region of interest is often sufficient to induce a therapeutic effect. In humans, however, brain regions are 800-2300 times larger than those of rodents, and neuroanatomic targets frequently exceed 1 cm^3.

When greater injection volume is indicated, increased cell concentration and modified injection parameters often result in reflux and compromised viability of the transplanted cells.[22,23] As flow rate and acceleration of the delivery device are increased, disruptive mechanical forces are amplified, injuring cell membranes. Not surprisingly, decreased viability of NSCs is associated with higher rates of device acceleration or flow.[22] For mesenchymal stem cells, higher cell concentrations were similarly linked to a decrease in the number of viable cells shortly after infusion, potentially related to a depletion of nutrients in the carrier media, local hypoxia, or both.[22]

When cell volume and injection rate are increased, unintended reflux often occurs along the cannula tract. In studies using prelabeled NSCs and real-time MRI, low rates of infusion (i.e., 1 μL/min) distributed cells in intermittent bursts to form an ovoid cloud of distribution at the tip of the cannula. When the infusion rate was increased (i.e., 5 μL/min), there was less dispersion of cells and significant backflow around the cannula.[24] Other groups experienced in the practice of convection-enhanced delivery—a technique that takes advantage of continuous positive pressure in addition to passive diffusion to facilitate dispersion of small molecules or viral vectors—have confirmed reflux with higher infusion rates, and have also documented tissue damage at infusion rates exceeding 5 μL/min.[27] With a standard straight cannula, reflux may result in deposition of approximately 75% (range, 62%–99%) of transplanted cells outside the intended target.[25] To help limit reflux and tissue injury, slower injection rates are therefore preferred and smaller diameter cannulas with a stepped taper design (i.e., the cannula diameter is decreased along a series of steps as it courses distally) have been used with variable success.[26,27]

With larger cellular engraftments, central areas of necrosis may develop as a result of impaired diffusion and delivery of both oxygen and nutrients to the core of the graft.[39] In solid organs, such as skeletal muscle, myocardium, and brain, multiple smaller cellular deposits are superior to larger cell transplants.[18,39-41] In the rodent striatum, small engraftments of fetal nigral cell suspensions led to both enhanced survival and greater integration with host neurons.[18,41]

THE NEED FOR MULTIPLE INJECTIONS

To overcome the technical shortcomings of straight cannula–syringe platforms, human trials have often relied on multiple transcortical injections to better distribute cells in the intended neural target (Fig. 1). For example, one group developed a template mounted to a stereotactic frame that was designed with two columns of nine holes each, spaced at 4-mm intervals. This template was used to guide cannula insertion into the putamen in human patients with PD,[42] and it resulted in up to 16 independent transcortical punctures in human subjects.[7,42] As one can extrapolate, an even greater number of punctures would be required if this approach were applied to larger cortical or subcortical targets. With each added transcortical puncture, the risk of neurologic morbidity from inadvertent vascular injury and hemorrhagic stroke increases.

Insights into risks with repeated transcortical penetration may be gained with review of established neurosurgical procedures for microelectrode placement—namely, stereoelectroencephalography (SEEG) and DBS. SEEG is a procedure in which multiple depth electrodes are placed to monitor cortical and subcortical structures for epileptiform discharges. The rates of intracerebral hemorrhage in patients who undergo SEEG range from 0.3% to 11.1%, and the rate of intracerebral hemorrhage per SEEG electrode has been reported between 0.2% and 0.7%.[43-45] When targeting deeper subcortical structures with DBS, such as the globus pallidus interna or the subthalamic nucleus, reported rates of hemorrhage depend on location, but often range between 1% and 5% per lead insertion.[46] Even when trajectories are closely monitored in real time with iMRI, hemorrhage rates of 2.4% per electrode have been reported.[33] Although this incidence may seem relatively low, up to half of procedure-related hemorrhages may result in permanent neurologic deficits.[47,48] Therefore, it is essential to select cell delivery strategies or devices that reduce the number of transcortical punctures, and trajectory planning must emphasize avoidance of vascular structures, such as eliminating transgression of brain sulci.

Radial Distribution Cannulas

In an effort to improve 3D distribution of engrafted cells in larger, more complex neural structures, investigators have developed new radial delivery cannulas that allow for delivery of stem cells along an angulated trajectory.[6,17,25,49,50] These delivery platforms were designed to accomplish several goals: (1) to minimize the number of transcortical penetrations required for cell transplantation, (2) to minimize local trauma and improve viability of engrafted cells, and (3) to improve accuracy and uniformity with both cannula targeting and cell delivery. The two available cannulas differ in their individual components but share several key features, which may be used to help understand their nuances. In simplified form, each is composed of a specialized, stereotactically placed guide cannula with an angulated distal port. Once in place, advancement of a cell delivery catheter on an angular trajectory allows divergence from the axis of the guide cannula. By varying the depth, angular trajectory, or radial displacement of the cell delivery catheter, multiple small cell injections are performed through a single transcortical puncture and allow "seeding" of the target of interest. Initial results have been encouraging, and suggest that radial distribution can place smaller deposits of cells throughout large, complex neuroanatomical structures without inducing brain hemorrhage, significant parenchymal injury, or inflammation.[6,17,25,49-51]

Intracerebral Microinjection Instrument

The intracerebral microinjection instrument (IMI) consists of two nested tubes (a guide cannula and delivery catheter), a stereotactic drive system, an electrophysiologic recording system, and either an automated infusion pump or manual delivery syringe.[17,49,51] The outer guide is a rigid, stainless steel tube specifically designed with a distal exit port angulated at 25 degrees. An inner tube translates along the outer guide cannula and is composed of several components. The proximal end is equipped with a Luer lock to allow attachment to a motorized infusion system or a standard delivery syringe. The inner tube continues as a series of polyimide-lined stainless steel tubes and terminates distally as a 20-mm-long polyimide-insulated microcannula made of flexible synthetic quartz polymer. The delivery microcannula also contains an uninsulated, low-impedance platinum wire to facilitate electrophysiological recordings to identify and refine trajectories to the target of interest.[17,49,51] The microcannula may be advanced up to 2 cm from the exit port at a fixed angle of 25 degrees. Repeated small cellular injections with variation in rotation of the guide cannula and displacement of the microcannula achieve a "cone-like" or "funnel-like" distribution of transplanted cells. In preclinical models, the IMI has been shown to accurately transplant human stem cells (NSI-566, NeuralStem, Inc.) along a radial trajectory with less reflux, decreased operative time, and less gliosis than standard straight cannulas.[49] The IMI is not MRI compatible, but instead relies on electrophysiology to refine targeting. The IMI lacks the resolution of standard microelectrode recordings used in DBS, but allows the operator to distinguish between white and gray matter and to accurately target regions with distinct electrophysiological patterns (e.g., the substantia nigra pars compacta).[51] The IMI has also been adapted to permit a range of microcannula diameters that allow the smallest, least invasive cannula to be used for a given therapeutic application, such as ~50 μm microcannulas for small molecules and ~200 μm microcannulas for cell suspensions.

Radially Branched Deployment Device

The radially branched deployment (RBD) device is a modular cannula system in which a cell-delivery catheter can be deployed at various angles and depths from a single longitudinal cannula. This facilitates delivery of multiple small cellular grafts in a "tree-like" pattern. The RBD device is made up of three nested tubes: an outer guide tube, an inner guide tube, and a cell delivery catheter. The rigid outer tube has a side port 4 mm proximal from the distal end, which is orthogonally oriented to the radial axis. The port may be opened or closed with rotation of the inner guide tube. The inner guide tube has a deflected tip that aligns with the side port of the outer guide tube when turned to the "open" position. In the initial prototype, both were constructed with stainless steel and resembled a standard stereotactic biopsy needle when in the closed configuration. The innermost cell delivery catheter is constructed from flexible nylon-12 and translates linearly within the inner guide tube. Unlike the fixed angulation of the IMI, the distal end of the cell-delivery catheter may be precurved to varying angles up to 90°. When fully deployed, the catheter can reach distances up to 2 cm from the guide cannula. By varying the depth and rotational angle of the side port and the radial deployment of the cell delivery catheter, multiple small cellular deposits can be transplanted into larger, anatomically complex target regions.

The RBD device also addresses some of the limitations of standard syringe-based delivery. Many other delivery cannulas are equipped with a Luer lock that attaches to a manual syringe. With many syringes, relatively small movements of the plungers dispense large volumes, introducing variability when trying to deliver small, precise cell doses. Estimates of variation range from 4.3% to 24% with repeat injection.[38] Also, the site of connection between the syringe and the cannula can produce an abrupt change in diameter. As previously described, this can lead to dramatic increases in linear velocity and greater extensional force across cell membranes, ultimately compromising viability.[20] Variations in the manual rate of displacement of the plunger introduces added mechanical stress on cells, unintended reflux, or both. These findings have led others to rely on automated syringe-infusion systems to improve precision and reliability with syringe-coupled cannulas.[24,36,38]

Although automated infusion devices may better control the volume of infusate delivered, cell dose may still vary due to sedimentation. With multiple or prolonged injections, cells in suspension begin to settle under the influence of gravity, creating a gradient of cellular density. With syringes kept in a vertical orientation, the greatest concentration of cells is found at the most dependent position—usually at the site of connection between the cannula and the syringe. Depending on the viscosity of the carrier media, up to threefold difference in cell concentration between the first and last fractions administered have been reported.[23]

To address these shortcomings, the cell delivery cannula of the RBD device is made of a modular, easy-to-interchange component that lacks abrupt transition points within its inner diameter. The lumen of the RBD device cell delivery catheter is directly fitted with a titanium wire of a flexible coil design, with a gas-tight seal that allows it to serve as a plunger. Before insertion into the RBD device guide cannula system, the RBD device cell delivery catheter is preloaded with the cell suspension intended for each individual injection location. In addition to eliminating potential dead space, this integrated plunger–catheter design allows small, precise delivery of the cellular suspension (1.36 ± 0.13 μL per cm of plunger travel)—a level of precision similar to that of an automated Hamilton syringe (Hamilton Company). Being fully interchangeable, the RBD device cell delivery catheter and plunger system allow different cell doses (or even distinct cell types) to be delivered at different injection targets. With each catheter deployment, the surgeon delivers the entire cell dose which has been preloaded in the catheter, thus avoiding dose variability due to sedimentation in the delivery device. After cell delivery, the expended catheter can be fully removed, and the RBD device guide side port is closed. The guide tube may then be redeployed to a new radial trajectory and reloaded with a new preloaded catheter-plunger combination (Fig. 3).

Through variations in depth by raising or lowering the RBD device, rotational angle of the side port, and radial extension of the cell delivery catheter, this technology can distribute small cell deposits over target regions $>4\ cm^3$ through a single transcortical penetration. The viability of engrafted NPCs or human embryonic stem cell-derived dopamine neuronal precursor cells is essentially unchanged by passage through the RBD device, perhaps due to the lack of transition points in the catheter–plunger design.[25,50] Unintended cell reflux is also impeded at the transition point between the cell catheter and the side port of the outer guide tube. Together, these design elements allow for the delivery of cells to a wide range of 3D "patterns" in anatomically complex target regions via a single transcortical penetration.[25,50]

Adaptation to Interventional MRI

To further improve the potential application of RBD to human patients, a second-generation RBD device uses only nonparamagnetic materials, which enables its use in the high-field MRI environment (Fig. 3). In this iteration, polyether ether ketone, nylon, and nickel titanium are used to construct the outer guide tube, inner guide tube, and plunger, respectively. The base of the device has also been adapted to allow mounting on the single-use, disposable ClearPoint SMARTframe (MRI Interventions) to permit target selection and accurate cannula placement with a commercially available iMRI platform. Initial results with this technology have been encouraging. Using an established neurosurgical work flow,[28,31,32] multiple deposits of human embryonic stem cell-derived dopaminergic neuronal precursor cells were safely implanted into the striatum of live swine without hemorrhage or adverse events.[50] With detailed surgical planning with iMRI, super paramagnetic iron oxide beads were distributed throughout the putamen in human cadaveric heads via a single guide cannula insertion, raising optimism for potential future applications in human patients.[50]

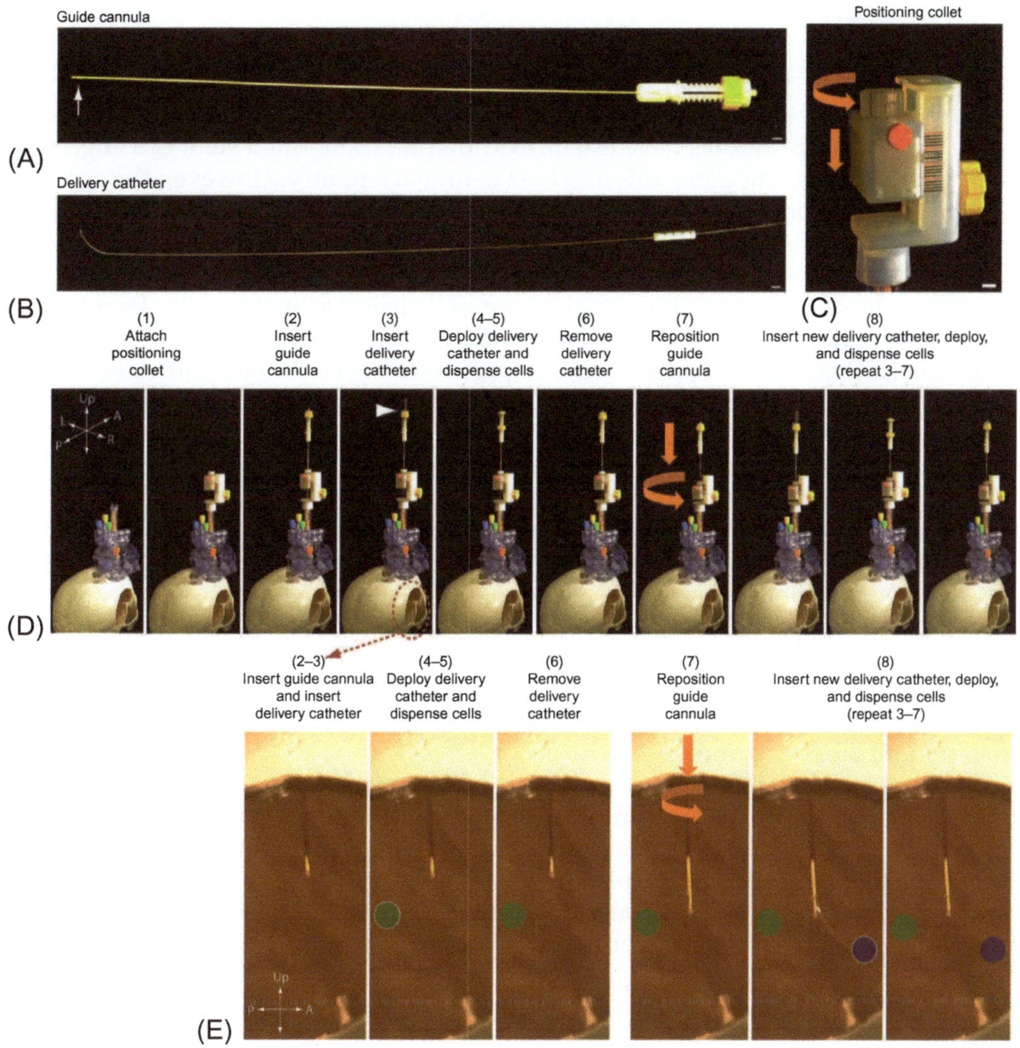

FIG. 3 The interventional magnetic resonance imaging (iMRI)-guided radially branched deployment (RBD) platform. The RBD platform consists of (A) a guide cannula with a distal side port *(white arrow)* and (B) a flexible cell-delivery catheter with a curved distal end and proximal locking hub. (C) The guide cannula is placed in a positioning collet. The *orange arrows* demonstrate rotational control by the *white hub*. The *yellow wheel* controls depth. Horizontal millimeter markings *(black and red lines)* indicate changes in depth. Scale bars, 5 mm. (D) RBD use illustrated on a human skull model. The RBD platform is mounted over a burr hole. (1) The positioning collet is attached to an MRI-compatible, skull-mounted aiming device; (2) the guide tube is inserted through the collet; (3) a prebent cell-delivery catheter is locked into the guide cannula; (4) the catheter is deployed through the open side port on the guide cannula; (5) cells are delivered with the plunger wire; (6) the cell-delivery catheter is removed after completion of injection and the side port is closed; (7) the guide tube is repositioned to a new depth and radial angle; (8) cells are delivered to the new location via repetition of steps 3–7. (E) Intracranial view of the different steps required for cell delivery with the RBD platform (see steps outlined in D).

CONCLUSIONS

Despite continued innovation and the drawbacks associated with more conventional straight-cannula delivery, neither the IMI nor the iMRI-guided RBD device has been used in live human patients. Studies have established that multiple small stem cell engraftments are superior in incorporating the cells into host neuronal circuitry. However, the optimal cell dosage and distance between engraftments have yet to be determined and may differ based on target region and disease process. Cell delivery strategies and device selection should similarly be tailored to the neuroanatomic region of interest. For small, well-defined brain regions, a straight cannula may be sufficient for cell delivery; meanwhile, larger or more complex brain regions may require newer radial delivery strategies.

While the RBD device represents some improvement over straight catheters, its performance may still fall short of the cell distribution achieved in smaller rodents, and multiple pial entrance points may still be required. Further innovations, such as the simultaneous use of multiple delivery catheters very small in diameter (<50 μm), may overcome some of the remaining limitations. Initial targeting also remains dependent on placement of a rigid linear guide cannula. When eloquent neural structures and/or regional vascular anatomy prohibit a linear trajectory, suboptimal catheter placement may be unavoidable. In such scenarios, the development of "steerable" or more flexible guide cannulas may be beneficial. With increasing complexity of delivery strategies and devices, parallel efforts are also needed to develop computer algorithms to optimally seed complex 3D structures with small cell engraftments. Finally, delivery strategies will also need to account for the migratory properties of the injected cells. Adjuvant strategies to biologically harness such cellular properties remain important areas of future research.

Nonetheless, the relative successes in the development of radial delivery and iMRI-mediated visualization of cellular infusions provide investigators with new options for testing the clinical benefit of intraparenchymal cell transplantation to treat the diseased or injured brain. As our understanding of the biology underlying cell-based therapeutics continues to advance in the decades to come, continued research into the devices and surgical aspects of this therapeutic strategy will help ensure that delivery platforms no longer present a critical bottleneck to clinical translation.

References

1. Shen Y, Huang J, Liu L, et al. A compendium of preparation and application of stem cells in Parkinson's disease: current status and future prospects. *Front Aging Neurosci*. 2016;8:117.
2. Connor B. Concise review: the use of stem cells for understanding and treating Huntington's disease. *Stem Cells*. 2018;36(2):146-160.
3. Scolding NJ, Pasquini M, Reingold SC, et al. Cell-based therapeutic strategies for multiple sclerosis. *Brain*. 2017;140(11):2776-2796.
4. Azad TD, Veeravagu A, Steinberg GK. Neurorestoration after stroke. *Neurosurg Focus*. 2016;40(5):E2.
5. Levy M, Boulis N, Rao M, Svendsen CN. Regenerative cellular therapies for neurologic diseases. *Brain Res*. 2016;1638(Pt A):88-96.
6. Potts MB, Silvestrini MT, Lim DA. Devices for cell transplantation into the central nervous system: design considerations and emerging technologies. *Surg Neurol Int*. 2013;4(suppl 1):S22-S30.
7. Freed CR, Greene PE, Breeze RE, et al. Transplantation of embryonic dopamine neurons for severe Parkinson's disease. *N Engl J Med*. 2001;344(10):710-719.
8. Lindvall O, Bjorklund A. Cell therapy in Parkinson's disease. *NeuroRx*. 2004;1(4):382-393.

9. Olanow CW, Goetz CG, Kordower JH, et al. A double-blind controlled trial of bilateral fetal nigral transplantation in Parkinson's disease. *Ann Neurol.* 2003;54(3):403-414.
10. Winkler EA, Bell RD, Zlokovic BV. Central nervous system pericytes in health and disease. *Nat Neurosci.* 2011; 14(11):1398-1405.
11. Yu G, Borlongan CV, Stahl CE, et al. Systemic delivery of umbilical cord blood cells for stroke therapy: a review. *Restor Neurol Neurosci.* 2009;27(1):41-54.
12. Rodriguez A, Tatter SB, Debinski W. Neurosurgical techniques for disruption of the blood-brain barrier for glioblastoma treatment. *Pharmaceutics.* 2015;7(3):175-187.
13. Hynynen K, McDannold N, Sheikov NA, Jolesz FA, Vykhodtseva N. Local and reversible blood-brain barrier disruption by noninvasive focused ultrasound at frequencies suitable for trans-skull sonications. *Neuroimage.* 2005; 24(1):12-20.
14. Cohen-Pfeffer JL, Gururangan S, Lester T, et al. Intracerebroventricular delivery as a safe, long-term route of drug administration. *Pediatr Neurol.* 2017;67:23-35.
15. Lim JY, Jeong CH, Jun JA, et al. Therapeutic effects of human umbilical cord blood-derived mesenchymal stem cells after intrathecal administration by lumbar puncture in a rat model of cerebral ischemia. *Stem Cell Res Ther.* 2011;2(5):38.
16. Smith EJ, Stroemer RP, Gorenkova N, et al. Implantation site and lesion topology determine efficacy of a human neural stem cell line in a rat model of chronic stroke. *Stem Cells.* 2012;30(4):785-796.
17. Cunningham MG, Bolay H, Scouten CW, et al. Preclinical evaluation of a novel intracerebral microinjection instrument permitting electrophysiologically guided delivery of therapeutics. *Neurosurgery.* 2004;54(6):1497-1507 [discussion 1507].
18. Nikkhah G, Cunningham MG, Jodicke A, Knappe U, Bjorklund A. Improved graft survival and striatal reinnervation by microtransplantation of fetal nigral cell suspensions in the rat Parkinson model. *Brain Res.* 1994;633(1–2):133-143.
19. Tol M, Akar AR, Durdu S, Ayyildiz E, Ilhan O. Comparison of different needle diameters and flow rates on bone marrow mononuclear stem cell viability: an ex vivo experimental study. *Cytotherapy.* 2008;10(1):98-99.
20. Aguado BA, Mulyasasmita W, Su J, Lampe KJ, Heilshorn SC. Improving viability of stem cells during syringe needle flow through the design of hydrogel cell carriers. *Tissue Eng Part A.* 2012;18(7–8):806-815.
21. Gobbel GT, Kondziolka D, Fellows-Mayle W, Uram M. Cellular transplantation for the nervous system: impact of time after preparation on cell viability and survival. *J Neurosurg.* 2010;113(3):666-672.
22. Kondziolka D, Gobbel GT, Fellows-Mayle W, Chang YF, Uram M. Injection parameters affect cell viability and implant volumes in automated cell delivery for the brain. *Cell Transplant.* 2011;20(11 – 12):1901-1906.
23. Rossetti T, Nicholls F, Modo M. Intracerebral cell implantation: preparation and characterization of cell suspensions. *Cell Transplant.* 2016;25(4):645-664.
24. Malloy KE, Li J, Choudhury GR, et al. Magnetic resonance imaging-guided delivery of neural stem cells into the basal ganglia of nonhuman primates reveals a pulsatile mode of cell dispersion. *Stem Cells Transl Med.* 2017; 6(3):877-885.
25. Silvestrini MT, Yin D, Coppes VG, et al. Radially branched deployment for more efficient cell transplantation at the scale of the human brain. *Stereotact Funct Neurosurg.* 2013;91(2):92-103.
26. Chen MY, Lonser RR, Morrison PF, Governale LS, Oldfield EH. Variables affecting convection-enhanced delivery to the striatum: a systematic examination of rate of infusion, cannula size, infusate concentration, and tissue-cannula sealing time. *J Neurosurg.* 1999;90(2):315-320.
27. Krauze MT, Saito R, Noble C, et al. Reflux-free cannula for convection-enhanced high-speed delivery of therapeutic agents. *J Neurosurg.* 2005;103(5):923-929.
28. Southwell DG, Narvid JA, Martin AJ, Qasim SE, Starr PA, Larson PS. Comparison of deep brain stimulation lead targeting accuracy and procedure duration between 1.5- and 3-Tesla interventional magnetic resonance imaging systems: an initial 12-month experience. *Stereotact Funct Neurosurg.* 2016;94(2):102-107.
29. Bjartmarz H, Rehncrona S. Comparison of accuracy and precision between frame-based and frameless stereotactic navigation for deep brain stimulation electrode implantation. *Stereotact Funct Neurosurg.* 2007; 85(5):235-242.
30. Bot M, van den Munckhof P, Bakay R, Sierens D, Stebbins G, Verhagen Metman L. Analysis of stereotactic accuracy in patients undergoing deep brain stimulation using nexframe and the leksell frame. *Stereotact Funct Neurosurg.* 2015;93(5):316-325.
31. Ostrem JL, Ziman N, Galifianakis NB, et al. Clinical outcomes using ClearPoint interventional MRI for deep brain stimulation lead placement in Parkinson's disease. *J Neurosurg.* 2016;124(4):908-916.

32. Starr PA, Markun LC, Larson PS, Volz MM, Martin AJ, Ostrem JL. Interventional MRI-guided deep brain stimulation in pediatric dystonia: first experience with the ClearPoint system. *J Neurosurg Pediatr*. 2014;14(4):400-408.
33. Martin AJ, Starr PA, Ostrem JL, Larson PS. Hemorrhage detection and incidence during magnetic resonance-guided deep brain stimulator implantations. *Stereotact Funct Neurosurg*. 2017;95(5):307-314.
34. Larson PS, Starr PA, Bates G, Tansey L, Richardson RM, Martin AJ. An optimized system for interventional magnetic resonance imaging-guided stereotactic surgery: preliminary evaluation of targeting accuracy. *Neurosurgery*. 2012;70(suppl 1 Operative):95-103 [discussion 103].
35. Larson PS, Starr PA, Martin AJ. Deep brain stimulation: interventional and intraoperative MRI approaches. *Prog Neurol Surg*. 2018;33:187-197.
36. Vermilyea SC, Lu J, Olsen M, et al. Real-time intraoperative MRI intracerebral delivery of induced pluripotent stem cell-derived neurons. *Cell Transplant*. 2017;26(4):613-624.
37. Lindvall O, Brundin P, Widner H, et al. Grafts of fetal dopamine neurons survive and improve motor function in Parkinson's disease. *Science*. 1990;247(4942):574-577.
38. Gobbel GT, Kondziolka D, Fellows-Mayle W, Uram M. Manual vs automated delivery of cells for transplantation: accuracy, reproducibility, and impact on viability. *Neurosurgery*. 2010;67(6):1662-1668 [discussion 1668].
39. Skuk D, Paradis M, Goulet M, Tremblay JP. Ischemic central necrosis in pockets of transplanted myoblasts in nonhuman primates: implications for cell-transplantation strategies. *Transplantation*. 2007;84(10):1307-1315.
40. Ott HC, Kroess R, Bonaros N, et al. Intramyocardial microdepot injection increases the efficacy of skeletal myoblast transplantation. *Eur J Cardiothorac Surg*. 2005;27(6):1017-1021.
41. Nikkhah G, Olsson M, Eberhard J, Bentlage C, Cunningham MG, Bjorklund A. A microtransplantation approach for cell suspension grafting in the rat Parkinson model: a detailed account of the methodology. *Neuroscience*. 1994;63(1):57-72.
42. Breeze RE, Wells Jr TH, Freed CR. Implantation of fetal tissue for the management of Parkinson's disease: a technical note. *Neurosurgery*. 1995;36(5):1044-1047 [discussion 1047–1048].
43. De Almeida AN, Olivier A, Quesney F, Dubeau F, Savard G, Andermann F. Efficacy of and morbidity associated with stereoelectroencephalography using computerized tomography—or magnetic resonance imaging-guided electrode implantation. *J Neurosurg*. 2006;104(4):483-487.
44. Gonzalez-Martinez J, Mullin J, Vadera S, et al. Stereotactic placement of depth electrodes in medically intractable epilepsy. *J Neurosurg*. 2014;120(3):639-644.
45. Mullin JP, Shriver M, Alomar S, et al. Is SEEG safe? A systematic review and meta-analysis of stereo-electroencephalography-related complications. *Epilepsia*. 2016;57(3):386-401.
46. Zrinzo L, Foltynie T, Limousin P, Hariz MI. Reducing hemorrhagic complications in functional neurosurgery: a large case series and systematic literature review. *J Neurosurg*. 2012;116(1):84-94.
47. Binder DK, Rau G, Starr PA. Hemorrhagic complications of microelectrode-guided deep brain stimulation. *Stereotact Funct Neurosurg*. 2003;80(1–4):28-31.
48. Binder DK, Rau GM, Starr PA. Risk factors for hemorrhage during microelectrode-guided deep brain stimulator implantation for movement disorders. *Neurosurgery*. 2005;56(4):722-732 [discussion 722–732].
49. Glud AN, Bjarkam CR, Azimi N, Johe K, Sorensen JC, Cunningham M. Feasibility of three-dimensional placement of human therapeutic stem cells using the intracerebral microinjection instrument. *Neuromodulation*. 2016;19(7):708-716.
50. Silvestrini MT, Yin D, Martin AJ, et al. Interventional magnetic resonance imaging-guided cell transplantation into the brain with radially branched deployment. *Mol Ther*. 2015;23(1):119-129.
51. Bjarkam CR, Glud AN, Margolin L, et al. Safety and function of a new clinical intracerebral microinjection instrument for stem cells and therapeutics examined in the Gottingen minipig. *Stereotact Funct Neurosurg*. 2010;88(1):56-63.

CHAPTER

18

CRISPR–Cas Gene Editing for Neurological Disease

Bryan P. Simpson[*,†], *Beverly L. Davidson*[*,†]

[*]The Raymond G. Perelman Center for Cellular and Molecular Therapeutics, The Children's Hospital of Philadelphia, Philadelphia, PA, United States [†]The Perelman School of Medicine, The University of Pennsylvania, Philadelphia, PA, United States

INTRODUCTION

Many neurological diseases are caused by genetic mutations, which means they are amenable to various gene therapy approaches. Such approaches include gene augmentation to replace the mutant gene, or gene knockdown to inhibit expression of a dominantly acting disease gene. Therapeutic modalities also include going after the mutant protein, either by replacing it (e.g., enzyme replacement therapy) or by inhibiting it (e.g., introducing antibodies to remove toxic gene products). Recently, gene editing has been added to the toolbox of methodologies to reduce the disease gene burden. Among gene editing techniques is the clustered regularly interspaced short palindromic repeats (CRISPR)–CRISPR-associated nuclease (Cas) gene editing technology, referred to as CRISPR–Cas. This modality offers the flexibility to edit, replace, and modify specific loci in the genome with simplicity and ease in the laboratory, which makes it a useful tool to knockout and knock-in genes, to activate or silence gene expression, and to create model systems.[1] The CRISPR–Cas system has been well characterized since its rise to fame and is now moving toward therapeutic development, having shown promise in disease models.[2, 3] This chapter highlights the studies and findings supporting the feasibility and utility of therapeutic CRISPR–Cas gene editing as it applies to neurological disease.

CRISPR–Cas GENE EDITING IN MAMMALIAN SYSTEMS

CRISPR is a bacterial adaptive immune system that targets foreign nucleic acid sequences (Fig. 1).[4, 5] Bacteria use this system to defend against infections by incorporating short

Nervous System Drug Delivery
https://doi.org/10.1016/B978-0-12-813997-4.00018-9

© 2019 Elsevier Inc. All rights reserved.

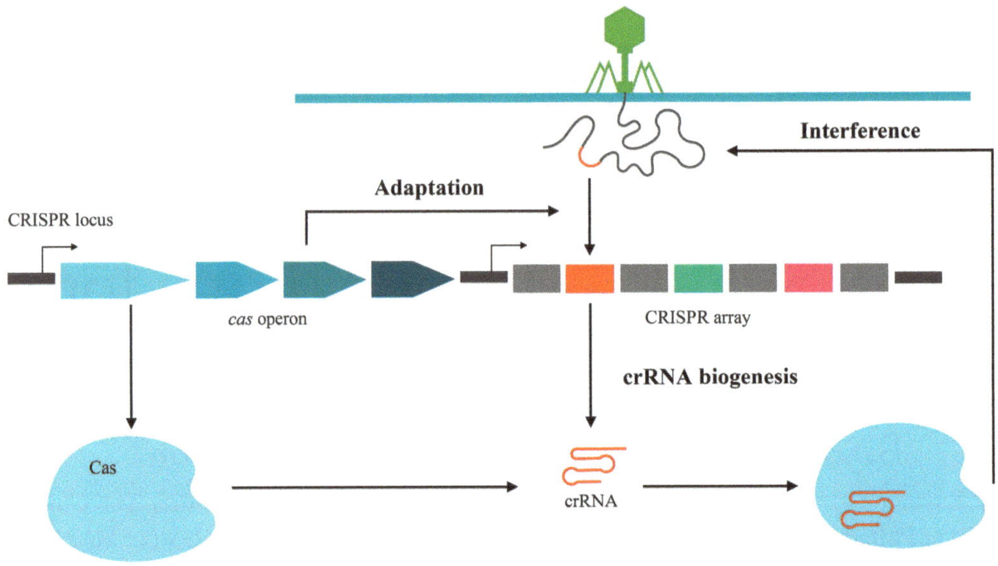

FIG. 1 The CRISPR–Cas adaptive immune system in prokaryotes. The clustered regularly interspaced short palindromic repeat (CRISPR) locus is a genomic region that provides adaptive immunity against viral infections in prokaryotes. When foreign nucleic acid is introduced into the host cell, the Cas operon expresses CRISPR-associated proteins (i.e., Cas) to select and integrate a protospacer sequence into the CRISPR array. The CRISPR array expresses pre-CRISPR RNA (crRNA) that is further processed into mature crRNA. The Cas nuclease protein complexes with the crRNA, which guides it to the sequence-specific target for cleavage.

fragments of the foreign DNA within the CRISPR region of its genome. The CRISPR region consists of short, repetitive, palindromic spacer sequences, and sequences encoding Cas proteins.[6] When the foreign DNA fragments are expressed, they behave as guide RNAs (gRNAs) to direct the Cas nuclease to the invading target, which results in cleavage of the foreign agent's DNA.[1]

The fusion of CRISPR RNA (crRNA) and the trans-activating CRISPR RNA (tracrRNA) into a chimeric single-guide RNA (sgRNA) allows for site-specific gene editing in eukaryote genomes (Fig. 2).[7, 8] The CRISPR type II system with the *Streptococcus pyogenes* Cas9 (SpCas9) nuclease is the most commonly used system, and scientists take advantage of it by designing an sgRNA that targets a specific locus, and expressing the sgRNA along with Cas9 in vitro or in vivo.[9] Cas9 associates with the sgRNA, and the complex is targeted to the desired DNA sequence.[2, 10, 11] The Cas9-sgRNA complex binds to the protospacer adjacent motif (PAM) sequence located next to the sgRNA target sequence, and the two enzymatic endonuclease domains of Cas9 (i.e., RuvC and HNH) cleave both strands of DNA to generate a double-strand break (DSB).[7, 12, 13] The DSB can be repaired by one of the two DNA repair mechanisms: nonhomologous end joining (NHEJ) or homology-directed repair (HDR).[14] NHEJ, the predominant repair pathway in mammalian cells, is prone to error and can result in insertion or deletion (i.e., indel) of nucleotides. Indels can create frameshift mutations and downstream premature stop codons, leading to inactivation of the target gene.[2, 11]

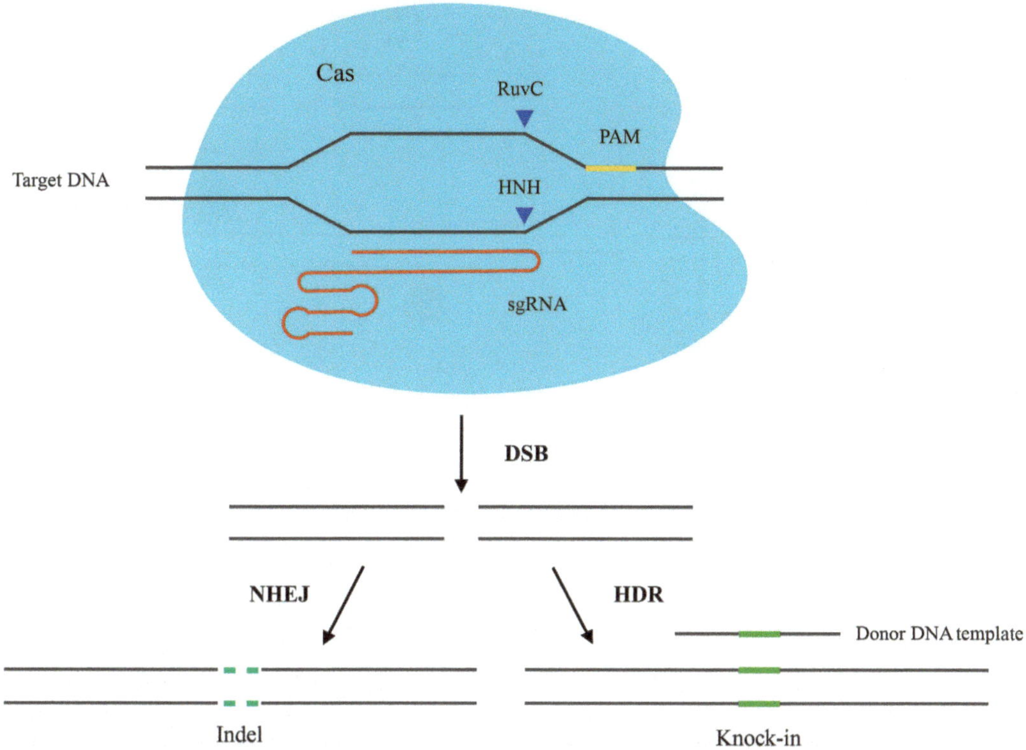

FIG. 2 CRISPR–Cas9 gene editing in eukaryotes. The sequence-specific Cas9-sgRNA complex binds the protospacer adjacent motif (PAM), the protospacer sequence of the sgRNA anneals; meanwhile, the catalytic nuclease domains (i.e., RuvC and HNH) cleave the target DNA sequence to create a double-strand break (DSB). The error-prone, nonhomologous end-joining (NHEJ) repair pathway can create insertion and deletion (i.e., indel) mutations that can result in a frameshift, premature stop codon, and, ultimately, gene knockout. Alternatively, the homology-directed repair pathway can modify or correct a gene by knock-in when a donor DNA template is provided.

In addition to gene knockout and inactivation, targeted gene insertion is possible by CRISPR–Cas9 HDR-mediated knock-in of donor DNA sequence templates,[15-17] which can be harnessed to correct loss-of-functions mutations.[18] HDR may be less efficient and somewhat limited in terminally differentiated postmitotic cells such as neurons.[16] However, NHEJ in neurons can occur, allowing for knock-in editing strategies. The CRISPR-Cpf1 system with RNA-guided Cpf1, a nuclease like Cas9, generates DSBs with overhangs; this in turn promotes NHEJ in neurons.[19-21] Finally, CRISPR–Cas9 multiplexing is possible when multiple sgRNAs are used to target different genes and pathways simultaneously, or when more than one guide is required to modify a gene.[15, 22-24]

The CRISPR–Cas9 system also allows for transcriptional regulation to modulate gene expression without permanently editing the genome (Fig. 3). For this, a mutated, dead Cas9 (dCas9) that lacks enzymatic endonuclease activity is used.[25] The dCas9-sgRNA complex binds the targeted DNA sequence without creating a DSB, and is used to silence or activate gene expression (processes known as CRISPR interference [CRISPRi] or CRISPR activation

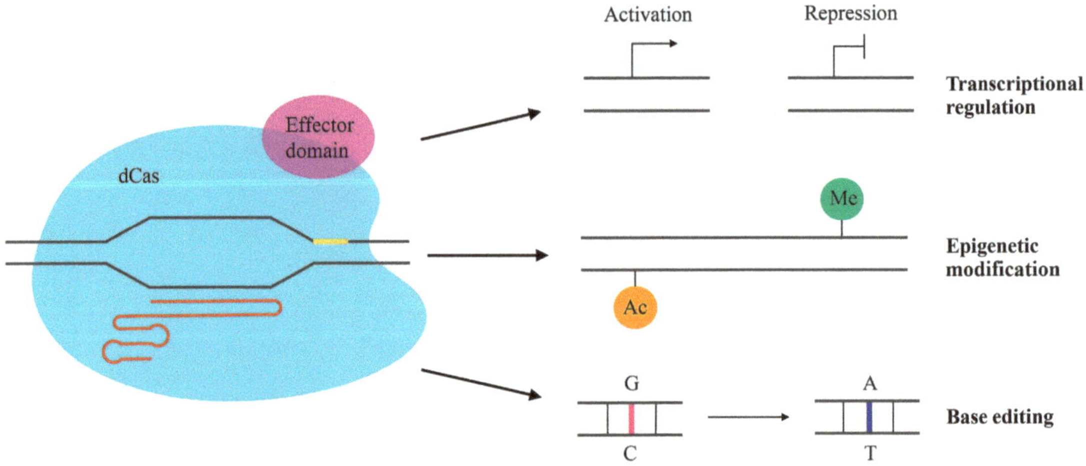

FIG. 3 dCas9-mediated gene manipulation. Catalytically inactive dead Cas9 (dCas9) can functionally bind target DNA to impart transcriptional downregulation. Additionally, when dCas9 is fused with different effector domains, it is possible to enhance gene repression, activate gene expression, target epigenetic modifications, or edit DNA bases.

[CRISPRa], respectively). This enhanced gene regulation is achieved with effector protein domains fused to dCas9 to recruit epigenetic modifying factors.[25-29] For CRISPRi, the transcriptional repressor domain Krüppel associated box (KRAB) is fused to dCas9; for CRISPRa, the transcriptional activator domain VP64 is fused to dCas9.[25, 30] Alternatively, sgRNA can be modified with an MS2 coat protein fused to an effector domain for robust transcriptional modulation.[31, 32]

CRISPR–Cas9 is delivered to cells using standard viral and nonviral methods. Currently, adeno-associated viral (AAV) vectors are the most widely used delivery method for in vivo delivery of CRISPR–Cas9; however, AAV packaging size is restricted to ~4.7 kb, so a dual-AAV vector system is required. One vector delivers the SpCas9, and the second vector delivers the sgRNA(s). Alternatively, *Staphylococcus aureus* Cas9 (SaCas9), a smaller Cas9 orthologue, can fit into a single AAV vector along with the sgRNAs.[33] Because viral delivery and constitutive expression of CRISPR–Cas9 may increase off-target effects and the risk of an immune response to the foreign bacterial protein, short-lived ribonucleoprotein (RNP) complexes of sgRNA-Cas9 are attractive alternatives for gene editing. RNP delivery has been shown to be effective after direct injections into mouse cortex, striatum, and hippocampus.[34] Additionally, in vivo RNP delivery of an allele-specific sgRNA-Cas9 complex resulted in correction of hearing loss in a dominant deafness-associated allele in the *Tmc1* Beethoven mouse model.[35] These studies support the potential for RNP delivery in neurological diseases to overcome the safety concern of sustained expression from viral delivery systems, or the need for more than one vector.

The CRISPR–Cas9 system provides advantages over alternative genome editing tools, such as zinc-finger nucleases and transcription activator-like effector nucleases.[36] Most notably, CRISPR–Cas9 is easily adaptable to any gene and does not require extensive optimization, although editing frequency can vary depending on the target sequence and the sgRNAs.[37] In addition, CRISPR–Cas9 overcomes the limitations of RNA-targeting strategies for gain-of-function mutations, such as RNA interference and antisense oligonucleotides. These latter approaches, if delivered via nonviral means, require repeated treatments, and both can be prone to high off-target effects and cytotoxicity.[38, 39]

One major drawback of the CRISPR–Cas9 system is off-target effects due to sgRNA sequence homology to other genomic sites.[40, 41] Cas9 can tolerate three to five mismatches, depending on the distribution and number of mismatched nucleotides.[2, 13, 30, 42] Designing sgRNAs with low GC content and minimal homology to off-target sites, reducing the size of the sgRNA, and limiting the amount of Cas9 protein are all ways to minimize off-targets.[12, 42, 43] The nickase Cas9, a mutated Cas9 with only one functionally enzymatic endonuclease domain, also reduces off-targets because it cuts only one strand of the DNA. However, the nickase Cas9 does require two sgRNAs targeting opposite DNA strands to create a DSB for generation of the desired indel.[44, 45]

Engineering advances in the CRISPR–Cas9 system have focused on optimizing editing efficiency, specificity, and Cas9 regulation. On-target specificity was improved by engineering an enhanced Cas9 and a high-fidelity Cas9; both were associated with reduced off-targeting.[46, 47] An evolved SpCas9 variant, xCas9, is associated with improved specificity, low off-target activity, and broader PAM compatibility, all of which facilitate broader applicability.[48] Another approach involves a synthetic crRNA to replace the wild-type crRNA, which reduces off-targets and enhances cleavage activity.[49] Additionally, split-Cas9 systems have been developed to control and regulate Cas9 expression. In this scenario, Cas9 is expressed by two expression systems that can be induced to dimerize, and has editing activity similar to that of wild-type SpCas9.[50-52] Finally, expression regulated by small molecules, light, or temperature have also been designed to control Cas9 activity.[53]

MOVING CRISPR–Cas GENE EDITING TOWARD THE CLINIC

Disease-modifying treatments for neurological diseases are currently lacking, with most drugs targeting symptom management. Early CRISPR–Cas9 genome editing studies were used to create model systems, but advances in gene delivery of CRISPR–Cas9 technology now support the clinical utility of this modality for genome editing.[54] For dominantly inherited neurological diseases, CRISPR–Cas9 gene editing could inactivate or transcriptionally downregulate disease-causing gain-of-function mutations. Additionally, CRISPR–Cas9 repair of mutant genes could correct autosomal recessive loss-of-function mutations.[55, 56] Even more exciting is the recent development of CRISPR–Cas9 single-base editing strategies for correcting pathogenic point mutations (Fig. 3).[57]

Not long after CRISPR–Cas9 gene editing was performed in eukaryotic cells, in vivo editing of genes was investigated in the mammalian brain. The initial CRISPR–Cas9 gene editing studies in postmitotic neurons and the mammalian brain support the technology as a viable therapeutic option.[58, 59] The first proof-of-principle study of CRISPR–Cas9 gene knockdown in neurons was done in embryonic mice,[58] and CRISPR–Cas9 knockout of the *N*-methyl-D-aspartate receptor (NMDAR) subunit protein GluN1 was done in mouse hippocampal pyramidal neurons after in utero electroporation delivery. NMDAR currents were abolished, demonstrating functional in vivo knockdown of *Grin1*. Additionally, the authors transfected the Cas9/gRNA complex into postmitotic neurons in hippocampal slice cultures from P6-P8 rats and showed similar reductions in NMDAR-dependent currents. Although the study did not demonstrate in vivo editing of neurons after CRISPR–Cas9 delivery into the brain, it was nonetheless a tremendous step forward.

In other work, *Mecp2* promoter-driven Cas9 and an sgRNA targeting the *Mecp2* gene demonstrated high editing efficiency with concomitant changes in learning and behavior.[59] The CRISPR–Cas9 modifications were stable after AAV delivery to adult mice hippocampal neurons. CRISPR–Cas9 multiplexing can also target multiple genes, suggesting the potential for treating polygenic disease such as psychiatric disorders;[60] in fact, Cas9 coupled with three sgRNAs targeting *Dnmt3a*, *Dnmt1*, and *Dnmt3b* in the hippocampus induced behavioral alterations, setting the stage for in vivo CRISPR–Cas9 interrogation of pathways to address the roles of pathways in complex behavior, with possible adaptation to clinical problems.

CRISPR–Cas9 application to correct a monogenic disease was first presented by three groups studying Duchenne muscular dystrophy in mice models.[61-63] They showed editing and improvement in phenotypes after delivery of the editing machinery. CRISPR–Cas9 gene therapy was also studied in the setting of hereditary liver disease.[64, 65] These pioneering researchers opened the door for ongoing in vivo CRISPR–Cas9 gene editing research and the application of this research toward treating neurological diseases. Currently, CRISPR–Cas9 technology is geared toward correcting monogenic diseases in patient cell lines and animal models of Alzheimer disease (AD), Parkinson disease, amyotrophic lateral sclerosis (ALS), and the repeat expansion disorders, such as fragile X syndrome (FXS), Huntington disease (HD), the autosomal dominant spinocerebellar ataxias (SCAs), and spinal and bulbar muscular atrophy (SBMA).

Alzheimer disease is a major cause of morbidity and mortality in elderly individuals. Mutations in the presenilin genes *PSEN1* and *PSEN2* and in the amyloid precursor protein (*APP*) gene result in amyloid-associated pathology that leads to early-onset familial AD, and variants of *APOE* are risk factors for late-onset AD.[43] Gyorgy and colleagues disrupted a mutant variant of APP in human fibroblasts, providing the first experimental evidence of CRISPR–Cas9 gene editing for treatment of familial forms of dominant AD.[66] *PSEN1* mutations were corrected in AD-patient-induced pluripotent stem cells (iPSCs) by HDR-mediated CRISPR–Cas9 when combined with homologous repair templates.[67]

Mutations in *LRRK2* and *SNCA* are causes of autosomal dominant Parkinson disease.[68] CRISPRi targeting the transcription start site of *SNCA* significantly reduced expression levels in vitro.[69] Additionally, studies using dCas9-KRAB repression and dCas9-VPR activation showed that *SNCA* expression could be modulated in human iPSC-derived neurons.

Polyglutamine (polyQ) repeat expansion disorders such as HD, Kennedy disease, and SCAs are caused by toxic gain-of-function mutations. HD is a neurodegenerative disease caused by an expanded CAG trinucleotide repeat in the huntingtin gene (i.e., *HTT*), which results in a mutant HTT protein with an expanded polyQ repeat. Several CRISPR–Cas9 approaches have been investigated to reduce expression of mutant HTT. Similar to findings in mouse models of muscular dystrophy, CRISPR-mediated HDR has been suggested as an exon-skipping strategy to target polyQ repeats.[70, 71] Another approach is to reduce the size of the expanded CAG repeat. Using the CRISPR–Cas9 D10A nickase resulted in CAG/CTG repeat contraction in a reporter vector.[72] The problem with this approach is that other repeat-containing genes in the genome may be targeted. Replacement of the expanded polyQ repeat with a normal polyQ repeat by homologous recombination in iPSCs from HD patients improved disease phenotypes, suggesting the potential for a similar HDR-mediated approach using CRISPR–Cas9.[73] However, the efficiency of an HDR-mediated approach may not be sufficient to correct disease in humans. Alternatively, CRISPRi targeting the transcription start site of *HTT* resulted in reduction of *HTT* mRNA and protein expression levels.[69] Meanwhile, CRISPR-NHEJ-mediated deletion reduced mutant *HTT* expression in mesenchymal stem cells from the YAC128 mouse model of HD.[74]

CRISPR–Cas9 targeted deletion or contraction of the HTT polyQ repeat might not restore normal protein function, and whether complete knockout of both the mutant and wild-type *HTT* alleles will be tolerated remains unknown. To overcome these problems, allele-specific editing of the mutant allele was demonstrated with CRISPR–Cas9 by using single-nucleotide polymorphisms (SNPs) found in PAMs of the mutant allele. In one approach, two allele-specific sgRNAs targeting SNPs within PAMs on the mutant allele created a large deletion including the promoter (i.e., TSS), and expanded CAG from the mutant allele in patient fibroblasts, iPSCs, and neural progenitor cells.[75] Ablation of mutant *HTT* mRNA and protein demonstrated the efficacy of allele-specific CRISPR–Cas9 without targeting the normal allele. A similar approach identified SNPs within PAMs flanking exon-1 of *HTT* in heterozygosity between the normal and mutant alleles.[76] Allele-specific deletion of *HTT* exon-1 resulted in targeted mutant *HTT* mRNA and protein knockdown in human HD fibroblasts. In vivo allele-specific editing in the BACHD transgenic mouse model reduced human mutant *HTT* expression to 40%, a level known to provide therapeutic benefit by RNA interference or antisense oligonucleotides, demonstrating CRISPR–Cas9 in vivo efficacy in an HD mouse model for the first time. A nonallele-specific approach with sgRNAs flanking the CAG repeat reduced human mutant HTT in striatal neurons and improved motor deficits in the HD140Q-knock-in mouse model of HD, which shows the behavioral effects of mutant HTT reduction by in vivo CRISPR–Cas9.[77]

To address constitutive Cas9 expression, an off-switch was engineered and tested in HD patient-derived neurons and in a mouse brain.[78] The "KamiCas9" contains an sgRNA that can reduce not only the target, but also the nuclease. Although this may reduce immune responses to the bacterial nuclease, further testing is required to determine the extent to which this also may alleviate unintended editing.

The expansion of GGGGCC hexanucleotide (G4C2) repeats in the *C9orf72* gene cause familial ALS and frontotemporal dementia.[79, 80] Mutihac and colleagues (cited in Ref 18) corrected an expanded G4C2 repeat with HDR-mediated CRISPR–Cas9 in iPSCs from an individual with a *C9orf72* mutation, and ameliorated disease-associated phenotypes such as protein aggregation and stress granule formation in differentiated motor neurons.[18] In other work, mutations in microtubule-associated protein tau were corrected in iPSCs derived from individuals with frontotemporal dementia.[81]

Autosomal dominant mutations in *SOD1* and *FUS* cause familial ALS. Single-nucleotide mutations in *SOD1* and *FUS* were corrected by HDR-mediated CRISPR–Cas9 in iPSCs from individuals with familial ALS.[82] Systemic AAV9 delivery of SaCas9 and an sgRNA targeting exon 2 of human *SOD1* reduced mutant SOD1 expression by more than 2.5-fold in the spinal cord of neonatal G93A-SOD1 model mice of ALS,[83] improving motor function and slowing muscle atrophy. Furthermore, the mice showed a delay in disease onset, more surviving neurons at end stage, and increased survival.

Fragile X syndrome is a neurodevelopmental disorder that causes learning disabilities and cognitive impairment. FXS is a trinucleotide repeat disorder caused by CGG repeat expansion in the 5′ UTR of the fragile X mental retardation 1 (*FMR1*) gene. This causes hypermethylation of *FMR1* with transcriptional silencing and reduced fragile X mental retardation protein (FMRP) expression. CRISPR–Cas9 and an sgRNA targeting upstream of the *FMR1* CGG repeat in patient FXS iPSCs resulted in deletion of CGG repeats and reactivation of *FMR1* expression.[84] In another study in FXS iPSCs, Cas9 was expressed along with two sgRNAs flanking the CGG repeat, resulting in precise, efficient removal of the repeat and increased FMR1 expression.[85] An alternative approach targeted the hypermethylation of *FMR1* with dCas9-Tet for demethylation of the CGG expansion at the *FMR1* promoter to reactivate expression of *FMR1* in FXS iPSCs.[86] The edited iPSCs that were subsequently differentiated into neurons had wild-type phenotypes, and maintained *FMR1* expression after engraftment into the mouse brain. Notably, in vivo CRISPR–Cas9 editing and reactivation of *FMR1* has not yet been shown in vivo. A related disorder, fragile X-associated tremor/ataxia (FXTAS), is caused by smaller repeat expansions in this locus, referred to as premutation of *FMR1*, and is a late-onset neurodegenerative disease caused by increased *FMR1* mRNA toxicity and decreased FMRP expression. Based on the CRISPR–Cas9 studies in FXS iPSCs, similar strategies to edit and correct *FMR1* in FXTAS are worth pursuing.

CRISPR–Cas RNA TARGETING

A new generation of CRISPR–Cas proteins is making it possible to study RNA biology and to develop RNA-targeting therapeutics. Programmable RNA-guided Cas13, an RNase protein, efficiently reduces target mRNAs, and a catalytically inactive dCas13 edited single-bases in mammalian cells.[87, 88] RNA-targeting CRISPR–Cas9 methods can reduce mRNAs associated with HD, *C9orf72*-linked ALS, FTXTAS, SBMA, the SCAs, and myotonic dystrophy types 1 and 2 (DM1 and DM2). An engineered RNA-targeting SpCas9 (RCas9), which can bind RNA, eliminated exogenous expression of CUG, CCUG, CAG, and GGGGCC repeat expansion RNAs in cells, and endogenous CUG and CCUG expansions in DM1 and DM2 patient cells.[89, 90] Targeting expanded CAG RNAs reduced disease-associated polyQ protein, supporting the approach in HD, SBMA, and the SCAs. The advantage of RNA targeting is that no permanent off-target effects at the DNA level have been observed, which makes it a potentially safer strategy. However, repeated dosing or sustained expression from vector systems will be required, raising the specter of off-targeting or immune responses to the foreign Cas proteins. Studying RNA-targeting CRISPR systems in vivo will be the next step toward an RNA-directed therapy for neurological diseases.

CRISPR TRANSCRIPTIONAL REGULATION IN THE MAMMALIAN BRAIN

Transcriptional regulation with dCas9 has been widely used in vitro to modulate gene expression. The first proof-of-principle use of in vivo CRISPRi in the mammalian brain was done with a lentivirus expressing dCas9-KRAB to knockdown *Syt1* expression in the dentate gyrus of the mouse hippocampus.[91] The CRISPRi system modulated the excitatory–inhibitory balance of the dentate gyrus, altering learning and behavior in mice. Finally, multiplex CRISPRi abolished the expression of five genes important for neurotransmitter release in the brain, promoting the utility of in vivo CRISPRi to transcriptionally regulate mutant genes for neurological disease therapy.

Gene editing with CRISPR–Cas allows for generally simple, precise gene modifications and regulation in the mammalian nervous system. The technology opens a new door in biomedical research for drug discovery, uncovering disease mechanisms, and personalized medicine and therapeutics. Both the recent successes of gene therapy in the clinic and in vivo CRISPR–Cas gene editing in animal models demonstrate tremendous potential for developing better models and for future treatment. Although further studies are needed to better understand in vivo off-target effects and to advance delivery systems for improved safety and reduced immunogenicity in the brain, CRISPR–Cas technology will continue to facilitate biomedical research and holds promise for application to a broad spectrum of neurological disorders.

References

1. Walters BJ, Azam AB, Gillon CJ, Josselyn SA, Zovkic IB. Advanced in vivo use of CRISPR/Cas9 and anti-sense DNA inhibition for gene manipulation in the brain. *Front Genet.* 2015;6:362.
2. Hsu PD, Lander ES, Zhang F. Development and applications of CRISPR-Cas9 for genome engineering. *Cell.* 2014;157(6):1262-1278.
3. Wright AV, Nunez JK, Doudna JA. Biology and applications of CRISPR systems: harnessing nature's toolbox for genome engineering. *Cell.* 2016;164(1-2):29-44.
4. Barrangou R, Fremaux C, Deveau H, et al. CRISPR provides acquired resistance against viruses in prokaryotes. *Science.* 2007;315(5819):1709-1712.
5. Jansen R, Embden JD, Gaastra W, Schouls LM. Identification of genes that are associated with DNA repeats in prokaryotes. *Mol Microbiol.* 2002;43(6):1565-1575.
6. Makarova KS, Grishin NV, Shabalina SA, Wolf YI, Koonin EV. A putative RNA-interference-based immune system in prokaryotes: computational analysis of the predicted enzymatic machinery, functional analogies with eukaryotic RNAi, and hypothetical mechanisms of action. *Biol Direct.* 2006;1:7.
7. Cong L, Ran FA, Cox D, et al. Multiplex genome engineering using CRISPR/Cas systems. *Science.* 2013; 339(6121):819-823.
8. Jinek M, Chylinski K, Fonfara I, Hauer M, Doudna JA, Charpentier E. A programmable dual-RNA-guided DNA endonuclease in adaptive bacterial immunity. *Science.* 2012;337(6096):816-821.
9. Wu Y, Liang D, Wang Y, et al. Correction of a genetic disease in mouse via use of CRISPR-Cas9. *Cell Stem Cell.* 2013;13(6):659-662.
10. Makarova KS, Aravind L, Wolf YI, Koonin EV. Unification of Cas protein families and a simple scenario for the origin and evolution of CRISPR-Cas systems. *Biol Direct.* 2011;6:38.
11. Doudna JA, Charpentier E. Genome editing. The new frontier of genome engineering with CRISPR-Cas9. *Science.* 2014;346(6213).
12. Hsu PD, Scott DA, Weinstein JA, et al. DNA targeting specificity of RNA-guided Cas9 nucleases. *Nat Biotechnol.* 2013;31(9):827-832.
13. Mali P, Yang L, Esvelt KM, et al. RNA-guided human genome engineering via Cas9. *Science.* 2013; 339(6121):823-826.
14. Fishman-Lobell J, Rudin N, Haber JE. Two alternative pathways of double-strand break repair that are kinetically separable and independently modulated. *Mol Cell Biol.* 1992;12(3):1292-1303.
15. Auer TO, Del Bene F. CRISPR/Cas9 and TALEN-mediated knock-in approaches in zebrafish. *Methods.* 2014; 69(2):142-150.
16. Ran FA, Hsu PD, Wright J, Agarwala V, Scott DA, Zhang F. Genome engineering using the CRISPR-Cas9 system. *Nat Protoc.* 2013;8(11):2281-2308.

17. Beumer KJ, Carroll D. Targeted genome engineering techniques in Drosophila. *Methods*. 2014;68(1):29-37.
18. Shin JW, Lee JM. The prospects of CRISPR-based genome engineering in the treatment of neurodegenerative disorders. *Ther Adv Neurol Disord*. 2018;11.
19. Maresca M, Lin VG, Guo N, Yang Y. Obligate ligation-gated recombination (ObLiGaRe): custom-designed nuclease-mediated targeted integration through nonhomologous end joining. *Genome Res*. 2013;23(3):539-546.
20. Zetsche B, Gootenberg JS, Abudayyeh OO, et al. Cpf1 is a single RNA-guided endonuclease of a class 2 CRISPR-Cas system. *Cell*. 2015;163(3):759-771.
21. Platt RJ, Chen S, Zhou Y, et al. CRISPR-Cas9 knockin mice for genome editing and cancer modeling. *Cell*. 2014; 159(2):440-455.
22. Liu Y, Ma S, Wang X, et al. Highly efficient multiplex targeted mutagenesis and genomic structure variation in *Bombyx mori* cells using CRISPR/Cas9. *Insect Biochem Mol Biol*. 2014;49:35-42.
23. Zhou J, Shen B, Zhang W, et al. One-step generation of different immunodeficient mice with multiple gene modifications by CRISPR/Cas9 mediated genome engineering. *Int J Biochem Cell Biol*. 2014;46:49-55.
24. Shalem O, Sanjana NE, Hartenian E, et al. Genome-scale CRISPR-Cas9 knockout screening in human cells. *Science*. 2014;343(6166):84-87.
25. Gilbert LA, Larson MH, Morsut L, et al. CRISPR-mediated modular RNA-guided regulation of transcription in eukaryotes. *Cell*. 2013;154(2):442-451.
26. Cheng AW, Wang H, Yang H, et al. Multiplexed activation of endogenous genes by CRISPR-on, an RNA-guided transcriptional activator system. *Cell Res*. 2013;23(10):1163-1171.
27. Maeder ML, Linder SJ, Cascio VM, Fu Y, Ho QH, Joung JK. CRISPR RNA-guided activation of endogenous human genes. *Nat Methods*. 2013;10(10):977-979.
28. Perez-Pinera P, Kocak DD, Vockley CM, et al. RNA-guided gene activation by CRISPR-Cas9-based transcription factors. *Nat Methods*. 2013;10(10):973-976.
29. Tanenbaum ME, Gilbert LA, Qi LS, Weissman JS, Vale RD. A protein-tagging system for signal amplification in gene expression and fluorescence imaging. *Cell*. 2014;159(3):635-646.
30. Hu J, Lei Y, Wong WK, et al. Direct activation of human and mouse Oct4 genes using engineered TALE and Cas9 transcription factors. *Nucleic Acids Res*. 2014;42(7):4375-4390.
31. Gilbert LA, Horlbeck MA, Adamson B, et al. Genome-scale CRISPR-mediated control of gene repression and activation. *Cell*. 2014;159(3):647-661.
32. Konermann S, Brigham MD, Trevino AE, et al. Genome-scale transcriptional activation by an engineered CRISPR-Cas9 complex. *Nature*. 2015;517(7536):583-588.
33. Ran FA, Cong L, Yan WX, et al. In vivo genome editing using *Staphylococcus aureus* Cas9. *Nature*. 2015; 520(7546):186-191.
34. Staahl BT, Benekareddy M, Coulon-Bainier C, et al. Efficient genome editing in the mouse brain by local delivery of engineered Cas9 ribonucleoprotein complexes. *Nat Biotechnol*. 2017;35(5):431-434.
35. Gao X, Tao Y, Lamas V, et al. Treatment of autosomal dominant hearing loss by in vivo delivery of genome editing agents. *Nature*. 2018;553(7687):217-221.
36. Carlson DF, Fahrenkrug SC, Hackett PB. Targeting DNA with fingers and TALENs. *Mol Ther Nucleic Acids*. 2012;1.
37. Ansai S, Kinoshita M. Targeted mutagenesis using CRISPR/Cas system in medaka. *Biol Open*. 2014;3(5):362-371.
38. Heidenreich M, Zhang F. Applications of CRISPR-Cas systems in neuroscience. *Nat Rev Neurosci*. 2016; 17(1):36-44.
39. Southwell AL, Skotte NH, Kordasiewicz HB, et al. In vivo evaluation of candidate allele-specific mutant huntingtin gene silencing antisense oligonucleotides. *Mol Ther*. 2014;22(12):2093-2106.
40. Wu X, Kriz AJ, Sharp PA. Target specificity of the CRISPR-Cas9 system. *Quant Biol*. 2014;2(2):59-70.
41. Wei C, Liu J, Yu Z, Zhang B, Gao G, Jiao R. TALEN or Cas9—rapid, efficient and specific choices for genome modifications. *J Genet Genomics*. 2013;40(6):281-289.
42. Fu Y, Sander JD, Reyon D, Cascio VM, Joung JK. Improving CRISPR-Cas nuclease specificity using truncated guide RNAs. *Nat Biotechnol*. 2014;32(3):279-284.
43. Kolli N, Lu M, Maiti P, Rossignol J, Dunbar GL. Application of the gene editing tool, CRISPR-Cas9, for treating neurodegenerative diseases. *Neurochem Int*. 2018;112:187-196.
44. Kim H, Kim JS. A guide to genome engineering with programmable nucleases. *Nat Rev Genet*. 2014;15(5):321-334.
45. Ran FA, Hsu PD, Lin CY, et al. Double nicking by RNA-guided CRISPR Cas9 for enhanced genome editing specificity. *Cell*. 2013;154(6):1380-1389.

46. Kleinstiver BP, Pattanayak V, Prew MS, et al. High-fidelity CRISPR-Cas9 nucleases with no detectable genome-wide off-target effects. *Nature*. 2016;529(7587):490-495.
47. Slaymaker IM, Gao L, Zetsche B, Scott DA, Yan WX, Zhang F. Rationally engineered Cas9 nucleases with improved specificity. *Science*. 2016;351(6268):84-88.
48. Hu JH, Miller SM, Geurts MH, et al. Evolved Cas9 variants with broad PAM compatibility and high DNA specificity. *Nature*. 2018;556(7699):57-63.
49. Rahdar M, McMahon MA, Prakash TP, Swayze EE, Bennett CF, Cleveland DW. Synthetic CRISPR RNA-Cas9-guided genome editing in human cells. *Proc Natl Acad Sci U S A*. 2015;112(51):E7110-E7117.
50. Truong DJ, Kuhner K, Kuhn R, et al. Development of an intein-mediated split-Cas9 system for gene therapy. *Nucleic Acids Res*. 2015;43(13):6450-6458.
51. Wright AV, Sternberg SH, Taylor DW, et al. Rational design of a split-Cas9 enzyme complex. *Proc Natl Acad Sci U S A*. 2015;112(10):2984-2989.
52. Zetsche B, Volz SE, Zhang F. A split-Cas9 architecture for inducible genome editing and transcription modulation. *Nat Biotechnol*. 2015;33(2):139-142.
53. Richter F, Fonfara I, Gelfert R, Nack J, Charpentier E, Moglich A. Switchable Cas9. *Curr Opin Biotechnol*. 2017;48:119-126.
54. Dai WJ, Zhu LY, Yan ZY, Xu Y, Wang QL, Lu XJ. CRISPR-Cas9 for in vivo gene therapy: promise and hurdles. *Mol Ther Nucleic Acids*. 2016;5:e349.
55. Anheim M, Tranchant C, Koenig M. The autosomal recessive cerebellar ataxias. *N Engl J Med*. 2012;366(7):636-646.
56. He X, Tan C, Wang F, et al. Knock-in of large reporter genes in human cells via CRISPR/Cas9-induced homology-dependent and independent DNA repair. *Nucleic Acids Res*. 2016;44(9):e85.
57. Gaudelli NM, Komor AC, Rees HA, et al. Programmable base editing of A*T to G*C in genomic DNA without DNA cleavage. *Nature*. 2017;551(7681):464-471.
58. Straub C, Granger AJ, Saulnier JL, Sabatini BL. CRISPR/Cas9-mediated gene knock-down in post-mitotic neurons. *PLoS One*. 2014;9(8).
59. Swiech L, Heidenreich M, Banerjee A, et al. In vivo interrogation of gene function in the mammalian brain using CRISPR-Cas9. *Nat Biotechnol*. 2015;33(1):102-106.
60. McCarroll SA, Hyman SE. Progress in the genetics of polygenic brain disorders: significant new challenges for neurobiology. *Neuron*. 2013;80(3):578-587.
61. Tabebordbar M, Zhu K, Cheng JKW, et al. In vivo gene editing in dystrophic mouse muscle and muscle stem cells. *Science*. 2016;351(6271):407-411.
62. Nelson CE, Hakim CH, Ousterout DG, et al. In vivo genome editing improves muscle function in a mouse model of Duchenne muscular dystrophy. *Science*. 2016;351(6271):403-407.
63. Long C, Amoasii L, Mireault AA, et al. Postnatal genome editing partially restores dystrophin expression in a mouse model of muscular dystrophy. *Science*. 2016;351(6271):400-403.
64. Yang Y, Wang L, Bell P, et al. A dual AAV system enables the Cas9-mediated correction of a metabolic liver disease in newborn mice. *Nat Biotechnol*. 2016;34(3):334-338.
65. Yin H, Song CQ, Dorkin JR, et al. Therapeutic genome editing by combined viral and non-viral delivery of CRISPR system components in vivo. *Nat Biotechnol*. 2016;34(3):328-333.
66. Gyorgy B, Loov C, Zaborowski MP, et al. CRISPR/Cas9 mediated disruption of the swedish APP allele as a therapeutic approach for early-onset Alzheimer's disease. *Mol Ther Nucleic Acids*. 2018;11:429-440.
67. Poon A, Schmid B, Pires C, et al. Generation of a gene-corrected isogenic control hiPSC line derived from a familial Alzheimer's disease patient carrying a L150P mutation in presenilin 1. *Stem Cell Res*. 2016;17(3):466-469.
68. Sundal C, Fujioka S, Uitti RJ, Wszolek ZK. Autosomal dominant Parkinson's disease. *Parkinsonism Relat Disord*. 2012;18(suppl 1):S7-10.
69. Heman-Ackah SM, Bassett AR, Wood MJ. Precision modulation of neurodegenerative disease-related gene expression in human iPSC-derived neurons. *Sci Rep*. 2016;6.
70. Long C, McAnally JR, Shelton JM, Mireault AA, Bassel-Duby R, Olson EN. Prevention of muscular dystrophy in mice by CRISPR/Cas9-mediated editing of germline DNA. *Science*. 2014;345(6201):1184-1188.
71. Ousterout DG, Kabadi AM, Thakore PI, Majoros WH, Reddy TE, Gersbach CA. Multiplex CRISPR/Cas9-based genome editing for correction of dystrophin mutations that cause Duchenne muscular dystrophy. *Nat Commun*. 2015;6:6244.
72. Cinesi C, Aeschbach L, Yang B, Dion V. Contracting CAG/CTG repeats using the CRISPR-Cas9 nickase. *Nat Commun*. 2016;7.

73. An MC, Zhang N, Scott G, et al. Genetic correction of Huntington's disease phenotypes in induced pluripotent stem cells. *Cell Stem Cell*. 2012;11(2):253-263.
74. Kolli N, Lu M, Maiti P, Rossignol J, Dunbar GL. CRISPR-Cas9 mediated gene-silencing of the mutant huntingtin gene in an in vitro model of Huntington's disease. *Int J Mol Sci*. 2017;18(4).
75. Shin JW, Kim KH, Chao MJ, et al. Permanent inactivation of Huntington's disease mutation by personalized allele-specific CRISPR/Cas9. *Hum Mol Genet*. 2016;25(20):4566-4576.
76. Monteys AM, Ebanks SA, Keiser MS, Davidson BL. CRISPR/Cas9 editing of the mutant huntingtin allele in vitro and in vivo. *Mol Ther*. 2017;25(1):12-23.
77. Yang S, Chang R, Yang H, et al. CRISPR/Cas9-mediated gene editing ameliorates neurotoxicity in mouse model of Huntington's disease. *J Clin Invest*. 2017;127(7):2719-2724.
78. Merienne N, Vachey G, de Longprez L, et al. The self-inactivating KamiCas9 system for the editing of CNS disease genes. *Cell Rep*. 2017;20(12):2980-2991.
79. DeJesus-Hernandez M, Mackenzie IR, Boeve BF, et al. Expanded GGGGCC hexanucleotide repeat in noncoding region of C9ORF72 causes chromosome 9p-linked FTD and ALS. *Neuron*. 2011;72(2):245-256.
80. Renton AE, Majounie E, Waite A, et al. A hexanucleotide repeat expansion in C9ORF72 is the cause of chromosome 9p21-linked ALS-FTD. *Neuron*. 2011;72(2):257-268.
81. Nimsanor N, Kitiyanant N, Poulsen U, et al. Generation of an isogenic, gene-corrected iPSC line from a symptomatic 57-year-old female patient with frontotemporal dementia caused by a P301L mutation in the microtubule associated protein tau (MAPT) gene. *Stem Cell Res*. 2016;17(3):556-559.
82. Wang L, Yi F, Fu L, et al. CRISPR/Cas9-mediated targeted gene correction in amyotrophic lateral sclerosis patient iPSCs. *Protein Cell*. 2017;8(5):365-378.
83. Gaj T, Ojala DS, Ekman FK, Byrne LC, Limsirichai P, Schaffer DV. In vivo genome editing improves motor function and extends survival in a mouse model of ALS. *Sci Adv*. 2017;3(12):eaar3952.
84. Park CY, Halevy T, Lee DR, et al. Reversion of FMR1 methylation and silencing by editing the triplet repeats in fragile X iPSC-derived neurons. *Cell Rep*. 2015;13(2):234-241.
85. Xie N, Gong H, Suhl JA, Chopra P, Wang T, Warren ST. Reactivation of FMR1 by CRISPR/Cas9-mediated deletion of the expanded CGG-repeat of the fragile X chromosome. *PLoS One*. 2016;11(10).
86. Liu XS, Wu H, Krzisch M, et al. Rescue of fragile X Syndrome neurons by DNA methylation editing of the FMR1 gene. *Cell*. 2018;172(5):979-992.e976.
87. Abudayyeh OO, Gootenberg JS, Essletzbichler P, et al. RNA targeting with CRISPR-Cas13. *Nature*. 2017; 550(7675):280-284.
88. Cox DBT, Gootenberg JS, Abudayyeh OO, et al. RNA editing with CRISPR-Cas13. *Science*. 2017; 358(6366):1019-1027.
89. Nelles DA, Fang MY, O'Connell MR, et al. Programmable RNA tracking in live cells with CRISPR/Cas9. *Cell*. 2016;165(2):488-496.
90. Batra R, Nelles DA, Pirie E, et al. Elimination of toxic microsatellite repeat expansion RNA by RNA-targeting Cas9. *Cell*. 2017;170(5):899-912 e810.
91. Zheng Y, Shen W, Zhang J, et al. CRISPR interference-based specific and efficient gene inactivation in the brain. *Nat Neurosci*. 2018;21(3):447-454.

SECTION III

CLINICAL APPLICATION OF NERVOUS SYSTEM DRUG DELIVERY

CHAPTER

19

Drug- and Disease-Specific Paradigms for Drug Delivery to the Central Nervous System

Kristin Huntoon, Russell R. Lonser, James B. Elder

Department of Neurological Surgery, The Ohio State University Wexner Medical Center, Columbus, OH, United States

CHALLENGES OF DELIVERING DRUGS TO THE CENTRAL NERVOUS SYSTEM

A major challenge in treating diseases of the central nervous system (CNS) is effective targeting and delivery of therapeutic agents into the brain, the spinal cord, or both. Physiologic obstacles include the blood–brain barrier (BBB) and the blood–cerebrospinal fluid (CSF) barrier. Multiple strategies are being investigated to overcome these hurdles. This chapter provides an overview of novel drug-delivery strategies as they relate to various CNS pathologies.

THE BLOOD–BRAIN BARRIER

The BBB is the physical barrier that separates brain interstitial fluid from the intravascular blood compartment and thus is vital in regulating the cerebral microenvironment. It comprises a microvascular endothelium, extracellular matrix, and neuroglial structures (i.e., astrocytes, pericytes, neurons, and microglia). The endothelial cells of the brain capillaries are connected by tight junctions (TJs). Lacking fenestrations, these cells restrict paracellular transport of large hydrophilic compounds. They also exhibit very low pinocytotic activity and endosomal transport. This leads to tightly regulated and limited transcellular transport of substances needed for brain functioning. The TJs are thought to be the major components

Nervous System Drug Delivery
https://doi.org/10.1016/B978-0-12-813997-4.00019-0

© 2019 Elsevier Inc. All rights reserved.

of the BBB that determine its permeability to hydrophilic molecules. The backbone of the TJ has several adhesive molecules, such as zonaoccludin-1 (ZO-1), ZO-2, ZO-3, occludin, claudin, and junctional adhesion molecules.[1]

The permeability of the BBB is also related to its high electrical resistance across the cell membrane (1000-8000 Ωcm^2).[2] The luminal surface of the endothelium presents a net negative charge, thereby creating an additional barrier to polar and ionic substances. This barrier also contains peptidases that inactivate proteins, as well as intracellular enzymes that inactivate neuroactive and neurotoxic substances. Interestingly, in comparison to their systemic counterpart, cerebral endothelial cells also possess a greater concentration of mitochondria to provide the energy needed for the maintenance of the barrier integrity as well as active transport.

Small and lipophilic molecules (O_2, CO_2, and ethanol) can diffuse freely across the BBB; generally, molecules of lipophilicity of ~1.1-1.6 Log$P$ and molecular weight of up to 180 Da are capable of crossing the BBB. Larger hydrophilic molecules require active transcellular transport to gain access to the CNS. Molecules with low solubility require facilitated transport, relying on mechanisms such as ion channels, specific transporters, energy-dependent pumps, or receptor-mediated endocytosis for transportation. However, other mechanisms such as plasma protein binding, efflux transporters such as P-glycoprotein, multidrug resistance protein, and organic anion transporters can also limit CNS entry. The BBB prevents up to 98% of all small-molecule therapeutics and essentially 100% of all unmodified large-molecule therapeutics from entering the brain, due to the closely sealed TJs. The final barrier after the microvascular endothelial cells are intervening pericytes, which form a mechanical barrier along with microglia—key immunologic guardians if the BBB is breached.[1]

Due to these barriers, drugs known to be effective for systemic diseases have often failed when used for CNS indications, due to various drug properties such as larger size (150-1200 Da), hydrophilic nature, and/or protein-bound status when delivered systemically. For example, the chemotherapy agent carboplatin is efficacious systemically against lung cancer and in vitro for glioblastoma, but is ineffective in humans with glioblastoma. Additionally, chemotherapy agents are often subjected to the active efflux transport proteins that can be highly expressed by the BBB, resulting in the transport of compounds from the brain back to the blood supply. In fact, most chemotherapy agents have been documented to have low concentrations in the CSF, with relevant permeability reported only for temozolomide, methotrexate, and topotecan.[3]

The Blood–brain Barrier in Disease States

The integrity of the BBB can be altered with age or development of certain diseases. For example, aged mice that underwent traumatic brain injury have been shown to have significantly slower repair of the barrier compared with younger mice.[4] Likewise, in stroke models in which the middle cerebral artery is occluded, mice had slower recovery of barrier integrity.[5] It is believed that there is a biphasic opening of the BBB with stroke or reperfusion injuries, in which the initial opening is due to the activation of MMP-2, while the second opening is due to activation of MMP-9 and MMP-3.[6] Similarly, in early stages of multiple sclerosis, there are increases in the activity of MMPs 1, 2, 3, 7, 9, and 12, which are known to disrupt the BBB and cause increased entry of leukocytes and degradation of myelin.[7]

Several studies argue that lapses in barrier function (i.e., "reversed" transport of A-beta [Aβ] across the BBB) precipitate the development of Alzheimer disease. New literature suggests that faulty BBB clearance of Aβ via LRP-1/RAGE-mediated transport may be the reason that Aβ accumulates in the brain.[8] Polymorphisms in P-gp and glutathione transferase in the BBB have been associated with risk of developing Parkinson disease (PD), suggesting a reduced capacity to detoxify neurotoxins. In patients diagnosed with malignancy of the CNS, the established thinking is that the integrity of the barrier is often compromised within the tumors of the brain and spinal cord. However, this alteration in permeability is highly variable and dependent on tumor type and size. Additionally, the core of a malignant brain tumor is often leaky, whereas the brain adjacent to the tumor has variable barrier integrity. The so-called *sink effect* is yet another mechanism that contributes to the failure of chemotherapy to effectively treat CNS neoplasms. The sink effect refers to the distribution of drug concentration—highest in the center of the lesion, with progressively decreasing concentration toward the periphery of the lesion. These alterations have significant implications for understanding disease pathology and progression as well as for optimizing drug-delivery strategies.

Blood–brain Barrier Modulation

Techniques to modulate BBB permeability have been explored for a variety of clinical applications. These strategies include use of pharmacologic agents such as mannitol, as well as targeted methods such as focused ultrasound and laser interstitial thermal therapy (LITT).

Frequency-tuned electric field therapy (also known as electrochemotherapy) is the transient use of an electric current to open the BBB for chemotherapeutic drugs. This technique uses a sublethal pulsed electric field to disrupt the endothelial membrane, allowing uptake of chemotherapeutic agents. Irreversible electroporation is achieved via the use of a series of electric pulses delivered through needle electrodes; these pulses induce a nonthermal focal ablation to the target. The mechanisms of action of this therapy are twofold: it exerts an antimicrotubule effect, and it creates a nonthermal ablative property of the electric field, which occurs as the pulse frequency increases, causing dielectric losses and developing friction between rapidly oscillating molecules. In vivo studies of two different glioma cell lines showed additional increased cytotoxicity by >50% for 1 ng/mL concentrations of either bleomycin or carboplatin at electric fields ≥400 V/cm.[9]

Laser therapy is an additional technology to modulate BBB permeability within the CNS. LITT is a minimally invasive therapy in which a thin laser probe is inserted into a lesion using magnetic resonance guidance. The laser probe then delivers hyperthermic ablation. LITT has been successfully used for treatment of primary or secondary tumors and deep seizure foci in epilepsy. This technique has also recently been demonstrated to sustain local disruption of the peritumoral BBB. In one case series,[10] LITT was shown to disrupt the BBB in the peritumoral region that extends outward 1-2 cm from the viable tumor rim. The disruption in the BBB persisted in all 14 patients for up to 4 weeks after LITT, as measured quantitatively by dynamic contrast-enhanced magnetic resonance imaging, and up to 6 weeks as measured by serum levels of the brain-specific factor. Importantly, this observation indicates that after LITT treatment, there is a time frame during which enhanced local delivery of therapeutic agents into the desired location, including the peritumoral region, can potentially be achieved.[10]

High-intensity Focused Ultrasound

High-intensity focused ultrasound (HIFU) is a relatively new methodology for disrupting the BBB that can provide a noninvasive, reversible, targetable strategy for drug delivery. For this technique, patients undergo HIFU with magnetic resonance imaging guidance and thermometry, and the drug (e.g., gene therapy-loaded microbubbles) is delivered via targeted intra-arterial injection. The ultrasound exposures used are 10-ms bursts at low-pressure amplitude (<1 MPa), repeated at the frequency of 1 Hz for a duration of 20 to 30 seconds. For some diseases or in certain trials, HIFU is administered in conjunction with injected microbubbles that can enhance the delivery of therapeutic compounds.[11] Microbubbles are made of gas (air or perfluorocarbon) covered by a shell of albumin or lipids, and are administered intra-arterially and intravenously within 90 minutes of HIFU. Their usual diameter (1–10 μm) allows them to pass through the capillary networks of the tissue exposed to the ultrasound, and the ultrasound wave causes microbubbles to expand and contract in the capillaries. The bubble oscillations also prompt the surrounding fluid to flow (a process known as *microstreaming*), thereby creating large shear forces around the bubbles as well. Above a certain threshold, the bubble oscillations become so large that the inertia of the surrounding fluid causes the bubbles to collapse. As the bubble collapses, the nearby vessel wall is punctured by fluid jets and mechanical opening of the TJs.[12]

Studies have shown successful delivery of gene therapy to the CNS by focused ultrasound plasmid-loaded microbubbles.[13,14] Plasmids can enter the cerebral parenchyma largely unhindered while the BBB remains open, but they are rapidly degraded in circulating blood upon release from microbubbles. There are preclinical and Phase I/II trials in place to study the effects on essential tremor, PD, brain tumors (both for ablation and for delivery of therapeutic agents), depression, obsessive–compulsive personality disorder, pain syndrome in pediatric and adult patient populations, epilepsy, Alzheimer disease, and stroke with thrombolysis and intracranial hematoma evacuation.[15]

THE BRAIN–CEREBROSPINAL FLUID BARRIER

The confining membrane of interconnected apical TJs of neuroependymal cells to adjacent CSF-containing spaces forms the brain–CSF barrier. The brain–CSF barrier inhibits paracellular diffusion of water-soluble molecules. The choroid plexus epithelial cells allow the directed transport of ions and nutrients into the CSF and the removal of toxic agents from the CSF.[16] The potential impact of the brain–CSF barrier is important when considering drug delivery via CSF routes, such as with an implanted CSF reservoir or lumbar puncture.

The structural mechanism that accounts for the permeability restriction between CSF and brain extracellular fluid has been suggested to be a membrane specialization (i.e., the "strap" junction), which is present between neuroependymal cells.[17] Foot processes of the pseudostratified neuroependymal layer form the outer surface of the cerebral wall. These foot processes are linked by strap junction-like elements. The ventricular system is the route by which many substances enter the brain, as the brain–CSF barrier is quantitatively leakier than the BBB. The bulk absorption by the arachnoid villi and the selective removal of molecules by the choroid plexus result in a concentration gradient for molecules reaching the interstitial fluid of the brain to diffuse into the CSF. Most of these molecules are then cleared by either bulk flow or active transport. This function of the brain–CSF barrier helps to maintain the low concentration of many substances in both the brain and the CSF, compared with plasma concentrations.

DRUG-DELIVERY ROUTES FOR THE CENTRAL NERVOUS SYSTEM

Intra-arterial Drug Delivery

Intra-arterial injections for various CNS diseases have been attempted since the 1950s.[18] In some disease states, such as treatment of vasospasm associated with aneurysmal subarachnoid hemorrhage or lysis of thrombosis in acute stroke management, intra-arterial therapy is performed without BBB disruption. One of the most profound examples of success with intra-arterial injection without BBB disruption has been the treatment of retinoblastoma.[19] In this procedure, a microcatheter is placed at the ostium of the ophthalmic artery under anesthesia and the drug is delivered selectively into the ophthalmic artery. Since 2006, when ophthalmic artery chemosurgery for retinoblastoma was introduced, enucleation rates have decreased from over 95% to <10% at Memorial Sloan Kettering Cancer Center in New York City.[20] Additionally, despite treating more advanced cases with intra-arterial chemotherapy rather than enucleation, patient survival has not been compromised.[19] However, for many CNS disease states, such as tumor, BBB disruption is performed in conjunction with intra-arterial drug delivery to enhance brain penetration of the therapeutic agent. This is because one of the major hurdles of intra-arterial injections is that, despite localized targeting, the infused agent cannot penetrate the BBB.[21]

Intra-arterial infusion of hyperosmotic agents is one of the most common methods of BBB disruption. Other agents that have been used include arabinose and urea. Mannitol reversibly opens the BBB by transiently shrinking endothelial cells and opening TJs between the cells. This approach temporarily increases the permeability of the BBB to large molecules such as chemotherapy agents, antibodies, and nanoparticles, potentially increasing their concentration in the targeted areas of the brain.[22] This increase in BBB permeabilization is transient, lasting between 30 minutes and 2 hours. During this time, drug permeability between the vascular system and the CNS interstitial fluid is potentially increased. Intra-arterial drug delivery with BBB disruption can lead to increased local exposure of the target tissue to the therapeutic agent, and has been shown to increase the dosage at the target area up to three to fivefold compared to systemic delivery.[21] Some clinical trials have investigated intra-arterial injection with BBB disruption for treatment of diseases of the CNS such as glioblastoma, radiation necrosis, and CNS lymphoma, with limited success.[23] In one example, a large series of patients with newly diagnosed primary CNS lymphoma were treated with intra-arterial methotrexate-based chemotherapy and BBB disruption. Overall survival was 3.1 years (25% estimated survival at 8.5 years), and durable remissions were achieved with stable or improved cognitive function for a median of 12 years (range, 2-26 years) after diagnosis.[24] Conversely, results for intra-arterial drug delivery for glioblastoma have been disappointing. Agents injected using selective intra-arterial injection have included BCNU,[25] carboplatin,[26] temozolomide,[27] and bevacizumab.[28] Examples of intra-arterial delivery used for nonneoplastic CNS disease entities include papaverine for ischemic stroke,[29] nimodipine for vasospasm,[30] and tPA for ischemic stroke.[31] Future clinical studies are needed to better clarify the role for this treatment modality.

Intrathecal/intraventricular Drug Delivery

Like other approaches, the intrathecal/intraventricular route is an alternative way to bypass the physical barriers of the brain. In this method, therapeutic agents are instilled directly into cerebral ventricles, CSF, or both. Due in part to the brain–CSF barrier, this route is perhaps best suited for pathologies such as meningiomas or leptomeningeal disease, as drugs for these conditions are mainly distributed into the ventricles and the subarachnoid areas of the brain.[32] One major advantage of intrathecal/intraventricular delivery over intracerebral delivery is the slower diffusion rate of the agent in the interstitial fluid that surrounds the parenchymal tissue of the brain. The drug therefore achieves a higher concentration in the brain compared to the extravascular concentration. The major disadvantage of the intrathecal/intraventricular approach is the risk of causing a subependymal astrogliatic reaction due to the high drug exposure at the ependymal surface of brain.

The drug concentration in the brain relies on the diffusion phenomenon, which determines how much of the drug penetrates the brain from the administration site. Diffusion decreases logarithmically with each millimeter of brain tissue; that is, from the injection site to the ependymal surface of the brain.[33] In general, slow intraventricular infusion is found to be very effective compared to bolus administration, as the slow intraventricular infusion is associated with sustained administration of the drug.[34] CSF turnover, which depends on the size of the patient and the volume of CSF, also affects drug concentration.[32] The intrathecal/intraventricular routes are now standard for the delivery of morphine for pain, baclofen for severe spasticity, and chemotherapeutics for some types of leptomeningeal carcinomatosis.[35] The intrathecal route has also emerged as a useful—and, in some cases, ideal—route of administration for certain targeted protein therapy for neuropathic lysosomal storage diseases, stroke, and brain metastases.[36] For example, enzymatic activity was detected in neurons and glia in all six neuronal layers of cortex and in white matter in nonhuman primates and beagles that received lumbar intrathecal administration of iduronate-2-sulfatase (i.e., the enzyme that is deficient in mucopolysaccharidosis II) via lumbar puncture.[35]

Direct Injection

The concept of direct injection into a brain lesion or brain target arose in the 1970s, primarily to administer chemotherapy into tumors either stereotactically or to the tumor cavity after surgical resection.[37] In many of these studies, the therapeutic agent evaluated was known to be efficacious in in vitro or in animal models, but poor penetration through the BBB limited the utility of this method in humans. For example, a one-time injection of lomustine into a mouse glioma model was found to be highly effective in increasing long-term survival, and had less systemic toxicity than when injected intraperitoneally.[38] However, in a small series of eight patients with glioblastoma who underwent one-time intracavitary methotrexate injection, no significant change in survival was observed.[37] Fig. 1 shows an example of direct injection for delivery of a therapeutic agent.

Another direct injection method uses osmotic minipumps for local delivery of therapeutic agents. The apparatus relies on an osmotic pressure difference between the local tissue environment and the minipump. This shifts water into the pump, driving the solution from the pump's reservoir. This method can be used to infuse materials into the cerebral ventricles or directly into brain tissue. An early study compared the extent of cellular uptake of 5-[125I] iodo-2′-deoxyuridine ([125I]IUdR) following either a single injection, slow-release PLGA polymer, or osmotic pump delivery in a C6 glioma rat model. This study showed a significantly increased distribution of [125I]IUdR in the target injection site with the Alzet osmotic pump compared to drug distribution via a polymer implant (mean labeling index, 34.3% vs 22.5% polymer implant vs 6.2% single injection).[39] Osmotic pumps are still being optimized and tested in in vivo models for neurological disease, and have not yet been investigated in human clinical trials.[40]

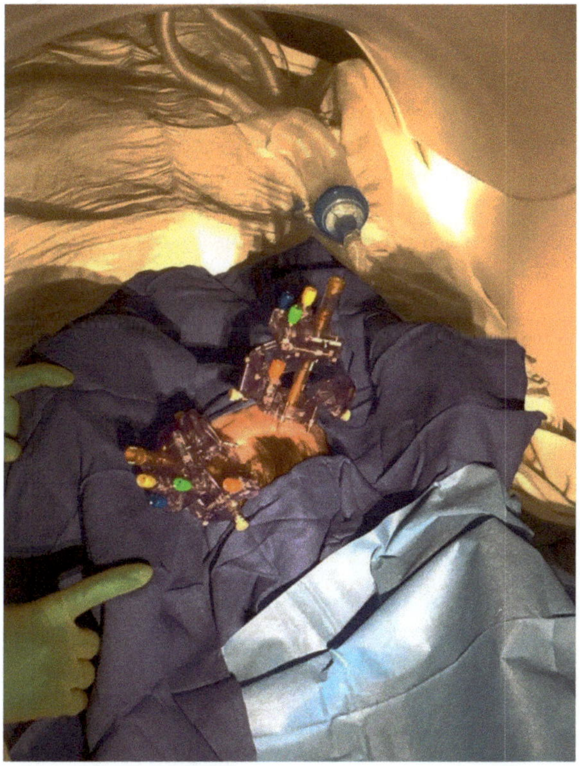

FIG. 1 Intraoperative photograph shows a patient positioned supine in the bore of the intraoperative magnetic resonance (MR) imaging suite. The patient underwent MR-guided injection of a clinical trial agent for recurrent glioblastoma; an MR-compatible scalp mounting integrated head fixation frame was used for injections monitored in real time.

Convection-enhanced Delivery

Since the early 1990s, convection-enhanced delivery (CED) has been used to deliver agents into the brain by direct infusion. This technique involves placement of an infusion catheter into the brain, followed by local delivery of a solute containing a therapeutic agent (Fig. 2). CED involves continuous infusion carried out at a low rate, under positive pressure over a period of time.[41] CED uses a pump connected to one or more catheters placed into a lesion, into the interstitial space of a lesion, or into the walls of a resection cavity. Multiple factors influence the volume of drug distribution with CED, including the catheter shape, size, and placement; rate and volume of infusion; physical properties of the drug (e.g., molecular weight, size, lipophilicity, charge, viscosity); and structural properties of the target tissue (e.g., interstitial fluid pressure within and around the tumor, tissue anisotropy). A major advantage of CED as a drug-delivery mechanism is its ability to bypass the BBB. This allows a more sustained delivery of drugs of varying molecular weights.

Convection-enhanced delivery has been used in humans, primarily in clinical trials for treatment of tumors of the CNS, especially GBM. In these trials, CED has been shown to produce markedly different drug distribution from that observed in normal brain.[42] As mentioned previously, the core of a GBM typically has poor vascularity, surrounded by heterogeneous tissue with rich vascularity and a pathologic interstitial composition. The leaky cytoarchitecture of the center coupled with the outward pressure gradient may result in higher infusate clearance rates from the lesion. Despite this observation, recent studies have shown that CED can deliver therapeutic doses to infiltrating tumor cells at the margin of the tumor.[43]

The first Phase III randomized clinical trial (PRECISE trial) to incorporate CED for treatment of GBM compared the use of carmustine wafers with CED infusion of cintredekin besudotox (IL13PE38QQR) in 296 patients with a first recurrence of GBM. Patients in the CED arm received an infusion starting 24 hours after resection, at a concentration of 0.5 μg/mL and at a total rate of 0.750 mL/hour for 96 hours. This trial did not show significant difference in overall survival outcomes between groups.[44] A number of other clinical trials involving CED for CNS disease states have been performed, with variable levels of success.

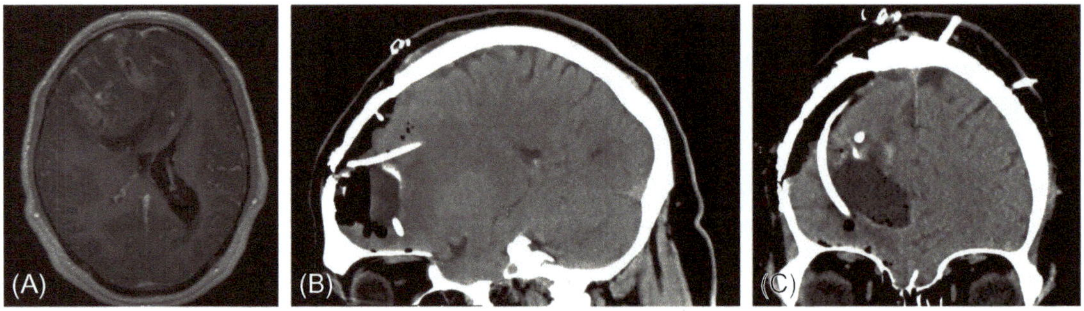

FIG. 2 A 29-year-old man underwent surgical resection of a right frontal glioblastoma multiforme, as well as stereotactic placement of two convection-enhanced delivery (CED) catheters into the brain surrounding the resection cavity. Postoperatively, the catheters were used for 72-hour CED of a clinical trial agent. (A) Preoperative axial magnetic resonance image shows a right frontal glioblastoma multiforme with associated mass effect distorting the frontal horn of the right lateral ventricle. (B) Sagittal and (C) coronal postoperative computed tomography show placement of CED catheters into the parenchyma surrounding the tumor resection cavity.

Many of the early clinical trials that used CED were plagued by underappreciation of lesional tissue characteristics and catheter-related backflow issues. However, changes in infusate and cannula characteristics, development of catheter placement guidelines, mathematical modeling to predict infusate spatial distribution, and monitoring of infusate delivery in real-time are currently being instituted in order to optimize this drug-delivery technique. Most studies have examined the use of CED in tumors, but there is increasing interest in the use of CED for PD (muscimol) and Alzheimer disease (vector-mediated neprilysin).[45]

Dissolvable Wafers

Polymer-based drug-delivery technology has been described for implantable devices and in carriers for systemic administration. Many types of implants are being studied, including wafers, gels, micro- and nanocarriers, and microchips.[46] Interstitial pressure gradients and drug concentration gradients favor the distribution of the agent enclosed in the implanted polymer from the center of the wafer to its periphery (i.e., the surrounding lesion). The controlled release of a therapeutic agent effectively results in a slower elimination from the target tissue and increased time of exposure to the agent. However, the major drawback of this approach is that local penetration of the agent is largely governed by the ability of the drug to diffuse into the local lesion microenvironment. These implants have been used to deliver a very large array of agents, including chemotherapeutic agents such as mitoxantrone, BCNU, 4-HC, paclitaxel, carboplatin, and doxorubicin in brain tumors.[47] BCNU wafers ((1,3-bis[2-chloroethyl]-1-nitrosourea), 3.85% carmustine) are probably the most recognized and most clinically used implantable device in neurosurgical oncology clinical practice. These wafers contain carmustine, a nitrosourea alkylating agent within polifeprosan, a biodegradable copolymer. The carmustine is typically released over a period of 5 days; the implant is completely degraded over 6 to 8 weeks.

Two randomized control trials compared carmustine wafers to placebo treatment in patients with newly diagnosed malignant glioma. Both reported a significant survival benefit for patients who received Gliadel (Eisai Inc. for Arbor Pharmaceuticals) compared with patients in the control groups.[48-50] However, in the case of recurrent malignant glioma, a subsequent randomized controlled trial demonstrated a significant survival benefit for Gliadel only after adjustment for prognostic factors, and a prospective cohort study reported no survival benefit for Gliadel compared with a historical control group.[24] In subsequent studies, median intent-to-treat overall survival was 13.9 months for BCNU wafers vs 11.6 months for placebo wafers ($P = .005$).[51] Another study is currently enrolling 500 patients for 3-year follow-up at over 30 sites, the prospective Vigilant ObservatIon of GlIadeL WAfer ImplaNT (VIGILANT) registry (NCT02684838), which will evaluate BCNU wafers for treatment of CNS malignancies.[52]

Intranasal Delivery

The nasal passageway connects to the brain using extracellular pathways (the olfactory and trigeminal channels) via the olfactory neuroepithelium or respiratory epithelium. The olfactory or trigeminal nerve systems could therefore potentially serve as external access points of the entrance to the brain, and could be exploited to bypass the BBB for intranasal drug delivery. Intranasal delivery of therapeutic agents is gaining popularity, as it is a much less invasive method of drug administration with potential to reduce systemic toxicity. However, some studies have shown poor efficiency of drug delivery using this method, with <0.1% of the drugs in the nasal cavity actually reaching the brain.[53,54] Additionally, due to the rapid clearance of molecules in the nasal cavity, retention at the nasal mucosa is limited. However, drug accumulation in the brain has been shown to spike within the first few minutes of intranasal delivery, compared with the sustained accumulation observed after oral delivery.[55,56] This may be beneficial in patients who require a quick burst of drug release (i.e., within minutes).Therefore, in an effort to increase brain concentrations and efficacy, nanoparticle and bioadhesion technologies have been explored in combination with intranasal delivery.

Intranasal agents have been successful in many studies, including neurotrophins (NGF and insulin-like growth factor [IGF]-1); neuropeptides (hypocretin-1 and exendin); cytokines (interferon β-1b and erythropoietin); polynucleotides (DNA plasmids and genes); and small molecules (chemotherapeutics and carbamazepine). Administration of intranasal neurotrophins provides a reliable mechanism to target neurotrophins to the CNS to treat neurodegenerative diseases. In vivo models of Alzheimer disease in which intranasal NGF was administered showed reduced neurodegeneration and improved performance in memory tasks.[57] In a rodent stroke model, intranasal IGF-1 administration correlated with reduced brain damage and neurologic deficits.[58] Clinical trials are underway for Alzheimer disease, cognitive impairment, attention deficit hyperactivity disorder, depression, and GBM.[59]

DRUG-BASED PARADIGMS

Chemotherapy

As noted previously, many forms of chemotherapy are barred from the CNS by the BBB. There are, however, a few classes of antiproliferative drugs called nitrosoureas (e.g., carmustine and lomustine) and alkylating agents (e.g., temozolomide) that have been shown to have some ability to cross the BBB and have been used clinically.[47] Researchers have developed several methods to circumvent the challenges of permeability, including chemical modification of existing drugs, production of a prodrug (e.g., cyclodextrin complexes, antibody-directed enzyme prodrug therapy, and biotin-avidin conjugated system),[32] receptor-mediated transportation of the agent, and coadministration of agents inhibiting drug efflux.[60] Drug delivery paradigms that directly introduce the agent into the brain, such as CED and direct injection (discussed in detail above and in subsequent chapters), are under active investigation. Ultimately, effective chemotherapy agents that are unable to cross the BBB via systemic delivery represent one of the challenges that investigators are attempting to address with technological innovations in drug delivery.

Immunotherapy

Immunotherapeutics were first used to treat malignancies in the early 1900s. The use of antibodies to target cancer was proposed by Ehrlich, who purported that they offered more specificity and less toxicity compared to the chemotherapy of the time.[61] Shortly thereafter, in 1904, Julius Donath and Karl Landsteiner reported the first concrete evidence that autoantibodies can cause disease; they did this by showing that autoantibodies caused paroxysmal cold hemoglobinuria.[62] The proof of concept was in the 1970s, when Moolten and Cooperband prepared a conjugation of antileukemia immunoglobins and diphtheria toxin with a specific toxicity against leukemia cells; the resulting product was termed an *immunotoxin*.[63] Since then, the number of immunotherapy options and indications for use has increased significantly.

In terms of cancer immunotherapy, several classes are in use or under further investigation. Many types of immunotherapy, such as those listed in Table 1, do not involve specialized CNS drug-delivery strategies. Instead, existing properties of the immune system are modulated, and access to the CNS remains governed by the properties of the BBB.[64,65] The ability of different types of immunotherapy to circumvent the BBB is the subject of ongoing investigation. Some forms of immunotherapy make use of drug-delivery techniques to access the CNS, such as oncolytic virus and viral vector systems. Oncolytic viruses (with up to 46% of the

TABLE 1 Types of Cancer Immunotherapy

Class of therapy	Example
Adjuvant therapy	Recombinant Listeria; STING agonists; TLR agonists
Adoptive T cell therapy	CAR T cells; TCR transgenic T cells
Altered peptide ligands	Glatiramer acetate for multiple sclerosis
Cell therapy	Treg cells for several neurological diseases
Checkpoint blockade therapy	Antagonistic antibody to CTLA-4 or PD-1
Cytokine therapy	IL-2, IFN-g, IL-15, IL-18, TNF-α
DNA vaccination	Myelin basic protein vaccination, T-cell receptor vaccination for multiple sclerosis
Gene therapy	Cytokine and peptides gene therapy for multiple sclerosis or trophic factor gene therapy for Parkinson disease and Huntington disease
Immunosuppression	Anti-VLA4 for multiple sclerosis, anti-amyloid β in Alzheimer disease
Macrophage activation	CD40 agonists; CD47 antagonists
NK cell therapy	NK cell lines; ex vivo expanded NK cells
Oncolytic virus therapy	Engineered HSV, measles virus, poliovirus
Recombinant protein	Interferon-beta for multiple sclerosis, neurotrophic factors for neurodegenerative diseases
RNA interference	Interfering pro-inflammatory cytokines
Vaccines: therapeutic	Amyloid β or Tau vaccination for Alzheimer disease, synapsin vaccination for Parkinson disease or bacille Calmette-Guerin vaccine in multiple sclerosis

Abbreviations: CAR, *chimeric antigen receptor;* CD, *cluster of differentiation;* CTLA-4, *cytotoxic T-lymphocyte-associated protein 4;* HSV, *herpes simplex virus;* IFN, *interferon;* IL, *interleukin;* NK, *natural killer;* PD-1, *programmed cell death protein 1;* STING, *stimulator of interferon genes;* TCR, *T-cell receptor;* TNF-α, *tumor necrosis factor alpha;* Treg, *regulatory T cell;* VLA4, *very late antigen-4.*

administered concentration within the lesion) have been shown to cross the BBB when delivered systemically; however, many current oncolytic virus therapy trials use direct-injection techniques into the tumor and surrounding brain to circumvent the BBB. For example, Delta-24-RGD (DNX-2401) is a tumor-selective, replication-competent oncolytic adenovirus, which was used in a recent dose-escalation, biologic-end-point clinical trial involving 37 patients with recurrent malignant glioma. Intratumoral injection of DNX-2401 was shown to improve survival >3 years from the time of treatment in 20% of patients, and 3/25 patients (12.0%) had >95% reduction in the enhancing tumor.[66,67] However, the BBB integrity is a crucial limiting factor for autologous cell delivery (adoptive T-cell therapy [e.g., chimeric antigen receptor (CAR) T cells; TCR transgenic T cells]), monoclonal antibodies, and checkpoint inhibitors. Thus, many of these agents are being engineered with liposomes and nanoparticles to facilitate BBB penetration.[68] Additionally, as will be discussed in later sections, many drug delivery systems (i.e., viral-based gene therapy) can activate immune responses.

Drug-coupling Agents

Nanoparticles

Nanoparticles are promising new instruments of noninvasive delivery systems. They enhance the permeability of therapeutic agents across the BBB, carry intracellular drugs, and increase cargo half-life. Nanoparticles have attracted much interest in the field of brain tumor drug delivery due to their proven ability to take full advantage of enhanced permeation and retention effects. This effect is especially noted in high-grade gliomas, in which high rates of angiogenesis produce leaky vasculature. This increased tumor permeability (compared with the permeability of normal tissue) allows the uptake of nanoparticles measuring between ~20 and ~100 nm, which are then retained in the tumor tissue rather than being returned to the systemic circulation. Lastly, their degree of surface coating or area per mass increases the dosage of the targeted agent delivered to lesional tissue. Currently, early preclinical studies in brain tumor models have shown superior efficacy of different therapeutic agents such temozolomide,[69] camptothecin[70], dithiazanine iodide, doxorubicin,[71] and aclarubicin and paclitaxel[72] when delivered in polymer poly(lactic-*co*-glycolic acid) (PLGA) nanoparticle formulations compared to free drugs.[73]

Nanoparticles or nanocarriers can be fabricated using a wide range of synthetic materials. Biocompatible and hydrolytically degradable polymers such as poly(glycolic acid) (PGA), poly(lactic acid) (PLA), PLGA, chitosan, Poly(beta-amino) esters (PBAE), and poly(amidoamines) have been used to deliver several therapeutic small macromolecules, such as proteins and nucleic acids. Polymers with a positive charge are being studied particularly for delivery of nucleic acids because their cationic character allows them to form nanocomplexes with negatively charged DNA and RNA.[74]

Nanoparticles that exhibit magnetic properties are called magnetic nanoparticles (MNPs) and are useful in their ability to create temporary pores in cell membranes (i.e., the BBB endothelium) to improve targeting and delivery of agents. They are often administered systemically, and their characteristics (i.e., size, shape, surface charge, coating, and magnetic properties) are specifically tailored for the planned clinical application and target. The size, shape, and coating can alter magnetic forces and resistance by order of magnitude:

Exceedingly small particles experience insufficient magnetic force; if they are overly large, they encounter too much resistance in the tissue. In general, the field strength should be 200-700 mT, with gradients of about 8-100 T/m, depending on blood flow (rates ranging from 10 cm/s in arteries to 0.05 cm/s in capillaries). In the human brain, the working distance (up to 30-50 cm from the magnet surface) and the safety restrictions of the magnetic field strength (8 T adults, 4 T children) have limited the applications of MNPs in clinical settings.[75] Some groups are attempting a combination of effects with nanoparticles; Tabatabaei et al.[76] provided preliminary evidence that magnetic heating of MNPs with a low radiofrequency source can transiently increase BBB permeability. This mechanism capitalizes on the low radiofrequency field magnetically exciting MNPs to release energy in the form of heat to their surroundings, a mechanism termed *Neel relaxation*.[76]

Liposomes

Lipid-based nanoparticles known as liposomes have attracted attention as drug carriers for a range of applications due to their biocompatibility and efficient packaging of drugs. Liposomes have a fluid lipid bilayer formulation with a hydrophilic core, and they often exhibit similar structures to a cell membrane but in the form of a sphere, encapsulating hydrophobic drugs with controlled-release capabilities. Liposomes also have good biocompatibility, low toxicity, and commercial availability. Additionally, liposomal encapsulation allows for a prolonged half-life for compounds while changes in size and surface modification by PEGylation can be used to further improve the compound's circulation and availability. The major disadvantage associated with liposomes is their relatively larger size compared to nanoparticles or other carriers.[77]

Liposomes have the potential to significantly improve CNS drug delivery, especially for the treatment of GBM. Compounds such as irinotecan, doxorubicin, and vincristine—formulations of chemotherapy agents that would not cross the BBB if unaltered—may have increased efficacy if delivered via liposomes.[78] Some preclinical data suggest that with liposomal formulation, treatment with those chemotherapy agents resulted in changes in tumor size as well as tumor vascular morphology and function in GBM grown subcutaneously and orthotopically, demonstrating the ability of liposomal formulations to circumvent the BBB, specifically by preventing drug efflux.[78] In some nontumor models, such as a transient middle cerebral artery occlusion rat model of ischemia, liposomes carrying asialoerythropoietin accumulated over three times greater in ischemic regions of the brain compared to controls. However, these liposomes also accumulated over twofold in nonischemic brain regions and in the liver.[79] In the realm of infectious disease, cytarabine liposomal formulation, Triolein, dipalmitoylphosphatidylcholine (DOPC), and 1,2-dipalmitoyl-*sn*-glycero-3-phosphorylglycerol sodium salt (DPPG) are used clinically to treat neoplastic meningitis and lymphomatous meningitis.[80]

Gene Therapy: Systemic and Direct Injection

The concept of gene therapy initially arose with the development of gene transfer technology in bacteria and later, with gene transfer into animals and humans using viral vectors and genetically modified cultured cells. Cloned genes became available in the 1970s and were used to demonstrate that foreign genes could indeed correct genetic defects and disease phenotypes in mammalian cells in vitro. Various viral vectors have been used and optimized for gene transfer and have demonstrated that gene therapy could have clinical efficacy.[81] Gene therapy offers some unique benefits, including the potential to correct disease-causing DNA mutations, eliminate toxic RNA/proteins, and increase the expression of therapeutic proteins. Moreover, it can ameliorate some aspects of debilitating neurological diseases, thereby providing a platform for functional recovery.[82]

Viral Vectors

In the early 2000s, Foust et al.[83] reported that adeno-associated virus serotype 9 (AAV9) vectors could cross the intact BBB in neonates and adult mice with high transmission. Although the rate of transmission depends on the strength of the promoter used, approximately 64% of glial fibrillary acidic protein (GFAP)-labeled astrocytes were green fluorescent protein (GFP)-positive in the lumbar spinal cord after intravenous administration of a high dose of 4×10^{12} genome copies of self-complementary AAV9-GFP.[83] Other vectors (e.g., adenovirus, lentivirus, and herpes simplex virus) exhibit local transduction around the local tract to varying degrees of efficiency both clinically and in in vivo studies.[74]

New advances in viral design have allowed researchers to produce two AAV variants: one that efficiently ferries genetic cargo past the BBB, and another that is efficiently picked up by peripheral neurons that reside outside the brain and spinal cord. Intravenous administration of 1×10^{11} vector genomes of AAV-PHP.eB transduced 69% of cortical and 55% of striatal neurons, while 1×10^{12} vg of AAV-PHP·S transduced 82% of dorsal root ganglion neurons, as well as cardiac and enteric neurons.[84] As of July 2018, a total of 121 clinical trials are listed on clinicaltrials.gov for the treatment of diseases of the CNS.

Nonviral Gene Therapy

Nonviral gene therapy is a way to transfer genetic material to target tissue without a viral backbone. Thus, these are usually naked DNA, particle-based and chemical-based gene therapy strategies. They are directly administered via a plasmid DNA or DNA fragment (i.e., naked DNA). Nonviral gene therapy has some advantages over viral-based gene therapy, including that nonviral gene therapy has no preexisting immunity to nonviral delivery systems. Moreover, they are not derived from pathogens and therefore have fewer safety concerns than viral vectors (e.g., replication competency and insertional oncogenesis). Unfortunately, nonviral systems are less efficient at in vivo gene delivery than their virus-based counterparts. Nonviral gene therapy includes delivery vehicles such as nanoparticles, liposomes, dendrimers, carbon nanotubes, MNPs, and cationic polymers.[85]

Extracellular vesicles are another field of neurotherapeutics for nonviral delivery of gene therapy. They are naturally secreted, membrane-encompassed structures that encapsulate many types of membrane particles shed from cells into the extracellular environment, including microparticles, microvesicles, exosomes, and apoptotic bodies. Extracellular vesicles can deliver proteins, RNA, miRNA, and DNA to neighboring cells. Their size (generally ranging from 50 to 200 nm), cell surface receptor expression, and ability to package nucleic acids and proteins makes these extracellular vesicles analogous to endogenous nanoparticles.[86] Extracellular vesicles have the advantages of originating from endogenous vesicles, efficient cell uptake, and low inherent toxicity. However, they are associated with disadvantages such as the low yields of current isolation techniques, no optimal purification method, and expense. In one preclinical example, vexosomes (i.e., extracellular vesicles with AAV capsid) have been shown to be a gene-delivery tool for ganglion cells, bipolar cells, Müller cells, and photoreceptors upon intravitreal injection in mice.[87]

Cell-based Gene Therapy

Cell-based gene therapy offers an increasingly promising delivery mechanism for targeted gene therapy. In terms of the CNS, cell-based gene therapy has primarily been used with neural stem cells and mesenchymal stem cells. Neural stem cells and mesenchymal stem cells have the innate ability to home to regions of the CNS that have pathology, such as genetic neurodegenerative disease, lysosomal storage disease, and metabolic disorders; this endogenously engineered property is an added advantage of this as a delivery system for neurotherapeutics. Interestingly, mesenchymal stem cells have immunosuppressive properties that lead to a more stable engraftment than neural stem cells.[88] Additionally, in the case of adoptive cell therapy, lymphocytes or peripheral blood mononuclear cells specific to tumor-associated antigens are expanded ex vivo and reintroduced to the patient. One type of adoptive transfer currently under investigation is CAR T-cell therapy, which has shown promise in directing antitumor immune responses.[89]

DISEASE-SPECIFIC PARADIGMS

Glioblastoma Multiforme

Many drug-delivery techniques have been investigated in targeting GBM, with varying degrees of success. These include CED, direct injection of oncolytic virus and gene therapy, immunotherapy, nanoparticles, or a combination of these therapies. As mentioned previously, oncolytic virus therapy is under investigation and is currently being used for GBM treatment. For example, a Phase I trial for patients with recurrent GBM involved inoculation with a single dose of oncolytic PVSRIPO virus delivered intratumorally via CED. PVSRIPO is a genetically engineered, nonpathogenic poliovirus:rhinovirus chimera. PVSRIPO naturally infects cancer cells due to the presence of poliovirus receptor (CD155/Necl-5), and the infection stimulates the host immune system to destroy other tumor cells. Safe and appreciable efficacy of the treatment was observed, and the overall survival rate was extended in half of treated patients.[90]

Another group has used a modified adenovirus Delta-24-RGD (Arg-Gly-Asp motif) to treat patients with recurrent GBM in which RGD binds to integrins (i.e., the surface receptors of cancer cells) for viral internalization. This virus contains a deletion of eight amino acids in the region for binding the Rb protein, which is highly expressed in cancer cells but prevents viral replication in normal cells. A Phase I trial of 25 patients with recurrent GBM was performed to assess the toxicity related to the treatment; it showed that even at the highest dose (i.e., 3×10^{10} viral particles) it was well tolerated. A 10-10,000-fold increased cytokine IL-12p70 was detected in the serum of these patients. Twenty percent of patients survived more than 3 years after treatment, and 3 patients had a $\geq 95\%$ reduction in the enhancing tumor.[67]

In terms of current immunotherapy, some peptide vaccines (mutated forms of EGFR protein [EGFRvIII]) dendritic cell vaccines, as well as immune checkpoint inhibitors, anti-(programmed death-ligand 1)-PD-1 and anti-CTLA-4, have been studied in clinical Phase II/III trials and demonstrated improved overall survival for the patients.[91] CAR T cells are modified to express receptors specific to a targeted alteration in tumor cells of a specific cancer; in the case of GBM, this is the EGFRvIII mutation. In vivo mouse models showed that CAR T cells in xenogeneic subcutaneous models of an orthotopic human EGFRvIII+ GBM showed that GBM CAR T cells inhibited GBM growth, and regressed to further degree with the addition of temodar.[92] A Phase I clinical trial was initiated after these results were found (NCT02209376). Two additional clinical trials utilizing CAR T cells to treat GBM are ongoing, one targeting HER2 antigen (NCT01109095), and another targeting EGFRvIII using different vector types and CAR design (NCT01454596). CAR T cells largely have been administered by regional stereotactic injection, CED, and nanoparticles.

Parkinson Disease

In patients with PD, the loss of nigrostriatal dopaminergic neurons alters striato-pallidal circuitry, reducing GABA input to the subthalamic nucleus and rendering this structure disinhibited. As PD progresses, the effectiveness of most of the current pharmacotherapies wanes. Aromatic amino acid decarboxylase (AADC) is the rate-limiting enzyme for the conversion of L-DOPA to dopamine. Clinical studies have investigated methods to increase AADC expression. In vivo studies in nonhuman primates in which an AAV vector containing the cDNA encoding human AADC (hAADC) to the putamen via CED of PD models have supported this theory by increasing dopamine in the striatum and providing functional correction of the motor deficit. In a recent Phase I trial, patients who received the hAADC via bilateral putaminal infusions had initial improvements based on the Unified PD Rating Scale off medication for 12 hours in the first 12 months; however, patients slowly deteriorated over subsequent years (Fig. 3).[93]

Another gene therapy target for PD is acid decarboxylase gene (GAD). GAD is the rate-limiting enzyme for GABA production, and the activity of both GABA efferents to the subthalamic nucleus and its targets within the basal ganglia circuitry are affected in patients with PD. In a double-blind, Phase II, randomized controlled trial, 21 patients were randomly assigned to placebo surgery and 16 received a one-time stereotactic intracerebral infusion of AAV2-GAD. At 6-month follow-up, the Unified PD Rating Scale score for the AAV2-GAD group decreased by 8.1 points, compared to a decrease of 4.7 points in the control group. However, many secondary outcome measures did not show improvement in the treatment group compared with the control group, in part due to the study being underpowered for outcomes that are more difficult to achieve (e.g., improvement in function, activities of daily living, and quality of life measures).[94]

Other Neurodegenerative Conditions

Huntington Disease

Huntington disease (HD) occurs from an expansion of a CAG tract in the huntingtin gene, resulting in ubiquitous expression of mutant huntingtin (HTT) protein, which is thought to be

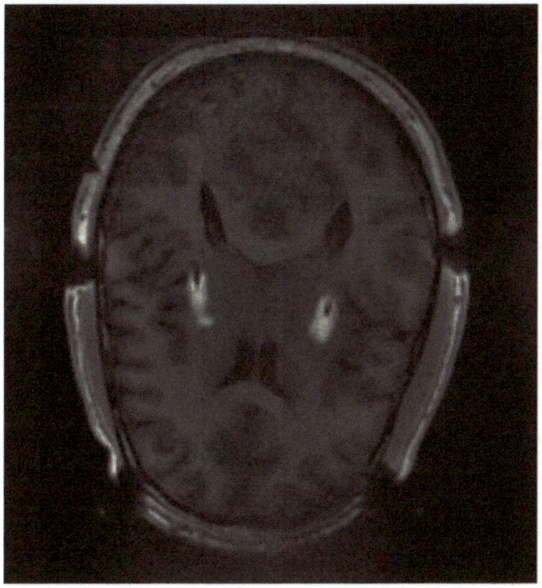

FIG. 3 T1-weighted magnetic resonance image (MRI) shows real-time stereotactic intracerebral infusion of AAV2-GAD in a patient with Parkinson disease. Two MRI-compatible stereotactic platforms were attached percutaneously to the skull. The infusion cannulae (*dark lines*) are noted within the infusate (virus co-infused with gadolinium) within bilateral putamen. Real-time monitoring of the infusion allows the surgeon to continuously assess the extent to which the agent has penetrated the target area.

the predominant toxic agent. Thus, clinical investigations have targeted development of therapies to correct HTT transcription and translation.[95] There are currently three main methods being investigated for targeting RNA: antisense oligonucleotides, RNA interference compounds, and small-molecule-splicing modulators. By acting on pre-mRNA, antisense oligonucleotides can target exons and introns, permitting an increased choice of target sequences; however, whether they can prevent the formation of truncated mRNA remains unclear. RNA interference compounds act on spliced mRNA; thus, they can reduce the formation of mRNA fragments only by directly targeting a sequence within that target exon, which may not prevent the formation of toxic HTT. For reasons that are unclear (but are perhaps related to immunity), single-stranded DNA diffuses well throughout the CNS and is readily taken up by neurons and other cells. Thus, the injection of antisense oligonucleotides into the CSF has been shown to result in widespread delivery of drug to the brain and a corresponding reduction in mRNA and protein.[74] In recent Phase I and II trials of antisense oligonucleotides given intrathecally, up to ~60% reduction (mean, ~40%) was seen in the mutant huntingtin protein (mHTT) in the CSF of patients treated for 13 weeks with four intrathecal injections of RG6042 at the two highest doses. mHTT levels were continuing to decline at the last reported measurement.[96,97]

Other DNA-targeting therapeutic agents may ameliorate all aspects of HD, including those mediated by alternative splicing that may be missed with RNA-targeted therapies. These included zinc finger proteins and clustered regularly interspaced short palindromic repeats (CRISPR)-Cas9. The use of zinc fingers protein has been limited, as the nonnative proteins can trigger inflammatory and immune reactions that result in neuronal death and limited duration of effect.[98] CRISPR and Cas have been found to form the basis of a prokaryote immune system that recognizes and subsequently destroys foreign DNA. Cas9 nuclease can be combined with synthetic "guide RNA" to produce a construct that can cut target DNA with high precision at any chosen site. In a rodent model of HD, treatment of the striatum with a

CRISPR-Cas9 construct that deleted the region containing the CAG repeat in the human HTT gene of Q140 knock-in mice showed attenuation of pathology and improved motor function, but it did not extended survival.[99] The success of CRISPR will dependent on ideal administration; so far, stereotactic injections have been studied, but additional studies are being designed to deliver the target CRISPR via liposomes and CED.

Spinal Muscular Atrophy

Spinal muscular atrophy is an autosomal recessive disease that causes weakness and muscle wasting via destruction of the anterior horn neurons of the spinal cord, which causes the loss of lower motor neurons that control movement. There is wide variability in age of onset, symptoms, and rate of progression. The disease spectrum is variously divided into 3-5 types, based on age at the time of symptom onset or on the highest-attained motor development milestone. Both gene therapy and stem cell therapies have shown promise in patients with spinal muscular atrophy, and clinical trials are ongoing. The first clinical success for such a therapy for spinal muscular atrophy was reported in 2016, when the Food and Drug Administration (FDA) approved nusinersen, an antisense oligonucleotide increases production of survival motor neuron2 protein. Nusinersen is administered intrathecally via lumbar puncture, and has resulted in extended survival in spinal muscular atrophy patients.[100]

Friedreich Ataxia

Friedreich ataxia is an incurable, inherited autosomal recessive progressive neurological disease that leads to gait disturbance, extremity weakness, vision and hearing impairments, dysarthria, scoliosis, diabetes mellitus, and heart disorders. The hallmark of the disease expansion of an intronic GAA triplet repeat in the FXN gene, which leads to reduced expression of the mitochondrial protein frataxin, resulting in the disease symptoms. Treatment in in vivo studies has ranged from cytokine treatments[101] to intrathecal stem cell transplantation,[102] both of which have shown some level of neuroprotection or disease stabilization. In a recent preclinical study, mice that received transplant of bone marrow cells expressing wild-type frataxin had improvements in several indicators of motor coordination and locomotor activity. Elevations in frataxin levels were also detected. Histologically, there was abrogation of disease pathology throughout the nervous system. This study reaffirms the value of cell-based gene therapy in individuals with neurodegenerative disorders.[103]

Genetic Disorders

Metachromatic Leukodystrophy

Metachromatic leukodystrophy is a fatal demyelinating lysosomal disease that arises from a deficiency of arylsulfatase A. Patients with metachromatic leukodystrophy exhibit progressive motor and cognitive impairment and die within a few years of symptom onset. Although the genetics of metachromatic leukodystrophy are known, its pathophysiology is not yet fully understood. Currently, no FDA-approved or effective treatments are available. Some case series and clinical trials have used umbilical cord blood transplantation from an unrelated donor[104] and, more recently, lentiviral vector-transduced cell infusion.[105,106] Intravenous infusion of autologous hematopoietic stem cells transduced with a lentiviral vector encoding arylsulfatase A cDNA resulted in sustained stable engraftment of gene-corrected hematopoietic stem cells in roughly 60% of patients (totally nine patients were treated) and remained stable during follow-up. This corresponded to a progressive reconstitution of arylsulfatase A activity in circulating hematopoietic cells and in the CSF, as well as a reduction of the toxic lysosomal storage material in peripheral nerve samples in 6/7 patients. Eight patients had prevention of disease onset or halted disease progression by clinical and instrumental assessment in comparison with the known natural history of untreated control patients with early-onset disease. Gross Motor Function Measure scores for six patients up to the last follow-up showed that gross motor performance was similar to that of children developing normally.[105]

Stroke

Many agents have been designed to combat thrombosis-related ischemic strokes, with the most widely known agent being intravenous or intra-arterial targeted recombinant tissue plasminogen activator (alteplase). Alteplase, although the only FDA-approved medication for ischemic stroke, is nonspecific for plasminogen activation and has a very limited treatment window after symptom onset. Antithrombotic aptamers have recently garnered interest due to their targeting of one or more proteins/targets participating in thrombosis, ability of extent or degree of inhibition by aptamer agent antidotes, and designed duration capabilities. The targets of aptamers include prothrombin, thrombin, Factor VII, Factor IX, Factor X, Factor XII, and vWF.[107]

A recently published randomized study attempted to determine whether intraventricular treatment of associated intraventricular hemorrhages would improve with the use of a thrombolytic agent. CLEAR III was a large, multiinstitutional, randomized, double-blind, placebo-controlled, Phase III trial that enrolled 500 patients with obstructing intraventricular hemorrhage from nontraumatic intracerebral hemorrhage (volume <30 mL) who qualified for placement of an extraventricular drain. The trial compared the intraventricular administration of 1 mg of alteplase or 0.9% saline (12 doses total, 8 hours apart). The study showed no significant difference with the use of alteplase for ventricular hemorrhage clearance radiographically. However, at 180 days, the treatment group had lower case fatality (18% vs saline 29%, $P=.006$). Nevertheless, these individuals had a greater modified Rankin Scale score 5 (17% vs 9%; $P=.007$; the modified Rankin Scale measures degree of disability/dependence after a stroke). Therefore, mortality may be improved with alteplase treatment but at the cost of severe disability among those who survive.[108]

Pain

Intrathecal analgesia has been used since the 1970s, initially with the use of intrathecal morphine for treatment of intractable cancer pain. Intrathecal morphine is now used for failed back syndrome and patients with spinal cord injuries. One novel drug agent currently approved in the United States is ziconotide, a synthetic equivalent of a neuroactive peptide originally found in the venom of the marine snail, *Conus magus*. In vivo models have shown antinociceptive properties with no development of pharmacologic tolerance or respiratory depression. Its mechanism of action appears to be through selective blockade of neuronal N-type voltage-sensitive calcium channels. In a randomized, double-blind, placebo-controlled clinical trial, the mean percent reduction in visual analog scale of pain intensity from baseline was 31.2% and 6.0% for intrathecal ziconotide- and placebo-treated patients respectively ($P < .01$; 169 treated with ziconotide, 86 placebo). However ziconotide-associated adverse effects (e.g., dizziness, blurred vision, nystagmus, and sedation) were observed; these were primarily related to the rapid titration and the high doses administered.[109]

FUTURE WORK

Stem Cell Therapies

The restorative capacity of isolated neural stem cells expanded in culture has already been demonstrated in several animal models of different neurologic diseases such as stroke, spinal cord injuries, PD, HD, multiple sclerosis, and amyotrophic lateral sclerosis.[110] Implantation of neural stem cells in vivo has been shown to attenuate deleterious inflammation, protect the CNS from degeneration, and enhance endogenous recovery processes.[110] Stem cell therapies with or without gene therapy or viral vectors still need to be optimized in terms of their delivery to the CNS, as many of the cells die before reaching the target. Likewise, their lack of migratory ability inside a lesion must be addressed, as this deficiency causes therapeutic failure in clinical trials of viruses and suicide genes delivered by cells. Therefore, these stem cell and cell-based therapies must be combined with targeting strategies such as CED, intrathecal injection, and osmotic pumps.

CRISPR

The CRISPR/Cas system is an adaptive bacterial immune system defense mechanism found in bacteria that inactivates foreign genetic material. The CRISPR-associated protein Cas9 is an endonuclease that uses a guide sequence within an RNA duplex, tracrRNA:crRNA, to form base pairs with DNA target sequences, enabling Cas9 to introduce a site-specific, double-strand break in the DNA. The dual tracrRNA:crRNA was engineered as a single guide RNA (sgRNA) that retains two critical features: a sequence at the 5′ side that determines the DNA target site by Watson-Crick base-pairing and a duplex RNA structure at the 3′ side that binds to Cas9. This finding has allowed changes in guide sequence of the sgRNA program Cas9 to target any DNA sequence of interest facilitating site-specific gene editing in mammalian cells.[111] CRISPR/Cas offers new hope for the development of effective treatments for neurological diseases with known causative mutations, and as discussed, advances have been seen HD. Ultimately, the mode for delivery of Cas proteins to the CNS may limit the potential of this therapy, but initial experiments have shown some encouraging strides with the use of viral vectors.[112]

CONCLUSIONS

Many therapeutic interventions are unsuccessful in treating disorders of the CNS because they cannot be effectively delivered to the brain. Physical and physiological barriers to effective CNS drug delivery include the BBB or disease-altered BBB, limited diffusion of therapy through parenchymal tissue, and the brain–CSF barrier. Various drug-delivery strategies are currently being used or are under investigation for their ability to circumvent these barriers. Further clinical work will clarify the role of drug-delivery techniques such as CED, coupling of HIFU with microbubbles, and viral-based therapeutics. Ultimately, some of these strategies may change the treatment paradigm and prognosis for patients with pathology of the CNS.

References

1. Pardridge WM. CSF, blood-brain barrier, and brain drug delivery. *Expert Opin Drug Deliv.* 2016;13(7):963-975.
2. Butt AM, Jones HC, Abbott NJ. Electrical resistance across the blood-brain barrier in anaesthetized rats: a developmental study. *J Physiol.* 1990;429:47-62.
3. Shapiro WR, Young DF, Mehta BM. Methotrexate: distribution in cerebrospinal fluid after intravenous, ventricular and lumbar injections. *N Engl J Med.* 1975;293(4):161-166.
4. Onyszchuk G, He YY, Berman NE, Brooks WM. Detrimental effects of aging on outcome from traumatic brain injury: a behavioral, magnetic resonance imaging, and histological study in mice. *J Neurotrauma.* 2008;25(2):153-171.
5. DiNapoli VA, Huber JD, Houser K, Li X, Rosen CL. Early disruptions of the blood-brain barrier may contribute to exacerbated neuronal damage and prolonged functional recovery following stroke in aged rats. *Neurobiol Aging.* 2008;29(5):753-764.
6. Rosenberg GA, Yang Y. Vasogenic edema due to tight junction disruption by matrix metalloproteinases in cerebral ischemia. *Neurosurg Focus.* 2007;22(5):E4.
7. Kurzepa J, Bartosik-Psujek H, Suchozebrska-Jesionek D, Rejdak K, Stryjecka-Zimmer M, Stelmasiak Z. Role of matrix metalloproteinases in the pathogenesis of multiple sclerosis. *Neurol Neurochir Pol.* 2005;39(1):63-67.
8. Wu Z, Guo H, Chow N, et al. Role of the MEOX2 homeobox gene in neurovascular dysfunction in Alzheimer disease. *Nat Med.* 2005;11(9):959-965.
9. Neal 2nd RE, Rossmeisl Jr JH, D'Alfonso V, et al. In vitro and numerical support for combinatorial irreversible electroporation and electrochemotherapy glioma treatment. *Ann Biomed Eng.* 2014;42(3):475-487.
10. Leuthardt EC, Duan C, Kim MJ, et al. Hyperthermic laser ablation of recurrent glioblastoma leads to temporary disruption of the peritumoral blood brain barrier. *PLoS One.* 2016;11(2).
11. Han H, Lee H, Kim K, Kim H. Effect of high intensity focused ultrasound (HIFU) in conjunction with a nanomedicines-microbubble complex for enhanced drug delivery. *J Control Release.* 2017;266:75-86.
12. Hynynen K. Ultrasound for drug and gene delivery to the brain. *Adv Drug Deliv Rev.* 2008;60(10):1209-1217.
13. Fan CH, Ting CY, Lin CY, et al. Noninvasive, targeted, and non-viral ultrasound-mediated gdnf-plasmid delivery for treatment of Parkinson's disease. *Sci Rep.* 2016;6:19579.
14. Lin CY, Hsieh HY, Pitt WG, et al. Focused ultrasound-induced blood-brain barrier opening for non-viral, non-invasive, and targeted gene delivery. *J Control Release.* 2015;212:1-9.
15. Meng Y, Suppiah S, Mithani K, Solomon B, Schwartz ML, Lipsman N. Current and emerging brain applications of MR-guided focused ultrasound. *J Ther Ultrasound.* 2017;5:26.
16. Engelhardt B, Sorokin L. The blood-brain and the blood-cerebrospinal fluid barriers: function and dysfunction. *Semin Immunopathol.* 2009;31(4):497-511.
17. Fossan G, Cavanagh ME, Evans CA, et al. CSF-brain permeability in the immature sheep fetus: a CSF-brain barrier. *Brain Res.* 1985;350(1–2):113-124.
18. Klopp CT, Alford TC, Bateman J, Berry GN, Winship T. Fractionated intra-arterial cancer; chemotherapy with methyl bis amine hydrochloride; a preliminary report. *Ann Surg.* 1950;132(4):811-832.
19. Abramson DH, Fabius AW, Issa R, et al. Advanced unilateral retinoblastoma: the impact of ophthalmic artery chemosurgery on enucleation rate and patient survival at MSKCC. *PLoS One.* 2015;10(12).
20. Abramson DH, Dunkel IJ, Brodie SE, Kim JW, Gobin YP. A phase I/II study of direct intraarterial (ophthalmic artery) chemotherapy with melphalan for intraocular retinoblastoma initial results. *Ophthalmology.* 2008; 115(8):1398-1404. 1404 e1391.
21. Joshi S, Ellis JA, Ornstein E, Bruce JN. Intraarterial drug delivery for glioblastoma mutiforme: will the phoenix rise again? *J Neurooncol.* 2015;124(3):333-343.
22. Doolittle ND, Muldoon LL, Culp AY, Neuwelt EA. Delivery of chemotherapeutics across the blood-brain barrier: challenges and advances. *Adv Pharmacol.* 2014;71:203-243.

23. Wong ET, Huberman M, Lu XQ, Mahadevan A. Bevacizumab reverses cerebral radiation necrosis. *J Clin Oncol*. 2008;26(34):5649-5650.
24. Angelov L, Doolittle ND, Kraemer DF, et al. Blood-brain barrier disruption and intra-arterial methotrexate-based therapy for newly diagnosed primary CNS lymphoma: a multi-institutional experience. *J Clin Oncol*. 2009; 27(21):3503-3509.
25. Figueiredo EG, Faria JW, Teixeira MJ. Treatment of recurrent glioblastoma with intra-arterial BCNU [1, 3-bis (2-chloroethyl)-1-nitrosourea]. *Arq Neuropsiquiatr*. 2010;68(5):778-782.
26. Newton HB. Intra-arterial chemotherapy of primary brain tumors. *Curr Treat Options Oncol*. 2005;6(6):519-530.
27. Muldoon LL, Pagel MA, Netto JP, Neuwelt EA. Intra-arterial administration improves temozolomide delivery and efficacy in a model of intracerebral metastasis, but has unexpected brain toxicity. *J Neurooncol*. 2016; 126(3):447-454.
28. Burkhardt JK, Hofstetter CP, Santillan A, et al. Orthotopic glioblastoma stem-like cell xenograft model in mice to evaluate intra-arterial delivery of bevacizumab: from bedside to bench. *J Clin Neurosci*. 2012;19(11):1568-1572.
29. Baltsavias G, Yella S, Al Shameri RA, Luft A, Valavanis A. Intra-arterial administration of papaverine during mechanical thrombectomy for acute ischemic stroke. *J Stroke Cerebrovasc Dis*. 2015;24(1):41-47.
30. Pala A, Schneider M, Brand C, et al. The evolution of invasive cerebral vasospasm treatment in patients with spontaneous subarachnoid hemorrhage and delayed cerebral ischemia-continuous selective intracarotid nimodipine therapy in awake patients without sedation. *Neurosurg Rev*. 2018.
31. Jayaraman MV, Haas RA, Ahn SH. Intra-arterial therapy for acute stroke: trials and tribulations. *Med Health R I*. 2011;94(12):357-359.
32. Alam MI, Beg S, Samad A, et al. Strategy for effective brain drug delivery. *Eur J Pharm Sci*. 2010;40(5):385-403.
33. Fung LK, Shin M, Tyler B, Brem H, Saltzman WM. Chemotherapeutic drugs released from polymers: distribution of 1,3-bis(2-chloroethyl)-1-nitrosourea in the rat brain. *Pharm Res*. 1996;13(5):671-682.
34. Murry DJ, Blaney SM. Clinical pharmacology of encapsulated sustained-release cytarabine. *Ann Pharmacother*. 2000;34(10):1173-1178.
35. Calias P, Banks WA, Begley D, Scarpa M, Dickson P. Intrathecal delivery of protein therapeutics to the brain: a critical reassessment. *Pharmacol Ther*. 2014;144(2):114-122.
36. Calias P, Papisov M, Pan J, et al. CNS penetration of intrathecal-lumbar idursulfase in the monkey, dog and mouse: implications for neurological outcomes of lysosomal storage disorder. *PLoS One*. 2012;7(1).
37. Garfield J, Dayan AD. Postoperative intracavitary chemotherapy of malignant gliomas. A preliminary study using methotrexate. *J Neurosurg*. 1973;39(3):315-322.
38. Tator CH. Intraneoplastic injection of CCNU for experimental brain tumor chemotherapy. *Surg Neurol*. 1977; 7(2):73-77.
39. Mairs RJ, Wideman CL, Angerson WJ, et al. Comparison of different methods of intracerebral administration of radioiododeoxyuridine for glioma therapy using a rat model. *Br J Cancer*. 2000;82(1):74-80.
40. http://www.alzet.com/research_applications/CHEMO/index.html.
41. Bobo RH, Laske DW, Akbasak A, Morrison PF, Dedrick RL, Oldfield EH. Convection-enhanced delivery of macromolecules in the brain. *Proc Natl Acad Sci U S A*. 1994;91(6):2076-2080.
42. Vogelbaum MA, Aghi MK. Convection-enhanced delivery for the treatment of glioblastoma. *Neuro Oncol*. 2015; 17(suppl 2):ii3-ii8.
43. Yang X, Saito R, Nakamura T, et al. Peri-tumoral leakage during intra-tumoral convection-enhanced delivery has implications for efficacy of peri-tumoral infusion before removal of tumor. *Drug Deliv*. 2016;23(3):781-786.
44. Kunwar S, Chang S, Westphal M, et al. Phase III randomized trial of CED of IL13-PE38QQR vs Gliadel wafers for recurrent glioblastoma. *Neuro Oncol*. 2010;12(8):871-881.
45. Mehta AM, Sonabend AM, Bruce JN. Convection-enhanced delivery. *Neurotherapeutics*. 2017;14(2):358-371.
46. Wolinsky JB, Colson YL, Grinstaff MW. Local drug delivery strategies for cancer treatment: gels, nanoparticles, polymeric films, rods, and wafers. *J Control Release*. 2012;159(1):14-26.
47. Sawyer AJ, Piepmeier JM, Saltzman WM. New methods for direct delivery of chemotherapy for treating brain tumors. *Yale J Biol Med*. 2006;79(3–4):141-152.
48. Perry J, Chambers A, Spithoff K, Laperriere N. Gliadel wafers in the treatment of malignant glioma: a systematic review. *Curr Oncol*. 2007;14(5):189-194.
49. Valtonen S, Timonen U, Toivanen P, et al. Interstitial chemotherapy with carmustine-loaded polymers for high-grade gliomas: a randomized double-blind study. *Neurosurgery*. 1997;41(1):44-48 [discussion 48–49].

50. Westphal M, Hilt DC, Bortey E, et al. A phase 3 trial of local chemotherapy with biodegradable carmustine (BCNU) wafers (Gliadel wafers) in patients with primary malignant glioma. *Neuro Oncol*. 2003;5(2):79-88.
51. Westphal M, Ram Z, Riddle V, Hilt D, Bortey E. Executive committee of the gliadel study G. Gliadel wafer in initial surgery for malignant glioma: long-term follow-up of a multicenter controlled trial. *Acta Neurochir (Wien)*. 2006;148(3):269-275 [discussion 275].
52. Lillehei KO, Kalkanis SN, Liau LM, et al. Rationale and design of the 500-patient, 3-year, and prospective vigilant observation of GlIadeL WAfer ImplaNT registry. *CNS Oncol*. 2018;7(2):CNS08.
53. Illum L. Is nose-to-brain transport of drugs in man a reality? *J Pharm Pharmacol*. 2004;56(1):3-17.
54. Landis MS, Boyden T, Pegg S. Nasal-to-CNS drug delivery: where are we now and where are we heading? An industrial perspective. *Ther Deliv*. 2012;3(2):195-208.
55. Lochhead JJ, Thorne RG. Intranasal delivery of biologics to the central nervous system. *Adv Drug Deliv Rev*. 2012;64(7):614-628.
56. Djupesland PG, Mahmoud RA, Messina JC. Accessing the brain: the nose may know the way. *J Cereb Blood Flow Metab*. 2013;33(5):793-794.
57. De Rosa R, Garcia AA, Braschi C, et al. Intranasal administration of nerve growth factor (NGF) rescues recognition memory deficits in AD11 anti-NGF transgenic mice. *Proc Natl Acad Sci U S A*. 2005;102(10):3811-3816.
58. Hanson LR, Frey II WH. Intranasal delivery bypasses the blood-brain barrier to target therapeutic agents to the central nervous system and treat neurodegenerative disease. *BMC Neurosci*. 2008;9(suppl 3):S5.
59. Sood S, Jain K, Gowthamarajan K. Intranasal therapeutic strategies for management of Alzheimer's disease. *J Drug Target*. 2014;22(4):279-294.
60. Groothuis DR. The blood-brain and blood-tumor barriers: a review of strategies for increasing drug delivery. *Neuro Oncol*. 2000;2(1):45-59.
61. Gedikoglu S. Immunotoxins. *Mikrobiyol Bul*. 1990;24(1):79-87.
62. Donath J, Landsteiner K. Uber paroxysmale haemoglobinurie. *Munch Med Wochenschr*. 1904;51:1590-1593.
63. Moolten FL, Cooperband SR. Selective destruction of target cells by diphtheria toxin conjugated to antibody directed against antigens on the cells. *Science*. 1970;169(3940):68-70.
64. Sampson JH, Maus MV, June CH. Immunotherapy for Brain Tumors. *J Clin Oncol*. 2017;35(21):2450-2456.
65. Villoslada P, Moreno B, Melero I, et al. Immunotherapy for neurological diseases. *Clin Immunol*. 2008; 128(3):294-305.
66. Cloughesy TF, Landolfi J, Hogan DJ, et al. Phase 1 trial of vocimagene amiretrorepvec and 5-fluorocytosine for recurrent high-grade glioma. *Sci Transl Med*. 2016;8(341):341ra375.
67. Lang FF, Conrad C, Gomez-Manzano C, et al. Phase I study of DNX-2401 (delta-24-RGD) oncolytic adenovirus: replication and immunotherapeutic effects in recurrent malignant glioma. *J Clin Oncol*. 2018;36(14):1419-1427.
68. Lyon JG, Mokarram N, Saxena T, Carroll SL, Bellamkonda RV. Engineering challenges for brain tumor immunotherapy. *Adv Drug Deliv Rev*. 2017;114:19-32.
69. Recinos VR, Tyler BM, Bekelis K, et al. Combination of intracranial temozolomide with intracranial carmustine improves survival when compared with either treatment alone in a rodent glioma model. *Neurosurgery*. 2010; 66(3):530-537 [discussion 537].
70. Weingart JD, Thompson RC, Tyler B, Colvin OM, Brem H. Local delivery of the topoisomerase I inhibitor camptothecin sodium prolongs survival in the rat intracranial 9L gliosarcoma model. *Int J Cancer*. 1995; 62(5):605-609.
71. Lesniak MS, Upadhyay U, Goodwin R, Tyler B, Brem H. Local delivery of doxorubicin for the treatment of malignant brain tumors in rats. *Anticancer Res*. 2005;25(6B):3825-3831.
72. Li KW, Dang W, Tyler BM, et al. Polilactofate microspheres for paclitaxel delivery to central nervous system malignancies. *Clin Cancer Res*. 2003;9(9):3441-3447.
73. Mangraviti A, Gullotti D, Tyler B, Brem H. Nanobiotechnology-based delivery strategies: new frontiers in brain tumor targeted therapies. *J Control Release*. 2016;240:443-453.
74. Maguire CA, Ramirez SH, Merkel SF, Sena-Esteves M, Breakefield XO. Gene therapy for the nervous system: challenges and new strategies. *Neurotherapeutics*. 2014;11(4):817-839.
75. Dilnawaz F, Sahoo SK. Therapeutic approaches of magnetic nanoparticles for the central nervous system. *Drug Discov Today*. 2015;20(10):1256-1264.
76. Tabatabaei SN, Girouard H, Carret AS, Martel S. Remote control of the permeability of the blood-brain barrier by magnetic heating of nanoparticles: a proof of concept for brain drug delivery. *J Control Release*. 2015;206:49-57.

77. Alyautdin R, Khalin I, Nafeeza MI, Haron MH, Kuznetsov D. Nanoscale drug delivery systems and the blood-brain barrier. *Int J Nanomedicine*. 2014;9:795-811.
78. Verreault M, Strutt D, Masin D, et al. Vascular normalization in orthotopic glioblastoma following intravenous treatment with lipid-based nanoparticulate formulations of irinotecan (Irinophore C), doxorubicin (Caelyx(R)) or vincristine. *BMC Cancer*. 2011;11:124.
79. Ishii T, Asai T, Oyama D, et al. Amelioration of cerebral ischemia-reperfusion injury based on liposomal drug delivery system with asialo-erythropoietin. *J Control Release*. 2012;160(1):81-87.
80. Fusco JP, Castanon E, Carranza OE, et al. Neurological and cytological response as potential early predictors of time-to-progression and overall survival in patients with leptomeningeal carcinomatosis treated with intrathecal liposomal cytarabine: a retrospective cohort study. *J Neurooncol*. 2013;115(3):429-435.
81. Friedmann T. A brief history of gene therapy. *Nat Genet*. 1992;2(2):93-98.
82. Tosolini AP, Smith GM. Editorial: gene therapy for the central and peripheral nervous system. *Front Mol Neurosci*. 2018;11:54.
83. Foust KD, Nurre E, Montgomery CL, Hernandez A, Chan CM, Kaspar BK. Intravascular AAV9 preferentially targets neonatal neurons and adult astrocytes. *Nat Biotechnol*. 2009;27(1):59-65.
84. Chan KY, Jang MJ, Yoo BB, et al. Engineered AAVs for efficient noninvasive gene delivery to the central and peripheral nervous systems. *Nat Neurosci*. 2017;20(8):1172-1179.
85. Singh S. Nanomaterials as non-viral siRNA delivery agents for cancer therapy. *Bioimpacts*. 2013;3(2):53-65.
86. van der Vos KE, Balaj L, Skog J, Breakefield XO. Brain tumor microvesicles: insights into intercellular communication in the nervous system. *Cell Mol Neurobiol*. 2011;31(6):949-959.
87. Rufino-Ramos D, Albuquerque PR, Carmona V, Perfeito R, Nobre RJ, Pereira de Almeida L. Extracellular vesicles: novel promising delivery systems for therapy of brain diseases. *J Control Release*. 2017;262:247-258.
88. Rossignol J, Fink K, Davis K, et al. Transplants of adult mesenchymal and neural stem cells provide neuroprotection and behavioral sparing in a transgenic rat model of Huntington's disease. *Stem Cells*. 2014; 32(2):500-509.
89. Rodriguez A, Brown C, Badie B. Chimeric antigen receptor T-cell therapy for glioblastoma. *Transl Res*. 2017;187:93-102.
90. Desjardins A, Sampson JH, Peters KB, et al. Oncolytic polio/rhinovirus recombinant (PVSRIPO) against recurrent glioblastoma (GBM): Optimal dose determination. *J Clin Oncol*. 2015;33(suppl 15):2068.
91. Dunn-Pirio AM, Vlahovic G. Immunotherapy approaches in the treatment of malignant brain tumors. *Cancer*. 2017;123(5):734-750.
92. Johnson LA, Scholler J, Ohkuri T, et al. Rational development and characterization of humanized anti-EGFR variant III chimeric antigen receptor T cells for glioblastoma. *Sci Transl Med*. 2015;7(275):275ra222.
93. Mittermeyer G, Christine CW, Rosenbluth KH, et al. Long-term evaluation of a phase 1 study of AADC gene therapy for Parkinson's disease. *Hum Gene Ther*. 2012;23(4):377-381.
94. LeWitt PA, Rezai AR, Leehey MA, et al. AAV2-GAD gene therapy for advanced Parkinson's disease: a double-blind, sham-surgery controlled, randomised trial. *Lancet Neurol*. 2011;10(4):309-319.
95. Wild EJ, Tabrizi SJ. Therapies targeting DNA and RNA in Huntington's disease. *Lancet Neurol*. 2017;16(10):837-847.
96. https://www.prnewswire.com/news-releases/ionis-htt-rx-rg6042-top-line-data-demonstrate-significant-reductions-of-disease-causing-mutant-huntingtin-protein-in-people-with-huntingtons-disease-300606962.html.
97. https://www.fiercebiotech.com/biotech/roche-ionis-hail-huntington-s-success-plan-pivotal-trial.
98. Agustin-Pavon C, Mielcarek M, Garriga-Canut M, Isalan M. Deimmunization for gene therapy: host matching of synthetic zinc finger constructs enables long-term mutant Huntingtin repression in mice. *Mol Neurodegener*. 2016;11(1):64.
99. Yang S, Chang R, Yang H, et al. CRISPR/Cas9-mediated gene editing ameliorates neurotoxicity in mouse model of Huntington's disease. *J Clin Invest*. 2017;127(7):2719-2724.
100. Mercuri E, Darras BT, Chiriboga CA, et al. Nusinersen versus sham control in later-onset spinal muscular atrophy. *N Engl J Med*. 2018;378(7):625-635.
101. Kemp KC, Cerminara N, Hares K, et al. Cytokine therapy-mediated neuroprotection in a Friedreich's ataxia mouse model. *Ann Neurol*. 2017;81(2):212-226.
102. Jones J, Estirado A, Redondo C, Martinez S. Stem cells from wildtype and Friedreich's ataxia mice present similar neuroprotective properties in dorsal root ganglia cells. *PLoS One*. 2013;8(5).

103. Kemp KC, Hares K, Redondo J, et al. Bone marrow transplantation stimulates neural repair in Friedreich's ataxia mice. *Ann Neurol*. 2018;83(4):779-793.
104. Chen X, Gill D, Shaw P, Ouvrier R, Troedson C. Outcome of early juvenile onset metachromatic leukodystrophy after unrelated cord blood transplantation: a case series and review of the literature. *J Child Neurol*. 2016; 31(3):338-344.
105. Sessa M, Lorioli L, Fumagalli F, et al. Lentiviral haemopoietic stem-cell gene therapy in early-onset metachromatic leukodystrophy: an ad-hoc analysis of a non-randomised, open-label, phase 1/2 trial. *Lancet*. 2016; 388(10043):476-487.
106. Biffi A, Montini E, Lorioli L, et al. Lentiviral hematopoietic stem cell gene therapy benefits metachromatic leukodystrophy. *Science*. 2013;341(6148):1233158.
107. Nimjee SM, Povsic TJ, Sullenger BA, Becker RC. Translation and clinical development of antithrombotic aptamers. *Nucleic Acid Ther*. 2016;26(3):147-155.
108. Hanley DF, Lane K, McBee N, et al. Thrombolytic removal of intraventricular haemorrhage in treatment of severe stroke: results of the randomised, multicentre, multiregion, placebo-controlled CLEAR III trial. *Lancet*. 2017;389(10069):603-611.
109. Wallace MS, Charapata SG, Fisher R, et al. Intrathecal ziconotide in the treatment of chronic nonmalignant pain: a randomized, double-blind, placebo-controlled clinical trial. *Neuromodulation*. 2006;9(2):75-86.
110. Einstein O, Ben-Hur T. The changing face of neural stem cell therapy in neurologic diseases. *Arch Neurol*. 2008; 65(4):452-456.
111. Doudna JA, Charpentier E. Genome editing. The new frontier of genome engineering with CRISPR-Cas9. *Science*. 2014;346(6213):1258096.
112. Murlidharan G, Sakamoto K, Rao L, et al. CNS-restricted transduction and CRISPR/Cas9-mediated gene deletion with an engineered AAV vector. *Mol Ther Nucleic Acids*. 2016;5(7).

CHAPTER

20

Dynamic Contrast-Enhanced Magnetic Resonance Imaging in Brain Tumors

James R. Ewing[*,†,‡,§], *Rasha Elmghirbi*[‡], *Tavarekere Nagaraja*[§]

[*]Department of Neurology, Henry Ford Hospital, Detroit, MI, United States [†]Department of Neurology, Wayne State Medical School, Detroit, MI, United States [‡]Department of Physics, Oakland University, Rochester, MI, United States [§]Department of Neurosurgery, Henry Ford Hospital, Detroit, MI, United States

CONVECTION-ENHANCED DELIVERY OF DRUGS TO THE BRAIN

The blood–brain barrier (BBB) is impermeable to many therapeutic agents that might otherwise prove effective for the treatment of brain tumors, and the blood tumor barrier (BTB) in the brain can also prove to be an impediment,[1,2] particularly to water-soluble agents. Even with a good (mostly passive) transfer of a therapeutic agent across the BBB and the BTB, the relative efficiency of drug delivery to the tumor shows that, at most, 1% of the administered drug will reach the brain tumor.[1] Convection-enhanced delivery (CED) of therapeutic agents to the brain[3,4] offers a number of important advantages for the treatment of cerebral tumors.[1] CED offers a means for delivering drugs to cerebral tumors, particularly those therapies that are systemically toxic, since they can be delivered in a locally targeted manner. Because of the toxicity of many of these agents, however, administration of these drugs should be planned to incorporate a best-effort prediction as to where, when, and in what concentration a therapeutic agent can be delivered via CED. This in turn requires a detailed knowledge of the mechanical and fluid properties of the tissue in which the CED of a therapeutic agent is to be performed.

Much progress has been made in this area.[5-7] Conventional magnetic resonance imaging (MRI) is often combined with maps of MRI apparent diffusion coefficients or MRI tractography to estimate tissue cellularity and tissue porosity, which facilitates estimates about tissue fluid

https://doi.org/10.1016/B978-0-12-813997-4.00020-7
© 2019 Elsevier Inc. All rights reserved.

conductivity and CED delivery patterns. A recent paper[8] demonstrated that incorporating a dynamic contrast-enhanced (DCE)-MRI measurement of extravascular, extracellular space generates a better estimate of tumor cellularity, and can thus can be expected to yield a more accurate estimate of local tissue fluid conductivity and CED drug delivery patterns, both spatial and temporal. This conclusion is supported by a recent demonstration of a noninvasive estimate of tumor fluid conductivity using DCE-MRI.[9] Those researchers concluded that DCE-MRI and arterial spin labeling contained sufficient information to noninvasively generate an estimate of tumor interstitial fluid pressure (TIFP).[9]

There is a relatively complete theoretical understanding of fluid flows in various types of tissue, including tumor tissue.[6, 7, 10-16] However, in and around tumors, few experimental studies have measured relevant physical constants such as mechanical moduli, tissue fluid conductivity, extracellular space, and sources and sinks of fluids using studies of perfusion, tumor exudate fluxes, and clearance rates. This chapter reviews recent developments in DCE-MRI with an emphasis on how tissue parameters estimated by DCE-MRI can be used to plan CED, particularly in cerebral tumors. Many of the demonstrations in this chapter are performed in model tumors in the brains of small animals; however, because MRI is essentially atraumatic and repeatable, the translation of these practices to humans is quite straightforward.

Convection-enhanced delivery planning demands best-effort estimates of mechanical and fluid properties of tissue; therefore, our emphasis will be on producing unbiased estimators of tissue physical quantities. The production of such unbiased estimates calls for the use of data-driven[17] models to describe the dynamic behavior of the data in DCE-MRI studies. As these data-driven models of DCE-MRI studies are a major product of our efforts,[17-25] they will be used throughout this chapter.

DCE-MRI FOR TUMORS

Before we consider MRI measures of tumor physiology per se, we should note that, as a method, DCE-MRI belongs to a broad class of procedures called *indicator-dilution studies*. These methods, which have been in use for more than a century,[26-36] are extensions of the Fick principle,[37] which is in turn an application of the principle of conservation of mass. Indicator-dilution studies use the temporal concentration of some agent in tissue, often in trace amounts, to quantitate such physiological parameters as perfusion,[38-41] vascular volume,[38, 39] vascular permeability,[42-45] indicator distribution volume, tracer binding potential,[46, 47] or some combination thereof. Whereas early studies were invasive, usually requiring sampling of both arterial and venous blood,[32,33,48] the introduction of methods for external monitoring of tissue concentration (e.g., single photon emission computed tomography, positron emission tomography, computed tomography, and eventually MRI) allowed the formation of a theory and practice[36, 49] of essentially noninvasive estimates of physiological quantities. Therefore, an extensive body of literature describing methods and results can be drawn upon when considering the development of DCE-MRI studies for measuring physiology in both normal and pathological tissue.

Indicator-Dilution Methods for Estimating Microvascular Permeability

As noted, in an indicator-dilution study, an indicator is introduced to the organ system to be studied and its concentration-time curve is followed so that the physiology (and possibly the pathology) of the organ system can be quantified. In DCE-MRI, the organ system is the vasculature and its associated perfused tissue, and the physiology of interest is often the permeability of the microvessels of a tumor, along with other measures of physiology that may be available. In DCE-MRI, the indicator, usually a lightweight (~1 kDa) magnetic resonance contrast agent (MRCA), is almost always administered as a venous bolus.

Among typical parameters to be estimated in an indicator-dilution experiment are the forward (i.e., vascular-to-interstitium) volumetric transfer constant (K^{trans}); the reverse (i.e., interstitium-to-vascular) transfer constant (k_{ep}); the extracellular, extravascular volume fraction (v_e), which is estimated by the ratio $\frac{K^{trans}}{k_{ep}}$ when the indicator used is neither bound nor metabolized in the tissue; the plasma volume (v_p); and the specific flow (F) as $\left[\frac{\text{mL/min}}{\text{mL}}\right]$. If both flow and K^{trans} are estimated, because K^{trans} is the product of F and the vascular permeability surface area (PS)-product, PS will be available. All MRCAs in use are excluded from intact cells; therefore, the extracellular distribution volume (V_D) can be estimated via a Logan Plot[47] as the tissue approaches equilibrium with the vascular concentration.[21] Given sufficient information, the exchange rates for protons of water between intracellular and extracellular pools, and between intravascular and extravascular pools might also be calculable.[50-52]

A number of approaches may address the challenge of modeling indicator dilution in the capillary bed of a tissue.[49] Among others in current use in MRI are the two-parameter Tofts model (which estimates K^{trans} and k_{ep}),[53] the two-parameter Patlak model (which estimates K^{trans} and the plasma volume, v_p),[44, 54] the three-parameter extended Tofts model, also known as the standard model (which estimates K^{trans}, k_{ep}, and v_p),[55] the four-parameter two-compartment exchange model,[56, 57] and the five-parameter shutter-speed model.[52] With very few exceptions, voxel-by-voxel estimates of model parameters in brain tumors have used models with two or three freely varying parameters, while possibly fixing one or two of the parameters of models with higher order,[17] so that there are three or fewer free parameters to be fit.

This is not the place to discuss ambiguities in the selection of models for analysis of DCE-MRI in cerebral tumors. Suffice it to say that despite the existence of clearly understood procedures for selection of models to balance variance and bias in parametric estimates,[58] there remains a lack of agreement among investigators as to what the best model is for such analyses; that different models will consider different parameters; and, even when the parameters of different models have the same names (e.g., K^{trans}), estimates of those parameters can differ by an order of magnitude when different models are applied to the same data. The field badly needs a consensus approach to the selection of models for fitting DCE-MRI data; this shortfall has slowed the use of DCE-MRI in places where it might otherwise provide valuable information on the physiological state of cerebral tumors.

In our experience of over 15 years of DCE-MRI in strokes and cerebral tumors of animals and humans, the most useful model is the standard model of Tofts et al.[55] A formal consideration of model selection demonstrated that the standard model was a member of a set of nested models,[17, 24] thus allowing a straightforward test of the standard model against a reduced (Patlak) model[44] and in turn, a test of the Patlak model against a model of no leakage. Test–retest studies demonstrated good stability in repeated measures[18, 22]; this stability allowed significant changes in tumor physiology to be measured in a relatively small numbers of animals.[19, 20, 23, 59] A further consideration of model *failures* in the periphery of model tumors led to an important result that related the fluid properties of embedded cerebral tumors to the compression of tissues in the periphery of the tumors.[19]

For the standard model, the time dependence of contrast agent (CA) in tissue can be presented as follows:

$$C(t) = K^{trans} \int_0^t C_p(\tau) e^{-k_{ep}(t-\tau)} d\tau + v_p C_p(t) \quad (1)$$

where t is time, $C(t)$ is the tissue concentration of CA, $C_p(t)$ is the plasma concentration, K^{trans} is the forward volumetric transfer rate constant, k_{ep} is the reverse transfer rate constant, and v_p is the vascular plasma volume fraction.

Derivations relating to Eq. (1) can be found in a number of early publications.[44, 45, 60-63] Of particular interest are the Crone-Renkin single-capillary model,[60] with its description of the *adiabatic assumption*, and Patlak's compartmental approach,[44] followed by a generalization[45] that produced Eq. (1) well in advance of its appearance in MRI studies. In terms of model presentation, our personal preference is the single-capillary model as presented by the great Danish neurologist Niels Lassen.[63] This model stays fairly close to the physical reality of microvessels in tissue, and its simplifying assumptions are clearly stated in the derivation.

As noted, k_{ep} is the exchange rate *from* the extravascular, extracellular space *to* the plasma, and equals K^{trans}/v_e. K^{trans} and k_{ep} are rate constants (units of inverse time), and thus the *change* in the measured tissue and plasma concentrations is the quantity that drives their estimations. Because K^{trans} is the product of the extraction fraction and flow, the interpretation of this rate constant can depend on whether the delivery of the indicator to the tissue is perfusion- or diffusion-limited. In the case of high flow (diffusion-limited delivery), $K^{trans} \approx \text{PS}$, where PS is the permeability-surface area product.

One notable feature of this two-compartment model and its associated derivation is that the vascular and extravascular spaces are the only spaces considered. Mass conservation occurs solely between these two spaces, and if, for instance, some MRCA leaks out of the extravascular space into another space, that event is not accounted for in this model. We shall return to this subject when we consider an instance of model *failure*, and inferences that might be drawn on the basis of that failure.

The Graphical Patlak Method and its Extension

As noted, in DCE-MRI the indicator is almost always administered as a venous bolus. If K^{trans} is very small, and v_{EES} is reasonably large, there are early time periods in many experiments during which k_{ep} is not measurable because a small amount of MRCA has exited the vasculature. In such cases, k_{ep} can be set to 0.

$$C_{tiss} \cong K^{trans} \int_0^t C_p(\tau) d\tau + v_p C_p(t) \quad (2)$$

In this case, both sides of the equation can be divided by $C_p(t)$:

$$\frac{C_{tiss}(t)}{C_p(t)} \cong \frac{K^{trans} \int_0^t C_p(\tau) d\tau}{C_p(t)} + v_p \quad (3)$$

When plotted with $\frac{C_{tiss}(t)}{C_p(t)}$ as the ordinate and $\frac{\int_0^t C_p(\tau)d\tau}{C_p(t)}$ as the abscissa, this is the equation of a straight line with a slope of K^{trans} and an intercept of v_p. The useful thing about this relationship is that it allows data to be plotted in such a way that the assumptions of the model (mainly, that reflux from the tissue to the vasculature is negligible) can be assessed. It can also be seen that, if a best estimate of k_{ep} can be found, dividing both sides of Eq. (1) by $C_p(t)$ will also yield a straight line whose slope is K^{trans}:

$$\frac{C_{tiss}}{C_p(t)} = \frac{K^{trans} \int_0^t C_p(\tau)e^{-k_{ep}(t-\tau)}d\tau}{C_p(t)} + v_p \tag{4}$$

Three nested models can be proposed: Model 1, where $K^{trans} = k_{ep} = 0$ with only v_p nonzero; Model 2, the Patlak model where $k_{ep} = 0$ and v_p and K^{trans} are nonzero; and Model 3, the standard model, with all three parameters to be estimated simultaneously. The linearization of these models allows an intuitive approach to model validity and a well-understood paradigm for comparing alternative models.[17, 24, 25]

MRI Contrast and Contrast Agent Concentration

Given the body of theory and practice in the use of indicators for other imaging modalities, the relationship between changes in the concentration of commonly used MRCAs and changes in MR signal becomes a primary consideration. If a reliable predictor of MRCA concentration can be established, many results of previous investigations are immediately applicable.

In non-MRI applications, the relationship between contrast and indicator concentration is usually straightforward; the signal change (counts detected, X-rays absorbed) is linear in the concentration of the indicator. However, MRI parameters that change with MRCA concentration in tissue—for example, changes in the longitudinal (T_1) and/or transverse (T_2^*) relaxation times—often have a complex relationship with MR contrast concentration. In particular, changes in both the measured (T_1) and (T_2^*) relaxation times of tissue can depend on where in the tissue the MRCA is at any particular time. If, for instance, the MRCA is primarily intravascular, T_2^*, depending as it does on local "static" field gradients, will be far shorter than if the same amount of MRCA were to be uniformly distributed in the vasculature and the surrounding extracellular space.

Signal changes with MRCA concentration in the MRI signals of real sequences never occur as pure (T_1) or (T_2^*) functions, but show mixed effects (as seen in Fig. 5 of Ewing et al.[17]). For instance, in conjunction with a bolus injection of MRCA, it is common practice to use a spoiled gradient-recalled echo (SPGRE) MRI sequence with a short TR and relatively high flip angle. In this case, the equilibrium signal response is.

$$S_n(t) = M_0 \frac{\left(1 - e^{TR\,R_1(t)}\right)}{\left(1 - \cos(\theta)e^{-TR\,R_1(t)}\right)} \sin(\theta)e^{-TE\,R_2^*(t)} \tag{5}$$

where $S_n(t)$ is the signal intensity of the nth image set in a T_1-weighted DCE-MRI procedure; t is time; M_0 is the magnetization of the protons in the voxel; θ is the local tip-angle; TR is the repetition time between pulses; TE is the echo time; and $R_1(t)$ and $R_2^*(t)$ are the longitudinal and the transverse relaxation rates in the voxel, respectively, as a function of time.

$R_1(t)=1/T_1(t)$, and $R_2^*(t)=1/T_2^*(t)$. For most achievable echo times in preclinical and clinical systems, $T_2^*(t)$ dephasing can be a significant effect at 3T and above, particularly for preclinical systems. $R_2^*(t)$ can be measured using a dual-echo SPGRE sequence, and R_2^* dephasing can be adjusted in Eq. (5), but it is rare to see such a sequence used, even though a loss of signal can usually be observed at the start of a DCE-MRI experiment when the bolus of MRCA first arrives.[17] The effect of loss of signal due to T_2^* dephasing can lead to a significant underestimation of MRCA concentration.

Even if T_2^* dephasing can be accounted for, there may be an additional impediment to estimating MRCA concentration via changes in $R_1(t)$. Longitudinal relaxation of the protons of water by paramagnetic compounds is a local phenomenon; the protons must be in the near vicinity of the MRCA to be relaxed by the agent. Intracellular water, for instance, becomes relaxed only by water exchange with spaces that contain MRCA. In the event that transcytolemmal water exchange rates are of the order of longitudinal relaxation rates in the interstitial spaces, changes in $C_{tiss}(t)$ will not be linear in $R_1(t)$.[52, 64-68] The importance of this effect, summarized in the "shutter-speed" model,[52, 69, 70] has been debated in the DCE-MRI community.[50, 71] Although the model is indisputably a fuller description of the relaxation processes in tissue, because the smallest possible shutter-speed model has at least five parameters, typical DCE-MRI studies lack the contrast to noise to fit more than three of the five parameters, thus yielding an inevitably biased estimate of the parameters and obviating its utility for further mechanical modeling. Mathematically, the shutter-speed model is a special case of a three-site chemical exchange system[72] in a nuclear magnetic resonance experimental measurement of T_1. Spencer and Fishbein[72] showed that short repetition times and relatively high tip-angles minimize exchange effects. This evidence provides assurance in DCE-MRI that, under most conditions, changes in $R_1(t)$ will be approximately linear in changes in $C_{tiss}(t)$. Our own work[50] and that of others[71] have reflected this judgment. In short, in most DCE-MRI studies, changes in $C_{tiss}(t)$ will be approximately linear in changes in $R_1(t)$, particularly if T_2^* dephasing can be compensated.

An Example Analysis in an Animal Model of Cerebral Tumor

This example is drawn from our previously published data.[19] Experimental details are included in that article, and in others from our laboratory,[19, 21-23, 59] including test–retest statistics on a 24-hour interval.[22]

Briefly, a U251 model of embedded cerebral tumor was studied in an athymic female rat. Fig. 1A shows a T1-weighted postcontrast image of the tumor in the brain of this animal. About 18 days after implantation, when tumors grew to about 6 mm in diameter, a DCE-MRI study was conducted, followed by a wick-in-needle measurement of TIFP (WIN-TIFP). A high-resolution T2-weighted MRI after TIFP measurement verified that the needle path went through the tumor.

The MRI study was performed in a 7T magnet in a Varian 20-cm bore system with a Direct Drive spectrometer and console (Varian Medical Systems, Inc.). Arterial spin labeling, DCE-MRI, Look-Locker (LL) sequence, T1- and T2-weighted image sets were acquired and used to derive the tumor parameters that were used in both the fluid-mechanical model and in multivariate empirical estimates of TIFP. Spin-echo arterial spin-labeled data were acquired to estimate cerebral blood flow in a single central slice, as previously described.[73]

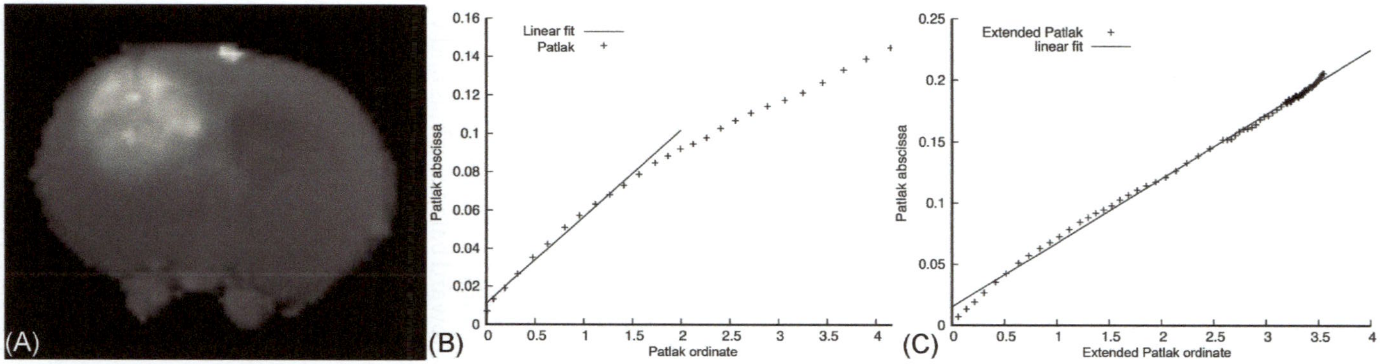

FIG. 1 (A) A T1-weighted, postcontrast image of a U251 model of cerebral tumor in an athymic rat. In this model, a halo of contrast is often observed around the tumor itself, around the time at which this image was taken (i.e., 15 min after contrast injection). (B) The Patlak plot (Model 2) using a magnetic resonance image estimate of magnetic resonance contrast agent concentration in the entire region of interest of the tumor. The approximate linearity of the first 15 data points (i.e., the first minute postinjection of data collection) is evident. The slope of the linear portion of the data is 0.0521 min^{-1}. (C) The linearized standard model considering 93 points, 6 minutes of data, with k_{ep} optimized by a nonlinear algorithm. K^{trans} estimated via this model is 0.0524 min^{-1}. For fitting the full 6 minutes of data, the extended model is clearly superior, but the close agreement of the two models in their domains of validity demonstrate that the nested models essentially agree in their estimates of model parameters that are held in common.

The DCE-MRI sequence was a dual-echo gradient-echo sequence that acquired three slices on 2-mm centers (1.8 mm slice, 0.2 mm gap). The slice set was centered on the tumor and 150 image sets at 4-second intervals were acquired with the following parameters: FA = 25 degrees, matrix = 128 × 64, NE = 2, NA = 1, $TE_1/TE_2/TR = 2.0/4.0/60$ ms. Total run time was about 10 minutes. At image 15 of the dual-echo gradient-echo sequence, bolus injection of the CA (Magnevist, Bayer Healthcare Pharmaceuticals), 0.25 mmol/kg at undiluted concentration, no flush, was performed by hand push.

Before and immediately after the DCE-MRI sequence, two LL sequences (FA = 15 degrees, matrix 128 × 64, five 2.0 mm slices, no gap, NE = 24 inversion-recovery echoes on 50 ms intervals, TE/TR = 4.0/2000 ms) were run so that a voxel-by-voxel estimate of longitudinal relaxation time (T_1) in the tissue could be made pre- and post-CA administration. Before the first LL sequence, a six-image progressive saturation set of dual-echo gradient-echo studies was run with varying repetition times and used, together with LL T_1 maps, to independently estimate the FA used in the subsequent DCE-MRI experiment. The sequence parameters were as follows: (nominal) FA = 25 degrees, matrix = 128 × 64, three 1.8 mm slices, 0.2 mm gap, NE = 2, NA = 1, $TE_1/TE_2/TR = 2.0/4.0/(60, 100, 200, 300, 400, 600)$ ms.

In this case, a time trace of R_1 was constructed using the data of the dual-echo sequence and Eq. (5), allowing correction for T_2^* dephasing and generating an approximately unbiased estimate of the concentration-time curve of the MRCA. A region of interest (ROI) in the tumor was established (see further), and the ROI-averaged time trace of R_1 was constructed. A group-averaged input function obtained from a series of studies using radiolabeled Gd-DTPA[74] was used throughout to approximate each study's arterial input function (AIF). The resultant Patlak graphical plot is shown in Fig. 1B, and the linearized standard model (Extended Patlak) is presented in Fig. 1C. It is clear from an inspection of the two graphs in Fig. 1 that the standard model is superior to the Patlak model when fitting 6 min of data, but that the two models tend to agree in their estimate of K^{trans} in the domain where each is valid.

Extending the practice of linearization and model comparison to a voxel-by-voxel analysis of DCE-MRI data generates the maps seen in Fig. 2. These maps are fairly standard parametric maps presenting the quantities that are usually estimated in DCE-MRI studies. Moreover, because they were constructed with an eye toward balancing bias and variance,[58] they have generated robust and repeatable estimates of model parameters.[18, 20, 22, 23] Notably, the same analytical procedures used in the production of DCE-MRI parametric maps in animal models of cerebral tumors can be used in the analysis of DCE-MRI data in human gliomas. For instance, Fig. 3, a DCE-MRI study conducted at 3 T, illustrates the same hierarchy of parametric estimates as that of the animal model shown in Fig. 2. It is reassuring that Model 3 parametric estimates in the cerebral tumors of animal model generally lie within the range of estimates of Model 3 estimates in human gliomas.[22, 24]

The Logan Plot in DCE-MRI

Following the work of Patlak, Logan developed a graphical approach to measure the tracer distribution volume for reversible systems in positron emission tomography data.[46, 47] In this graphical approach, after some time ($t > t^*$), when an equilibrium of the MRCA between the tissue and the vasculature is established, the plot becomes linear with a slope of distribution volume (V_D).

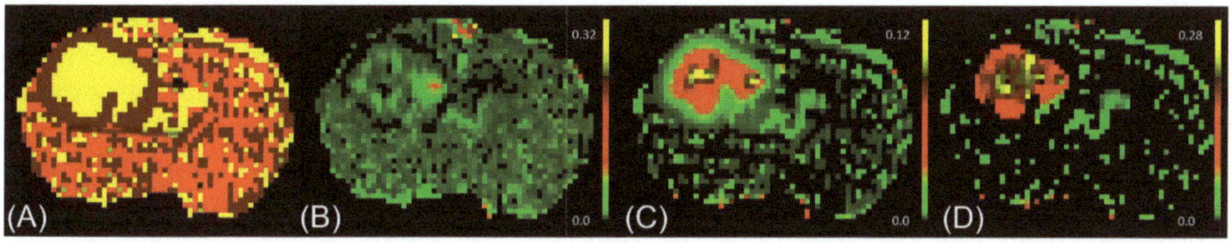

FIG. 2 Maps of model selection and dynamic contrast-enhanced magnetic resonance imaging parameters. (A) A map of data-driven model selection: Model 3 = *yellow*, Model 2 = *dark red*, Model 1 = *light red*. (B) Plasma volume fraction (v_p). (C) K^{trans} [min^{-1}]. (D) Extracellular, extravascular volume fraction (v_e). In a data-driven analysis, only models that meet the criterion for necessity are accepted. Therefore, v_p can be estimated almost anywhere in the brain; v_p and K^{trans} can be estimated in regions where Model 2 is necessary; and v_p, K^{trans}, and v_e can be estimated only in regions where it is necessary to account for an approach to equilibrium between the vascular and extravascular pools of magnetic resonance contrast agent.

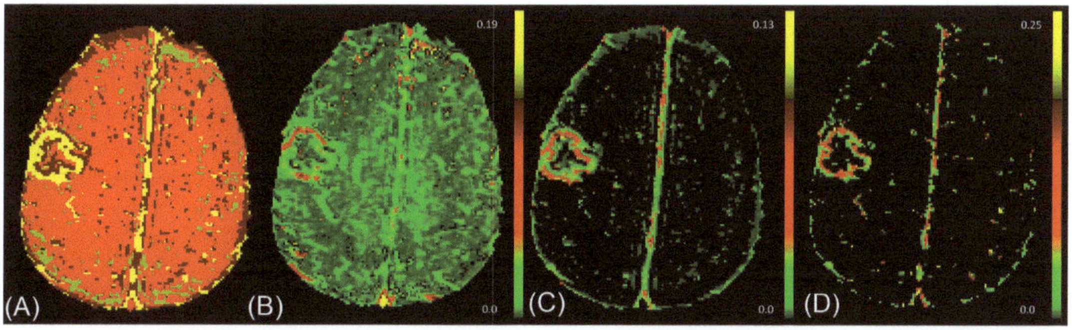

FIG. 3 As in Fig. 2, maps of model selection and dynamic contrast-enhanced magnetic resonance imaging parameters. (A) A map of data-driven model selection: Model 3 = *yellow*, Model 2 = *dark red*, Model 1 = *light red*. (B) Plasma volume fraction (v_p). (C) K^{trans} [min^{-1}]. (D) Extracellular, extravascular volume fraction (v_e).

Logan's expression describes the relationship of tissue indicator concentration to plasma indicator concentration as indicator concentrations in tissue and vasculature approach equilibrium:

$$\frac{\int_0^t C_{tiss}(\tau)d\tau}{C_{tiss}(t)} = V_D \frac{\int_0^t C_p(\tau)d\tau}{C_{tiss}(t)} + Constant \tag{6}$$

where $C_p(t)$ and $C_{tiss}(t)$ are the plasma and tissue concentrations of CA, respectively, and V_D is the distribution volume. In the case that an equilibrium is established between the vascular compartment and the tissue compartment, if $\frac{\int_0^t C_{tiss}(\tau)d\tau}{C_{tiss}(t)}$ is plotted on the ordinate, and $\frac{\int_0^t C_p(\tau)d\tau}{C_{tiss}(t)}$ is plotted on the abscissa, a plot of these quantities becomes linear with a slope of V_D, the distribution volume fraction. Because most MRCAs do not bind to any elements of the intra- or extravascular compartments, in DCE-MRI, V_D is a simple estimate of the extracellular volume fraction.

The advantages of a Logan plot for estimating V_D are threefold. First, because only two parameters (i.e., slope and intercept) are to be estimated, the estimate of V_D (i.e., the estimate of the total extracellular volume) is more stable than the equivalent quantity of the standard model, which is (v_e+v_p). We have shown a strong correlation between tumor cellularity (i.e., the number of cells in a given volume of tumor tissue) and V_D estimated via two independent methods—the standard model and the Logan graphical approach.[75, 76] Second, the regime in which the Logan plot becomes linear is useful for describing where the data becomes uninformative in estimating the parameters of the standard model, since the estimation of both K^{trans} and k_{ep} require tissue concentration to be changing with respect to plasma concentration. Third, unlike the kinematic parameters of K^{trans} and k_{ep}, a DCE-MRI estimate of V_D expressed as a volume fraction is independent of the size of the MRCA.

Many DCE-MRI studies have focused on vascular parameters such as blood volume, blood flow, and vascular permeability, particularly as measured by the forward volume transfer constant K^{trans}. The estimation of interstitial space and distribution volume has received less attention. However, it has been noted that distribution volume estimated via DCE-MRI and quantitative autoradiographic studies did not differ in the means and was highly correlated in rat models of stroke[77] and cerebral ischemia.[78]

Because distribution volume directly estimates the leakage volume available for MRCA distribution outside the cellular space, quantitative estimates of this volume can reveal important physiological changes in the tumor microenvironment. In particular, because V_D approximates the porosity (φ) of the tissue, and because porosity is strongly related to fluid conductivity, V_D is an important quantity in the consideration of interstitial fluid flow.

DCE-MRI Measures of Tumor Fluid and Mechanical Properties

This section, which describes investigations summarized elsewhere,[19, 21] demonstrates that a Logan plot of DCE-MRI data can be used to estimate tumor extracellular volume fraction, and that graphical methods can be extended to normal regions in the tumor boundary. Fig. 4A draws an ROI defined by the Model 3 region of the embedded tumor of the animal shown in Figs. 2 and 3. By applying Eq. (6) to the summed data of this ROI, using the same group-averaged AIF as was used to produce the Patlak maps of Fig. 2, the Logan plot of Fig. 5A was constructed. The linearity of the final portion of the curve is evident, demonstrating that MRCA equilibrated between intra- and extravascular spaces around point number

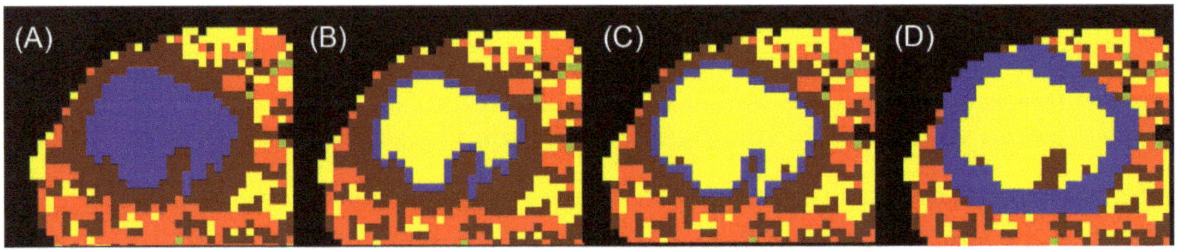

FIG. 4 In the animal of Fig. 3, regions of interest (ROIs) are seen in *blue*. ROIs defined by: (A) the Model 3 ROI, (B) the edge of the Model 3 ROI, (C) the pixels immediately adjacent to and outside of the Model 3 ROI, and (D) a wider region drawn to include most of the Model 2 ROI surrounding the Model 3 ROI.

77 (5 minutes after CA injection). The slope of the linear fitting from point 77 on was 0.32, meaning that 32% of the volume of the selected ROI was extracellular space. Using the same AIF as before, a Logan plot of data from the edge of the tumor (defined by the ROI of Fig. 4B) shows a smaller extracellular volume of about 13% of the total space, as might be expected of a tumor with a necrotic core and a rapidly growing rim. The other two ROIs of Fig. 4 (Fig. 4C and D) lie entirely outside the ROI defined by Model 3 and do not use the AIF of Fig. 5A and B, but rather use the MRCA concentration at the edge of the tumor (ROI, Fig. 4B) as an input function to the outer adjacent single-voxel region (ROI, Fig. 4C), or to a larger and still adjacent catchment region (ROI, Fig. 4D). The Logan-type plot of Fig. 5C (see Ewing et al.[19] for details) demonstrates an eventual equilibration around 5.5 min with an extracellular volume fraction in this mostly normal tissue region of 8.3%. These studies imply that the tumor rim has lower porosity and may be more compressed than the tumor itself, and that the mostly normal tissue in the tumor periphery is even more compressed. Because it is linear, the Patlak-type plot of Fig. 5D implies that, during the experiment, the catchment region is a final sink for

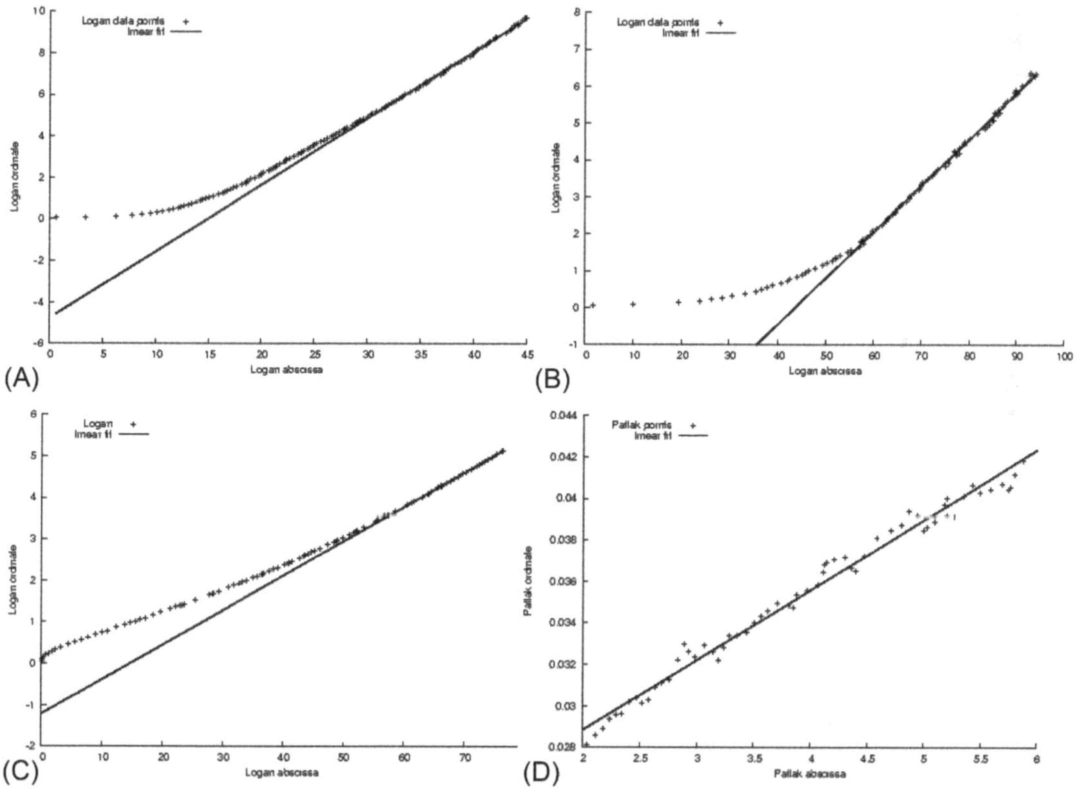

FIG. 5 (A, B, C) Logan plots drawn from the data of the ROIs of Fig. 4A-C, respectively. (D) A Patlak plot drawn from the data of the ROI of Fig. 4D. The Logan plots of (A) and (B) use the same group-averaged arterial input function that produced the Patlak plots of Fig. 1B and C. The Logan plot of (C) and the Patlak plot of (D) use the data of the ROI of Fig. 4B as its input function. That is to say, the input to ROIs of Fig. 4C and D is the tissue concentration of the outermost pixels of the Model 3 ROI.

the MRCA being carried out of the tumor by its exudate. This plot allows an estimate of the flux of tumor exudate across the rim of the tumor and can be correlated with the degree of compression in both the tumor rim and in the tumor periphery.

As described previously,[19] plotting the relation between tumor exudate flux and peritumoral compression leads to a remarkable result in this animal model (Fig. 6). In a sample of 18 animals (16 studied twice) with implanted cerebral tumors, tumor exudate flux was strongly related to the porosity of peritumoral tissue, with a simple R^2 of 0.9 for all 34 points. This study implies that the peritumoral region is under considerable mechanical stress; the measured distribution volume of the peritumoral region ranges from 4% to 18%, compared to that of 20% for normal brain across a wide range of species,[79-81] including the rat. Peritumoral compression as a control of tumor exudate directly relates to a very long line of work in the study of TIFP,[9, 11, 12, 14, 82-87] and more generally to the mechanical and fluid properties of embedded tumors.[13-16, 88-90]

Due to the typically abnormal structure and function of tumor blood vessels, tumor lymphatics, and interstitium,[91] TIFP is centrally elevated in most tumors.[82, 83, 92] The physical model of an embedded tumor in many ways echoes the conditions of CED of drugs, in that there is a centrally increased fluid pressure that demands a description of local stresses and fluxes. Thus, a mechanical and fluid-mechanical theory that correctly describes the local mechanical and fluidic stresses in and around an embedded tumor can also be expected to

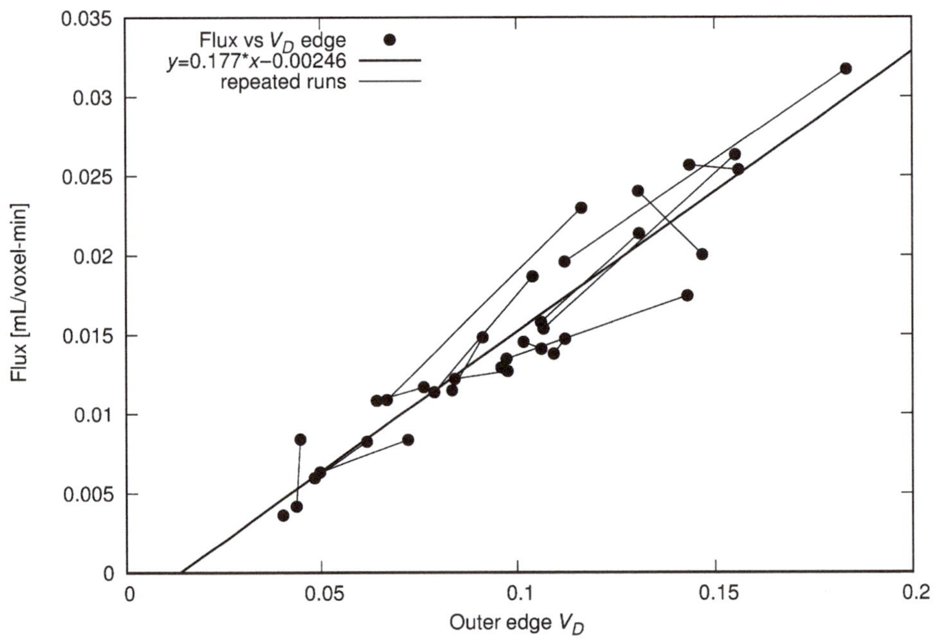

FIG. 6 Exudate flux from the rim of the tumor to the surrounding ring vs distribution volume in the mostly normal tissue surrounding the tumor in 18 animals. The 16 repeated studies are connected with *dashed lines*. This very significant covariance (R^2 ~0.9) implies that a knowledge of V_D in the rim of the tumor will yield a remarkably precise prediction of exudate flow, regardless of those other parameters, such as tumor interstitial fluid pressure and tumor porosity, which might otherwise be thought necessary to predict the exudate flow rate.

describe the delivery and distribution of drugs in an injected fluid, when CED is used in the context of treating a solid tumor.

Combined Measures of Tumor Physiology: Acute Changes after High-Dose Radiotherapy

An MRI study of the acute response of model tumors to high-dose radiotherapy[20] illustrates the range of noninvasive measures of tumor structure and physiology that have become available. In addition to the parameters available via DCE-MRI, there are most notably arterial spin labeling[93-97] to measure tumor blood flow (TBF), and the apparent diffusion coefficient of water[98-101] to probe tissue structures on the microscopic level.

The acute response to radiotherapy in embedded cerebral tumors was studied in 25 rats at 2, 4, 8, 12, and 24 hours after radiotherapy. The radiotherapy used in this study was 20 Gy, delivered using 6 MV photons, 8 Gy/min. Acute postradiation changes in the tissue-vascular interface in a rat model of implanted cerebral gliomas were studied. In addition to the DCE-MRI parameters (i.e., K^{trans}, v_e, V_D, v_p) produced in and around the tumors, TBF and apparent diffusion coefficient were assessed. TBF decreased at 2 hours postirradiation by 70%-80% of preirradiation levels, followed by a steady increase in later studies until, at 24 hours postirradiation, TBF exceeded preirradiation levels by 40% (Fig. 7).

In the tumor itself, the change in the interstitial volume fraction, v_e, which was negative throughout the study, pointed to a significant loss of interstitial volume (i.e., cell swelling). At 2 hours after radiation, v_e was 30% below preirradiated values. The tumor v_e temporarily increased to near preirradiated values 4 hours after radiation, but subsequently decreased to 30% below preirradiated levels at 24 hours postirradiation. V_D in the periphery of the tumor (mostly normal tissue) followed a similar pattern. The vascular permeability, K^{trans}, demonstrated a bimodal response, increasing at 4-8 hours, reaching 35% above preirradiated values,

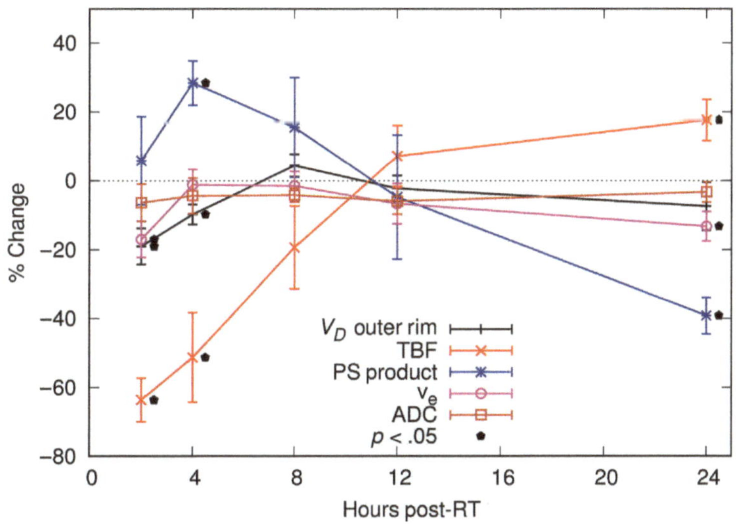

FIG. 7 Changes in tumor physiological parameters after single high-dose radiotherapy. Early and profound changes in tumor blood flow and tumor extracellular volume fraction are evident. Together with Fig. 6, these data imply that tumor blood flow is decreased in synchrony with acute cell swelling in the tumor.

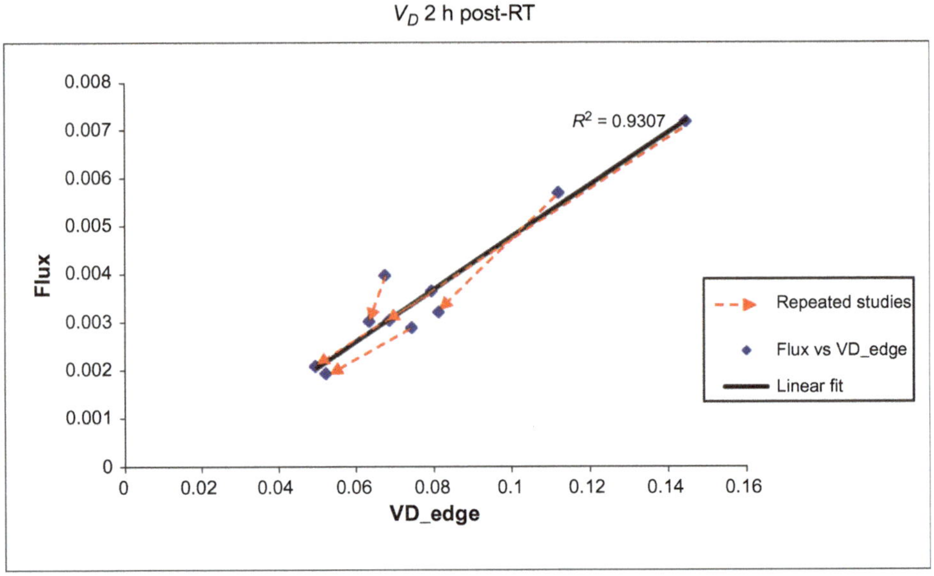

FIG. 8 In a group of five animals studied just before and 2 hours after single high-dose radiotherapy, tumor exudate flux is highly correlated with V_D in the tissue of the mostly normal peritumoral region. Both flux and peritumoral V_D move downward, suggesting that the tumor swells and further compresses the peritumoral tissue.

followed by a decrease at 24 hours to 20%-40% below the preirradiation level. Noting the profound decrease in TBF, it appears that the tumor cells swell (decrease in V_D in the periphery of the tumor, and of v_e in the tumor itself) and throttle TBF in the tumor (Fig. 8).

To our knowledge, this is the first measurement of changes in v_e and/or TBF in the hours immediately following 20 Gy irradiation. Since vascular changes have profound implications for the delivery of adjunct therapies including CED, these findings provide direction for the optimization of combined therapies.

MRI Estimates of Tissue Fluid Conductivity and TIFP

This section summarizes studies presented in a recent paper[9] that demonstrated that DCE-MRI studies could be used to make in vivo estimates of fluid conductivities in tissues and that MRI studies of tumor physiology contain enough information to make a noninvasive estimate of TIFP possible. In this study, experiments were conducted in 36 athymic rats with implanted cerebral tumor, studied either as controls, or after treatments. A direct measure of the central TIFP was conducted using an invasive WIN-TIFP technique, a method that has been used in both in animal[102] and human studies.[103] WIN-TIFP measurements are not easily repeated and are clinically applicable only in locations where the tumor can be easily reached. In the study presented further, MRI estimates of TIFP were compared to WIN-TIFP measures performed immediately after DCE-MRI.

Earlier studies suggested that TIFP could be sensed by dynamic studies. Pishko and colleagues[104] used a model of interstitial transport of contrast in porous media to estimate the flow and pressure of interstitial fluid, showing increased central TIFP and higher exudate streaming velocities at the tumor boundaries. Hassid et al.,[105] in a mouse model of non-small cell lung cancer, used a slow intravenous infusion of MRCA blood for about 2 hours, showing that parametric images of steady-state tissue CA concentration in the tumors and their surroundings reflected the spatial distribution of the TIFP. A study of this type is clearly impractical in human subjects, both because of the length of the study and the amount of MRCA required.

Other MRI studies[106, 107] have evaluated TIFP using parameters that related to tumor perfusion and permeability. However, the resulting estimates were not well correlated with needle-based TIFP measurements. Using a different approach, in a DCE-MRI experiment, the velocity of interstitial fluid flow along a line drawn normal to the tumor surface was proposed as a measure of central TIFP,[108] but our experience with this approach suggests that it may strongly depend on which normally drawn line is used.

Two approaches using MRI data to model TIFP were used, one based on a well-recognized fluid-mechanical model of flow in porous media, and the other an empirical multivariate analysis of MRI vascular parameters as predictors of TIFP. The fluid-mechanics approach demonstrated that, given good estimates for fluid-mechanical constants, the underlying physical mechanical system could be understood in the framework of classical poro-elastic theory. The empirical approach confirmed that MRI studies contain sufficient information to predict TIFP noninvasively.

The fluid-mechanical model for estimating TIFP depended on a measurement of the mean flux of an MRCA at the tumor periphery. By Darcy's law,[109] this flux is driven by the interstitial fluid pressure gradient at the tumor surface, and by such characteristics of the interstitium as porosity and interstitial hydraulic conductivity. The Kozeny–Carman equation[110] was used to estimate fluid conductivity, and TIFP was then estimated in each individual tumor via Darcy's law.

To our knowledge, this research presented the first in vivo estimate of hydraulic conductivity in an embedded tumor. In control studies ($n=13$), the mean estimated hydraulic conductivity (MRI-K) in U251 tumors was $(2.3 \pm 3.1) \times 10^{-5}$ [mm^2/mmHg s]. This value is in approximate agreement with previously reported hydraulic conductivities and porosities in other tumors.[90, 111] Significant positive correlations were found between WIN-TIFP and MRI-TIFP in both mechanical and empirical models. For instance, in the control group of the fluid-mechanical model, MRI-TIFP was a strong predictor of WIN-TIFP ($R^2=0.76$; $P<.0001$).

As we have noted, DCE-MRI provides a number of important measures of a tumor's mechanical and fluid-mechanical status, among which are K^{trans} in the tumor and its rim, V_D in the tumor, rim, and peritumoral region; and flux at the rim of the tumor. To these measures, we can add an arterial spin-labeled measure of TBF and a measure of the apparent diffusion coefficient in the tumor and its peritumoral region. These parameters were estimated in the three different ROIs, as described in Fig. 4 (i.e., whole tumor, tumor rim, tumor periphery), that are believed to have different physiological characteristics and may affect TIFP differently. Using an Akaike information criterion test, four parameters emerged that best described TIFP. In their order of importance, these parameters were V_D in the tumor, the

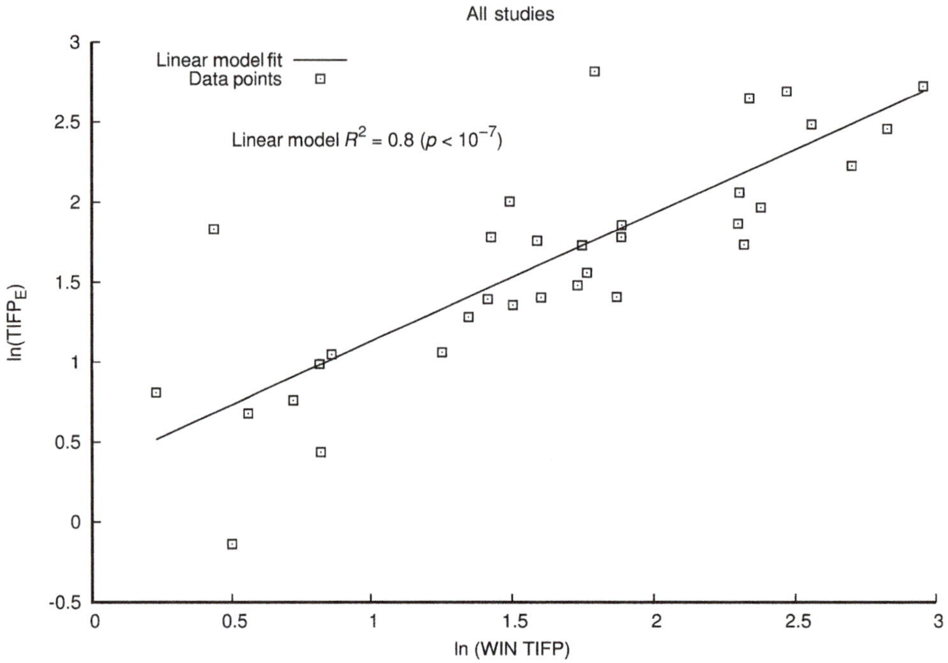

FIG. 9 An empirical prediction ($TIFP_E$) of the measured value of the central tumor interstitial fluid pressure (wick-in-needle measurement of tumor interstitial fluid pressure). The predictor was constructed from four magnetic resonance imaging-measured parameters.

mean interstitial fluid velocity, v, at the tumor surface, TBF_{-peri} in the adjacent normal tissue surrounding the tumor, and TBF in the tumor.

Based on the result of this multivariate analysis, an empirical model for estimating log TIFP was established (MRI-$TIFP_E$; Fig. 9), consisting of linear regression of the above-mentioned four parameters. The R^2 for the linear regression analysis was 0.8 ($P<10^{-6}$). Thus, it can be said with some confidence that noninvasive DCE-MRI and perfusion measures combined contain sufficient information to predict TIFP.

CONCLUSIONS

This research implies that MRI dynamic studies contain enough information to noninvasively estimate TIFP in this, and possibly other, tumor models. More importantly for the purposes of planning CED of drugs, it demonstrates that descriptive measurements of the fluid-mechanical properties of tissue are available via an essentially noninvasive, and therefore repeatable, technology.

APPENDIX 1 THE CRONE-RENKIN SINGLE-CAPILLARY MODEL

This model stays fairly close to the physical reality of microvessels in tissue, and its simplifying assumptions are perforce clearly stated in the derivation. Moreover, there are few places where this derivation is presented in full. We consider that the capillary bed is composed of identical straight tubes of length, L, total surface area, S, and uniform permeability, P, and that flow rate, F, and mean plasma velocity, v, are constant. Consider the schematic representation shown here, along with a differential element and the mass-flow balances drawn around the differential element. Note that conservation of mass requires that the sum of outflows and inflows of tracer sum to zero (Fig. A.1).

We see the following mass balance in the differential element

$$FC(x) = FC(x+dx) + PS[C(x) - C_{EES}(x)]dx \quad \text{(A.1)}$$

where $C(x)$ is the blood concentration of indicator at x, $C_{EES}(x)$ is the concentration of indicator outside the capillary in the extravascular extracellular space, and $x = l/L$, is the fractional distance by which the element of blood has traversed the capillary at velocity v. Note that the leakage out is PS times the concentration at x. The quantity PS is known as the permeability-surface area product, or PS-product. It has the same dimensions as flow, [mL/(mm^3 min)], that is [min^{-1}]. Rearranging the equation and forming the differential.

$$\frac{dC(x)}{dx} + \frac{PS}{F}[C(x) - C_{EES}(x)] = 0 \quad \text{(A.2)}$$

We make the *adiabatic assumption* that tracer concentration outside the capillary is essentially unchanged during the passage of the disk down the capillary; $C_{EES}(x)$ is equal to a constant C_{EES}. This is true if the leakage rate out of the capillary is small, so that very little happens during a single passage. If C_{EES} is a constant, we can solve the equation for the capillary concentration.

$$C(x) = [C_{in} - C_{EES}]e^{-\left(\frac{PS}{F}\right)x} + C_{EES} \quad \text{(A.3)}$$

where C_{in} is the concentration at the input to the capillary. We are interested in the net flux from the capillary to the extravascular, extracellular space. Solving for $x = 1$ to find the venous concentration.

$$C_v = C(x=1) = [C_{in} - C_{EES}]e^{-\frac{PS}{F}} + C_{EES}$$

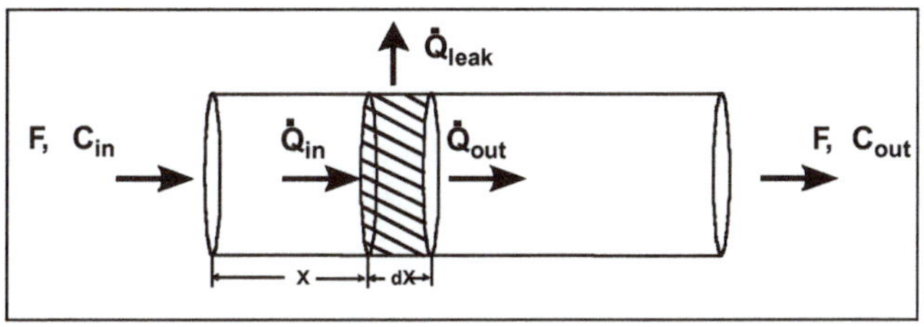

FIG. A.1 A schematic of the Crone-Renkin single-capillary model, with a mass-flow balance for a differential disk of the capillary.

where C_v is the venous concentration at the outlet of the capillary. The arterial concentration is C_{in}, and the arteriovenous difference is

$$C_p - C_v = (C_{in} - C_{EES})\left[1 - e^{-\left(\frac{PS}{F}\right)}\right] \tag{A.4}$$

where $C_p(t)$ is the plasma concentration of indicator as a function of time. This is the first result that we need.

An expression for the arteriovenous concentration difference provides the quantity of indicator leaking out of the capillary per unit time. The *rate* of leakage of indicator is the flow times the arteriovenous concentration difference

$$\frac{dQ}{dt} = F(C_p - C_v) = F(C_{in} - C_{EES})\left[1 - e^{-\left(\frac{PS}{F}\right)}\right] \tag{A.5}$$

where Q is the quantity (moles) of MRCA leaking from the vasculature to the extravascular space. We now change over into units that are *specific*, that is, per-unit-mass, or per-unit-volume. In the case of MRI, we use per-unit-volume, which is massively simpler than the old radiological units, which are per-unit-mass. dQ/dt (leakage) becomes change in tissue concentration, dC_{tiss}/dt, F becomes $f = F/V$, that is, flow per unit volume of tissue, PS becomes a specific quantity

$$\frac{dC_{tiss}}{dt} = f(C_p - C_v) = f(C_{in} - C_{EES})\left[1 - e^{-\left(\frac{PS}{f}\right)}\right] \tag{A.6}$$

In this differential equation, the extravascular, extracellular concentration is just the tissue concentration divided by the extravascular, extracellular volume, v_{EES}, and defining the extraction fraction, E, as the fraction of indicator that leaves the capillary on one pass, we then get a differential equation in C_{tiss} and time

$$\frac{dC_{tiss}}{dt} = fE\left(C_p - \frac{C_{tiss}}{v_{EES}}\right) \tag{A.7}$$

The quantity fE is known as the transfer constant, K^{trans}

$$\frac{dC_{tiss}}{dt} + \frac{K^{trans}}{v_{EES}} C_{tiss} = K^{trans} C_p(t) \tag{A.8}$$

This has a very familiar form—the arterial input function is the forcing function, the tissue concentration is the response, and K^{trans} is the time constant characteristic of the system. The quantity $\frac{K^{trans}}{v_{EES}}$ is a time constant, notated k_{ep} in the standard model, which measures the transfer rate of indicator returning to the vasculature from the extravascular space. This is what is known as a "low-pass" system, which means that, given a constant forcing function, the response will eventually reach some constant level.

By virtue of the linearity of the differential equation, there is an equivalent integral expression to that of Eq. (A.8). Taking account of vascular volume also, we arrive at Eq. (A.1), the working equation that is typically used in the analysis of what Patlak calls 'concentration–time' experiments

$$C(t) = K^{trans} \int_0^t C_p(\tau) e^{-k_{ep}(t-\tau)} d\tau + v_p C_p(t) \tag{A.9}$$

APPENDIX 2 LOGAN PLOT THEORY

The time course of the tissue concentration of MRCA, $C_{tiss}(t)$, can be expressed by a differential equation[44-47]

$$\frac{dC_{tiss}(t)}{dt} = K^{trans}\left(C_p(t) - \frac{C_{tiss}(t)}{v_e}\right) \tag{B.1}$$

where K^{trans} is forward transfer constant, $C_p(t)$ is plasma volume concentration at time t, and v_e is interstitial volume fraction. This equation can be rewritten as

$$C_{tiss}(t) = v_e C_p(t) - \left(\frac{v_e}{K^{trans}}\right)\frac{dC_{tiss}(t)}{dt} \tag{B.2}$$

Following Patlak et al.,[44,45] the total concentration of the tracer in the region of interest (ROI) contains the contribution from the blood volume. Thus,

$$ROI(t) = C_{tiss}(t) + v_p C_p(t) \tag{B.3}$$

where v_p is the plasma volume of the tracer within the ROI. From Eqs. (B.2), (B.3)

$$ROI(t) = (v_e + v_p)C_{p(t)} - \left(\frac{v_e}{K^{trans}}\right)\frac{dC_{tiss}}{dt} \tag{B.4}$$

Inserting $(v_e + v_p) = V_D$, the distribution volume, Eq. (B.4) reduces to

$$ROI(t) = V_D C_{p(t)} - \left(\frac{v_e}{K^{trans}}\right)\frac{dC_{tiss}}{dt}$$

Integrating on both sides of this equation, we get

$$\int_0^t ROI(\tau)d\tau = V_D\int_0^t C_p(\tau)d\tau - \left(\frac{v_e}{K^{trans}}\right)C_{tiss}(t) \tag{B.5}$$

Dividing both sides by $ROI(t)$ and rearranging by inserting $ROI(t)$ from Eq. (B.3), the Eq. (B.5) becomes

$$\frac{\int_0^t ROI(\tau)d\tau}{ROI(t)} = \frac{V_D\int_0^t C_p(\tau)d\tau}{ROI(t)} - \left(\frac{v_e}{K^{trans}}\right)\frac{C_{tiss}(t)}{C_{tiss}(t) + v_p C_p(t)} \tag{B.6}$$

Assume that, at some point in time after the administration of MRCA, there exists a steady-state condition in which the tissue concentration follows the plasma concentration. In essence, this criterion means that the time constant of the input function is much longer than the time constant of the exchange of MRCA across the microvasculature. When this equilibrium condition is valid, the last term in Eq. (B.6) becomes a constant, and the equation reduces to

$$\frac{\int_0^t ROI(\tau)d\tau}{ROI(t)} = V_D\frac{\int_0^t C_p(\tau)d\tau}{ROI(t)} + Constant \tag{B.7}$$

Since the ROI to be examined is the always the tissue region, Eq. (B.7) can simply be rewritten in terms of plasma and tissue concentration of CA.

$$\frac{\int_0^t ROI(\tau)d\tau}{C_{tiss}(t)} = V_D \frac{\int_0^t C_p(\tau)d\tau}{C_{tiss}(t)} + Constant \tag{B.8}$$

where $C_p(t)$ and $C_{tiss}(t)$ are the plasma and tissue concentrations of CA, respectively, and V_D is the distribution volume. In the case that an equilibrium has been established between the vascular compartment and the tissue compartment, if $\frac{\int_0^t C_{tiss}(\tau)d\tau}{C_{tiss}(t)}$ is plotted on the ordinate, and $\frac{\int_0^t C_p(\tau)d\tau}{C_{tiss}(t)}$ is plotted on the abscissa, a plot of these quantities becomes linear with a slope of V_D, the distribution volume fraction.

References

1. Groothuis DR. The blood-brain and blood-tumor barriers: a review of strategies for increasing drug delivery. *Neuro-Oncology*. 2000;2(1):45-59.
2. Schlageter KE, Molnar P, Lapin GD, Groothuis DR. Microvessel organization and structure in experimental brain tumors: microvessel populations with distinctive structural and functional properties. *Microvasc Res*. 1999;58(3):312-328.
3. Bobo RH, Laske DW, Akbasak A, Morrison PF, Dedrick RL, Oldfield EH. Convection-enhanced delivery of macromolecules in the brain. *Proc Natl Acad Sci U S A*. 1994;91(6):2076-2080.
4. Lieberman DM, Laske DW, Morrison PF, Bankiewicz KS, Oldfield EH. Convection-enhanced distribution of large molecules in gray matter during interstitial drug infusion. *J Neurosurg*. 1995;82(6):1021-1029.
5. Rosenbluth KH, Eschermann JF, Mittermeyer G, Thomson R, Mittermeyer S, Bankiewicz KS. Analysis of a simulation algorithm for direct brain drug delivery. *Neuroimage*. 2012;59(3):2423-2429.
6. Dai W, Astary GW, Kasinadhuni AK, Carney PR, Mareci TH, Sarntinoranont M. Voxelized model of brain infusion that accounts for small feature fissures: comparison with magnetic resonance tracer studies. *J Biomech Eng*. 2016;138(5).
7. Kim JH, Astary GW, Kantorovich S, Mareci TH, Carney PR, Sarntinoranont M. Voxelized computational model for convection-enhanced delivery in the rat ventral hippocampus: comparison with in vivo MR experimental studies. *Ann Biomed Eng*. 2012;40(9):2043-2058.
8. Meyer HJ, Garnov N, Surov A. Comparison of two mathematical models of cellularity calculation. *Transl Oncol*. 2018;11(2):307-310.
9. Elmghirbi R, Nagaraja TN, Brown SL, et al. Toward a noninvasive estimate of interstitial fluid pressure by dynamic contrast-enhanced MRI in a rat model of cerebral tumor. *Magn Reson Med*. 2018;5:2040-2052.
10. Netti PA, Baxter LT, Boucher Y, Skalak R, Jain RK. Time-dependent behavior of interstitial fluid pressure in solid tumors: implications for drug delivery. *Cancer Res*. 1995;55(22):5451-5458.
11. Liu LJ, Brown SL, Ewing JR, Ala BD, Schneider KM, Schlesinger M. Estimation of tumor interstitial fluid pressure (TIFP) noninvasively. *PLoS One*. 2016;11(7).
12. Liu LJ, Brown SL, Ewing JR, Schlesinger M. Phenomenological model of interstitial fluid pressure in a solid tumor. *Phys Rev E*. 2011;84(2). 021919–021911 to 021919–021919.
13. Sarntinoranont M. *A Mechanical Model of a Growing Solid Tumor With Applications to Vessel Collapse and Drug Transport* [Dissertation]. Berkeley, CA: Engineering—Mechanical Engineering, University of California, Berkeley; 1999.
14. Sarntinoranont M, Rooney F, Ferrari M. Interstitial stress and fluid pressure within a growing tumor. *Ann Biomed Eng*. 2003;31(3):327-335.
15. Stylianopoulos T, Martin JD, Snuderl M, Mpekris F, Jain SR, Jain RK. Coevolution of solid stress and interstitial fluid pressure in tumors during progression: implications for vascular collapse. *Cancer Res*. 2013;73(13):3833-3841.
16. Jain RK, Martin JD, Stylianopoulos T. The role of mechanical forces in tumor growth and therapy. *Annu Rev Biomed Eng*. 2014;16:321-346.

17. Ewing JR, Bagher-Ebadian H. Model selection in measures of vascular parameters using dynamic contrast-enhanced MRI: experimental and clinical applications. *NMR Biomed*. 2013;26(8):1028-1041.
18. Nagaraja TN, Elmghirbi R, Brown SL, et al. Reproducibility and relative stability in magnetic resonance imaging indices of tumor vascular physiology over a period of 24 h in a rat 9L gliosarcoma model. *Magn Reson Imaging*. 2017;44:131-139.
19. Ewing JR, Nagaraja TN, Aryal MP, et al. Peritumoral tissue compression is predictive of exudate flux in a rat model of cerebral tumor: an MRI study in an embedded tumor. *NMR Biomed*. 2015, 1557-1559.
20. Brown SL, Nagaraja TN, Aryal MP, et al. MRI-tracked tumor vascular changes in the hours after single-fraction irradiation. *Radiat Res*. 2015;183(6):713-721.
21. Aryal MP, Nagaraja TN, Keenan KA, et al. Dynamic contrast enhanced MRI parameters and tumor cellularity in a rat model of cerebral glioma at 7 T. *Magn Reson Med*. 2014;71(6):2206-2214.
22. Aryal MP, Nagaraja TN, Brown SL, et al. Intratumor distribution and test-retest comparisons of physiological parameters quantified by dynamic contrast-enhanced MRI in rat U251 glioma. *NMR Biomed*. 2014; 27(10):1230-1238.
23. Nagaraja TN, Aryal MP, Brown SL, et al. Cilengitide-induced temporal variations in transvascular transfer parameters of tumor vasculature in a rat glioma model: identifying potential MRI biomarkers of acute effects. *PLoS One*. 2013;8(12).
24. Bagher-Ebadian H, Jain R, Nejad-Davarani SP, et al. Model selection for DCE-T1 studies in glioblastoma. *Magn Reson Med*. 2012;68(1):241-251.
25. Ewing JR, Brown SL, Lu M, et al. Model selection in magnetic resonance imaging measurements of vascular permeability: Gadomer in a 9L model of rat cerebral tumor. *J Cereb Blood Flow Metab*. 2006;26(3):310-320.
26. Stewart G. Researches on the circulation time and on the influences which affect it. The output of the heart. *J Physiol*. 1897;XXII:159-183.
27. Stewart G. The pulmonary circulation time, the quantity of blood in the lungs and the output of the heart. *Am J Phys*. 1921;58:20-44.
28. Stewart G. The output of the heart in dogs. *Am J Phys*. 1921;57:27-50.
29. Hamilton W, Moore J, Kingsman J, Spurling R. Simultaneous determination of the greater and lesser circulation times, of the mean velocity of blood flow through the heart and luns, of the cardiac output and an approximation of the amount of blood actively circulating. *Am J Physiol*. 1928;IXXXV:377.
30. Moore J, Kinsman J, Hamilton W, Spurling R. Studies on the circulation. *Am J Physiol*. 1929;IXXXIX:331-339.
31. Hamilton WF, Moore JW, Kinsman JM, Spurling RG. Studies on the circulation. IV. Further analysis of the injection method, and of changes in hemodynamics under physiological and pathological conditions. *Am J Physiol*. 1932;99(3):534-551.
32. Kety S, Schmidt C. The determination of cerebral blood flow in man by the use of nitrous oxide in low concentrations. *Am J Phys*. 1945;143:53-66.
33. Kety S, Schmidt C. The nitrous oxide method for the quantitative determination of cerebral blood flow in man: theory, procedure, and normal values. *J Clin Invest*. 1948;27:476-483.
34. Meier P, Zierler KL. On the theory of the indicator-dilution method for measurement of blood flow and volume. *J Appl Physiol*. 1954;6:731-743.
35. Zierler K. Theory of the use of arteriovenous concentration differences for measuring metabolism in steady and non-steady states. *Circ Res*. 1961;40:2111-2125.
36. Zierler K. Equations for measuring blood flow by external monitoring of radioisotopes. *Circ Res*. 1965;16:309-321.
37. Fick A. Uber die messung des blutquantums in der herzventrikeln. *Sitz Der Physik-Med Ges Wurzburg*. 1870;2:16-28.
38. Grubb R, Raichle M, Eichling J, Ter-Pogossian M. The effects of changes in paco2 on cerebral blood volume, blood flow, and vascular mean transit time. *Stroke*. 1974;5:630-639.
39. Grubb R, Raichle M, Gado M, Eichling J, Hughes C. Cerebral blood flow, oxygen utilization, and blood volume in dementia. *Neurology*. 1977;27:905-910.
40. Phelps M, Hoffman E, Coleman R, et al. Tomographic images of blood pool and perfusion in brain and heart. *J Nucl Med*. 1977;17:603-612.
41. Herscovitch P, Markham J, Raichle ME. Brain blood flow measured with intravenous $H_2^{15}O$. I Theory and error analysis. *J Nucl Med*. 1983;24:782-789.

42. Raichle M, Eichling J, Straatmann M, Welch M, Larson K, Ter-Pogossian M. Blood-brain barrier permeability of 11C-labeled alcohols and 15O-labeled water. *Am J Phys*. 1976;230:543-552.
43. Raichle ME, Eichling JO, Grubb RL. Brain permeability of water. *Arch Neurol*. 1974;30:319-321.
44. Patlak CS, Blasberg RG, Fenstermacher JD. Graphical evaluation of blood-to-brain transfer constants from multiple-time uptake data. *J Cereb Blood Flow Metab*. 1983;3:1-7.
45. Patlak C, Blasberg R. Graphical evaluation of blood to brain transfer constants from multiple time up take data. Generalizations. *J Cereb Blood Flow Metab*. 1985;5:584-590.
46. Logan J. Graphical analysis of PET data applied to reversible and irreversible tracers. *Nucl Med Biol*. 2000; 27(7):661-670.
47. Logan J, Fowler JS, Volkow ND, et al. Graphical analysis of reversible radioligand binding from time-activity measurements applied to [N-11C-methyl]-(−)-cocaine PET studies in human subjects. *J Cereb Blood Flow Metab*. 1990;10(5):740-747.
48. Sokoloff L. *Cerebral Circulatory and Metabolic Changes Associated With Aging*. Paper presented at: Research Publications, Association for Research in Nervous & Mental Diseases, 1960.
49. Sourbron SP, Buckley DL. Classic models for dynamic contrast-enhanced MRI. *NMR Biomed*. 2013; 26(8):1004-1027.
50. Paudyal R, Bagher-Ebadian H, Nagaraja TN, Fenstermacher JD, Ewing JR. Modeling of look-locker estimates of the magnetic resonance imaging estimate of longitudinal relaxation rate in tissue after contrast administration. *Magn Reson Med*. 2011;66(5):1432-1444.
51. Huang W, Tudorica LA, Li X, et al. Discrimination of benign and malignant breast lesions by using shutter-speed dynamic contrast-enhanced MR imaging. *Radiology*. 2011;261(2):394-403.
52. Li X, Rooney WD, Springer Jr CS. A unified magnetic resonance imaging pharmacokinetic theory: intravascular and extracellular contrast reagents. *Magn Reson Med*. 2005;54(6):1351-1359.
53. Tofts P, Kermode A. Measurement of the blood-brain barrier permeability and leakage space using dynamic MR imaging. 1. Fundamental concepts. *Magn Reson Med*. 1991;17:357-367.
54. Patlak CS, Blasberg RG, Fenstermacher JD. An evaluation of errors in the determination of blood flow by the indicator fractionation and tissue equilibration (Kety) methods. *J Cereb Blood Flow Metab*. 1984;4:47-60.
55. Tofts PS, Brix G, Buckley DL, et al. Estimating kinetic parameters from dynamic contrast-enhanced T(1)-weighted MRI of a diffusable tracer: standardized quantities and symbols. *J Magn Reson Imaging*. 1999;10 (3):223-232.
56. Brix G, Kiessling F, Lucht R, et al. Microcirculation and microvasculature in breast tumors: pharmacokinetic analysis of dynamic MR image series. *Magn Reson Med*. 2004;52(2):420-429.
57. Ingrisch M, Sourbron S. Tracer-kinetic modeling of dynamic contrast-enhanced MRI and CT: a primer. *J Pharmacokinet Pharmacodyn*. 2013;40(3):281-300.
58. Burnham KP, Anderson DR. *Model Selection and Multimodel Inference: A Practical Information-Theoretic Approach*. 2nd ed. New York: Springer; 2002.
59. Elmghirbi R, Nagaraja TN, Brown SL, et al. Acute temporal changes of MRI-tracked tumor vascular parameters after combined anti-angiogenic and radiation treatments in a rat glioma model: identifying signatures of synergism. *Radiat Res*. 2017;187(1):79-88.
60. Crone C. The permeability of capillaries in various organs as determined by use of the 'indicator diffusion' method. *Acta Physiol Scand*. 1963;58:292-305.
61. Renkin EM. Exchangeability of tissue potassium in skeletal muscle. *Am J Physiol*. 1959;6(197):1211-1215.
62. Johnson JA, Wilson T. A model for capillary exchange. *Am J Physiol*. 1966;210(6):1299-1303.
63. Lassen NA, Perl W. *Tracer Kinetic Methods in Medical Physiology*. 1st ed. New York, NY: Raven Press; 1979.
64. Huang W, Li X, Morris EA, et al. The magnetic resonance shutter speed discriminates vascular properties of malignant and benign breast tumors in vivo. *Proc Natl Acad Sci U S A*. 2008;105(46):17943-17948.
65. Landis CS, Li X, Telang FW, et al. Determination of the MRI contrast agent concentration time course in vivo following bolus injection: effect of equilibrium transcytolemmal water exchange. *Magn Reson Med*. 2000; 44(4):563-574.
66. Landis CS, Li X, Telang FW, et al. Equilibrium transcytolemmal water-exchange kinetics in skeletal muscle in vivo. *Magn Reson Med*. 1999;42:467-478.
67. Li X, Huang W, Morris EA, et al. Dynamic NMR effects in breast cancer dynamic-contrast-enhanced MRI. *Proc Natl Acad Sci U S A*. 2008;105(46):17937-17942.
68. Li X, Priest RA, Woodward WJ, et al. Feasibility of shutter-speed DCE-MRI for improved prostate cancer detection. *Magn Reson Med*. 2013;69(1):171-178.

69. Li X, Springer Jr CS, Jerosch-Herold M. First-pass dynamic contrast-enhanced MRI with extravasating contrast reagent: evidence for human myocardial capillary recruitment in adenosine-induced hyperemia. *NMR Biomed*. 2009;22(2):148-157.
70. Yankeelov TE, Rooney WD, Huang W, et al. Evidence for shutter-speed variation in CR bolus-tracking studies of human pathology. *NMR Biomed*. 2005;18(3):173-185.
71. Buckley DL, Kershaw LE, Stanisz GJ. Cellular-interstitial water exchange and its effect on the determination of contrast agent concentration in vivo: dynamic contrast-enhanced MRI of human internal obturator muscle. *Magn Reson Med*. 2008;60(5):1011-1019.
72. Spencer RG, Fishbein KW. Measurement of spin-lattice relaxation times and concentrations in systems with chemical exchange using the one-pulse sequence: breakdown of the Ernst model for partial saturation in nuclear magnetic resonance spectroscopy. *J Magn Reson*. 2000;142(1):120-135.
73. Ewing JR, Wei L, Knight RA, et al. Direct comparison of local cerebral blood flow rates measured by MRI arterial spin-tagging and quantitative autoradiography in a rat model of experimental cerebral ischemia. *J Cereb Blood Flow Metab*. 2003;23(2):198-209.
74. Nagaraja TN, Karki K, Ewing JR, et al. The MRI-measured arterial input function resulting from a bolus injection of Gd-DTPA in a rat model of stroke slightly underestimates that of Gd-[(14)C]DTPA and marginally overestimates the blood-to-brain influx rate constant determined by Patlak plots. *Magn Reson Med*. 2010;63(6):1502-1509.
75. Aryal MP, Nagaraja TN, Keenan KA, et al. Dynamic contrast enhanced MRI parameters and tumor cellularity in a rat model of cerebral glioma at 7T. *Mag Reson Med*. 2012;71:2206-2214.
76. Aryal MP, Nagaraja TN, Keenan KA, et al. *Logan Plot Estimates of Tracer Distribution Volume from Dynamic Contrast Enhanced MRI Data and Tumor Cellularity in a Rat Model of Cerebral Glioma at 7T*. Salt Lake City, UT: International Society for Magnetic Resonance in Medicine; 2013.
77. Nagaraja TN, Ewing JR, Karki K, et al. MRI and quantitative autoradiographic studies following bolus injections of unlabeled and (14)C-labeled gadolinium-diethylenetriaminepentaacetic acid in a rat model of stroke yield similar distribution volumes and blood-to-brain influx rate constants. *NMR Biomed*. 2011;24(5):547-558.
78. Knight RA, Karki K, Ewing JR, et al. Estimating blood and brain concentrations and blood-to-brain influx by magnetic resonance imaging with step-down infusion of Gd-DTPA in focal transient cerebral ischemia and confirmation by quantitative autoradiography with Gd-[(14)C]DTPA. *J Cereb Blood Flow Metab*. 2009;29(5):1048-1058.
79. Levin VA, Fenstermacher JD, Patlak CS. Sucrose and inulin space measurements of cerebral cortex in four mammalian species. *Am J Phys*. 1970;219(5):1528-1533.
80. Rosenberg G, Kyner W, Fenstermacher J, Patlak C. Effect of vasopressin on ependymal and capillary permeability to tritiated water in cat. *Am J Phys*. 1986;251:f485-f489.
81. Sykova E, Nicholson C. Diffusion in brain extracellular space. *Physiol Rev*. 2008;88(4):1277-1340.
82. Jain RK. Transport of molecules in the tumor interstitium: a review. *Cancer Res*. 1987;47(12):3039-3051.
83. Boucher Y, Baxter LT, Jain RK. Interstitial pressure gradients in tissue-isolated and subcutaneous tumors: implications for therapy. *Cancer Res*. 1990;50(15):4478-4484.
84. Nathanson SD, Nelson L. Interstitial fluid pressure in breast cancer, benign breast conditions, and breast parenchyma. *Ann Surg Oncol*. 1994;1(4):333-338.
85. Boucher Y, Salehi H, Witwer B, Harsh GR, Jain RK. Interstitial fluid pressure in intracranial tumours in patients and in rodents. *Br J Cancer*. 1997;75(6):829-836.
86. Milosevic M, Fyles A, Hedley D, et al. Interstitial fluid pressure predicts survival in patients with cervix cancer independent of clinical prognostic factors and tumor oxygen measurements. *Cancer Res*. 2001;61(17):6400-6405.
87. Jain RK. Normalization of tumor vasculature: an emerging concept in antiangiogenic therapy. *Science*. 2005;307(5706):58-62.
88. Tse JM, Cheng G, Tyrrell JA, et al. Mechanical compression drives cancer cells toward invasive phenotype. *Proc Natl Acad Sci U S A*. 2012;109(3):911-916.
89. Voutouri C, Mpekris F, Papageorgis P, Odysseos AD, Stylianopoulos T. Role of constitutive behavior and tumor-host mechanical interactions in the state of stress and growth of solid tumors. *PLoS One*. 2014;9(8).
90. Pishko GL, Astary GW, Mareci TH, Sarntinoranont M. Sensitivity analysis of an image-based solid tumor computational model with heterogeneous vasculature and porosity. *Ann Biomed Eng*. 2011;39(9):2360-2373.
91. Jain RK. Normalizing tumor vasculature with anti-angiogenic therapy: a new paradigm for combination therapy. *Nat Med*. 2001;7(9):987-989.

92. Wiig H, Rubin K, Reed RK. New and active role of the interstitium in control of interstitial fluid pressure: potential therapeutic consequences. *Acta Anaesthesiol Scand*. 2003;47(2):111-121.
93. Ewing JR, Jiang Q, Boska M, et al. *A Fast Method for Measuring Cerebral Perfusion Using Arterial Spin Labeling and Variable Flip-Angle FLASH Imaging*, 1995 Paper presented at: Society of Magnetic Resonance, Nice.
94. Roberts DA, Detre JA, Bolinger L, Insko EK, Leigh Jr. JS. Quantitative magnetic resonance imaging of human brain perfusion at 1.5 T using steady-state inversion of arterial water. *Proc Natl Acad Sci U S A*. 1994;91(1):33-37.
95. Williams DS, Detre JA, Leigh JS, Koretsky AP. Magnetic resonance imaging of perfusion using spin inversion of arterial water. *Proc Natl Acad Sci U S A*. 1992;89(1):212-216.
96. Ewing JR, Cao Y, Fenstermacher JA. Single-coil arterial spin-tagging for estimating cerebral blood flow as viewed from the capillary—the relative contributions of intra- and extra-vascular signal. *Magn Reson Med*. 2001;46:465-475.
97. Alsop DC, Detre JA, Golay X, et al. Recommended implementation of arterial spin-labeled perfusion MRI for clinical applications: a consensus of the ISMRM perfusion study group and the European consortium for ASL in dementia. *Magn Reson Med*. 2014;73:102-116.
98. Tsien C, Cao Y, Chenevert T. Clinical applications for diffusion magnetic resonance imaging in radiotherapy. *Semin Radiat Oncol*. 2014;24(3):218-226.
99. Lemasson B, Galban CJ, Boes JL, et al. Diffusion-weighted MRI as a biomarker of tumor radiation treatment response heterogeneity: a comparative study of whole-volume histogram analysis versus voxel-based functional diffusion map analysis. *Transl Oncol*. 2013;6(5):554-561.
100. Chenevert TL, Stegman LD, Taylor JMG, et al. Diffusion magnetic resonance imaging: an early surrogate marker of therapeutic efficacy in brain tumors. *J Natl Cancer Inst*. 2000;92(24):2029-2036.
101. Chenevert T, Branberg J, Pipe J. Anisotropic diffusion in human white matter: demonstration with MR techniques in vivo. *Radiology*. 1990;177:401-405.
102. Wiig H, Reed RK, Aukland K. Measurement of interstitial fluid pressure in dogs: evaluation of methods. *Am J Phys*. 1987;253(2 Pt 2):H283-H290.
103. Boucher Y, Kirkwood JM, Opacic D, Desantis M, Jain RK. Interstitial hypertension in superficial metastatic melanomas in humans. *Cancer Res*. 1991;51(24):6691-6694.
104. Pishko GL, Astary GW, Zhang J, Mareci TH, Sarntinoranont M. Role of convection and diffusion on DCE-MRI parameters in low leakiness KHT sarcomas. *Microvasc Res*. 2012;84(3):306-313.
105. Hassid Y, Furman-Haran E, Margalit R, Eilam R, Degani H. Noninvasive magnetic resonance imaging of transport and interstitial fluid pressure in ectopic human lung tumors. *Cancer Res*. 2006;66(8):4159-4166.
106. Ellingsen C, Walenta S, Hompland T, Mueller-Klieser W, Rofstad EK. The microenvironment of cervical carcinoma xenografts: associations with lymph node metastasis and its assessment by DCE-MRI. *Transl Oncol*. 2013; 6(5):607-617.
107. Haider MA, Sitartchouk I, Roberts TP, Fyles A, Hashmi AT, Milosevic M. Correlations between dynamic contrast-enhanced magnetic resonance imaging-derived measures of tumor microvasculature and interstitial fluid pressure in patients with cervical cancer. *J Magn Reson Imaging*. 2007;25(1):153-159.
108. Hompland T, Ellingsen C, Ovrebo KM, Rofstad EK. Interstitial fluid pressure and associated lymph node metastasis revealed in tumors by dynamic contrast-enhanced MRI. *Cancer Res*. 2012;72(19):4899-4908.
109. Nield DA, Bejan A. *Convection in Porous Media*. 3rd ed. New York, NY: Springer Science+Business Media, Inc.; 2006.
110. Truskey GA, Yuan F, Katz DF. *Transport Phenomena in Biological Systems*. Pearson Prentice Hall; 2010.
111. Netti PA, Berk DA, Swartz MA, Grodzinsky AJ, Jain RK. Role of extracellular matrix assembly in interstitial transport in solid tumors. *Cancer Res*. 2000;60(9):2497-2503.

CHAPTER

21

Clinical Methods of Nervous System Drug Delivery for Tumors

Lee S. Hwang, Daria Krivosheya, Michael A. Vogelbaum

Department of Neurosurgery, The Cleveland Clinic, Cleveland, OH, United States

DRUG DELIVERY FOR TREATMENT OF GLIOBLASTOMA

Malignant primary brain tumors are uncommon and are typically aggressive and difficult to treat. Glioblastoma is the most common (47.1%) primary malignant tumor of the central nervous system (CNS).[1] The current standard of care for patients with newly diagnosed glioblastoma consists of surgical debulking followed by adjuvant temozolomide (TMZ) and radiation. Despite advances in surgery, radiation therapy, and chemotherapy, the overall prognosis of patients with glioblastoma remains dismal, with the median survival of just 12-18 months with maximal treatment.[2,3] The 5-year survival rate is less than 10%.[4,5] Furthermore, recurrence is inevitable even if the confluent tumor is totally resected, because the infiltrative nature of the tumor cells makes it impossible to completely obliterate microscopic disease. Choucair et al.[6] reported that more than 90% of patients with glioblastoma presented with recurrence at the original tumor location, and multiple lesions developed in 5% of patients after treatment.

The blood–brain barrier (BBB) limits the delivery and efficacy of chemotherapeutic agents that are administered systemically. The BBB, which restricts access of molecules from the bloodstream to the CNS, is primarily composed of endothelial cells linked by tight junctions. The structural bases of the BBB also consist of pericytes and astrocyte foot processes, which enhance protection of the CNS by wrapping around endothelial cells (Fig. 1).[7] Molecules in the circulation gain access to the brain interstitial fluid through simple/facilitated diffusion or carrier/receptor-mediated transport, favoring the passage of small lipophilic molecules, water, amino acids, and peptides.[8] As a result, approximately 98% of small-molecule drugs and almost all large-molecule drugs fail to permeate the BBB.[9,10]

When considering drug delivery to a brain tumor, one must decide whether the drug can be delivered effectively as-is, or whether it requires special packaging for selective targeting.

Nervous System Drug Delivery
https://doi.org/10.1016/B978-0-12-813997-4.00021-9

© 2019 Elsevier Inc. All rights reserved.

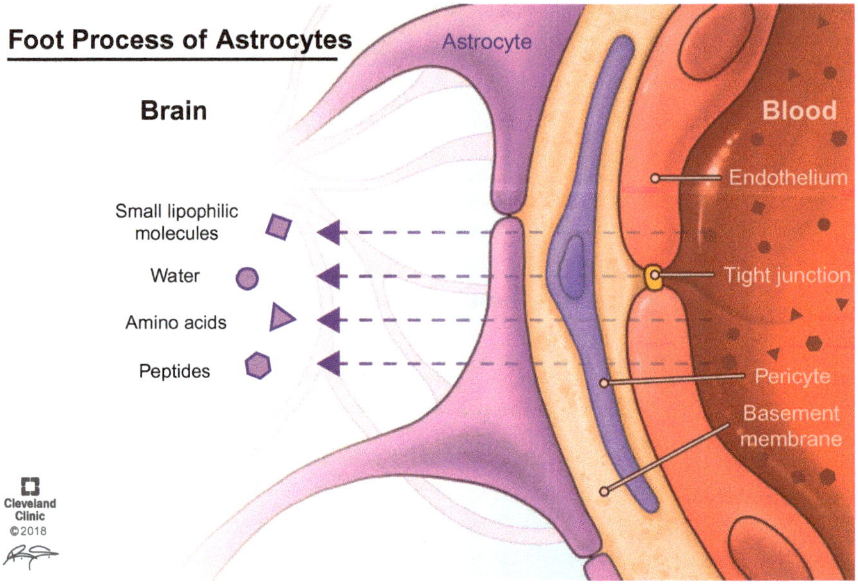

FIG. 1 The blood–brain barrier, which restricts access of molecules from the bloodstream to the central nervous system, primarily comprises endothelial cells linked by tight junctions. The structural basis of the blood–brain barrier also consists of pericytes and astrocyte foot processes, which enhance protection of the central nervous system by wrapping around endothelial cells. Molecules in the circulation gain access to the brain interstitial fluid through simple/facilitated diffusion or carrier/receptor-mediated transport, favoring the passage of small lipophilic molecules, water, amino acids, and peptides.

Very few chemotherapeutic drugs can cross the BBB and therefore require a means, or a vector, to ensure drug delivery to tumor tissue in the CNS. Nature provides a great example of horizontal delivery of genetic material in the form of viruses. Using these self-propagation machines in science for drug delivery is being actively explored and is effective for transfer of genetic material. This active area of research will be discussed in a separate chapter of this book.

Although the virus strains used in medicine have decreased virulence (either naturally or by design), the potential for acute toxicity or latent infection and chronic disease still exists. Therefore, exploration of other means of drug packaging continues. This chapter first discusses different vectors currently used for drug delivery in treatment of glioblastoma. We will then discuss various methods of therapeutic agent delivery to the tumor. Finally, we will consider the importance of matching the therapeutic drug to the ideal mode of delivery, to ensure that the drug successfully reaches its target and exerts its therapeutic effect.

METHODS OF DRUG PACKAGING

Nanovectors

Nanotechnology-based devices have been extensively studied to improve systemic delivery of chemotherapeutic agents for brain tumor. These nanovectors are a promising approach to targeted drug delivery and include nanoparticles, liposomes, carbon nanotubes (CNTs), microcapsules, micelles, and dendrimers (Fig. 2A).[11]

Many anticancer drugs have been conjugated to nanomaterials for local administration and have demonstrated significant success in preclinical models targeting malignant brain tumors, including glioblastoma.[12-14] Positron emission tomography (PET) showed that nanoparticles can penetrate the rodent brain by convection-enhanced delivery (CED) and can be successfully distributed in large intracranial volumes.[10] Furthermore, CED administration of anticancer agents loaded into nanoparticles significantly increases the survival rate of animals with glioblastoma.[10] One challenge is that these polymeric devices degrade quickly and are rapidly cleared from the blood when administered by intravenous injection.[15] Liposomes, which are small artificial vesicles composed of lipid bilayers, also have a short half-life in blood.[15] However, surface-modified liposomes and nanoparticles, coated with specific ligands or proteins to increase the circulation time, have been more therapeutically effective for brain tumors.[16,17]

CNTs are another class of nanotechnology-based devices extensively studied for the treatment of brain tumors. They are large, cylindrical molecules consisting of a hexagonal

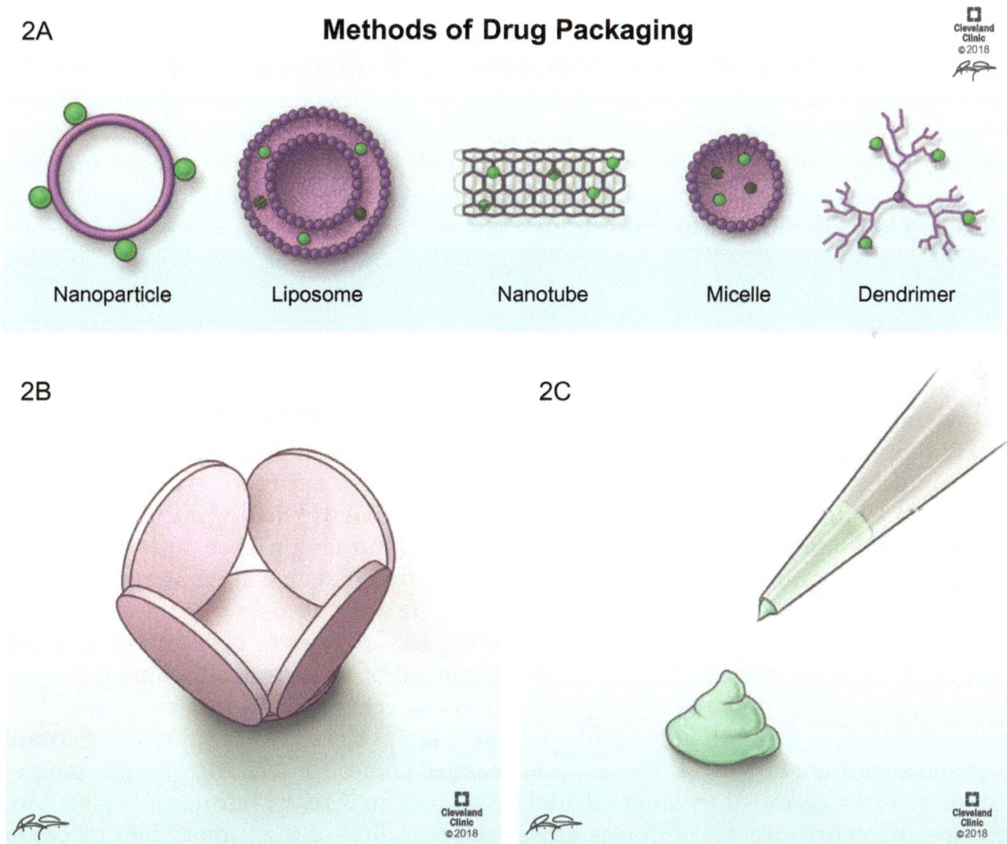

FIG. 2 Methods of drug packaging. (A) Nanovectors, which can carry and deliver drugs, include nanoparticles, liposomes, carbon nanotubes, microcapsules, micelles, and dendrimers. (B) Polymers are placed on the surface of the brain surrounding the resection cavity, where the drug can be slowly released. (C) Hydrogels are three-dimensional polymeric and hydrophilic networks that also allow for controlled release of drugs.

arrangement of carbon atoms.[18] Unlike other nanovectors, CNTs persist in the brain for several months, allowing continuous release of drugs.[19,20] They can become potentially toxic to the brain and other organs because they are capable of crossing membrane barriers.[21] On the other hand, CNTs can be conjugated to ligands for targeted drug delivery and can penetrate tumor cells through direct diffusion or endocytosis without causing damage to noncancerous cells.

Another promising nanotechnology for targeted drug delivery is the use of cell encapsulation, which involves the formation of small microcapsules in a spherical shape that can be loaded with drugs.[22,23] Microcapsules typically have a semipermeable membrane composed of polymers, protecting the encapsulated material from the extracellular environment while allowing their release for therapeutic effect.[24] Microcapsules have been used to successfully deliver endostatin,[25] an antiangiogenic agent, as well as TMZ to glioma cells, significantly prolonging survival in animal models.[22]

Polymeric micelles have also recently emerged as a promising tool for delivery of therapeutic compounds to brain tumors. They can cross the BBB and persist in the brain tissue for prolonged periods of time.[26] Another advantage is that they are very flexible in terms of design modification, allowing incorporation of a wide variety of ligands into their structures.[27]

Dendrimers provide another avenue for delivering drugs to specific regions of the brain. Their unique tree-like structure allows for reduced toxicity, low immune response, and ability to diffuse into tissues for an extended period.[28,29] The most commonly used form is the polyamidoamine dendrimer, which is associated with low cytotoxicity and can incorporate a high density of functional groups.[30,31] However, one challenge is that the amount of drug released is relatively difficult to control.[32]

Polymers

Controlled drug-delivery systems may enhance the efficacy of therapeutic agents that failed to penetrate the BBB and are rapidly cleared or metabolized. Controlled-release devices for direct delivery of chemotherapy to brain tumors were first approved by the Food and Drug Administration (FDA) in 1996.[33,34] The rationale for this approach is that biocompatible materials can be introduced directly into the brain, particularly during surgical resection of the tumor. If the materials are loaded with drugs and are engineered to release the drug slowly after implantation, then long-term chemotherapy is feasible at the tumor site without the need for BBB penetration (Fig. 2B). Furthermore, the drug diffuses into the tissue adjacent to the site of tumor resection, presumably targeting residual tumor cells and generating the highest drug concentrations in the region most in need of postsurgical treatment.

Both nondegradable polymers (that persist after delivery) and degradable polymers have been investigated. In general, degradable polymers are preferred to avoid chronic complications at the site of implantation. The main implant of choice for treating glioblastoma using this strategy is the carmustine wafer Gliadel (Eisai Inc. for Arbor Pharmaceuticals).[35] In this procedure, the neurosurgeon performs gross total resection of the tumor, then places small polymer drug wafers at the surface of the brain surrounding the resection cavity. The drug is then slowly released from these wafers for approximately 3 weeks. Because the drug is delivered locally rather than systemically, adverse effects are minimized. However, there may be more local effects with this treatment, such as issues with wound healing and an increased risk of treatment-related necrosis. One clinical trial demonstrated that after 2 years, 31% of patients treated with Gliadel wafers were alive, whereas only 6% of patients who underwent standard brain tumor therapy survived, although these results were questioned due to the small study size.[36] When the FDA approved this local therapy in 1996 for patients with recurrent glioblastoma, it was the first new brain cancer therapy approved in over 20 years. Two subsequent randomized controlled trials led to the 2003 FDA extension of approval to include initial surgery for newly diagnosed malignant glioma.[37]

Preclinical testing of local delivery of chemotherapeutic agents using the same polymer technology followed the initial clinical success of the carmustine wafers. Paclitaxel, a microtubule-binding agent that can arrest mitosis, has been shown to be safe and effective after local interstitial delivery in animal models of malignant glioma.[38] Another chemotherapeutic agent, cyclophosphamide, has had limited systemic use due to the need for high dosing for its active metabolite, 4-hydroperoxycyclophosphamide (4-HC), to cross the BBB. Two studies demonstrated that local delivery of 4-HC incorporated into a polymer matrix extended the median survival of rats with intracranial F98 gliomas.[39,40] Other drugs that have been investigated for local delivery using the polymer system include 5-fluorouracil,[41] adriamycin,[42] methotrexate,[43,44] camptothecin,[45-47] and minocycline.[48]

However, therapeutic limitations have been identified in the use of polymers to deliver these agents. Most notably, a concentration gradient cannot drive a large influx of molecules into the surrounding brain tissue. As the drug diffuses, it is degraded, released into the vasculature, internalized by cells, and bound to the extracellular matrix—further hindering effective transport. Small molecules are able to avoid some of these mechanisms by moving more rapidly through brain tissue, but large molecules are significantly impaired and limited in distribution. Furthermore, there are some intrinsic disadvantages associated with polymer drug delivery. The use of an adequate number of polymers (preferably eight carmustine wafers) requires a large surgical cavity, which is not always feasible with needle biopsies and eloquent tumor locations.[35,49-52] They also cannot be placed beyond the resection cavity, limiting distribution in peritumoral areas.

The net result is that there have been no additional therapeutics developed clinically in this manner since the FDA approved carmustine wafers, and the small clinical benefit associated with this therapy has been surpassed by the introduction of an oral chemotherapy that provides greater clinical benefit when combined with radiation therapy.

Hydrogels

Hydrogels are three-dimensional polymeric and hydrophilic networks that can assimilate large amounts of water or biological fluid without dissolution of the polymer and can swell in aqueous media.[53] They are emerging as excellent candidates for controlled-release and targeted drug delivery, as they are able to encapsulate biomacromolecules, including proteins and DNA as well as hydrophilic or hydrophobic drugs (Fig. 2C).[54] Drug-loaded hydrogels can be administered intratumorally or within the surgical resection cavity.[55]

Hydrophobic polymeric networks can be constructed with poly-lactic acid (PLA) or poly-lactide-co-glycolide (PLGA), which are both endogenous and easily metabolized by the body and are therefore associated with minimal systemic toxicity. A biodegradable gel matrix incorporating PLGA and TMZ was developed by Akbar et al.[56] Another form of hydrogel with PLGA and paclitaxel was shown to be effective for local delivery in a 9L rodent model of glioblastoma.[57,58] In addition, Rahman et al.[59] developed a novel thermosensitive formulation of chemotherapeutic drug-loaded PLGA microparticles that could form matrices that mold around the walls of the resection cavity.

Photopolymerization uses light (either visible or ultraviolet) to initiate and to propagate a polymerization reaction to form a linear or cross-linked polymer structure. Photopolymerized polymer networks can be used in tissue engineering due to their capacity to entrap a wide range of substances and cells.[60] For instance, a photopolymerizable hydrogel for local delivery of TMZ has been developed for treatment of glioblastoma.[61] When this solution is irradiated with a light at 400 nm for over 15 seconds, the hydrogel rapidly forms and presents a viscous modulus. Furthermore, tumor growth in mice treated with the photopolymerized TMZ hydrogel significantly decreased compared to mice in the control group.

Lipid nanocapsules are nanotechnology-based hydrogels composed of an oily core of triglycerides surrounded by a shell of surfactants.[62] They are relatively cost effective and simple to prepare, are stable for prolonged periods of time, and are compatible for encapsulation of a variety of drugs.[63] Recently, a hydrogel incorporating gemcitabine-loaded lipid nanocapsules was developed for local treatment of glioblastoma.[64] The advantage of this system is that the degradation of the gel corresponds to the release of the gemcitabine-loaded lipid nanocapsules, as no other components (e.g., synthetic or natural polymers, gelling agents, external stimuli) are present in the formulation, which also reduces the risk of adverse effects.[65] This system was also well tolerated in the mouse brain and reduced the tumor growth, in comparison to the free drug, in a subcutaneous human glioblastoma model.[65]

In addition to drug delivery, hydrogels are ideal candidates for theranostic applications combining treatment with an imaging platform. They can noninvasively assess the biodistribution and target-site accumulation of the drug, control the drug release, enhance the therapeutic efficacy via triggered drug release, and predict the therapeutic response.[66] For instance, Kim et al.[67,68] designed an injectable, magnetic resonance imaging-monitored, long-term therapeutic hydrogel incorporating a hydrophobic magnetic core and an active metabolite of irinotecan. Another example of a theranostic hydrogel is the pH- and temperature-sensitive magnetic nanogel developed by Jiang et al.[69]

In this manner, drug-loaded hydrogels can fill the gap between tumor resection and administration of chemoradiation, sustainably release the drug over a prolonged period, and allow diffusion of the drug from the resection cavity borders to the brain parenchyma to obliterate the infiltrating tumor cells responsible for recurrence. However, none of these hydrogels have yet reached the stage of clinical investigation.[70]

Microchips

Drug-impregnated microchips can overcome some limitations of polymer technology.[22,71] Microchips comprise pumps, valves, and channels at the micrometer scale and are controlled by time-dependent biodegradation[72] or electrochemical dissolution.[73] They can be controlled remotely and can release single or multiple agents.[72] Scott et al.[22] demonstrated the use of such microchips to release TMZ in a rodent gliosarcoma model. Furthermore, they showed that the TMZ flow rates from the microchips were predictable, and rodents receiving this treatment survived longer than those that received oral TMZ therapy. On the other hand, the potential disadvantages of this approach are that the devices need to be refilled, and are subject to possible electronic malfunction and alterations in magnetic fields. No clinical trials have been reported for this method of local drug delivery in patients diagnosed with malignant gliomas.

METHODS OF DRUG DELIVERY

Direct Injection

Chemotherapeutic agents can be directly injected into the tumor resection cavity, the surrounding brain parenchyma, the ventricle, or some combination thereof. This method can involve repeated needle-based injections and/or catheter implants that are connected to a reservoir (e.g., an Ommaya reservoir) for continued injection of drugs, radioactive compounds, viruses, antibodies, lymphocytes, and other therapies.[74-83] The distribution of chemotherapeutic drugs after direct injection relies on a concentration gradient and permeability of the agent into the tumor and surrounding brain parenchyma (Fig. 3A).

This therapeutic approach is simple and easily repeated. A large volume can be delivered with minimal systemic toxicity, and the reservoir can also be refilled for continued delivery.[84] On the other hand, repeated injections are associated with increased risk of intracranial hemorrhage, infection, and a malpositioned catheter.[75,79,85] Due to its dependence on a concentration gradient, the depth of distribution is often limited to approximately 3 mm, with an exponential decay in concentration from the injection site.[76,86] Furthermore, this method relies on a bolus-based approach, which makes it difficult to predict drug concentration and distribution.[76]

Multiple studies report intermittent bolus injections of both chemotherapeutic[84,87-91] and biological agents.[74-83] However, no successful large-scale clinical trials have demonstrated significant efficacy.[92-94] Gaspar et al.[75] placed permanent catheters containing ^{125}I seeds in 59 patients with recurrent malignant astrocytomas (37 glioblastoma and 22 anaplastic glioma), delivering a radiation dose of 0.05 Gy/hour to the periphery of the contrast-enhancing tumor. The median survival for these patients was 1.34 years (0.9 years for patients with glioblastoma and 2.04 years for patients with anaplastic glioma). Similarly, Riva et al.[74] conducted a Phase I study in which ^{131}I radio-conjugated antibodies against a stromal antigen tenascin were directly injected into malignant gliomas. There was only a 17.8% response rate for bulky tumors and a 66% response rate for small tumors. Torres et al.[76] also performed a Phase I study, in which they placed intracavitary catheters attached to Ommaya reservoirs in nine patients with recurrent malignant astrocytomas (eight glioblastoma and one anaplastic astrocytoma) loaded with variable concentrations of 188 relabeled humanized monoclonal antibody nimotuzumab against the epidermal growth factor receptor. They reported that 85% of the antibody was retained in the surgical cavity after injection; however, no survival analyses were performed. Other studies have assessed the efficacy of directly injecting viral

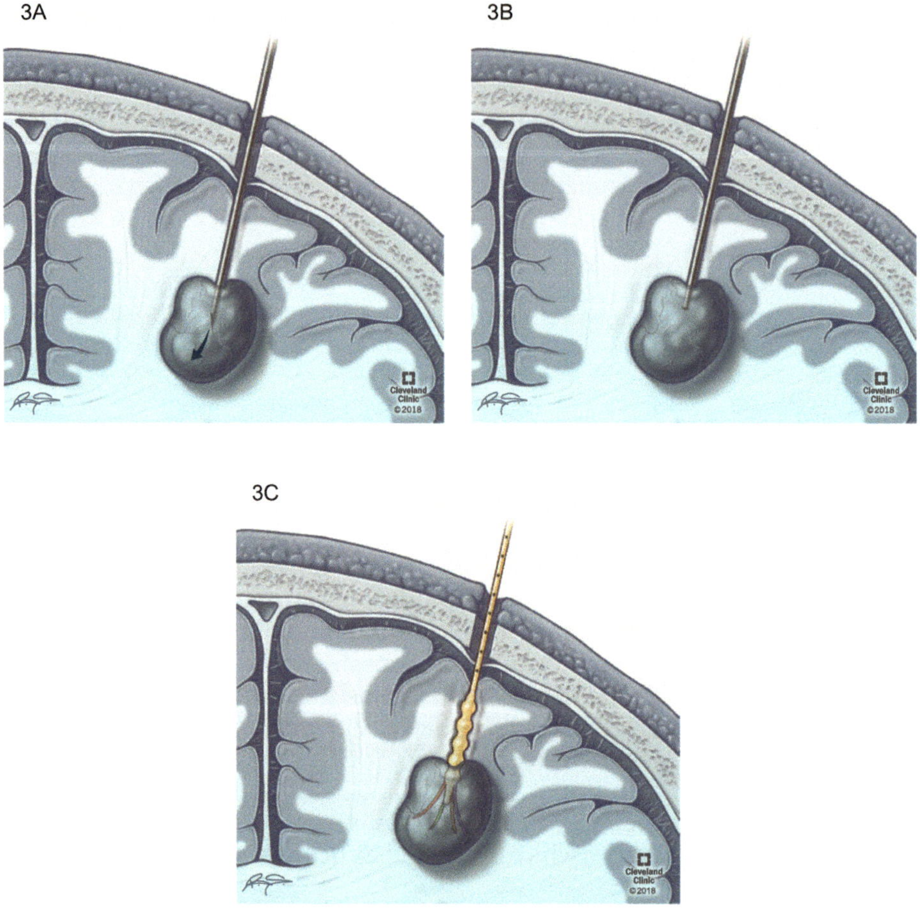

FIG. 3 Methods of drug delivery. Drugs can be locally delivered to the intracranial mass via (A) direct injection; (B) single-port convection-enhanced delivery; or (C) multi-port convection-enhanced delivery (e.g., the Cleveland Multiport Catheter).

agents[79] and autologous lymphocytes with monoclonal antibodies.[81,82,95] While local delivery via direct injection initially showed promising results, its use in clinical trials has decreased dramatically.

Convection-Enhanced Delivery

Convection-enhanced delivery is a strategy for direct drug delivery into the brain powered by bulk flow kinetics from pressure gradients, as opposed to concentration gradients in standard diffusion-based delivery. CED uses implanted intracranial microcatheters through which drugs are infused at precisely controlled infusion rates (Fig. 3B).[96] This method of delivery can homogenously distribute drugs through large volumes of brain tissue, almost an eightfold increase in distribution relative to direct injection,[96] regardless of the size of the infused molecule.[97] Factors affecting infusate distribution include: (1) infusion rate, volume, and concentration; (2) tumor tissue architecture and interstitial fluid pressure; (3) infusate characteristics, including half-life and metabolism of the drug; (4) cannula size, shape, and number; and (5) catheter position and actual volume of distribution.[98]

Since the initial conceptualization by Bobo et al. (1994),[96] CED for treatment of glioblastoma continues to be investigated in preclinical models. Kaiser et al.[99] demonstrated that CED of topotecan increased overall survival in a rat glioma model. Similarly, other studies evaluating CED of the chemotherapies gemcitabine and carboplatin also showed promising results for treatment of glioblastoma in animal models.[100] Furthermore, CED of tumor necrosis factor-related apoptosis-inducing ligand (TRAIL) combined with oral TMZ[101] as well as CED of bevacizumab, a humanized vascular endothelial growth factor antibody,[102] prolonged overall survival in animal models. Irinotecan, carboplatin, and cetuximab as nanoparticles and liposomes have also been developed to allow gradual release after CED.[103-105]

The first clinical trial of CED for patients with glioblastoma was carried out in 1997,[106] and there have been 13 to date.[106-118] Most clinical trials used flexible, single-lumen catheters that were not designed specifically for CED—rather, they were available "off the shelf." Eight studies involved conjugated toxins specifically taken up by high-grade glioma cells,[106-109, 112-115] and two studies incorporated conventional chemotherapies unable to penetrate the BBB if administered systemically.[110,117] A select few will be discussed in this section.

After promising results from animal models, the first clinical trial used the targeted toxin TF-CRM107, a human transferrin conjugated to a diphtheria toxin. Of the 15 patients, 9 had significant reduction in their tumor sizes with limited toxicity, demonstrating feasibility and safety.[106] The phase II study reported slightly less encouraging results, with only 31 of 44 patients completing treatment and just 12 of those 31 patients showing complete or partial radiographic response. The median survival was 37 weeks. Although well tolerated, the most common toxicity was cerebral edema, which occurred in 14% of patients. The Phase III study, conducted at 40 centers in the United States and in Europe, was aborted because the response rate was only 39%, according to an intermediate analysis of 44 patients.[106]

A subsequent therapeutic approach using CED attempted to take advantage of the interleukin-4 receptor overexpression in malignant glioma cells. A chimeric recombinant fusion protein incorporating interleukin-4 and Pseudomonas exotoxins was developed for a clinical trial in 2000.[108] In a cohort of nine patients with glioblastoma, six demonstrated tumor necrosis after treatment without significant systemic toxicity.[109] Epidermal growth factor receptor overexpression is another feature of glioblastoma that can be therapeutically targeted. TP-38 is a chimeric protein containing a Pseudomonas exotoxin and a TGF-alpha binding domain with a strong affinity for epidermal growth factor receptor. Its binding delivers the exotoxin and then induces apoptosis. In a study of 20 patients with recurrent or progressive malignant brain tumors (17 glioblastomas), the median overall survival was 28 weeks; however, many patients experienced leaks into the ventricles or the subarachnoid space, which impaired intraparenchymal distribution.[115,119]

Paclitaxel, a chemotherapeutic agent that cannot cross the BBB, was administered via CED for treatment of high-grade gliomas by Lidar et al.[110] Of the 15 patients, 11 showed radiographic response, and the median overall survival was 7.5 months.[110] Another chemotherapeutic agent, topotecan (a topoisomerase I inhibitor), was also used in a CED clinical trial with 16 patients (10 with glioblastoma and 6 with anaplastic astrocytoma).[117] Topotecan was not detected in the serum, and patients receiving the highest dosing experienced only minor neurological deficits. The median overall survival was 60 weeks, but this needs to be confirmed in a prospective, randomized trial.

The only randomized Phase III clinical trial of CED that has been completed is the PRECISE trial, which evaluated IL13-PE38QQR (i.e., cintredekin besudotox), a chimeric protein composed of IL-13 and a truncated form of Pseudomonas exotoxin A.[114] This trial enrolled 296 patients, with one arm receiving IL13-PE38QQR via CED through catheters implanted after craniotomy for resection 96 hours earlier, and the control arm receiving carmustine wafers placed on the walls of the resection cavity after craniotomy. No significant difference was observed in overall survival; however, the study was powered to detect greater than 50% survival benefit over the control group with carmustine wafers. In addition, a follow-up study discovered that only 68% of catheter placements were performed per protocol, which may have skewed the results.[120]

Although CED appears to be a promising technique for local drug delivery, clinical trials have had limited success. A challenge associated with CED is the possibility of infusate reflux around the delivery cannula, which has been reported in several trials.[121] Reflux usually occurs when the pressure gradient between the cannula and the tumor region equalizes, impeding the flow of the infusate into the tumor region. Tissue disruption at the tip of the cannula, infusion rate, and cannula diameter may contribute to reflux. Tissue disruption can cause reflux when a cannula is inserted too deeply into the tissue, if there is a shift of the brain after cannula insertion, or if biopsy is performed prior to infusion. A cavity is subsequently formed and filled, creating a pressure reservoir that ultimately causes backflow up the catheter. Softer cannulas and catheters as well as quick catheter placement may reduce brain-shift-induced tissue disruption.[96,122,123] Furthermore, high infusion rates increase the risk of reflux when the pressure created by the infusion exceeds the interstitial pressure.[124] An infusion rate of 3 μL/min or higher can drive the infusate into the subarachnoid and ventricular spaces, causing adverse events.[125] In designing a safe and efficacious CED system, infusion rate and cannula diameter must both be considered, given that each cannula diameter and flow rate is associated with a specific and predictable backflow distance along the cannula.[121]

Catheter design is key to preventing unwanted reflux of infusate. Several modifications in CED catheter construction were implemented to minimize the likelihood of backflow. The original catheter structure consisted of a hollow tube. Later it was noted that decreasing the gauge size of the lead catheter reduces the incidence of reflux.[125] Further modification of this design consists of several stepwise reductions in tube diameter along the catheter length and was shown to reduce the amount of backflow.[126] Another modification is to use a porous tube at the end of the catheter, resulting in infusate delivery over the length of the porous segment rather than release at a single outlet.[127] Yet another strategy is to use microfabricated catheter tips that can be attached to the end of the standard stepped catheter. These microfabricated tips make use of the micro-electromechanical systems technology that enables fluid delivery via microfluidic circuits.[128] One such system, the Alcyone microcathether (Alcyone Lifesciences, Inc.) was shown to provide superior distribution at low level of backflow that was demonstrated in a pig brain using magnetic resonance imaging of infused tracers.[129] This catheter is currently being used in a Phase II study of virus-based immunotherapy for glioblastoma developed by DNAtrix, Inc., DNX-2401, with pembrolizumab (CAPTIVE).[130]

The drive to improve drug distribution while minimizing backflow led to the development of the Cleveland Multiport Catheter. This system contains four microcatheters that allow chemotherapeutic drug delivery in four different directions from the catheter tip, which significantly enhances the volume of drug distribution (Fig. 3C). The results from the phase I study of administration of intratumoral topotecan in conjunction with a gadolinium tracer into newly diagnosed or recurrent glioblastoma showed excellent drug distribution with no evidence of backflow.[131] A series of pilot trials using this device are currently enrolling patients with recurrent glioblastoma to investigate delivery of topotecan via CED.[132,133]

MATCHING DRUGS WITH METHODS

Many drug-delivery systems have been designed to maximize the concentration of the drug in the tumor in the most selective way possible, and to ensure effective drug delivery to the tumor, an appropriate method of delivery must be chosen. In other words, the mechanism of action of the administered drug as well as its inherent properties must be taken into consideration when choosing the delivery method. The various methods of systemic and local delivery, along with examples of each type of drug and vector, are listed in Table 1.

A large number of chemotherapeutics that have shown in vitro efficacy against gliomas cannot cross the BBB, and yet they are routinely delivered systemically via an oral or intravenous route. It is illogical to expect these drugs to show clinical efficacy given this limitation. On the other hand, local or regional delivery may also have limitations that need to be considered. Delivery into the tumor resection cavity may be accomplished by placing polymers or hydrogels at the end of surgery. However, studies have demonstrated the limited distribution of the drug administered in this manner, traveling only a few millimeters through the surrounding tissue.[134]

Local drug delivery via a cannula or a catheter may also be limited, depending on the device and infusion technique used. The extent of tissue penetration when the drug is delivered locally via injection or Ommaya reservoir is bound by passive diffusion and amounts to 2–3 mm of penetration. Although this extent of distribution is not likely to be useful for a conventional chemotherapy or targeted agent, it may be effective for some types of agents (e.g., replication-competent viral vectors). In this case, the vector infects tumor cells locally and releases more virus particles that are capable of infecting neighboring tumor cells, thereby extending its therapeutic effect beyond the area of original delivery. In addition, vectors may potentiate an innate antitumor response, resulting in a synergistic tumor-control mechanism.

Most chemotherapeutic agents currently being used or under investigation require additional mechanisms to enhance their distribution within the tumor and in the surrounding tissue. This is where the CED method, using continuous positive-pressure drug delivery, may be able to overcome the limitation of passive diffusion and deliver a greater volume of effective drugs. Small agents and nanoparticles may penetrate the tumor more effectively via CED.[112,135,136] Several improvements in cannula design allow for increased infusion rate, reduced reflux, and increased infusate distribution.[112,127,137-140]

TABLE 1 Matching the Drug Delivery Mechanism to the Type of Drug or Vector to be Delivered

Delivery mechanism	Type of drug/vector	Examples
Systemic		
Oral	Chemotherapy Targeted small molecule drugs	Temozolomide
Intravenous	Chemotherapy Targeted small molecule drugs Nanoparticles Liposomes Monoclonal antibodies	Carmustine Liposomal doxorubicin Bevacizumab
Intranasal	Nanoparticles Stem cells	Carboplatin nanoparticles
Local		
Intracavitary release	Polymers Hydrogels Microchips	Carmustine wafers
Direct injection	Chemotherapy Monoclonal antibodies Viral vectors	Methotrexate Nitrosoureas Nimousumab Toca 511
Local/Regional		
Intrathecal	Chemotherapy	Methotrexate
Convection enhanced delivery	Chemotherapy Nanoparticles Viruses	Topotecan Cintredekin besudotox

In addition to characteristics of the delivered drug and the cannula, the heterogeneous nature of glioblastoma makes homogeneous delivery of therapeutic agents challenging. One physiological barrier is that certain tumor zones metabolize the delivered agent more effectively than others. Physical barriers include aberrant blood vessel growth and variable interstitial spaces, creating differential rates of drug clearance.[141-143] Whether delivery should target enhancing tumor or tumor-infiltrated brain, which has an intact BBB and more uniform tissue architecture, is a question being actively investigated.[131]

CONCLUSIONS

Despite significant advances in drug therapy, glioblastoma remains one of the most aggressive, fatal malignant brain tumors in humans. Even after gross total resection and chemoradiation, the probability of recurrence is extremely high. New strategies for drug delivery, particularly in the local peritumoral microenvironment of the resection cavity, are rapidly emerging and already affecting the treatment of glioblastoma.

With the capability of specific molecular targeting and improved pharmacokinetic properties of chemotherapeutic agents, it will be necessary to be equally attentive to the development of more efficacious methods of packaging and delivering these agents. As described in this chapter, the most commonly used drug packaging strategies include nanovectors, polymers, hydrogels, and microchips. In addition to drug formulation, effective delivery and distribution of the drug at the site of tumor cells are crucial. Local drug delivery is facilitated by techniques such as direct injection, CED, and the more recently developed Cleveland Multiport Catheter.

Although challenges persist in preclinical testing and clinical trials, further modifications of current local drug delivery techniques and incorporation of new technological applications will become powerful tools in developing more therapeutic modalities for patients with aggressive brain tumors such as glioblastoma.

References

1. Ostrom QT, Gittleman H, Liao P, et al. CBTRUS statistical report: primary brain and other central nervous system tumors diagnosed in the United States in 2010-2014. *Neuro-Oncology*. 2017;19(S5):1-88.
2. Davis FG, Freels S, Grutsch J, Barlas S, Brem S. Survival rates in patients with primary malignant brain tumors stratified by patient age and tumor histological type: an analysis based on surveillance, epidemiology, and end results (SEER) data, 1973-1991. *J Neurosurg*. 1998;88:1-10.
3. McLendon RE, Halperin EC. Is the long-term survival of patients with intracranial glioblastoma multiforme overstated? *Cancer*. 2003;98:1745-1748.
4. Deen DF, Chiarodo A, Grimm EA, et al. Brain tumor working group report on the 9th international conference on brain tumor research and therapy. Organ system program, National Cancer Institute. *J Neuro-Oncol*. 1993;16:243-272.
5. Kleihues P, Sobin LH. World Health Organization classification of tumors. *Cancer*. 2000;88:2887.
6. Choucair AK, Levin VA, Gutin PH, et al. Development of multiple lesions during radiation therapy and chemotherapy in patients with gliomas. *J Neurosurg*. 1986;65(5):654-658.
7. Serwer LP, James CD. Challenges in drug delivery to tumors of the central nervous system: an overview of pharmacological and surgical considerations. *Adv Drug Deliv Rev*. 2012;64:590-597.
8. Fakhoury M, Takechi R, Al-Salami H. Drug permeation across the blood-brain barrier: applications of nanotechnology. *Br J Med Med Res*. 2015;6:547-556.
9. Pardrige WM. The blood-brain barrier and neurotherapeutics. *NeuroRx*. 2005;2:1-2.
10. Zhou J, Patel TR, Sirianni RW, et al. Highly penetrative, drug-loaded nanocarriers improve treatment of glioblastoma. *Proc Natl Acad Sci U S A*. 2013;110(29):11751-11756.
11. Benny O, Pakneshan P. Novel technologies for antiangiogenic drug delivery in the brain. *Cell Adhes Migr*. 2009;3:224-229.
12. Koziara JM, Lockman PR, Allen DD, Mumper RJ. Paclitaxel nanoparticles for the potential treatment of brain tumors. *J Control Release*. 2004;99:259-269.
13. Inoue T, Yamashita Y, Nishihara M, et al. Therapeutic efficacy of a polymeric micellar doxorubicin infused by convection-enhanced delivery against intracranial 9L brain tumor models. *Neuro-Oncology*. 2009; 11(2):151-157.
14. Tripathi RM, Shrivastav A, Shrivastav BR. Biogenic gold nanoparticles: as a potential candidate for brain tumor directed drug delivery. *Artif Cells Nanomed Biotechnol*. 2015;43(5):311-317.
15. Laquintana V, Trapani A, Denora N, Wang F, Gallo JM, Trapani G. New strategies to deliver anticancer drugs to brain tumors. *Expert Opin Drug Deliv*. 2009;6:1017-1032.
16. Huwyler J, Drewe J, Krahenbuhl S. Tumor targeting using liposomal antineoplastic drugs. *Int J Nanomedicine*. 2008;3:21-29.
17. Jin Y, Li J, Rong LF, Lu XW, Huang Y, Xu SY. Pharmacokinetics and tissue distribution of 5-fluorouracil encapsulated by galactosyl-ceramide liposomes in mice. *Acta Pharmacol Sin*. 2005;26:250-256.

18. Zhang W, Zhang Z, Zhang Y. The application of carbon nanotubes in target drug delivery systems for cancer therapies. *Nanoscale Res Lett*. 2011;6:555.
19. Jain AK, Das M, Swarnakar NK, Jain S. Engineered PLGA nanoparticles: an emerging delivery tool in cancer therapeutics. *Crit Rev Ther Drug Carrier Syst*. 2011;28:1-45.
20. Rastogi V, Yadav P, Bhattacharya SS, et al. Carbon nanotubes: an emerging drug carrier for targeting cancer cells. *J Drug Deliv*. 2014;2014:670815.
21. Kolosnjaj J, Szwarc H, Moussa F. Toxicity studies of carbon nanotubes. *Adv Exp Med Biol*. 2007;620:181-204.
22. Scott AW, Tyler BM, Masi BC, et al. Intracranial microcapsule drug delivery device for the treatment of an experimental gliosarcoma model. *Biomaterials*. 2011;32(10):2532-2539.
23. Negrulj R, Mooranian A, Chen-Tan N, et al. Swelling, mechanical strength, and release properties of probucol microcapsules with and without a bile acid, and their potential oral delivery in diabetes. *Artif Cells Nanomed Biotechnol*. 2016;44(5):1290-1297.
24. Bhujbal SV, de Vos P, Niclou SP. Drug and cell encapsulation: alternative delivery options for the treatment of malignant brain tumors. *Adv Drug Deliv Rev*. 2014;67–68:142-153.
25. Joki T, Machluf M, Atala A, et al. Continuous release of endostatin from microencapsulated engineered cells for tumor therapy. *Nat Biotechnol*. 2001;19:35-39.
26. Chen PY, Liu HL, Hua MY, et al. Novel magnetic/ultrasound focusing system enhances nanoparticle drug delivery for glioma treatment. *Neuro-Oncology*. 2010;12(10):1050-1060.
27. Morshed RA, Cheng Y, Auffinger B, Wegscheid ML, Lesniak MS. The potential of polymeric micelles in the context of glioblastoma therapy. *Front Pharmacol*. 2013;4:157.
28. Albertazzi L, Gherardini L, Brondi M, et al. In vivo distribution and toxicity of PAMAM dendrimers in the central nervous system depend on their surface chemistry. *Mol Pharm*. 2013;10:249-260.
29. Hernández-Pedro NY, Rangel-López E, Magaña-Maldonado R, et al. Application of nanoparticles on diagnosis and therapy in gliomas. *Biomed Res Int*. 2013;2013:351031.
30. Bai CZ, Choi S, Nam K, An S, Park JS. Arginine modified PAMAM dendrimer for interferon beta gene delivery to malignant glioma. *Int J Pharm*. 2013;445:79-87.
31. Wu G, Barth RF, Yang W, Kawabata S, Zhang L, Green-Church K. Targeted delivery of methotrexate to epidermal growth factor receptor-positive brain tumors by means of cetuximab (IMC-C225) dendrimer bioconjugates. *Mol Cancer Ther*. 2006;5:52-59.
32. Singh R, Lillard JW. Nanoparticle-based targeted drug delivery. *Exp Mol Pathol*. 2009;86:215-223.
33. Fleming AB, Saltzman WM. Pharmacokinetics of the carmustine implant. *Clin Pharmacokinet*. 2002;41:403-419.
34. Sawyer AJ, Piepmeier JM, Saltzman WM. New methods for direct delivery of chemotherapy for treating brain tumors. *Yale J Biol Med*. 2007;79:141-152.
35. Brem H, Piantadosi S, Burger PC, et al. Placebo-controlled trial of safety and efficacy of intraoperative controlled delivery by biodegradable polymers of chemotherapy for recurrent gliomas. The polymer-brain tumor treatment group. *Lancet*. 1995;345(8956):1008-1012.
36. Valtonen S, Timonen U, Toivanen P, et al. Interstitial chemotherapy with carmustine-loaded polymers for high-grade gliomas: a randomized double-blind study. *Neurosurgery*. 1997;41:44-49.
37. Westphal M, Hilt DC, Bortey E, et al. A phase 3 trial of local chemotherapy with biodegradable carmustine (BCNU) wafers (Gliadel wafers) in patients with primary malignant glioma. *Neuro-Oncology*. 2003;5(2):79-88.
38. Walter KA, Cahan MA, Gur A, et al. Interstitial taxol delivered from a biodegradable polymer implant against experimental malignant glioma. *Cancer Res*. 1994;54:2207-2212.
39. Domb A, Bogdansky S, Olivi A, et al. Controlled delivery of water soluble and hydrolytically unstable anticancer drugs from polymeric implants. *Polym Prepr*. 1992;32:219-222.
40. Judy KD, Olivi A, Buahin KG, et al. Effectiveness of controlled release of a cyclophosphamide derivative with polymers against rat gliomas. *J Neurosurg*. 1995;82(3):481-486.
41. Menei P, Boisdron-Celle M, Croué A, Guy G, Benoit JP. Effect of stereotactic implantation of biodegradable 5-fluorouracil-loaded microspheres in healthy and C6 glioma-bearing rats. *Neurosurgery*. 1996;39(1):117-123.
42. Lin SY, Cheng LF, Lui WY, Chen CF, Han SH. Tumoricidal effect of controlled-release polymeric needle devices containing adriamycin HCl in tumor-bearing mice. *Biomater Artif Cells Artif Organs*. 1989;17(2):189-203.
43. Rama B, Mandel T, Jansen J, Dingeldein E, Mennel HD. The intraneoplastic chemotherapy in a rat brain tumour model utilizing methotrexate-polymethylmethacrylate pellets. *Acta Neurochir*. 1987;87(1–2):70-75.
44. Dang W, Colvin OM, Brem H, Saltzman WM. Covalent coupling of methotrexate to dextran enhances the penetration of cytotoxicity into a tissue-like matrix. *Cancer Res*. 1994;54(7):1729-1735.

45. Sampath P, Amundson E, Wall ME, et al. Camptothecin analogs in malignant gliomas: comparative analysis and characterization. *J Neurosurg*. 2003;98(3):570-577.
46. Storm PB, Moriarity JL, Tyler B, Burger PC, Brem H, Weingart J. Polymer delivery of camptothecin against 9L gliosarcoma: release, distribution, and efficacy. *J Neuro-Oncol*. 2002;56(3):209-217.
47. Weingart JD, Thompson RC, Tyler B, Colvin OM, Brem H. Local delivery of the topoisomerase I inhibitor camptothecin sodium prolongs survival in the rat intracranial 9L gliosarcoma model. *Int J Cancer*. 1995; 62(5):605-609.
48. Weingart JD, Sipos EP, Brem H. The role of minocycline in the treatment of intracranial 9L glioma. *J Neurosurg*. 1995;82:635-640.
49. Raza SM, Pradilla G, Legnani FG, et al. Local delivery of antineoplastic agents by controlled-release polymers for the treatment of malignant brain tumours. *Expert Opin Biol Ther*. 2005;5(4):477-494.
50. Brem H, Mahaley Jr MS, Vick NA, et al. Interstitial chemotherapy with drug polymer implants for the treatment of recurrent gliomas. *J Neurosurg*. 1991;74(3):441-446.
51. Olivi A, Ewend MG, Utsuki T, et al. Interstitial delivery of carboplatin via biodegradable polymers is effective against experimental glioma in the rat. *Cancer Chemother Pharmacol*. 1996;39(1–2):90-96.
52. Olivi A, Grossman SA, Tatter S, et al. New approaches to brain tumor therapy CNS consortium. Dose escalation of carmustine in surgically implanted polymers in patients with recurrent malignant glioma: a new approaches to brain tumor therapy CNS consortium trial. *J Clin Oncol*. 2003;21(9):1845-1849.
53. Peppas NA, Bures P, Leobandung W, Ichikawa H. Hydrogels in pharmaceutical formulations. *Eur J Pharm Biopharm*. 2000;50(1):27-46.
54. Lin CC, Metters AT. Hydrogels in controlled release formulations: network design and mathematical modeling. *Adv Drug Deliv Rev*. 2006;58:1379-1408.
55. Karim R, Palazzo C, Evrard B, Piel G. Nanocarriers for the treatment of glioblastoma multiforme: current state-of-the-art. *J Control Release*. 2016;227:23-37.
56. Akbar U, Jones T, Winestone J, et al. Delivery of temozolomide to the tumor bed via biodegradable gel matrices in a novel model of intracranial glioma with resection. *J Neuro-Oncol*. 2009;94:203-212.
57. Vellimana AK, Recinos VR, Hwang L, et al. Combination of paclitaxel thermal gel depot with temozolomide and radiotherapy significantly prolongs survival in an experimental rodent glioma model. *J Neuro-Oncol*. 2013;111:229-236.
58. Tyler B, Fowers KD, Li KW, et al. A thermal gel depot for local delivery of paclitaxel to treat experimental brain tumors in rats. *J Neurosurg*. 2010;113:210-217.
59. Rahman CV, Smith SJ, Morgan PS, et al. Adjuvant chemotherapy for brain tumors delivered via a novel intra-cavity moldable polymer matrix. *PLoS One*. 2013;8.
60. Nguyen KT, West JL. Photopolymerizable hydrogels for tissue engineering applications. *Biomaterials*. 2002;23:4307-4314.
61. Fourniols T, Randolph LD, Staub A, et al. Temozolomide-loaded photopolymerizable PEG-DMA-based hydrogel for the treatment of glioblastoma. *J Control Release*. 2015;210:95-104.
62. Huynh NT, Passirani C, Saulnier P, Benoit JP. Lipid nanocapsules: a new platform for nanomedicine. *Int J Pharm*. 2009;379:201-209.
63. Aparicio-Blanco J, Torres-Suarez AI. Glioblastoma multiforme and lipid nanocapsules: a review. *J Biomed Nanotechnol*. 2015;11:1283-1311.
64. Moysan E, Gonzalez-Fernandez Y, Lautram N, Bejaud J, Bastiat G, Benoit JP. An innovative hydrogel of gemcitabine-loaded lipid nanocapsules: when the drug is a key player of the nanomedicine structure. *Soft Matter*. 2014;10:1767-1777.
65. Bastiancich C, Vanvarenberg K, Ucakar B, et al. Lauroyl-gemcitabine-loaded lipid nanocapsules hydrogel for the treatment of glioblastoma. *J Control Release*. 2016;225:283-293.
66. Schleich N, Danhier F, Preat V. Iron oxide-loaded nanotheranostics: major obstacles to in vivo studies and clinical translation. *J Control Release*. 2015;198:35-54.
67. Kim JI, Kim B, Chun C, Lee SH, Song SC. MRI-monitored long-term therapeutic hydrogel system for brain tumors without surgical resection. *Biomaterials*. 2012;33:4836-4842.
68. Kim JI, Chun C, Kim B, et al. Thermosensitive/magnetic poly(organophosphazene) hydrogel as a long-term magnetic resonance contrast platform. *Biomaterials*. 2012;33:218-224.
69. Jiang L, Zhou Q, Mu K, et al. pH/temperature sensitive magnetic nanogels conjugated with Cy5.5-labled lactoferrin for MR and fluorescence imaging of glioma in rats. *Biomaterials*. 2013;34:7418-7428.

70. Bastiancich C, Danhier P, Preat V, Danhier F. Anticancer drug-loaded hydrogels as drug delivery systems for the local treatment of glioblastoma. *J Control Release*. 2016;243:29-42.
71. Farra R, Sheppard Jr NF, McCabe L, et al. First-in-human testing of a wirelessly controlled drug delivery microchip. *Sci Transl Med*. 2012;4(122):122ra21.
72. Richards Grayson AC, Choi IS, Tyler BM, et al. Multi-pulse drug delivery from a resorbable polymeric microchip device. *Nat Mater*. 2003;2(11):767-772.
73. Santini Jr JT, Cima MJ, Langer R. A controlled-release microchip. *Nature*. 1999;397(6717):335-338.
74. Riva P, Franceschi G, Frattarelli M, et al. 131I radioconjugated antibodies for the locoregional radioimmunotherapy of high-grade malignant glioma—phase I and II study. *Acta Oncol*. 1999;38(3):351-359.
75. Gaspar LE, Zamorano LJ, Shamsa F, Fontanesi J, Ezzell GE, Yakar DA. Permanent 125iodine implants for recurrent malignant gliomas. *Int J Radiat Oncol Biol Phys*. 1999;43(5):977-982.
76. Torres LA, Coca MA, Batista JF, et al. Biodistribution and internal dosimetry of the 188Re-labelled humanized monoclonal antibody anti-epidermal growth factor receptor, nimotuzumab, in the locoregional treatment of malignant gliomas. *Nucl Med Commun*. 2008;29(1):66-75.
77. Bartolomei M, Mazzetta C, Handkiewicz-Junak D, et al. Combined treatment of glioblastoma patients with locoregional pre-targeted 90Y-biotin radioimmunotherapy and temozolomide. *Q J Nucl Med Mol Imaging*. 2004;48(3):220-228.
78. Paganelli G, Bartolomei M, Ferrari M, et al. Pre-targeted locoregional radioimmunotherapy with 90Y-biotin in glioma patients: phase I study and preliminary therapeutic results. *Cancer Biother Radiopharm*. 2001;16(3):227-235.
79. Prados MD, McDermott M, Chang SM, et al. Treatment of progressive or recurrent glioblastoma multiforme in adults with herpes simplex virus thymidine kinase gene vector-producer cells followed by intravenous ganciclovir administration: a phase I/II multi-institutional trial. *J Neuro-Oncol*. 2003;65(3):269-278.
80. Goetz C, Riva P, Poepperl G, et al. Locoregional radioimmunotherapy in selected patients with malignant glioma: experiences, side effects and survival times. *J Neuro-Oncol*. 2003;62(3):321-328.
81. Jung G, Brandl M, Eisner W, et al. Local immunotherapy of glioma patients with a combination of 2 bispecific antibody fragments and resting autologous lymphocytes: evidence for in situ t-cell activation and therapeutic efficacy. *Int J Cancer*. 2001;91(2):225-230.
82. Quattrocchi KB, Miller CH, Cush S, et al. Pilot study of local autologous tumor infiltrating lymphocytes for the treatment of recurrent malignant gliomas. *J Neuro-Oncol*. 1999;45(2):141-157.
83. Bigner DD, Brown MT, Friedman AH, et al. Iodine-131-labeled antitenascin monoclonal antibody 81C6 treatment of patients with recurrent malignant gliomas: phase I trial results. *J Clin Oncol*. 1998;16(6):2202-2212.
84. Patchell RA, Regine WF, Ashton P, et al. A phase I trial of continuously infused intratumoral bleomycin for the treatment of recurrent glioblastoma multiforme. *J Neuro-Oncol*. 2002;60(1):37-42.
85. Boiardi A, Eoli M, Salmaggi A, et al. New approach in delivering chemotherapy: locoregional treatment for recurrent glioblastoma (rGBM). *J Exp Clin Cancer Res*. 2003;22(4):123-127.
86. Lang FF, Bruner JM, Fuller GN, et al. Phase I trial of adenovirus-mediated p53 gene therapy for recurrent glioma: biological and clinical results. *J Clin Oncol*. 2003;21(13):2508-2518.
87. Ferroli P, Broggi M, Franzini A, et al. Surgifoam and mitoxantrone in the glioblastoma multiforme postresection cavity: the first step of locoregional chemotherapy through an ad hoc-placed catheter: technical note. *Neurosurgery*. 2006;59(2):E433-E434.
88. Boiardi A, Silvani A, Eoli M, et al. Treatment of recurrent glioblastoma: can local delivery of mitoxantrone improve survival? *J Neuro-Oncol*. 2008;88(1):105-113.
89. Voulgaris S, Partheni M, Karamouzis M, Dimopoulos P, Papadakis N, Kalofonos HP. Intratumoral doxorubicin in patients with malignant brain gliomas. *Am J Clin Oncol*. 2002;25(1):60-64.
90. Boiardi A, Eoli M, Salmaggi A, et al. Efficacy of intratumoral delivery of mitoxantrone in recurrent malignant glial tumours. *J Neuro-Oncol*. 2001;54(1):39-47.
91. Boiardi A, Silvani A, Pozzi A, Fariselli L, Broggi G, Salmaggi A. Interstitial chemotherapy plus systemic chemotherapy for glioblastoma patients: improved survival in sequential studies. *J Neuro-Oncol*. 1999;41(2):151-157.
92. Boiardi A, Eoli M, Pozzi A, Salmaggi A, Broggi G, Silvani A. Locally delivered chemotherapy and repeated surgery can improve survival in glioblastoma patients. *Ital J Neurol Sci*. 1999;20(1):43-48.
93. Oshiro S, Tsugu H, Komatsu F, et al. Evaluation of intratumoral administration of tumor necrosis factor-alpha in patients with malignant glioma. *Anticancer Res*. 2006;26(6A):4027-4032.
94. Koch D, Hundsberger T, Boor S, Kaina B. Local intracerebral administration of O(6)-benzylguanine combined with systemic chemotherapy with temozolomide of a patient suffering from a recurrent glioblastoma. *J Neuro-Oncol*. 2007;82(1):85-89.

95. Nakagawa K, Kamezaki T, Shibata Y, Tsunoda T, Meguro K, Nose T. Effect of lymphokineactivated killer cells with or without radiation therapy against malignant brain tumors. *Neurol Med Chir (Tokyo)*. 1995;35(1):22-27.
96. Bobo RH, Laske DW, Akbasak A, Morrison PF, Dedrick RL, Oldfield EH. Convection-enhanced delivery of macromolecules in the brain. *Proc Natl Acad Sci U S A*. 1994;91(6):2076-2080.
97. Mehta AI, Choi BD, Ajay D, et al. Convection enhanced delivery of macromolecules for brain tumors. *Curr Drug Discov Technol*. 2012;9(4):305-310.
98. Healy AT, Vogelbaum MA. Convection-enhanced drug delivery for gliomas. *Surg Neurol Int*. 2015;6(1):S59-S67.
99. Kaiser MG, Parsa AT, Fine RL, Hall JS, Chakrabarti I, Bruce JN. Tissue distribution and antitumor activity of topotecan delivered by intracerebral clysis in a rat glioma model. *Neurosurgery*. 2000;47:1391-1399.
100. Degen JW, Walbridge S, Vortmeyer AO, Oldfield EH, Lonser RR. Safety and efficacy of convection-enhanced delivery of gemcitabine or carboplatin in a malignant glioma model in rats. *J Neurosurg*. 2003;99:893-898.
101. Saito R, Bringas JR, Panner A, et al. Convection-enhanced delivery of tumor necrosis factor-related apoptosis-inducing ligand with systemic administration of temozolomide prolongs survival in an intracranial glioblastoma xenograft model. *Cancer Res*. 2004;64:6858-6862.
102. Wang W, Sivakumar W, Torres S, et al. Effects of convection-enhanced delivery of bevacizumab on survival of glioma-bearing animals. *Neurosurg Focus*. 2015;38(3).
103. Arshad A, Yang B, Bienemann AS, et al. Convection-enhanced delivery of carboplatin PLGA nanoparticles for the treatment of glioblastoma. *PLoS One*. 2015;10.
104. Kaluzova M, Bouras A, Machaidze R, Hadjipanayis CG. Targeted therapy of glioblastoma stem-like cells and tumor non-stem cells using cetuximab-conjugated iron-oxide nanoparticles. *Oncotarget*. 2015;6:8788-8806.
105. Mendiburu-Eliçabe M, Gil-Ranedo J. Combination therapy of intraperitoneal rapamycin and convection-enhanced delivery of nanoliposomal CPT-11 in rodent orthotopic brain tumor xenografts. *Curr Cancer Drug Targets*. 2015;15:352-362.
106. Laske DW, Youle RJ, Oldfield EH. Tumor regression with regional distribution of the targeted toxin TF-CRM107 in patients with malignant brain tumors. *Nat Med*. 1997;3:1362-1368.
107. Weaver M, Laske DW. Transferrin receptor ligand-targeted toxin conjugate (Tf-CRM107) for therapy of malignant gliomas. *J Neuro-Oncol*. 2003;65:3-13.
108. Rand RW, Kreitman RJ, Patronas N, Varricchio F, Pastan I, Puri RK. Intratumoral administration of recombinant circularly permuted interleukin-4-Pseudomonas exotoxin in patients with high-grade glioma. *Clin Cancer Res*. 2000;6:2157-2165.
109. Weber F, Asher A, Bucholz R, et al. Safety, tolerability, and tumor response of IL4- Pseudomonas exotoxin (NBI-3001) in patients with recurrent malignant glioma. *J Neuro-Oncol*. 2003;64:125-137.
110. Lidar Z, Mardor Y, Jonas T, et al. Convection-enhanced delivery of paclitaxel for the treatment of recurrent malignant glioma: a phase I/II clinical study. *J Neurosurg*. 2004;100:472-479.
111. Patel SJ, Shapiro WR, Laske DW, et al. Safety and feasibility of convection-enhanced delivery of Cotara for the treatment of malignant glioma: initial experience in 51 patients. *Neurosurgery*. 2005;56:1243-1253.
112. Vogelbaum MA, Sampson JH, Kunwar S, et al. Convection-enhanced delivery of cintredekin besudotox (interleukin 13 PE38QQR) followed by radiation therapy with and without temozolomide in newly diagnosed malignant gliomas: phase 1 study of final safety results. *Neurosurgery*. 2007;61:1031-1038.
113. Kunwar S, Prados MD, Chang SM, et al. Direct intracerebral delivery of cintredekin besudotox (IL13-PE38QQR) in recurrent malignant glioma: a report by the Cintredekin Besudotox Intraparenchymal Study Group. *J Clin Oncol*. 2007;25:837-844.
114. Kunwar S, Chang S, Westphal M, et al. Phase III randomized trial of CED of IL13-PE38QQR vs Gliadel wafers for recurrent glioblastoma. *Neuro-Oncology*. 2010;12:871-881.
115. Sampson JH, Akabani G, Archer GE, et al. Intracerebral infusion of an EGFR-targeted toxin in recurrent malignant brain tumors. *Neuro-Oncology*. 2008;10:320-329.
116. Bogdahn U, Hau P, Stockhammer G, et al. Targeted therapy for high-grade glioma with the TGF-b2 inhibitor trabedersen: results of a randomized and controlled phase IIb study. *Neuro-Oncology*. 2011;13:132-142.
117. Bruce JN, Fine RL, Canoll P, et al. Regression of recurrent malignant gliomas with convection-enhanced delivery of topotecan. *Neurosurgery*. 2011;69:1272-1280.
118. Voges J, Reszka R, Gossmann A, et al. Imaging-guided convection-enhanced delivery and gene therapy of glioblastoma. *Ann Neurol*. 2003;54:479-487.
119. Sampson JH, Reardon DA, Friedman AH, et al. Sustained radiographic and clinical response in patient with bifrontal recurrent glioblastoma multiforme with intracerebral infusion of the recombinant targeted toxin TP-38: case study. *Neuro-Oncology*. 2005;7:90-96.

120. Mueller S, Polley MY, Lee B, et al. Effect of imaging and catheter characteristics on clinical outcome for patients in the PRECISE study. *J Neuro-Oncol*. 2011;101(2):267-277.
121. Morrison PF, Lonser RR, Oldfield EH. Convective delivery of glial cell line-derived neurotrophic factor in the human putamen. *J Neurosurg*. 2007;107:74-83.
122. Guarnieri M, Carson BS, Khan A, Penno M, Jallo GI. Flexible versus rigid catheters for chronic administration of exogenous agents into central nervous system tissues. *J Neurosci Methods*. 2005;144:147-152.
123. Raghavan R, Brady ML, Rodríguez-Ponce MI, Hartlep A, Pedain C, Sampson JH. Convection-enhanced delivery of therapeutics for brain disease, and its optimization. *Neurosurg Focus*. 2006;20(4).
124. Morrison PF, Chen MY, Chadwick RS, Lonser RR, Oldfield EH. Focal delivery during direct infusion to brain: role of flow rate, catheter diameter, and tissue mechanics. *Am J Phys*. 1999;277:R1218-R1229.
125. Fiandaca MS, Forsayeth JR, Dickinson PJ, Bankiewicz KS. Image-guided convection-enhanced delivery platform in the treatment of neurological diseases. *Neurotherapeutics*. 2008;5:123-127.
126. Krauze MT, Saito R, Noble C, et al. Reflux-free cannula for convection-enhanced high-speed delivery of therapeutic agents. *J Neurosurg*. 2005;103(5):923-929.
127. Oh S, Odland R, Wilson SR, et al. Improved distribution of small molecules and viral vectors in the murine brain using a hollow fiber catheter. *J Neurosurg*. 2007;107(3):568-577.
128. Olbricht WL, Neeves KB, Foley CP. Microfluidic probes in the treatment of brain-related diseases. *Drug News Perspect*. 2010;23(8):491-497.
129. Brady ML, Raghavan R, Singh D, et al. In vivo performance of a microfabricated catheter for intraparenchymal delivery. *J Neurosci Methods*. 2014;229:76-83.
130. DNAtrix, Inc. Combination Adenovirus + Pembrolizumab to Trigger Immune Virus Effects (CAPTIVE). Available from: https://clinicaltrials.gov/ct2/show/NCT02798406. NLM identifier: NCT02798406 (Accessed 6 April 2018), 2019.
131. Vogelbaum MA, Brewer C, Barnett GH, et al. First-in-human evaluation of the Cleveland multiport catheter for convection-enhanced delivery of topotecan in recurrent high-grade glioma: results of pilot trial 1. *J Neurosurg*. 2018;1-10.
132. Vogelbaum MA. Intratumorally Administered Topotecan Using CED in High Grade Glioma Undergoing Stereotactic Biopsy. Available from: https://clinicaltrials.gov/ct2/show/NCT03193463. NLM identifier: NCT03193463 (Accessed 6 April 2018), 2019.
133. Vogelbaum MA. Topotecan in Glioma Undergoing a Clinically-Indicated Surgical Resection. Available from: https://clinicaltrials.gov/ct2/show/NCT02500459. NLM identifier: NCT02500459 (Accessed 6 April 2018), 2019.
134. Fung LK, Ewend MG, Sills A, et al. Pharmacokinetics of interstitial delivery of carmustine, 4-hydroperoxycyclophosphamide, and paclitaxel from a biodegradable polymer implant in the monkey brain. *Cancer Res*. 1998;58(4):672-684.
135. Salegio EA, Streeter H, Dube N, et al. Distribution of nanoparticles throughout the cerebral cortex of rodents and non-human primates: implications for gene and drug therapy. *Front Neuroanat*. 2014;8:9.
136. Valles F, Fiandaca MS, Bringas J, et al. Anatomic compression caused by high-volume convection-enhanced delivery to the brain. *Neurosurgery*. 2009;65(3):579-585.
137. White E, Bienemann A, Malone J, et al. An evaluation of the relationships between catheter design and tissue mechanics in achieving high-flow convection-enhanced delivery. *J Neurosci Methods*. 2011;199(1):87-97.
138. Yin D, Forsayeth J, Bankiewicz KS. Optimized cannula design and placement for convection-enhanced delivery in rat striatum. *J Neurosci Methods*. 2010;187:46-51.
139. Olson JJ, Zhang Z, Dillehay D, Stubbs J. Assessment of a balloon-tipped catheter modified for intracerebral convection-enhanced delivery. *J Neuro-Oncol*. 2008;89:159-168.
140. Sonabend AM, Stuart RM, Yun J, et al. Prolonged intracerebral convection-enhanced delivery of topotecan with a subcutaneously implantable infusion pump. *Neuro-Oncology*. 2011;13(8):886-893.
141. Groothuis DR, Ward S, Itskovich AC, et al. Comparison of 14C-sucrose delivery to the brain by intravenous, intraventricular, and convection-enhanced intracerebral infusion. *J Neurosurg*. 1999;90(2):321-331.
142. Jain RK. Physiological barriers to delivery of monoclonal antibodies and other macromolecules in tumors. *Cancer Res*. 1990;50(3):814s-819s.
143. Smith JH, Humphrey JA. Interstitial transport and transvascular fluid exchange during infusion into brain and tumor tissue. *Microvasc Res*. 2007;73:58-73.

CHAPTER

22

Delivery Methods for Treatment of Genetic Disorders

Kunal Gupta, Nathan R. Selden

Department of Neurological Surgery, Oregon Health & Science University, Portland, OR, United States

GENE THERAPY IN THE 21ST CENTURY

There has never been a more exciting time for gene therapy. With the advent of novel technologies and multiple human applications of gene therapy technologies, the hope for permanent cures for monogenic disorders has increased substantially. Recent studies have attempted to address disorders such as hemophilia,[1] epidermolysis bullosa,[2] and Hunter syndrome (ClinicalTrials.gov NCT03041324).[3] Hemophilia A and B are coagulopathies characterized by a deficiency of clotting factors VIII and IX, respectively, causing recurrent hemorrhages and requiring repeat clotting factor transfusions. George et al.[1] significantly reduced hemorrhage rates in 10 adult men with hemophilia by using peripheral infusion of adeno-associated virus (AAV) vector carrying a modified clotting factor IX gene. Hemorrhage rates fell from 11.1 events per year to 0.4 at 1-year follow-up. Moreover, reduced factor infusions among the treated patients saved an estimated $3.6 million over the course of the study.[1]

Junctional epidermolysis bullosa is a dermatological disorder characterized by severe mechanical fragility of the skin, causing blistering, sloughing, and superinfection. It is caused by a series of mutations in the genes encoding laminin 5 (also known as laminin 332). In one recent, particularly severe case of a 7-year-old boy with 60% total body epidermis loss, a group at Ruhr University Bochum in Germany performed biopsies on samples of the patient's tissue. They transfected the harvested tissue with a non-pathological version of the laminin 3B gene (*LAMB3*) using the Moloney murine leukemia virus (MMLV) retroviral vector, expanded the subsequent epidermal stem cell colonies on transplantable sheets, and replaced the child's epidermis. At 21-month follow-up, the patient's modified skin appeared histologically normal and the patient demonstrated no evidence of autoantibodies to the basement

Nervous System Drug Delivery
https://doi.org/10.1016/B978-0-12-813997-4.00022-0

© 2019 Elsevier Inc. All rights reserved.

membrane.[2] Successful treatment of these peripheral systemic disorders depends on accessibility of the pathological tissue as well as delivery of the therapeutic gene. In contrast, disorders of the central nervous system (CNS) require unique adaptations of new and existing technologies for safe and effective trials of gene therapies. In this chapter, we will discuss gene delivery methods and their adaptations to the CNS.

The goal of gene therapy is stable, continuous expression of a gene product that targets a specific disorder. Therapeutic gene products for the treatment of human disease are often defective or missing protein molecules, such as clotting factor IX in hemophilia B or laminin 3B in junctional epidermolysis bullosa, as described above. Other forms of gene therapy involve the exogenous expression of trophic gene products (e.g., growth factors) or suppression of deleterious gene products (e.g., oncogenes) by negative regulator genes. In order for gene therapy to be delivered effectively, the therapy must: (1) target a specific organ, region, or cell type; (2) deliver exogenous genetic material to the cell nucleus, where it is either integrated or remains as an extrachromosomal unit; and (3) the gene product must be transcribed and translated, and must be sufficiently efficacious to affect the specific disorder.

There are many ways to target gene therapy efficiently and effectively. Reducing the required load of exogenous material and off-target effects improves clinical outcomes. Gene therapy can be performed at the population level, using germline genetic editing, or at the individual level. Germline genetic editing is currently prohibited in most Western European countries and is not approved for use in the United States—the ethical ramifications are vast and are well-considered elsewhere, and therefore will not be discussed further.[4] Individually targeted gene therapy can be applied in vivo, with direct application of the vector to the individual; and ex vivo, with modification of explanted or laboratory-cultured cells and subsequent implantation of these cells. The ex vivo approach has been used effectively in patients with immunological disease; for example, it has been used in the development of chimeric antigen receptor-modified T cells targeting acute and chronic lymphoid leukemia.[5, 6] In these trials, patients' T cells were removed, amplified, and modified to target CD19-expressing tumor cells; these cells were subsequently reinfused into patients. The results of these trials showed early success in bringing a medically refractory disease into remission. A trial examining neural stem cell implantation in stroke patients has also been carried out; in this case, the neural stem cell line was modified to express c-Myc by MMLV-type retroviral transduction to allow for clonal expansion ex vivo, and implanted into infarcted areas by stereotactic injection.[7] Because ex vivo and in vivo therapies share common features (e.g., mechanisms for gene modification, targeting methods), the technologies underpinning these applications will be discussed together.

Specific targeting can be considered at every stage of gene therapy design, from route of administration to vector choice to aspects of the genetic sequence delivered. Organ systems can be anatomically targeted by route of administration—for example, intravenous infusion can be used to target the hematopoietic/lymphoreticular system, including the liver, spleen, and endothelial cells[1]; inhalational approaches can target the pulmonary epithelium,[8] etc. Specific approaches to the CNS are discussed in greater detail below. In addition to delivery route, the vector and the genetic material can be targeted to specific cell populations. Cellular targeting can be dramatically altered by vector choice and modification.

NON-INTEGRATING VIRAL VECTORS

Gene therapy vectors can be largely divided into viral and nonviral subgroups. Viral vectors can be further subdivided into those that integrate into the host/patient genome and those that do not. Non-integrating viral vectors include adenovirus and AAV, which in their native forms integrate into the host genome at a specific locus on chromosome 19, but in their recombinant forms maintain their genome as an extrachromosomal episome. Gene expression by adenoviral transduction is transient and requires expression of the coxsackie adenovirus receptor (CAR) for cellular transduction. Transduction does occur in mammalian motoneurons, for example, using the CAV-2-type adenovirus.[9] Adenoviruses bind the CAR with fibers expressed on the capsid surface; modification of the fiber knob domain permits modification of the cellular tropism of the adenovirus.[10, 11] Adenoviruses replicate and disperse by cell lysis, which makes them particularly attractive as an oncolytic therapy.[10] In the CNS, however, replication-deficient viruses can permit cell survival.[12]

The AAV is a single-strand DNA parvovirus that also infects replicating and nonreplicating cells. In the recombinant form, it does not integrate into the host genome due to removal of the required *rep* genes. AAVs instead form extrachromosomal viral genomes within the nucleus and allow temporary transcription of the transduced genes. Although the episome is lost during cell replication, AAV may produce relatively more stable transduction in CNS neuronal disorders due to low cellular turnover. AAVs have low intrinsic immunogenicity and do not appear to contribute to human disease. A range of AAV serotypes with differential tissue-tropism have been evaluated; for example, AAV9 preferentially transduces myocardium, AAV2 has broad tissue-tropism, and AAV1, 6, and 7 preferentially transduce skeletal muscle.[13] AAV-vector-based gene therapy has been applied to neurological disorders including Canavan disease, demonstrating a reduction in *N*-acetylaspartate accumulation and seizure frequency.[14] AAV has also been applied in a number of trials for the treatment of Parkinson disease, in which early neuronal loss is concentrated in a defined population of dopaminergic neurons in the substantia nigra pars compacta, making it a popular therapy for anatomically and biologically targeted CNS therapy. AAV-based therapies of Parkinson disease have generally been found safe, but their clinical efficacy has not been established—perhaps due in part to the multiplicity and heterogeneity of the relevant genetic targets, as well as uncertainty about appropriate dosing.[15, 16] The overall safety profile of AAV application in the CNS, however, is encouraging. Further refinement of anatomical and biological disease targets as well as therapeutic transgenes may improve clinical efficacy in future studies.

The benefits of the adenovirus include high transduction efficiency, non-integrative mechanism, and ability to modify the virus. However, host immunity to the adenovirus (up to 80% in the human population) remains a critical factor in limiting efficacy, as does the ubiquitous cellular expression of the CAR and subsequent off-target sequestration (mostly hepatic) of the adenovirus. Cellular immunity of the host against transduced cells (due to viral protein expression) limits adenoviral gene therapy, both by direct cellular toxicity and clearance of transduced cells, which in turn limits the duration of gene expression. Hepatic sequestration, for example, can result in host cellular immunity against transduced liver cells and cause hepatitis, as demonstrated in early trials of gene therapy for hemophilia

B.[1] The transient nature of gene expression with adenovirus vectors also limits the durability of gene therapy, and may require either higher viral titers or repeat administration that can exacerbate host immunity.

Compared with adenovirus, AAV demonstrates lower pathogenicity and often more stable expression in long-lived post-mitotic cells (such as those of the CNS) due to episome formation. Replication of AAV, however, requires use of adenovirus helper virus, as the AAV itself is replication-deficient. The helper adenovirus, however, can in turn drive an immune response to AAV transduction. In order to circumvent this immune response, AAV can be generated using adenoviral-free mechanisms.[17] Any remaining immune response to the AAV capsid itself can be mitigated using various approaches, including low titer doses, immune suppression, capsid modification, and plasma exchange.[13] Despite these limitations, AAV-mediated therapy has been effectively applied to patients with a range of disorders, such as factor IX deficiency in hemophilia B. For this therapy, low viral titers were used in order to avoid an AAV-mediated immune reaction in conjunction with a highly active transgene promoter to maintain efficacy.[1]

INTEGRATING VIRAL VECTORS

Integrating viral vectors include retroviruses and lentiviruses. Retroviruses are RNA viruses carrying a gene for reverse transcriptase, which transcribes the viral RNA genome into double-stranded DNA. This DNA is then integrated into the host genome, allowing stable expression of viral genes or recombinant transgenes. Examples of *Retroviridae* include the murine leukemia virus, bovine leukemia virus, avian leucosis virus, and human T-cell leukemia virus. The best-known, most-studied example of a lentivirus is the human immunodeficiency virus; however, others include the feline, equine, and simian immunodeficiency viruses. The minimal genomic requirement for a replication-competent retrovirus includes three open reading frames; group-specific antigen genes that encode structural capsid and matrix proteins; polymerase genes that encode the reverse transcriptase, protease, and integrase; envelope genes that encode the surface glycoproteins, bound by long terminal repeat regions that contain transcriptional enhancer sequences; and a Psi sequence that directs virion packaging. Retroviruses differ from lentiviruses in that they only transduce replicating cells, thereby limiting their application to the CNS, whereas lentiviruses transduce dividing and latent cells.

Retroviruses and lentiviruses are modified for use as recombinant vectors by removal of the viral genome and replacement with a therapeutic transgene cassette. This also eliminates the replication competency of the virus, which is paramount for preventing iatrogenic viral illness. Elimination of the viral genome, however, poses a barrier to viral reproduction for clinical use. Retroviruses and lentiviruses therefore also require helper viruses for replication. Recombination between the transgenic virus and packaging viruses can and does occur. Using a combination of helper viruses with the relevant genes split between them, however, reduces the risk of recreating a replication-competent retrovirus. The packaging Psi sequence is maintained on the transgenic virus so that only the transgenic genome is packaged into viruses. As a result, generating sufficient titers and quantity of virus is technically demanding and may limit large-scale clinical therapies.[18, 19]

Additional genetic modification may be used to affect the tissue tropism of therapeutic viruses. For example, the viral glycoprotein coat can be modified or pseudotyped to alter tissue tropism for a therapeutic purpose. Pseudotyping involves the addition of heterologous envelope genes that encode viral vector surface glycoproteins; most often, the vesicular stomatitis virus glycoprotein is added due to its broad tropism. Examples of neurotropic pseudotyping include the addition of Mokola virus glycoproteins for cerebral white matter tracts, and lyssavirus glycoproteins that allow retrograde viral transport from peripheral muscle across the neuromuscular junction, into motor neurons.[20]

The ability of retroviral and lentiviral vectors to stably integrate into the host genome is key for continued transgene expression and potential continuous treatment of disease in humans. However, retroviral approaches are associated with considerable technical and biological issues. Retroviruses also have low transduction efficiency and only infect replicating cells, which is beneficial for certain systems such as epidermal stem cells, but not for post-mitotic neurons in the CNS. In contrast, lentiviral vectors can transduce replicating and post-mitotic cells. With either vector, random insertion site mutagenesis may cause significant long-term complications. For example, in a trial of retroviral treatment for X-linked severe combined immune deficiency (SCID-X1), autologous hematopoietic stem cells were transduced with a transgenic Gammaretrovirus derived from the murine leukemia virus carrying the *IL-2R gamma* gene and transfused into human patients.[21] This treatment restored T cell immunity in 9 of 10 patients, but was associated with leukemia occurrence in 4 patients 3 to 5 years later. In the patients who developed leukemia, gammaretroviral vector insertions were found close to a variety of proto-oncogenes, including *LMO2*, *BMI1*, and *CCND2*, indicating that this virus preferentially inserted at gene start sites and allowed the long terminal repeat promoter elements to drive the expression of nearby proto-oncogenes.[22, 23]

The benefits of stable transgene expression after retroviral or lentiviral infection must therefore be weighed against the risks of random insertion. To overcome these risks, directed insertional techniques have been developed that exploit the specific insertion site of the AAV at chromosome 19. AAV itself has low transgene capacity when used as a recombinant virus; the integrating *rep* genes are typically removed to increase transgene capacity, however this removes the AAV virus' ability for site-specific transgene integration. The ability of the AAV virus to perform site-specific transgene integration may be combined with the neurotropism and high transgene capacity of the herpes simplex virus genome, overcoming the transient expression seen with herpes simplex virus vectors as well as the low transgene capacity of the AAV virus while preserving its ability for site-specific integration. The herpes simplex virus/AAV hybrid amplicon with transgene expression stably integrates in certain mammalian systems up to a year after treatment.[24, 25]

NONVIRAL GENE EDITING

The development of a novel technology, clustered regularly interspaced short palindromic repeats (CRISPR), has revolutionized gene modification and clinical therapy. In nature, CRISPR/Cas9 (originally isolated from bacteria) provides a form of adaptive immunity against bacteriophages.[26, 27] CRISPR is a gene locus present in bacteria comprising an array of repeat-spacer sequences that store the genetic sequences of past invasions by bacteriophages (i.e., protospacers). The CRISPR locus is accompanied by an operon of CRISPR-associated sequence (Cas) genes that encode proteins that modify the transcribed CRISPR-RNA (crRNA) into a mature crRNA. There are three classes of CRISPR/Cas systems. Type II uses Cas9 as a single effector protein and has been the most studied for gene modification. Types I and III require multiple downstream effector molecules and therefore have more barriers to translation in eukaryotic systems.[28] The mature crRNA (also termed a *guide RNA*), which is specific for a bacteriophage gene sequence, can bind the Cas nuclease and guide it to cleave the invading bacteriophage genome in a sequence-specific manner. The guide RNA is specific to the target gene both in terms of the sequence and in the presence of an obligatory protospacer adjacent motif sequence, which ensures that the Cas nuclease does not bind and cleave the host's own genetic material (Fig. 1).[26]

CRISPR/Cas9 is currently being adapted for use in gene therapy. The ability to easily re-encode novel guide-RNAs for alternative sequences allows targeting of a wide range of genes. Cas9 alone introduces sequence-specific, double-stranded DNA breaks. The genomic engineering aspect comes from the subversion of native gene repair mechanisms. These double-strand breaks are repaired either by nonhomologous end joining or homology-directed repair. Nonhomologous end joining introduces insertions or deletions (i.e., indels) that disrupt the gene, whereas homology-directed repair relies on a preexisting template gene to repair the cut gene.[29] CRISPR/Cas9 technology has been applied to human cells in vitro and has been shown to target specific sequences and even multiple sequences simultaneously. However, efficiency was low, ranging from just 2% to 25%.[30] CRISPR/Cas9 technology is limited in its ability to introduce genetic components into cells, which are still dependent on existing viral transduction and nonviral transfection techniques. An additional challenge is the occurrence of off-target binding. In human cell lines, Cas9-mediated gene editing results in a mean of 40% off-target deletions, with rates up to 125% of the intended target site, suggesting that off-target sites can be cleaved by Cas9 at a higher rate than the target site itself.[31] Management of off-target effects will be critical for safe and widespread adoption of CRISPR/Cas9-mediated gene therapy.

To reduce off-site targeting, the Cas9/guide-RNA ribonucleoprotein complex (Cas9-RNP) has been injected directly into target cells. In a mouse feasibility study, Cas9-RNP was injected into the mouse brain and shown to efficiently disinhibit expression of a fluorescent marker in neurons.[32] The Cas9-RNP itself does not penetrate cells. To penetrate cells, it must be conjugated to additional structures, such as nanoparticles, polyarginine peptides, or a viral nuclear localization signal (e.g., from the simian vacuolating virus 40).[32, 33] This exciting approach does not use replication-competent techniques, and could be adapted to target genetic disorders in the CNS. The use of Cas9-RNP has also been applied to human embryos to edit out the mutated MYBPC3 gene, which causes hypertrophic cardiomyopathy. In that study, off-target effects were mitigated by use of the RNP complex, and embryo mosaicism was mitigated by injecting the RNP complex into the zygote prior to cell division.[34] Mutation in the paternal genome may have been corrected using the maternal template; however, the exact mechanism is still under investigation.[35]

Other nonviral-mediated gene editing mechanisms include cationic polymers[36] and cationic liposomal therapy,[37] which deliver oligonucleotides intracellularly. These positively charged molecules bind negatively charged genetic material for delivery and target cell membranes, causing endocytosis of the therapeutic sequence. Cationic polymers and liposomes have the benefits of larger sequence capacity than viruses, plus increased targeting specificity; however, they are also associated with lower efficiency and significant toxicity.[38, 39] Protein-based nuclease systems such as zinc-finger nucleases and transcription activator-like effector nucleases induce sequence-specific double-strand breaks, which are subsequently repaired by homology-directed repair or nonhomologous end joining, similar to CRISPR/Cas9. These nucleases comprise a Fok1 nuclease conjugated to zinc finger and transcription

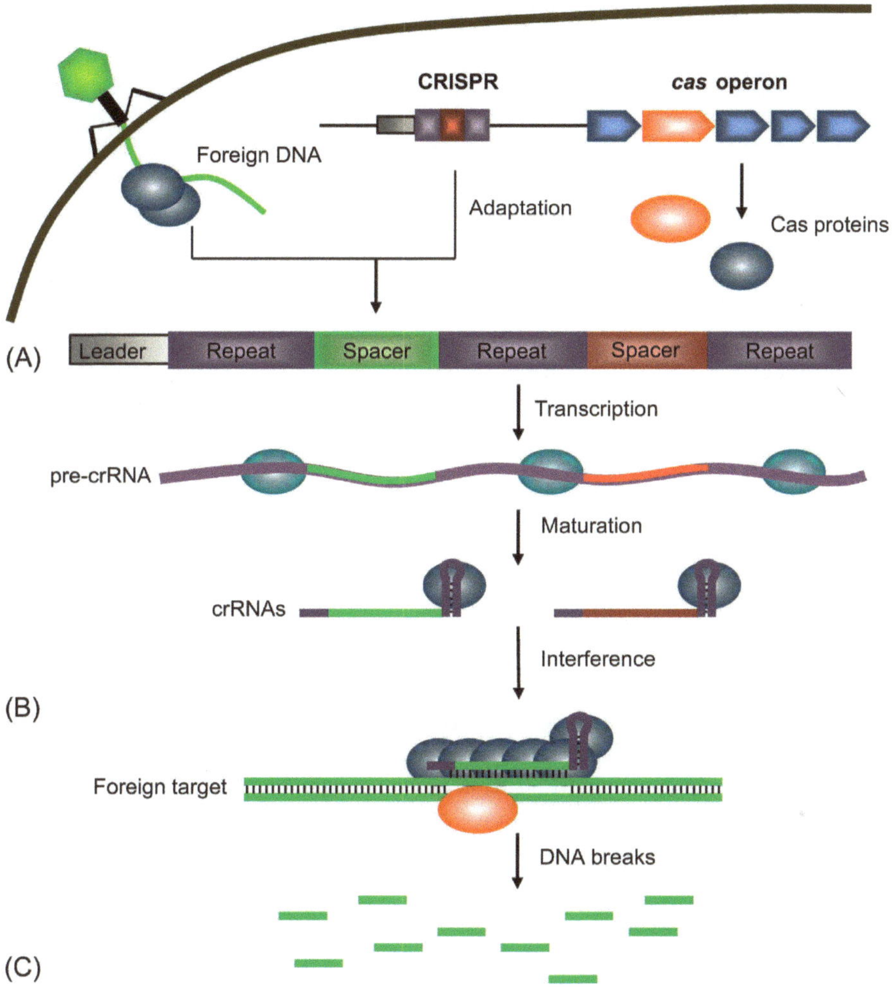

FIG. 1 Overview of CRISPR/Cas immunity to viruses of bacteria and archaea based on the type I system. (A) In adaptation, phage infection is recognized by Cas proteins (encoded by the cas operon) and a short sequence of the phage DNA (protospacer) is added to the leader end of the CRISPR array, resulting in a new spacer sequence and a duplicated repeat. (B) Transcription of the CRISPR array from a promoter within the leader sequence results in a precursor CRISPR RNA (pre-crRNA) transcript. The pre-crRNA is matured into individual crRNAs by a process involving Cas proteins. (C) The mature crRNAs form a ribonucleoprotein complex, which targets nucleic acids that are complementary to the spacer sequence in the crRNAs. In some cases, a separate Cas nuclease (orange) is recruited, resulting in interference and destruction of the nucleic acid target. *Reproduced with permission from Fineran PC, Charpentier E. Memory of viral infections by CRISPR-Cas adaptive immune systems: acquisition of new information. Virology 2012;434(2):202-209.*

activator-like effector nuclease domains that confer sequence specificity.[39, 40] A key limitation of protein-based genome editing is the requirement for protein sequence engineering to confer genomic sequence specificity. As a result, RNA-based genome editing therapies such as CRISPR/Cas9 have sprung to the forefront of gene editing therapies, due to the ease of modification for a range of different target sequences.

CELLULAR THERAPY

The techniques discussed previously use gene therapy to modify the functions of existing cells. Cellular therapy approaches CNS genetic disorders with delivery of specific gene products or trophic factors by cell replacement. Cells used for treatment of genetic disorders in the CNS can take multiple forms; they can take the form of prespecified terminally differentiated cells, such as substantia nigra neurons derived from human fetal tissue, or immature cells. Immature cells can be pluripotent stem cells or neuroectodermal lineage specified, and can be derived from laboratory-maintained human embryonic stem cells (HESCs) or patient-specific inducible pluripotent stem cells (iPSCs). Outside the CNS, patient autologous cells can be extracted, expanded, and potentially modified and replaced—for example, in the treatment of junctional epidermolysis bullosa[2] or leukemia.[5, 6] Each system has potential benefits and caveats associated with its therapeutic use.

Terminally differentiated cells have the benefit of known cell identity and defined functions, lending credibility to the hypothesis that transplanting these cells in regions of loss may restore function. Terminally differentiated cells are also less likely than HESCs to display tumorgenicity. Terminally differentiated cells can be removed stably and maintained from human adult and fetal cadaveric tissue, though their shelf-life may be short. Typically, however, these cells can be neither maintained in continuous culture nor expanded in vitro, potentially limiting widespread application to large-scale clinical patient cohorts.

Immature cells include HESCs and iPSCs. HESCs were initially described in 1998, with further lines described in 2004.[41-43] iPSCs were described by Shinya Yamanaka in 2007,[44] who used a combination of oncogenes to turn human fibroblasts into iPSCs. Human stem cells, regardless of whether HESC or iPSC, have the benefit of near-infinite expansion in chemically defined media, providing the scale required for clinical treatment programs. There is, however, significant teratogenicity risk with immature cells,[45] which limits their clinical use. Furthermore, differentiation in the post-injury environment may occur along nonneural lineages and provide little benefit in neurological disorders.

Human embryonic stem cells and iPSCs can be differentiated into lineage-specific precursors, including human neural precursor cells (NPCs). NPCs can also be maintained and expanded in culture, similarly allowing the magnitude of scale required for clinical treatment programs. It is possible for these cells to continue to replicate posttransplant, providing an increased number of functional cells and expanding the impact of smaller transplants, perhaps even providing a stable niche for therapeutic cell replacement. Compared to HESCs, NPCs have the additional benefit of following defined pathways of terminal differentiation into glia and neurons, which are the cell types of principal clinical benefit in neurodegenerative disorders. Furthermore, the transplantation of neural precursors rather than terminally differentiated cells may permit the microenvironment to dictate the differentiation of the precursors into cell subtypes that replace the acquired deficit. In theory, this allows for a single technology—the generation and transplantation of enriched clinical-grade neural precursor cells—to treat many disorders. In contrast, terminally differentiated cells or viral therapies require unique individual technologies engineered specifically for each genetic or degenerative disorder.

Several trials of human stem cell implantation specifically for the treatment of human genetic disorders of the CNS have been described. Human NPCs were transplanted in six children with neuronal ceroid lipofuscinosis (i.e., Batten disease). In a first human trial, 6 patients with infantile or late-infantile neuronal ceroid lipofuscinosis underwent stereotactic intracerebral and intraventricular injection of allogeneic human NPCs with concomitant immunosuppression for 12 months. No adverse events were attributed to the procedure or to the immunosuppression, demonstrating the safety of this approach.[46] This clinical study was supported by in vitro data demonstrating production of the deficient enzymes, and animal in vivo data demonstrating self-renewal, cell migration, and efficacy in an animal model of neuronal ceroid lipofuscinosis.[47, 48] Clinical efficacy in this Phase I safety trial was limited by the small number of enrolled patients and the advanced state of their disease. The trial was, however, critical to subsequent testing aimed at the development of large-scale therapeutic programs. For example, a similar open-label trial was subsequently performed in four patients with Pelizaeus-Merzbacher disease, an inherited leukodystrophy caused by a *proteolipid protein 1* gene mutation. Human NPCs were surgically transplanted into the frontal lobes of four male patients and immunosuppression was continued for 9 months. This procedure was also well tolerated. Posttransplant imaging suggested increased myelination, and modest neurological improvements were noted in three patients.[49] Interestingly, non-CNS transplantation of hematopoietic stem cells has been demonstrated to stabilize cerebral adrenoleukodystrophy (the most severe CNS variant of X-linked adrenoleukodystrophy) and to improve survival. The mechanism of this therapeutic effect is unclear, but may involve cross-correction of the defective peroxisomal membrane-bound protein (Fig. 2).[50, 51]

Cellular replacement therapies may be combined with specific gene modifications to alter the activity of transplanted cells and target specific disorders, providing increased trophic support to native cells or providing missing gene products. It is possible that viral transduction of existing cells alone may be insufficient to provide functional clinical benefit in patients with neurodegenerative disorders associated with significant cell loss; in these cases, cellular therapies also may directly replace missing cells and, combined with gene modification, may increase the possibility of functional clinical benefit.

Irrespective of whether using immature precursor cells with functional integration guided by the disease microenvironment, or cells genetically engineered to provide a specific therapy, significant barriers remain. In general, both cellular therapy strategies for genetic disorders require a cross-correction strategy, in which the transplanted cells secrete, and host cells absorb, the deficient gene product. The source of transplantable cells also remains controversial, with significant technical and ethical limitations on cadaveric adult and fetal cells, as well as financial and ethical limits on the use of HESCs.[52]

Inducible pluripotent stem cells are expensive, as they necessitate creation of stable cell lines and clinical-grade transplantable cells for each patient; however, the use of patient-specific tissue to create stem cells may circumvent the ethical issues posed by embryonic or fetal tissue procurement. In patients with genetic disorders, autologous cell sources generally require substantial ex vivo modification of the genetic defect they also carry, unless allogeneic cells are substituted. For allogeneic CNS transplants, it is currently unclear what duration of immunosuppression is required. Although the use of autologous patient-derived iPSCs may abrogate the requirement for immunosuppression, this may not be a feasible therapeutic strategy in all cases. In some cases, allograft NPC transplants into the CNS appear to survive well after posttransplant immunosuppression is discontinued.[46] It is possible that the immunological privilege of the CNS can be exploited; however, clinical studies are needed to determine the functional longevity of transplanted tissue. Future challenges include the

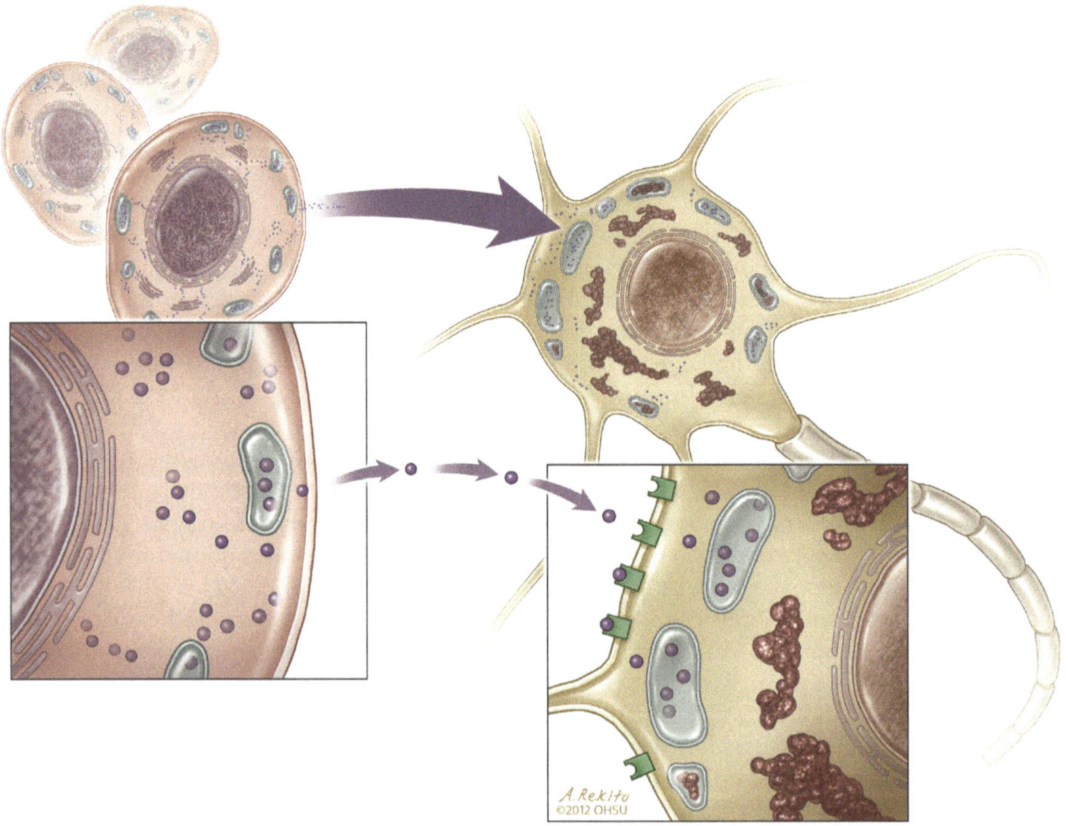

FIG. 2 Cellular transplantation therapy allows cross-correction by transfer of missing or defective proteins from transplanted cells to autologous cells, functionally correcting the genetic defect. *Reproduced with permission from A. J. Rekito, Oregon Health & Science University.*

development of human in vivo monitoring strategies; the ability to track cell survival, engraftment, and function, by direct imaging and serological testing would be of tremendous benefit in guiding trial outcomes as well as clinical efficacy.[53, 54]

PRINCIPLES OF APPROACHES TO THE CNS

Gene therapy in the CNS will require specific adaptation of existing technologies to its unique biology. The blood–brain barrier (BBB) holds the CNS largely impenetrable to a range of systemically applied therapies. In animal studies, it has been shown that certain adenovirus serotypes are able to penetrate the CNS after systemic administration; however, these phenomena remain untested in humans as of yet.[55]

Disruption of the BBB is used therapeutically to apply systemic chemotherapy agents that typically do not penetrate the BBB to access tumors in the CNS. The technique uses arterial catheterization of the major cranial vessels and administration of intra-arterial mannitol to temporarily open the BBB.[56] The benefit of BBB disruption is that it can affect a large volume of brain and does not require neurological surgery. However, its limitations include reduced specificity for small discrete brain regions and invasiveness.

Surgical access to the CNS using open and stereotactic techniques is safe and accurate for a range of disorders, including tumors, movement disorders, epilepsy, and neurodegenerative disease. Stereotactic techniques allow more precise anatomical targeting of therapies, to even submillimeter accuracy. They also allow avoidance of particularly sensitive or eloquent areas in the planning of injection sites. Cranial approaches are typically frame-based (the most common frames in use are the Leksell Stereotactic System [Elekta] and Cosman-Roberts-Wells frames, Fig. 3), as well as computer-based frameless stereotactic systems. Stereotactic systems have been used in the application of AAV-glutamic acid decarboxylase therapy to globus pallidus neurons. Stereotactic administration of the virus was safe and resulted in improved motor scores.[16, 57] Interestingly, in these studies, there was minimal change in antiviral titers

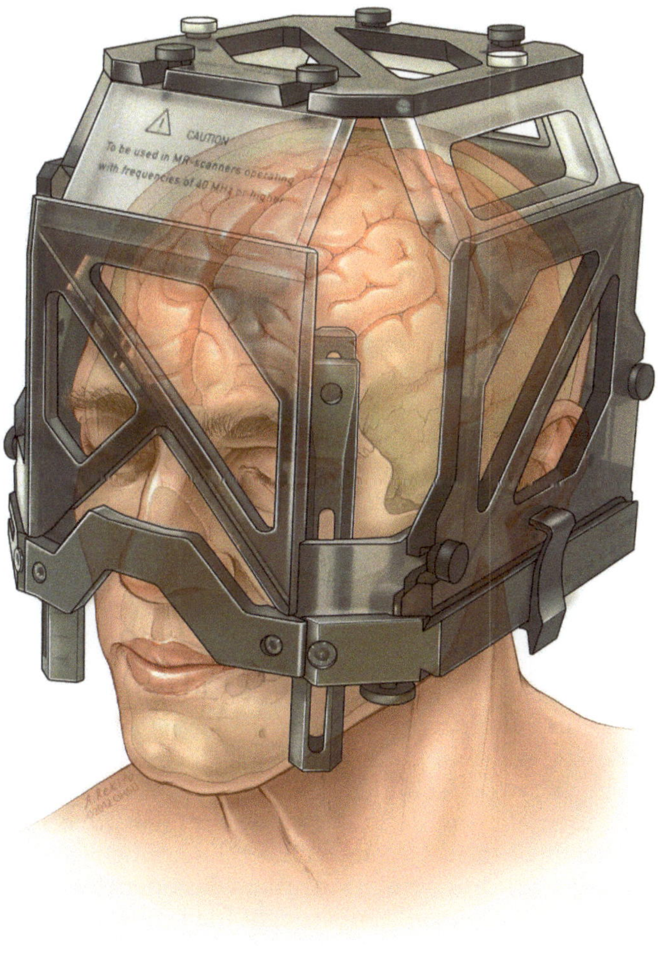

FIG. 3 Image of frame-based Leksell cranial targeting system for stereotactic surgery. *Reproduced with permission from A. J. Rekito, Oregon Health & Science University.*

after stereotactic intracerebral viral injection. Immunological privilege may make the CNS particularly receptive to viral-mediated gene therapies.

Stereotactic targeting techniques also exist for spinal surgery. Experimental platforms have been developed for targeted spinal cord injection, using either physiological microelectrode recordings[58] or magnetic resonance imaging technologies.[59] Spinal cord injection techniques have been used safely in human patients.[60] Targeting systems with real-time localization, either by physiological or radiographic mapping, will be crucial for safely cannulating specific tracts within the spinal cord.

Intrathecal injection of virally mediated gene therapies has been performed in a number of experimental models. Lumbar puncture, while invasive, is a routine bedside procedure with low morbidity and does not involve penetration of neural parenchyma. In animal models, cisterna magna and lumbar cistern infusion of AAV transduces brain and spinal cord neurons diffusely, with greater levels of transduction closer to the injection site.[61] However, other studies using lentiviruses show greater transduction in the meninges.[62] Interestingly, intrathecal virus has been shown to transduce peripheral nerves and Schwann cells, suggesting that viruses applied intrathecally can transgress the Redlich-Obersteiner zones and can be used to treat peripheral nerve disorders.[63, 64] It will be of critical importance to characterize the relative transduction efficiencies and specificities of different viruses, as each may be suited to varying pathologies. An ongoing European trial for patients with Huntington disease is intrathecally administering antisense Huntingtin to bind mutant Huntingtin gene transcripts (ClinicalTrials.gov NCT02519036).[65]

Viral transduction, nonviral gene transfection, and cellular replacement modalities share common challenges for CNS administration. The high tissue volume of the CNS suggests that a systemic administration method may be the most effective way to treat genetic disorders. However, peripheral venous administration is hampered by the BBB, although this limitation may be spontaneously modified in some inflammatory conditions or subject to therapeutic manipulation. It is also unclear how effectively material infused into the cerebrospinal fluid penetrates the brain and spinal cord parenchyma. Conversely, the compartmentalization of the CNS into defined areas with specific functional affiliations allows targeted administration of therapies by surgical means in specific disease states that have some degree of anatomic focality (e.g., Parkinson disease). Appropriate therapeutic approaches will therefore need to be tailored to specific disorders.

CONCLUSIONS

Delivery of gene therapy can be accomplished by viral-mediated, nonviral-mediated, and cell-mediated methods. A considerable number of variables influence the efficacy of gene therapy, including vector choice, gene design, and anatomical delivery method. These factors must be individually tailored for each specific disorder and gene therapy application. Initial hopes that gene therapy would provide dramatic and definitive cures with few adverse effects were sharply curtailed after the death of Jesse Gelsinger in 1999. Gelsinger was an 18-year-old man who received viral therapy for ornithine transcarbamylase deficiency and died from a severe immunological reaction to the vector.[66] Subsequent trials, however, have demonstrated the safety of viral-mediated gene therapies, and their clinical applications are growing.

In this chapter, we have discussed viral-mediated therapies, including integrating and non-integrating viruses; nonviral-mediated gene therapies, such as CRISPR/Cas9; and cellular therapies for genetic disorders and specific adaptations of gene therapy to the CNS. Application of these approaches may result in rapid diminution or even eradication of certain monogenic disorders, in the same way that vaccines have eliminated certain communicable diseases. Nonetheless, these technologies are currently in their infancy, and further refinement and continued safety studies will be required for optimal safe application at a population level.

References

1. George LA, Sullivan SK, Giermasz A, et al. Hemophilia B gene therapy with a high-specific-activity factor IX variant. *N Engl J Med*. 2017;377(23):2215-2227.
2. Hirsch T, Rothoeft T, Teig N, et al. Regeneration of the entire human epidermis using transgenic stem cells. *Nature*. 2017;551(7680):327-332.
3. Faul M, Xu L, Wald MM, Coronado V, Dellinger AM. Traumatic brain injury in the United States: national estimates of prevalence and incidence, 2002–2006. *Inj Prev*. 2010;16(Suppl 1):A268.
4. Lanphier E, Urnov F, Haecker SE, Werner M, Smolenski J. Don't edit the human germ line. *Nature*. 2015; 519(7544):410-411.
5. Grupp SA, Kalos M, Barrett D, et al. Chimeric antigen receptor-modified T cells for acute lymphoid leukemia. *N Engl J Med*. 2013;368(16):1509-1518.
6. Porter DL, Levine BL, Kalos M, Bagg A, June CH. Chimeric antigen receptor-modified T cells in chronic lymphoid leukemia. *N Engl J Med*. 2011;365(8):725-733.
7. Sinden JD, Hicks C, Stroemer P, Vishnubhatla I, Corteling R. Human neural stem cell therapy for chronic ischemic stroke: charting progress from laboratory to patients. *Stem Cells Dev*. 2017;26(13):933-947.
8. Kim J, Hall RR, Lesniak MS, Ahmed AU. Stem cell-based cell carrier for targeted oncolytic virotherapy: translational opportunity and open questions. *Viruses*. 2015;7(12):6200-6217.
9. Salinas S, Bilsland LG, Henaff D, et al. CAR-associated vesicular transport of an adenovirus in motor neuron axons. *PLoS Pathog*. 2009;5(5).
10. Panek WK, Kane JR, Young JS, et al. Hitting the nail on the head: combining oncolytic adenovirus-mediated virotherapy and immunomodulation for the treatment of glioma. *Oncotarget*. 2017;8(51):89391-89405.
11. Wickham TJ, Roelvink PW, Brough DE, Kovesdi I. Adenovirus targeted to heparan-containing receptors increases its gene delivery efficiency to multiple cell types. *Nat Biotechnol*. 1996;14(11):1570-1573.
12. Wold WS, Toth K. Adenovirus vectors for gene therapy, vaccination and cancer gene therapy. *Curr Gene Ther*. 2013;13(6):421-433.
13. Mingozzi F, High KA. Immune responses to AAV vectors: overcoming barriers to successful gene therapy. *Blood*. 2013;122(1):23-36.
14. Leone P, Shera D, McPhee SW, et al. Long-term follow-up after gene therapy for canavan disease. *Sci Transl Med*. 2012;4(165):165ra163.
15. Bartus RT, Weinberg MS, Samulski RJ. Parkinson's disease gene therapy: success by design meets failure by efficacy. *Mol Ther*. 2014;22(3):487-497.
16. Kaplitt MG, Feigin A, Tang C, et al. Safety and tolerability of gene therapy with an adeno-associated virus (AAV) borne GAD gene for Parkinson's disease: an open label, phase I trial. *Lancet*. 2007;369(9579):2097-2105.
17. Matsushita T, Elliger S, Elliger C, et al. Adeno-associated virus vectors can be efficiently produced without helper virus. *Gene Ther*. 1998;5(7):938-945.
18. Ichim CV, Wells RA. Generation of high-titer viral preparations by concentration using successive rounds of ultracentrifugation. *J Transl Med*. 2011;9:137.
19. Miller DL, Meikle PJ, Anson DS. A rapid and efficient method for concentration of small volumes of retroviral supernatant. *Nucleic Acids Res*. 1996;24(8):1576-1577.

20. Cronin J, Zhang XY, Reiser J. Altering the tropism of lentiviral vectors through pseudotyping. *Curr Gene Ther*. 2005;5(4):387-398.
21. Cavazzana-Calvo M, Hacein-Bey S, de Saint Basile G, et al. Gene therapy of human severe combined immunodeficiency (SCID)-X1 disease. *Science*. 2000;288(5466):669-672.
22. Hacein-Bey-Abina S, Garrigue A, Wang GP, et al. Insertional oncogenesis in 4 patients after retrovirus-mediated gene therapy of SCID-X1. *J Clin Invest*. 2008;118(9):3132-3142.
23. Hacein-Bey-Abina S, Von Kalle C, Schmidt M, et al. LMO2-associated clonal T cell proliferation in two patients after gene therapy for SCID-X1. *Science*. 2003;302(5644):415-419.
24. Heister T, Heid I, Ackermann M, Fraefel C. Herpes simplex virus type 1/adeno-associated virus hybrid vectors mediate site-specific integration at the adeno-associated virus preintegration site, AAVS1, on human chromosome 19. *J Virol*. 2002;76(14):7163-7173.
25. Saydam O, Glauser DL, Fraefel C. Construction and packaging of herpes simplex virus/adeno-associated virus (HSV/AAV) hybrid amplicon vectors. *Cold Spring Harb Protoc*. 2012;2012(3):352-356.
26. Fineran PC, Charpentier E. Memory of viral infections by CRISPR-Cas adaptive immune systems: acquisition of new information. *Virology*. 2012;434(2):202-209.
27. Pourcel C, Salvignol G, Vergnaud G. CRISPR elements in Yersinia pestis acquire new repeats by preferential uptake of bacteriophage DNA, and provide additional tools for evolutionary studies. *Microbiology*. 2005;151(Pt 3):653-663.
28. Makarova KS, Haft DH, Barrangou R, et al. Evolution and classification of the CRISPR-Cas systems. *Nat Rev Microbiol*. 2011;9(6):467-477.
29. Wilkinson R, Wiedenheft B. A CRISPR method for genome engineering. *F1000Prime Rep*. 2014;6:3.
30. Mali P, Yang L, Esvelt KM, et al. RNA-guided human genome engineering via Cas9. *Science*. 2013;339(6121):823-826.
31. Fu Y, Foden JA, Khayter C, et al. High-frequency off-target mutagenesis induced by CRISPR-Cas nucleases in human cells. *Nat Biotechnol*. 2013;31(9):822-826.
32. Staahl BT, Benekareddy M, Coulon-Bainier C, et al. Efficient genome editing in the mouse brain by local delivery of engineered Cas9 ribonucleoprotein complexes. *Nat Biotechnol*. 2017;35(5):431-434.
33. Mout R, Ray M, Yesilbag Tonga G, et al. Direct cytosolic delivery of CRISPR/Cas9-ribonucleoprotein for efficient gene editing. *ACS Nano*. 2017;11(3):2452-2458.
34. Ma H, Marti-Gutierrez N, Park SW, et al. Correction of a pathogenic gene mutation in human embryos. *Nature*. 2017;548(7668):413-419.
35. Egli D, Zuccaro M, Kosicki M, Church G, Bradley A, Jasin M. Inter-homologue repair in fertilized human eggs? bioRxiv, *Nature*. 2017;560:E5-E7.
36. Boussif O, Lezoualc'h F, Zanta MA, et al. A versatile vector for gene and oligonucleotide transfer into cells in culture and in vivo: polyethylenimine. *Proc Natl Acad Sci U S A*. 1995;92(16):7297-7301.
37. Felgner PL, Gadek TR, Holm M, et al. Lipofection: a highly efficient, lipid-mediated DNA-transfection procedure. *Proc Natl Acad Sci U S A*. 1987;84(21):7413-7417.
38. El-Aneed A. An overview of current delivery systems in cancer gene therapy. *J Control Release*. 2004;94(1):1-14.
39. Yin H, Kanasty RL, Eltoukhy AA, Vegas AJ, Dorkin JR, Anderson DG. Non-viral vectors for gene-based therapy. *Nat Rev Genet*. 2014;15(8):541-555.
40. Urnov FD, Rebar EJ, Holmes MC, Zhang HS, Gregory PD. Genome editing with engineered zinc finger nucleases. *Nat Rev Genet*. 2010;11(9):636-646.
41. Cowan CA, Klimanskaya I, McMahon J, et al. Derivation of embryonic stem-cell lines from human blastocysts. *N Engl J Med*. 2004;350(13):1353-1356.
42. Shamblott MJ, Axelman J, Wang S, et al. Derivation of pluripotent stem cells from cultured human primordial germ cells. *Proc Natl Acad Sci U S A*. 1998;95(23):13726-13731.
43. Thomson JA, Itskovitz-Eldor J, Shapiro SS, et al. Embryonic stem cell lines derived from human blastocysts. *Science*. 1998;282(5391):1145-1147.
44. Takahashi K, Tanabe K, Ohnuki M, et al. Induction of pluripotent stem cells from adult human fibroblasts by defined factors. *Cell*. 2007;131(5):861-872.
45. Merkle FT, Ghosh S, Kamitaki N, et al. Human pluripotent stem cells recurrently acquire and expand dominant negative P53 mutations. *Nature*. 2017;545(7653):229-233.
46. Selden NR, Al-Uzri A, Huhn SL, et al. Central nervous system stem cell transplantation for children with neuronal ceroid lipofuscinosis. *J Neurosurg Pediatr*. 2013;11(6):643-652.

47. Tamaki S, Eckert K, He D, et al. Engraftment of sorted/expanded human central nervous system stem cells from fetal brain. *J Neurosci Res*. 2002;69(6):976-986.
48. Tamaki SJ, Jacobs Y, Dohse M, et al. Neuroprotection of host cells by human central nervous system stem cells in a mouse model of infantile neuronal ceroid lipofuscinosis. *Cell Stem Cell*. 2009;5(3):310-319.
49. Gupta N, Henry RG, Strober J, et al. Neural stem cell engraftment and myelination in the human brain. *Sci Transl Med*. 2012;4(155):155ra137.
50. Mahmood A, Raymond GV, Dubey P, Peters C, Moser HW. Survival analysis of haematopoietic cell transplantation for childhood cerebral X-linked adrenoleukodystrophy: a comparison study. *Lancet Neurol*. 2007; 6(8):687-692.
51. Miller WP, Rothman SM, Nascene D, et al. Outcomes after allogeneic hematopoietic cell transplantation for childhood cerebral adrenoleukodystrophy: the largest single-institution cohort report. *Blood*. 2011;118(7):1971-1978.
52. Selden NR. Stem cells and central nervous system therapeutics. *Neurosurg Focus*. 2008;24(3–4):E2.
53. Guillaume DJ, Huhn SL, Selden NR, Steiner RD. Cellular therapy for childhood neurodegenerative disease. Part I: rationale and preclinical studies. *Neurosurg Focus*. 2008;24(3–4):E22.
54. Selden NR, Guillaume DJ, Steiner RD, Huhn SL. Cellular therapy for childhood neurodegenerative disease. Part II: clinical trial design and implementation. *Neurosurg Focus*. 2008;24(3–4):E23.
55. Bourdenx M, Dutheil N, Bezard E, Dehay B. Systemic gene delivery to the central nervous system using adeno-associated virus. *Front Mol Neurosci*. 2014;7:50.
56. Neuwelt EA, Goldman DL, Dahlborg SA, et al. Primary CNS lymphoma treated with osmotic blood-brain barrier disruption: prolonged survival and preservation of cognitive function. *J Clin Oncol*. 1991;9(9):1580-1590.
57. LeWitt PA, Rezai AR, Leehey MA, et al. AAV2-GAD gene therapy for advanced Parkinson's disease: a double-blind, sham-surgery controlled, randomised trial. *Lancet Neurol*. 2011;10(4):309-319.
58. Riley J, Butler J, Baker KB, et al. Targeted spinal cord therapeutics delivery: stabilized platform and microelectrode recording guidance validation. *Stereotact Funct Neurosurg*. 2008;86(2):67-74.
59. Lamanna JJ, Urquia LN, Hurtig CV, et al. Magnetic resonance imaging-guided transplantation of neural stem cells into the porcine spinal cord. *Stereotact Funct Neurosurg*. 2017;95(1):60-68.
60. Riley J, Glass J, Feldman EL, et al. Intraspinal stem cell transplantation in amyotrophic lateral sclerosis: a phase I trial, cervical microinjection, and final surgical safety outcomes. *Neurosurgery*. 2014;74(1):77-87.
61. Bey K, Ciron C, Dubreil L, et al. Efficient CNS targeting in adult mice by intrathecal infusion of single-stranded AAV9-GFP for gene therapy of neurological disorders. *Gene Ther*. 2017;24(5):325-332.
62. Fedorova E, Battini L, Prakash-Cheng A, Marras D, Gusella GL. Lentiviral gene delivery to CNS by spinal intrathecal administration to neonatal mice. *J Gene Med*. 2006;8(4):414-424.
63. Kagiava A, Sargiannidou I, Theophilidis G, et al. Intrathecal gene therapy rescues a model of demyelinating peripheral neuropathy. *Proc Natl Acad Sci U S A*. 2016;113(17):E2421-E2429.
64. Vulchanova L, Schuster DJ, Belur LR, et al. Differential adeno-associated virus mediated gene transfer to sensory neurons following intrathecal delivery by direct lumbar puncture. *Mol Pain*. 2010;6:31.
65. Cassidy JD, Carroll LJ, Peloso PM, et al. Incidence, risk factors and prevention of mild traumatic brain injury: results of the WHO collaborating centre task force on mild traumatic brain injury. *J Rehabil Med*. 2004;43 (43 Suppl):28-60.
66. Sibbald B. Death but one unintended consequence of gene-therapy trial. *CMAJ*. 2001;164(11):1612.

CHAPTER

23

Direct Convective Nervous System Drug Delivery for Patients with Neurodegenerative Disorders

R. Mark Richardson, Paul Larson*[†]

[*]Department of Neurological Surgery, University of Pittsburgh, Pittsburgh, PA, United States
[†]Department of Neurological Surgery, University of California–San Francisco, San Francisco, CA, United States

INTRAOPERATIVE MAGNETIC RESONANCE IMAGING-GUIDED CONVECTION-ENHANCED DELIVERY

Neurodegenerative diseases affect a large portion of our increasingly elderly population. Idiopathic Parkinson disease (PD) alone affects over 1 million people in the United States, and 1% of individuals over the age of 60 are likely to develop PD. Current therapies for neurodegenerative disorders are aimed at symptomatic relief and do not address the underlying pathology. As neurodegeneration progresses, the efficacy of pharmacological treatments is reduced. Because the blood–brain barrier prevents significant amounts of most systemically administered agents from reaching therapeutic parenchymal levels without producing toxicity, significant adverse effects typically increase with escalating medication doses. Direct intracerebral delivery of therapeutics circumvents the blood–brain barrier and targets the disease process with functional-anatomic specificity. Therapeutic molecules, such as growth factors, can be delivered directly into the brain parenchyma; alternatively, the genes that code for these types of molecules can be delivered. The former approach requires chronic readministration of the therapeutic agent, whereas a single gene therapy infusion may last a decade or more.[1]

Convection-enhanced delivery (CED) uses a pressure gradient established at the tip of an infusion cannula to distribute the therapeutic agent within the brain parenchyma by bulk flow. This technique has become the accepted standard for focal intracerebral delivery, rather than relying on diffusion, which results in nonhomogeneous distribution restricted to a few millimeters from the source. CED can uniformly deliver high concentrations of macromolecules over significant volumes, although the distribution is dependent on the structural

Nervous System Drug Delivery
https://doi.org/10.1016/B978-0-12-813997-4.00023-2

© 2019 Elsevier Inc. All rights reserved.

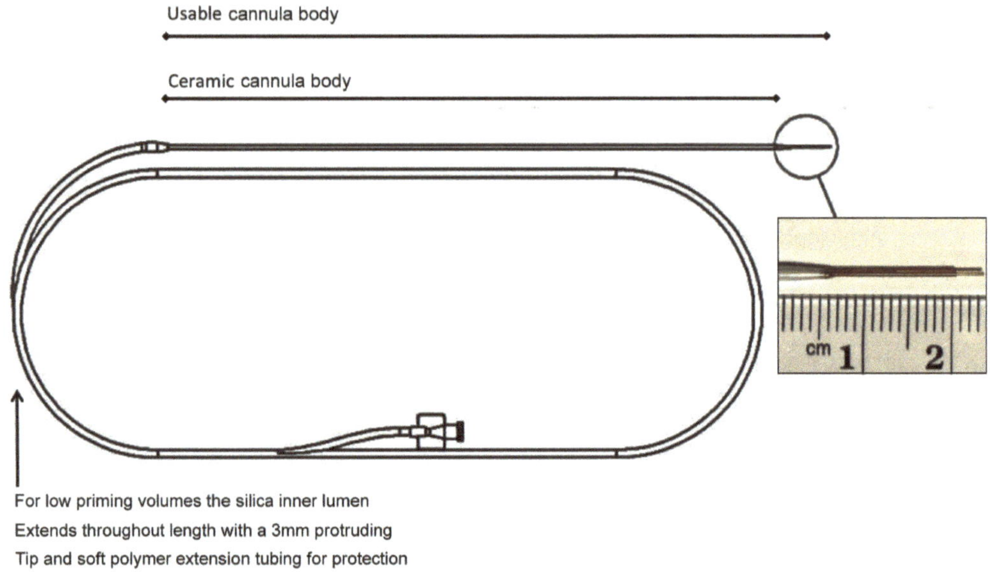

FIG. 1 A magnetic resonance imaging-compatible infusion cannula for clinical convection-enhanced delivery. A two-step design cannula with a ceramic body and inner silica sleeve allow low priming volumes. The ceramic body is encased with a protective outer polymer sleeve.

properties of the tissue and the parameters of the infusion procedure. Development of an infusion cannula with a step close to the cannula tip now allows positive pressure to be exerted without significant reflux (Fig. 1).

Historically, stereotactic infusions used in clinical trials have occurred without a way to verify the extent to which the infusate is distributed within the target during the infusion, versus leaking into neighboring regions or refluxing up the cannula track. For this reason, the failure of protein infusion and gene therapy trials in PD is thought to be largely due to inadequate delivery. The development of real-time magnetic resonance imaging (MRI)-guided CED using interventional or intraoperative MRI (iMRI-CED) seeks to overcome these challenges by enabling monitoring of the infusion via coadministration of an MRI contrast agent mixed with the therapeutic drug. Studies in nonhuman primates have allowed for quantification of CED dynamics and the intraoperative detection of leakage and reflux,[2, 3] definition of structure-specific "green zones" delineated as reliable cannula placement zones,[4-6] and verification that gadoteridol distributions visible on MRI correspond well with subsequent transgene expression patterns.[7] These results set the stage for planning cannula placement with respect to individual patients' anatomy,[8, 9] leading to the clinical development of the iMRI-CED method described in this chapter.

THE CLEARPOINT SYSTEM

Currently, the sole commercially available MRI-compatible drug delivery platform includes a skull-mounted aiming device (SmartFrame), a reflux-resistant CED cannula (SmartFlow), and an MRI-integrated software package (ClearPoint) that communicates with both the console and the operating neurosurgeon in the MRI suite (all products from MRI Interventions, Inc.). The platform is based on the concept of prospective stereotaxy, the alignment of a skull-mounted trajectory guide within an MRI system.[10] The key features of this strategy are: (1) patient positioning on the MRI gantry under general anesthesia; (2) definition of target coordinates with respect to the MRI isocenter, rather than to a separate stereotactic space; (3) trajectory alignment and cannula insertion via a burr hole-mounted trajectory guide instead of a traditional stereotactic frame-based system; and (4) integration of planning, insertion, and real-time MRI confirmation of cannula placement and subsequent infusate distribution during a single procedure. The accuracy of the ClearPoint system was first validated preclinically in nonhuman primate infusion studies,[11] and has since been validated extensively for the clinical placement of deep brain stimulation leads, with average targeting errors on the order of 1 mm.[12-15]

iMRI ENVIRONMENT

Depending on the institution, iMRI-CED can be performed in intraoperative MRI or standard diagnostic MRI suites. In either case, there are several other factors to consider in the iMRI environment. First, there must be adequate space for the anesthetists to carry out their duties. This starts with a separate area outside the MRI suite in which patients can be intubated and lines can be placed. Most diagnostic scanner rooms in large hospitals can accommodate ventilators to handle intubated patients, but not all have adequate space for an anesthesia machine. In those settings, anesthesiologists may need to monitor the patient outside the scanner room. In addition, the scanner room must have adequate space on the opposite end of the MRI bore to allow for the sterile table and space for the surgeon to operate. The scanner room must also be outfitted with adequate lighting, and proper connections for suction and a pneumatic, MRI-compatible drill. Finally, a large-bore scanner is critical for adequate clearance of the stereotactic frame during the alignment procedure. The ClearPoint platform is compatible with 1.5 and 3T scanners. At higher field strengths, the potential for image distortion should be carefully evaluated; however, equivalent targeting accuracy using ClearPoint in the 3T environment has been reported.[16] Nonetheless, involvement of an MR physicist in establishing initial scan parameters on a site-specific 3 T scanner is recommended.

ClearPoint Workflow

The ClearPoint system has a specially designed trajectory guide (the SmartFrame) and integrated planning software, which enables imaging to guide each of these steps in the procedure: (1) acquisition of an initial volumetric MRI and trajectory planning, (2) placement of the skull-mounted SmartFrame and burr hole, (3) definition of target coordinates with respect to the MRI isocenter, (4) trajectory alignment, dural puncture, and infusion cannula insertion, and (5) monitoring of infusate distribution.

After intubation, arterial line placement, and shaving of the scalp around the planned entry point, the head is secured on the MRI gantry with skull pins, in a head holder that allows positioning of two loop coils on either side of the head. If available, a portable MRI-compatible board that allows attachment of a head fixation device, before the patient is moved onto the MRI gantry, may be used (e.g., the AtamA System, Monteris Medical). The patient's head is then brought through the bore of the scanner and is prepped and draped; a custom elasticized drape that allows the head to move in and out of the bore of the scanner without contaminating the operative field is available. An initial volumetric scan is then obtained after placing a temporary gadolinium-impregnated marking grid on the scalp. MRI sequences for iMRI-CED should provide clear visualization of the target of interest. T1-weighted spoiled gradient-echo sequences (BRAVO on GE consoles, MPRAGE on Siemens consoles) and the 3 T-MRI FGATIR (fast gray matter acquisition T1 inversion recovery)[17] sequence provide good gray–white tissue differentiation. A contrasted sequence is required to delineate vascular structures.

The ClearPoint planning software exists on a stand-alone workstation in the MRI control room and communicates with the MR console via a network link. The general workflow is divided into four different stages in the software: trajectory planning (entry), target verification and trajectory visualization (target), alignment of trajectory guide (navigate), and insertion monitoring (evaluate). In the first stage, the surgeon selects the target and entry point, defining the expected trajectory. The trajectory should be aligned down the longest axis of the target structure if possible (Fig. 2), with the tip remaining >3 mm from the edge of any gray–white junction or cerebrospinal fluid space at all times to avoid the potential for the infusate to be carried away from the target via white matter or cerebrospinal fluid pathways. It is advisable to avoid large Virchow-Robins spaces when possible, given their increased potential to divert the infusate through perivascular spread away from the target.

The software then provides grid coordinates at which the burr holes should be created for the planned entry sites. Two options exist for mounting the SmartFrame: a base that mounts directly to the skull after retraction of a formal incision, or a base that mounts to the skull through percutaneous screws, with additional screws countersunk into the scalp to maximize stability. The choice of mounting may depend on the location and number of planned trajectories. For instance, multiple trajectories through a single burr hole may be achieved more easily using the skull-mounted base, whereas the percutaneously mounted base is favored when only one trajectory is anticipated.

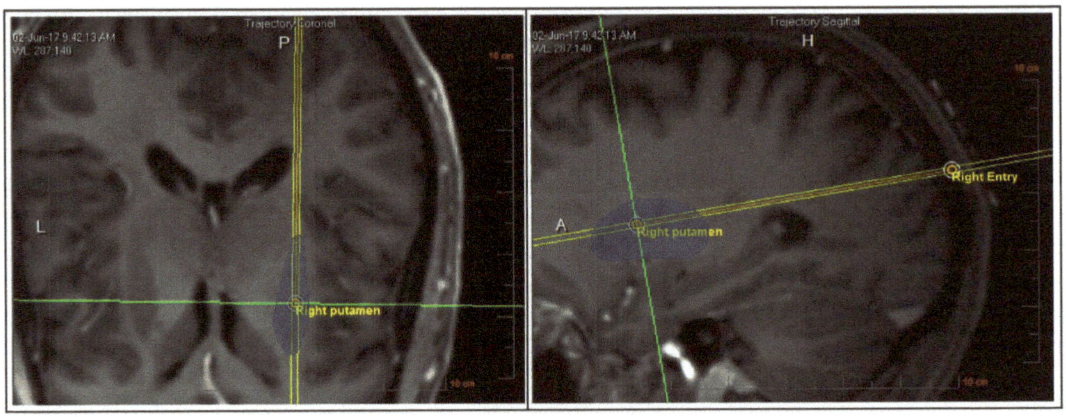

FIG. 2 Planned posterior-to-anterior trajectory targeting the putamen. The trajectory traverses the long axis of the putamen (*blue*-shaded regions) from a transparietal/occipital entry point.

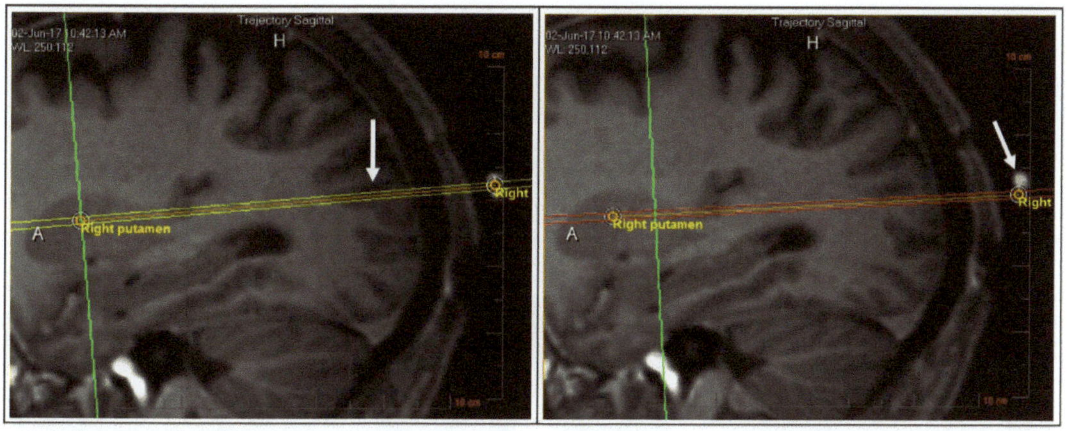

FIG. 3 Offset of entry point after frame placement. The arrow in the left panel shows a planned trajectory traversing a sulcus. The *red* lines in the right panel delineate the corrected planned trajectory, achieved by translation of the X-Y stage, rather than repositioning of the SmartFrame.

If using the skull-mounted bases, the scalp is opened and the SmartFrames are mounted over the skull markings that were previously made through the grid with an awl. A pneumatic drill is then used to make burr holes centered on these awl marks. It is recommended to leave the dura closed at this step and puncture it later with a fine-pointed ceramic probe, just prior to cannula insertion, to provide the best stability for the infusion cannula. This maneuver requires careful planning of the entry site, to ensure that cortical vessels are >3 mm from the planned insertion point. If the percutaneously mounted frame is used, the incision and burr hole are made in a single maneuver using a hand drill, after frame alignment.

Once the SmartFrame is mounted, the final trajectory depends on the pivot point of the trajectory guide, which now is fixed in space. Targeting scans are obtained and transferred to the ClearPoint software, where the target is reselected based on real-time anatomic visualization, in order to confirm that the planned target can be reached via an acceptable trajectory. If the SmartFrame is not mounted ideally with respect to the intended cortical entry point (e.g., entry too close to a cortical vein or trajectory too close to a sulcus), the software allows the surgeon to offset the pivot point using the X-Y stage, up to 2.5 mm (Fig. 3). If this maneuver is undertaken, one must anticipate having to rely on the pitch-and-roll adjustments in subsequent steps, as little room may remain on the X-Y stage. Following subsequent serial MRI-based instructions for aligning the SmartFrame, a predicted vector error of <0.5 mm should be achieved. During alignment, the infusion cannula should be attached to the Luer Lok syringe (Becton, Dickinson and Company) containing the infusate. The syringe is then mounted on an MRI-compatible infusion pump, and should be held in a vertical position during attachment of the infusion line to reduce the chance of trapping air. The infusion line and cannula are then primed to expel air from the circuit.

Once both SmartFrames are aligned, the initial insertion depth is calculated, which is different than the target depth. The depth for initial implantation should render the tip of the infusion cannula in the target structure, with the first step on the cannula positioned just proximal to the border of the target structure (Fig. 4). In this way, one can implement a strategy of

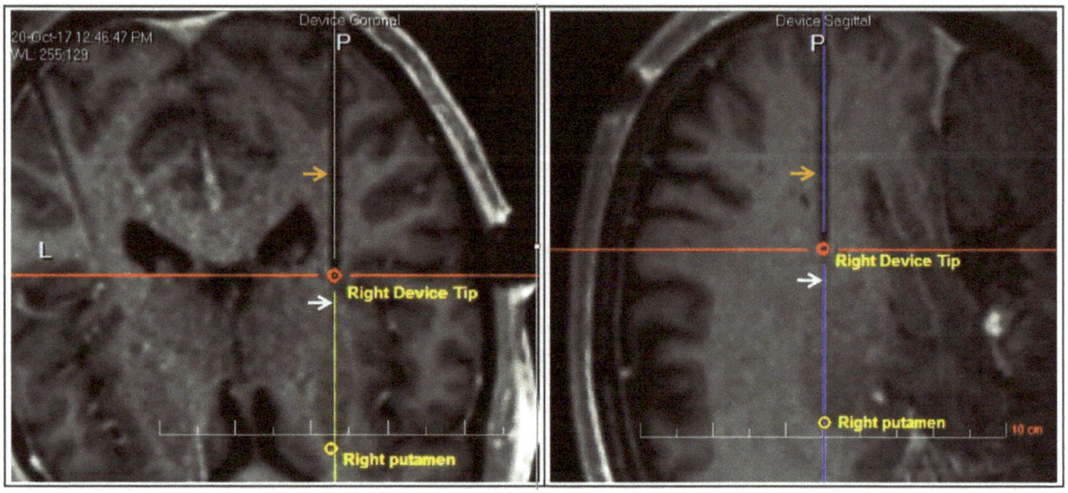

FIG. 4 Initial position of infusion cannula. *Orange* arrows indicate the artifact from the infusion cannula and *white* arrows indicate the border of the putamen, in *(left)* axial and *(right)* sagittal views. The red circle indicates the location of the second step, which is approximately 8 mm from the border of the putamen, placing the tip approximately 5 mm into the target structure (using this 13-mm tip length SmartFlow cannula).

progressive advancement of the cannula as the infusion fills a given region of the target structure. Before inserting the infusion cannulas, a fine-pointed ceramic probe is used to pierce the dura and pia sharply, in one maneuver. A fresh probe is used on each side, in the case of simultaneous bilateral infusions, and inserted approximately 1–2 cm into the brain parenchyma to ensure subsequent smooth entry of the infusion cannula. Again, the surgeon should plan the entry such that the trajectory is not expected to pass too close to cortical surface vessels. When passing the fine-pointed probe, the surgeon should be vigilant in confirming the absence of any unusual resistance that would indicate potential contact with and deflection off bone at the edge of the burr hole. If a complication is suspected, imaging should be obtained and a decision can be made whether to continue, modify, or abort the procedure.

Next, with the infusion pump set at a rate of 1–3 μL/min, flow from the cannula tip is confirmed visually. After verifying that the patient's systolic and mean arterial blood pressures are well controlled, the cannula is then inserted to the first planned infusion depth. Care must be taken during these steps to avoid movement of the brain around the cannula, which can create space for reflux to occur, where the path of least resistance from the tip becomes the space around the cannula itself. Steps to minimize the risk for reflux include insertion of the cannula without jostling the SmartFrame, securing the infusion line so that no undue torque is placed on the device, and smooth operation of the gantry.

A volumetric scan is obtained to verify accurate implantation and to confirm no complications have occurred. If targeting accuracy is unacceptable, the infusion should be halted, and the source of error determined. In this case, the surgeon has the option to remove the cannula, make a software-guided trajectory adjustment, and reinsert. Reinserting in close proximity to the existing tract is not recommended, given the risk for reflux. The ongoing infusion is visualized using serial repetition of volumetric T1-weighted scans, each taking 3 to 6 minutes to complete, with minimal to no delay between scans. When the cannula is accurately placed in the desired target, the infusion is expected to grow concentrically after reaching the first step (3 mm from the tip; Fig. 5A), after which the volume tends to elongate to a greater degree along the long axis of the cannula. When the volume begins to exceed the proximal boundary of the target structure through which the cannula was inserted, the cannula is advanced in 2- to 3-mm increments, to fill the structure moving proximally to distally (Fig. 5B). A second step, when reached (usually a distance of 13–18 mm from the cannula tip, based on designs currently available), acts as an additional temporary barrier to infusion volume growth along the cannula axis (Fig. 5C). Infusion rates are slowly increased during the procedure, based on review of the ongoing infusion, without stopping the pump. Perivascular leakage is inevitable (Fig. 5D), but in some instances subsequent leakage outside the target may be attenuated by advancing the cannula a few millimeters, past the presumed vascular offender. High infusion rates increase the risk of backflow and should not be

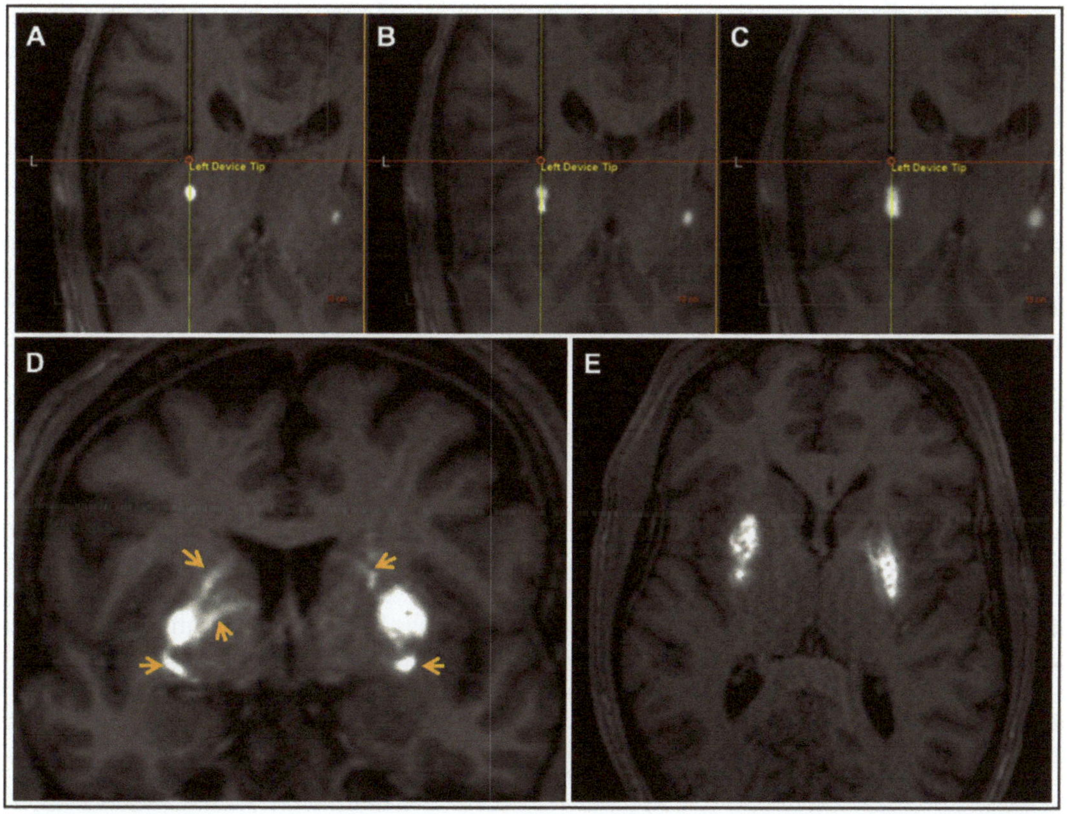

FIG. 5 Real-time visualization of drug infusion. The infusion is expected to grow concentrically after reaching the first step (A). The cannula is advanced in increments of approximately 2–3 mm, with the goal of filling the structure in a proximal-to-distal fashion (B). A second step acts as an additional temporary barrier to infusion volume growth along the cannula axis (C). Perivascular leakage is always evident but should be carefully assessed (D, *orange* arrows show areas of infusate leakage) until target structures are covered with adequate volume of infusate (E).

attempted unless the cannula tip is deep within the target structure. The infusion continues with simultaneous evaluation of serial MRI scans and progressive advancement of the cannula, until the desired infusion volume and distribution have been achieved (Fig. 5E). The cannula should not be pulled back at any stage in the infusion, as this would lead to loss of infusion pressure at the cannula tip, significant reflux, and inability to carry out further CED. The infusion is considered complete when the target infusion volume has been achieved, or when further spread within the target becomes limited secondary to significant leakage outside the target.

OTHER CONSIDERATIONS IN INTRACEREBRAL DRUG DELIVERY

The method described above is appropriate for therapeutic strategies that require single rather than chronic delivery, such as gene therapy. This method is currently being investigated in two ongoing clinical trials involving iMRI-CED of gene therapy for PD in the United States (NCT01621581 and NCT03562494). It should be noted that an alternative therapeutic strategy, intermittent glial cell line-derived neurotrophic factor infusions administered via CED, is being used in a PD clinical trial in the United Kingdom (EudraCT# 2011–003866-34). This trial tests a recently developed platform consisting of a transcutaneous bone-anchored port and implantable multi-catheter system for chronic intermittent CED.[18] MRI-based techniques for assessing and optimizing infusate distribution during chronic intermittent drug delivery have not yet been reported.

Although the delivery of viral vectors is focal and subsequent gene expression overlaps with gadoteridol distribution visualized during intraoperative MRI,[7, 8] it is important to consider that gene expression also occurs in both retrograde and anterograde fashion, depending on vector serotype. Vectors derived from adeno-associated viruses (AAVs) have been used almost exclusively in clinical trials for CNS disorders. AAVs can be subdivided into serotypes, based on their capsid profiles. Viral vector particles are transported within the brain from one region to another, either in an anterograde manner (i.e., transduction of cells in a brain region that receives projections from the infusion site), retrograde manner (i.e., transduction of cells in a brain region that sends projections to the injection site), or both, depending on the serotype and transduced species.[19] Understanding how a specific serotype distributes gene products from the targeted region for vector delivery is crucial for creating a predictable and reproducible gene therapy paradigm.[20-25] Due to its relative phenotype specificity for neurons and its clinical safety profile, AAV2 has been the gold standard for neurosurgical gene therapy and has been used in multiple Phase I[26-29] and Phase II studies.[30, 31] Given that AAV2 has a robust preference for anterograde transportation of the gene product,[22] when using AAV2 vectors one must assume that downstream targets of projection neurons also will express the transgene.

CONCLUSIONS

The field of nervous system drug delivery for neurodegenerative disorders has recently entered a new technological era, one in which iMRI-CED has established a means for standardization of accurate delivery. Clinical research is ongoing to optimize the direct intracerebral delivery of therapeutics to the brain. The critical challenge is to achieve the most complete coverage of the target structure while avoiding leakage into neighboring regions or perivascular spaces.

References

1. Hadaczek P, Eberling JL, Pivirotto P, Bringas J, Forsayeth J, Bankiewicz KS. Eight years of clinical improvement in MPTP-lesioned primates after gene therapy with AAV2-hAADC. *Mol Ther*. 2010;18(8):1458-1461.
2. Varenika V, Dickinson P, Bringas J, et al. Detection of infusate leakage in the brain using real-time imaging of convection-enhanced delivery. *J Neurosurg*. 2008;109(5):874-880.
3. Varenika V, Kells AP, Valles F, Hadaczek P, Forsayeth J, Bankiewicz KS. Controlled dissemination of AAV vectors in the primate brain. *Prog Brain Res*. 2009;175:163-172.
4. Yin D, Richardson RM, Fiandaca MS, et al. Cannula placement for effective convection-enhanced delivery in the nonhuman primate thalamus and brainstem: implications for clinical delivery of therapeutics. *J Neurosurg*. 2010;113(2):240-248.
5. Yin D, Valles FE, Fiandaca MS, et al. Optimal region of the putamen for image-guided convection-enhanced delivery of therapeutics in human and non-human primates. *Neuroimage*. 2011;54(Suppl 1):S196-S203.
6. Yin D, Valles FE, Fiandaca MS, et al. Striatal volume differences between non-human and human primates. *J Neurosci Methods*. 2009;176(2):200-205.
7. Su X, Kells AP, Salegio EA, et al. Real-time MR imaging with Gadoteridol predicts distribution of transgenes after convection-enhanced delivery of AAV2 vectors. *Mol Ther*. 2010;18(8):1490-1495.
8. Richardson RM, Kells AP, Rosenbluth KH, et al. Interventional MRI-guided putaminal delivery of AAV2-GDNF for a planned clinical trial in Parkinson's disease. *Mol Ther*. 2011;19(6):1048-1057.
9. San Sebastian W, Richardson RM, Kells AP, et al. Safety and tolerability of magnetic resonance imaging-guided convection-enhanced delivery of AAV2-hAADC with a novel delivery platform in nonhuman primate striatum. *Hum Gene Ther*. 2012;23(2):210-217.
10. Truwit CL, Liu H. Prospective stereotaxy: a novel method of trajectory alignment using real-time image guidance. *J Magn Reson Imaging*. 2001;13(3):452-457.
11. Richardson RM, Kells AP, Martin AJ, et al. Novel platform for MRI-guided convection-enhanced delivery of therapeutics: preclinical validation in nonhuman primate brain. *Stereotact Funct Neurosurg*. 2011;89(3):141-151.
12. Larson PS, Starr PA, Bates G, Tansey L, Richardson RM, Martin AJ. An optimized system for interventional magnetic resonance imaging-guided stereotactic surgery: preliminary evaluation of targeting accuracy. *Neurosurgery*. 2012;70(1 Suppl Operative):95-103. discussion 103.
13. Lee PS, Weiner GM, Corson D, et al. Outcomes of Interventional-MRI versus microelectrode recording-guided subthalamic deep brain stimulation. *Front Neurol*. 2018;9:241
14. Ostrem JL, Ziman N, Galifianakis NB, et al. Clinical outcomes using ClearPoint interventional MRI for deep brain stimulation lead placement in Parkinson's disease. *J Neurosurg*. 2016;124(4):908-916.
15. Sidiropoulos C, Rammo R, Merker B, et al. Intraoperative MRI for deep brain stimulation lead placement in Parkinson's disease: 1 year motor and neuropsychological outcomes. *J Neurol*. 2016;263(6):1226-1231.
16. Southwell DG, Narvid JA, Martin AJ, Qasim SE, Starr PA, Larson PS. Comparison of deep brain stimulation lead targeting accuracy and procedure duration between 1.5- and 3-Tesla interventional magnetic resonance imaging systems: an initial 12-month experience. *Stereotact Funct Neurosurg*. 2016;94(2):102-107.
17. Sudhyadhom A, Haq IU, Foote KD, Okun MS, Bova FJ. A high resolution and high contrast MRI for differentiation of subcortical structures for DBS targeting: the Fast Gray Matter Acquisition T1 Inversion Recovery (FGATIR). *Neuroimage*. 2009;47(Suppl 2):T44-T52.
18. Barua N, Gill S. Drug delivery for movement disorders. *Prog Neurol Surg*. 2018;33:243-252.
19. Salegio EA, Samaranch L, Kells AP, et al. Axonal transport of adeno-associated viral vectors is serotype-dependent. *Gene Ther*. 2013;20(3):348-352.
20. Ciesielska A, Mittermeyer G, Hadaczek P, Kells AP, Forsayeth J, Bankiewicz KS. Anterograde axonal transport of AAV2-GDNF in rat basal ganglia. *Mol Ther*. 2011;19(5):922-927.
21. Green F, Samaranch L, Zhang HS, et al. Axonal transport of AAV9 in nonhuman primate brain. *Gene Ther*. 2016; 23(6):520-526.

22. Kells AP, Hadaczek P, Yin D, et al. Efficient gene therapy-based method for the delivery of therapeutics to primate cortex. *Proc Natl Acad Sci U S A*. 2009;106(7):2407-2411.
23. Samaranch L, Blits B, San Sebastian W, et al. MR-guided parenchymal delivery of adeno-associated viral vector serotype 5 in non-human primate brain. *Gene Ther*. 2017;24(4):253-261.
24. Samaranch L, Salegio EA, San Sebastian W, et al. Strong cortical and spinal cord transduction after AAV7 and AAV9 delivery into the cerebrospinal fluid of nonhuman primates. *Hum Gene Ther*. 2013;24(5):526-532.
25. San Sebastian W, Samaranch L, Heller G, et al. Adeno-associated virus type 6 is retrogradely transported in the non-human primate brain. *Gene Ther*. 2013;20(12):1178-1183.
26. Christine CW, Starr PA, Larson PS, et al. Safety and tolerability of putaminal AADC gene therapy for Parkinson disease. *Neurology*. 2009;73(20):1662-1669.
27. Eberling JL, Jagust WJ, Christine CW, et al. Results from a phase I safety trial of hAADC gene therapy for Parkinson disease. *Neurology*. 2008;70(21):1980-1983.
28. Marks Jr WJ, Ostrem JL, Verhagen L, et al. Safety and tolerability of intraputaminal delivery of CERE-120 (adeno-associated virus serotype 2-neurturin) to patients with idiopathic Parkinson's disease: an open-label, phase I trial. *Lancet Neurol*. 2008;7(5):400-408.
29. Muramatsu S, Fujimoto K, Kato S, et al. A phase I study of aromatic L-amino acid decarboxylase gene therapy for Parkinson's disease. *Mol Ther*. 2010;18(9):1731-1735.
30. LeWitt PA, Rezai AR, Leehey MA, et al. AAV2-GAD gene therapy for advanced Parkinson's disease: a double-blind, sham-surgery controlled, randomised trial. *Lancet Neurol*. 2011;10(4):309-319.
31. Marks Jr WJ, Bartus RT, Siffert J, et al. Gene delivery of AAV2-neurturin for Parkinson's disease: a double-blind, randomised, controlled trial. *Lancet Neurol*. 2010;9(12):1164-1172.

CHAPTER

24

Central Nervous System Drug Delivery After Ischemic or Hemorrhagic Stroke

Nguyen Hoang, Shahid M. Nimjee

Department of Neurosurgery, The Ohio State University Medical Center, Columbus, OH, United States

DRUG DELIVERY AFTER STROKE

Stroke is the leading cause of disability and the fourth most common cause of death in the United States.[1-3] Additionally, the associated costs from stroke morbidity are expected to triple over the next decade, rising from $71.55 billion in 2012 to a projected $184.13 billion in 2030.[1] Ischemic stroke, which results from both vessel thrombosis and thromboembolism, represents 85% of all strokes. Hemorrhagic stroke accounts for the remaining 15%; 5% of these start with subarachnoid hemorrhage (SAH) from aneurysmal rupture.[2]

The mechanism of brain injury after stroke depends on the type of stroke. Ischemic injury from thrombosis results in a loss of blood supply and deprivation of oxygen and nutrients, thereby altering metabolic function and leading to a cascade of cell death. Intracranial hemorrhage causes regional mechanical injury from the dissecting space-occupying lesion, resulting in increased intracranial pressure and subsequent ischemia, peri-infarct edema, and a robust inflammatory response. These reactions create secondary injury that results in significant morbidity. In patients with acute SAH, aneurysmal rupture leads to blood sequestration within the subarachnoid space. After the initial insult from SAH, subsequent morbidity and mortality are largely due to rehemorrhage, hydrocephalus, and delayed cerebral ischemia secondary to vasospasm. Emerging stroke therapeutics may play a key role in mitigating stroke mortality and decreasing long-term disability. Targeted central nervous system (CNS) therapies are focused on prevention of brain damage by mitigating primary and secondary brain injury; these therapies include early revascularization and therapeutics for neuroprotection and promotion of functional recovery.

Treatment can be delivered to the CNS systemically or it can be directed locally. Systemic delivery routes involve intravenous or more targeted intra-arterial (IA) administration.

https://doi.org/10.1016/B978-0-12-813997-4.00024-4

© 2019 Elsevier Inc. All rights reserved.

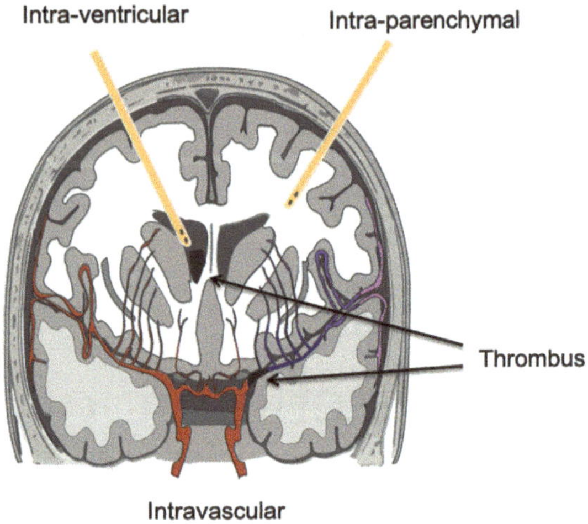

FIG. 1 Administration routes of therapeutics for central nervous system delivery in patients who have had a stroke. Systemic delivery of therapeutics is less invasive and allows for localization to the central nervous system and across the blood–brain barrier (BBB) through parenteral, intramuscular, intravenous, and subcutaneous routes. An acute thrombus can be broken down through fibrinolytic therapy intravenously or intra-arterially. Intrathecal administration of therapeutics allows for bypassing the BBB and access to the ventricular system and subarachnoid space. Intraparenchymal delivery is invasive, but allows bypassing the BBB to deliver therapeutics that may not cross the BBB in higher and more localized concentrations. Intranasal route (not shown) is a noninvasive delivery strategy that can bypass the BBB at the level of the nasal cribriform plate and olfactory system.

Systemic delivery must bypass the microvascular endothelial barrier of the brain, known as the blood–brain barrier (BBB). Local delivery to the CNS can bypass the BBB, but may be invasive in nature (Fig. 1).

PATHOPHYSIOLOGY OF ISCHEMIC STROKE

Initial neuronal damage in patients who experience acute ischemic stroke depends on cerebral blood flow and perfusion. Reduction below the normal cerebral blood flow range (i.e., 40-60 mL per 100 g of brain tissue per minute) results in inadequate tissue oxygenation and disruption of metabolic homeostasis.[4-6] Cell death occurs within minutes once cerebral blood flow drops below 10 mL per 100 g of brain tissue per minute.[7, 8] However, cerebral blood flow between 10 and 18 mL per 100 g of brain tissue per minute is called the *ischemic penumbra*, which is tissue without neuronal function and electrical activity at risk for permanent structural injury, but that could be salvaged if reperfused.[8-10] Collateral circulation in arterial occlusion or hypotension helps preserve ischemic penumbra, limiting ischemic damage.

A complex cascade of molecular events occur after brain ischemia (Fig. 2). The depletion of oxygen impairs brain energy metabolism by shunting energy pathways to anaerobic metabolism and decreasing glycolysis and cellular respiration.[11] The loss of adenosine triphosphate (ATP) production is coupled with accumulation of lactic acid, which decreases regional pH and causes critical functional and structural changes that lead to cell damage. Loss of ATP from diversion of oxidative phosphorylation causes failure to maintain the ATP-dependent transmembrane electrochemical gradient established by the sodium-potassium adenosine triphosphatase (Na^+, K^+-ATPase) pump, as well as initiation of free radical production by the mitochondrial chain. Failure of ionic pumps of the plasma membrane causes the alteration of the membrane gradient potential, leading to regional depolarization associated with influx of Na^+, Ca^{2+}, and H_2O, as well as efflux of K^+.[12] This phenomenon contributes to increased membrane permeability and causes a destructive process that leads to cellular swelling,

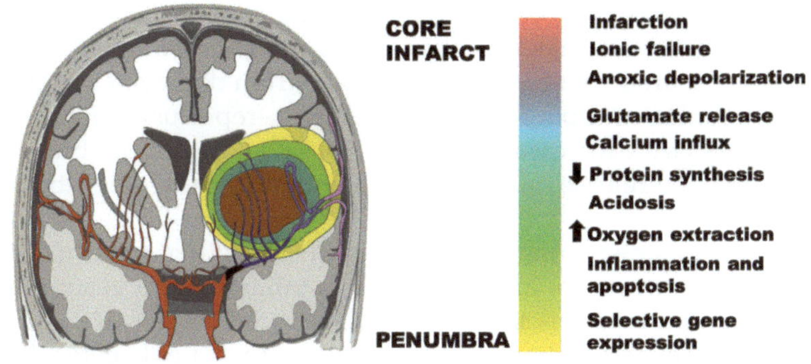

FIG. 2 Overview of pathophysiological changes in focally ischemic brain. Initial neuronal damage in acute ischemic stroke depends on cerebral blood flow and perfusion thresholds where certain biochemical functions are ill-maintained. The depletion of oxygen impairs brain energy metabolism by shunting energy pathways to anaerobic metabolism and decreasing glycolysis and cellular respiration. Energy failure leads to the depolarization of neurons and activation of specific glutamate receptors that increases intracellular Na^+, Ca^{2+}, H_2O, Cl^- levels, and efflux of K^+. Intracellular Ca^{2+} activates various enzyme systems (e.g., proteases, lipases, endonucleases, etc.) that alter cellular biochemistry. Mitochondrial depolarization and failure increases free radical formation, leading to apoptosis and formation of inflammatory mediators.

clearing of the cytosol, endoplasmic reticulum and Golgi organelle swelling, and cytoplasmic blebbing in a process termed *oncosis* (ischemic cell death).[13, 14] This membrane failure contributes to release of excitatory amino acids into the extracellular environment, particularly glutamate-release. Glutamate disinhibited activation of *N*-methyl-D-aspartate (NMDA), a-amino-3-hydroxy-5-methyl-4-isoxazole propionate (AMPA), and kainate glutamate receptors contribute to downstream excitotoxic neuronal injury.[11, 15] Overactivation of NMDA and AMPA receptors contributes to large Ca^{2+} influx, and is a major contributor to mitochondrial Ca^{2+} accumulation and resultant damage.[16-18] Mitochondrial depolarization and failure increases reactive oxygen species and reactive nitrogen species and release of cytochrome *c*, leading to apoptosis.[19]

Mitochondrial failure initiates free radical formation and increases formation of superoxide ($O_2^{\bullet\bullet}$), hydrogen peroxide (H_2O_2), hydroxyl radical (OH^-), nitric oxide ($NO^{\bullet\bullet}$), or peroxynitrite ($OONO^-$).[20] In particular, there is an increase in production of reactive oxygen species and reactive nitrogen species within surrounding endothelial cells (ECs) of the cerebral microvasculature and surrounding glial cells (e.g., astrocytes, oligodendrocytes, and microglia) in the penumbra after stroke. Free radical formation overwhelms endogenous antioxidant reserves and is exacerbated by concomitant reduction of free radical scavenger enzymes such as catalase, glutathione peroxidase, and superoxide dismutase.[21] Subsequent reperfusion exacerbates and fuels oxidative and nitrosative stress, promoting protein nitration and oxidation, lipid peroxidation, mitochondrial damage, DNA damage, and activation of signaling pathways that lead to apoptosis and upregulation of inflammatory responses, ultimately causing reperfusion injury.[20, 22, 23]

Free radical-mediated reperfusion induces release of pro-inflammatory cytokines, including interleukin-1, interleukin-6, and tumor necrosis factor-α, which are involved in upregulation of endothelial adhesion proteins, including intercellular adhesion molecule-1, vascular cell adhesion molecule-1, and selectins, thereby promoting neutrophil accumulation into ischemic brain parenchyma.[24-28] The release of damage-associated molecular pattern molecules, (e.g., chromatin-associated protein [termed *high mobility group protein B1*], heat shock proteins, ATP, heparan sulfate, DNA, and RNA) from damaged ischemic cells triggers postischemic inflammation and plays a role in ischemia–reperfusion injury. In response to these cytokines, leukocytes extravasate from capillaries and infiltrate brain tissue, causing proinflammatory cytokine release, which eventually results in deterioration of the penumbra.[29, 30]

The role of the heightened immune response in mediating reperfusion injury is not completely understood, but secondary injury due to inflammation is supported by several studies that show that immunodeficient mice have reduced infarct size and improved outcomes after transient occlusion of the middle cerebral artery.[31] Leukocyte adhesion molecule (intercellular adhesion molecule-1, CD11b/CD18, P-selectin) inhibition by antibodies leads to smaller cerebral infarcts.[32, 33] Ischemic brain injury is exacerbated by leukocyte infiltration and development of brain edema. These damaging events ultimately lead to ischemic cell death and perpetuate positive-feedback loops that cause structural and functional changes, contributing to secondary injury after stroke. Mitigating these secondary injury processes is critical in neuroprotection after stroke.

BARRIERS TO CENTRAL NERVOUS SYSTEM DRUG DELIVERY

The Blood–Brain Barrier

The BBB separates brain interstitial fluid from intravascular blood, and is made up of a neurovascular unit of brain microvascular ECs and supporting astrocytes, pericytes, microglia, neurons, and extracellular matrix.[34] Transport and diffusion across this barrier are particularly selective due to the biological importance of cerebral parenchyma and the risk of neurotoxicity. The BBB is formed at the plasma membrane of microvascular ECs that form a cell monolayer lumen of capillaries within brain parenchyma and separate brain interstitial fluid from blood.

The selective permeability of the BBB is the result of EC-to-EC tight junctions (TJs) and adherens junctions, which form tightly packed ECs supported by astrocytes end feet that border the abluminal membrane of endothelial vessels and form a continuous sleeve surrounding TJs between intervening ECs. TJs are junctional complexes that play a key role in maintaining the BBB, and are composed of claudins and occludins that restrict paracellular diffusion of ions between ECs and maintain the polarity of the BBB. Permeability of the BBB is a function of this polarity and of the electrical resistance across the cell membrane. Electrical resistance across brain parenchyma vessels is estimated to be as high as 8000 $\Omega\,cm^2$, compared with the 100-300 $\Omega\,cm^2$ systemic vessel resistance in arterial and venous vessels.[35] Astrocytes regulate BBB permeability and maintain BBB integrity by the synthesis and secretion of soluble growth factors. The basal lamina surrounding the ECs also encloses intervening pericytes, which are undifferentiated contractile connective tissue cells that form a secondary barrier and have a role in phagocytosis if the BBB is breached.[36]

Pathologic Changes in Stroke that Alter Drug Delivery

Stroke alters the BBB neurovascular unit by affecting neuroglial support elements and changing the microvascular ECs.[37] The changes and response of the BBB to focal ischemia are complex and follow a heterogeneous spatial and temporal pattern.[38] The mechanisms of BBB breakdown are manifold. They begin with initial ischemic response and are exacerbated by reperfusion injury, inflammation, and activation of cytokine cascade, and remodeling. During stroke, both vasogenic edema and cytotoxic edema are present; in vasogenic edema, there is parenchyma brain swelling due to disruption of the BBB, whereas in cytotoxic edema the BBB remains intact.[36]

The breakdown of the BBB after focal ischemia enables considerable movement of vascular fluid across the microvascular endothelium and leads to vasogenic edema. Early cytotoxic edema formation is mediated by increased activity of Na^+–K^+–Cl^- cotransporter, resulting in the influx of Na^+ and H_2O.[39, 40] In a rat middle cerebral artery occlusion (MCAO) model, Na^+–K^+–Cl^- cotransporter inhibition resulted in reduced edema and infarct volume.[41] The first signs of increased BBB permeability occur early, within 4-6 hours of ischemia onset, and can be seen as vasogenic and cytoxic edema.[41] There is also a stepwise impairment of transcellular transcytosis transport that contributes to early peak in enhanced BBB permeability, followed by disruption of paracellular barrier mechanisms, which involves remodeling of TJ protein complexes in a later phase, leading to progressive peri-infarct edema after 48 hours.[42] The early breakdown of the BBB involves increased permeability from increased transcellular transport. After acute stroke, transmission electron microscopy of the BBB shows upregulation in the number of endothelial caveolae and endocytotic vesicles within ECs.[42, 43] At the molecular level, there is an increase in expression of Caveolin1, a protein involved in endocytotic transport, in its phosphorylated state as early as 2 hours after ischemic injury.[44] In a rodent MCAO model, decreased mRNA expression of claudin-5, occludin, and zonula occludens-1 was seen as early as 3 hours after ischemia.[45] The structural changes in TJs can be seen at 2 days after stroke. In a mouse transgenic model in which claudin-5 was attached with an enhanced green fluorescent protein under 2-photon microscopy, gaps or protrusions were seen in endothelial blood vessels of Tg-enhanced green fluorescent protein-claudin-5 transgenic mice, suggesting increased paracellular permeability and dynamic disassembly and reassembly of endothelial TJs.[42] Reperfusion and subsequent oxidative stress lead to a heightened inflammatory response, contributing to breakdown of the BBB. Leukocyte infiltration leads to disruption of basement membranes by leukocyte-derived matrix metalloproteinases that degrade the extracellular matrix of the BBB.[38] The presence of major inflammatory cytokines in the stroke region upregulates production of angiogenic growth factors (e.g., vascular endothelial growth factor) that lead to secondary TJ remodeling.[38] BBB breakdown leads to vascular edema and risk for hemorrhagic transformation, both of which contribute to high stroke morbidity and mortality.

Blood -brain barrier changes are not a static phenomenon, which complicates CNS drug delivery in the acute injury phase. Enhanced BBB permeability after acute injury can follow a heterogeneous spatial and temporal pattern, as discussed above. This variability in BBB disruption can result in incomplete drug penetration, depending on the timing after initial ischemia, with optimal BBB disruption at 4-6and 48-72 hours after ischemia onset.[37, 46]

The diminished cerebral blood flow to the necrotic core and penumbra decreases drug CNS bioavailability to the desired target tissues, thereby lowering efficacious drug concentrations. Similarly, regional vasogenic edema, cytotoxic edema, and mass effect from hematomas in hemorrhagic stroke exert regional mass effect, limiting blood flow and CNS bioavailability. The increase in global intracranial pressure after stroke causes diminished cerebral perfusion; at the regional level, increased interstitial pressures may affect CNS drug diffusion in the acute setting.[47] Increased intracranial pressure and edema diminish CNS drug penetration from an intraventricular route, as the drug concentration in cerebrospinal fluid (CSF) distributes into brain parenchyma via diffusion. Lastly, alterations in regional metabolism after acute stroke increase activation of proteolytic enzymes and downregulation of protein synthesis. This may limit prodrugs and the delivery of genes, which require active transcription and translation to be effective. Although the ischemic core is difficult to target due to these changes, the penumbra remains a site for therapeutic intervention.

SYSTEMIC DELIVERY

There are several mechanisms by which molecules can cross the BBB (Fig. 3).[36] Passive diffusion of molecules can occur through a paracellular aqueous pathway or through a transcellular lipophilic pathway. Paracellular transport is mediated by TJs, and is limited to small, water-soluble molecules. Transcellular diffusion across the EC membrane is restricted to lipophilic molecules. Carrier-mediated transport and receptor-mediated

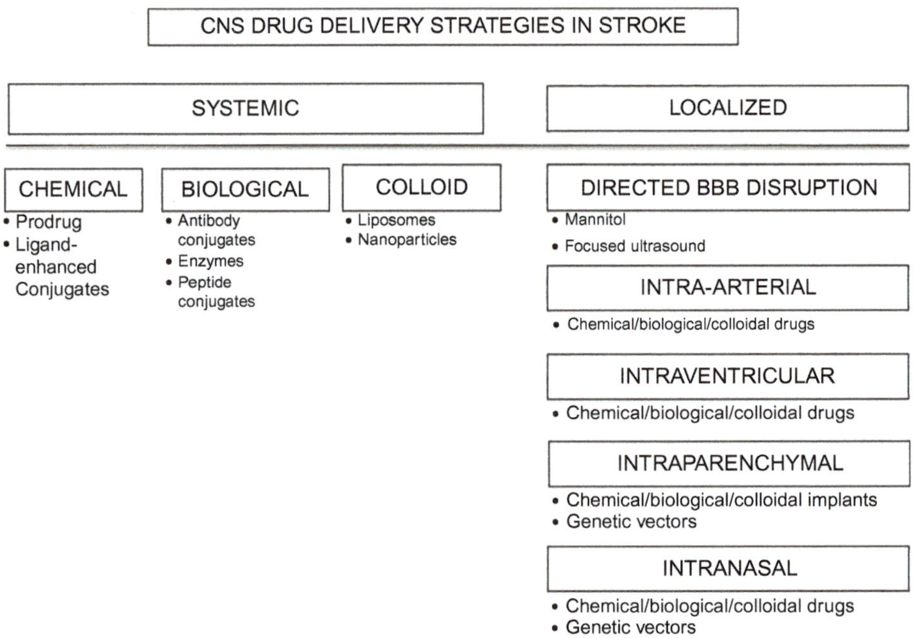

FIG. 3 Approaches to delivery of drugs to the central nervous system.

transcytosis are saturable modes of transport across the BBB. In carrier-mediated transport, molecular carriers present at the luminal and abluminal membranes of the BBB selectively bind and transport small molecules such as ions, sugars, and amino acids. Carrier-mediated transport is typically restricted to small molecules within stereospecific pores. The transport of larger macromolecules primarily occurs through receptor-mediated transcytosis. The influx of nutrients such as insulin, transferrin, and leptin takes place by transcellular receptor binding and subsequent endocytosis. Lastly, adsorptive transcytosis is an endocytotic process mediated by electrostatic interactions between the positively charged cation moieties on proteins and the negatively charged membrane surface regions on the brain ECs by clathrin-coated pits or caveolae.[48]

Systemic delivery of CNS drugs is dependent on intravascular transport of drugs across the BBB or blood–CSF barrier, and these barriers are major obstacles to the delivery of therapeutics into the CNS after stroke. CNS drugs can passively diffuse through a paracellular aqueous pathway or a transcellular lipophilic pathway. There is low permeability to hydrophilic compounds due to the restrictive nature of paracellular TJs. Lipophilic compounds can freely diffuse through a transcellular route, but those larger than 500-600 kDa have low penetration due to low solubility or due to expression a high number of drug efflux transporters (e.g., P-glycoprotein).[49] Many drug delivery strategies have been explored to bypass this unique barrier; enhanced delivery techniques that have been considered to include chemical conjugation (i.e., prodrugs and lipophilic analogues), biological conjugation (i.e., antibodies, receptor/vector-mediated, and peptide-conjugates), and colloidal (i.e., liposomal, nanoparticles, and micelles) carriers.[37, 48, 50-52]

Chemical Ligand-enhanced Conjugation

The addition of a lipophilic moiety, such as a fatty acid, cholesterol, glyceride, or a phospholipid, to a CNS drug/prodrug can aid in passive transport across the BBB by increasing its lipophilicity. Modification of a large-lipid moiety can render a biologically active drug inactive. The increase in lipophilicity can inadvertently increase the volume of distribution and alter bioavailability by decreasing a drug's solubility and increasing its plasma-protein binding, deposition in other tissues, and rate of oxidative metabolism by cytochrome P-450 upregulation.[53] The increase in molecular weight can lead to lower CNS penetration and higher efflux due to drug efflux transporters (e.g., P-glycoprotein) but can alternatively help to evade export as well by reducing the affinity for the MDR1 binding site.

Biological Conjugation

Receptor-mediated transcytosis CNS drug delivery can be used to enhance CNS drug penetration by conjugating drugs with bioactive molecules. Biological conjugation takes advantage of endogenous nutrient transport systems present at the BBB that transport compounds necessary for CNS function, such as sugars, ions, amino acids, vitamins, and proteins. These systems exploit receptor-mediated transcytosis processes rather than carrier-mediated transport system, which is typically restricted to smaller molecules through stereospecific pores or adsorptive-mediated transcytosis, which can be nonspecific.[48, 51, 54]

Drugs are conjugated to a molecule that can target a receptor-mediated transport system; typically, antibodies or peptides conjugated display an affinity to a specific receptor on the BBB luminal membrane. Upon binding to its natural receptor, the vector-conjugated drug cargo undergoes endocytosis and gains access to the brain interstitium. A variety of vectors and drug-vector chemical conjugation methods are available (not discussed here), but conjugation must ensure functionality of the drug-vector and biological moiety.

Erythropoietin (EPO) is glycoprotein hematopoietic growth factor long recognized as the hormone that regulates erythropoiesis. EPO is also expressed endogenously in astrocytes and neurons, especially after hypoxia, and serves as a neuroprotectant. It also plays a role in: limiting production of reactive oxygen species and glutamate, modulating neurotransmission, reversing vasospasm, promoting angiogenesis, preventing apoptosis, and reducing inflammation.[55] In a mouse model of permanent MCAO, EPO was targeted to ischemic tissues using cTfRMAb–EPO fusion protein, in which EPO is fused to the heavy chain of a chimeric monoclonal antibody (MAb) against the mouse transferrin receptor (TfR). Intravenous administration of either low- or high-dose EPO compared to treatment with low- or high-dose cTfRMAb–EPO fusion protein led to 81% reduction in hemispheric stroke and 78% reduced neurological deficit in the mice that received high-dose cTfRMAb–EPO fusion protein.[56]

Similarly, brain-derived neurotrophic factor (BDNF), a neurotrophin essential for neuronal development, synaptic plasticity, and neurogenesis, was coupled to biotin (B) via polyethylene glycol (PEG)-hydrazide and paired with streptavidin coupled to OX26, a rat transferrin receptor antibody, yielding an OX26-BDNF conjugate. Unconjugated BDNF and OX26-BDNF conjugate were intravenously administered into rats subjected to permanent MCAO. Mice in the OX26-BDNF conjugate group had a 243% increase in motor performance relative to BDNF alone showing the CNS targetability of the OX26-BDNF and the neuroprotective effects of BDNF in stroke.[57] Glial-derived neurotrophic factor (GDNF), another potent neuroprotective neurotrophin, was engineered and delivered to the CNS through conjugation to MAb to the human insulin receptor (HIR), forming a HIRMAb–GDNF fusion protein that could bind to the HIR and trigger receptor-mediated endocytosis. Once transported across the BBB, the HIRMAb–GDNF fusion protein was able to bind GDNF receptor (GFR)-a1 to activate GDNF neuroprotection pathways.[58]

Colloidal Drug Carriers

Colloidal drug carriers include a broad range of drug delivery systems that encapsulate a therapeutic agent within a nanoparticle shell. Colloidal drug carriers include micelles, liposomes, polymeric particles, hydrogels, inorganic/solid particles, dendrimers, nanospheres/nanocapsules, and quantum dots.[37, 52, 53] They are composed of biocompatible polymers that can be lipid-based compounds (e.g., dipalmitoyl-phosphatidylcholine), or polymer-based compounds, including natural polymers such as proteins, polysaccharides, and biocompatible synthetic polymers (e.g., poly[e-caprolactone], poly[ethylene oxide], poly[alkyl cyanoacrylate]).[52] Colloidal carriers can vary in size from 10 nm to 5 μm, but are typically 10 to 100 nm to evade systemic clearance.[52] Their larger size allows encapsulation of drugs and larger molecules, including proteins, peptides, and DNA/RNA. Colloid carriers have been engineered to have multiple properties based on different targeting ligands, surface chemistry, and polymer qualities that improve clearance and controlled release of drug, and increase drug stability, circulation time, and targetable delivery.[52, 59] Surface modifications help target colloids to the BBB and prevent clearance through the kidney, reticuloendothelial system, and a macrophage-mediated process called *opsonization*. In opsonization, serum proteins bind electrostatically to circulating colloids, marking them for clearance by macrophages. Sterically stabilized colloids prevent opsonization and loss via the reticuloendothelial system, through incorporation of PEG moieties into the liposome bilayer that prevent serum protein adherence.[52]

Applications of colloid drug carriers in stroke are seminal but evolving. Asialoerythropoietin (AEPO), an enzymatically-desialylated derivative of recombinant human EPO, was delivered to the CNS via a novel AEPO-modified PEGylated liposome (AEPO-liposome) in a transient MCAO rat model after reperfusion. AEPO-liposomes were found to accumulate preferentially in ischemic parenchyma, seen through [^{60}I] labeling. Treatment with AEPO-liposomes significantly reduced cerebral infarct lesions after cerebral ischemia–reperfusion injury.[61] Another application involves encapsulating tacrolimus, a potent immunomodulator, to suppress the inflammatory response secondary to ischemia–reperfusion injury after transient MCAO in a rat model. Tacrolimus entrapped in liposomal membranes was found to accumulate in the ischemic brain parenchyma of these mice after reperfusion. Histopathology showed that treatment with the liposomes suppressed neutrophil infiltration and apoptosis.[62] Other groups have used colloid drug carrier systems to target reactive oxygen species,[63, 64] inflammation,[65] vasospasm,[66, 67] angiogenesis,[68] and thrombolysis.[69-71]

LOCALIZED CNS DELIVERY

Localized CNS drug delivery is a direct method of delivering therapeutic agents to targeted CNS sites (Fig. 3). Within the CNS, localized delivery may be more invasive but can bypass the BBB, prevent unwanted systemic delivery, and increase therapeutic concentrations to a targeted site. Targeted localized delivery includes IA, intraventricular/intrathecal, intranasal, and direct local injection.

Intra-arterial Delivery

Intra-arterial drug delivery is advantageous when a drug has high systemic clearance or high regional drug extraction with low regional flow. IA therapy is less effective when a drug requires long infusion times and is prone to mixing with high cerebral flow.[72] In terms of CNS delivery, IA therapy is invasive, requiring arterial access and selective catheterization into cervical carotid arteries, intracranial vessels, or both. The risks associated with selective catheterization include transient ischemic attack (TIA)/stroke and non-neurological complications. One meta-analysis reported TIA/stroke transient and permanent neurological complication risks ranging from 1.8% to 3.6% and 0.0% to 1.8%, respectively.[73] IA drug delivery in stroke includes applications for thrombolysis with recombinant tissue plasminogen activator (rtPA) and treatments for vasospasm, which will be discussed later in this chapter.

Intraventricular/Intrathecal Delivery

The blood–CSF barrier is composed of the lining of the ECs and TJs that make up the choroid plexus. The barrier is profoundly different from the BBB, and is much more permeable, as seen by the 300-fold difference in greater electrical resistance across the parenchymal endothelium (8000 Ωcm^2) than the choroid plexus (26 Ωcm^2).[74] The choroid plexus lines the cavities of the two lateral ventricles and the midline third and fourth ventricles, and CSF is produced by the choroid plexus. Thus, drug distribution into the CSF is a function of transport across the choroid plexus.

There are three main pathways for a drug in the CSF to pass through the brain parenchyma: circulatory, diffusion, or convective routes.[74] The circulatory route relies on CSF turnover and entry into the systemic blood circulation through absorption across the valves of the arachnoid villi into systemic venous circulation and then reentering across the endothelial BBB. CSF turnover is constant, and its volume in humans is replaced every 4-5 hours. Drugs in the CSF can also diffuse into the brain parenchyma, but because diffusion is dependent on exposure time and diffusion distance, drug penetration is slow. This leads to a logarithmic decrease in drug concentration within brain matter with each millimeter of distance removed from the CSF surface.[75] Drugs in the CSF can also travel by bulk flow or convection via penetration into perivascular spaces (i.e., Virchow-Robin spaces), which originate at the CSF ependymal surface of the brain. The volume of flow through these perivascular spaces is small compared to that of the CSF circulation; therefore, CNS delivery of a drug to a parenchymal target lesion is not always effective. Intraventricular drug delivery in stroke may be warranted to deliver treatment to a specific pathology in the arachnoid space or ventricular system, such as intraventricular hemorrhage (IVH) and vasospasm.

Intranasal

Intranasal administration is a noninvasive way to bypass the BBB and can be used to deliver a wide variety of biologics such as peptides, proteins, oligonucleotides, viral vectors, and stem cells.[76] Intranasal delivery is attractive for its ease of use and administration, but is limited by its low permeability to hydrophilic drugs, low CNS delivery efficiency for proteins, and low delivery volume, which requires higher drug potency. In addition, the nasal environment can easily clear therapeutic drugs by mucociliary clearance, and can deactivate drugs by low pH or enzymatic degradation by nasal cytochrome P450/peptidases/proteases in the nasal mucosa.[77] Intranasal transport to the CNS requires three sequential steps. First, drugs must be transported across the epithelial barrier that lines the nasal passages. This is achieved by intracellular transport across epithelial supporting cells, sensory nerve fibers, olfactory sensory nerves, or peripheral trigeminal nerves; it can also be achieved via paracellular transport around epithelial supporting cells to cross into the lamina propria. Second, once through the epithelial barrier, drugs are transported from the nasal mucosa to the brain entry sites near the pial–brain interface. Drugs that have been transported into sensory nerves are trafficked via intraneuronal transport to the olfactory bulb or brainstem. Drugs contained within the lamina propria must enter the CNS through diffusion within perineural, perivascular, or lymphatic channels associated with olfactory and trigeminal nerve bundles that extend from the lamina propria to the brain. Finally, drugs are transported from brain entry sites to other CNS areas by intracellular transport to second-order neurons, synapsing with peripheral olfactory sensory neurons or trigeminal ganglion cells through diffusion from perivascular spaces into the parenchyma.[76]

Therapeutic peptides, proteins, gene vectors, and stems cells have been delivered to the CNS for stroke applications. Calcitonin gene-related peptide (CGRP) is a potent microvascular vasodilator and is released from sensory fibers originating in the trigeminal ganglia to dilate cerebral vessels.[78] Sun et al.[79] delivered CGRP intranasally, after induced SAH by injection of autologous blood into the cisterna magna of rats. Intranasal CGRP ameliorated vasospasm, improved cerebral blood flow, and reduced cortical and EC death. Moreover, CGRP increased the levels of vascular endothelial growth factor and stimulated angiogenesis.[79] Growth factors such as insulin-like growth factor 1, fibroblast growth factor, EPO, and vascular endothelial growth factor have also been delivered to ischemic brain via an intranasal route after MCAO in mice, and have shown efficacy in protecting the brain from ischemia, significantly reducing infarct volume and brain edema, and improving neurologic function.[80-90] Likewise, the intranasal route can be used to delivery of genetic vectors and stem cell-based therapies with good CNS penetrance.[76, 91] The ability to deliver DNA can lead to a large variety of targets and applications, including viable neuroprotective and neurorestorative therapy for stroke. Wei et al.[92] delivered hypoxic, preconditioned bone marrow-derived mesenchymal stem cells (BMSCs) intranasally after induced ischemia in a mouse MCAO and carotid ligation stroke model. Hypoxia-treated BMSCs decreased cell death in the peri-infarct region, with significant neuroprotection and reduced infarct volume. Furthermore, hypoxic-conditioned BMSCs performed better on sensorimotor functional test 3 days after transplantation than vehicle-treated controls.[92]

Direct Local Injection

Direct drug delivery can be achieved by local parenchymal injections or by an implanted device. Direct delivery bypasses the BBB and can administer larger amounts of therapeutics to reach efficacious levels, and can deliver a wider variety of drugs and biologics without the restrictions associated with intravenous, intraventricular, or intranasal routes. Direct delivery of therapeutics is invasive and necessitates surgical access with associated morbidity and mortality. It unavoidably causes mechanical disruption of the parenchyma from a needle, catheter, or implanted device, which may cause neurological complications. Direct injection can be performed stereotactically with a needle or micro-injector. Multiple implantable devices (e.g., convection delivery catheters, biodegradable polymers, and hydrogels) can be placed to control the release of therapeutics.

Drug penetration into target sites is determined by diffusion coefficients of compounds or by transfection rates when gene therapy is delivered. For example, transforming growth factor β1, TGF-β1, 25 kDa protein, with a diffusion coefficient of $\sim 10^{-6}$ cm^2/s, would diffuse approximately 1 mm into surrounding brain parenchyma only after roughly 3 days.[47] This underscores that slow diffusive process of local direct injection. In stroke, where there is a large area of infarction, delivery of therapeutics via local injection may be ineffective given the large area requiring therapy. Convection-enhanced delivery (CED) has been developed to overcome this slow diffusion process.[93, 94] CED can enhance diffusion through positive pressure and constant bulk flow to disperse therapeutics within the interstitium along perivascular spaces. Developed at National Institutes of Health, Edward Oldfield and his group demonstrated the efficacy of CED with CNS delivery of radiolabeled transferrin (^{111}In-Tf; mw = 80 kDa) and [^{14}C] sucrose (mw = 342 Da). CED supplemented simple diffusion and enhanced the distribution of macromolecules in the brain while achieving drug concentrations on orders of magnitude greater than systemic levels.[93] CED can be complicated by infusate reflux and delivery dynamics, which are influenced by interstitial parenchymal edema.[94] Vasogenic edema, which is associated with malignant brain tumors due to increased fluid permeability in surrounding tissue, presents an obstacle for CED, resulting in lower desired target volume. There are currently no applications of CED in stroke; however, the cytotoxic edema present in ischemic tissue may enhance CED to desired target tissue.[95] Haar et al.[95] studied CED in a rat MCAO model and infused Gd-DTPA into ischemic tissue. Cytotoxic edema decreased the fluid in the extracellular space, and the interstitial pressure from the volume of infused fluid was distributed to a larger region of brain tissue in the setting of ischemia than when the same volume was infused into normal brain tissue. However, due to this larger region of tissue distribution, tissue received a lower concentration of the infused agent. This could theoretically be overcome by increasing the concentration of the infusate.

Applications of direct drug delivery in stroke include delivering genes or restorative cells to ischemic brain tissue through viral vectors or colloid/nanoparticles and stem cell therapy. These strategies will be discussed in subsequent sections.

GOALS OF DRUG DELIVERY TO THE CNS AFTER STROKE

The principal goal in patients with acute ischemic stroke is early revascularization, as early restoration of cerebral blood flow reduces ischemia, attenuates neurological injury, and improves clinical outcomes and survival.[96] Early stroke therapy seeks to salvage penumbra and prevent progression of the ischemic core. If revascularization fails, secondary goals include mitigating cerebral edema, reducing oxidative stress, and promoting recovery through angiogenesis, neurogenesis, and synaptogenesis.

Various thrombolytic enzymes, including streptokinase, urokinase, rtPA, and tenecteplase have been tested. Thrombolytic drugs activate plasminogen into plasmin, which serves as a proteolytic enzyme critical in dissolving fibrin clots.

Streptokinase and urokinase are non-fibrin-specific plasminogen activators that can activate plasmin systemically. rtPA and tenecteplase are specific for plasmin associated with fibrin clots and have reduce systemic activation.[97] The standard of care for intravenous thrombolytics involves the of rtPA or tenecteplase largely because randomized placebo-controlled trials have not shown benefit from intravenous administration of streptokinase within 6 hours after stroke onset in patients with ischemic penumbra, large intracranial artery occlusion, or severe stenosis.[98] Currently, alteplase is the only intravenous treatment for acute ischemic stroke that has been approved by the Food and Drug Administration. The first 10% of the drug dose is administered as a bolus, and the remaining 90% is infused over 1 hour. Tenecteplase is given as a single intravenous bolus.[99, 100] In superiority trials, tenecteplase appears to be similarly safe; however, it is unclear whether it is as effective or more effective than alteplase.

Although revascularization methods have evolved over time, thrombolytic therapy remains the standard of care, and consists of alteplase administered within a narrow time window to a small population who meet criteria. Trials such as NINDS, ECASS, ATLANTIS-A, and ATLANTIS-B have demonstrated a safety window for intravenous thrombolytic administration. These trials found a strong association between favorable outcome if the thrombolytic was given within a 3-hour timeframe and increased risk of hemorrhage >3 hours after stroke onset.[101-103] The therapeutic window for intravenous thrombolytic therapy was expanded in hopes of increasing its benefit from 3 hours of stroke onset to 4.5 hours after the ECASS-III trial showed clear benefit for patients treated between 3 and 4.5 hours after stroke onset.[104] Localized endovascular IA thrombolytic therapy extended the window to 6 hours, but with the advent of stent retrievers, IA thrombolytic therapy is no longer used as monotherapy. Instead, it is used as salvage therapy in conjunction with mechanical thrombectomy.[105]

Intra-arterial Thrombolytic Therapy

Initially, intravenous thrombolytic therapy initiated >3 hours after stroke showed no benefit; instead, it carried high risk of symptomatic hemorrhage.[101, 104] Endovascular IA fibrinolysis emerged due to the need for extended administration—that is, administration of thrombolytic therapy >3 hours after stroke onset—and the need to limit the risk of hemorrhage. PROACT II was a Phase III, multicenter, open-label, randomized controlled clinical trial that tested the efficacy and safety of IA r-pro-urokinase in patients with acute MCAO who presented within 6 hours of stroke onset.[106] Functional outcomes were favorable: 40% of patients who received IA r-pro-urokinase had a modified Rankin Scale (mRS) score of ≤ 2, compared to just 25% of controls ($P = .04$). Recanalization rates were greater in the group that received IA r-pro-urokinase compared to controls (66% vs. 18%; $P < .001$), but rates of intracranial hemorrhage within 24 hours of treatment were also increased in treatment groups (10% vs. 2%; $P = .06$).[106]

Endovascular stent retrievers and aspiration devices have supplanted IA fibrinolytics as monotherapy for large vessel occlusion stroke, and IA fibrinolytics have become adjunctive or salvage therapy in patients with acute stroke. The IMS III and SYNTHESIS Expansion trials studied the effectiveness of IA fibrinolysis with first-generation mechanical embolectomy devices. Due to the poor efficacy of early mechanical embolectomy devices and low usage rates, both IMS III and SYNTHESIS Expansion allowed for better comparison of IA fibrinolytics as monotherapy for patients.

IMS III was a Phase III, multicenter, prospective, randomized, open-label superiority trial that enrolled 656 patients who received either intravenous rtPA within 3 hours of stroke symptom onset or combined intravenous rtPA and endovascular therapy (2:1 ratio). Endovascular treatment was not standardized and varied, with all patients receiving IA rtPA with or without EKOS (EKOS Corporation), Merci Retrieval System (Concentric Medical Inc.), Penumbra (Penumbra Inc.), or Solitaire Revascularization Device (Medtronic). Of note, stent retrievers were rarely used in this study population, with only two patients treated with the Trevo Retriever (Stryker) and 12 patients with the Solitaire device (Medtronic) due to evolving regulatory approval of these devices at that time. The IMS III trial was stopped early due to futility, after the randomization of 656 participants. Of the 334 patients who received endovascular therapy, the majority was treated with IA tPA alone or in combination with mechanical thrombectomy. No significant difference was seen between the endovascular therapy group and the intravenous tPA group in overall functional independence, as measured by mRS score of 0-2 at 90 days (40.8% vs. 38.7%, respectively).

SYNTHESIS Expansion was a Phase III, prospective, randomized, open-label, blinded-end point, 2-arm superiority trial that enrolled 362 eligible patients who experienced ischemic stroke and received intravenous rtPA within 4.5 hours of onset. Patients who received intravenous rtPA were compared to those who received endovascular therapy (i.e., IA rtPA, mechanical clot disruption or retrieval, or combination of these approaches). Endovascular treatment varied, with 66% of patients undergoing infusion of IA tPA and thrombus fragmentation with a guidewire only. Stent retrievers were used in 14% of endovascular cases. Data on rates and efficacy of recanalization were not published. No difference was observed in the primary end point of "good outcome," defined as mRS score ≤ 2 at 3 months or symptomatic intracerebral hemorrhage at 7 days.[105, 107] When taken in context, these studies showed no benefit of endovascular IA delivery of fibrinolytics compared to intravenous tPA within the 3-hour window. They also showed the potential of second-generation endovascular thrombectomy devices that help establish the role of mechanical thrombectomy in acute stroke care. The use of IA fibrinolysis for acute ischemic stroke has become salvage therapy due to the established superiority of endovascular mechanical thrombectomy for recanalization. Current guidelines show that salvage IA fibrinolytics may be used in some clinical circumstances.[108]

Six randomized controlled trials (MR CLEAN, ESCAPE, REVASCAT, SWIFT PRIME, EXTEND-IA, and THRACE), show superiority of mechanical thrombectomy in patients with acute stroke within a 6- to 8-hour window of symptom onset, with endovascular thrombectomy leading to significantly reduced disability at 90 days compared with controls.[109-114] In addition, with the completion of DAWN and DEFUSE-3 trials, mechanical thrombectomy in select patients with clinical perfusion imaging mismatch shows resulted in large-area salvageable penumbra, which was associated with overall benefit in functional outcome at 90 days in the treatment group (Fig. 4).[115, 116] An in-depth discussion of mechanical thrombectomy is beyond the scope of this chapter, but future considerations will explore the potential for drug delivery post-intervention to deliver antispasmodics, antiplatelet

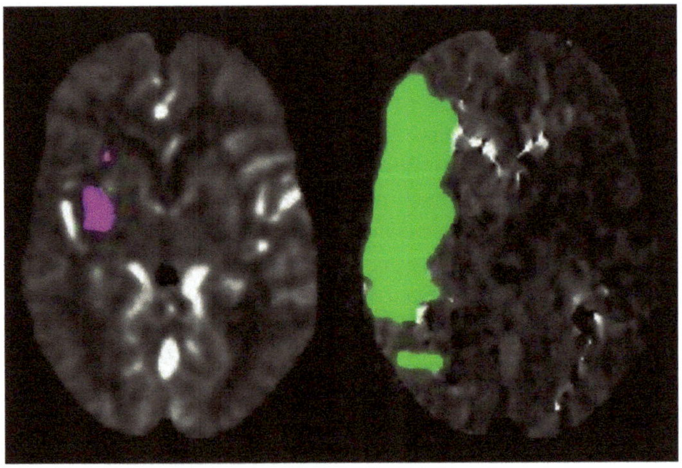

FIG. 4 Computed tomography perfusion (RAPID software) shows an area of salvageable penumbra. Using perfusion imaging, the early infarct (ischemic core and *pink*) can be subtracted from the region of perfusion delay of >6s (hypoperfusion and *green*) to determine size of infarct and penumbra, allowing clinicians to direct treatment after acute stroke.

therapy, fibrinolytics, anti-inflammatory agents, neuroprotectants, antiapoptotics, and cell therapy in hyperacute settings for stroke.

Three Phase I clinical trials are currently underway to explore the potential administration of neuroprotectants, including verapamil and magnesium in the hyperacute stroke setting after recanalization. These trials include Superselective Administration of VErapamil During Recanalization in Acute Ischemic Stroke (SAVER-I), NCT02235558; Verapamil for Neuroprotection in Stroke, NCT03347786; and Magnesium And Verapamil After Recanalization in Ischemia of the Cerebrum: a Clinical and Translational Study (MAVARIC), NCT02912663.

Intra-arterial Vasodilator Therapy For Cerebral Vasospasm

Intra-arterial therapies have been used to treat cerebral vasospasm in patients with SAH. Randomized controlled trials of oral nimodipine have shown a reduction in delayed neurological deterioration due to vasospasm, as well as infarct frequency and poor outcome.[117] No randomized controlled trials have examined IA delivery of vasodilators in patients with cerebral vasospasm, but this therapy has become the standard of care for patients with delayed cerebral ischemia. IA verapamil, a calcium channel blocker, has been the mainstay for IA treatment of cerebral vasospasm. IA delivery of papaverine,[118, 119] nicardipine,[119] nimodipine,[120] amrinone,[121] mannitol, and colforsin daropate hydrochloride[122] have also been used to treat vasospasm.

The Intra-arterial Vasospasm Trial (iVAST)- A Multi-center Randomized Study (NCT01996436) is a Phase IV randomized, double-blind, clinical trial underway that aims to determine whether monotherapy or dual-therapy antispasmodics improve outcomes in medically refractory vasospasm. Patients are randomized into three groups based on treatment: Group 1, nicardipine; Group 2, verapamil; and Group 3, nicardipine, verapamil, and nitroglycerin.

Intraventricular Applications in Patients with Hemorrhagic Stroke

Due to poor diffusion of drugs from the CSF into the brain parenchyma, intraventricular delivery of therapeutics to the subarachnoid or intraventricular spaces is more efficacious in stroke treatment. Intraventricular treatment thus far has been used to treat IVH caused by hemorrhagic stroke and to treat vasospasm secondary to SAH.

The CLEAR III trial examined patients with spontaneous intracerebral hemorrhage with IVH to determine whether faster clearance of IVH improved survival and long-term functional outcomes by relieving acute obstructive hydrocephalus and reducing neurotoxicity. CLEAR III was a randomized, double-blinded, placebo-controlled, multiregional trial that enrolled 500 patients with obstructing IVH from nontraumatic intracranial hemorrhage (volume < 30 mL) who had an extraventricular drain. Treatment included intraventricular administration of 12 doses, 8 hours apart, of 1 mg alteplase or 0.9% saline. Investigators found no significant advantage to using alteplase for clot lysis and ventricular hemorrhage clearance. Good functional outcomes were achieved in 48% of the group that received the drug, compared to 45% of the group that received saline ($P = .554$). At 180 days, the treatment group had lower case fatality (18% vs. saline 29%, $P = .006$), but a greater proportion with mRS score of 5 (17% vs. 9%; $P = .007$). Therefore, administering intraventricular tPA to treat hemorrhage IVH may save lives, but the increased survival might come at the expense of severe disability.[123]

Likewise, intraventricular administration of thrombolytics has been used in aneurysmal SAH to accelerate IVH and cisternal clot clearance. Four randomized controlled trials have suggested that intracisternal thrombolytic therapy may prevent chronic hydrocephalus, attenuate angiographic vasospasm, reduce delayed cerebral ischemia, and improve neurological recovery.[60, 124-126] Kramer et al.[60] assessed the use of low-dose intraventricular rtPA in patients with SAH who were managed with endovascular coiling and an extraventricular drain. This was a prospective, parallel group, placebo-controlled, randomized, Phase II trial. Patients were included if they were older than 18 years, had a modified Fisher score of 4, had aneurysmal SAH with an extraventricular drain, and were treated by endovascular coil embolization. Treatment with rtPA at a dose of 2-mg delivered intraventricularly every 12 hours for a maximum of 5 doses (10 mg) was compared to a placebo group that received saline. After drug or placebo administration, computed tomography of the head was carried out to confirm hemorrhage stability and adequate extraventricular drain placement. In total, just 12 patients were enrolled in this pilot study; therefore, the study was not sufficiently powered to determine treatment benefit in terms of functional outcome. Toward their primary end point of safety, no patients in the rtPA treatment group developed new intracranial hemorrhage, and complication rates between groups were similar. No differences in rates of mortality, delayed cerebral ischemia, vasospasm, or shunt dependences were observed.[60]

Etminan et al.[125] carried out a larger randomized, open-label Phase II study with a cohort of 60 patients who were randomized to either intraventricular rtPA and lateral rotational therapy, or standard of care therapy. The primary end point was functional outcome, measured by the Glasgow Outcome Scale at discharge and at 3-month follow-up. Clot clearance rate, radiographic features of delayed cerebral ischemia, and posthemorrhagic hydrocephalus were secondary end points. Their results showed no difference in the incidence of adverse or severe adverse events between the two groups, with the exception of increased incidence of subgaleal hematoma in the rtPA treatment group. Despite significantly higher clot clearance rates, there was no significant beneficial effect on the incidence of delayed cerebral ischemia, poor functional outcome, and posthemorrhagic hydrocephalus after experimental treatment. Etminan et al.,[125] however, reported a nonsignificant trend of improved vasospasm, with 40% of patients treated with rtPA having moderate or severe angiographic vasospasm compared with 57% of patients in the control group ($P = .438$), thus warranting further study for therapeutic intraventricular treatment in patients with SAH. The Intraventricular Fibrinolysis for Aneurysmal Subarachnoid Hemorrhage (FIVHeMA) trial (NCT03187405) is a Phase III, randomized, parallel assignment, single-blinded clinical trial underway that hopes to enroll 440 patients to examine this treatment.

Intraventricular nicardipine injection has also been explored for treatment of vasospasm. Currently, this treatment has not been analyzed in any randomized controlled trials, but many prospective and retrospective series detail its use in patients with aneurysmal SAH and refractory vasospasm for whom conventional therapies were ineffective or contraindicated.[127-132] Nicardipine, a second-generation L-type calcium channel blocker, had been deemed effective as an intracranial implant to decrease the incidence of cerebral vasospasm in patients with aneurysmal SAH.[129, 133] When intraventricular nicardipine was administered to patients with refractory vasospasm, transcranial Doppler ultrasound showed a decrease in radiographic vasospasm (evidenced by lower mean cerebral blood flow velocity and radiographic angiography) but clinical outcomes assessed at 90 days were not clearly established.[127, 128] Differences in intrathecal nicardipine are not established as seen in clinical outcomes at 90 days. This may be a result of biased patient selection, as intrathecal nicardipine may be reserved as salvage therapy.

Sonolysis

Focused ultrasound has been used to temporally open the BBB to deliver drugs to the CNS. In patients with stroke, focused ultrasound has been used alternatively for clot disruption. Ultrasound-enhanced drug delivery of echogenic liposomes is an evolving thrombolytic treatment for ischemic stroke currently under development.[69, 134] Ultrasound-enhanced thrombolytic therapy in patients with acute stroke was shown to be effective in vitro; however, results in clinical trials have been modest.[135, 136] In CLOTBUST, patients with acute ischemic stroke due to MCAO were treated with combined 2 MHz transcranial Doppler and rtPA within 3 hours of stroke onset. Ultrasound-enhanced thrombolytic therapy augmented rtPA-induced arterial recanalization, with a nonsignificant trend toward increased rate of recovery from stroke.[135]

A second Phase II trial, CLOTBUST-HF, used a hands-free ultrasound device to deliver 2 hours of therapeutic exposure to 2 MHz of pulsed-wave ultrasound after rtPA was administered to patients with middle cerebral artery large vessel occlusion. This trial showed safety without increased risk of intracranial hemorrhage, but had a modest recanalization rate of 40%.[136] The mechanism of ultrasound-enhanced thrombolysis may be related to microbubble cavitation. As small microbubbles are produced under ultrasound, they oscillate in response to the ultrasound acoustic field that may disrupt or enhance rtPA thrombolysis. Although recent studies of mechanical thrombectomy have shown superiority in recanalization and overall survival, ultrasound-enhanced thrombolytic therapy has poor clinical equipoise but may be beneficial in more remote settings (i.e., clinics without access to a Level 1 Comprehensive Stroke Center). The use of echogenic liposomes with rtPA or plasmin has been explored for focal delivery of rtPA to the site of the thrombus and to enhance thrombolysis without systemic exposure of thrombolytic therapies.[70, 134] In vitro experiments found that the lytic efficacy of tPA-containing echogenic liposomes is comparable to that of rtPA alone. The addition of 120-kHz ultrasound significantly enhanced lytic treatment efficacy for both tPA (71%, confidence interval: 56%-86%) and t-ELIP (89%, confidence interval: 76%-100%).[133]

Neuroprotection

Stroke is the leading cause of disability in the United States, and the healthcare-associated costs of stroke are on the rise.[1-3] Stroke-associated disability is so common in part because so few patients are eligible for and actually receive treatment. In fact, less than 4% of people who experience stroke receive tPA.[137] The seriousness of stroke-associated disability underscores the need for additional treatments aimed at neuroprotection and functional recovery.

Neuroprotection involves measures taken to minimize the harmful effects of ischemia. Optimally, neuroprotective strategies promote neurogenesis while lessening the damage that results during ischemia and reperfusion, including excitotoxicity, oxidative and nitrosative stress, and inflammation.[138-140]

Excitotoxicity refers to the overactivation of neuronal receptors, particularly NMDA and AMPA receptors, from the rapid release of the excitatory amino acid glutamate after acute ischemia.[11] Extracellular glutamate initiates Ca^{2+} cell influx, which leads to the cascade of mitochondrial injury. Although preclinical models show that attenuation of NMDA and AMPA receptor-activity by inhibitors improves stroke outcomes in small animal models, this has not demonstrated in large clinical trials.[141] Magnesium sulfate reduces presynaptic release of the neurotransmitter glutamate and blocks NMDA receptors and, in preclinical trials, significantly reduced infarct volume.[142] In clinical trials, however, there was increased mortality in the magnesium-treated group than in the placebo group.[143] Gavestinel (GV150526), Selfotel (CGS 19755), Licostinel (ACEA 1021), and Aptiganel (CNS 1102) are NMDA-receptor antagonists with varying competitive and noncompetitive inhibitor mechanisms acting at glutamate- or glycine-binding sites of NMDA receptors. All four drugs failed to improve mortality and morbidity and, at times, even exacerbated it.[144-147] NMDA receptor antagonism causes psychomimetic side effects, which are a major cause of clinical failure. Given its large role in neuronal function, nonselective inhibition worsened clinical outcomes. Ifenprodil is a selective NMDA inhibitor specifically at the GluN2BR subunit that does not have neurological side effects and has demonstrated therapeutic efficacy in small, double-blinded, placebo-controlled Phase II studies.[148]

The neutralization of oxidative and nitrosative stresses caused by ischemia–reperfusion has become a potential therapeutic strategy. Endogenous free radical scavengers are depleted after stroke, and the prevention of oxidative and nitrosative stress with antioxidants has proven to be effective in limiting the expansion of the ischemic lesion in preclinical stroke models.[138] Disufenton sodium (NXY-059) is a peroxynitrite scavenger that has shown mixed results. In the SAINT I trial, the administration of NXY-059 within 6 hours after onset of acute ischemic stroke significantly reduced disability at 90 days[149]; however, in a larger second study (SAINT II), NXY-059 was ineffective.[150] The limited solubility and BBB penetrance of NXY-059 may have reduced its effectiveness.[151] A randomized controlled trial of edaravone, a free radical scavenger that readily crosses the BBB, showed significant benefit with patients with mRS scores of 0 and 1 after treatment.[152] A meta-analysis of 496 patients in three trials showed overall neurological improvement in patients treated with edaravone, but there was a moderate risk of bias.[153]

Immune-mediated inflammatory response is a hallmark of ischemia and is therefore a potential target for neuroprotection. Fingolimod is a sphingosine-1-phosphate receptor modulator that prevents leukocyte infiltration by limiting the egress of lymphocytes from lymph nodes. Rodent preclinical data show that fingolimod can reduce infarct size and edema with improved functional neurological behavior.[154] In an open-label, evaluator-blinded, parallel-group clinical pilot trial, fingolimod was given to 11 patients after acute ischemic stroke and 11 patients were used as controls. Patients treated with fingolimod had better neurological function recovery, lower circulating lymphocyte counts, and less severe neurological deficits, especially in the first week after treatment, when reduction of National Institutes of Health Stroke Scale (NIHSS) scores were 4 in the treatment group vs. −1 in the control group ($P = .0001$). Patients who received fingolimod also had speedier neurological rehabilitation and less edema on magnetic resonance imaging at 7 days. In this study, treatment with fingolimod was not associated with any drug-related serious events.[155] A combination of fingolimod and rtPA administered within 4.5 hours of stroke onset resulted in lower circulating lymphocytes, smaller lesion volumes, less hemorrhage, and improved NIHSS scores.[156]

Restorative Stem Cell Therapies

Currently, stroke therapies are focused on revascularization to prevent cell death and on neuroprotection to mitigate cellular damage. However, an emerging cell-based strategy still in its infancy has tremendous potential for restoration of neurological function after stroke. Stroke is a heterogenous disease that afflicts many different cell types. Stem cell-based strategies can potentially replace neurons and the supporting milieu, and may trophically stimulate endogenous repair mechanisms to promote neurogenesis, angiogenesis, and remodeling.

Stem cell therapies can be delivery intravenously, intra-arterially, intranasally, intraventricularly, or via direct injection. When administered, stem cells/progenitor cells exhibit the ability to migrate to pathological sites, a mechanism called *pathotrophism*. These cells express chemokine receptors, such as chemokine receptor type 4, and undergo chemotaxis toward cytokines secreted from damaged brain, including stromal cell-derived factor 1, monocyte chemotactic protein-1, chemokine ligand 2, and other cytokines.[157, 158] The targeting efficiency to sites of damage when exogenous cells are delivered intravascularly is 1%-10%. Cells infused intra-arterially have greater CNS uptake and localization to damaged targets compared to cells infused intravenously, which can become trapped in filtering reticuloendothelial system organs.[159, 160] In direct injection to either the contralateral hemisphere or an adjacent ischemic site, approximately 33% of cells migrate to target infarct areas.[3, 161-164] Likewise, during intrathecal transplantation of stem cells, cells migrate from the CSF space into the subependymal ischemic region and have been found to improve behavioral deficits in mice after stroke.[165]

Multiple stem cell lineages exist, but they have not been created equal. Embryonic stem cells (ESCs) are pluripotent stem cells derived from a blastocyst that can give rise to all cell types as well as functional neurons, astrocytes, and oligodendrocytes. Induced pluripotent stem cells (iPSCs) are established from skin fibroblasts transduced with Oct3/4, Sox2, Klf4, and c-Myc that result in stemness with proliferative abilities, like ESCs without the associated ethical issues.[166, 167] The use of undifferentiated ESCs and iPSCs can also be complicated by teratoma formation or even highly malignant teratocarcinomas. This tumorigenic potential, however, can be minimized by in vitro neural pre-differentiation methods.[168-170] Neural stem cells (NSCs) are multipotent, self-renewing cells that primarily differentiate into neurons, astrocytes, and oligodendrocytes, and can be found endogenously in subventricular zones. NSCs can be derived from differentiated ESCs, iPSCs, or from a fetal brain source. The most common progenitor cells used in current clinical trials are mesenchymal stem cells (MSCs) or marrow stromal cells, which are mesoderm in origin and can be derived from bone marrow, placenta, umbilical cord, muscle, skin, adipose tissue, and Wharton's jelly.[158] MSCs have poor potential to differentiate into neurons; upon transplantation, only 2%-20% differentiate to become neurons, whereas NSCs, iPSCs, and ESCs differentiate into neural lineage 34%-60%, 40%-66%, and 30% of the time, respectively.[158, 161, 171, 172]

Because only a small number of MSCs become neurons when transplanted, MSCs primarily promote neurological recovery by stimulating endogenous repair via the release of soluble trophic factors, including neurotrophins and angiogenic growth factors like BDNF, nerve growth factor, vascular endothelial growth factor, insulin-like growth factor 1, hepatocyte growth factor, and GDNF.[173-175] Thus, exogenous stem cells do not simply replace lost neuronal circuits. Instead, functional recovery is attributed to the angiogenesis and growth factor secretion from these cells that promote neurogenesis by inducing migration of endogenous progenitor cell from subventricular zones to sites of injury, and by promoting synaptogenesis through axonal remodeling to adapt the existing neuronal circuits to injured circuits.

In rat MCAO models, intravenous and intracranial stem cell replacement improves neurological function, ameliorates behavioral impairments, and decreases infarction size.[172, 176-180] To date, Phase I and Phase II clinical trials have shown that intravenous and IA administration of BMSCs are safe and feasible. Bang et al.[181] first reported on autologous MSC infusion in five patients who had acute stroke onset with average NIHSS scores of 10, and mean time from stroke onset to infusion of 46 days. No serious adverse events occurred within 1 year. Functional outcomes at 3, 6, and 12 months (measured by Barthel index scores and mRS scores) showed improved performance scores in the treatment group compared to the placebo group.[181] Suárez-Monteagudo et al.[182] showed that stereotactic implantation of autologous BMSCs in five patients after ischemic and hemorrhagic stroke was safe and well-tolerated, without adverse events. Sites of injection included affected thalamus, striatum, and motor cortex. At 1 year, when compared to preoperative baseline testing, treated patients had improved motor function and functional capacity.[182] Early studies of IA and intravenous delivery of MSCs or mononuclear cells (MNCs) were unblended and studied chronic stroke for up to 10 years after onset. The results showed good safety and feasibility, and had either modest functional improvements or no appreciable difference.[181-187]

Savitz et al.,[188] in an open-label study, administered MNCs and the acute setting 24-72 hours after stroke. They showed safety and feasibility of harvesting bone marrow in acute stroke settings, and demonstrated safety with MNC administration. At 6 months, all surviving patients had shifted down by at least 1 point on the mRS compared to day 7. Seven of ten patients achieved a Barthel Index score over 90.[188] Further studies of cell replacement in acute settings showed improvement in neurological function when given in acute and subacute timeframes.[186, 189] Interestingly, higher plasma levels of beta nerve growth factor were found in treated patients compared with control patients.[190] Clinical improvement in the MSC treatment groups was associated with higher serum levels of stromal cell-derived factor-1 after administration,[185] thereby supporting the theory that stem cell therapy induces activation of neurogenesis from inactive neural progenitor cells.

CONCLUSIONS

To date, the only randomized, blinded, placebo-controlled Phase II trial with published results involves administration of a proprietary multipotent adult progenitor cell product called MultiStem (Athersys, Inc.) to acute stroke patients 24 and 48 hours after symptom onset. Of the 129 patients enrolled in the study, 67 were randomized to receive multipotent adult progenitor cells and the remaining 62 patients were given a placebo treatment. The safety end point was met, but no difference was found between groups in global stroke recovery at 90 days.[191] There are 26 clinical trials currently underway to explore the use of stem cell therapy in stroke. These include randomized controlled trials exploring various delivery routes and cell types to discern maximum benefit.

References

1. Ovbiagele B, Goldstein LB, Higashida RT, et al. Forecasting the future of stroke in the United States: a policy statement from the American Heart Association and American Stroke Association. *Stroke*. 2013;44(8):2361-2375.
2. Go AS, Mozaffarian D, VLL R, et al. Heart disease and stroke statistics—2013 update: a report from the American Heart Association. *Circulation*. 2013;127(1):e6-e245.
3. North American Symptomatic Carotid Endarterectomy Trial (NASCET) Investigators. Clinical alert: benefit of carotid endarterectomy for patients with high-grade stenosis of the internal carotid artery. National Institute of Neurological Disorders and Stroke stroke and trauma division. *Stroke*. 1991;22(6):816-817.
4. Kety SS, Schmidt CF. The nitrous oxide method for the quantitative determination of cerebral blood flow in man: theory, procedure and normal values. *J Clin Invest*. 1948;27(4):476-483.
5. Sundt TM, Sharbrough FW, Anderson RE, Michenfelder JD. Cerebral blood flow measurements and electroencephalograms during carotid endarterectomy. *J Neurosurg*. 1974;41(3):310-320.
6. Waltz AG, Wanek AR, Anderson RE. Comparison of analytic methods for calculation of cerebral blood flow after intracarotid injection of 133 Xe. *J Nucl Med*. 1972;13(1):66-72.
7. Morawetz RB, Crowell RH, DeGirolami U, Marcoux FW, Jones TH, Halsey JH. Regional cerebral blood flow thresholds during cerebral ischemia. *Fed Proc*. 1979;38(11):2493-2494.
8. Astrup J, Symon L, Branston NM, Lassen NA. Cortical evoked potential and extracellular K+ and H+ at critical levels of brain ischemia. *Stroke*. 1977;8(1):51-57.
9. Branston NM, Strong AJ, Symon L. Extracellular potassium activity, evoked potential and tissue blood flow. Relationships during progressive ischaemia in baboon cerebral cortex. *J Neurol Sci*. 1977;32(3):305-321.
10. Astrup J, Blennow G, Nilsson B. Effects of reduced cerebral blood flow upon EEG pattern, cerebral extracellular potassium, and energy metabolism in the rat cortex during bicuculline-induced seizures. *Brain Res*. 1979; 177(1):115-126.
11. Dirnagl U, Iadecola C, Moskowitz MA, Dirnagl U, Iadecola C, Moskowitz MA. Pathobiology of ischaemic stroke: an integrated view. *Trends Neurosci*. 1999;22(9):391-397.
12. Stokum JA, Gerzanich V, Simard MJ. Molecular pathophysiology of cerebral edema. *J Cereb Blood Flow Metab*. 2016;36(3):513-538.
13. Weerasinghe P, Buja ML. Oncosis: an important non-apoptotic mode of cell death. *Exp Mol Pathol*. 2012; 93(3):302-308.
14. Majno G, Joris I. Apoptosis, oncosis, and necrosis. An overview of cell death. *Am J Pathol*. 1995;146(1):3-15.
15. Mergenthaler P, Dirnagl U, Meisel A. Pathophysiology of stroke: lessons from animal models. *Metab Brain Dis*. 2004;19(3/4):151-167.
16. Hartley DM, Kurth MC, Bjerkness L, Weiss JH, Choi DW. Glutamate receptor-induced 45Ca2+ accumulation in cortical cell culture correlates with subsequent neuronal degeneration. *J Neurosci*. 1993;13(5):1993-2000.
17. Choi DW. Glutamate neurotoxicity in cortical cell culture is calcium dependent. *Neurosci Lett*. 1985;58(3):293-297.
18. Choi DW. Ionic dependence of glutamate neurotoxicity. *J Neurosci*. 1987;7(2):369-379.
19. Broughton B, Reutens DC, Sobey CG. Apoptotic mechanisms after cerebral ischemia. *Stroke*. 2009;40(5):e331-e339.
20. Pacher P, Beckman JS, Liaudet L. Nitric oxide and peroxynitrite in health and disease. *Physiol Rev*. 2007; 87(1):315-424.
21. Ishige K, Schubert D, Sagara Y. Flavonoids protect neuronal cells from oxidative stress by three distinct mechanisms. *Free Radic Biol Med*. 2001;30(4):433-446.
22. Walford GA, Moussignac R-LL, Scribner AW, Loscalzo J, Leopold JA. Hypoxia potentiates nitric oxide-mediated apoptosis in endothelial cells via peroxynitrite-induced activation of mitochondria-dependent and -independent pathways. *J Biol Chem*. 2004;279(6):4425-4432.

23. Zhu C, Xu F, Wang X, et al. Different apoptotic mechanisms are activated in male and female brains after neonatal hypoxia-ischaemia. *J Neurochem*. 2006;96(4):1016-1027.
24. Gelderblom M, Leypoldt F, Steinbach K, et al. Temporal and spatial dynamics of cerebral immune cell accumulation in stroke. *Stroke*. 2009;40(5):1849-1857.
25. Barone FC, Arvin B, White RF, et al. Tumor necrosis factor-alpha. A mediator of focal ischemic brain injury. *Stroke*. 1997;28(6):1233-1244.
26. Jean WC, Spellman SR, Nussbaum ES, Low WC. Reperfusion injury after focal cerebral ischemia: the role of inflammation and the therapeutic horizon. *Neurosurgery*. 1998;43(6):1382.
27. Loddick SA, Rothwell NJ. Neuroprotective effects of human recombinant interleukin-1 receptor antagonist in focal cerebral ischaemia in the rat. *J Cereb Blood Flow Metab*. 1996;16(5):932-940.
28. Yamasaki Y, Matsuura N, Shozuhara H, Onodera H, Itoyama Y, Kogure K. Interleukin-1 as a pathogenetic mediator of ischemic brain damage in rats. *Stroke*. 1995;26(4):676.
29. Huang J, Upadhyay UM, Tamargo RJ. Inflammation in stroke and focal cerebral ischemia. *Surg Neurol*. 2006;66 (3):232-245.
30. Gelderblom M, Leypoldt F, Steinbach K, et al. Temporal and spatial dynamics of cerebral immune cell accumulation in stroke. *Stroke*. 2009;40(5):1849-1857.
31. Hum PD, Subramanian S, Parker SM, et al. T- and B-cell-deficient mice with experimental stroke have reduced lesion size and inflammation. *J Cereb Blood Flow Metab*. 2007;27(11):1798-1805.
32. Connolly ES, Winfree CJ, Springer TA, et al. Cerebral protection in homozygous null ICAM-1 mice after middle cerebral artery occlusion. Role of neutrophil adhesion in the pathogenesis of stroke. *J Clin Invest*. 1996; 97(1):209-216.
33. Soriano SG, Coxon A, Wang YF, et al. Mice deficient in mac-1 (CD11b/CD18) are less susceptible to cerebral ischemia/reperfusion injury. *Stroke*. 1999;30(1):134-139.
34. Zoppo DGJ. The neurovascular unit in the setting of stroke. *J Intern Med*. 2010;267(2):156-171.
35. Butt AM, Jones HC, Abbott NJ. Electrical resistance across the blood-brain barrier in anaesthetized rats: a developmental study. *J Physiol*. 1990;429:47-62.
36. Prager BCN, Shahid M, Grant GA, Ghosh C, Janigro D. The blood-brain barrier. In: *Youmans and Winn Neurological Surgery*. Philadelphia, PA; Elsevier; 2017:372.
37. Thompson BJ, Ronaldson PT. Drug delivery to the ischemic brain. *Adv Pharmacol*. 2014;71:165-202.
38. Arai K, Lok J, Guo S, Hayakawa K, Xing C, Lo EH. Cellular mechanisms of neurovascular damage and repair after stroke. *J Child Neurol*. 2011;26(9):1193-1198.
39. Wallace BK, Foroutan S, O'Donnell ME. Ischemia-induced stimulation of Na-K-cl cotransport in cerebral microvascular endothelial cells involves AMP kinase. *Am J Phys Cell Phys*. 2011;301(2):26.
40. Shah K, Abbruscato T. The role of blood-brain barrier transporters in pathophysiology and pharmacotherapy of stroke. *Curr Pharm Des*. 2014;20(10):1510-1522.
41. O'Donnell ME, Lam TI, Tran LQ, Foroutan S, Anderson SE. Estradiol reduces activity of the blood–brain barrier Na–K–cl cotransporter and decreases edema formation in permanent middle cerebral artery occlusion. *J Cereb Blood Flow Metab*. 2006;26(10):1234-1249.
42. Knowland D, Arac A, Sekiguchi KJ, et al. Stepwise recruitment of transcellular and paracellular pathways underlies blood-brain barrier breakdown in stroke. *Neuron*. 2014;82(3):603-617.
43. Lossinsky AS, Shivers RR. Structural pathways for macromolecular and cellular transport across the blood-brain barrier during inflammatory conditions. Review. *Histol Histopathol*. 2004;19(2):535-564.
44. Nag S, Venugopalan R, Stewart DJ. Increased caveolin-1 expression precedes decreased expression of occludin and claudin-5 during blood-brain barrier breakdown. *Acta Neuropathol*. 2007;114(5):459-469.
45. Jiao H, Wang Z, Liu Y, Wang P, Xue Y. Specific role of tight junction proteins claudin-5, occludin, and ZO-1 of the blood–brain barrier in a focal cerebral ischemic insult. *J Mol Neurosci*. 2011;44(2):130-139.
46. Ronaldson PT, Davis TP. Targeting blood-brain barrier changes during inflammatory pain: an opportunity for optimizing CNS drug delivery. *Ther Deliv*. 2011;2(8):1015-1041.
47. Lo EH, Singhal AB, Torchilin VP, Abbott NJ. Drug delivery to damaged brain. *Brain Res Rev*. 2001;38(1–2):140-148.
48. Hervé F, Ghinea N, Scherrmann J-M. CNS delivery via adsorptive transcytosis. *AAPS J*. 2008;10(3):455-472.
49. Banks WA. Characteristics of compounds that cross the blood-brain barrier. *BMC Neurol*. 2009;9(Suppl 1).
50. O'Reilly MA, Hynynen K. Ultrasound enhanced drug delivery to the brain and central nervous system. *Int J Hyperth*. 2012;28(4):386-396.
51. Jones AR, Shusta EV. Blood–brain barrier transport of therapeutics via receptor-mediation. *Pharm Res*. 2007;24 (9):1759-1771.

52. Beija M, Salvayre R, Viguerie N, Marty J-D. Colloidal systems for drug delivery: from design to therapy. *Trends Biotechnol*. 2012;30(9):485-496.
53. Pathan SA, Iqbal Z, Zaidi SM, et al. CNS drug delivery systems: novel approaches. *Recent Pat Drug Deliv Formul*. 2009;3(1):71-89.
54. Lee HJ, Pardridge WM. Pharmacokinetics and delivery of tat and tat-protein conjugates to tissues in vivo. *Bioconjug Chem*. 2001;12(6):995-999.
55. Marti HH, Wenger RH, Rivas LA, et al. Erythropoietin gene expression in human, monkey and murine brain. *Eur J Neurosci*. 1996;8(4):666-676.
56. Fu A, Hui E, Lu J, Boado RJ, Pardridge WM. Neuroprotection in stroke in the mouse with intravenous erythropoietin–Trojan horse fusion protein. *Brain Res*. 2011;1369:203-207.
57. Zhang Y, Pardridge WM. Blood-brain barrier targeting of BDNF improves motor function in rats with middle cerebral artery occlusion. *Brain Res*. 2006;1111:227-229.
58. Boado RJ, Zhang Y, Zhang Y, Wang Y, Pardridge WM. GDNF fusion protein for targeted-drug delivery across the human blood–brain barrier. *Biotechnol Bioeng*. 2008;100(2):387-396.
59. Singh R, Lillard JW. Nanoparticle-based targeted drug delivery. *Exp Mol Pathol*. 2009;86(3):215-223.
60. Kramer AH, Roberts DJ, Holodinsky J, et al. Intraventricular tissue plasminogen activator in subarachnoid hemorrhage patients: a prospective, randomized, placebo-controlled pilot trial. *Neurocrit Care*. 2014;21(2):275-284.
61. Ishii T, Asai T, Oyama D, et al. Amelioration of cerebral ischemia-reperfusion injury based on liposomal drug delivery system with asialo-erythropoietin. *J Control Release*. 2012;160(1):81-87.
62. Ishii T, Asai T, Oyama D, et al. Treatment of cerebral ischemia-reperfusion injury with PEGylated liposomes encapsulating FK506. *FASEB J*. 2013;27(4):1362-1370.
63. Jin Q, Cai Y, Li S, et al. Edaravone-encapsulated agonistic micelles rescue ischemic brain tissue by tuning blood-brain barrier permeability. *Theranostics*. 2017;7(4):884-898.
64. Yun X, Maximov VD, Yu J, Zhu H, Vertegel AA, Kindy MS. Nanoparticles for targeted delivery of antioxidant enzymes to the brain after cerebral ischemia and reperfusion injury. *J Cereb Blood Flow Metab*. 2013;33(4):583-592.
65. Hyun H, Lee J, Hwang D, et al. Combinational therapy of ischemic brain stroke by delivery of heme oxygenase-1 gene and dexamethasone. *Biomaterials*. 2011;32(1):306-315.
66. Basalious EB, Shamma RN. Novel self-assembled nano-tubular mixed micelles of Pluronics P123, Pluronic F127 and phosphatidylcholine for oral delivery of nimodipine: in vitro characterization, ex vivo transport and in vivo pharmacokinetic studies. *Int J Pharm*. 2015;493(1–2):347-356.
67. Huang S-L, Kim H, Britton G, Peng T, Holland CK, McPherson D. Nitric oxide-loaded echogenic liposomes for treatment of vasospasm following subarachnoid hemorrhage. *Int J Nanomedicine*. 2013;9(1):155-165.
68. Kim D, Seo Y, Thambi T, et al. Enhancing neurogenesis and angiogenesis with target delivery of stromal cell derived factor-1α using a dual ionic pH-sensitive copolymer. *Biomaterials*. 2015;61:115-125.
69. Shekhar H, Bader KB, Huang S, et al. In vitro thrombolytic efficacy of echogenic liposomes loaded with tissue plasminogen activator and octafluoropropane gas. *Phys Med Biol*. 2016;62(2):517-538.
70. Kandadai MA, Meunier JM, Hart K, Holland CK, Shaw GJ. Plasmin-loaded echogenic liposomes for ultrasound-mediated thrombolysis. *Transl Stroke Res*. 2015;6(1):78-87.
71. Zhang N, Li C, Zhou D, et al. Cyclic RGD functionalized liposomes encapsulating urokinase for thrombolysis. *Acta Biomater*. 2018;70:227-236.
72. Joshi S, Emala CW, Pile-Spellman J. Intra-arterial drug delivery: a concise review. *J Neurosurg Anesthesiol*. 2007;19 (2):111-119.
73. Cloft HJ, Joseph GJ, Dion JE. Risk of cerebral angiography in patients with subarachnoid hemorrhage, cerebral aneurysm, and arteriovenous malformation: a meta-analysis. *Stroke*. 1999;30(2):317-320.
74. Pardridge WM. CSF, blood-brain barrier, and brain drug delivery. *Expert Opin Drug Deliv*. 2016;13(7):963-975.
75. Covell DG, Narang PK, Poplack DG. Kinetic model for disposition of 6-mercaptopurine in monkey plasma and cerebrospinal fluid. *Am J Phys*. 1985;248(2 Pt. 2):56.
76. Lochhead JJ, Thorne RG. Intranasal delivery of biologics to the central nervous system. *Adv Drug Deliv Rev*. 2012;64(7):614-628.
77. Costantino HR, Illum L, Brandt G, Johnson PH, Quay SC. Intranasal delivery: physicochemical and therapeutic aspects. *Int J Pharm*. 2007;337(1–2):1-24.
78. Kokkoris S, Andrews P, Webb DJ. Role of calcitonin gene-related peptide in cerebral vasospasm, and as a therapeutic approach to subarachnoid hemorrhage. *Front Endocrinol (Lausanne)*. 2012;3:135.
79. Sun B-L, Shen F-P, Wu Q-J, et al. Intranasal delivery of calcitonin gene-related peptide reduces cerebral vasospasm in rats. *Front Biosci*. 2010;2:1502-1513.

80. Liu XF, Fawcett JR, Thorne RG, Frey WH. Non-invasive intranasal insulin-like growth factor-I reduces infarct volume and improves neurologic function in rats following middle cerebral artery occlusion. *Neurosci Lett*. 2001;308(2):91-94.
81. Wang Z-L, Cheng S-M, Ma M-M, et al. Intranasally delivered bFGF enhances neurogenesis in adult rats following cerebral ischemia. *Neurosci Lett*. 2008;446(1):30-35.
82. Yu Y-P, Xu Q-Q, Zhang Q, Zhang W-P, Zhang L-H, Wei E-Q. Intranasal recombinant human erythropoietin protects rats against focal cerebral ischemia. *Neurosci Lett*. 2005;387(1):5-10.
83. Lioutas V-A, Alfaro-Martinez F, Bedoya F, Chung C-C, Pimentel DA, Novak V. Intranasal insulin and insulin-like growth factor 1 as neuroprotectants in acute ischemic stroke. *Transl Stroke Res*. 2015;6(4):264-275.
84. Ma M, Ma Y, Yi X, et al. Intranasal delivery of transforming growth factor-beta1 in mice after stroke reduces infarct volume and increases neurogenesis in the subventricular zone. *BMC Neurosci*. 2008;9(1):1-10.
85. Fletcher L, Kohli S, Sprague SM, et al. Intranasal delivery of erythropoietin plus insulin-like growth factor–I for acute neuroprotection in stroke. *J Neurosurg*. 2009;111(1):164-170.
86. Ma Y-P, Ma M-M, Cheng S-M, et al. Intranasal bFGF-induced progenitor cell proliferation and neuroprotection after transient focal cerebral ischemia. *Neurosci Lett*. 2008;437(2):93-97.
87. Liu XF, Fawcett JR, Thorne RG, DeFor TA, Frey WH. Intranasal administration of insulin-like growth factor-I bypasses the blood-brain barrier and protects against focal cerebral ischemic damage. *J Neurol Sci*. 2001;187 (1–2):91-97.
88. Yang J-P, Liu H-J, Wang Z-L, et al. The dose-effectiveness of intranasal VEGF in treatment of experimental stroke. *Neurosci Lett*. 2009;461(3):212-216.
89. Yang J-P, Liu H-J, Cheng S-M, et al. Direct transport of VEGF from the nasal cavity to brain. *Neurosci Lett*. 2009;449(2):108-111.
90. Thorne RG, Pronk GJ, Padmanabhan V, Frey WH. Delivery of insulin-like growth factor-I to the rat brain and spinal cord along olfactory and trigeminal pathways following intranasal administration. *Neuroscience*. 2004;127 (2):481-496.
91. Zhao Q, Hu J, Xiang J, et al. Intranasal administration of human umbilical cord mesenchymal stem cells-conditioned medium enhances vascular remodeling after stroke. *Brain Res*. 2015;1624:489-496.
92. Wei N, Yu SP, Gu X, et al. Delayed intranasal delivery of hypoxic-preconditioned bone marrow mesenchymal stem cells enhanced cell homing and therapeutic benefits after ischemic stroke in mice. *Cell Transplant*. 2013;22 (6):977-991.
93. Bobo RH, Laske DW, Akbasak A, Morrison PF, Dedrick RL, Oldfield EH. Convection-enhanced delivery of macromolecules in the brain. *Proc Natl Acad Sci U S A*. 1994;91(6):2076-2080.
94. Raghavan R, Brady ML, Rodríguez-Ponce MII, Hartlep A, Pedain C, Sampson JH. Convection-enhanced delivery of therapeutics for brain disease, and its optimization. *Neurosurg Focus*. 2006;20(4):E12.
95. Haar PJ, Broaddus WC, Z-j C, Fatouros PP, Gillies GT, Corwin FD. Quantification of convection-enhanced delivery to the ischemic brain. *Physiol Meas*. 2010;31(9):1075-1089.
96. Lee K-YY, Han SW, Kim SH, et al. Early recanalization after intravenous administration of recombinant tissue plasminogen activator as assessed by pre- and post-thrombolytic angiography in acute ischemic stroke patients. *Stroke*. 2007;38(1):192-193.
97. Lijnen HR, Collen D. Fibrinolytic agents: mechanisms of activity and pharmacology. *Thromb Haemost*. 1995;74 (1):387-390.
98. Powers WJ, Rabinstein AA, Ackerson T, et al. Guidelines for the early management of patients with acute ischemic stroke: a guideline for healthcare professionals from the American Heart Association/American Stroke Association. *Stroke*. 2018;49(3):e46-e110.
99. Huang X, Cheripelli B, Lloyd SM, et al. Alteplase versus tenecteplase for thrombolysis after ischaemic stroke (ATTEST): a phase 2, randomised, open-label, blinded endpoint study. *Lancet Neurol*. 2015;14(4):368-376.
100. Parsons M, Spratt N, Bivard A, et al. A randomized trial of tenecteplase versus alteplase for acute ischemic stroke. *N Engl J Med*. 2012;366(12):1099-1107.
101. Hacke W, Donnan G, Fieschi C, et al. Association of outcome with early stroke treatment: pooled analysis of ATLANTIS, ECASS, and NINDS rt-PA stroke trials. *Lancet*. 2004;363(9411):768-774.
102. Hacke W, Kaste M, Fieschi C, et al. Randomised double-blind placebo-controlled trial of thrombolytic therapy with intravenous alteplase in acute ischaemic stroke (ECASS II). *Lancet*. 1998;352(9136):1245-1251.

103. Albers GW, Clark WM, Madden KP, Hamilton SA. ATLANTIS trial: results for patients treated within 3 hours of stroke onset. Alteplase thrombolysis for acute noninterventional therapy in ischemic stroke. *Stroke.* 2002; 33(2):493-495.
104. Hacke W, Kaste M, Bluhmki E, et al. Thrombolysis with alteplase 3 to 4.5 hours after acute ischemic stroke. *N Engl J Med.* 2008;359(13):1317-1329.
105. Ciccone A, Valvassori L, Nichelatti M, et al. Endovascular treatment for acute ischemic stroke. *N Engl J Med.* 2013;368(10):904-913.
106. Furlan A, Higashida R, Wechsler L, et al. Intra-arterial prourokinase for acute ischemic stroke. The PROACT II study: a randomized controlled trial. Prolyse in acute cerebral thromboembolism. *JAMA.* 1999; 282(21):2003-2011.
107. Investigators III T. The interventional management of stroke (IMS) II study. *Stroke.* 2007;38(7):2127-2135.
108. Powers WJ, Derdeyn CP, Biller J, et al. 2015 American Heart Association/American Stroke Association focused update of the 2013 guidelines for the early Management of Patients with acute ischemic stroke regarding endovascular treatment a guideline for healthcare professionals from the American Heart Association/American Stroke Association. *Stroke.* 2015;46(10):3020-3035.
109. Bracard S, Ducrocq X, Mas J, et al. Mechanical thrombectomy after intravenous alteplase versus alteplase alone after stroke (THRACE): a randomised controlled trial. *Lancet Neurol.* 2016;15(11):1138-1147.
110. Campbell BC, Mitchell PJ, Kleinig TJ, et al. Endovascular therapy for ischemic stroke with perfusion-imaging selection. *N Engl J Med.* 2015;372(11):1009-1018.
111. Berkhemer OA, PSS F, Beumer D, et al. A randomized trial of intraarterial treatment for acute ischemic stroke. *N Engl J Med.* 2015;372(1):11-20.
112. Goyal M, Demchuk AM, Menon BK, et al. Randomized assessment of rapid endovascular treatment of ischemic stroke. *N Engl J Med.* 2015;372(11):1019-1030.
113. Jovin TG, Chamorro A, Cobo E, et al. Thrombectomy within 8 hours after symptom onset in ischemic stroke. *N Engl J Med.* 2015;372(24):2296-2306.
114. Saver JL, Goyal M, Bonafe A, et al. Solitaire™ with the intention for thrombectomy as primary endovascular treatment for acute ischemic stroke (SWIFT PRIME) trial: protocol for a randomized, controlled, multicenter study comparing the solitaire revascularization device with IV tPA with IV tPA alone in acute ischemic stroke. *Int J Stroke.* 2015;10(3):439-448.
115. Nogueira RG, Jadhav AP, Haussen DC, et al. Thrombectomy 6 to 24 hours after stroke with a mismatch between deficit and infarct. *N Engl J Med.* 2018;378(1):11-21.
116. Albers GW, Marks MP, Kemp S, et al. Thrombectomy for stroke at 6 to 16 hours with selection by perfusion imaging. *N Engl J Med.* 2018;378(8):708-718.
117. Rinkel GJ, Feigin VL, Algra A, Vermeulen M, van Gijn J. Calcium antagonists for aneurysmal subarachnoid haemorrhage. *Cochrane Database Syst Rev.* 2002;4.
118. Kassell NF, Helm G, Simmons N, Phillips CD, Cail WS. Treatment of cerebral vasospasm with intra-arterial papaverine. *J Neurosurg.* 1992;77(6):848-852.
119. Kaku Y, Yonekawa Y, Tsukahara T, Kazekawa K. Superselective intra-arterial infusion of papaverine for the treatment of cerebral vasospasm after subarachnoid hemorrhage. *J Neurosurg.* 1992;77(6):842-847.
120. Böker DK, Solymosi L, Wassmann H. Immediate postangiographic intraarterial treatment of cerebral vasospasm after subarachnoid hemorrhage with nimodipine. Report on 3 cases. *Neurochirurgia (Stuttg).* 1985;28(Suppl 1):118-120.
121. Yoshida K, Watanabe H, Nakamura S. Intraarterial injection of amrinone for vasospasm induced by subarachnoid hemorrhage. *AJNR Am J Neuroradiol.* 1997;18(3):492-496.
122. Suzuki S, Ito O, Sayama T, Yamaguchi S, Goto K, Sasaki T. Intraarterial injection of colforsin daropate hydrochloride for the treatment of vasospasm after aneurysmal subarachnoid hemorrhage: preliminary report of two cases. *Neuroradiology.* 2006;48(1):50-53.
123. Hanley DF, Lane K, McBee N, et al. Thrombolytic removal of intraventricular haemorrhage in treatment of severe stroke: results of the randomised, multicentre, multiregion, placebo-controlled CLEAR III trial. *Lancet.* 2017;389(10069):603-611.
124. Litrico S, Almairac F, Gaberel T, et al. Intraventricular fibrinolysis for severe aneurysmal intraventricular hemorrhage: a randomized controlled trial and meta-analysis. *Neurosurg Rev.* 2013;36(4):523.

125. Etminan N, Beseoglu K, Eicker SO, Turowski B, Steiger H-JJ, Hänggi D. Prospective, randomized, open-label phase II trial on concomitant intraventricular fibrinolysis and low-frequency rotation after severe subarachnoid hemorrhage. *Stroke*. 2013;44(8):2162-2168.
126. Findlay JM, Kassell NF, Weir BK, et al. A randomized trial of intraoperative, intracisternal tissue plasminogen activator for the prevention of vasospasm. *Neurosurgery*. 1995;37(1):168.
127. Goodson K, Lapointe M, Monroe T, Chalela JA. Intraventricular nicardipine for refractory cerebral vasospasm after subarachnoid hemorrhage. *Neurocrit Care*. 2008;8(2):247-252.
128. Lu N, Jackson D, Luke S, Festic E, Hanel RA, Freeman W. Intraventricular nicardipine for aneurysmal subarachnoid hemorrhage related vasospasm: assessment of 90 days outcome. *Neurocrit Care*. 2012;16(3):368-375.
129. Shibuya M, Suzuki Y, Enomoto H, Okada T, Ogura K, Sugita K. Effects of prophylactic intrathecal administrations of nicardipine on vasospasm in patients with severe aneurysmal subarachnoid haemorrhage. *Acta Neurochir*. 1994;131(1–2):19-25.
130. Kasuya H, Onda H, Sasahara A, Takeshita M, Hori T. Application of nicardipine prolonged-release implants: analysis of 97 consecutive patients with acute subarachnoid hemorrhage. *Neurosurgery*. 2005;56(5):895-902.
131. Ko S-B, Choi AH, Helbok R, et al. Acute effects of intraventricular nicardipine on cerebral hemodynamics: a preliminary finding. *Clin Neurol Neurosurg*. 2016;144:48-52.
132. Kassell NF, Torner JC, Haley EC, Jane JA, Adams HP, Kongable GL. The international cooperative study on the timing of aneurysm surgery. Part 1: overall management results. *J Neurosurg*. 1990;73(1):18-36.
133. Barth M, Capelle H-H, Weidauer S, et al. Effect of nicardipine prolonged-release implants on cerebral vasospasm and clinical outcome after severe aneurysmal subarachnoid hemorrhage: a prospective, randomized, double-blind phase iia study. *Stroke*. 2007;38(2):330-336.
134. Shaw GJ, Meunier JM, Huang S-L, Lindsell CJ, McPherson DD, Holland CK. Ultrasound-enhanced thrombolysis with tPA-loaded echogenic liposomes. *Thromb Res*. 2009;124(3):306-310.
135. Alexandrov AV, Molina CA, Grotta JC, et al. Ultrasound-enhanced systemic thrombolysis for acute ischemic stroke. *N Engl J Med*. 2004;351(21):2170-2178.
136. Barreto AD, Alexandrov AV, Shen L, et al. CLOTBUST-hands free. *Stroke*. 2013;44(12):3376-3381.
137. Benjamin EJ, Virani SS, Callaway CW, et al. Heart disease and stroke statistics-2018 update: a report From the American Heart Association. *Circulation*. 2018;137(12):e67-e492.
138. Chamorro Á, Dirnagl U, Urra X, Planas AM. Neuroprotection in acute stroke: targeting excitotoxicity, oxidative and nitrosative stress, and inflammation. *Lancet Neurol*. 2016;15(8):869-881.
139. Karsy M, Brock A, Guan J, Taussky P, Kalani YMS, Park MS. Neuroprotective strategies and the underlying molecular basis of cerebrovascular stroke. *Neurosurg Focus*. 2017;42(4):E3.
140. Ginsberg MD. Neuroprotection for ischemic stroke: past, present and future. *Neuropharmacology*. 2008;55(3):363-389.
141. Lai TW, Zhang S, Wang YT. Excitotoxicity and stroke: identifying novel targets for neuroprotection. *Prog Neurobiol*. 2014;115:157-188.
142. Yang Y, Li Q, Ahmad F, Shuaib A. Survival and histological evaluation of therapeutic window of post-ischemia treatment with magnesium sulfate in embolic stroke model of rat. *Neurosci Lett*. 2000;285(2):119-122.
143. Muir KW, Lees KR, Ford I, Davis S, In investigators I. Magnesium for acute stroke (intravenous magnesium efficacy in stroke trial): randomised controlled trial. *Lancet*. 2004;363(9407):439-445.
144. Albers GW, Goldstein LB, Hall D, Lesko LM, Investigators A. Aptiganel hydrochloride in acute ischemic stroke: a randomized controlled trial. *JAMA*. 2001;286(21):2673-2682.
145. Albers GW, Clark WM, Atkinson RP, Madden K, Data JL, Whitehouse MJ. Dose escalation study of the NMDA glycine-site antagonist licostinel in acute ischemic stroke. *Stroke*. 1999;30(3):508-513.
146. Sacco RL, DeRosa JT, Haley EC, et al. Glycine antagonist in neuroprotection for patients with acute stroke: GAIN Americas: a randomized controlled trial. *JAMA*. 2001;285(13):1719-1728.
147. Davis SM, Albers GW, Diener HC, Lees KR, Norris J. Termination of acute stroke studies involving selfotel treatment. ASSIST Steering Committed. *Lancet*. 1997;349(9044):32.
148. Bullock MR, Merchant RE, Carmack CA, et al. An open-label study of CP-101,606 in subjects with a severe traumatic head injury or spontaneous intracerebral hemorrhage. *Ann N Y Acad Sci*. 1999;890:51-58.
149. Lees KR, Zivin JA, Ashwood T, et al. NXY-059 for acute ischemic stroke. *N Engl J Med*. 2006;354(6):588-600.
150. Shuaib A, Lees KR, Lyden P, et al. NXY-059 for the treatment of acute ischemic stroke. *N Engl J Med*. 2007;357(6):562-571.

151. Ginsberg MD. Life after cerovive: a personal perspective on ischemic neuroprotection in the post-NXY-059 era. *Stroke*. 2007;38(6):1967-1972.
152. Edaravone Acute Infarction Study Group. Effect of a novel free radical scavenger, edaravone (MCI-186), on acute brain infarction. Randomized, placebo-controlled, double-blind study at multicenters. *Cerebrovasc Dis*. 2003; 15(3):222-229.
153. Feng S, Yang Q, Liu M, et al. Edaravone for acute ischaemic stroke. *Cochrane Database Syst Rev*. 2011;12.
154. Liu J, Zhang C, Tao W, Liu M. Systematic review and meta-analysis of the efficacy of sphingosine-1-phosphate (S1P) receptor agonist FTY720 (fingolimod) in animal models of stroke. *Int J Neurosci*. 2013;123(3):163-169.
155. Fu Y, Zhang N, Ren L, et al. Impact of an immune modulator fingolimod on acute ischemic stroke. *Proc Natl Acad Sci U S A*. 2014;111(51):18315-18320.
156. Zhu Z, Fu Y, Tian D, et al. Combination of the immune modulator fingolimod with alteplase in acute ischemic stroke: a pilot trial. *Circulation*. 2015;132(12):1104-1112.
157. Guzman R, Bliss T, Angeles A, Moseley M, Palmer T, Steinberg G. Neural progenitor cells transplanted into the uninjured brain undergo targeted migration after stroke onset. *J Neurosci Res*. 2008;86(4):873-882.
158. Liu X, Ye R, Yan T, et al. Cell based therapies for ischemic stroke: from basic science to bedside. *Prog Neurobiol*. 2014;115:92-115.
159. Fischer UM, Harting MT, Jimenez F, et al. Pulmonary passage is a major obstacle for intravenous stem cell delivery: the pulmonary first-pass effect. *Stem Cells Dev*. 2009;18(5):683-692.
160. Pendharkar AV, Chua JY, Andres RH, et al. Biodistribution of neural stem cells after intravascular therapy for hypoxic–ischemia. *Stroke*. 2010;41(9):2064-2070.
161. Darsalia V, Kallur T, Kokaia Z. Survival, migration and neuronal differentiation of human fetal striatal and cortical neural stem cells grafted in stroke-damaged rat striatum. *Eur J Neurosci*. 2007;26(3):605-614.
162. Kelly S, Bliss TM, Shah AK, et al. Transplanted human fetal neural stem cells survive, migrate, and differentiate in ischemic rat cerebral cortex. *Proc Natl Acad Sci U S A*. 2004;101(32):11839-11844.
163. Li Y, Chen J, Chen XG, et al. Human marrow stromal cell therapy for stroke in rat: neurotrophins and functional recovery. *Neurology*. 2002;59(4):514-523.
164. Li Y, Chen J, Wang L, Lu M, Chopp M. Treatment of stroke in rat with intracarotid administration of marrow stromal cells. *Neurology*. 2001;56(12):1666-1672.
165. Jafari SS, Aghaei AA, Asadi-Shekaari M, Nematollahi-mahani SN, Sheibani V. Investigating the effects of adult neural stem cell transplantation by lumbar puncture in transient cerebral ischemia. *Neurosci Lett*. 2011;495(1):1-5.
166. Takahashi K, Tanabe K, Ohnuki M, et al. Induction of pluripotent stem cells from adult human fibroblasts by defined factors. *Cell*. 2007;131(5):861-872.
167. Takahashi K, Yamanaka S. Induction of pluripotent stem cells from mouse embryonic and adult fibroblast cultures by defined factors. *Cell*. 2006;126(4):663-676.
168. Eiraku M, Watanabe K, Matsuo-Takasaki M, et al. Self-organized formation of polarized cortical tissues from ESCs and its active manipulation by extrinsic signals. *Cell Stem Cell*. 2008;3(5):519-532.
169. Reubinoff BE, Pera MF, Fong C-Y, Trounson A, Bongso A. Embryonic stem cell lines from human blastocysts: somatic differentiation in vitro. *Nat Biotechnol*. 2000;18(4):399-404.
170. Thomson JA, Itskovitz-Eldor J, Shapiro SS, et al. Embryonic stem cell lines derived from human blastocysts. *Science*. 1998;282(5391):1145-1147.
171. Chen J, Li Y, Wang L, Lu M, Zhang X, Chopp M. Therapeutic benefit of intracerebral transplantation of bone marrow stromal cells after cerebral ischemia in rats. *J Neurol Sci*. 2001;189(1–2):49-57.
172. Chen J, Li Y, Wang L, et al. Therapeutic benefit of intravenous administration of bone marrow stromal cells after cerebral ischemia in rats. *Stroke*. 2001;32(4):1005-1011.
173. Chen J, Zhang ZG, Li Y, et al. Intravenous administration of human bone marrow stromal cells induces angiogenesis in the ischemic boundary zone after stroke in rats. *Circ Res*. 2003;92(6):692-699.
174. Chen X, Katakowski M, Li Y, et al. Human bone marrow stromal cell cultures conditioned by traumatic brain tissue extracts: growth factor production. *J Neurosci Res*. 2002;69(5):687-691.
175. Shen LH, Li Y, Chopp M. Astrocytic endogenous glial cell derived neurotrophic factor production is enhanced by bone marrow stromal cell transplantation in the ischemic boundary zone after stroke in adult rats. *Glia*. 2010;58(9):1074-1081.
176. Chen J, Sanberg PR, Li Y, et al. Intravenous administration of human umbilical cord blood reduces behavioral deficits after stroke in rats. *Stroke*. 2001;32(11):2682-2688.

177. Chen J, Shehadah A, Pal A, et al. Neuroprotective effect of human placenta-derived cell treatment of stroke in rats. *Cell Transplant*. 2013;22(5):871.
178. Kang SK, Lee DH, Bae YC, Kim HK, Baik SY, Jung JS. Improvement of neurological deficits by intracerebral transplantation of human adipose tissue-derived stromal cells after cerebral ischemia in rats. *Exp Neurol*. 2003;183(2):355-366.
179. Borlongan CV, Hadman M, Sanberg C, Sanberg PR. Central nervous system entry of peripherally injected umbilical cord blood cells is not required for neuroprotection in stroke. *Stroke*. 2004;35(10):2385-2389.
180. van Velthoven C, Sheldon AR, Kavelaars A, et al. Mesenchymal stem cell transplantation attenuates brain injury after neonatal stroke. *Stroke*. 2013;44(5):1426-1432.
181. Bang O, Lee J, Lee P, Lee G. Autologous mesenchymal stem cell transplantation in stroke patients. *Ann Neurol*. 2005;57(6):874-882.
182. Suárez-Monteagudo C, Hernández-Ramírez P, Alvarez-González L, et al. Autologous bone marrow stem cell neurotransplantation in stroke patients. An open study. *Restor Neurol Neurosci*. 2009;27(3):151-161.
183. Battistella V, de Freitas GR, da Fonseca L, et al. Safety of autologous bone marrow mononuclear cell transplantation in patients with nonacute ischemic stroke. *Regen Med*. 2011;6(1):45-52.
184. Savitz SI, Dinsmore J, Wu J, Henderson GV, Stieg P, Caplan LR. Neurotransplantation of fetal porcine cells in patients with basal ganglia infarcts: a preliminary safety and feasibility study. *Cerebrovasc Dis*. 2005; 20(2):101-107.
185. Lee J, Hong J, Moon G, Lee P, Ahn Y, Bang O. A long-term follow-up study of intravenous autologous mesenchymal stem cell transplantation in patients with ischemic stroke. *Stem Cells*. 2010;28(6):1099-1106.
186. Bhasin A, Srivastava MV, Kumaran SS, et al. Autologous mesenchymal stem cells in chronic stroke. *Cerebrovasc Dis Extra*. 2011;1(1):93-104.
187. Honmou O, Houkin K, Matsunaga T, et al. Intravenous administration of auto serum-expanded autologous mesenchymal stem cells in stroke. *Brain*. 2011;134(6):1790-1807.
188. Savitz SI, Misra V, Kasam M, et al. Intravenous autologous bone marrow mononuclear cells for ischemic stroke. *Ann Neurol*. 2011;70(1):59-69.
189. Friedrich MAG, Martins MP, Araújo MD, et al. Intra-arterial infusion of autologous bone marrow mononuclear cells in patients with moderate to severe middle cerebral artery acute ischemic stroke. *Cell Transplant*. 2012; 21(Suppl 1):S13-S21.
190. Moniche F, Gonzalez A, Gonzalez-Marcos J-RR, et al. Intra-arterial bone marrow mononuclear cells in ischemic stroke: a pilot clinical trial. *Stroke*. 2012;43(8):2242-2244.
191. Hess DC, Wechsler LR, Clark WM, et al. Safety and efficacy of multipotent adult progenitor cells in acute ischaemic stroke (MASTERS): a randomised, double-blind, placebo-controlled, phase 2 trial. *Lancet Neurol*. 2017;16(5):360-368.

CHAPTER

25

Intrathecal Drug Delivery for Cancer Pain

*Philip S. Kim**, *Peter S. Staats*[†], *Timothy R. Deer*[‡], *Michael J. Iadarola*[§], *Andrew J. Mannes*[§]

[*]Center for Interventional Pain & Spine, LLC, Helen Graham Center, Newark, DE, United States [†]National Spine and Pain Centers, Shrewsbury, NJ, United States [‡]Center for Pain Relief, INC, Charleston, WV, United States [§]Department of Perioperative Medicine, National Institutes of Health, Bethesda, MD, United States

PAIN MANAGEMENT IN CANCER PATIENTS

Interventional, neurosurgical, and soon, molecular neurosurgical procedures can supplement pharmacological and complementary approaches to treat cancer pain. The primary indications for sophisticated pain treatment (e.g., an implantable drug delivery system [IDDS]) are cancer pain that does not respond to systemic analgesic therapy. Advanced pain treatment techniques may also be appropriate for patients with unmanageable severe adverse effects that cannot be treated due to dose-limiting side effects (e.g., systemic opioids causing refractory constipation, nausea, vomiting, or sedation). Of the approximately 500,000 patients who die each year from cancer, 5%-15% suffer from poor pain control.[1-3] IDDSs may optimize pain management while minimizing side effects, enhancing quality of life, improving survival, and potentially reducing costs.[4]

INITIAL EVALUATION

An initial evaluation for an IDDS must include evaluation of general medical condition along with the primary disease, as well as the patient's prognosis and associated comorbidities. It is also important to consider the expectations of both the patient and the family. A complete general medical history and disease-specific oncologic disorder evaluation

https://doi.org/10.1016/B978-0-12-813997-4.00025-6
© 2019 Elsevier Inc. All rights reserved.

are required; these must include details on local recurrence and metastasis, as well as psychosocial and pain history. The physical examination should be general, with additional focus on neurological findings. Examination of the site of pain and the surrounding anatomic regions is critical (e.g., evaluating whether significant motor and sensory deficits are present). Assessment of the spine and the abdominal region is critical to determine placement of the catheter and implantable device. The usual location for the intrathecal pump device is the abdomen, but other locations are possible (e.g., the buttock or torso). Specific pain history to determine suitability of intrathecal therapy includes the following: quality of pain, pain intensity, alleviating or exacerbating factors, temporal characteristics, duration of pain, and associated features (e.g., numbness, weakness, and vasomotor changes). Psychosocial evaluation should assess the presence of any psychiatric disorders, such as uncontrolled depression or anxiety. The nature and meaning of the presenting pain must be distinguished from anxiety and suffering affecting the patient (to distinguish emotional vs physical suffering). The ability to cope and the availability of psychosocial support systems must be assessed and reinforced with proper health and social professionals. Each of these elements factors into the success of intrathecal treatment.

Appropriate selection of an IDDS is based on the therapeutic goal. If the patient's pain is severe, uncontrolled, and strongly diminishing quality of life, IDDS may be an appropriate therapy. Life expectancy must be considered when selecting IDDS over externalized neuraxial infusion. An IDDS becomes more cost-effective with a life expectancy of 3 months or more.[5] Econometric analysis supports IDDS with intrathecal morphine or ziconotide compared to conventional medical management with brand pain medication at 7 and 10 months, respectively.[6]

Intraspinal delivery of drugs via IDDSs has been available since the early 1980s. These devices allow for precise, compartmental delivery of drugs for treatment of pain and spasticity. The delivery of medications in the intraspinal or intrathecal space allows direct access to their site of receptor binding in the spinal column at much smaller doses than needed for systemic administration. The result is a potent analgesic effect with lower systemic side effects. IDDS may also be considered as a route of administration of non-opioid agents such as ziconotide, clonidine, and local anesthetics. Local anesthetics (e.g., bupivacaine) and neuronal calcium channel blockers (e.g., ziconotide) can produce profound analgesic effects on many intractable pain conditions unresponsive to opioid treatment.

Neuraxial delivery systems have two components: a subarachnoid or epidural catheter and a delivery mode (e.g., syringe pump, internal port, or internal or external pump) capable of continuous infusion or bolus dosing. For short-term use and cost, these external devices can be used. For long-term needs, a completely implantable pump system is recommended for infection control, safety, and reduced cost.

Intrathecal Drug Delivery Systems

An intrathecal drug delivery system consists of a pump implanted in a subcutaneous pocket, usually in the abdomen, which is attached to a catheter tunneled under the skin and inserted in the intrathecal space of the spinal cord at the appropriate level (Fig. 1). Pumps can be programmable or nonprogrammable. Programmable pumps are set up to deliver the

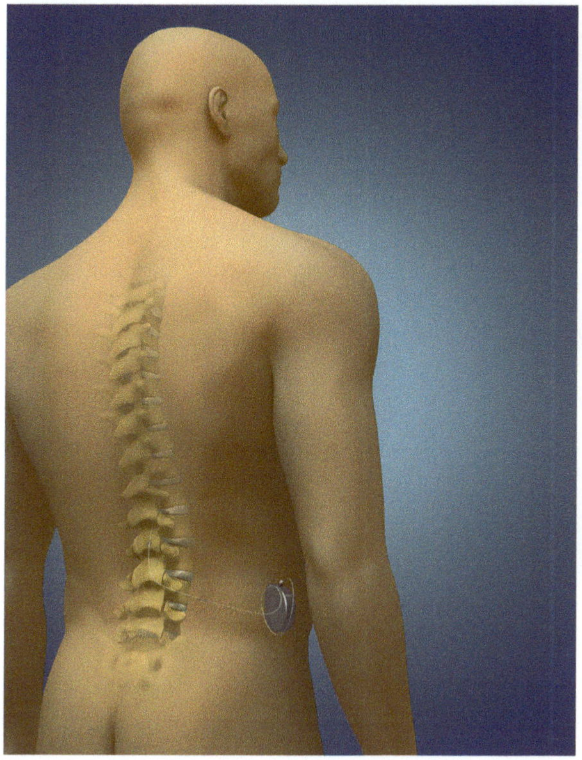

FIG. 1 Implantable pump and catheter. *Reprinted with the permission of Medtronic, Inc.*

drug or admixture at a controlled rate, based on the drug concentration and dose requirements of the patient. Nonprogrammable pumps run at a constant flow rate, with dose adjustments made by altering the concentration of medications.

Several IDDS are commercially available, with options for programmable and nonprogrammable pumps. The programmable pumps include SynchroMed II (Medtronic, Fig. 2), Prometra (Flowonix, Fig. 3), and MedStream (Codman). The SynchroMed II pump is approved by Food and Drug Administration (FDA) for morphine and ziconotide for chronic intractable pain and baclofen for severe spasticity. The pump uses a peristaltic roller system to move the drug from the reservoir to the implanted intrathecal catheter. It has flexible dosing and patient-administered dosing capabilities, and is available in volumes of 20 and 40 cc. The Prometra pump has FDA approval for morphine to treat chronic intractable pain. This device uses a pressurized gas chamber as the driving force, with a programmable flow-metering valve to allow variable drug delivery. The pump is available in a volume of 20 cc. The advantage of the Prometra pump is that it is associated with much lower energy requirements, because a battery is used to control the electronics only and not to pump the drug (as in the SynchroMed II). The Prometra battery has a 10-year lifespan. The Prometra pump is not compatible with magnetic resonance imaging (MRI), and requires removal of drug before MRI to prevent uncontrolled overdosing of the drug. The SynchroMed II pump is MRI-compatible, but it may temporarily stall during a scan. The MedStream pump is FDA

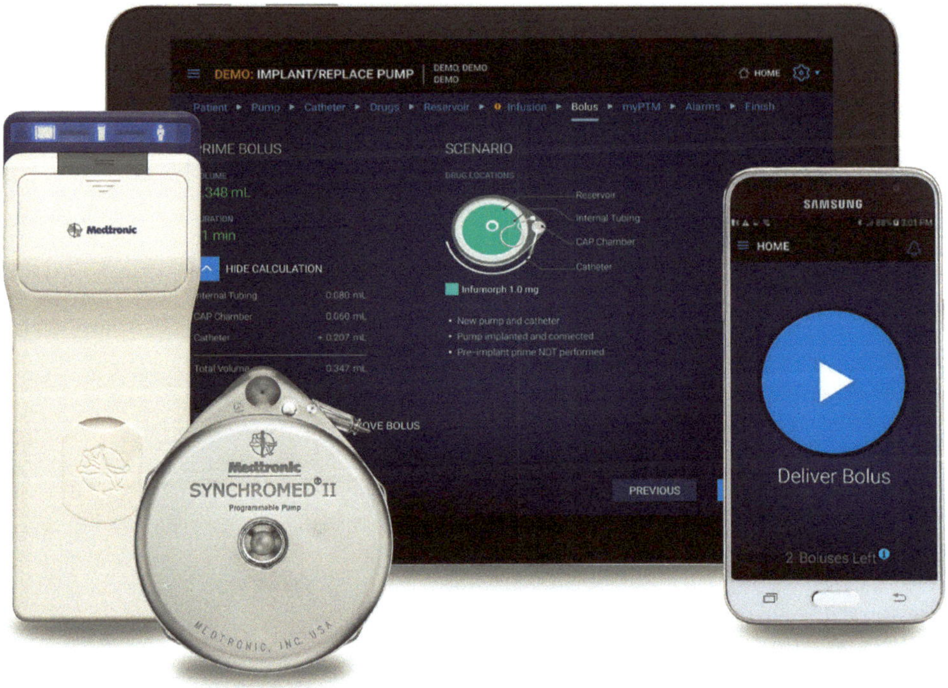

FIG. 2 Medtronic SynchroMed II pump with programmer. *Reprinted with the permission of Medtronic, Inc.*

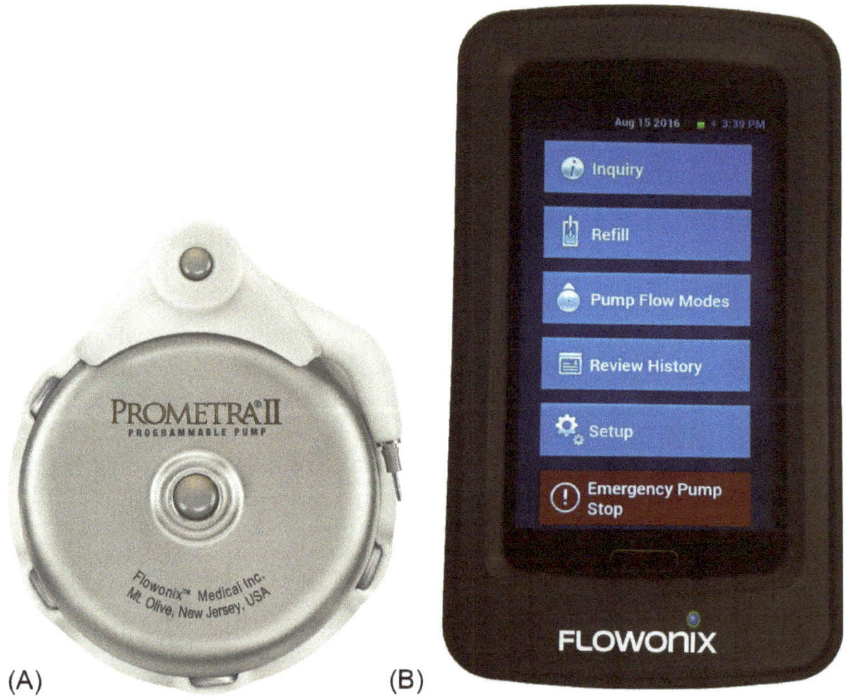

FIG. 3 (A) Prometra II programmable pump and (B) programmer. *Reproduced with permission from Flowonix.*

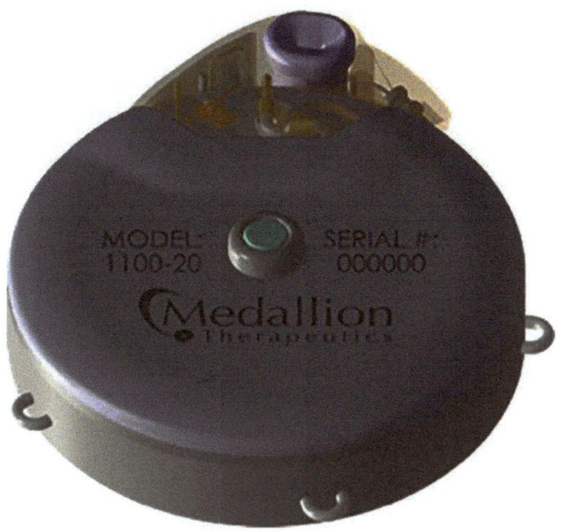

FIG. 4 Medallion P100 programmable pump. *Reproduced with permission from Medallion Therapeutics, The Alfred Mann Foundation.*

approved for baclofen to treat severe spasticity. The MedStream uses compressed gas as the driving force and a ceramic drive flow valve system to maintain a constant infusion rate from either a 20- or 40-cc reservoir.

The Medallion (Medallion Therapeutics) is an IDDS currently under investigation in clinical trials. The Medallion system offers improved safety with a negative-pressure reservoir. The negative pressure draws medication from the syringe during pump refills rather than requiring positive pressure from the syringe plunger, which reduces the risk of inadvertent injection into the pump pocket rather than the reservoir. The Medallion will be available with a 20-cc reservoir (Fig. 4).

The only constant-flow pump currently available is the Codman 3000 (Johnson & Johnson). This nonprogrammable pump does not use a battery; instead, it is based on a self-contained propellant that maintains a constant infusion rate. It is more commonly used in Europe than in the United States. All dose adjustments must be made with changes in concentration, and frequent dose adjustments and bolus dosing are not available with this model.

DRUG SELECTION

Once a patient has been deemed a good candidate for intrathecal drug therapy, the next step in the process is selecting the appropriate drug. The Polyanalgesic Consensus Conference (PACC) panel developed recommendations on the rational use of intrathecal analgesics based on reviews of preclinical and clinical research and experience in 2000, 2003, 2007, 2011, and 2017.[7] Findings from the published literature are combined with the clinical experience of a large panel of scientists and clinicians, and the resulting body of knowledge forms recommendations regarding the use of intrathecal analgesics to treat chronic pain. It is our belief that these recommendations should be a guide in clinical practice.

Panel Recommendations from the 2017 PACC

The most recent PACC panel recommendations,[7] published in 2017, contain new algorithm tracks for neuropathic and nociceptive pain states in both cancer and noncancer conditions as new elements for improving patient care (Tables 1 and 2).[8] Previous and updated drug selection recommendations provide clinical practice guidelines for the optimization of intrathecal therapy with single and multidrug combinations in a rational and prioritized order. In each arm of the algorithm, the medications are arranged in a hierarchy of efficacy and safety. First-line medications are supported by extensive clinical experience as well as published clinical and preclinical literature. Typically, these medications are used to initiate intrathecal therapies. Note that morphine and ziconotide are the 2 FDA-approved agents for intrathecal therapies. Non-approved agents are commonly used among pain practitioners who manage

TABLE 1 2017 Polyanalgesic Algorithm for Intrathecal Therapies for Cancer and Other Terminal Condition-related Pain with Localized Nociceptive or Neuropathic Pain

Line 1	Morphine	Ziconotide		Morphine + bupivacaine
Line 2	Hydromorphone	Hydromorphone + bupivacaine or Hydromorphone + clonidine		Morphine + clonidine
Line 3	Clonidine	Ziconotide + opioid	Fentanyl	Fentanyl + bupivacaine or Fentanyl + clonidine
Line 4	Opioid + clonidine + bupivacaine		Bupivacaine + clonidine	
Line 5	Baclofen			

Line 1: Morphine and ziconotide are approved by the US Food and Drug Administration for intrathecal therapy and are recommended as first-line therapy for neuropathic pain. The combination of morphine and bupivacaine is recommended for neuropathic pain on the basis of clinical use and apparent safety.
Line 2: Hydromorphone, alone or in combination with bupivacaine or clonidine is recommended. Alternatively, the combination of morphine and clonidine may be used.
Line 3: Third-line recommendations for neuropathic pain include clonidine, ziconotide plus an opioid, and fentanyl alone or in combination with bupivacaine or clonidine.
Line 4: The combination of bupivacaine and clonidine (with or without an opioid drug) is recommended.
Line 5: Baclofen is recommended on the basis of safety, although reports of efficacy are limited.
Content reproduced with permission from Deer TR, Pope JE, Hayek SM, et al. The Polyanalgesic Consensus Conference (PACC): recommendations on intrathecal drug infusion systems best practices and guidelines. Neuromodulation 2017;20(2):96-132.

TABLE 2 2017 Polyanalgesic Algorithm for Intrathecal Therapies in Cancer or Other Terminal Condition-related Pain with Diffuse Nociceptive or Neuropathic pain

Line 1	Morphine	Hydromorphone	Ziconotide	Fentanyl
Line 2	Morphine + bupivacaine	Ziconotide + opioid	Hydromorphone + bupivacaine	Fentanyl + bupivacaine
Line 3	Opioid (morphine, hydromorphone, or fentanyl) + clonidine			Sufentanil
Line 4	Opioid + clonidine + bupivacaine		Sufentanil + bupivacaine OR clonidine	
Line 5	Sufentanil + bupivacaine + clonidine			

Content reproduced with permission from Deer TR, Pope JE, Hayek SM, et al. The Polyanalgesic Consensus Conference (PACC): recommendations on intrathecal drug infusion systems best practices and guidelines. Neuromodulation 2017;20(2):96-132.

TABLE 3 2017 Polyanalgesic Consensus Conference Recommendations for Maximum Daily Drug Doses and Concentrations

Drug	Maximum concentration	Maximum daily dose
Morphine	20 mg/mL	15 mg
Hydromorphone	15 mg/mL	10 mg
Fentanyl	10 mg/mL	1000 μg
Sufentanil	5 mg/mL	500 μg
Bupivacaine	30 mg/mL	15-20 mg
Clonidine	1000 μg/day	600 μg/day
Ziconotide	100 >μg/mL	19.2 μg/day

Content reproduced with permission from Deer TR, Pope JE, Hayek SM, et al. The Polyanalgesic Consensus Conference (PACC): recommendations on intrathecal drug infusion systems best practices and guidelines. Neuromodulation 2017;20(2):96-132.

patients with an intrathecal pump; the safety and efficacy of these agents have been thoroughly discussed. Specific recommendations are provided for ziconotide for neuropathic pain and baclofen for pain from spasticity. Best practices for improved patient care and outcomes with intrathecal infusion are recommended to minimize the potential for morbidity and mortality.

Starting and maximum doses recommended by the PACC panel are summarized in Tables 3 and 4.[7] Appropriate starting doses may vary based on a patient's baseline oral intake at the time that intrathecal therapy is initiated. Recommended bolus trial doses are listed in Table 5.[9]

Compounding of Medications

The FDA regards traditional "pharmacy compounding" as the combining or altering of ingredients by a pharmacist in response to a licensed practitioner's prescription to treat a patient's special medical needs.[8] When prescribing a drug for intrathecal administration, special

TABLE 4 Recommended Starting Drug Dose Ranges for Medications Administered Intrathecally

Drug	Recommended starting dose
Morphine	0.1-0.5 mg/day
Hydromorphone	0.02-0.5 mg/day
Ziconotide	0.5-2.4 μg/day
Fentanyl	25-75 μg/day
Bupivacaine	1-4 mg/day
Clonidine	40-100 μg/day
Sufentanil	10-20 μg/day

TABLE 5 2017 Polyanalgesic Consensus Conference Recommendations for Intrathecal Bolus Trial Dose[7]

Drug	Recommended starting dose
Morphine	0.1-0.5 mg
Hydromorphone	0.025-0.1 mg
Ziconotide	1-5 μg
Fentanyl	15-75 μg
Bupivacaine	0.5-2.5 mg
Clonidine	5-20 μg
Sufentanil	5-20 μg

and strict requirements are needed to ensure patient safety and protection of the practice. Intrathecal drug preparation requires strict adherence to sterile practices to ensure the preparation is free of contaminants so that potentially catastrophic effects may be avoided.

The compounding pharmacist must have written policies and procedures established and updated to remain compliant with USP 797 standards. This means compounding sterile products under the manufacturer's labeled instructions and with any manipulations that expose the original contents to potential contamination. Other considerations in evaluating and selecting a compounding pharmacy for intrathecal drug preparation include: aseptic training of personnel, a segregated sterile compounding area, environmental monitoring of air quality, calibration of equipment, cleaning and disinfection programs, and quality assurance protocols. This requires a pharmacist's adherence to the following guidelines:

- use a pH that is physiologically appropriate and is consistent with the drug solubility and delivery system, generally in the range of pH 4-8;
- use solutions that are isotonic with normal cerebrospinal fluid (CSF) (approximately 300 mOsm/L);
- prepare the solution in a manner that does not alter the solubility of the constituents to minimize neurotoxicity or incompatibility with the pump system being used;
- verify the chemical and physical stability, and sterility of the preparation in accordance with the USP and ASHP publications; and
- ensure appropriate control of bacterial endotoxins (pyrogens).

Drug Flow Rates and Intermittent Bolusing

Continuous intrathecal delivery of medications is a standard mechanism. No studies have investigated the best rate of medication delivery. Some PACC members theorize that lower flow rates may result in higher concentration of the drug at the catheter tip and may increase the risk of granuloma, but this has not been demonstrated clinically.[8, 10]

Constant-flow pumps cannot be used to change the flow rate or to administer boluses. Intermittent bolusing and patient-controlled analgesia are now available with programmable pumps. Bolus doses can be given by programming the pump to give doses at set times and, when available, the patient has the option of self-administering boluses as needed (i.e., patient-controlled analgesia or personal therapy manager). Generally speaking, boluses are 5%-20% of the daily dose that is administered at a continuous rate. This manner of drug administration is often advantageous given that many patients' pain is intermittent and can be affected by certain activities of daily living and by the time of day. Caution must be used as baseline doses climb with clonidine and bupivacaine. To avoid hypotension or motor block, the dose of clonidine should not exceed 200 mcg, and the dose of bupivacaine should not exceed 4 mg.

Trial Period for Determination of Permanent Intrathecal Therapy

The concept of carrying out a trial before implanting an IDDS is based on the assumption that the trial will be predictive of the efficacy of the proposed therapy.[9] Trialing was thought to be critical prior to implantation; however, this is now a debatable issue. Trialing may be considered in patients with already-implanted pumps, for whom a change in intrathecal medication is being considered.[8, 9] Trialing may also be performed before consideration of pump implantation. The duration of the trial is also an important element. For example, the issue of opioid-induced hyperalgesia cannot be addressed during a single-shot trial or during a short-term, 72-hour drug infusion.

Many PACC members felt that trialing was unnecessary in end-of-life cancer patients if those patients have previously tolerated the same drug administered by another route. However, we feel that trialing should still be considered for most patients. The trial of various medications (e.g., ziconotide) allows a patient to assess potential side effects and toxicities, and to determine the overall benefit in pain relief measured by quality-of-life assessments and activities scales. Realistic expectations for the ability of long-term IDDS to reduce pain and improve quality of life can be established during the trial period. Trialing can be performed by intrathecal or epidural injection infusion for most drugs.

Adverse Effects and Complications

A number of complications can occur with IDDS.[11] These complications can be subdivided into early complications—that is, initial technical and surgical complications, and late complications—in other words, complications that arise over the long term related to the device or the drug (Table 6). With appropriate workup, good surgical technique, and postoperative vigilance, surgical complications of bleeding and infection are rare.

TABLE 6 Early and Late Complications after Implantation of an Intrathecal Pump[11]

Early complications	Late complications
Infection	Granuloma
Dural leak	Pump refill pocket fill
Pain at incision site	Drug effects (e.g., low testosterone)
Spinal cord or nerve injury	Catheter break or failure (i.e., withdrawal)
Intrathecal drug overdose	Pain at implant site, pump overdose

Complications related to the pump and catheter system exist and need to be addressed before intrathecal drug medications are chosen or adjusted.[11, 12] The most common operational problems are related to either pump programming or motor stalling. Pump programming errors are most likely to occur when drug concentrations are changed. It is necessary to verify drug concentration, daily dose, volume, and refill date. A motor stall is typically caused by gear shift wear. The incidence is low (0.5%-2.2%) and can be detected with radiographic examination.[13] Medtronic SynchroMed II pumps are equipped with electronic logs to detect and report motor stall issues and include other advances (e.g., corrosion-resistant motor and wiring). Catheter malfunctions may include kinks, holes, or blockages that prevent accurate delivery of medications.[12] Other complications include catheter fibrosis or inflammation, and granuloma formation.

Granuloma development of a noninfectious inflammatory mass or granuloma has been reported in 0.1%-5.0% of patients with implantable pumps.[11, 14, 15] Physicians should be suspicious of possible granuloma if a patient complains of increased pain that was previously controlled, neurological deficits of weakness and sensory loss, or a combination of these symptoms. Granulomas have been reported most frequently with morphine, followed by hydromorphone and fentanyl.

CURRENT EVIDENCE FOR INTRATHECAL THERAPY

Intrathecal therapy is an established and accepted alternative to standard medical management for cancer patients with severe pain. By using a wide variety of agents, IDDS can provide effective analgesia with potentially fewer adverse effects. Numerous prospective and randomized controlled trials have evaluated the use of IDDS in cancer patients (Table 7). Results from these trials have demonstrated reduction in severity of pain in cancer patients, improvement in quality of life, and reduction in drug-related toxicities. Similar results were noted in a prospective randomized study by Staats et al.,[17] which reported significant pain relief in patients with AIDS and patients with cancer who received ziconotide via IDDS.

Thimineur et al.[19] carried out a prospective study that showed improvements in pain, mood, and function from baseline to 36 months in participants treated with IDDS compared to patients who received conservative therapy. Other prospective studies have been published for refractory cancer pain using morphine and bupivacaine as the intrathecal agent, with significant reduction in pain over 3-6 months.[20-23] One study by Mitchell et al.[21] demonstrated pain relief in 22 patients over 6 months with levobupivacaine and morphine. In that study, the IDDS provided long-term pain control in patients with difficult-to-control cancer-related pain. A recent retrospective review of 160 patients from a single center demonstrated significant pain reduction over 3 months.[24] The median pain score was 7.1 at the time of implantation; however, available pain scores showed a median decrease in pain to 2.5 ($P < .0001$). At a comprehensive cancer care center in France, 93 patients diagnosed with pancreatic cancer were treated with IDDS and observed over an 11-year period.[25] In this largest study of the use of IDDS in pancreatic cancer patients, pain was reported to be significantly reduced with a low rate of complications.[25] Finally, a recent retrospective study of health service utilization and payments for cancer patients with IDDS compared to conservative

TABLE 7 Prospective and Randomized Controlled Trials of Intrathecal Drug Delivery Systems for Patients with Cancer-related Pain

Study, year	Study design	Participants, follow-up	Intervention	Results	Conclusions
Rauck et al. (2003)[16]	Prospective, open-label study	119 patients with refractory cancer pain and/or intolerable side effects. 17 US and international sites Follow-up: up to 4.7 years over 16 months postimplantation	Patient-activated bolus IDDS with morphine	Overall success, >50% reduction in numeric analog scale, use of systemic opioids	Use of patient-activated drug delivery system resulted in clinical effective analgesia and fewer adverse effects
Staats et al. (2004)[17]	Double-blind, placebo-controlled, randomized controlled trial	111 patients with cancer or AIDS randomized to receive ziconotide or placebo treatment. Average Visual Analogue Scale of Pain Intensity score: >50 mm Follow-up: 11 days	Intrathecal ziconotide titrated over 5-6 days, then maintenance phase for responders and crossover of nonresponders to the opposite cohort	Mean VASPI improved 53.1% in the ziconotide group compared to 18.1% in the placebo group ($P < .001$) 52.9% of patients in ziconotide group reported moderate to complete pain relief compared to 17.5% in the placebo group ($P < .001$)	Intrathecal ziconotide showed significant improvements in pain in patients with cancer- or AIDS-related pain
Smith et al. (2005)[18]	Double-blind, placebo-controlled, randomized controlled trial	200 patients with intractable cancer pain randomized to CMM or IDDS Follow-up: 4 weeks	IDDS with intrathecal morphine; other intrathecal agents used if morphine yielded no results	84.5% of patients with IDDS achieved clinical success compared to 70.8% relief in the CMM group ($P = .05$) Decreased composite toxicity: 17.1% CMM vs 50.3% IDDS. Fatigue and reduced consciousness in IDDS group compared to CMM group ($P < .05$)	Reduction in pain in crossover group; reduced toxicity, especially in fatigue and reduced consciousness; positive trend in quality of life improvements

IDDS, *implantable drug delivery system;* VASPI, *Visual Analogue Scale of Pain Intensity;* CMM, *comprehensive medical management.*

medical management showed a trend of lower medical utilization and payments during the first year postimplantation.[26]

Emerging Intrathecal Pain Control Agents

Two new approaches designed for interventional pain control are currently in development and are being evaluated in Phase I clinical trials. Both use the principle of selective neuronal chemoablation and, with certain considerations for disease progression, require just 1 administration of the therapeutic agent. The chemoneutralization process removes the entire neuron or part of a neuron (e.g., nerve endings) and this removal can be permanent or reversible, depending on the route of administration. The first of the new approaches is based on the restricted expression of the heat, capsaicin, and chemo-responsive ionotropic receptor TRPV1 in a subpopulation of dorsal root ganglion neurons. The therapeutic agent is a TRPV1 agonist called resiniferatoxin (RTX). The second of the new approaches is based on a biochemical process known as *agonist-mediated receptor endocytosis*. This process moves the receptor from the plasma membrane to the interior of the cell, and occurs when an agonist ligand binds to its cognate G-protein-coupled receptor. In the case of pain control, the ligand is the 11-amino-acid neuropeptide substance P, which is found in high concentrations in noci-responsive primary afferent neurons. During the process of tissue damage, substance P is released into the dorsal spinal cord, where it binds to its receptor on the surface of second-order spinal neurons, and the ligand receptor complex is internalized in a vesicle in the cytoplasm. This process constitutes a route of entry for therapeutic macromolecules into a set of spinal neurons that are critical for pain transmission to the brain. The neuroablative effect is achieved by coupling substance P with a plant enzyme or a bioengineered bacterial enzyme, both of which stop protein synthesis. Several days after exposure, the neuron dies; this treatment obviously is permanent.

Substance P-saporin and Substance P-pseudomonas Exotoxin-35 for Pain Control

BASIC BIOCHEMISTRY

The "substance P" portions of substance P-saporin (SP-SAP) and substance P-pseudomonas exotoxin-35 (SP-PE35) are basically identical. Substance P is an 11-amino-acid, C-terminally amidated peptide. It is released from nociceptive primary afferents and binds to the neurokinin 1 receptor on postsynaptic neurons in the dorsal horn. Thus, pain originating in any dermatome can be targeted by localized application. In both cases, the neurotoxic moiety has enzymatic activity. Pseudomonas endotoxin catalyzes the addition of an adenosine diphosphate-ribose group to elongation factor 2, and saporin cleaves the ribosomal RNA. Taken together, both of these actions block ribosomal translation of messenger RNAs. Consequently, if ribosome inactivation is effective enough, protein synthesis ends and the cell dies. This takes several days, a timeframe notably longer than RTX, which acts in a matter of minutes to roughly 2 hours.

Saporin is about half as large as PE (~600 amino acids), and PE has been N-terminally truncated to 35,000 MW. Despite the size of these two molecules, once conjugated to substance P, the entire conjugate is internalized into an endosome, and through a series of steps the active enzyme is delivered to the cytoplasm, where it has access to the neuron's complement of ribosomes.[27] A third SP-protein toxin, Substance P-diphtheria toxin, has also been evaluated for analgesic activity. However, this is a fusion protein, and the C-terminal substance P portion is difficult to amidate in vitro, which is a prerequisite for receptor recognition, and has not been pursued. Consideration of the amount of enzyme taken up vs potential compensating resiliency of neuronal protein synthetic capacity are two major factors thought to govern susceptibility of NK1 receptor neurons, compared to "off-target" neurons that do not make the NK1 receptor (or neurons that do make the NK1 receptor, but are not sensory neurons).

PRECLINICAL TRIALS WITH SP-SAP AND SP-PE35

Despite the somewhat elaborate underlying biochemistry, the therapeutic concept is fairly straightforward—treating pain with SP-SAP or SP-PE35 involves selective removal of a key neural link in the pain pathway, a process often referred to as *molecular neurosurgery*. The internalization of substance P after binding to the NK1 receptor was demonstrated in the mid-1990s,[28, 29] and the analgesic actions of SP-SAP and substance P-diphtheria toxin were described in 1997 and 1999, respectively.[30, 31] An investigation of the preclinical analgesic actions of SP-PE35 in rats was recently published.[32] This molecule seems to be more efficacious than SP-SAP, as it produces strong analgesia against several nociceptive sensory modalities as well as an antihyperalgesic action, whereas the action of SP-SAP is mainly antihyperalgesic.

Before it was investigated in human clinical trials, SP-SAP was evaluated in two different canine clinical trials. One trial was a safety study performed in purpose-bred dogs and did not evaluate analgesic activity[33]; the other was performed on client-owned dogs with pain from hindlimb or forelimb osteosarcoma.[34] In both studies, SP-SAP was administered intrathecally, via a catheter or a needle placed such that the tip was near the lumbar enlargement (or, in the case of forepaw tumors, into the CSF space of the cisterna magna). The main analgesia-related finding from this large cancer study was that, within 6 weeks of SP-SAP administration, significantly fewer SP-SAP-treated dogs required unblinding for adjustment of their standard analgesic regimen compared to the vehicle-treated (i.e., control) dogs. Furthermore, unblinding occurred much sooner for the controls compared to the SP-SAP-treated animals. The original primary and secondary outcomes were acquired at 2 weeks postrandomization and, although trends toward improvement in pain intensity ratings, interference with activities of daily living, and activity levels were noted, none of these was statistically significant. At the highest doses, both studies encountered motor deficit problems. The default would be to attribute this to the saporin moiety acting to compromise motor neurons, but the routes of administration may also be contributing factors. In our rat study of SP-PE35, we observed no motor deficits, even with an ED_{100} dose. However, based on the dog studies, further experiments with higher doses are needed to examine toxicity and to estimate the protective index.

HUMAN CLINICAL TRIAL EXPERIENCE WITH SP-SAP AND SP-PE35

It will likely be another 5 years before a human clinical trial is initiated for SP-SAP as palliative pain control in human cancer patients. The first patients to receive this therapy were treated in 2014. An intrathecal catheter was inserted at the S1-L5 vertebral level, advanced roughly 5 cm, and doses of 1, 2, and 4 μg were examined in one patient each. While only abstracts are available, the investigators report no adverse toxic, neurological, or cardiac effects. At the same time, the researchers observed no distinct signs of reduced pain in participants.[35] Unfortunately, because neither analgesia nor toxicity occurred, very few concrete conclusions can be drawn other than that this human trial was in the "no observable effect level" category of dose response. It is worth noting that these doses are far-removed from the 150-μg dose that caused motor impairment in dogs.

Two other factors might be contributing to the lack of effect in this small human clinical trial. First, based on the investigators' description, the drug appears to have been administered into the region of the cauda equina, rather than at the level of the lumbar spinal cord itself. This likely imposes a large dilution effect on SP-SAP, reducing the likelihood of obtaining a clinically meaningful dose at the level of the lumbar dorsal horn. In most rat and dog studies with SP-SAP or SP-PE35, the conjugates were given at the level of the spinal cord lumbar enlargement and testing was done on the hind paw. The dorsal horn is where the NK1 receptors are highest and where the target cells located. Therefore, administration at the relevant spinal cord levels would likely improve the chances of obtaining an analgesic action. Second, the potential for metabolism of the substance P moiety by peptidases in the CSF-filled lumbar cistern is a real consideration. CSF contains high levels of dipeptidyl-carboxypeptidase (an angiotensin-converting enzyme). This enzyme can remove the last two amino acids from the C-terminus of substance P, including the methionine amide needed for receptor recognition.[36, 37]

It is still too early to discern whether SP-SAP or SP-PE35 will be effective in treating pain in humans. SP-PE35 appears to be more potent than SP-SAP, and this may improve the capacity for obtaining on-target analgesia and enhance the protective index. As is standard with interventional approaches, the nuances of the administration procedure are critical variables. This applies to the above peptide-directed toxins and to resiniferatoxin (described in detail in the following section): These compounds perform best when the drug is placed close to the site of action. In the hands of a skilled interventionist, the use of a single-administration agent can provide long-lasting pain relief and obviate the need for excessive opioids.

Resiniferatoxin for Pain Control

Our group and another set of investigators have provided two detailed reviews of the mechanisms of action, analgesic activity, and spectrum of pain indications for RTX.[38, 39] As a versatile interventional analgesic, RTX has been analyzed in a variety of sites in the body, and many routes of administration have been explored for different pain indications.

Basic Neurobiology

The concept of RTX is similar to the above-described substance P conjugates; however, the site of action within the nociceptive circuit is vastly different. RTX is an agonist at the TRPV1 heat and ligand-gated Na^+/Ca^{++} cation channel; it is also 600 MW, which is much different than the conjugates (~35,000 MW). This allows for different physicochemical strategies for purification from the natural plant sources responsible for its biosynthesis.[40] In vivo, binding of RTX is pseudo-irreversible. When bound, the ion channel remains open, allowing an overload of calcium into the neuronal cell body, axon, or nerve ending.[41] Because TRPV1 is only made in very high quantities in a select population of dorsal root ganglion neurons,[42] the cellular toxicity of the calcium overload is confined to the TRPV1 dorsal root ganglion neurons. Other neurons in the ganglion are unaffected; therefore, sensations such as a pinprick, pressure sense, vibration, and proprioception are unaffected.[43] In terms of clinical usage, the site of action (either the cell body, axon, or nerve ending) must be accessible to RTX administration, and most sites are. This allows the clinician to tailor RTX administration to the individual pain problem. Only transient inactivation of the nerve ending occurs when RTX is administered into peripheral sites. This includes subcutaneous, peripheral, perineural, corneal surface, and intra-articular sites.[43-48] The recovery time is generally proportional to the distance between the affected nerve ending and the body surface. The cornea regains sensitivity in a matter of 4-5 days, because only a few fibers have to regenerate through the thin corneal epithelial layer to restore sensitivity in the capsaicin eye-wipe test. The fiber-type-specificity of TRPV1 and RTX are emphasized by the fact that the blink reflex remained intact throughout the study.[44]

Clinical Indications and Administration Routes for Resiniferatoxin

Clinical indications for RTX administration by intrathecal, epidural, or periganglionic routes include palliative care for pain in advanced cancer (currently in clinical trial), or localized osteosarcoma or postherpetic neuralgia (ganglionic), as well as various back or spinal column problems associated with pain. In contrast to the reversible, peripheral actions of RTX on TRPV1-containing neural structures, injection beside the ganglion or directly into the ganglion will kill the TRPV1-positive neuronal cell bodies, and the analgesia is permanent. Injections into the CSF selectively target the centrally projecting axons of the TRPV1 neurons. This, too, is permanent, as the central axons of dorsal root ganglion neurons do not regenerate. A unique characteristic of CSF administration is that the neuronal perikarya in the dorsal root ganglion and associated peripheral axons are largely spared (Fig. 5). Remarkably, the ability of these dorsal root ganglion neurons to communicate with peripheral end organs is left intact.[49] Nonetheless, the disconnect with the spinal cord is quite thorough, producing profound analgesic effects when properly administered.

Whether administration is peripheral or central, two modalities of pain sensation are removed by RTX. One is the sensation of heat pain; the other is referred to as "loss of clinical pain." This idea assumes that pain from tissue damage or inflammation is distinct from epicritic pain. Sharp pain from a pinprick is intact, as are other somatosensory sensitivities. The extraordinary potency and selectivity of RTX and limited repertoire of cell types that express TRPV1, yield clinically relevant analgesia yet with preservation of non-pain modalities, translate RTX into a nearly ideal analgesic action.

Preclinical Trials with Resiniferatoxin

The intrathecal analgesic actions of RTX have been examined mainly in rat models and in companion dogs with osteosarcoma. Three trials have been conducted in companion dogs with osteosarcoma: two open studies and one double-blinded study.[43, 50, 51] For these palliative care cases, the drug is administered intrathecally, either into the lumbar intrathecal space for hindlimb osteosarcomas or into the cisterna magna for forelimb bone tumors. Initial intrathecal dose ranging used block of thermal nociception to demonstrate target engagement and amounts of 0.1, 1.2, or 3.0 μg/kg were given intracisternally and fore paws were tested. It is important to remember that RTX is a TRPV1 agonist and, when injected, it activates pain

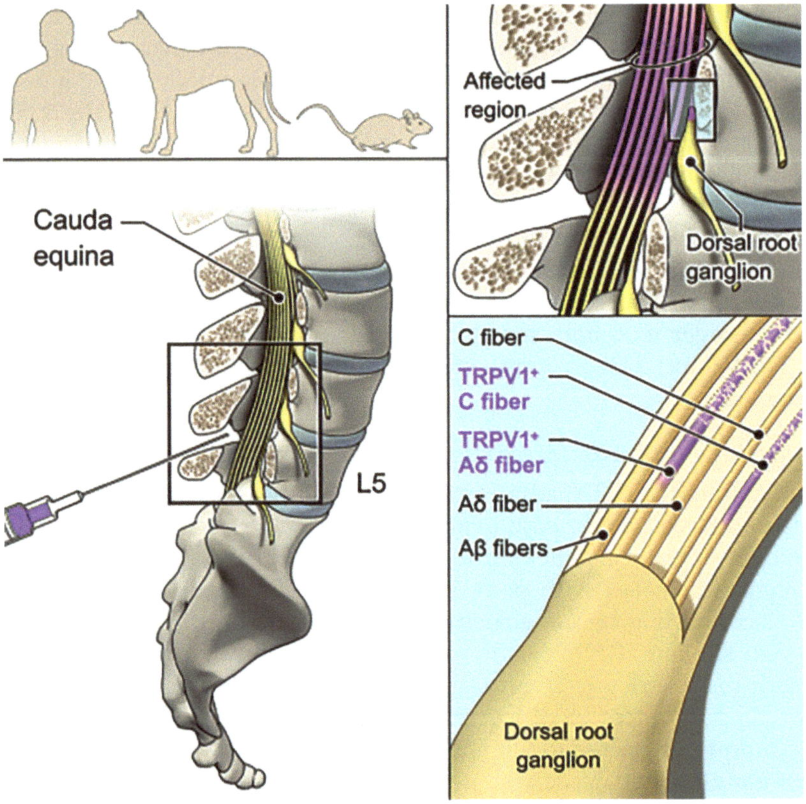

FIG. 5 Model of resiniferatoxin (RTX) mechanism after intrathecal injection. RTX acts on TRPV1-containing axons in the dorsal roots of the cauda equina and inactivates or ablates only the centrally projecting axonal portion of the TRPV1$^+$ C- and A-delta primary afferent neurons. The TRPV1$^+$ axons are only a subpopulation of the somatosensory and proprioceptive afferent populations. Only the TRPV1$^+$ axons are susceptible to RTX. *Reproduced with permission from Sapio MR, Neubert JK, LaPaglia DM, et al. Pain control through selective chemo-axotomy of centrally projecting TRPV1$^+$ sensory neurons. J Clin Invest 2018;128(4):1657-1667.*

fibers first, followed by axonal inactivation and analgesia. To counteract the algesic effect, general anesthesia was maintained for 2 hours, which allowed the lesion/inactivation process to reach completion. Intrathecal administration also causes a transient increase in blood pressure and heart rate, which peak 5 minutes after administration, then rapidly decreases over 60 minutes and slowly returns toward baseline over the next 240 minutes. Behavioral testing showed that thermal pain was ablated at the 1.2 μg/kg dose, and doses of 1.0-1.2 μg/kg were used in dogs with osteosarcoma. Recovery after extubation was generally uneventful, and dogs were discharged home the next day. Of the 20 dogs treated, 18 survived to 2-week follow-up. All 18 showed marked decreases in pain evaluations (according to the owners) on a brief canine pain inventory similar to the human Brief Pain Inventory short form. The analgesia lasted until the animals died or were euthanized due to progression of their primary disease (as much as 14 weeks later, or beyond that for 4 animals). Additionally, analgesic medications were decreased or discontinued in most dogs.

In the larger, blinded study, the primary end points were: (1) the time of unblinding; that is, the point at which dogs were experiencing unacceptable levels of discomfort and required additional intervention to their standard of care pain management regimen, or for controls offering of RTX injection; and (2) the number of animals that required unblinding within 6 weeks of entering the study. Intrathecal administration of 1.2 μg/kg produced significant improvement in both end points, as well as significant improvement in lameness as evaluated by a blinded veterinarian from video recordings obtained at 2 weeks after administration. No dog in any study developed a deafferentation pain syndrome, as evidence by a lack of self-injurious behavior. In terms of pain and progression of disease, many parallels exist between the dog's primary bone cancer and metastatic progression to bone in humans. The pain from bone destruction and the occurrence of breakthrough pain are similar, with persistently increasing frequency and intensity of pain. RTX appears to reduce the severity of these problems because, when this pain is severe or prolonged, it often leads to owners requesting euthanasia for their dog.

Human Clinical Trial Experience with Resiniferatoxin

To date, clinical experience with intrathecal RTX is limited. One ongoing Phase I study, Resiniferatoxin to Treat Severe Pain Associated with Advanced Cancer (NCT00804154), is being conducted. In this study, RTX has been administered intrathecally to 12 people. Thus far, three doses (3.3, 13, and 26 μg) have been examined and in all cases, the compound was administered via lumbar puncture in a volume of 1-2 mL under propofol anesthesia and supplemented with intravenous fentanyl where appropriate. This route allows RTX to bind to TRPV1 receptors on axons in the cauda equina, cutting the connection between the ganglion and spinal cord.[49] Patients are eligible for study if they have severe cancer-related pain in the lower half of the body and if standard treatments have failed to reduce their pain. The goal is to treat their pain, but also to retain the capacity to sense hot thermal stimuli in the face and hands to prevent burn injury. In terms of dose justification, autonomic activation, and the need to extend the duration of postinjection anesthesia, the previously described canine trials formed a strong translational basis for the human study. The predictive value withstood the fact that, although canine osteosarcoma affects different limbs and different species were represented, taken together the dogs were more uniform than the human patients who had great heterogeneity in terms of site of involvement and course of their disease. Nonetheless, RTX produced analgesia and opioid use was frequently reduced as well.

As part of a larger investigation of the mechanisms of intrathecal RTX, we analyzed spinal cord and dorsal root ganglion tissues at several levels from 1 treated patient from whom consent for autopsy had been previously obtained. Alterations in gene expression were performed using RNA-Seq and immunohistochemistry. We also obtained pre- and postinjection assessments of this patient's pain level and opioid consumption.[49] Intrathecal RTX decreased ganglionic TRPV1 transcript levels, but overall, there were minimal alterations in the patient's dorsal root ganglion gene expression profile. This is consistent with findings from the aforementioned animal studies, indicating that although some TRPV1+ neurons are deleted, many survive the exposure. This suggests that the axons in the dorsal roots are the structures most susceptible to RTX rather than the cell bodies in the dorsal root ganglion. In rats, dogs, and humans, it appears that the lesion of TRPV1 axons is an important component of the analgesic action of RTX. Postinjection, this person exhibited a 50% decrease in numerical pain rating (i.e., from a score of 8 to a 4), and also showed significant decreases in both immediate and slow-release opioids on average during the 14 days of observation. These data, though limited, are consistent with the analgesic efficacy of RTX in dogs; additional studies are in progress.

Future Research

To date, we have examined intrathecal route of administration of several pain medications. Additional studies are underway to conduct palliative care treatments using an epidural route (NCT03226574) and a periganglionic route (NCT02522611). These two routes offer several advantages over intrathecal administration, allowing for more precise targeting of pain. For example, in the case of a unilateral osteosarcoma, RTX can be given next to the relevant dorsal root ganglion only on the affected side of the body. This level of precision targeting is quite different from treating the whole lower half of the body with the intrathecal route. Exploration of all of these routes in various patient populations will be a part of the clinical development process of this intervention. The therapeutic molecule is extraordinarily potent and specific, making the administration procedure one of the more critical variables in determining the thoroughness of the analgesia obtained.

CONCLUSIONS

Numerous prospective and retrospective studies demonstrate significant benefit of IDDS for severe intractable pain. With proper patient selection and consideration of a patient's medical condition, successful IDDS can be performed with limited complications. Communication and education are needed from the oncologist and the care team to recognize the utility of IDDS and its implementation. Working closely with the oncologist may allow seamless integration of IDDSs. Additionally, the two new approaches to analgesia represent an unprecedented opportunity to dramatically reduce the use of morphine and other prescription opioids by replacing them with highly specific, efficacious, nonaddictive, long-acting pain treatments.

References

1. de Leon-Casasola OA. Interventional procedures for cancer pain management: when are they indicated? *Cancer Investig*. 2004;22(4):630-642.
2. Hanks GW, Justins DM. Cancer pain: management. *Lancet*. 1992;339(8800):1031-1036.
3. Meuser T, Pietruck C, Radbruch L, Stute P, Lehmann KA, Grond S. Symptoms during cancer pain treatment following WHO-guidelines: a longitudinal follow-up study of symptom prevalence, severity and etiology. *Pain*. 2001;93(3):247-257.
4. Practice guidelines for cancer pain management. A report by the American Society of Anesthesiologists Task Force on pain management, cancer pain section. *Anesthesiology*. 1996;84(5):1243-1257.
5. Bedder MD, Burchiel K, Larson A. Cost analysis of two implantable narcotic delivery systems. *J Pain Symptom Manag*. 1991;6(6):368-373.
6. Deer TR, Smith HS, Burton AW, et al. Comprehensive consensus based guidelines on intrathecal drug delivery systems in the treatment of pain caused by cancer pain. *Pain Physician*. 2011;14(3):E283-E312.

7. Deer TR, Pope JE, Hayek SM, et al. The Polyanalgesic Consensus Conference (PACC): recommendations on intrathecal drug infusion systems best practices and guidelines. *Neuromodulation.* 2017;20(2):96-132.
8. Deer T, Levy R, Prager J, et al. Polyanalgesic Consensus Conference 2011: Recommendations for the Management of Pain by Intrathecal (Intraspinal) Drug Delivery: Report of an Interdisciplinary Expert Panel, 2011.
9. Deer TR, Hayek SM, Pope JE, et al. The Polyanalgesic consensus conference (PACC): Recommendations for trialing of intrathecal drug delivery infusion therapy. *Neuromodulation.* 2017;20(2):133-154.
10. Kim P. Case series of distal catheter obstruction. In: *NANS Annual Meeting, Las Vegas, Nevada*; 2011.
11. Deer TR, Pope JE, Hayek SM, et al. The Polyanalgesic consensus conference (PACC): Recommendations for intrathecal drug delivery: guidance for improving safety and mitigating risks. *Neuromodulation.* 2017;20(2):155-176.
12. Follett KA, Burchiel K, Deer T, et al. Prevention of intrathecal drug delivery catheter-related complications. *Neuromodulation.* 2003;6(1):32-41.
13. Fluckiger B, Knecht H, Grossmann S, Felleiter P. Device-related complications of long-term intrathecal drug therapy via implanted pumps. *Spinal Cord.* 2008;46(9):639-643.
14. Deer T, Levy R, Prager J, et al., PACC 2011: Consensus on Diagnosis, Detection, and Treatment of Catheter-Tip Inflammatory Masses (Granulomas), 2011.
15. Langsam A. A case of spinal cord compression syndrome by a fibrotic mass presenting in a patient with an intrathecal pain management pump system. *Pain.* 1999;83(1):97-99.
16. Rauck RL, Cherry D, Boyer MF, Kosek P, Dunn J, Alo K. Long-term intrathecal opioid therapy with a patient-activated, implanted delivery system for the treatment of refractory cancer pain. *J Pain.* 2003;4(8):441-447.
17. Staats PS, Yearwood T, Charapata SG, et al. Intrathecal ziconotide in the treatment of refractory pain in patients with cancer or AIDS: a randomized controlled trial. *JAMA.* 2004;291(1):63-70.
18. Smith TJ, Coyne PJ. Implantable drug delivery systems (IDDS) after failure of comprehensive medical management (CMM) can palliate symptoms in the most refractory cancer pain patients. *J Palliat Med.* 2005;8(4):736-742.
19. Thimineur MA, Kravitz E, Vodapally MS. Intrathecal opioid treatment for chronic non-malignant pain: a 3-year prospective study. *Pain.* 2004;109(3):242-249.
20. Mercadante S, Intravaia G, Villari P, et al. Intrathecal treatment in cancer patients unresponsive to multiple trials of systemic opioids. *Clin J Pain.* 2007;23(9):793-798.
21. Mitchell A, McGhie J, Owen M, McGinn G. Audit of intrathecal drug delivery for patients with difficult-to-control cancer pain shows a sustained reduction in pain severity scores over a 6-month period. *Palliat Med.* 2015;29(6):554-563.
22. Sjoberg M, Appelgren L, Einarsson S, et al. Long-term intrathecal morphine and bupivacaine in "refractory" cancer pain. I. Results from the first series of 52 patients. *Acta Anaesthesiol Scand.* 1991;35(1):30-43.
23. Sjoberg M, Nitescu P, Appelgren L, Curelaru I. Long-term intrathecal morphine and bupivacaine in patients with refractory cancer pain. Results from a morphine:Bupivacaine dose regimen of 0.5:4.75 mg/ml. *Anesthesiology.* 1994;80(2):284-297.
24. Sayed D, Monroe F, Orr WN, et al. Retrospective analysis of intrathecal drug delivery: outcomes, efficacy, and risk for Cancer-related pain at a high volume Academic Medical Center. *Neuromodulation.* 2018;21:660-664.
25. Carvajal G, Dupoiron D, Seegers V, et al. Intrathecal drug delivery systems for refractory pancreatic cancer pain: observational follow-up study over an 11-year period in a comprehensive cancer center. *Anesth Analg.* 2018.
26. Stearns LJ, Hinnenthal JA, Hammond K, Berryman E, Janjan NA. Health services utilization and payments in patients with Cancer pain: a comparison of intrathecal drug delivery vs. conventional medical management. *Neuromodulation.* 2016;19(2):196-205.
27. Weidle UH, Tiefenthaler G, Schiller C, Weiss EH, Georges G, Brinkmann U. Prospects of bacterial and plant protein-based immunotoxins for treatment of cancer. *Cancer Genomics Proteomics.* 2014;11(1):25-38.
28. Grady EF, Garland AM, Gamp PD, Lovett M, Payan DG, Bunnett NW. Delineation of the endocytic pathway of substance P and its seven-transmembrane domain NK1 receptor. *Mol Biol Cell.* 1995;6(5):509-524.
29. Mantyh PW, DeMaster E, Malhotra A, et al. Receptor endocytosis and dendrite reshaping in spinal neurons after somatosensory stimulation. *Science.* 1995;268(5217):1629-1632.
30. Benoliel R, Eliav E, Mannes AJ, Caudle RM, Leeman S, Iadarola MJ. Actions of intrathecal diphtheria toxin-substance P fusion protein on models of persistent pain. *Pain.* 1999;79(2–3):243-253.
31. Mantyh PW, Rogers SD, Honore P, et al. Inhibition of hyperalgesia by ablation of lamina I spinal neurons expressing the substance P receptor. *Science.* 1997;278(5336):275-279.
32. Iadarola MJ, Sapio MR, Wang X, et al. Analgesia by deletion of spinal neurokinin 1 receptor expressing neurons using a bioengineered substance P-Pseudomonas exotoxin conjugate. *Mol Pain.* 2017;13. 1744806917727657.

33. Wiese AJ, Rathbun M, Butt MT, et al. Intrathecal substance P-saporin in the dog: distribution, safety, and spinal neurokinin-1 receptor ablation. *Anesthesiology*. 2013;119(5):1163-1177.
34. Brown DC, Agnello K. Intrathecal substance P-saporin in the dog: efficacy in bone cancer pain. *Anesthesiology*. 2013;119(5):1178-1185.
35. Frankel AE, Nymeyer H, Lappi DA, Higgins D, Ahn C, Noe C. Preliminary results from a phase I study of substance P-saporin in terminal cancer patients with intractable pain. *J Clin Oncol*. 2014;32(Suppl 31):191.
36. Mellstrom B, Iadarola MJ, Yang HY, Costa E. Inhibition of Met5-enkephalin-Arg6-Phe7 degradation by inhibitors of dipeptidyl carboxypeptidase. *J Pharmacol Exp Ther*. 1986;239(1):174-178.
37. Yokosawa H, Endo S, Ogura Y, Ishii S. A new feature of angiotensin-converting enzyme in the brain: hydrolysis of substance P. *Biochem Biophys Res Commun*. 1983;116(2):735-742.
38. Iadarola MJ, Gonnella GL. Resiniferatoxin for pain treatment: an interventional approach to personalized pain medicine. *Open Pain J*. 2013;6:95-107.
39. Iadarola MJ, Mannes AJ. The vanilloid agonist resiniferatoxin for interventional-based pain control. *Curr Top Med Chem*. 2011;11(17):2171-2179.
40. Appendino G, Szallasi A. Euphorbium: modern research on its active principle, resiniferatoxin, revives an ancient medicine. *Life Sci*. 1997;60(10):681-696.
41. Karai LJ, Russell JT, Iadarola MJ, Olah Z. Vanilloid receptor 1 regulates multiple calcium compartments and contributes to Ca2+–induced Ca2+ release in sensory neurons. *J Biol Chem*. 2004;279(16):16377-16387.
42. Goswami SC, Mishra SK, Maric D, et al. Molecular signatures of mouse TRPV1-lineage neurons revealed by RNA-Seq transcriptome analysis. *J Pain*. 2014;15(12):1338-1359.
43. Karai L, Brown DC, Mannes AJ, et al. Deletion of vanilloid receptor 1-expressing primary afferent neurons for pain control. *J Clin Invest*. 2004;113(9):1344-1352.
44. Bates BD, Mitchell K, Keller JM, et al. Prolonged analgesic response of cornea to topical resiniferatoxin, a potent TRPV1 agonist. *Pain*. 2010;149(3):522-528.
45. Iadarola MJ, Sapio MR, Raithel SJ, Mannes AJ, Brown DC. Long-term pain relief in canine osteoarthritis by a single intra-articular injection of resiniferatoxin, a potent TRPV1 agonist. *Pain*. 2018;159(10):2105-2114.
46. Neubert JK, Karai L, Jun JH, Kim HS, Olah Z, Iadarola MJ. Peripherally induced resiniferatoxin analgesia. *Pain*. 2003;104(1–2):219-228.
47. Neubert JK, Mannes AJ, Karai LJ, et al. Perineural resiniferatoxin selectively inhibits inflammatory hyperalgesia. *Mol Pain*. 2008;4:3.
48. Raithel SJ, Sapio MR, LaPaglia DM, Iadarola MJ, Mannes AJ. Transcriptional changes in dorsal spinal cord persist after surgical incision despite preemptive analgesia with peripheral Resiniferatoxin. *Anesthesiology*. 2018; 128(3):620-635.
49. Sapio MR, Neubert JK, LaPaglia DM, et al. Pain control through selective chemo-axotomy of centrally projecting TRPV1+ sensory neurons. *J Clin Invest*. 2018;128(4):1657-1670.
50. Brown DC, Agnello K, Iadarola MJ. Intrathecal resiniferatoxin in a dog model: efficacy in bone cancer pain. *Pain*. 2015;156(6):1018-1024.
51. Brown DC, Iadarola MJ, Perkowski SZ, et al. Physiologic and antinociceptive effects of intrathecal resiniferatoxin in a canine bone cancer model. *Anesthesiology*. 2005;103(5):1052-1059.

Index

Note: Page numbers followed by *f* indicate figures and *t* indicate tables.

FSC
www.fsc.org
MIX
Papier | Fördert gute Waldnutzung
FSC® C083411

Kolibri 360 GmbH
Ferdinand-Jühlke-Straße 7
99095 Erfurt, Deutschland
produktsicherheit@kolibri360.de